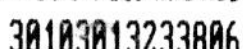

1323380.6

30103013233806

AF442674

Pavol Quittner

Philippe Souplet

Superlinear Parabolic Problems
Blow-up, Global Existence and Steady States

Birkhäuser
Basel · Boston · Berlin

Authors:

Pavol Quittner
Department of Applied Mathematics
and Statistics
Comenius University
Mlynská dolina
84248 Bratislava
Slovakia
e-mail: quittner@fmph.uniba.sk

Philippe Souplet
Laboratoire Analyse Géométrie et Applications
UMR CNRS 7539 - Institut Galilée
Université Paris-Nord
99 av. J.-B. Clément
93430 Villetaneuse
France
souplet@math.univ-paris13.fr

2000 Mathematics Subject Classification: primary 35-01, 35-02, 35B (35B05, 35B30, 35B33, 35B35, 35B40, 35B45, 35B50, 35B60, 35B65), 35J (35J55, 35J60, 35J65), 35K (35K50, 35K55, 35K60); secondary 35A07, 35A20, 35B10, 25B20, 35B37, 35B41, 35C15, 35D05, 35D10, 35J20, 35H25, 35K05, 35K15, 35K20, 35K40, 35K45, 35P30, 45K05, 46B70, 46E30, 46E35, 46N20, 47D06, 47D60, 78A55, 80A20, 80A25, 92C15, 92D25, 92E20, 93C20

Library of Congress Control Number: 2007929012

Bibliographic information published by Die Deutsche Bibliothek
Die Deutsche Bibliothek lists this publication in the Deutsche Nationalbibliografie;
detailed bibliographic data is available in the Internet at <http://dnb.ddb.de>.

ISBN 978-3-7643-8441-8 Birkhäuser Verlag AG, Basel · Boston · Berlin

© 2007 Birkhäuser Verlag AG
Basel · Boston · Berlin
P.O. Box 133, CH-4010 Basel, Switzerland
Part of Springer Science+Business Media
Printed on acid-free paper produced of chlorine-free pulp. TCF ∞
Printed in Germany

ISBN 978-3-7643-8441-8   e-ISBN 978-3-7643-8442-5

9 8 7 6 5 4 3 2 1   www.birkhauser.ch

Contents

Introduction

This book is devoted to the qualitative study of solutions of superlinear elliptic and parabolic partial differential equations and systems. Here "superlinear" means that the problems involve nondissipative terms which grow faster than linearly for large values of the solutions. This class of problems contains, in particular, a number of reaction-diffusion systems which arise in various mathematical models, especially in chemistry, physics and biology.

For parabolic problems of this type it is known that a solution may cease to exist in a finite time as a consequence of its L^∞-norm becoming unbounded: The solution blows up. On the other hand, in many of these problems there exist also global solutions (in particular, stationary solutions). Both global and blowing-up solutions may be very unstable and they may exhibit a rather complicated asymptotic behavior.

Concerning elliptic problems, we consider questions of existence and nonexistence, multiplicity, regularity, singularities and a priori estimates. Special emphasis is put on those results which are useful in the investigation of the corresponding parabolic problems. As for parabolic problems, we study the questions of local and global existence, a priori estimates and universal bounds, blow-up, asymptotic behavior of global and nonglobal solutions.

The study of superlinear parabolic and elliptic equations and systems has attracted the attention of many mathematicians during the past decades. Although a lot of challenging problems have already been solved, there are still many open questions even in the case of the simplest possible model problems. Unfortunately, most of the material, including many of the fundamental ideas, is scattered throughout hundreds of research articles which are not always easily readable for non-specialists. One of the main purposes of this book is thus to give an up-to-date and, as much as possible, self-contained account of the most important results and ideas of the field. In particular we try to find a balance between fundamental ideas and current research. Special effort is made to describe in a pedagogical way the main methods and techniques used in the study of these problems and to clarify the connections between several important results. Moreover, a number of the original proofs have been significantly simplified. In this way, the topic should be accessible to a larger audience of non-specialists.

The book contains five chapters. The first two are intended to be an introduction to the field and to enable the reader to get acquainted with the main ideas by studying simple model problems, respectively of elliptic and parabolic type. These model problems are of the form

$$\left.\begin{aligned}
-\Delta u &= f(u), && x \in \Omega, \\
u &= 0, && x \in \partial\Omega,
\end{aligned}\right\} \tag{0.1}$$

and

$$\left.\begin{aligned}
u_t - \Delta u &= f(u), & x \in \Omega,\ t > 0,\\
u &= 0, & x \in \partial\Omega,\ t > 0,\\
u(x,0) &= u_0(x), & x \in \Omega,
\end{aligned}\right\} \qquad (0.2)$$

where $\Omega \subset \mathbb{R}^n$ and f is a superlinear function, typically $f(u) = |u|^{p-1}u$ for some $p > 1$. The subsequent three chapters are devoted to problems with more complex structure; namely, elliptic and parabolic systems, equations with gradient depending nonlinearities, and nonlocal equations. They include several problems arising in biological or physical contexts. These chapters contain many developments which reflect several aspects of current research. Although the techniques introduced in Chapters I and II provide efficient tools to attack some aspects of these problems, they often display new phenomena and specifically different behaviors, whose study requires new ideas. Many open problems are mentioned and commented.

For the reader's convenience we have collected a number of frequently used results in several appendices. These include estimates of solutions of linear elliptic and parabolic equations, maximum principles, and basic notions from dynamical systems. Also, in one of the appendices, we give an account of the local theory of semilinear parabolic problems based on the abstract framework of interpolation-extrapolation spaces. However, this material is not essential for the understanding of the main contents of the book and can be left for a second reading. In particular, for the case of the model problem (0.2), the most useful results on local existence-uniqueness are proved by more elementary methods in the main text. On the other hand, we assume knowledge of the fundamentals of ordinary differential equations, of measure theory, of functional analysis (distributions, self-adjoint and compact operators in Hilbert spaces, Sobolev-Slobodeckii spaces and their embeddings, interpolation, Nemytskii mapping) and of the calculus of variations (minimizing of coercive, weakly lower semicontinuous functionals). Finally, a section of methodological notes and an index are provided.

We would like to stress that, due to the broadness of the field of superlinear problems, our list of results and methods is of course not complete and is influenced in part by the interests of the authors. For reasons of space, many interesting topics and results could not be mentioned in this book (and we also apologize for any omission.) In particular, we do not touch degenerate problems with superlinear source (involving for instance porous medium, fast diffusion, or p-Laplace operators), nor higher order equations (where the maximum principle does not generally apply). We do not consider superlinear problems involving nonlinear boundary conditions, nor parabolic systems with convection (chemotaxis, Navier-Stokes). These are very interesting and intensively studied topics, but would require a book on their own. Finally, let us mention that there exist several textbooks and monographs dealing, at least in part, with certain aspects of superlinear problems; see [460], [466], [63], [113], [405], [504], [372], [222], for example.

We would like to express our gratitude to several colleagues for their careful and critical reading of (some parts of) the manuscript, particularly H. Amann, M. Balabane, M. Chipot, M. Fila, Ph. Laurençot, P. Poláčik, A. Rodríguez-Bernal, J. Rossi, F.B. Weissler and M. Winkler. Our special thanks go to H. Amann for his stimulating encouragements to this project. We also thank T. Hempfling from Birkhäuser for his helpfulness and the first author thanks the Slovak Literary Fund for providing financial support.

$$\varphi_j : U_j \rightarrow B_1 \cap (R^{n-1} \times R^+)$$

$$\left|\varphi_j\right|, \left|\frac{\partial \varphi_j}{\partial x}\right|, \left|\varphi_j\right| < M$$

$$\bigcap_{\substack{k_0+1}} U_j = \varphi$$

1. Preliminaries

General

We denote by $B_R(x)$ or $B(x, R)$ the open ball in $\mathbb{R}^n$ with center x and radius R. We set $B_R := B_R(0)$. The $(n-1)$-dimensional unit sphere is denoted by S^{n-1}. The characteristic function of a given set M is denoted by χ_M. We write $D' \subset\subset D$ for $D', D \subset \mathbb{R}^n$ if the closure of D' is a compact subset of D. For any real number s, we set $s_+ := \max(s, 0)$ and $s_- := \max(-s, 0)$. We also denote $\mathbb{R}_+ := [0, \infty)$.

Domains

Let Ω be a **domain**, i.e. a nonempty, connected, open subset of $\mathbb{R}^n$ and let $k \in \mathbb{N}$. We shall say that Ω is **uniformly regular of class** C^k (cf. [13, p. 642]), if either $\Omega = \mathbb{R}^n$ or there exists a countable family (U_j, φ_j), $j = 1, 2, \ldots$ of coordinate charts with the following properties:

(i) Each φ_j is a C^k-diffeomorphism of U_j onto the open unit ball B_1 in $\mathbb{R}^n$ mapping $U_j \cap \Omega$ onto the "upper half-ball" $B_1 \cap (\mathbb{R}^{n-1} \times (0, \infty))$ and $U_j \cap \partial\Omega$ onto the flat part $B_1 \cap (\mathbb{R}^{n-1} \times \{0\})$. In addition, the functions φ_j and the derivatives of φ_j and φ_j^{-1} up to the order k are uniformly bounded on U_j and B_1, respectively.

(ii) The set $\bigcup_j \varphi_j^{-1}(B_{1/2})$ contains an ε-neighborhood of $\partial\Omega$ in $\overline{\Omega}$ for some $\varepsilon > 0$.

(iii) There exists $k_0 \in \mathbb{N}$ such that any $k_0 + 1$ distinct sets U_j have an empty intersection.

In an analogous way we define a uniformly regular domain of class $C^{2+\alpha}$ (shortly domain of class $C^{2+\alpha}$). Unless explicitly stated otherwise[1], we will always assume that

$$\Omega \subset \mathbb{R}^n \text{ is a uniformly regular domain of class } C^{2+\alpha} \text{ for some } \alpha \in (0, 1).$$

On the other hand, we do not assume Ω to be bounded unless this is explicitly mentioned.

We denote the distance to the boundary function by

$$\delta(x) := \text{dist}(x, \partial\Omega).$$

The exterior unit normal on $\partial\Omega$ at a point $x \in \partial\Omega$ is denoted by $\nu(x)$, and the outer normal derivative by ∂_ν or $\partial/\partial\nu$. The surface measure (on e.g. $\partial\Omega$ or S^{n-1}) will be denoted by $d\sigma$ or $d\omega$.

For a given domain Ω and $0 < T < \infty$, we set

$$Q_T := \Omega \times (0, T),$$
$$S_T := \partial\Omega \times (0, T) \qquad \text{(lateral boundary)},$$
$$\mathcal{P}_T := S_T \cup (\overline{\Omega} \times \{0\}) \qquad \text{(parabolic boundary)}.$$

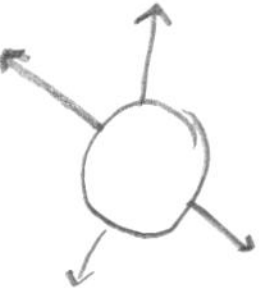

---

[1] In fact, if we want to allow nonsmooth domains, we will refer to an *arbitrary* domain.

Functions of space and time

Let $u = u(x, t)$ be a real function of the space variable $x \in \Omega$ and the time variable t. Without fearing confusion we will also consider u as a function of a single variable t with values in a space of functions defined in Ω, hence $u(t)(x) = u(x, t)$.

By a solution of a PDE being **positive** we usually mean that $u(x) > 0$ or $u(x, t) > 0$ in the domain under consideration. Note that, due to the strong maximum principles in Appendix F, positive is often equivalent to nontrivial nonnegative.

Radial functions. We say that a domain $\Omega \subset \mathbb{R}^n$ is symmetric if either $\Omega = \mathbb{R}^n$, or $\Omega = B_R = \{x \in \mathbb{R}^n : |x| < R\}$, or $\Omega = \{x \in \mathbb{R}^n : R < |x| < R'\}$, where $0 < R < R' \leq \infty$ (an annulus if $R' < \infty$). Denote $r = |x|$ and let $J \subset \mathbb{R}$ be an interval. A function u defined on a symmetric domain Ω (resp., on $\Omega \times J$) is said to be **radially symmetric**, or simply **radial**, if it can be written in the form $u = u(r)$ (resp., $u = u(r, t)$ for each $t \in J$). The function u is said to be **radial nonincreasing** if it is radial and if, moreover, u is nonincreasing as a function of r.

Banach spaces and linear operators

If X is a Banach space and $p \geq 1$, then X' and p' denote the (topological) dual space and dual exponent $(1/p + 1/p' = 1)$, respectively. We write $X \hookrightarrow Y$ or $X \hookrightarrow\hookrightarrow Y$ if X is continuously or compactly embedded in Y, respectively. If both $X \hookrightarrow Y$ and $Y \hookrightarrow X$ (that is X and Y coincide and carry equivalent norms), then we write $X \doteq Y$. We denote by $\mathcal{L}(X, Y)$ the space of continuous linear operators $A : X \to Y$, $\mathcal{L}(X) = \mathcal{L}(X, X)$. If A is a linear operator in X with the domain of definition $D(A)$ and $Y \subset X$, then the operator A_Y, the **Y-realization of** A, is defined by $A_Y u = Au$, $D(A_Y) := \{u \in D(A) \cap Y : Au \in Y\}$.

Function spaces

We denote by $\mathcal{D}(\Omega)$ the space of C^∞-functions with compact support in Ω. The norms in the Sobolev space $W^{k,p}(\Omega)$ (or the Sobolev-Slobodeckii space $W^{k,p}(\Omega)$ if k is not an integer) and the Lebesgue space $L^p(\Omega)$ will be denoted by $\| \cdot \|_{k,p}$ and $\| \cdot \|_p$, respectively. We denote by $W_0^{1,2}(\Omega)$ the closure of $\mathcal{D}(\Omega)$ in $W^{1,2}(\Omega)$. The spaces $W^{k,2}(\Omega)$, $k \in \mathbb{N}$, and $W_0^{1,2}(\Omega)$ will also be denoted as $H^k(\Omega)$ and $H_0^1(\Omega)$, respectively. The functions in these spaces are usually understood to be real valued. If no confusion is likely, we shall use the same notation for similar spaces of functions with values in $\mathbb{R}^n$. Otherwise we shall use the notation $L^p(\Omega, \mathbb{R}^n)$, for example.

Let Ω be a bounded domain in $\mathbb{R}^n$ (not necessarily smooth). The weighted Lebesgue spaces $L_\delta^p(\Omega)$ are defined as follows. Denoting as before

$$\delta(x) = \mathrm{dist}(x, \partial\Omega), \qquad x \in \Omega,$$

we put, for all $1 \leq p \leq \infty$,

$$L^p_\delta = L^p_\delta(\Omega) := L^p(\Omega; \delta(x)\,dx).$$

For $1 \leq p < \infty$, L^p_δ is endowed with the norm

$$\|u\|_{p,\delta} = \left(\int_\Omega |u(x)|^p\,\delta(x)\,dx\right)^{1/p}.$$

Remark 1.1. Let us note that $L^\infty_\delta(\Omega) = L^\infty(\Omega)$, with same norm. Indeed, $L^\infty_\delta(\Omega)$ consists, by definition, of those measurable functions that are essentially bounded with respect to the measure $\delta(x)\,dx$. $\quad\square$

For any $1 \leq p < \infty$, the uniformly local Lebesgue space (cf. [297], [253]) L^p_{ul} is defined by

$$L^p_{ul} = L^p_{ul}(\mathbb{R}^n) = \left\{\phi \in L^p_{loc}(\mathbb{R}^n) : \|\phi\|_{p,ul} < \infty\right\},$$

where

$$\|\phi\|_{p,ul} := \sup_{a\in\mathbb{R}^n} \left(\int_{|y-a|<1} |\phi(y)|^p\,dy\right)^{1/p}.$$

These are Banach spaces with the norm $\|.\|_{p,ul}$. Also, for $p = \infty$, we define $L^\infty_{ul} := L^\infty = L^\infty(\mathbb{R}^n)$. We note that $L^r_{ul} \hookrightarrow L^p_{ul}$ whenever $1 \leq p \leq r \leq \infty$.

In what follows X denotes a Banach space.

Let M be a metric space. Then $B(M, X)$, $BC(M, X)$, $BUC(M, X)$ denote the Banach spaces of bounded, bounded and continuous, bounded and uniformly continuous functions $u : M \to X$, respectively, all endowed with the sup-norm

$$\|u\|_\infty = \|u\|_{\infty,M} := \sup_{t\in M} \|u(t)\|_X.$$

We denote by $C(M, X)$ the space of continuous functions endowed with the topology of locally uniform convergence. If M is locally compact, then we denote by $C_0(M, X)$ the space of functions $u \in BUC(M, X)$ with the following property: Given $\varepsilon > 0$, there exists a compact set $K \subset M$ such that $\|u(t)\|_X < \varepsilon$ for all $t \in M \setminus K$. We also set $B(M) := B(M, \mathbb{R})$, $BC(M) := BC(M, \mathbb{R})$, etc.

Let $M \subset \mathbb{R}^n$. A function $u : M \to X$ is said to be locally Hölder continuous if, for each point $t \in M$, there exist $\alpha \in (0, 1)$, $C > 0$ and a neighborhood V of t, such that

$$\lfloor u \rfloor_{\alpha, M\cap V} := \sup_{x,y\in M\cap V,\ x\neq y} \frac{\|u(x) - u(y)\|_X}{|x - y|^\alpha} < \infty. \tag{1.1}$$

If α in (1.1) can be chosen independent of $t \in M$, then u is said to be locally α-Hölder continuous. The space of such functions is denoted by $C^\alpha(M, X)$ (or $C^\alpha(M)$ if $X = \mathbb{R}$) and endowed with the family of seminorms $\|\cdot\|_{\infty,K} + \lfloor\cdot\rfloor_{\alpha,K}$, where K runs over all compact subsets of M. By $UC^\alpha(M, X)$, $\alpha \in (0, 1)$, we denote the set of functions $u : M \to X$ such that

$$\lfloor u \rfloor_\alpha := \lfloor u \rfloor_{\alpha, M} < \infty.$$

The norm in the Banach space $BUC^\alpha(M, X) = B(M, X) \cap UC^\alpha(M, X)$ is the sum of the sup-norm and the seminorm $\lfloor \cdot \rfloor_\alpha$. Note that if M is compact, then any locally Hölder continuous function $u : M \to X$ belongs to $BUC^\alpha(M, X)$ for some α and $C^\alpha(M, X) = BUC^\alpha(M, X)$.

If Ω is an arbitrary domain in $\mathbb{R}^n$, then $BC^1(\overline{\Omega})$ denotes the space of functions $u \in BC(\overline{\Omega})$ whose first derivatives in Ω are bounded, continuous and can be continuously extended to $\overline{\Omega}$. The norm of a function u in this space is defined as the sum of sup-norms of u and its first-order derivatives. The spaces $BC^k(\overline{\Omega})$ and $BUC^k(\Omega)$, $k \geq 1$ integer, are defined in an obvious way. If no confusion is likely, we shall denote their norms by $\| \cdot \|_{BC^k}$. The spaces $C^{k+\alpha}(\Omega)$, $UC^{k+\alpha}(\Omega)$, $BUC^{k+\alpha}(\Omega)$, where $k \geq 1$ is an integer and $\alpha \in (0, 1)$ are defined similarly.

Let Ω be a bounded domain in $\mathbb{R}^n$. Then $\overline{\Omega}$ is compact, hence any function in $C(\overline{\Omega})$ is bounded and uniformly continuous. On the other hand, the functions in $BUC(\Omega)$ can be uniquely extended to functions in $C(\overline{\Omega})$. Identifying the function $u \in BUC(\Omega)$ with its extension and endowing the space $C(\overline{\Omega})$ with the sup-norm, we can write $BUC(\Omega) = C(\overline{\Omega})$. Similarly, $BUC^\alpha(\Omega) = C^\alpha(\overline{\Omega})$.

If $Q \subset \mathbb{R}^n \times \mathbb{R}$ is a domain in space and time, then $C^{2,1}(Q)$ is the space of functions which are twice continuously differentiable in the spatial variable x and once in the time variable t. The space $BC^{2,1}(\overline{Q})$ has obvious meaning. If $u \in L^p(Q)$, then u_t, $D_x u$ and $D_x^2 u$ denote the time derivative and first and second spatial derivatives of u in the sense of distributions. Alternatively, we shall also use the notation $\nabla u, D^2 u$ instead of $D_x u, D_x^2 u$. We denote by $W^{2,1;p}(Q)$ the space of functions $u \in L^p(Q)$ satisfying $u_t, D_x u, D_x^2 u \in L^p(Q)$, endowed with the norm

$$\|u\|_{2,1;p} = \|u\|_{2,1;p;Q} := \|u\|_{p;Q} + \|D_x u\|_{p;Q} + \|D_x^2 u\|_{p;Q} + \|u_t\|_{p;Q}.$$

Let $Q = Q_T = \Omega \times (0, T)$ where Ω is an arbitrary domain in $\mathbb{R}^n$ and $T > 0$. Given $\alpha \in (0, 1]$ set

$$[f]_{\alpha;Q} = \sup\left\{ \frac{|f(x,t) - f(y,s)|}{|x - y|^\alpha + |t - s|^{\alpha/2}} : x, y \in \Omega, \ t, s \in (0, T), \ (x,t) \neq (y,s) \right\}.$$

Let k be a nonnegative integer, $\alpha \in (0, 1)$ and $a = k + \alpha$. Then we put

$$|f|_{a;Q} = \sum_{|\beta| + 2j \leq k} \sup_Q |D_x^\beta D_t^j f| + \sum_{|\beta| + 2j = k} [D_x^\beta D_t^j f]_{\alpha;Q}$$

and $BUC^{a, a/2}(Q) := \{f : |f|_{a;Q} < \infty\}$. The spaces $UC^{a, a/2}(Q)$ and $C^{a, a/2}(Q)$ are defined analogously as in the case of functions defined in R^n. Note that if $p > n+2$,

$a < 2 - (n+2)/p$ and Ω is smooth enough (for example, if Ω satisfies a uniform interior cone condition), then

$$W^{2,1;p}(Q) \hookrightarrow BUC^{a,a/2}(Q); \tag{1.2}$$

see [320, Lemmas II.3.3, II.3.4], [399, Theorem 6.9] and the references therein for this statement and more general embedding and trace theorems for anisotropic spaces. Embedding (1.2) can also be derived by using the interpolation embedding in Proposition 51.3 and embeddings for isotropic spaces.

Eigenvalues and eigenfunctions

If Ω is bounded, then we denote by $\lambda_1, \lambda_2, \ldots$ the eigenvalues of $-\Delta$ in $W_0^{1,2}(\Omega)$ and by $\varphi_1, \varphi_2, \ldots$ the corresponding eigenfunctions. Recall that $\lambda_1 < \lambda_2 \le \lambda_3 \le \cdots$, $\lambda_k \to \infty$ as $k \to \infty$, that

$$\frac{1}{\lambda_1} = \sup\left\{ \int_\Omega u^2 \, dx : u \in W_0^{1,2}(\Omega), \ \int_\Omega |\nabla u|^2 \, dx = 1 \right\}, \tag{1.3}$$

and that we can choose $\varphi_1 > 0$. Unless explicitly stated otherwise, we shall assume that φ_1 is normalized by

$$\int_\Omega \varphi_1 \, dx = 1.$$

We shall often use the fact that if Ω is of class C^2, then there exist constants $c_1, c_2 > 0$ such that

$$c_1 \delta(x) \le \varphi_1(x) \le c_2 \delta(x), \qquad x \in \Omega \tag{1.4}$$

(this is a consequence of $u \in C^1(\overline{\Omega})$ and of Hopf's lemma; cf. Proposition 52.1(iii)).

Further frequent notation

We denote by $G(x, y, t) = G_\Omega(x, y, t)$ the Dirichlet heat kernel; $G_t(x) = G(x, t)$ is the Gaussian heat kernel in $\mathbb{R}^n$. The (elliptic) Dirichlet Green kernel is denoted by $K(x, y) = K_\Omega(x, y)$. We implicitly mean by e^{-tA} the Dirichlet heat semigroup in Ω.

The Dirac distribution at point y will be denoted by δ_y.

We shall use the symbols C, C_1, etc. to denote various positive constants. The dependence of these constants will be made precise whenever necessary.

Definitions of various critical exponents ($p_F, p_{BT}, p_{sg}, p_S, p_{JL}, p_L, 2_*, 2^*, q_c$) and other symbols can be found via the List of Symbols.

Chapter I

Model Elliptic Problems

2. Introduction

In Chapter I, we study the problem

$$\left.\begin{array}{ll} -\Delta u = f(x, u), & x \in \Omega, \\ u = 0, & x \in \partial\Omega, \end{array}\right\} \tag{2.1}$$

where $f : \Omega \times \mathbb{R} \to \mathbb{R}$ is a Carathéodory function (i.e. $f(\cdot, u)$ is measurable for any $u \in \mathbb{R}$ and $f(x, \cdot)$ is continuous for a.e. $x \in \Omega$). Of course, the boundary condition in (2.1) is not present if $\Omega = \mathbb{R}^n$. We will be mainly interested in the model case

$$f(x, u) = |u|^{p-1}u + \lambda u, \qquad \text{where } p > 1 \text{ and } \lambda \in \mathbb{R}. \tag{2.2}$$

Denote by p_S the critical Sobolev exponent,

$$p_S := \begin{cases} \infty & \text{if } n \leq 2, \\ (n+2)/(n-2) & \text{if } n > 2. \end{cases}$$

We shall refer to the cases $p < p_S$, $p = p_S$ or $p > p_S$ as to (Sobolev) subcritical, critical or supercritical, respectively.

3. Classical and weak solutions

Let u be a solution of (2.1) and $\tilde{f}(x) := f(x, u(x))$. Then u solves the linear problem

$$\left.\begin{array}{ll} -\Delta u = \tilde{f} & \text{in } \Omega, \\ u = 0 & \text{on } \partial\Omega. \end{array}\right\} \tag{3.1}$$

In what follows we define several types of solutions of the linear problem (3.1) (and, consequently, of (2.1)).

Definition 3.1. (i) We call u a **classical solution** of (3.1) if $\tilde{f} \in C(\Omega)$, $u \in C^2(\Omega) \cap C(\overline{\Omega})$ and u satisfies the equation and the boundary condition in (3.1) pointwise.

(ii) We call $u \in W_0^{1,2}(\Omega)$ a **variational solution** of (3.1) if $\tilde{f} \in \left(W_0^{1,2}(\Omega)\right)'$ and

$$\int_\Omega \nabla u \cdot \nabla \varphi \, dx = \int_\Omega \tilde{f} \varphi \, dx \quad \text{for all } \varphi \in W_0^{1,2}(\Omega). \tag{3.2}$$

(iii) Let Ω be bounded, $u \in L^1(\Omega)$. Set

$$\delta(x) := \text{dist}\,(x, \partial\Omega) \qquad \text{and} \qquad L_\delta^1(\Omega) := L^1(\Omega, \delta(x)dx).$$

We call u an L^1-**solution** of (3.1) if $\tilde{f} \in L^1(\Omega)$ and

$$\int_\Omega u(-\Delta\varphi) \, dx = \int_\Omega \tilde{f} \varphi \, dx \quad \text{for all } \varphi \in C^2(\overline{\Omega}), \ \varphi = 0 \text{ on } \partial\Omega. \tag{3.3}$$

More generally, we call u an L_δ^1-**solution**, or a **very weak solution**, of (3.1) if $\tilde{f} \in L_\delta^1(\Omega)$ and (3.3) is satisfied. Note that the definition makes sense since $|\varphi| \leq C\delta$ hence $\tilde{f}\varphi \in L^1(\Omega)$. Existence-uniqueness and properties of L_δ^1 solutions of the linear problem (3.1) are studied in Appendix C.

(iv) If $\Omega = \mathbb{R}^n$, then $u \in L_{loc}^1(\Omega)$ is called a **distributional solution** of (3.1) if the integral identity in (3.3) is true for all $\varphi \in \mathcal{D}(\mathbb{R}^n)$. $\square$

Remarks 3.2. (i) If we assume that $\tilde{f}$ is a bounded Radon measure in Ω (instead of $\tilde{f} \in L^1(\Omega)$), then the definition of an L^1-solution still makes sense and we refer to [20] and the references therein for properties of such solutions.

(ii) If $\tilde{f} \in L^\infty(\Omega)$, then any classical solution of (3.1) satisfies $u \in W^{2,q}(K)$ for any $K \subset\subset \overline{\Omega}$ and any $q < \infty$. This is a consequence of Remark 47.4(iii). If we further assume that $\tilde{f}$ is locally Hölder continuous in $\overline{\Omega}$, then $u \in C^2(\overline{\Omega})$.

(iii) Assume Ω bounded. If $\tilde{f} \in C(\overline{\Omega})$, for example, then any classical solution of (3.1) is also a variational solution (this follows from Remark (ii) and integration by parts). If $\tilde{f} \in L^2(\Omega)$, then any variational solution is an L^1-solution. Some other relations between various types of solutions defined above will be mentioned below (see also Lemma 47.7 in Appendix A). $\square$

In the following sections we shall often use variational methods in order to prove the solvability of (2.1). Therefore, we derive now a sufficient condition on f which guarantees that any variational solution of (2.1) is classical.

If $n \geq 3$ we set $2^* := p_S + 1 = 2n/(n-2)$, $2_* := (2^*)' = 2n/(n+2)$. Assume that the Carathéodory function f satisfies the following growth assumption

$$|f(x,u)| \leq \alpha(x) + C_f(|u| + |u|^p), \quad \alpha \in L^{(p+1)'}(\Omega) + L^2(\Omega), \ C_f > 0, \ p \leq p_S. \tag{3.4}$$

This growth condition can be significantly weakened if $n \leq 2$ but (3.4) will be sufficient for our purposes; cf. (2.2). Denote

$$F(x,u) := \int_0^u f(x,s) \, ds$$

and

$$E(u) := \frac{1}{2}\int_\Omega |\nabla u(x)|^2\, dx - \int_\Omega F\big(x, u(x)\big)\, dx. \qquad (3.5)$$

Since $p \le p_S$ we have $W^{1,2}(\Omega) \hookrightarrow L^{p+1}(\Omega)$ and the embedding is compact provided $p < p_S$ and Ω is bounded. In addition, the energy functional E is C^1 (continuously Fréchet differentiable) in $W^{1,2}(\Omega)$ and

$$E'(u)\varphi = \int_\Omega \nabla u \cdot \nabla \varphi\, dx - \int_\Omega f(\cdot, u)\varphi\, dx$$

for all $u, \varphi \in W^{1,2}(\Omega)$. In particular, each critical point of E in $W_0^{1,2}(\Omega)$ is a variational solution of (2.1).

The following proposition is essentially due to [96]; our proof closely follows the proof of [505, Lemma B.3].

Proposition 3.3. *Assume (3.4). If $n \ge 3$ assume also $\alpha \in L^{n/2}(\Omega)$. Let u be a variational solution of (2.1). Then $u \in L^q(\Omega)$ for all $q \in [2, \infty)$.*

Proof. Since the assertion is obviously true if $n \le 2$ due to $W^{1,2}(\Omega) \hookrightarrow L^q(\Omega)$, we may assume $n \ge 3$.

Denote $\tilde{f}(x) := f\big(x, u(x)\big)$. Then

$$|\tilde{f}| \le \alpha + C_f(|u| + |u|^p) \le a + b + 2C_f(|u| + |u|^{p_S}),$$

where $a := \alpha\chi_{|u|>1} \in L^{n/2}(\Omega)$, $b := \alpha\chi_{|u|\le 1}$ and α can be written in the form $\alpha = \alpha_1 + \alpha_2$ with $\alpha_1 \in L^{(p+1)'}(\Omega)$, $\alpha_2 \in L^2(\Omega)$.

Choose $s \ge 0$ such that $u \in L^{2(s+1)}(\Omega)$. We shall prove that $u \in L^{2^*(s+1)}(\Omega)$ so that an obvious bootstrap argument proves the assertion.

Choose $L > 0$ and set

$$\psi := \min\big(|u|^s, L\big), \qquad \varphi := u\psi^2, \qquad \Omega_L := \{x \in \Omega : |u|^s \le L\}.$$

In what follows we denote by C, C_1, C_2 various positive constants which may vary from step to step and which may depend on u, s, α, C_f but which are independent of L. We have

$$\nabla(u\psi) = (1 + s\chi_{\Omega_L})(\nabla u)\psi,$$
$$\nabla\varphi = (1 + 2s\chi_{\Omega_L})(\nabla u)\psi^2,$$

and $\varphi \in W_0^{1,2}(\Omega)$. Therefore, we obtain

$$\int_\Omega |\nabla u|^2\psi^2\, dx \le \int_\Omega \nabla u \cdot \nabla\varphi\, dx = \int_\Omega \tilde{f}\varphi\, dx = \int_\Omega \tilde{f}u\psi^2\, dx$$

$$\le C\int_\Omega \big[(a+b)|u|\psi^2 + u^2\psi^2 + |u|^{2^*}\psi^2\big]\, dx$$

$$\le C\int_\Omega \big[au^2\psi^2 + b|u| + |u|^{2s+2} + |u|^{2^*}\psi^2\big]\, dx$$

$$\le C\Big(1 + \int_\Omega (a + |u|^{2^*-2})u^2\psi^2\, dx\Big),$$

where we have used

$$\int_\Omega b|u|\,dx \le \int_\Omega \alpha|u|\,dx \le \int_\Omega (|\alpha_1| + |\alpha_2|)|u|\,dx$$
$$\le \|\alpha_1\|_{(p+1)'}\|u\|_{p+1} + \|\alpha_2\|_2\|u\|_2 = C.$$

Consequently, denoting $v := a + |u|^{2^*-2} \in L^{n/2}(\Omega)$, we obtain

$$\int_\Omega |\nabla(u\psi)|^2\,dx \le C\int_\Omega |\nabla u|^2\psi^2\,dx \le C\Big(1 + \int_\Omega vu^2\psi^2\,dx\Big)$$
$$\le C\Big(1 + K\int_{|v|\le K} u^2\psi^2\,dx + \int_{|v|>K} v(u\psi)^2\,dx\Big)$$
$$\le C\Big(1 + K\int_\Omega |u|^{2s+2}\,dx + \Big(\int_{|v|>K} v^{n/2}\,dx\Big)^{2/n}\Big(\int_\Omega |u\psi|^{2^*}\,dx\Big)^{(n-2)/n}\Big)$$
$$\le C_1(1+K) + C_2\varepsilon_K\int_\Omega |\nabla(u\psi)|^2\,dx,$$

where $\varepsilon_K := \big(\int_{|v|>K} v^{n/2}\,dx\big)^{2/n} \to 0$ as $K \to +\infty$. Choosing K such that $C_2\varepsilon_K < 1/2$ we arrive at

$$\int_{\Omega_L} |\nabla(|u|^{s+1})|^2\,dx = \int_{\Omega_L} |\nabla(u\psi)|^2\,dx \le 2C_1(1+K).$$

Letting $L \to +\infty$ we get $|u|^{s+1} \in W^{1,2}(\Omega)$, hence $u \in L^{2^*(s+1)}(\Omega)$. □

Corollary 3.4. *If f has the form (2.2) with $p \le p_S$, then any variational solution u of (2.1) is also a classical solution. Moreover, $u \in C^2(\overline{\Omega})$.*

Proof. The assertion is a consequence of standard regularity results for linear elliptic equations. More precisely, for any $2 \le q < \infty$, since $\tilde{f} := f(u) \in L^q(\Omega)$, Theorem 47.3(i) implies the existence of $\tilde{u} \in W^{2,q} \cap W_0^{1,q}(\Omega)$ such that $-\Delta\tilde{u} = \tilde{f}$. Since $u, \tilde{u} \in H_0^1(\Omega)$, the maximum principle in Proposition 52.3(i) yields $u = \tilde{u}$. Due to the embedding $W^{2,q}(\Omega) \subset C^1(\overline{\Omega})$ for $q > n$, we deduce that $\tilde{f} \in C^1(\overline{\Omega})$. Applying now Theorem 47.3(ii), and Proposition 52.3(i) again, we deduce that $u \in C^2(\overline{\Omega})$. □

As for L^1-solutions, we have the following regularity result (we shall see in Remarks 3.6 below that the growth conditions in Propositions 3.3 and 3.5 are optimal).

Proposition 3.5. *Assume Ω bounded. Let the Carathéodory function f satisfy the growth assumption*

$$|f(x,u)| \le C(1 + |u|^p), \quad p < p_{sg}, \tag{3.6}$$

where p_{sg} is defined in (3.8). Let u be an L^1-solution of (2.1). Then $u \in C_0 \cap W^{2,q}(\Omega)$ for all finite q.

Proof. It is based on a simple bootstrap argument. Fix $\rho \in (1, n/(n-2)p)$ and put $\tilde{f}(x) = f(x, u(x))$. Assume that there holds

$$\tilde{f} \in L^{\rho^i}(\Omega) \tag{3.7}$$

for some $i \geq 0$ (this is true for $i = 0$ by assumption). Since

$$\frac{1}{\rho^i} - \frac{1}{p\rho^{i+1}} = \frac{1}{\rho^i}\left(1 - \frac{1}{p\rho}\right) < \frac{2}{n},$$

by using Proposition 47.5(i), we obtain $u \in L^{p\rho^{i+1}}(\Omega)$, hence $\tilde{f} \in L^{\rho^{i+1}}(\Omega)$ due to (3.6). By induction, it follows that (3.7) is true for all integers i. In particular $\tilde{f} \in L^k(\Omega)$ for some $k > n/2$ and we may apply Proposition 47.5(i) once more to deduce that $u \in L^\infty(\Omega)$. The conclusion then follows similarly as in the proof of Corollary 3.4 (using the uniqueness part of Theorem 49.1 instead of Proposition 52.3). $\square$

Remarks 3.6. (i) **Singular solution**. Define the exponent

$$p_{sg} := \begin{cases} \infty & \text{if } n \leq 2, \\ n/(n-2) & \text{if } n > 2. \end{cases} \tag{3.8}$$

For $p > p_{sg}$ (hence $n \geq 3$), we let

$$U_*(r) := c_p r^{-2/(p-1)}, \quad r > 0, \qquad \text{where } c_p^{p-1} := \frac{2}{(p-1)^2}\big((n-2)p - n\big). \tag{3.9}$$

One can easily check that $u_*(x) := U_*(|x|)$ is a positive, radial distributional solution of the equation $-\Delta u = u^p$ in $\mathbb{R}^n$. This singular solution (hence the notation p_{sg}) plays an important role in the study of the parabolic problem (0.2) with $f(u) = |u|^{p-1}u$ (see for example Theorems 20.5, 22.4 and 23.10).

On the other hand, if we set $u(x) := u_*(x) - c_p$ for $0 < |x| \leq 1$, $\Omega := B_1(0) = \{x \in \mathbb{R}^n : |x| < 1\}$, then it is easy to verify that u is an L^1-solution of (2.1) with $f(x, u) = (u + c_p)^p$. Moreover, u is a variational solution of this problem if $p > p_S$. Hence the condition $p \leq p_S$ in Proposition 3.3 is necessary.

(ii) Let $n \geq 3$ and let Ω be bounded, $f \in C^1$, $|f(x, u)| \leq C(1 + |u|^p)$. The example in (i) shows that an L^1-solution need not be classical if $p > p_{sg}$. In fact, it was proved in [44], [394] that the problem

$$\left. \begin{aligned} -\Delta u &= |u|^{p-1}u && \text{in } \Omega, \\ u &= 0 && \text{on } \partial\Omega, \end{aligned} \right\} \tag{3.10}$$

has a positive unbounded radial L^1-solution $u \in C^2(\overline{\Omega} \setminus \{0\})$ provided $p \in [p_{sg}, p_S)$ and $\Omega = B_1(0)$. See also [404] and the references therein for related nonradial results.

(iii) For the case of L^1_δ-solutions, we shall see in Section 11 that the critical exponent is different, namely $(n+1)/(n-1)$. $\quad\square$

Remark 3.7. Classical vs. very weak solutions for the nonlinear eigenvalue problem. Another type of relations between different notions of solutions appears when one considers the nonlinear eigenvalue problem

$$\left.\begin{array}{ll} -\Delta u = \lambda f(u), & x \in \Omega, \\ u = 0, & x \in \partial\Omega. \end{array}\right\} \tag{3.11}$$

Here we assume that $f : [0, \infty) \to (0, \infty)$ is a C^1 nondecreasing, convex function, and $\lambda > 0$. Namely, it was proved in [94] (see also [233] for earlier related results) that if there exists a *very weak* solution of (3.11) for some $\lambda_0 > 0$, then there exists a *classical* solution for all $\lambda \in (0, \lambda_0)$. The proof is based on a perturbation argument relying on a variant of Lemma 27.4 below. As a consequence of this and of results from [305], [142], assuming in addition that $\lim_{u\to\infty} f(u)/u = \infty$, there exists $\lambda^* \in (0, \infty)$ such that:

(i) for $0 < \lambda < \lambda^*$, problem (3.11) has a (unique minimal) classical solution u_λ, and the map $\lambda \mapsto u_\lambda$ is increasing;

(ii) for $\lambda = \lambda^*$, problem (3.11) has a very weak solution defined by $u_{\lambda^*} = \lim_{\lambda \uparrow \lambda^*} u_\lambda$;

(iii) for $\lambda > \lambda^*$, problem (3.11) has no very weak solution.

On the other hand, the solution u_{λ^*} may be classical or singular, depending on the nonlinearity. For instance, in the case $f(u) = (u+1)^p$ with $\Omega = B_R$, (3.11) has a classical solution for $\lambda = \lambda^*$ if and only if $p < p_{JL}$, where p_{JL} is defined in (9.3); in the case $f(u) = e^u$, the condition is replaced with $n \leq 9$ (see [293], [369]). Illustrations of these facts appear on the bifurcation diagram in Remark 6.10(ii) (see Figure 3). $\quad\square$

4. Isolated singularities

In this section we study the question of isolated singularities of positive classical solutions to the equation $-\Delta u = u^p$. The following result classifies the possible singular behaviors for subcritical or critical p.

Theorem 4.1. *Let $n \geq 3$ and $1 < p \leq p_S$. Assume that u is a positive classical solution of*

$$-\Delta u = u^p \quad in \ B_1 \setminus \{0\} \tag{4.1}$$

and that u is unbounded at 0. Then there exist constants $C_2 \geq C_1 > 0$ such that

$$C_1 \psi(x) \leq u(x) \leq C_2 \psi(x), \quad 0 < |x| < 1/2,$$

where

$$\psi(x) = \begin{cases} |x|^{2-n} & \text{if } 1 < p < p_{sg}, \\ |x|^{2-n}(-\log|x|)^{(2-n)/2} & \text{if } p = p_{sg}, \\ |x|^{-2/(p-1)} & \text{if } p_{sg} < p \leq p_S. \end{cases}$$

Moreover, if $p < p_S$, then we have $C_2 \leq \tilde{C}_2$ with $\tilde{C}_2 = \tilde{C}_2(n,p) > 0$.

Furthermore, for all $p > 1$, we have the following result, which explains in what sense the equation can be extended to the whole unit ball.

Theorem 4.2. *Let $p > 1$ and $n \geq 3$. Assume that u is a positive classical solution of*

$$-\Delta u = u^p \quad \text{in } B_1 \setminus \{0\}.$$

(i) *Then $u^p \in L^1_{loc}(B_1)$ and there exists $a \geq 0$ such that u is a solution of*

$$-\Delta u = u^p + a\delta_0 \quad \text{in } \mathcal{D}'(B_1),$$

where δ_0 denotes the Dirac delta distribution. Moreover, we have $a \leq \tilde{a}$ with $\tilde{a} = \tilde{a}(n,p) > 0$.

(ii) *If $p < p_{sg}$ and $a = 0$, then the singularity is removable, i.e. u is bounded near $x = 0$.*

(iii) *If $p \geq p_{sg}$, then $a = 0$.*

Remarks 4.3. (i) Theorem 4.1 follows from [340], [44], [240] (see also [83]), and [108], for the cases $p < p_{sg}$, $p = p_{sg}$, $p_{sg} < p < p_S$ and $p = p_S$ respectively. Theorem 4.2 follows from [97] and [340]. See also the book [523] for further results and references.

(ii) Under the assumptions of Theorem 4.1 with $p_{sg} < p < p_S$, it can be shown more precisely that $|x|^{2/(p-1)}u(x) \to c_p$, as $x \to 0$, where c_p is given by (3.9) (cf. [240], [83]). If $1 < p < p_{sg}$, then actually $|x|^{n-2}u(x) \to C > 0$, as $x \to 0$ (see the proof below). Examples in [340] show that singular solutions do exist for $1 < p < p_{sg}$ and that the constant C may depend on the solution u.

(iii) If $p > p_S$, then the upper estimate $u(x) \leq C|x|^{-2/(p-1)}$ is still true in the radial case (cf. [213], [397]). In fact, as a consequence of $-(r^{n-1}u_r)_r = r^{n-1}u^p > 0$, for $r > 0$ small, we have either $u_r > 0$, hence u bounded, or $u_r \leq 0$. In this second case, by integration, we get $-u_r \geq r^{1-n} \int_0^r s^{n-1}u^p(s)\,ds \geq (r/n)u^p$ for $r > 0$ small, hence $(u^{1-p})_r \geq Cr$, and the upper estimate follows by a further integration. The estimate is unknown in the nonradial case for $p > p_S$, but related integral estimates of solutions can be found in e.g. [238] and [89].

(iv) A result similar to Theorem 4.1 is true for $n = 2$, with $\psi(x)$ given by the fundamental solution $\log|x|$ instead of $|x|^{2-n}$. These results are related to the fact that the (H^1-) capacity of a point is 0 when $n \geq 2$. When the origin is replaced by a closed subset of 0 capacity, related results can be found in [170]. On the other hand, the upper estimate $u(x) \leq C|x|^{-2/(p-1)}$, from the case $p_{sg} < p < p_S$, can be generalized to sets other than a single point (see Theorem 8.7 in Section 8). $\square$

We shall first prove Theorem 4.2. Theorem 4.1 in the case $1 < p < p_{sg}$ will then follow as a consequence of Theorem 4.2 and of a bootstrap argument. In the case $p_{sg} < p < p_S$, the upper estimate will be a consequence of the more general result Theorem 8.7 in Section 8. For the cases $p = p_{sg}$, $p = p_S$, and for the lower estimate when $p_{sg} < p < p_S$, see the above mentioned references.

In view of the proofs, we introduce the following notation. We denote by $\Gamma(x) = c_n|x|^{2-n}$ the fundamental solution of the Laplacian (Newton potential), i.e. $-\Delta\Gamma = \delta_0$ in $\mathcal{D}'(\mathbb{R}^n)$. We let $\omega = \{x \in \mathbb{R}^n : |x| < 1/2\}$ and fix $\chi \in \mathcal{D}(B_1)$ such that $\chi = 1$ on ω and $0 \leq \chi \leq 1$. For each positive integer j, denote $\chi_j(x) = \chi(jx)$. By a straightforward calculation using $n \geq 3$, we see that $\chi_j \to 0$ in $H^1(B_1)$ as $j \to \infty$. For any $\varphi \in \mathcal{D}(B_1)$, we put $\varphi_j := (1 - \chi_j)\varphi$. Observe that $\varphi_j \to \varphi$ in $H^1(B_1)$.

We need the following lemma.

Lemma 4.4. *Let $n \geq 3$. Assume that $u \in C^2(B_1 \setminus \{0\})$ satisfies $u \geq 0$ and*

$$-\Delta u \geq 0 \quad in\ B_1 \setminus \{0\}.$$

Then $u \in L^1_{loc}(B_1)$ and

$$-\Delta u \geq 0 \quad in\ \mathcal{D}'(B_1).$$

Proof. For each $k > 0$, we take a function $G_k \in C^2([0,\infty))$ such that $G_k(s) = s$ for $0 \leq s \leq k$, $G_k(s) = k + 1$ for s large, $G_k' \geq 0$ and $G_k'' \leq 0$. Define $u_k := G_k(u)$ and note that the sequence $\{u_k\}_k$ is monotone increasing and converges to u pointwise in $B_1 \setminus \{0\}$. The function u_k satisfies

$$-\Delta u_k = -G_k'(u)\Delta u - G_k''(u)|\nabla u|^2 \geq 0 \quad in\ B_1 \setminus \{0\}. \tag{4.2}$$

Fix $\alpha > 0$ and $\varphi \in \mathcal{D}(B_1)$. Multiplying inequality (4.2) by the test-function $\varphi_j^2(1 + u_k)^{-\alpha}$ and integrating by parts, we obtain

$$0 \leq \int_{B_1} \nabla u_k \cdot \nabla(\varphi_j^2(1 + u_k)^{-\alpha})$$

$$= -\alpha \int_{B_1} |\nabla u_k|^2 \varphi_j^2(1 + u_k)^{-1-\alpha} + 2 \int_{B_1} \nabla u_k \cdot \nabla \varphi_j (1 + u_k)^{-\alpha}\varphi_j.$$

It follows that

$$\alpha \int_{B_1} |\nabla u_k|^2 \varphi_j^2(1 + u_k)^{-1-\alpha}$$

$$\leq \frac{\alpha}{2} \int_{B_1} |\nabla u_k|^2 \varphi_j^2(1 + u_k)^{-1-\alpha} + C(\alpha) \int_{B_1} |\nabla \varphi_j|^2(1 + u_k)^{1-\alpha},$$

hence

$$\int_{B_1} |\nabla u_k|^2 \varphi_j^2 (1+u_k)^{-1-\alpha} \le C(\alpha) \int_{B_1} |\nabla \varphi_j|^2 (1+u_k)^{1-\alpha}.$$

Since $|\nabla \varphi_j|^2 \to |\nabla \varphi|^2$ in $L^1(B_1)$ and $(1+u_k)^{1-\alpha} \in L^\infty(B_1)$, we may pass to the limit $j \to \infty$ (using Fatou's lemma on the LHS) and we obtain

$$\int_{B_1} |\nabla u_k|^2 \varphi^2 (1+u_k)^{-1-\alpha} \le C(\alpha) \int_{B_1} |\nabla \varphi|^2 (1+u_k)^{1-\alpha}.$$

First taking $\alpha = 1$ and using $1 + u_k \le k+2$, we deduce that $u_k \in H^1(\omega)$, hence $u_k \in H^1_{loc}(B_1)$.

Next take $\alpha = 2/n$. Consider φ such that $\varphi = 1$ for $|x| \le 1/4$ and with support in ω. Applying the Sobolev and Hölder inequalities, we get, for any $\rho \in (0, 1/2)$,

$$\left(\int_{|x|<1/4} (1+u_k) \right)^{\frac{n-2}{n}} \le C \int_{|x|<1/4} \left| \nabla \left[(1+u_k)^{\frac{n-2}{2n}} \right] \right|^2 + C \int_{|x|<1/4} (1+u_k)^{\frac{n-2}{n}}$$

$$\le C \int_\omega (1+u_k)^{\frac{n-2}{n}}$$

$$\le C \int_{\rho<|x|<1/2} (1+u_k)^{\frac{n-2}{n}} + C\rho^2 \left(\int_{|x|\le\rho} (1+u_k) \right)^{\frac{n-2}{n}}.$$

Since u is bounded on $\{\rho < |x| < 1/2\}$ and $u_k \le u$, by taking $\rho \in (0, 1/4)$ small enough, we deduce that $\int_{|x|<1/4} u_k \le C$ independent of k. Consequently $u \in L^1(\omega)$, hence $u \in L^1_{loc}(B_1)$, and $u_k \to u$ in $L^1_{loc}(B_1)$.

Now assuming $\varphi \ge 0$, we multiply inequality (4.2) by φ_j and integrate by parts. We obtain

$$\int_{B_1} \nabla u_k \cdot \nabla \varphi_j = \int_{B_1} (-\Delta u_k)\varphi_j \ge 0.$$

Since $u_k \in H^1_{loc}(B_1)$, we may pass to the limit $j \to \infty$ to get $\int_{B_1} \nabla u_k \cdot \nabla \varphi \ge 0$, hence $\int_{B_1} (-\Delta\varphi)u_k \ge 0$. Since $u_k \to u$ in $L^1_{loc}(B_1)$, we conclude that $\int_{B_1} (-\Delta\varphi)u \ge 0$ and the proof of the lemma is complete. $\square$

Proof of Theorem 4.2. (i) By Lemma 4.4, we know that $u \in L^1_{loc}(B_1)$ and that $-\Delta u \ge 0$ in $\mathcal{D}'(B_1)$. It follows that Δu is a Radon measure (in other words, a 0-order distribution) on ω. Indeed, for each $\varphi \in \mathcal{D}(B_1)$ with $\mathrm{supp}(\varphi) \subset \omega$, using $\|\varphi\|_\infty \chi \pm \varphi \ge 0$, we obtain

$$\langle -\Delta u, \|\varphi\|_\infty \chi \pm \varphi \rangle \ge 0$$

hence

$$|\langle -\Delta u, \varphi \rangle| \le |\langle -\Delta u, \chi \rangle| \, \|\varphi\|_\infty =: C\|\varphi\|_\infty. \tag{4.3}$$

We next claim that $u^p \in L^1_{loc}(B_1)$. To this end, we assume $\varphi \geq 0$, we multiply (4.1) by φ_j and we integrate by parts. We obtain

$$\int_{B_1} u^p \varphi_j = \langle -\Delta u, \varphi_j \rangle \leq C\|\varphi_j\|_\infty \leq C\|\varphi\|_\infty,$$

due to (4.3), and the claim follows from Fatou's lemma.

Now a classical argument in distribution theory allows us to conclude: Denote $T = \Delta u + u^p \in \mathcal{D}'(B_1)$ and let $\varphi \in \mathcal{D}(B_1)$. Since $T = 0$ in $\mathcal{D}'(B_1 \setminus \{0\})$ and $(1 - \chi_j)\varphi = 0$ in the neighborhood of 0, we have $\langle T, (1 - \chi_j)\varphi \rangle = 0$. Consequently,

$$\langle T, \varphi \rangle - \langle T, \chi_j \rangle \varphi(0) = \langle T, \varphi \chi_j \rangle - \langle T, \chi_j \rangle \varphi(0) = \langle T, (\varphi - \varphi(0))\chi_j \rangle. \tag{4.4}$$

But since $\|(\varphi - \varphi(0))\chi_j\|_\infty \to 0$ as $j \to \infty$, it follows that the LHS of (4.4) converges to 0 as $j \to \infty$. We first deduce that $\ell = \lim_{j \to \infty} \langle T, \chi_j \rangle$ exists (take a φ such that $\varphi(0) \neq 0$). Moreover, since $-\Delta u \geq 0$ in $\mathcal{D}'(B_1)$, we have $\ell \leq \lim_{j \to \infty} \int_\omega u^p \chi_j = 0$ by dominated convergence. Returning to (4.4), we obtain

$$\Delta u + u^p = -a\delta_0 \tag{4.5}$$

with $a = -\ell \geq 0$. Now let $\psi \in \mathcal{D}(B_1)$ satisfy $-\Delta \psi \leq \mu \psi^{1/p}$ in B_1 for some $\mu > 0$ and $\psi \geq C > 0$ for $|x| < 2/3$ (such function is given for instance by $\psi(x) = \exp[-(1 - 2|x|^2)^{-1}]$ for $|x|^2 < 1/2$ and $\psi(x) = 0$ otherwise). Testing equation (4.5) with ψ, we get

$$a\psi(0) + \int_{B_1} u^p \psi = -\int_{B_1} u\Delta \psi \leq \mu \int_{B_1} u\psi^{1/p} \leq \frac{1}{2} \int_{B_1} u^p \psi + C(p, n).$$

It follows that

$$a + \int_{\{|x| < 2/3\}} u^p < \tilde{a}(n, p). \tag{4.6}$$

In particular, assertion (i) is proved.

For further reference, we also observe that

$$u \geq a\Gamma - C \quad \text{in } \omega. \tag{4.7}$$

To show this, we first note that $v := u - a\Gamma$ satisfies $-\Delta v = u^p$ in $\mathcal{D}'(B_1)$. By Lemma 47.7, $w := \chi v$ is an L^1-solution of

$$\left. \begin{aligned} -\Delta w = g &:= u^p \chi - h && \text{in } B_1, \\ w &= 0 && \text{on } \partial B_1, \end{aligned} \right\} \tag{4.8}$$

where $h := 2\nabla u \cdot \nabla \chi + u\Delta \chi \in L^\infty(B_1)$. At this point, let us introduce the function $\Theta \in C^2(\overline{B_1})$, $\Theta \geq 0$, classical solution of the problem

$$\left. \begin{aligned} -\Delta \Theta &= 1 && \text{in } B_1, \\ \Theta &= 0 && \text{on } \partial B_1. \end{aligned} \right\}$$

(This is the so-called "torsion" function, which will be useful as a comparison or test-function later again.) Then $w + \|h\|_\infty \Theta$ is an L^1-solution of (4.8) with g replaced by $g + \|h\|_\infty \geq 0$. By the maximum principle part of Theorem 49.1, we deduce that $w + \|h\|_\infty \Theta \geq 0$, hence (4.7).

(ii) Let $1 < p < p_{sg}$ and assume that $a = 0$. We have seen that $w = \chi u$ is an L^1-solution of (4.8). Moreover, since $\chi = 1$ near $x = 0$, we may write $g = w^p + \tilde{h}$ in (4.8) for some $\tilde{h} \in L^\infty(B_1)$. It then follows from Proposition 3.5 that $w \in L^\infty(B_1)$, hence $u \in L^\infty(\omega)$.

(iii) Assume $p \geq p_{sg}$. If we had $a > 0$, then (4.7) would imply $u^p \geq C|x|^{-(n-2)p}$ as $x \to 0$ for some $C > 0$. Since $u^p \in L^1_{loc}(B_1)$ due to (i), we conclude that $a = 0$. $\square$

Proof of Theorem 4.1 for $1 < p < p_{sg}$. By Theorem 4.2, we know that

$$-\Delta u = u^p + a\delta_0 \quad \text{in } \mathcal{D}'(B_1)$$

with $a > 0$. Denote $v_0 = u$, $\alpha_1 = n - 2$, and put $v_1 := u - a\Gamma = v_0 - C_1|x|^{-\alpha_1}$. Then we have

$$-\Delta v_1 = u^p \quad \text{in } \mathcal{D}'(B_1).$$

On the other hand, an easy calculation shows that $-\Delta(|x|^{-\alpha}) = C(\alpha)|x|^{-\alpha-2}$ in $\mathcal{D}'(B_1)$ for all $\alpha \in (0, n - 2)$ and some $C(\alpha) > 0$. Set $\alpha_2 := p\alpha_1 - 2$ if $p\alpha_1 > 2$ and choose $\alpha_2 \in (0, \alpha_1)$ otherwise. Notice that $\alpha_2 \in (0, \alpha_1) = (0, n - 2)$ in both cases due to $p\alpha_1 < n$. Since $u^p \leq C(v_1)_+^p + C|x|^{-p\alpha_1} \leq C(v_1)_+^p + C|x|^{-\alpha_2-2}$, there exists $C_2 > 0$ such that $v_2 := v_1 - C_2|x|^{-\alpha_2}$ satisfies

$$-\Delta v_2 \leq C(v_1)_+^p \quad \text{in } \mathcal{D}'(B_1).$$

Since $(v_1)_+^p \leq C(v_2)_+^p + C|x|^{-p\alpha_2}$, we can iterate this procedure and we obtain functions v_i $(i = 0, 1, \dots)$ satisfying $v_{i+1} = v_i - C_{i+1}|x|^{-\alpha_{i+1}}$, with $\alpha_{i+1} \in (0, \alpha_i)$, and

$$-\Delta v_{i+1} \leq C_i'(v_i)_+^p \quad \text{in } \mathcal{D}'(B_1).$$

Moreover, due to $0 < a \leq \tilde{a}(n, p)$, the constants C_i, C_i' may be chosen to depend only on n, p, i.

To conclude, we apply a bootstrap argument similar to that in the proof of Proposition 3.5: Fix $\rho \in (1, n/(n - 2)p)$, let $\Omega_1 = \{|x| < 2/3\}$, and assume that $(v_i)_+ \in L^{p\rho^i}_{loc}(\Omega_1)$ for some $i \geq 0$ (this is true for $i = 0$ in view of (4.6)). Since $(-\Delta v_{i+1})_+ \in L^{\rho^i}_{loc}(\Omega_1)$ and

$$\frac{1}{\rho^i} - \frac{1}{p\rho^{i+1}} = \frac{1}{\rho^i}\left(1 - \frac{1}{p\rho}\right) < \frac{2}{n},$$

we may apply Proposition 47.6(ii) to deduce that $(v_{i+1})_+ \in L^{p\rho^{i+1}}_{loc}(\Omega_1)$. By iterating, we get $(v_i)_+ \in L^k_{loc}(\Omega_1)$ for some sufficiently large i and some $k > n/2$. We may then apply Proposition 47.6(ii) once more to deduce that $(v_{i+1})_+ \in L^\infty(\omega)$. This implies

$$u - a\Gamma = v_1 = v_{i+1} + \sum_{j=2}^{i+1} C_j |x|^{-\alpha_j} \le C(1 + |x|^{-\alpha_2}), \quad |x| < 1/2.$$

Moreover, starting from (4.6), it is easy to check that the constant C depends only on n, p. This along with (4.7) yields the conclusion. $\square$

5. Pohozaev's identity and nonexistence results

In this section we prove the nonexistence of nontrivial solutions of (2.1) provided f satisfies (2.2) with $p \ge p_S$, $\lambda \le 0$ and Ω is a bounded starshaped domain. The following identity [419] plays a crucial role in the proof.

Theorem 5.1. *Let u be a classical solution of (2.1) with $f = f(u)$ being locally Lipschitz and Ω bounded. Then*

$$\frac{n-2}{2} \int_\Omega |\nabla u|^2 \, dx - n \int_\Omega F(u) \, dx + \frac{1}{2} \int_{\partial\Omega} \left| \frac{\partial u}{\partial \nu} \right|^2 x \cdot \nu \, d\sigma = 0, \qquad (5.1)$$

where $F(u) = \int_0^u f(s)ds$.

Proof. First notice that $u \in C^2(\overline{\Omega})$ (see Remark 3.2(ii)). Using integration by parts we obtain

$$\int_\Omega \left[\nabla u \cdot \nabla(x \cdot \nabla u) - |\nabla u|^2 \right] dx$$

$$= \int_\Omega \sum_{i,j} \frac{\partial u}{\partial x_i} x_j \frac{\partial}{\partial x_i} \left(\frac{\partial u}{\partial x_j} \right) dx = \int_\Omega \sum_{i,j} \frac{\partial u}{\partial x_i} x_j \frac{\partial}{\partial x_j} \left(\frac{\partial u}{\partial x_i} \right) dx$$

$$= \int_{\partial\Omega} \sum_{i,j} \frac{\partial u}{\partial x_i} x_j \frac{\partial u}{\partial x_i} \nu_j \, d\sigma - n \int_\Omega |\nabla u|^2 \, dx - \int_\Omega \sum_{i,j} \frac{\partial}{\partial x_j} \left(\frac{\partial u}{\partial x_i} \right) x_j \frac{\partial u}{\partial x_i} \, dx,$$

hence

$$\int_\Omega \sum_{i,j} \frac{\partial u}{\partial x_i} x_j \frac{\partial}{\partial x_j} \left(\frac{\partial u}{\partial x_i} \right) dx = \frac{1}{2} \left(\int_{\partial\Omega} \left| \frac{\partial u}{\partial \nu} \right|^2 x \cdot \nu \, d\sigma - n \int_\Omega |\nabla u|^2 \, dx \right)$$

and

$$\int_\Omega \nabla u \cdot \nabla(x \cdot \nabla u) \, dx = \frac{1}{2} \int_{\partial\Omega} \left| \frac{\partial u}{\partial \nu} \right|^2 x \cdot \nu \, d\sigma - \frac{n-2}{2} \int_\Omega |\nabla u|^2 \, dx.$$

Multiplying the equation in (2.1) by $x \cdot \nabla u$, integrating over Ω and denoting the left- or the right-hand side of the resulting equation by (LHS) or (RHS), respectively, we obtain

$$-(\text{LHS}) = \int_\Omega \Delta u (x \cdot \nabla u)\, dx = \int_{\partial\Omega} \frac{\partial u}{\partial\nu}(x \cdot \nabla u)\, d\sigma - \int_\Omega \nabla u \cdot \nabla(x \cdot \nabla u)\, dx$$

$$= \frac{1}{2}\int_{\partial\Omega} \left|\frac{\partial u}{\partial\nu}\right|^2 x \cdot \nu\, d\sigma + \frac{n-2}{2}\int_\Omega |\nabla u|^2\, dx,$$

$$(\text{RHS}) = \int_\Omega f(u)(x \cdot \nabla u)\, dx$$

$$= \int_{\partial\Omega} F(u)(x \cdot \nu)\, d\sigma - n \int_\Omega F(u)\, dx = -n \int_\Omega F(u)\, dx,$$

where we have used that, on $\partial\Omega$, $\nabla u = C\nu$ for suitable $C \in \mathbb{R}$ and $F(u) = F(0) = 0$. The comparison of (LHS) and (RHS) yields now the assertion. $\qquad\square$

Corollary 5.2. *Assume Ω bounded and starshaped with respect to some point $x_0 \in \Omega$ (i.e. the segment $[x_0, x]$ is a subset of Ω for any $x \in \Omega$), $n \geq 3$. Assume that*

$$F(u) \leq \frac{n-2}{2n} f(u)u \qquad \text{for all } u. \tag{5.2}$$

Then (2.1) does not possess classical positive solutions. If, in addition, $f(0) = 0$, then (2.1) does not possess classical nontrivial solutions.

Condition (5.2) is satisfied if, for example, $f(u) = |u|^{p-1}u + \lambda u$, $p \geq p_S$ and $\lambda \leq 0$.

Proof. We proceed by contradiction. We can assume that Ω is starshaped with respect to $x_0 = 0$. Then $x \cdot \nu \geq 0$ on $\partial\Omega$ and

$$\int_{\partial\Omega} x \cdot \nu\, d\sigma = \int_\Omega \Delta\left(\frac{x^2}{2}\right) dx > 0,$$

hence $x \cdot \nu > 0$ on a set of positive surface measure in $\partial\Omega$.

If u is a positive solution of (2.1), then $\partial u/\partial\nu < 0$ on $\partial\Omega$ by the maximum principle and we obtain

$$\frac{1}{2}\int_{\partial\Omega} \left|\frac{\partial u}{\partial\nu}\right|^2 x \cdot \nu\, d\sigma > 0. \tag{5.3}$$

Multiplication of the equation in (2.1) by u and integration by parts yields

$$\int_\Omega |\nabla u|^2\, dx = \int_\Omega f(u)u\, dx \tag{5.4}.$$

Using (5.1), (5.3), (5.4) we arrive at

$$\int_\Omega \left[\frac{n-2}{2n} f(u)u - F(u)\right] dx < 0,$$

which contradicts (5.2).

If $f(0) = 0$ and u is a sign-changing solution of (2.1), then the assertion follows from the unique continuation property. In fact, let $x_1 \in \partial\Omega$ be such that $x \cdot \nu > 0$ in a neighborhood Γ_1 of x_1 in $\partial\Omega$ (recall that $\partial\Omega$ is smooth). Then the above arguments guarantee $\partial u/\partial\nu = 0$ on Γ_1. Since $u = 0$ and $\Delta u = f(u) = 0$ on $\partial\Omega$, all the second derivatives of u have to vanish on Γ_1. Set $u(x) := 0$ for $x \notin \overline{\Omega}$. Then u is a solution of (2.1) in a neighborhood of Γ_1, hence $u \equiv 0$ in this neighborhood due to [272, Satz 2]. Using the same result one can easily show $u \equiv 0$ in Ω. $\square$

Remark 5.3. The idea of considering the multiplier $x \cdot \nabla u$ was used before in [455] in the linear case $f(u) = \mu u$ (for a different purpose, namely an integral representation of the eigenvalues of the Laplacian). Identities similar to (5.1) (see Lemma 31.4 for the case of systems and see also [431]) are sometimes called Rellich-Pohozaev type identities in the literature. $\square$

6. Homogeneous nonlinearities

In this section we use variational methods in order to study the problem

$$\left.\begin{array}{ll} -\Delta u = |u|^{p-1}u + \lambda u, & x \in \Omega, \\ u = 0, & x \in \partial\Omega. \end{array}\right\} \qquad (6.1)$$

The energy functional E has the form $E(u) = \Psi(u) - \Phi(u)$, where

$$\Psi(u) := \frac{1}{2}\int_\Omega \left[|\nabla u(x)|^2 - \lambda u^2\right] dx \quad \text{and} \quad \Phi(u) := \frac{1}{p+1}\int_\Omega |u|^{p+1}\, dx. \qquad (6.2)$$

Notice that Ψ is quadratic and Φ is positively homogeneous of order $p + 1 \neq 2$. Therefore, if

$$\Psi'(w) = \mu\Phi'(w) \qquad (6.3)$$

for some $\mu > 0$, then, setting $t := \mu^{1/(p-1)}$, we get

$$E'(tw) = \Psi'(tw) - \Phi'(tw) = t\left[\Psi'(w) - t^{p-1}\Phi'(w)\right] = 0. \qquad (6.4)$$

Consequently, tw is a critical point of E, hence a classical solution of (6.1) if $p \leq p_S$. A nontrivial function w satisfying (6.3) will be found by minimizing the functional Ψ with respect to the set $M := \{u : \Phi(u) = 1\}$ and using the following well-known Lagrange multiplier rule.

Theorem 6.1. *Let X be a real Banach space, $w \in X$ and let $\Psi, \Phi_1, \ldots, \Phi_k : X \to \mathbb{R}$ be C^1 in a neighborhood of w. Denote $M := \{u \in X : \Phi_i(u) = \Phi_i(w) \text{ for } i = 1, \ldots, k\}$ and assume that w is a local minimizer of Ψ with respect to the set M. If $\Phi_1'(w), \ldots, \Phi_k'(w)$ are linearly independent, then there exist $\mu_1, \ldots, \mu_k \in \mathbb{R}$ such that*

$$\Psi'(w) = \sum_{i=1}^{k} \mu_i \Phi_i'(w).$$

Our proofs of the main results of this section (Theorem 6.2 and Theorem 6.7(i)) follow those in [505, Theorem I.2.1 and Lemma III.2.2]. Let us first consider the subcritical case.

Theorem 6.2. *Assume Ω bounded. Let $1 < p < p_S$ and $\lambda < \lambda_1$. Then there exists a positive classical solution of (6.1).*

Proof. Set $X := W_0^{1,2}(\Omega)$ and define Ψ, Φ as in (6.2). Since

$$\Psi''(u)[h, h] = 2\Psi(h) \geq c_\lambda \int_\Omega |\nabla h|^2 \, dx, \quad c_\lambda := 1 - \frac{\lambda}{\lambda_1} > 0,$$

the functional Ψ is convex and coercive. Let $u_k \in M := \{u \in X : \Phi(u) = 1\}$, $u_k \rightharpoonup u$ in X. Then $u_k \to u$ in $L^{p+1}(\Omega)$ due to $X \hookrightarrow\hookrightarrow L^{p+1}(\Omega)$, hence $u \in M$. Consequently, the set M is weakly sequentially closed in the reflexive space X and there exists $w \in M$ such that $\Psi(w) = \inf_M \Psi$. Since $|w| \in M$ and $\Psi(|w|) = \Psi(w)$, we may assume $w \geq 0$. Moreover, $\Phi'(w)w = (p+1)\Phi(w) = p+1$, hence $\Phi'(w) \neq 0$. Theorem 6.1 guarantees the existence of $\mu \in \mathbb{R}$ such that $\Psi'(w) = \mu \Phi'(w)$, hence

$$0 < 2\Psi(w) = \Psi'(w)w = \mu\Phi'(w)w = \mu(p+1)\Phi(w) = \mu(p+1).$$

Consequently, $\mu > 0$ and we deduce from (6.4) that $u := \mu^{1/(p-1)} w$ is a nonnegative variational solution of (6.1), $u \not\equiv 0$. Corollary 3.4 guarantees that u is a classical solution and the strong maximum principle shows $u > 0$ in Ω. $\square$

Remarks 6.3. (i) **Annulus.** Assume that $\Omega = \{x \in \mathbb{R}^n : 1 < |x| < 2\}$, $\lambda < \lambda_1$ and let X denote the space all of radial functions in $W_0^{1,2}(\Omega)$. It is easily seen that X is compactly embedded into the space Y of all radial functions in $L^{p+1}(\Omega)$ for any $p > 1$ (in fact, X and Y are isomorphic to $W_0^{1,2}((1,2))$ and $L^{p+1}((1,2))$, respectively). Moreover, any critical point of E in X is obviously a classical solution of (6.1). Hence the proof of Theorem 6.2 guarantees the existence of a positive classical solution of (6.1) for all $p > 1$.

(ii) **Nonexistence for $\lambda \geq \lambda_1$.** If Ω is bounded, $\lambda \geq \lambda_1$ and $p > 1$ is arbitrary, then (6.1) does not have positive stationary solutions. To see this, it is sufficient to multiply the equation in (6.1) by the first eigenfunction φ_1 to obtain

$$0 = \int_\Omega |u|^{p-1} u \varphi_1 \, dx + (\lambda - \lambda_1) \int_\Omega u \varphi_1 \, dx > 0$$

provided u is a positive solution. $\square$

Remark 6.4. Unbounded domains. Let $\Omega = \mathbb{R}^n$, $1 < p < p_S$ and $\lambda < 0$ (notice that 0 is the minimum of the spectrum of $-\Delta$ in $W^{1,2}(\mathbb{R}^n)$). Let X and Y denote the space of radial functions in $W^{1,2}(\mathbb{R}^n)$ and $L^{p+1}(\mathbb{R}^n)$, respectively. If $n \geq 2$, then X is compactly embedded in Y (see [76, Theorem A.I']) so that we may use the approach above in order to get a positive solution of (6.1). Moreover, using Schwarz symmetrization it is easy to see that the minimizer of $\Psi(u) = \frac{1}{2} \int_\Omega \left(|\nabla u|^2 - \lambda u^2 \right) dx$ in $M_X := \{u \in X : \frac{1}{p+1} \int_\Omega |u|^{p+1}\, dx = 1\}$ is also a minimizer in the larger set $M := \{u \in W^{1,2}(\mathbb{R}^n) : \frac{1}{p+1} \int_\Omega |u|^{p+1}\, dx = 1\}$.

In the case $\Omega = \mathbb{R}^n$ one can use a similar approach to that used in Theorem 6.2 for functions $f = f(u)$ (or $f = f(|x|, u)$) which need not be homogeneous. In fact, if one is able to find a minimizer u of $\frac{1}{2} \int_\Omega |\nabla u|^2\, dx$ in the set $N := \{u \in X : \int_\Omega F(u)\, dx = 1\}$, then there exists $\sigma > 0$ such that the function $u_\sigma(x) := u(x/\sigma)$ solves (6.1). This idea was used in [76], for example. For more recent results on existence and uniqueness of positive solutions of this problem with $f = f(u)$ we refer to [235], [420] and the references therein.

If f depends on x (and not only on $|x|$) or if Ω is unbounded and not symmetric, then the situation is more delicate. In some cases, one can use the concentration compactness arguments in order to get a solution (see [50] and the references therein). $\quad\square$

Let us now turn to the critical case $p = p_S$. In view of Corollary 5.2 and the proof of Theorem 6.2, the functional Ψ cannot attain its infimum over the set M if Ω is starshaped and $\lambda = 0$. In other words, denoting

$$S_\lambda(u, \Omega) := \frac{\int_\Omega \left[|\nabla u|^2 - \lambda |u|^2 \right] dx}{\|u\|_{2^*}^2},$$

$$S_\lambda(\Omega) := \inf\{S_\lambda(u, \Omega) : u \in W_0^{1,2}(\Omega),\ u \neq 0\}$$

$$= \inf\{S_\lambda(u, \Omega) : u \in W_0^{1,2}(\Omega),\ \|u\|_{2^*} = 1\},$$

the value $S_0(\Omega)$ cannot be attained if Ω is starshaped. The following proposition shows that the same is true for any $\Omega \neq \mathbb{R}^n$. In particular, this means that the solution from Remark 6.3(i) (for $p = p_S$ and $\lambda = 0$) is not a minimizer of $S_0(\cdot, \Omega)$.

Proposition 6.5. *We have $S_0(\Omega_1) = S_0(\Omega_2)$ for any open sets $\Omega_1, \Omega_2 \subset \mathbb{R}^n$. If $\Omega \neq \mathbb{R}^n$, then $S_0(\Omega)$ is not attained.*

Proof. Let $\Omega_1, \Omega_2 \subset \mathbb{R}^n$ be open. Since $S_0(\Omega) = S_0(x + \Omega)$ for any $x \in \mathbb{R}^n$, we may assume $0 \in \Omega_1 \cap \Omega_2$. Denote $w^R(x) := w(Rx)$.

Let $\varepsilon > 0$ and $0 \neq u_1 \in W_0^{1,2}(\Omega_1)$, $S_0(u_1, \Omega_1) < S_0(\Omega_1) + \varepsilon$. Setting $\tilde{u}_1(x) := u(x)$ if $x \in \Omega_1$, $\tilde{u}_1(x) = 0$ if $x \notin \Omega_1$, we have $\tilde{u}_1 \in W_0^{1,2}(\mathbb{R}^n) = W^{1,2}(\mathbb{R}^n)$ and $\operatorname{supp}(\tilde{u}_1^R) \subset \Omega_2$ if R is sufficiently large. Let u_2 be the restriction of $\tilde{u}_1^R$ to Ω_2. Then $u_2 \in W_0^{1,2}(\Omega_2)$, $u_2 \neq 0$, and

$$S_0(\Omega_2) \leq S_0(u_2, \Omega_2) = S_0(\tilde{u}_1^R, \mathbb{R}^n) = S_0(\tilde{u}_1, \mathbb{R}^n)$$

$$= S_0(u_1, \Omega_1) < S_0(\Omega_1) + \varepsilon.$$

Letting $\varepsilon \to 0$ we obtain $S_0(\Omega_2) \leq S_0(\Omega_1)$. Exchanging the role of Ω_1 and Ω_2 we obtain the reversed inequality.

Now assume $\Omega \neq \mathbb{R}^n$, $u \in W_0^{1,2}(\Omega)$ and $S_0(u, \Omega) = S_0(\Omega)$. We may assume $u \geq 0$, $u \neq 0$. Set $\tilde{u}(x) := u(x)$ for $x \in \Omega$, $\tilde{u}(x) := 0$ otherwise. Then $S_0(\tilde{u}, \mathbb{R}^n) = S_0(\Omega) = S_0(\mathbb{R}^n)$, hence $\tilde{u}$ is a minimizer of $S_0(\cdot, \mathbb{R}^n)$ and the proof of Theorem 6.2 shows the existence of $\mu > 0$ such that $\tilde{u}$ is a classical positive solution of the equation $-\Delta u = \mu|u|^{p-1}u$ in $\mathbb{R}^n$. But this is a contradiction with $u = 0$ outside Ω. $\quad\square$

Remark 6.6. Best constant in Sobolev's inequality. The function $S_0(\cdot, \mathbb{R}^n)$ attains its minimum $S := S_0(\mathbb{R}^n) = \big(n(n-2)\pi\big)^{-1/2}\big(\Gamma(n)/\Gamma(n/2)\big)^{1/n}$ at any function of the form $u_\varepsilon(x - x_0)$, where $\varepsilon > 0$, $x_0 \in \mathbb{R}^n$ and

$$u_\varepsilon(x) := (\varepsilon^2 + |x|^2)^{-(n-2)/2}.$$

This was proved by symmetrization techniques in [43] and [508] (for more general results of this kind see [111] and the references therein). If we set

$$C_\varepsilon := \big[n(n-2)\varepsilon^2\big]^{(n-2)/4},$$

then the functions $C_\varepsilon u_\varepsilon(\cdot - x_0)$ are the only positive classical solutions of (6.1) with $\Omega = \mathbb{R}^n$, $p = p_S$ and $\lambda = 0$: This follows from Theorems 8.1 and 9.1 below. $\quad\square$

Theorem 6.7. *Let $n \geq 3$ and $p = p_S$. Assume Ω bounded, $0 < \lambda < \lambda_1$. Let S be the constant from Remark* 6.6.

(i) *If $S_\lambda(\Omega) < S$, then there exists $u \in W_0^{1,2}(\Omega)$ such that $u > 0$ in Ω and $S_\lambda(\Omega) = S_\lambda(u, \Omega)$.*

(ii) *If λ is close to λ_1, then $S_\lambda(\Omega) < S$.*

Proof. (i) Let $\{u_k\}$ be a minimizing sequence for $S_\lambda(\cdot, \Omega)$, $\|u_k\|_{2^*} = 1$. Replacing u_k by $|u_k|$ we may assume $u_k \geq 0$. Since

$$\left(1 - \frac{\lambda}{\lambda_1}\right) \int_\Omega |\nabla u_k|^2 \, dx \leq \int_\Omega \big[|\nabla u_k|^2 - \lambda u_k^2\big] \, dx = S_\lambda(u_k, \Omega) \to S_\lambda(\Omega),$$

the sequence $\{u_k\}$ is bounded in $W_0^{1,2}(\Omega)$ and we may assume $u_k \rightharpoonup u$ in $W_0^{1,2}(\Omega)$. Due to the embeddings $W_0^{1,2}(\Omega) \hookrightarrow L^{2^*}(\Omega)$ and $W_0^{1,2}(\Omega) \hookrightarrow\hookrightarrow L^2(\Omega)$ we obtain $u_k \rightharpoonup u$ in $L^{2^*}(\Omega)$ and $u_k \to u$ in $L^2(\Omega)$. Passing to a subsequence we may assume $u_k(x) \to u(x)$ a.e. Given $t \in [0,1]$, denote

$$\psi_k = \psi_k(t) := 2^*\big(u_k + (t-1)u\big)\big|u_k + (t-1)u\big|^{2^*-2}, \qquad \psi = \psi(t) := 2^*tu|tu|^{2^*-2}.$$

Then $\psi_k \to \psi$ a.e. in Ω and ψ_k, ψ are uniformly bounded in $L^{2_*}(\Omega)$, where $2_* := (2^*)' = 2n/(n+2)$. Using Vitali's convergence theorem we obtain

$$\int_\Omega \left[|u_k|^{2^*} - |u_k - u|^{2^*} \right] dx = \int_\Omega \int_0^1 \frac{d}{dt} |u_k + (t-1)u|^{2^*} \, dt \, dx$$

$$= \int_0^1 \int_\Omega \psi_k u \, dx \, dt \to \int_0^1 \int_\Omega \psi u \, dx \, dt = \int_\Omega |u|^{2^*} dx \quad \text{as } k \to \infty,$$

hence

$$\|u\|_{2^*}^{2^*} = 1 - \|u_k - u\|_{2^*}^{2^*} + o(1),$$

where $o(1) \to 0$ as $k \to \infty$. The weak convergence $u_k \rightharpoonup u$ in $W_0^{1,2}(\Omega)$ implies

$$\int_\Omega |\nabla u_k|^2 \, dx = \int_\Omega |\nabla(u_k - u)|^2 \, dx + \int_\Omega |\nabla u|^2 \, dx + o(1),$$

hence

$$S_\lambda(\Omega) = S_\lambda(u_k, \Omega) + o(1)$$

$$= \int_\Omega |\nabla(u_k - u)|^2 \, dx + \int_\Omega \left[|\nabla u|^2 - \lambda u^2 \right] dx + o(1)$$

$$\geq S\|u_k - u\|_{2^*}^2 + S_\lambda(\Omega)\|u\|_{2^*}^2 + o(1)$$

$$\geq S\|u_k - u\|_{2^*}^{2^*} + S_\lambda(\Omega)\|u\|_{2^*}^{2^*} + o(1)$$

$$= \left(S - S_\lambda(\Omega) \right) \|u_k - u\|_{2^*}^{2^*} + S_\lambda(\Omega) + o(1).$$

Now $S > S_\lambda(\Omega)$ implies $u_k \to u$ in $L^{2^*}(\Omega)$, hence $\|u\|_{2^*} = 1$. The weak lower semi-continuity of the norm in $W_0^{1,2}(\Omega)$ guarantees

$$S_\lambda(u, \Omega) \leq \liminf_{k \to \infty} S_\lambda(u_k, \Omega) = S_\lambda(\Omega),$$

thus $S_\lambda(u, \Omega) = S_\lambda(\Omega)$. Similarly as in the proof of Theorem 6.2, a suitable positive multiple of u is a classical positive solution of (6.1) with $p = p_S$, hence $u > 0$ in Ω.

(ii) Let φ_1 be the first eigenfunction, $\|\varphi_1\|_{2^*} = 1$. Then

$$S_\lambda(\Omega) \leq S_\lambda(\varphi_1, \Omega) = (\lambda_1 - \lambda) \int_\Omega \varphi_1^2 \, dx < S$$

if λ is close to λ_1. $\quad \square$

Corollary 6.8. *Let $n \geq 3$ and $p = p_S$. Assume Ω bounded, $0 < \lambda < \lambda_1$. If λ is close to λ_1, then problem (6.1) has a classical positive solution.*

Remarks 6.9. (i) **Positive solutions in the critical case** [98]. Let Ω be bounded, $p = p_S$,

$$\lambda^* := \inf\{\lambda \in (0, \lambda_1) : S_\lambda(\Omega) < S\}.$$

Set $u_\varepsilon(x) := (\varepsilon + |x|^2)^{-(n-2)/2}$ (cf. Remark 6.6) and assume $0 \in \Omega$. If $n \geq 4$ and $\lambda > 0$, then careful estimates show $S_\lambda(u_\varepsilon\varphi, \Omega) < S$ provided $\varphi \in \mathcal{D}(\Omega)$ is nonnegative, $\varphi = 1$ in a neighborhood of 0 and ε is small enough. Consequently, $\lambda^* = 0$ in this case and problem (6.1) possesses a positive solution for any $\lambda \in (0, \lambda_1)$.

Now let $n = 3$ and $\Omega = B_1(0)$. If $\lambda > \lambda_1/4$, then $S_\lambda(u_\varepsilon\varphi, \Omega) < S$ provided $\varphi(x) = \cos(\pi|x|/2)$ and ε is small enough. On the other hand, one can use a Pohozaev-type identity for radial functions in order to prove that (6.1) does not have positive radial solutions if $\lambda \leq \lambda_1/4$. Since any positive solution of (6.1) is symmetric due to [239] we have $\lambda^* = \lambda_1/4$ in this case and the problem possesses positive solutions if and only if $\lambda \in (\lambda_1/4, \lambda_1)$.

Another proof of the above results for $\Omega = B_1(0)$ based on the ODE techniques can be found in [40]. The authors use the symmetry of positive solutions $u = u(|x|)$ of (6.1) and the substitution $y(t) = u(|x|)$, $t = (n-2)^{n-2}|x|^{-(n-2)}$, which transforms the problem into the ODE $y'' + t^{-k}(\lambda y + y^{p_S}) - 0$ with $k := 2(n-1)/(n-2)$.

(ii) **Uniqueness for $p \leq p_S$.** Uniqueness of positive solutions of (6.1) in the case $\Omega = B_1(0)$, $p \leq p_S$, was established in [239] (if $\lambda = 0$), [393] (if $\lambda \geq 0$, $p \leq p_{sg}$), [310] (if $\lambda < 0$, $p < p_S$) and [544], [503] (if $\lambda > 0$, $p \leq p_S$). Some of these articles contain also uniqueness results for more general functions $f(|x|, u)$ and for Ω being an annulus.

Uniqueness fails for general bounded domains (see (iii) and (iv) below). On the other hand if Ω satisfies some convexity and symmetry properties, then uniqueness (and non-degeneracy) for positive solutions of (6.1) is true, at least for some values of p and/or λ (see [147], [118], [256], for example).

(iii) **Nonradial minimizers.** Let $\Omega = \{x : 1 < |x| < 2\}$, $n \geq 3$, $\lambda = 0$ and $p > 1$. Set

$$S(u, \Omega, p) := \frac{\int_\Omega |\nabla u|^2 \, dx}{\|u\|_{p+1}^2},$$

$$S(\Omega, p) := \inf\{S(u, \Omega, p) : u \in W_0^{1,2}(\Omega) \ u \neq 0\},$$

$$S^r(\Omega, p) := \inf\{S(u, \Omega, p) : u \in W_0^{1,2}(\Omega) \ u \neq 0, \ u \text{ is radial}\}.$$

By Remark 6.3(i), problem (6.1) with $\lambda = 0$ has a positive radially symmetric solution u which minimizes $S(\cdot, \Omega, p)$ in the class of radial functions. Since $S(\Omega, p_S)$ is not attained (see Proposition 6.5), we have $S(\Omega, p_S) < S^r(\Omega, p_S)$. It is easy to see that the functions $p \mapsto S(\Omega, p)$ and $p \mapsto S^r(\Omega, p)$ are continuous. Consequently, $S(\Omega, p) < S^r(\Omega, p)$ also for $p < p_S$, p close to p_S. Since $S(\Omega, p)$ is attained in the subcritical case, the corresponding (positive) minimizer is not radially symmetric.

(iv) **Effect of the topology of domain.** Let Ω be bounded, $n \geq 3$, $p = p_S$ and $\lambda = 0$. The above considerations show that (6.1) has a positive solution if Ω is an annulus but it does not possess positive solutions if Ω is starshaped. It was proved in [47] that this problem has positive solutions whenever the homology of dimension d of Ω with $\mathbb{Z}_2$ coefficients is nontrivial for some positive integer d. In particular, this is true when $n = 3$ and Ω is not contractible. On the other hand, there are several examples showing that positive solutions do exist even if Ω is contractible (see [149], for example).

Let Ω be bounded and let its Ljusternik-Schnirelman category be bigger than 1. If $p < p_S$, then problem (6.1) admits multiple positive solutions whenever p is close to p_S or $\lambda < 0$ and $|\lambda|$ is large enough (see [68]); the same is true if $p = p_S$, $\lambda > 0$ is small and $n \geq 4$ (see [456], [323]). Again, this topological condition on Ω is not necessary (see [148], where multiple positive solutions are constructed for any $p < p_S$, $\lambda = 0$ and Ω being starshaped, and see [408] for the critical case).

(v) **Critical case in the unit ball.** Let $\Omega = B_1(0)$, $n \geq 3$, $p = p_S$ and consider radial (classical) solutions of (6.1).

Due to Corollary 5.2, nontrivial solutions do not exist if $\lambda \leq 0$. Denote by X the space of all radial functions in $W_0^{1,2}(\Omega)$ and let λ_k^r denote the k-th eigenvalue of $-\Delta$ in X ($\lambda_k^r = k^2\pi^2$ if $n = 3$). The corresponding radial eigenfunction φ_k^r (considered as a function of $r := |x|$) has $(k-1)$ zeros in $(0,1)$ and each point $(0, \lambda_k^r) \in X \times \mathbb{R}$ is a bifurcation point for (6.1) (see [451]). The corresponding bifurcation branch $\mathcal{B}_k$ of nontrivial solutions is an unbounded continuous curve and u has $(k-1)$ zeros for any $(u, \lambda) \in \mathcal{B}_k$. Moreover, there exists $\mu_k := \lim\{\lambda : (u, \lambda) \in \mathcal{B}_k, \|u\|_X \to \infty\}$, $k = 1, 2, \ldots$, and we have $\mu_k = (k - \frac{1}{2})^2\pi^2$ if $n = 3$, $\mu_1 = 0$ if $n \geq 4$, $\mu_{k+1} = \lambda_k^r$ if $n = 4, 5$, $\mu_{k+1} \in (0, \lambda_k^r)$ if $n = 6$, $\mu_k = 0$ if $n \geq 7$ (see Figure 1 and [40], [41], [42], [39]).

Denote $\tilde{\mu}_k := \inf\{\lambda : (u, \lambda) \in \mathcal{B}_k\}$. The results mentioned in (i) and (ii) imply $\tilde{\mu}_1 = \mu_1 = \lambda_1/4$ if $n = 3$, $\tilde{\mu}_1 = \mu_1 = 0$ if $n \geq 4$. Similarly, [34] and [234] imply $\tilde{\mu}_2 = \mu_2$ if $n = 4$, $\tilde{\mu}_2 < \mu_2$ if $n = 5$ but the relation between $\tilde{\mu}_2$ and μ_2 for $n \in \{3, 6\}$ seems to be an open problem.

Denote also $\lambda_* := \inf\{\tilde{\mu}_k : k \geq 2\}$. Then $\lambda_* > 0$ provided $n \leq 6$ (see [39]). On the other hand, problem (6.1) with $\Omega = B_1(0)$, $n \geq 4$, $p = p_S$ and $\lambda > 0$ has infinitely many nontrivial solutions in $W_0^{1,2}(\Omega)$ (see [212]). Consequently, if $n \in \{4, 5, 6\}$ and $\lambda < \lambda_*$, then all these solutions (except for $\pm u_1$ where u_1 denotes the unique positive solution) have to be nonradial. The existence of (nonradial sign-changing) solutions for $\Omega = B_1(0)$, $n = 3$ and $\lambda \in (0, \lambda_1/4]$ seems to be open.

Many interesting results on singular radial solutions of (6.1) for $\Omega = B_1(0)$ and $p > 1$ can be found in [73]. $\square$

Remarks 6.10. Supercritical case. Let $n \geq 3$, $p > p_S$.

(i) If $\lambda = 0$, then the analogue of the result of [47] mentioned in Remark 6.9(iv) does not hold (see [406], [407]).

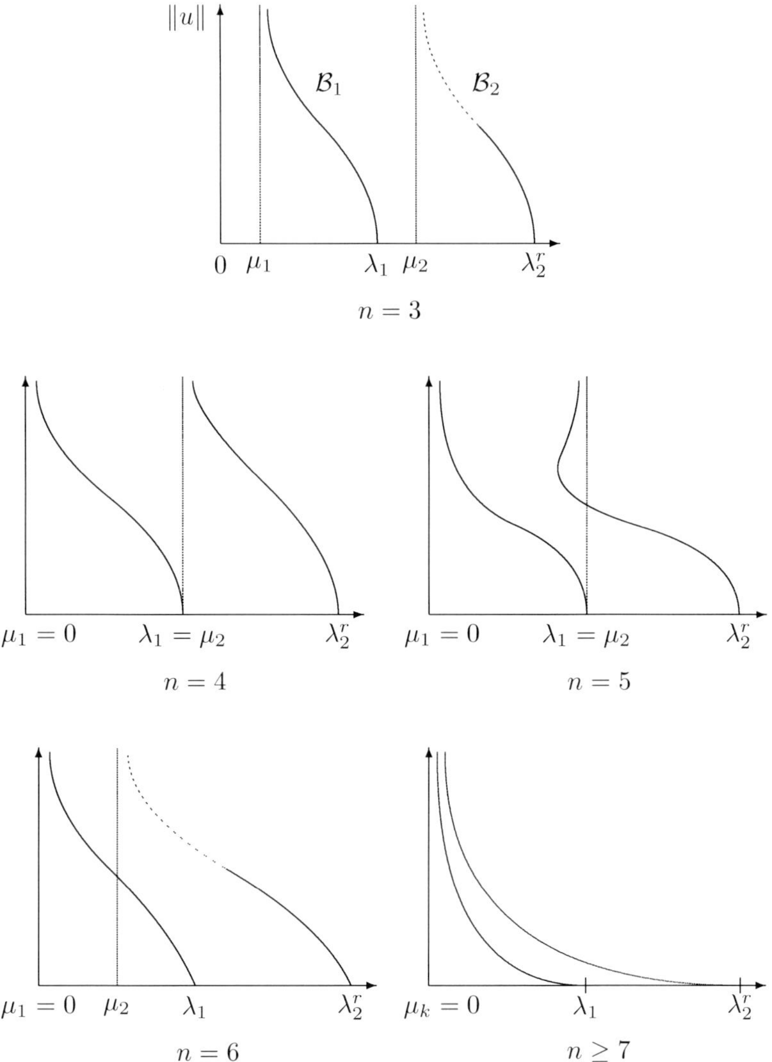

Figure 1: Bifurcation diagrams for radial solutions of (6.1)
with $p = p_S$ and $\Omega = B_1(0)$.

(ii) Let $\Omega = B_1(0)$. Then the points $(0, \lambda_k^r)$ from Remark 6.9(v) are bifurcation points for (6.1) also in this case. Let $\mathcal{B}_k(p)$ denote the corresponding bifurcation branch and let $\mu_k(p), \tilde{\mu}_k(p)$ have similar meaning as in Remark 6.9(v). If $n > 6$, assume also $p < p_{ZZ} := \left(n + 1 - \sqrt{2n - 3}\right)/\left(n - 3 - \sqrt{2n - 3}\right)$. Then

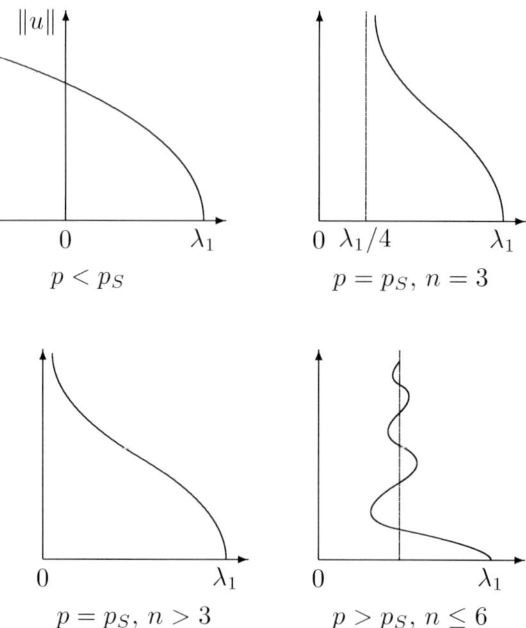

Figure 2: Bifurcation diagrams for positive solutions
of (6.1) with $\Omega = B_1(0)$.

$0 < \tilde{\mu}_1(p) < \mu_1(p) < \lambda_1$ and problem (6.1) has infinitely many radial positive (classical) solutions if $\lambda = \mu_1(p)$ (see Figure 2 and [546], [104], [365]). It is not clear whether the condition $p < p_{ZZ}$ for $n > 6$ is optimal, but some restrictions on n or p for this behavior of $\mathcal{B}_1(p)$ may be expected. In fact, bifurcation diagrams for positive solutions of the related problem

$$\left.\begin{aligned} -\Delta u &= \lambda(1+u)^p, & x &\in B_1(0), \\ u &= 0, & x &\in \partial B_1(0), \end{aligned}\right\} \tag{6.5}$$

in the supercritical case are completely different for $p < p_{JL}$ and $p \geq p_{JL}$, where p_{JL} is defined in (9.3) (see Figure 3 and [293]).

Note also that the same diagrams as in Figure 3 are true for the problem

$$\left.\begin{aligned} -\Delta u &= \lambda e^u, & x &\in B_1(0), \\ u &= 0, & x &\in \partial B_1(0), \end{aligned}\right\} \tag{6.6}$$

and the three cases I, II and III correspond to $n \leq 2$, $3 \leq n \leq 9$ and $n \geq 10$, respectively. $\quad\square$

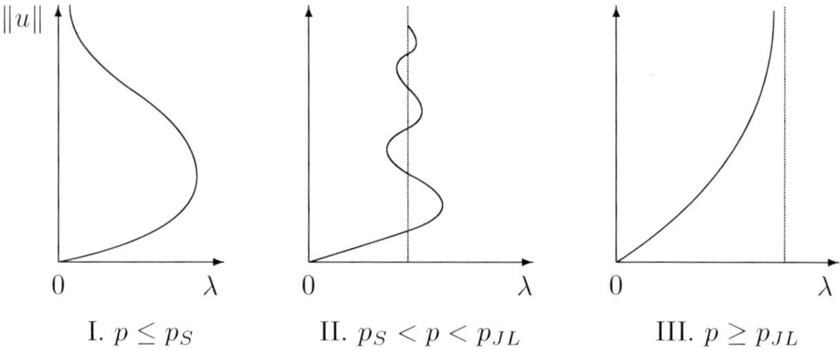

Figure 3: Bifurcation diagrams for positive solutions of (6.5).

7. Minimax methods

In this section we look for saddle points of the energy functional E defined in (3.5) by minimax methods. Throughout this section we assume that f satisfies the growth assumption (3.4) so that E is a C^1-functional in the Hilbert space $W_0^{1,2}(\Omega)$ and its critical points correspond to (variational) solutions of (2.1).

Even if we considered a finite-dimensional space $X = \mathbb{R}^2$ and a smooth functional $E : X \to \mathbb{R}$, then (looking at the graph of E as the earth's surface) existence of a saddle (mountain pass) on a mountain range between two valleys is not clear, in general. For example, if $E : \mathbb{R}^2 \to \mathbb{R} : (x, y) \mapsto e^x - y^2$, $A_0 = (0, -2)$, $A_1 = (0, 2)$, then any path from A_1 to A_2 in $\mathbb{R}^2$ has to cross the line $\{y = 0\}$ where $E > 0 > \max\{E(A_1), E(A_2)\}$, but the functional E does not possess critical points at all. If one looks for a point with a minimal height on the "mountain range" described by the graph of E on $\{(x, y) : y = 0\}$, then any minimizing sequence has the form $(x_k, 0)$, where $x_k \to -\infty$. In particular, it is not compact and we cannot choose a subsequence converging to the desired saddle point. Therefore, dealing with abstract functionals E in a real Banach space X, we shall need additional information on E which will prevent the problem mentioned above.

Definition 7.1. A sequence $\{u_k\}$ in X is called a **Palais-Smale sequence** if the sequence $\{E(u_k)\}$ is bounded and $E'(u_k) \to 0$. We say that E satisfies condition (PS) if any Palais-Smale sequence is relatively compact. We say that E satisfies condition $(PS)_\beta$ (Palais-Smale condition at level β) if any sequence $\{u_k\}$ satisfying $E(u_k) \to \beta$, $E'(u_k) \to 0$, is relatively compact. A real number β is called a **critical value** of E if there exists $u \in X$ with $E'(u) = 0$ and $E(u) = \beta$. $\square$

The following mountain pass theorem is due to [23]. Our proofs of this theorem and Theorems 7.4, 7.8 below closely follow those in [505, Chapter II].

Theorem 7.2. *Suppose that $E \in C^1(X)$ satisfies* (PS). *Let $u_0, u_1 \in X$,*

$$M := \max\{E(u_0), E(u_1)\},$$
$$P := \{p \in C([0,1], X) : p(0) = u_0, \ p(1) = u_1\}, \tag{7.1}$$
$$\beta := \inf_{p \in P} \max_{t \in [0,1]} E(p(t)).$$

If $\beta > M$, then β is a critical value of E.

Given $\beta \in \mathbb{R}$ and $\delta > 0$, denote

$$N_\delta = N_\delta(\beta) := \{u \in X : |E(u) - \beta| \leq \delta, \ \|E'(u)\| \leq \delta\}$$

and $E_\beta := \{u \in X : E(u) < \beta\}$.

In the proof of Theorem 7.2 we shall need the following deformation lemma.

Lemma 7.3. *Suppose that $E \in C^1(X)$ and let $N_\delta(\beta) = \emptyset$ for some $\delta < 1$. Choose $\varepsilon = \delta^2/2$. Then there exists a continuous mapping $\Phi : X \times [0,1] \to X$ such that*
(i) $\Phi(u,t) = u$ whenever $t = 0$ or $|E(u) - \beta| \geq 2\varepsilon$,
(ii) $t \mapsto E(\Phi(u,t))$ is nonincreasing for all u,
(iii) $\Phi(E_{\beta+\varepsilon}, 1) \subset E_{\beta-\varepsilon}$.
In addition, $\Phi(\cdot, t)$ is odd if E is even.

Proof. In order to avoid all technicalities we shall assume, in addition, that $E \in C^2(X)$ and X is a Hilbert space. Notice that these assumptions are satisfied in our applications if f has the form (2.2), for example (and see e.g. [505] for the proof in the general case).

Choose functions $\varphi, \psi : \mathbb{R} \to [0,1]$ such that φ is smooth, $\varphi(t) = 1$ for $|t-\beta| \leq \varepsilon$, $\varphi(t) = 0$ for $|t - \beta| \geq 2\varepsilon$, $\psi(t) = 1$ for $t \leq 1$ and $\psi(t) = 1/t$ for $t > 1$. The vector field

$$\mathcal{F} : X \to X : u \mapsto -\varphi\big(E(u)\big)\psi\big(\|E'(u)\|\big)\nabla E(u)$$

is bounded and locally Lipschitz. Consequently, the initial value problem

$$\Phi_t(u,t) = \mathcal{F}\big(\Phi(u,t)\big), \qquad \text{for } t \in [0,1],$$
$$\Phi(u,0) = u$$

has a unique solution for any $u \in X$. The function Φ defined in this way is obviously continuous and satisfies (i). Denoting $v := \Phi(u,t)$ we have

$$\frac{d}{dt}E\big(\Phi(u,t)\big) = \frac{d}{dt}E(v) = E'(v)\mathcal{F}(v) = -\varphi\big(E(v)\big)\psi\big(\|E'(v)\|\big)\|E'(v)\|^2 \leq 0,$$

thus (ii) is true.

Assertion (iii) will be proved by a contradiction argument. Let $u \in E_{\beta+\varepsilon}$ and assume $\Phi(u,1) \notin E_{\beta-\varepsilon}$. Then (ii) implies $|E(\Phi(u,t))-\beta| \leq \varepsilon < \delta$ for $t \in [0,1]$, hence $N_\delta = \emptyset$ implies $\|E'(\Phi(u,t))\| \geq \delta$ for $t \in [0,1]$. Using this estimate and the properties of the functions φ, ψ we get

$$E(\Phi(u,1)) = E(u) + \int_0^1 \frac{d}{dt} E(\Phi(u,t))\, dt$$

$$= E(u) - \int_0^1 \underbrace{\varphi(\ldots)}_{=1} \underbrace{\psi(\ldots)\|E'(\Phi(u,t))\|^2}_{\geq \delta^2}\, dt$$

$$< \beta + \varepsilon - \delta^2 \leq \beta - \varepsilon,$$

a contradiction. $\square$

Proof of Theorem 7.2. Assume that β is not a critical value of E. Then it is easy to use condition (PS) in order to find $\delta > 0$ such that $N_\delta(\beta) = \emptyset$. We may assume $\delta < 1$, $\delta^2 < \beta - M$. Let $\varepsilon := \frac{1}{2}\delta^2$ be from Lemma 7.3. By the definition of β there exists $p \in P$ such that $\max_{t \in [0,1]} E(p(t)) < \beta + \varepsilon$. Since $E(u_i) \leq M < \beta - \delta^2 = \beta - 2\varepsilon$ for $i = 0, 1$, Lemma 7.3(i) guarantees that $p_1 : t \mapsto \Phi(p(t), 1)$ is an element of P. Now Lemma 7.3(iii) implies $\max_{t \in [0,1]} E(p_1(t)) \leq \beta - \varepsilon$, which contradicts the definition of β. $\square$

The next theorem is again due to [23]. It represents a symmetric variant of Theorem 7.2 and we will use it for the proof of existence of infinitely many solutions of problem (2.1).

Theorem 7.4. *Suppose that $E \in C^1(X)$ is even and satisfies* (PS). *Let X^+, X^- be closed subspaces of X with $\dim X^- = \operatorname{codim} X^+ + 1 < \infty$. Let $E(0) = 0$ and let there exist $\alpha, \rho, R > 0$ such that $E(u) \geq \alpha$ for all $u \in S_\rho^+ := \{u \in X^+ : \|u\| = \rho\}$ and $E(u) \leq 0$ for all $u \in X^-$, $\|u\| \geq R$. Set*

$$\Gamma := \{h \in C(X, X) : h \text{ is odd}, \ h(u) = u \text{ if } E(u) \leq 0\},$$

$$\beta := \inf_{h \in \Gamma} \max_{u \in X^-} E(h(u)).$$

Then β is a critical value of E, $\beta \geq \alpha$.

The proof of the above theorem will be almost the same as the proof of Theorem 7.2 provided we prove the following Intersection Lemma.

Lemma 7.5. *If $\rho > 0$ and $h \in \Gamma$, then $h(X^-) \cap S_\rho^+ \neq \emptyset$.*

Proof of Theorem 7.4. Lemma 7.5 implies $\beta \geq \alpha$. Assume that β is not a critical value of E. Then $N_\delta(\beta) = \emptyset$ for some $\delta > 0$ and we may assume $\delta < 1$, $\delta^2 < \alpha$. Let $\varepsilon := \delta^2/2$ and Φ be from Lemma 7.3 and choose $h \in \Gamma$ such that

$E\big(h(u)\big) < \beta + \varepsilon$ for all $u \in X^-$. Set $h_1(u) := \Phi\big(h(u), 1\big)$. Then $h_1 \in \Gamma$ and $E\big(h_1(u)\big) = E\big(\Phi(h(u), 1)\big) < \beta - \varepsilon$, due to Lemma 7.3(iii). But this contradicts the definition of β. $\quad\square$

In the proof of Lemma 7.5 we shall need the notion of **Krasnoselskii genus**.

Definition 7.6. Let $\mathcal{A}$ be the set of all closed subsets of X satisfying $A = -A$. If $A \in \mathcal{A}$, then we set $\gamma(A) := 0$ if $A = \emptyset$ and

$$\gamma(A) := \inf\{m : \exists h \in C(A, \mathbb{R}^m \setminus \{0\}),\ h \text{ odd}\}$$

otherwise. $\quad\square$

The following proposition is proved in [505, Propositions II.5.2 and II.5.4]:

Proposition 7.7. *Suppose that* $A, A_1, A_2 \in \mathcal{A}$ *and* $h \in C(X, X)$ *is odd. Then the following is true:*
(1) $\gamma(A) \geq 0$, $\gamma(A) = 0$ *if and only if* $A = \emptyset$,
(2) *if* $A_1 \subset A_2$, *then* $\gamma(A_1) \leq \gamma(A_2)$,
(3) $\gamma(A_1 \cup A_2) \leq \gamma(A_1) + \gamma(A_2)$,
(4) $\gamma(A) \leq \gamma\big(\overline{h(A)}\big)$,
(5) *if* A *is compact and* $0 \notin A$, *then* $\gamma(A) < \infty$ *and there exists a symmetric neighborhood* U *of* A *such that* $\overline{U} \in \mathcal{A}$ *and* $\gamma(A) = \gamma(\overline{U})$.
(6) *Let* D *be a bounded symmetric neighborhood of zero in* Y, *where* Y *is a subspace of* X *with* $m := \dim(Y) < \infty$, *and let* $\partial_Y D$ *denote the boundary of* D *in* Y. *Then* $\gamma(\partial_Y D) = m$.

Proof of Lemma 7.5. Let $\rho > 0$ and $h \in \Gamma$. Set $R_1 := \max\{R, \rho\}$, $B_{R_1}^- := \{u \in X^- : \|u\| < R_1\}$ and $S_\rho := \{u \in X : \|u\| = \rho\}$. Since $E(u) \leq 0$ for $u \in X^-$, $\|u\| \geq R$, we have $\|h(u)\| = \|u\| > \rho$ for all $u \in X^-$, $\|u\| > R_1$, hence $h(X^-) \cap S_\rho = h\big(\overline{B_{R_1}^-}\big) \cap S_\rho$ is compact. In particular, $A := h(X^-) \cap S_\rho^+$ fulfills the assumptions of Proposition 7.7(5), thus there exists its symmetric neighborhood U with $\gamma(\overline{U}) = \gamma(A)$. By (2) and (3) in Proposition 7.7 we obtain

$$\gamma(A) = \gamma(\overline{U}) \geq \gamma\big(h(X^-) \cap S_\rho \cap \overline{U}\big) \geq \gamma\big(h(X^-) \cap S_\rho\big) - \gamma(B), \qquad (7.2)$$

where $S_\rho := \{u \in X : \|u\| = \rho\}$ and $B := h(X^-) \cap S_\rho \setminus U$. Let Z be a direct complement of X^+ in X and let $\pi : X \to Z$ denote the projection along X^+. Since U is a neighborhood of $h(X^-) \cap S_\rho^+$, we get $B \cap X^+ = \emptyset$, hence $0 \notin \pi(B)$ and the definition of γ implies

$$\gamma(B) \leq \dim Z = \operatorname{codim} X^+. \qquad (7.3)$$

Now (2) and (4) in Proposition 7.7 guarantee $\gamma\big(h(X^-) \cap S_\rho\big) \geq \gamma\big(h^{-1}(S_\rho) \cap X^-\big)$. Since $h(0) = 0$ and $h(u) = u$ for $u \in X^-$, $\|u\| > R$, the set $h^{-1}(S_\rho) \cap X^-$ contains the relative boundary of $\{u \in X^- : \|h(u)\| < \rho\}$ which is a symmetric bounded neighborhood of zero in X^-. Consequently, using (2) and (6) in Proposition 7.7 we arrive at

$$\gamma\big(h(X^-) \cap S_\rho\big) \geq \dim X^- = \operatorname{codim} X^+ + 1. \tag{7.4}$$

Now (7.2)–(7.4) imply $\gamma(A) \geq 1$, hence $A \neq \emptyset$. $\quad\square$

Theorems 7.2 and 7.4 guarantee the following solvability result.

Theorem 7.8. *Assume Ω bounded. Let f be a Carathéodory function, and let there exist $p < p_S$, $R > 0$ and $\mu > 2$ such that $|f(x, u)| \leq C(1 + |u|^p)$ for all $x \in \Omega$, $u \in \mathbb{R}$ and $f(x, u)u \geq \mu F(x, u) > 0$ for all $x \in \Omega$ and $|u| > R$.*

(i) If there exist $c < \lambda_1$ and $\rho \in (0, 1)$ such that $f(x, u)/u \leq c$ for all $x \in \Omega$ and $|u| < \rho$, then there exists a positive solution of (2.1).

(ii) If $f(x, -u) = -f(x, u)$ for all $x \in \Omega$ and $u \in \mathbb{R}$, then there exists a sequence $\{u_k\}$ of solutions of (2.1) with $E(u_k) \to \infty$ as $k \to \infty$.

Proof. The energy functional E associated with (2.1) is C^1. Let us first verify that E satisfies condition (PS). Let $\{u_k\}$ be a Palais-Smale sequence. Denote $|u|_{1,2} := \left(\int_\Omega |\nabla u|^2 \, dx\right)^{1/2}$ and notice that this is an equivalent norm in $X := W_0^{1,2}(\Omega)$. Then

$$o(1 + |u_k|_{1,2}) = -E'(u_k)u_k = -|u_k|_{1,2}^2 + \int_\Omega f(x, u_k)u_k \, dx$$

$$= \left(\frac{\mu}{2} - 1\right)|u_k|_{1,2}^2 + \int_\Omega \big[f(x, u_k)u_k - \mu F(x, u_k)\big] \, dx - \mu E(u_k)$$

$$\geq \left(\frac{\mu}{2} - 1\right)|u_k|_{1,2}^2 - C_1,$$

where $C_1 > 0$ is independent of k. Consequently, the sequence $\{u_k\}$ is bounded in X. We have $\nabla E(u) = u + \mathcal{F}_1(u)$, where $\mathcal{F}_1$ is compact.[2] Since $\{u_k\}$ is bounded in X, we may assume (passing to a subsequence if necessary) $\mathcal{F}_1(u_k) \to w$ in X for some $w \in X$. Since $o(1) = \nabla E(u_k) = u_k - \mathcal{F}_1(u_k)$, we obtain $u_k \to w$, hence $\{u_k\}$ is relatively compact.

(i) We will use Theorem 7.2. In order to get a positive solution, let us define $\tilde{f}(x, u) := f(x, u)$ if $u \geq 0$, $\tilde{f}(x, u) = 0$ otherwise, $\tilde{F}(x, u) := \int_0^u \tilde{f}(x, s) \, ds$, $\tilde{E}(u) := \frac{1}{2}\int_\Omega |\nabla u|^2 \, dx - \int_\Omega \tilde{F}(x, u) \, dx$, and notice that $\tilde{E}$ is C^1 and satisfies condition (PS). Set $u_0 := 0$, then $\tilde{E}(u_0) = 0$. The assumption $f(x, u)/u \leq c$ for $|u| < \rho$ guarantees $|\tilde{F}(x, u)| \leq (c/2)u^2$ for $|u| < \rho$. If $|u| \geq \rho$, then the growth assumption $|f(x, u)| \leq C(1 + |u|^p)$ implies

$$|\tilde{F}(x, u)| \leq C(|u| + |u|^{p+1}) \leq (c/2)u^2 + C_2|u|^{p+1},$$

---

[2] The Nemytskii mapping $\mathcal{F} : L^{p+1}(\Omega) \to L^{(p+1)'}(\Omega) : u \mapsto f(\cdot, u)$ is continuous. The embedding $I_p : X \hookrightarrow L^{p+1}(\Omega)$ is compact, hence the dual mapping $I_p' : (L^{p+1}(\Omega))' \to X'$ is compact as well. Let $R : X' \to X$ denote the Riesz isomorphism in the Hilbert space X (thus $RE'(u) = \nabla E(u)$) and let $J : L^{(p+1)'}(\Omega) \to (L^{p+1}(\Omega))'$ be the isomorphism defined by $(Jw)u = \int_\Omega uw \, dx$ for $u \in L^{p+1}(\Omega)$. Then $\nabla E(u) = u + \mathcal{F}_1(u)$, where $\mathcal{F}_1 : X \to X : u \mapsto RI_p'J\mathcal{F}I_p(u)$ is compact.

where $C_2 := C(1 + \rho^{-p})$. Consequently, if C_p denotes the norm of the embedding $X \hookrightarrow L^{p+1}(\Omega)$, then

$$\tilde{E}(u) \geq \frac{1}{2} \int_\Omega |\nabla u|^2 \, dx - \frac{c}{2} \int_\Omega u^2 \, dx - C_2 \int_\Omega |u|^{p+1} \, dx$$

$$\geq \left(\frac{1}{2} - \frac{c}{2\lambda_1} - C_2 C_p^{p+1} |u|_{1,2}^{p-1} \right) |u|_{1,2}^2 \geq \alpha > 0$$

provided $|u|_{1,2} = \delta$ is small enough. Now the assumption $f(x,u)u \geq \mu F(x,u) > 0$ implies $\frac{d}{du}\left(u^{-\mu} F(x,u) \right) \geq 0$ for $u > R$, hence $F(x,u) \geq b(x)u^\mu$ for $u > R$, where $b(x) := R^{-\mu} F(x,R) > 0$. Hence, fixing $u \in X$, $u > 0$ in Ω, denoting

$$A(u) := \frac{1}{2} \int_\Omega |\nabla u|^2 \, dx, \qquad B(u) := \int_\Omega b(x)u^\mu \, dx > 0, \qquad (7.5)$$

and taking $t > 0$, we obtain

$$\tilde{E}(tu) = E(tu) \leq t^2 A(u) - t^\mu B(u) + C_3 \to -\infty \qquad \text{as } t \to \infty,$$

where we used the estimate

$$\int_{0 < tu \leq R} \left[b(x)(tu)^\mu - F(x,tu) \right] dx \leq C_3$$

with C_3 independent of t and u. Hence, choosing $u_1 := tu$ with t large enough we have $E(u_1) < 0$. Let β be the number defined in Theorem 7.2. Since any path joining u_0 and u_1 has to intersect the sphere $S_\delta := \{u : |u|_{1,2} = \delta\}$, we have $\beta \geq \alpha > 0$ and Theorem 7.2 guarantees the existence of a solution u with $\tilde{E}(u) \geq \alpha$. Since $\tilde{f}(x,u) = 0$ for $u \leq 0$, the maximum principle implies $u \geq 0$. Now $E(u) = \tilde{E}(u) > 0$, hence $u \neq 0$ and using the maximum principle again we obtain $u > 0$ in Ω.

(ii) Choose a positive integer k. Let X^- denote the linear hull of $\varphi_1, \varphi_2, \ldots, \varphi_k$, and X^+ be the closure of the linear hull of $\varphi_k, \varphi_{k+1}, \ldots$. The growth condition on f guarantees $|F(x,u)| \leq C_F(1 + |u|^{p+1})$ for suitable $C_F > 0$. Set $q := p_S$ if $n \geq 3$ and choose $q > p$ otherwise. Let $C_4 := C_F C_q^{p+1-r}$ and $C_5 := C_F |\Omega|$, where C_q denotes the norm of the embedding $I_q : W_0^{1,2}(\Omega) \hookrightarrow L^{q+1}(\Omega)$, $r \in (0, p+1)$ is defined by $r/2 + (p+1-r)/q = 1$ and $|\Omega|$ denotes the measure of Ω. If $u \in X^+$ and $\|u\| = \rho := \rho_k := \left(\lambda_k^{r/2}/(4C_4) \right)^{1/(p-1)}$, then

$$E(u) \geq \frac{1}{2} \int_\Omega |\nabla u|^2 \, dx - C_F \int_\Omega |u|^{p+1} \, dx - C_5$$

$$\geq \frac{1}{2} |u|_{1,2}^2 - C_F \|u\|_2^r \|u\|_{q+1}^{p+1-r} - C_5$$

$$\geq \left(\frac{1}{2} - C_4 \lambda_k^{-r/2} |u|_{1,2}^{p-1} \right) |u|_{1,2}^2 - C_5$$

$$= \left(\frac{1}{2} - C_4 \lambda_k^{-r/2} \rho^{p-1} \right) \rho^2 - C_5 = C_6 \lambda_k^{r/(p-1)} - C_5,$$

where $C_6 = (4^p C_4)^{-1/(p-1)}$. Denote $\alpha = \alpha_k := \inf\{E(u) : u \in X^+,\ |u|_{1,2} = \rho\}$. Since $\lambda_k \to \infty$ as $k \to \infty$, we have $\alpha_k \to \infty$.

On the other hand, estimates in (i) show $E(tu) \le t^2 A(u) - t^\mu B(u) - C_3$, where A, B are defined in (7.5). Since $A(u) = 1/2$ for $|u|_{1,2} = 1$ and $C_7 := \inf\{B(u) : u \in X^-,\ |u|_{1,2} = 1\} > 0$, we have

$$E(u) \le \frac{1}{2}|u|_{1,2}^2 - C_7|u|_{1,2}^\mu + C_3 \quad \text{for all } u \in X^-,$$

hence the assumptions of Theorem 7.4 are satisfied for any k large enough and we obtain a sequence of critical points u_k of E satisfying $E(u_k) \ge \alpha_k$. (In fact, a more careful choice of ρ above enables one to use Theorem 7.4 for any k.) $\square$

Remarks 7.9. (i) **Linking.** Let f be differentiable in u, $f(x,0) = 0$, $f(x,u)/u \ge f_u(x,0)$ for all $x \in \Omega$ and $u \in \mathbb{R}$. If the assumption $f(x,u)/u \le c < \lambda_1$ for u small in Theorem 7.8(i) fails, then one can use a modification of the mountain pass theorem, so called "linking", in order to prove the existence of a nontrivial solution of (2.1) (see [505, Section II.8] and the references therein).

(ii) **Perturbation results.** Consider the problem

$$\left.\begin{array}{ll} -\Delta u = |u|^{p-1}u + \varphi, & x \in \Omega, \\[2mm] u = 0, & x \in \partial\Omega, \end{array}\right\} \tag{7.6}$$

where $\Omega \subset \mathbb{R}^n$ is bounded, $1 < p < p_S$ and $\varphi \in W^{-1,2}(\Omega) := \left(W_0^{1,2}(\Omega)\right)'$. Theorem 7.8(ii) guarantees existence of infinitely many solutions of (7.6) provided $\varphi = 0$. The same result is known to be true for φ belonging to a residual set in $W^{-1,2}(\Omega)$ (see [45]) and for all $\varphi \in W^{-1,2}(\Omega)$ provided $p(n-2) < n$ (see [300, Théorème V.4.6.]; see also [506], [46], [452] and [48]). On the other hand, if $n > 2$, $p \in [p_{sg}, p_S)$ and φ is a general (smooth) function, then even the solvability of (7.6) seems to be open.

(iii) **Unbounded domains.** If $\Omega = \mathbb{R}^n$, then the existence of infinitely many solutions of (2.1) is known in many cases as well. We refer to [76], [140], [139], [9] and the references therein.

(iv) **Critical case.** Let $\Omega \subset \mathbb{R}^n$ be bounded, $p = p_S$ and $\lambda > 0$. If $n \ge 7$, then problem (6.1) possesses infinitely many solutions, see [162]. Such a result is known for any $n \ge 4$ if the domain Ω exhibits suitable symmetries (see [212]) but not for general domains (cf. also the results for Ω being a ball mentioned in Remark 6.9(v)). If $n = 6$ and $\lambda \in (0, \lambda_1)$, then (6.1) has at least two (pairs of) solutions for any bounded Ω, see [117]. Recall also that if $\lambda \le 0$, $p \ge p_S$ and Ω is starshaped, then (6.1) does not possess nontrivial classical solutions due to Corollary 5.2. $\square$

8. Liouville-type results

In order to prove a priori bounds for positive solutions of (2.1) with $f(x, u) \sim u^p$ as $u \to +\infty$, $1 < p < p_S$ (see the rescaling method in Section 12), it will be important to know that the problems

$$-\Delta u = u^p, \qquad x \in \mathbb{R}^n \tag{8.1}$$

and

$$\left. \begin{array}{ll} -\Delta u = u^p, & x \in \mathbb{R}^n_+, \\ u = 0, & x \in \partial\mathbb{R}^n_+, \end{array} \right\} \tag{8.2}$$

do not possess positive bounded (classical) solutions. Here $\mathbb{R}^n_+$ denotes the half-space $\{x \in \mathbb{R}^n : x_n > 0\}$. In fact, we shall see in Chapter II that these Liouville-type results have important applications for parabolic problems as well. In this section we even prove that these problems do not possess any positive classical solution. The following two results are due to [240], [241], except for Theorem 8.1(ii) which was proved in [108].

Theorem 8.1. *Let $\Omega = \mathbb{R}^n$ and $p > 1$.*

(i) *If $p < p_S$, then equation (8.1) does not possess any positive classical solution.*

(ii) *If $p = p_S$, then any positive classical solution of (8.1) is radially symmetric with respect to some point.*

Theorem 8.2. *Let $1 < p \leq p_S$. Then problem (8.2) does not possess any positive classical solution.*

We will see in the next section that the condition $p < p_S$ is optimal for nonexistence in $\mathbb{R}^n$. However, in the case of a half-space and if we consider only *bounded* positive solutions, nonexistence is known for a larger range of exponents, namely $p < p'_S := (n+1)/(n-3)_+$ (note that p'_S is the Sobolev exponent in $(n-1)$ dimensions). This result is due to [150].

Theorem 8.3. *Let $1 < p < p'_S$, where*

$$p'_S := \begin{cases} \infty & \text{if } n \leq 3, \\ (n+1)/(n-3) & \text{if } n > 3. \end{cases}$$

Then problem (8.2) does not possess any positive, bounded classical solution.

On the other hand, under a stronger assumption on p, one can extend the nonexistence result in $\mathbb{R}^n$ to elliptic inequalities. The following result is due to [238].

Theorem 8.4. *Let $1 < p \le p_{sg}$. Then the inequality*

$$-\Delta u \ge u^p, \qquad x \in \mathbb{R}^n \tag{8.3}$$

does not possess any positive classical solution.

Remarks 8.5. (i) It seems unknown if the condition $p \le p_S$ is optimal for the nonexistence of positive solutions of (8.2). In the case of positive bounded solutions, the results recently announced in [178] indicate that the condition $p < p'_S$ can be improved, but the optimal exponent seems to remain unknown.

(ii) The condition $p \le p_{sg}$ in Theorem 8.4 is optimal, as shown by the explicit example $u(x) = k(1 + |x|^2)^{-1/(p-1)}$ with $n \ge 3$, $p > p_{sg}$ and $k > 0$ small enough.

(iii) Consider the inequality $-\Delta u \ge u^p$ in the half-space $\mathbb{R}^n_+$ (no boundary conditions required). Then nonexistence of positive solutions holds whenever $p \le (n+1)/(n-1)$ (see [74]).

(iv) Consider "quasi-solutions" of (8.1), i.e. (nonnegative) functions satisfying the double inequality

$$au^p \le -\Delta u \le u^p, \quad x \in \mathbb{R}^n, \tag{8.4}$$

for some $a \in (0, 1)$. It is shown in [509] that if $1 < p < p_S$ and $a \in (0, 1)$ is close enough to 1, then (8.4) has no positive solution $u \in C^2(\mathbb{R}^n)$ (see also Remark 8.8(ii)). On the other hand, if $p > p_{sg}$ and $a \in (0, 1)$ is small enough, then (8.4) possesses positive solutions $u \in C^2(\mathbb{R}^n)$. Note that a simple example is provided by the function $u(x) = k(1 + |x|^2)^{-1/(p-1)}$ with $k > 0$ large enough. $\quad\square$

We start by proving Theorem 8.4, which is much easier than Theorems 8.1 and 8.2. The following proof (cf. [74], [501], [372]) is based on a rescaled test-function argument, and it is different and simpler than the original proof of [238].

Proof of Theorem 8.4. Take $\xi \in \mathcal{D}(B_1)$, $0 \le \xi \le 1$, with $\xi = 1$ for $|x| \le 1/2$, and let $m = 2p/(p-1)$. Fix $R > 0$ and define $\varphi_R(x) = \xi^m(x/R)$. We observe that

$$\Delta \varphi_R = mR^{-2}\big[\xi^{m-1}\Delta\xi + (m-1)\xi^{m-2}|\nabla\xi|^2\big](x/R)$$

hence

$$|\Delta\varphi_R| \le CR^{-2}\xi^{m-2}(x/R)\chi_{\{|x|>R/2\}} = CR^{-2}\varphi_R^{1/p}\chi_{\{|x|>R/2\}}.$$

Multiplying (8.3) by φ_R, integrating by parts, and using Hölder's inequality, we obtain

$$\int_{\mathbb{R}^n} u^p \varphi_R \le -\int_{\mathbb{R}^n} u\,\Delta\varphi_R \le CR^{-2}\int_{R/2<|x|<R} u\,\varphi_R^{1/p}$$

$$\le CR^{n(p-1)/p-2}\bigg(\int_{R/2<|x|<R} u^p\varphi_R\bigg)^{1/p}. \tag{8.5}$$

In particular, it follows that

$$\int u^p \varphi_R \le C R^{n - 2p/(p-1)}. \tag{8.6}$$

If $p < p_{sg}$, i.e. $n - 2p/(p-1) < 0$, then this implies $u \equiv 0$ upon letting $R \to \infty$. If $p = p_{sg}$, then (8.6) implies $\int_{\mathbb{R}^n} u^p < \infty$. Therefore, the RHS of (8.5) goes to 0 as $R \to \infty$ and we again conclude that $u \equiv 0$. $\square$

Theorem 8.1 is much more delicate. Note that, in view of Theorem 8.4, we may restrict ourselves to $n \ge 3$. We will give a first proof of Theorem 8.1(i) which, like the original proof of [240], is based on integral estimates for (local) positive solutions (cf. Proposition 8.6 below). Here we essentially follow the (simplified) treatment of [83]. Next, we will prove Theorem 8.2 by using moving plane arguments, following [241]. We will then give a second, completely different proof of Theorem 8.1(i), also based on moving planes arguments, which is due to [123], [78] and allows us to prove Theorem 8.1(ii) at the same time. We point out that the techniques of both proofs of Theorem 8.1(i) are important and can be extended to some other problems (see e.g. Section 21 and [123], [78], respectively). Finally, we will prove Theorem 8.3 following [150]. Note that, although the proofs of both Theorems 8.2 and 8.3 are based on moving planes arguments, they use different ideas: reduction to the one-dimensional problem on a half-line for the former, monotonicity and reduction to the $(n-1)$-dimensional problem in the whole space for the latter.

Proposition 8.6. *Let $1 < p < p_S$ and let B_1 be the unit ball in $\mathbb{R}^n$. There exists $r = r(n, p) > \max(n(p-1)/2, p)$ such that if $0 < u \in C^2(B_1)$ is a solution of*

$$-\Delta u = u^p \tag{8.7}$$

in B_1, then

$$\int_{|x|<1/2} u^r \le C(n, p). \tag{8.8}$$

Let us assume for the moment that Proposition 8.6 is proved and deduce some consequences of it. To prove Theorem 8.1(i) it suffices to apply a simple homogeneity argument.

Proof of Theorem 8.1(i). Assume that u is a positive solution of (8.1). Then, for each $R > 0$, $v(x) := R^{2/(p-1)} u(Rx)$ solves (8.7) in B_1. It follows from Proposition 8.6 that

$$\int_{|y|<R/2} u^r(y)\, dy = R^n \int_{|x|<1/2} u^r(Rx)\, dx$$

$$= R^{n - 2r/(p-1)} \int_{|x|<1/2} v^r(x)\, dx \le C(n, p) R^{n - 2r/(p-1)}.$$

By letting $R \to \infty$, we conclude that $\int_{\mathbb{R}^n} u^r(y)\, dy = 0$, a contradiction. $\square$

As another important consequence of Proposition 8.6, we have the following result (cf. [151]) concerning singularities of local solutions to (8.7) in arbitrary domains. Note that when $p_{sg} < p < p_S$, the upper estimate in Theorem 4.1 concerning isolated singularities follows as a special case.

Theorem 8.7. *Let* $1 < p < p_S$ *and let* Ω *be an arbitrary domain in* $\mathbb{R}^n$. *There exists* $C = C(n, p) > 0$ *such that if* $0 < u \in C^2(\Omega)$ *is a solution of*

$$-\Delta u = u^p, \qquad x \in \Omega, \tag{8.9}$$

then
$$u(x) \le C(n, p)[\operatorname{dist}(x, \partial\Omega)]^{-2/(p-1)}. \tag{8.10}$$

Proof. It relies on Proposition 8.6 and a bootstrap argument. Let

$$r > \max(n(p-1)/2, p)$$

be given by Proposition 8.6. We may fix $\rho > 1$ such that

$$p - \frac{1}{\rho} < \frac{2r}{n}. \tag{8.11}$$

Assume that v is a solution of

$$-\Delta v = v^p \quad \text{in } B := \{x \in \mathbb{R}^n : |x| < 1\}. \tag{8.12}$$

Let i be a nonnegative integer and assume that, for all $\omega \subset\subset B$, there exists a constant $C_i(n, p, \omega) > 0$ (independent of v) such that

$$\|v\|_{L^{r\rho^i}(\omega)} \le C_i(n, p, \omega). \tag{8.13}$$

Note that (8.13) is true for $i = 0$ by Proposition 8.6. Since $r\rho^i/p > 1$ and

$$\frac{p}{r\rho^i} - \frac{1}{r\rho^{i+1}} = \frac{1}{r\rho^i}\left(p - \frac{1}{\rho}\right) < \frac{2}{n}$$

due to (8.12), we may apply Proposition 47.6(ii) to deduce that (8.13) is true with i replaced by $i+1$. After a finite number of steps, we obtain $\|v\|_{L^k(\omega)} \le C(n, p, \omega)$ for some $k > n/2$. We may then apply Proposition 47.6(ii) once more to deduce that

$$v(0) \le C(n, p). \tag{8.14}$$

Now assume that u is a solution of (8.9), fix $x_0 \in \Omega$ and let $R := \operatorname{dist}(x_0, \partial\Omega)$. Then $v(x) := R^{2/(p-1)}u(x_0 + Rx)$ solves (8.12) and the conclusion follows from (8.14). $\square$

Remarks 8.8. (i) **More general nonlinearities.** Results similar to Theorem 8.7 for more general nonlinearities can be found in [240], [83], [473], [424]. In particular, universal singularity estimates of the type of (8.10) are established in [424] when the nonlinearity u^p is replaced by any $f(x, u)$ such that $f(x, u) \sim u^p$, as $u \to \infty$, with $1 < p < p_S$. The method of proof is different: The estimate is directly deduced from the Liouville-type Theorem 8.1(i) by using rescaling and doubling arguments (see Theorem 26.8 and Lemma 26.11 below for a similar approach in the parabolic case).

(ii) **Singularities of quasi-solutions.** For "quasi-solutions" of (8.1) (cf. Remark 8.5(iv)), the local behavior near an isolated singularity was studied in [509]. Let $\Omega = B(0, 1) \setminus \{0\}$. If $p_{sg} < p < p_S$ and $a \in (0, 1)$ is close enough to 1, then any positive classical solution of

$$au^p \leq -\Delta u \leq u^p, \qquad x \in \Omega, \tag{8.15}$$

satisfies $\limsup_{x \to 0} |x|^{2/(p-1)} u(x) < \infty$. On the contrary, if $p > p_{sg}$ and $a \in (0, 1)$ is small enough, then there exist solutions of (8.15) with arbitrarily large growth rates as $x \to 0$.

On the other hand, by a straightforward modification of the proof of [424, Theorem 2.1], one can show the following uniform and global property: For each $p \in (1, p_S)$, there exist $a = a(n, p) \in (0, 1)$ and $C(n, p) > 0$ such that, for any domain $\Omega \subset \mathbb{R}^n$, estimate (8.10) is true for any positive solution $u \in C^2(\Omega)$ of (8.15). Note that, as a consequence of this estimate, one recovers the nonexistence result in Remark 8.5(iv).

(iii) **Radial supercritical case.** When $p \geq p_S$, $\Omega = B_R$ and u is a radial positive classical solution of (8.9), a similar argument as in Remark 4.3(iii) shows that $u(r) \leq C(R-r)^{-2/(p-1)}$, $0 \leq r < R$, for some $C > 0$. However the constant C cannot depend only on n, p, since otherwise this would imply nonexistence of radial positive classical solutions of (8.9) for $\Omega = \mathbb{R}^n$ and $p \geq p_S$, hence contradicting Theorem 9.1 below. $\square$

We now turn to the proof of Proposition 8.6. It is based on a key gradient estimate for local solutions of (8.7) (see (8.22) below). To establish this estimate, we prepare the following lemma, which provides a family of integral estimates relating any C^2-function with its gradient and its Laplacian. The proof relies on the Bochner identity (8.18), on the change of variable $v = u^{k+1}$, and on test-functions of the form φv^m.

In the rest of this section, we use the notation $\int = \int_\Omega$ for simplicity.

Lemma 8.9. *Let Ω be an arbitrary domain in $\mathbb{R}^n$, $0 \leq \varphi \in \mathcal{D}(\Omega)$, and $0 < u \in C^2(\Omega)$. Fix $q \in \mathbb{R}$ and denote*

$$I = \int \varphi\, u^{q-2} |\nabla u|^4, \quad J = \int \varphi\, u^{q-1} |\nabla u|^2 \Delta u, \quad K = \int \varphi\, u^q (\Delta u)^2.$$

Then, for any $k \in \mathbb{R}$ with $k \neq -1$, there holds

$$\alpha I + \beta J + \gamma K \leq \frac{1}{2} \int u^q |\nabla u|^2 \Delta\varphi + \int u^q \big[\Delta u + (q-k)u^{-1}|\nabla u|^2\big]\nabla u \cdot \nabla\varphi, \quad (8.16)$$

where

$$\alpha = -\frac{n-1}{n}k^2 + (q-1)k - \frac{q(q-1)}{2}, \quad \beta = \frac{n+2}{n}k - \frac{3q}{2}, \quad \gamma = -\frac{n-1}{n}.$$

Proof. *Step 1.* We first claim that for all $v \in C^2(\Omega)$, $v > 0$ and any $m \in \mathbb{R}$, there holds

$$\frac{m(1-m)}{2}\int \varphi\, v^{m-2}|\nabla v|^4 - \frac{3m}{2}\int \varphi\, v^{m-1}|\nabla v|^2\Delta v - \frac{n-1}{n}\int \varphi\, v^m(\Delta v)^2$$

$$\leq \frac{1}{2}\int v^m|\nabla v|^2\Delta\varphi + \int \big[v^m\Delta v + mv^{m-1}|\nabla v|^2\big]\nabla v \cdot \nabla\varphi.$$

$$(8.17)$$

First note that, by density, it suffices to prove (8.17) for $v \in C^3(\Omega)$. To prove the claim, we start from the identity

$$\frac{1}{2}\Delta|\nabla v|^2 = \nabla(\Delta v) \cdot \nabla v + |D^2 v|^2, \quad (8.18)$$

where $|D^2 u|^2 = \displaystyle\sum_{1 \leq i,j \leq n} (u_{x_i x_j})^2$. Multiplying by $\varphi\, v^m$ and integrating over Ω, we obtain

$$T_1 + T_2 := \int \varphi\, v^m \nabla(\Delta v) \cdot \nabla v + \int \varphi\, v^m |D^2 v|^2 = \frac{1}{2}\int \varphi\, v^m \Delta|\nabla v|^2 =: T_3. \quad (8.19)$$

Integrating by parts and using $\varphi \in \mathcal{D}(\Omega)$, we can rewrite the first and third terms as follows:

$$T_1 = -\int (\Delta v)\nabla \cdot (\varphi\, v^m \nabla v)$$

$$= -\int v^m(\Delta v)\,\nabla v \cdot \nabla\varphi - m\int \varphi\, v^{m-1}|\nabla v|^2\Delta v - \int \varphi\, v^m(\Delta v)^2$$

and

$$T_3 = \int |\nabla v|^2\Big[\frac{1}{2}v^m\Delta\varphi + mv^{m-1}\nabla v \cdot \nabla\varphi + \frac{m}{2}\varphi(v^{m-1}\Delta v + (m-1)v^{m-2}|\nabla v|^2)\Big]$$

$$= \frac{1}{2}\int v^m|\nabla v|^2\Delta\varphi + m\int v^{m-1}|\nabla v|^2\nabla v \cdot \nabla\varphi$$

$$+ \frac{m}{2}\int \varphi\, v^{m-1}|\nabla v|^2\Delta v + \frac{m(m-1)}{2}\int \varphi\, v^{m-2}|\nabla v|^4.$$

Moving the first term of T_1 to the right of (8.19) and the last two terms of T_3 to the left, it follows that

$$\frac{m(1-m)}{2}\int \varphi\, v^{m-2}|\nabla v|^4 - \frac{3m}{2}\int \varphi\, v^{m-1}|\nabla v|^2 \Delta v + \int \varphi\, v^m |D^2 v|^2$$

$$= \int \varphi\, v^m (\Delta v)^2 + \frac{1}{2}\int v^m |\nabla v|^2 \Delta\varphi + \int \left[v^m \Delta v + m v^{m-1}|\nabla v|^2\right] \nabla v \cdot \nabla\varphi.$$

$$(8.20)$$

By Cauchy-Schwarz' inequality (applied with the inner product $(A,B) = \operatorname{tr}(AB^*)$ on matrices), we have

$$(\Delta v)^2 = \left(\operatorname{tr}(D^2 v)\right)^2 \le \operatorname{tr}\left[(D^2 v)(D^2 v)^*\right]\operatorname{tr}(I_n) = n|D^2 v|^2. \qquad (8.21)$$

Due to $\varphi \ge 0$, Claim (8.17) follows by combining (8.20) and (8.21).

Step 2. We set $v = u^{k+1}$, $m = (k+1)^{-1}(q-2k)$, that is $q = (k+1)m+2k$, and we compute

$$\int \varphi\, v^{m-2}|\nabla v|^4 = (k+1)^4 \int \varphi\, u^{(k+1)(m-2)+4k}|\nabla u|^4 = (k+1)^4 I,$$

$$\int \varphi\, v^{m-1}|\nabla v|^2 \Delta v = (k+1)^3 \int \varphi\, u^{(k+1)(m-1)+3k}|\nabla u|^2(\Delta u + ku^{-1}|\nabla u|^2)$$

$$= (k+1)^3 (kI + J),$$

$$\int \varphi\, v^m (\Delta v)^2$$

$$= (k+1)^2 \int \varphi\, u^{(k+1)m+2k}\left[(\Delta u)^2 + 2k(\Delta u)u^{-1}|\nabla u|^2 + k^2 u^{-2}|\nabla u|^4\right]$$

$$= (k+1)^2 (k^2 I + 2kJ + K),$$

$$\int v^m (\Delta v)\nabla v \cdot \nabla\varphi = (k+1)^2 \int u^{(k+1)m+2k}\left[\Delta u + ku^{-1}|\nabla u|^2\right]\nabla u \cdot \nabla\varphi,$$

and

$$\int v^{m-1}|\nabla v|^2 \nabla v \cdot \nabla\varphi = (k+1)^3 \int u^{(k+1)m+2k-1}|\nabla u|^2 \nabla u \cdot \nabla\varphi.$$

Substituting in (8.17) and dividing by $(k+1)^2$, we get

$$\left[\frac{m(1-m)}{2}(k+1)^2 - \frac{3m}{2}k(k+1) - \frac{n-1}{n}k^2\right]I + \left[-\frac{3m}{2}(k+1) - 2k\frac{n-1}{n}\right]J$$

$$- \frac{n-1}{n}K \le \frac{1}{2}\int u^q |\nabla u|^2 \Delta\varphi + \int u^q \left[\Delta u + (k+m(k+1))u^{-1}|\nabla u|^2\right]\nabla u \cdot \nabla\varphi,$$

which readily implies the lemma. $\quad\square$

Lemma 8.10. (i) *Let Ω be an arbitrary domain in $\mathbb{R}^n$, and $0 \leq \varphi \in \mathcal{D}(\Omega)$. Let $0 < u \in C^2(\Omega)$ be a solution of (8.7) in Ω. Fix $q, k \in \mathbb{R}$ with $q > -p$, $k \neq -1$ and denote*

$$I = \int \varphi\, u^{q-2} |\nabla u|^4, \quad K = \int \varphi\, u^{2p+q}.$$

Then there holds

$$\alpha I + \delta K \leq \frac{1}{2} \int u^q |\nabla u|^2 \Delta\varphi + C \int \left[u^{p+q} + u^{q-1}|\nabla u|^2 \right] |\nabla u \cdot \nabla\varphi|, \qquad (8.22)$$

where $C = C(n, p, q, k) > 0$ and

$$\alpha = -\frac{n-1}{n} k^2 + (q-1)k - \frac{q(q-1)}{2}, \quad \delta = \frac{1}{p+q}\left(\frac{3q}{2} - \frac{n+2}{n} k \right) - \frac{n-1}{n}. \quad (8.23)$$

(ii) *Assume that $1 < p < p_S$. Then there exist $q, k \in \mathbb{R}$, with $q \neq -p$, $k \neq -1$, such that the constants α, δ defined in (8.23) satisfy*

$$\alpha, \delta > 0, \qquad 2p + q > n(p-1)/2. \qquad (8.24)$$

Proof. (i) Since $-\Delta u = u^p$, we have

$$(p+q)J = -\int \varphi\,(p+q) u^{p+q-1} |\nabla u|^2 = -\int \varphi\, \nabla u \cdot \nabla(u^{p+q})$$

$$= \int \varphi\,(\Delta u) u^{p+q} + \int (\nabla\varphi \cdot \nabla u) u^{p+q},$$

where J is defined in Lemma 8.9, hence

$$(p+q)J = -\int \varphi\, u^{2p+q} + \int (\nabla\varphi \cdot \nabla u) u^{p+q}.$$

Substituting in (8.16), we obtain (8.22).

(ii) A simple computation shows that $\delta > 0$ and $2p + q > n(p-1)/2$ is equivalent to

$$k < k_0(q) := \frac{q}{2} - \frac{(n-1)p}{n+2} \quad \text{and} \quad q > q_0(p) := \frac{(n-4)p - n}{2}.$$

For $k = k_0(q)$, we have

$$\alpha = \alpha(k_0(q)) = \frac{n-1}{n}\left(-\frac{q^2}{4} + \frac{(n-1)pq}{n+2} - \frac{(n-1)^2 p^2}{(n+2)^2} \right)$$

$$+ (q-1)\left(\frac{q}{2} - \frac{(n-1)p}{n+2} \right) - \frac{q(q-1)}{2}$$

$$= \frac{n-1}{n}\left(-\frac{q^2}{4} + \frac{(n-1)pq}{n+2} - \frac{(n-1)^2 p^2}{(n+2)^2} - \frac{np(q-1)}{n+2} \right)$$

$$= \frac{n-1}{n}\left(-\frac{q^2}{4} - \frac{pq}{n+2} + \frac{n(n+2)p - (n-1)^2 p^2}{(n+2)^2} \right).$$

The discriminant of the above polynomial in q is given by

$$D = \frac{p^2 + n(n+2)p - (n-1)^2 p^2}{(n+2)^2} = \frac{np[(n+2) - (n-2)p]}{(n+2)^2} > 0.$$

Therefore we have $\alpha(k_0(q)) > 0$ for $q \in (-\frac{2p}{n+2} - 2\sqrt{D}, -\frac{2p}{n+2} + 2\sqrt{D})$. Moreover, $-\frac{2p}{n+2} > q_0(p)$ is equivalent to $n(n+2) > (n^2 - 2n - 4)p$, which is true due to $p < p_S$. Choosing

$$q = -\frac{2p}{n+2} \quad \text{and} \quad k = k_0(q)^- = \left(-\frac{np}{n+2}\right)^- \quad \text{(with } k \neq -1\text{)},$$

we see that (8.24) is fulfilled. $\square$

Proof of Proposition 8.6. Take q, k as in Lemma 8.10(ii) and $\Omega = B_1$. We shall estimate the terms on the RHS of (8.22). Let $\xi \in \mathcal{D}(\Omega)$, be such that $\xi = 1$ for $|x| \leq 1/2$ and $0 \leq \xi \leq 1$. Put $\theta = (3p + 1 + 2q)/2(2p + q) \in (0, 1)$. By taking $\varphi = \xi^m$ with $m = 2/(1 - \theta)$, we have

$$|\nabla\varphi| \leq C\xi^{m-1} \leq C\varphi^\theta, \quad |\Delta\varphi| \leq C\xi^{m-2} \leq C\varphi^\theta. \tag{8.25}$$

Fix $\varepsilon > 0$. Using Young's inequality under the form

$$xyz \leq \varepsilon x^a + \varepsilon y^b + C(\varepsilon)z^c, \qquad a^{-1} + b^{-1} + c^{-1} = 1,$$

and (8.25), we obtain

$$\int u^q |\nabla u|^2 \Delta\varphi = \int \left(\varphi^{1/2} u^{(q-2)/2} |\nabla u|^2\right) \left(\varphi^{(q+2)/2(2p+q)} u^{(q+2)/2}\right)$$
$$\times \left(\varphi^{-(p+1+q)/(2p+q)} \Delta\varphi\right) \leq \varepsilon \int \varphi\, u^{q-2} |\nabla u|^4 + \varepsilon \int \varphi\, u^{2p+q} + C(\varepsilon),$$

$$C \int u^{p+q} |\nabla u \cdot \nabla\varphi| \leq \int \left(\varphi^{1/4} u^{(q-2)/4} |\nabla u|\right) \left(\varphi^{(4p+3q+2)/4(2p+q)} u^{(4p+3q+2)/4}\right)$$
$$\times \left(\varphi^{-(3p+1+2q)/2(2p+q)} |\nabla\varphi|\right) \leq \varepsilon \int \varphi\, u^{q-2} |\nabla u|^4 + \varepsilon \int \varphi\, u^{2p+q} + C(\varepsilon),$$

and

$$C \int u^{q-1} |\nabla u|^2 |\nabla u \cdot \nabla\varphi| \leq \int \left(\varphi^{3/4} u^{3(q-2)/4} |\nabla u|^3\right) \left(\varphi^{(q+2)/4(2p+q)} u^{(q+2)/4}\right)$$
$$\times \left(\varphi^{-(3p+1+2q)/2(2p+q)} |\nabla\varphi|\right) \leq \varepsilon \int \varphi\, u^{q-2} |\nabla u|^4 + \varepsilon \int \varphi\, u^{q+2p} + C(\varepsilon).$$

Combining this with (8.22), we obtain

$$\alpha I + \delta K \leq C(n, p, q, k)\varepsilon(I + K) + C(\varepsilon).$$

Since $\alpha, \delta > 0$, by choosing ε sufficiently small, we conclude that $I, K \leq C$, hence (8.8) with $r = 2p + q > \max(n(p-1)/2, p)$. $\square$

Proof of Theorem 8.2. Let u be a positive solution of (8.2).

Assume $n \geq 2$ and denote $x' = (x_1, \ldots, x_{n-1})$. Choose $\bar{x}, \tilde{x} \in \mathbb{R}^n_+$ with $\bar{x}_n = \tilde{x}_n$. We will show $u(\bar{x}) = u(\tilde{x})$ so that u depends only on x_n.

Choose the origin to be the point $\left(\left(\frac{\bar{x}+\tilde{x}}{2}\right)', 0\right)$. Given $x \in \overline{\mathbb{R}^n_+}$, set

$$z = \frac{x + e_n}{|x + e_n|^2}, \qquad v(z) = |x + e_n|^{n-2} u(x) = \frac{u(x)}{|z|^{n-2}}.$$

The function v is the **Kelvin transform** of u. It solves the problem

$$\left. \begin{array}{rl} \Delta v + |z|^\gamma v^p = 0 & \text{in } D, \\[2mm] v = 0 & \text{on } \partial D \setminus \{0\}, \end{array} \right\} \tag{8.26}$$

where $D := B_{1/2}(e_n/2)$ and $\gamma := (n-2)p - (n+2) \leq 0$. We want to show that v is axisymmetric about the z_n axis, i.e. $v = v(|z'|, z_n)$. Choose any direction e perpendicular to the z_n-axis. Without loss of generality we may assume $e = e_1$.

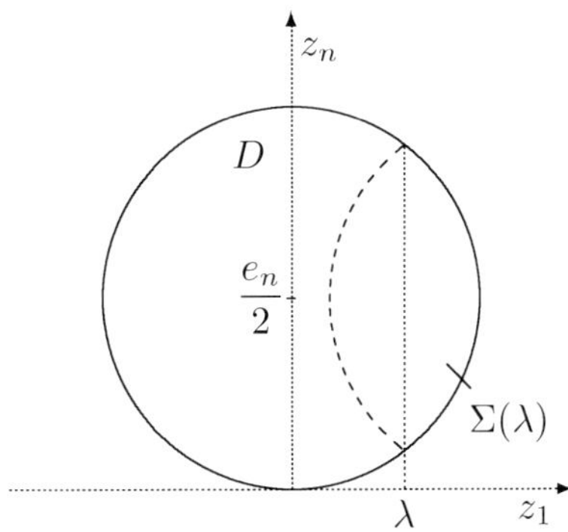

Figure 4: Moving planes.

We shall apply the **moving planes method** to problem (8.26). Given $\lambda \in [0, 1/2)$, set $\Sigma(\lambda) := \{z \in D : z_1 > \lambda\}$, $z^\lambda := (2\lambda - z_1, z_2, \ldots, z_n)$. The point z^λ is the **reflection** of z with respect to the hyperplane $\{z_1 = \lambda\}$ and $\Sigma(\lambda)$ is called a **cap**. We next define

$$w(z; \lambda) := v(z^\lambda) - v(z) \qquad \text{for } z \in \overline{\Sigma}(\lambda)$$

(the parameter λ will be omitted in w when no risk of confusion arises). Then

$$\Delta w = \Delta v(z^\lambda) - \Delta v(z) = -|z^\lambda|^\gamma v^p(z^\lambda) + |z|^\gamma v^p(z)$$
$$= \left(|z|^\gamma - |z^\lambda|^\gamma\right) v^p(z^\lambda) - |z|^\gamma \left(v^p(z^\lambda) - v^p(z)\right).$$

Since $v^p(z^\lambda) - v^p(z) = p\xi^{p-1} w(z;\lambda)$ for some $\xi = \xi(z,\lambda)$ lying between $v(z^\lambda)$ and $v(z)$, we obtain

$$\Delta w + |z|^\gamma p\xi^{p-1} w = \left(|z|^\gamma - |z^\lambda|^\gamma\right) v^p(z^\lambda) \geq 0 \quad \text{in } \Sigma(\lambda) .$$

The maximum principle (see Proposition 52.1) implies $v > 0$ in D and $\partial v / \partial \nu < 0$ on $\partial D \setminus \{0\}$, hence $w \geq 0$ on $\Sigma(\lambda)$ for λ close to $1/2$.

Set

$$\bar{\mu} := \inf\{\mu > 0 : w \geq 0 \text{ in } \Sigma(\lambda) \text{ for all } \lambda \geq \mu\}$$

and assume $\bar{\mu} > 0$. Then $w \geq 0$ on $\Sigma(\bar{\mu})$ and there exist $\lambda_i \in (0, \bar{\mu})$, $\lambda_i \to \bar{\mu}$, such that $\inf\{w(z;\lambda_i) : z \in \Sigma(\lambda_i)\} < 0$. Since $w(\cdot;\lambda_i) \geq 0$ on $\partial \Sigma(\lambda_i)$, this infimum is attained at some $q_i \in \Sigma(\lambda_i)$ and $\nabla w(q_i, \lambda_i) = 0$. Since $w(\cdot;\lambda_i) \geq 0$ in an ε-neighborhood of $\partial D \cap \overline{\Sigma(\lambda_i)}$ (with ε being independent of i), we may assume $q_i \to \bar{q} \in \overline{\Sigma(\bar{\mu})} \setminus \partial D$. Continuity arguments and $w \geq 0$ on $\Sigma(\bar{\mu})$ guarantee $w(\bar{q};\bar{\mu}) = 0$ and $\nabla w(\bar{q};\bar{\mu}) = 0$, hence $w(\cdot;\bar{\mu}) \equiv 0$ by the maximum principle. This contradicts $w > 0$ on $\{z \in \partial \Sigma(\bar{\mu}) : z_1 > \bar{\mu}\}$. Consequently, $\bar{\mu} = 0$ and $w(\cdot;0) \geq 0$ on $\Sigma(0)$. A symmetric argument shows $w(\cdot;0) \leq 0$ on $\Sigma(0)$, hence v is symmetric with respect to the hyperplane $\{e_1 = 0\}$. Since this holds for any hyperplane containing the z_n-axis, v is axially symmetric. Hence, $u = u(|x'|, x_n)$ and, consequently, $u(\bar{x}) = u(\tilde{x})$.

Thus we have reduced the problem to the case $n = 1$. Assume that u is a positive solution of

$$u''(t) + u^p(t) = 0, \qquad t > 0,$$
$$u(0) = 0.$$

Since u is concave and positive for $t > 0$, it must fulfill $u' \geq 0$. Fix $t_1 > 0$ and consider $t > t_1$. Then

$$u(t) = u(t_1) + (t - t_1)u'(t_1) + \int_{t_1}^{t} (t - s)u''(s)\, ds.$$

Since $u''(s) = -u^p(s) \leq -u^p(t_1)$, we obtain

$$0 < u(t) \leq u(t_1) + (t - t_1)u'(t_1) - \frac{1}{2}(t - t_1)^2 u^p(t_1),$$

hence

$$u^p(t_1) < \frac{2u(t_1)}{(t - t_1)^2} + \frac{2u'(t_1)}{t - t_1} \to 0 \qquad \text{as } t \to +\infty,$$

a contradiction. $\square$

We now turn to the proof of Theorem 8.1 based on moving planes.

Proof of Theorem 8.1. Due to Theorem 8.4, we may assume $n \geq 3$. Let $p \leq p_S$ and let u be a positive classical solution of (8.1). Set

$$v(z) := \frac{1}{|z|^{n-2}} u\left(\frac{z}{|z|^2}\right), \qquad z \in \mathbb{R}^n \setminus \{0\}$$

(v is the Kelvin transform of u). We have $v \in C(\mathbb{R}^n \setminus \{0\})$, $v > 0$,

$$v(z) \leq C|z|^{2-n} \qquad \text{as } |z| \to \infty, \tag{8.27}$$

and v solves the equation

$$\Delta v + |z|^\gamma v^p = 0 \quad \text{in } \mathbb{R}^n \setminus \{0\}, \tag{8.28}$$

where $\gamma := (n-2)p - (n+2) \leq 0$. Due to (8.28) and $n \geq 3$, we infer from Lemma 4.4 that $\Delta v \leq 0$ in $\mathcal{D}'(\mathbb{R}^n)$. It follows from the maximum principle in Proposition 52.3(ii) that, for each $R > 0$,

$$v \geq \eta(R) := \min_{|x|=R} v > 0 \quad \text{in } B_R(0) \setminus \{0\}. \tag{8.29}$$

Given $\lambda \leq 0$, set $z^\lambda := (2\lambda - z_1, z_2, \ldots, z_n)$, $\Sigma(\lambda) := \{z \in \mathbb{R}^n : z_1 < \lambda\}$, $\Sigma'(\lambda) := \Sigma(\lambda) \setminus \{0^\lambda\}$ and

$$w(z; \lambda) := v(z^\lambda) - v(z), \quad z \in \overline{\Sigma(\lambda)} \setminus \{0^\lambda\}$$

(the parameter λ will be omitted in w when no risk of confusion arises). As in the preceding theorem we obtain

$$\Delta w + |z|^\gamma p \xi^{p-1} w \leq 0 \quad \text{in } \Sigma'(\lambda), \tag{8.30}$$

where $\xi = \xi(z, \lambda)$ lies between $v(z^\lambda)$ and $v(z)$. Set $\alpha := (n-2)/2$ and $\tilde{w}(z; \lambda) = |z|^\alpha w(z; \lambda)$. Then

$$\Delta \tilde{w} - \frac{n-2}{|z|^2} z \cdot \nabla \tilde{w} + c(z, \lambda)\tilde{w} \leq 0 \quad \text{in } \Sigma'(\lambda), \tag{8.31}$$

where

$$c(z, \lambda) := -\frac{(n-2)^2}{4|z|^2} + |z|^\gamma p \xi^{p-1}(z, \lambda).$$

Let us first show that

$$\tilde{w} \geq 0 \quad \text{in } \Sigma'(\lambda), \quad \text{for } \lambda \ll -1. \tag{8.32}$$

We shall argue by contradiction. Assume that $\lambda_i \to -\infty$ and $\inf_{\Sigma'(\lambda_i)} \tilde{w}(\cdot; \lambda_i) < 0$. By (8.27) and (8.29) with $R = 1$, we have $\tilde{w}(z; \lambda_i) \geq 0$ if $|z - 0^{\lambda_i}| < 1$ and i is large enough. Since also, for each i,

$$\tilde{w}(z; \lambda_i) = 0 \text{ on } \partial\Sigma(\lambda_i) \quad \text{and} \quad \tilde{w}(z; \lambda_i) \to 0, \ |z| \to \infty, \tag{8.33}$$

we see that the infimum of $\tilde{w}(\cdot; \lambda_i)$ over $\Sigma'(\lambda_i)$ is attained at some $q_i \in \Sigma'(\lambda_i)$ and $|q_i - 0^{\lambda_i}| \geq 1$. We have $|q_i| \to \infty$, thus $v(q_i) \to 0$. If the sequence $\{q_i^{\lambda_i}\}$ were bounded, then (8.29) would imply $v(q_i^{\lambda_i}) \geq c_1 > 0$, hence $w(q_i) > 0$ for i large, a contradiction. Therefore $|q_i^{\lambda_i}| \to \infty$. Now the definition of v implies $v(z)|z|^{n-2} \to u(0)$ if $|z| \to \infty$, so that we cannot have $|q_i^{\lambda_i}|/|q_i| \to 0$ (otherwise $w(q_i) > 0$ for large i). Thus both $v(q_i)$ and $v(q_i^{\lambda_i})$ can be estimated above by Cq_i^{2-n} for some fixed $C > 0$ and the same is true for $\xi(q_i, \lambda_i)$. Hence,

$$c(q_i, \lambda_i) \leq -\frac{(n-2)^2}{4q_i^2} + \frac{Cp}{q_i^4} < 0 \qquad \text{for } i \text{ large enough.} \tag{8.34}$$

Since $\tilde{w} = \tilde{w}(\cdot; \lambda_i)$ attains an interior minimum at q_i, we have $\Delta\tilde{w}(q_i) \geq 0$, $\nabla\tilde{w}(q_i) = 0$ and $\tilde{w}(q_i) < 0$ so that (8.31) and (8.34) yield a contradiction. This proves (8.32).

Now denote

$$\bar{\mu} := \sup\{\mu \leq 0 : \tilde{w}(\cdot; \lambda) \geq 0 \text{ in } \Sigma'(\lambda) \text{ for all } \lambda \leq \mu\}$$

and assume $\bar{\mu} < 0$. Then $\tilde{w}(\cdot, \bar{\mu}) \geq 0$ in $\Sigma'(\bar{\mu})$ by continuity, and there exist $\lambda_i > \bar{\mu}$, $\lambda_i \to \bar{\mu}$, such that $\inf\{\tilde{w}(z; \lambda_i) : z \in \Sigma'(\lambda_i)\} < 0$. Assume that $\tilde{w}(\cdot, \bar{\mu})$ is not identically zero. Since $\Delta w(\cdot, \bar{\mu}) \leq 0$ in $\Sigma'(\bar{\mu})$, the maximum principle (see Proposition 52.1) implies $w(\cdot, \bar{\mu}) > 0$ in $\Sigma'(\bar{\mu})$. Arguing similarly as for (8.29), we deduce that $w(\cdot, \bar{\mu}) \geq c_2 > 0$ in $U := B_{\bar{\mu}/2}(0^{\bar{\mu}}) \setminus \{0^{\bar{\mu}}\}$. Due to the continuity of v in U and

$$w(z; \lambda_i) = w\big(z - 2(\lambda_i - \bar{\mu})e_1; \bar{\mu}\big) + v\big(z - 2(\lambda_i - \bar{\mu})e_1\big) - v(z), \qquad e_1 := (1, 0, \ldots, 0),$$

we obtain $w(\cdot; \lambda_i) \geq 0$ (hence $\tilde{w}(\cdot; \lambda_i) \geq 0$) in $B_{\bar{\mu}/4}(0^{\lambda_i}) \setminus \{0^{\lambda_i}\}$ for i large. Consequently, in view of (8.33), the infimum of $\tilde{w}(\cdot; \lambda_i)$ over $\Sigma'(\lambda_i)$ has to be attained at some $q_i \in \Sigma'(\lambda_i)$, with $|q_i - 0^{\lambda_i}| \geq \bar{\mu}/4$. Assume $|q_i| \to \infty$. Then $|q_i^{\lambda_i}|/|q_i| \to 1$ and we obtain a contradiction as above (cf. (8.34)). Therefore we may assume that $\{q_i\}$ is bounded and $q_i \to \bar{q} \in \overline{\Sigma(\bar{\mu})} \setminus \{0^{\bar{\mu}}\}$. By continuity and $\tilde{w}(\cdot, \bar{\mu}) \geq 0$, we obtain $\tilde{w}(\bar{q}, \bar{\mu}) = 0$ and $\nabla\tilde{w}(\bar{q}, \bar{\mu}) = 0$, hence $w(\bar{q}, \bar{\mu}) = 0$ and $\nabla w(\bar{q}, \bar{\mu}) = 0$. Applying the maximum principle in Proposition 52.1(ii) and (iii) to equation (8.30), it follows that $w(\cdot, \bar{\mu}) \equiv 0$, hence $\tilde{w}(\cdot, \bar{\mu}) \equiv 0$, a contradiction. Consequently, $\tilde{w}(\cdot, \bar{\mu}) \equiv 0$, which means that v is symmetric with respect to $\{z_1 = \bar{\mu}\}$. Now using (8.28) we see that $(-\Delta v)/v^p = |z|^\gamma$ has the same symmetry, which is not possible unless $p = p_S$.

If $p < p_S$, then we get $\bar{\mu} = 0$, so that $w(\cdot, 0) \geq 0$ and $v(z^0) \geq v(z)$ provided $z_1 \leq 0$. Considering the function $\tilde{v}(z) := v(z^0)$ instead of v we obtain the reversed inequality, hence $v(z_1, z_2, \ldots, z_n) = v(-z_1, z_2, \ldots, z_n)$. Repeating this procedure with any given direction instead of e_1 we see that v, hence u, are radially symmetric (about zero). If we repeat this procedure with $\tilde{u}(x) = u(x - x_0)$ for a given $x_0 \neq 0$ instead of u, we show that u is radially symmetric about the point x_0. Since this is true for any x_0, the function u has to be constant. But the only constant solution of (8.1) is the trivial solution.

If $p = p_S$ and $\bar{\mu} < 0$, then v is symmetric with respect to $\{z_1 = \bar{\mu}\}$. If $\bar{\mu} = 0$, then we can repeat the procedure with $\tilde{v}(z) := v(z^0)$ and in any case we obtain the symmetry of v with respect to $\{z_1 = \tilde{\mu}\}$ for suitable $\tilde{\mu}$. Now we can repeat the above proof with directions $e_2, e_3, \ldots, e_n$ instead of e_1 and we obtain the existence of $\bar{z} \in \mathbb{R}^n$ such that v is symmetric with respect to $\{z_k = \bar{z}_k\}$ for $k = 1, 2, \ldots, n$, hence $v(\bar{z} + z) = v(\bar{z} - z)$ for all z. Rotating the coordinate system and repeating the procedure we find $\tilde{z} \in \mathbb{R}^n$ such that $v(\tilde{z} + z) = v(\tilde{z} - z)$ for all z. Assume $\bar{z} \neq \tilde{z}$. Without loss of generality, we may assume $\bar{z} \neq 0$. The symmetry relations for v imply

$$v(\bar{z}) = v(2\tilde{z} - \bar{z}) = v(3\bar{z} - 2\tilde{z}) = v(4\tilde{z} - 3\bar{z}) = \cdots \to 0,$$

hence $v(\bar{z}) = 0$, a contradiction. Consequently, $\bar{z} = \tilde{z}$ and we obtain the rotational symmetry of v (hence of u) about $\bar{z}$. $\square$

Proof of Theorem 8.3. Assume that (8.2) admits a positive, bounded classical solution u. As a special case of Theorem 21.10 below (which we shall prove by using moving planes arguments), it follows that u is nondecreasing in x_n:

$$\partial_{x_n} u(x) \geq 0, \qquad x \in \mathbb{R}_+^n.$$

Therefore, for each $x' \in \mathbb{R}^{n-1}$,

$$U(x') := \lim_{x_n \to \infty} u(x', x_n)$$

is well defined and is a bounded positive function. Take now $\varphi \in \mathcal{D}(\mathbb{R}^{n-1})$ and $\psi \in \mathcal{D}(\mathbb{R})$, with $\operatorname{supp} \psi \subset (0, 1)$ and $\int_0^1 \psi = 1$. Let $k > 0$. Testing the equation with $\varphi(x')\psi(x_n - k)$, we have

$$-\int_{\mathbb{R}^{n-1}} \int_{\mathbb{R}} u^p\, \varphi(x')\, \psi(x_n - k)\, dx_n\, dx' = \int_{\mathbb{R}^{n-1}} \int_{\mathbb{R}} \varphi(x')\, \psi(x_n - k)\, \Delta u\, dx_n\, dx'$$

$$= \int_{\mathbb{R}^{n-1}} \int_{\mathbb{R}} u\, \Delta\big(\varphi(x')\, \psi(x_n - k)\big)\, dx_n\, dx',$$

hence

$$-\int_{\mathbb{R}^{n-1}} \int_{\mathbb{R}} u^p(x', s + k)\, \varphi(x')\, \psi(s)\, ds\, dx'$$

$$= \int_{\mathbb{R}^{n-1}} \int_{\mathbb{R}} u(x', s + k)\, \Delta\big(\varphi(x')\, \psi(s)\big)\, ds\, dx'.$$

By dominated convergence, letting $k \to \infty$, it follows that

$$-\int_{\mathbb{R}^{n-1}} U^p(x')\,\varphi(x')\,dx' = -\int_{\mathbb{R}^{n-1}} U^p(x')\,\varphi(x')\int_0^1 \psi(s)\,ds\,dx'$$

$$= \int_{\mathbb{R}^{n-1}}\int_{\mathbb{R}} U(x')\,\Delta\big(\varphi(x')\,\psi(s)\big)\,ds\,dx'.$$

But the RHS is equal to

$$\int_{\mathbb{R}^{n-1}} U(x')\,\Delta_{x'}\varphi(x')\,dx'\int_0^1 \psi(s)\,ds + \int_{\mathbb{R}^{n-1}} U(x')\,\varphi(x')\,dx'\int_0^1 \psi''(s)\,ds$$

$$= \int_{\mathbb{R}^{n-1}} U(x')\,\Delta_{x'}\varphi(x')\,dx'.$$

It follows that U solves (8.1) in $\mathbb{R}^{n-1}$ in the distribution sense, hence in the classical sense (this is a consequence of the boundedness of U and of Remark 47.4). The result is then a consequence of Theorem 8.1(i). $\square$

9. Positive radial solutions of $\Delta u + u^p = 0$ in $\mathbb{R}^n$

In this section we study positive radial classical solutions of the equation

$$-\Delta u = u^p, \qquad x \in \mathbb{R}^n. \tag{9.1}$$

Since this problem does not possess positive classical solutions if $1 < p < p_S$ due to Theorem 8.1, we restrict ourselves to the case $p \geq p_S$. Consequently, $n \geq 3$.

Positive radial classical solutions of (9.1) can be written in the form $u(x) = U(r)$, where $r = |x|$ and $U \in C^2([0, \infty))$ is a positive classical solution of

$$U'' + \frac{n-1}{r}U' + U^p = 0, \quad r \in (0, \infty), \qquad U'(0) = 0. \tag{9.2}$$

It is easily seen that prescribing initial values $U(0) = \alpha > 0$, $U'(0) = 0$, the equation in (9.2) has a unique solution for r small enough. In fact, this equation can be written in the form $(r^{n-1}U')' = -r^{n-1}U^p$ and, by integration we obtain the equivalent integral equation

$$U(r) = \alpha - \int_0^r\int_0^s \left(\frac{t}{s}\right)^{n-1} U^p(t)\,dt\,ds,$$

which can be solved by the Banach fixed point theorem.

Let $U_*(r) = c_p r^{-2/(p-1)}$ be the singular solution defined in (3.9) and set

$$p_{JL} := \begin{cases} +\infty & \text{if } n \leq 10, \\ 1 + 4\frac{n-4+2\sqrt{n-1}}{(n-2)(n-10)} & \text{if } n > 10. \end{cases} \tag{9.3}$$

The main result of this section is the following theorem.

Theorem 9.1. *Let $p \geq p_S$. Given $\alpha > 0$, problem (9.2) possesses a unique positive solution $U_\alpha \in C^2([0,\infty))$ satisfying $U_\alpha(0) = \alpha$. This solution is decreasing and we have*

$$U_\alpha(r) = \alpha U_1(\alpha^{(p-1)/2} r). \tag{9.4}$$

If $p > p_S$, then $r^{2/(p-1)} U_\alpha(r) \to c_p$ as $r \to \infty$. If $p = p_S$, then

$$U_1(r) = \left(\frac{n(n-2)}{n(n-2) + r^2} \right)^{(n-2)/2}. \tag{9.5}$$

Let $\alpha_1 > \alpha_2 > 0$. If $p \geq p_{JL}$, then $U_*(r) > U_{\alpha_1}(r) > U_{\alpha_2}(r)$ for all $r > 0$. If $p_S < p < p_{JL}$, then U_{α_1} and U_{α_2} intersect infinitely many times and U_{α_1}, U_* intersect infinitely many times as well. If $p = p_S$, then $U_{\alpha_1}, U_{\alpha_2}$ intersect once and U_{α_1}, U_* intersect twice.

Proof. Using the transformation

$$w(s) = r^{2/(p-1)} U(r), \qquad s = \log r, \tag{9.6}$$

problem (9.2) becomes

$$w'' + \beta w' + w^p - \gamma w = 0, \qquad s \in \mathbb{R}, \tag{9.7}$$

where

$$\beta := \frac{1}{p-1}\big((n-2)p - (n+2)\big) \geq 0, \qquad \gamma := c_p^{p-1} = \frac{2}{(p-1)^2}\big((n-2)p - n\big) > 0,$$

and we are looking for solutions w satisfying $w(s), w'(s) \to 0$ as $s \to -\infty$. Set

$$\mathcal{E}(w) = \mathcal{E}(w, w') := \frac{1}{2}|w'|^2 - \frac{\gamma}{2}w^2 + \frac{1}{p+1}w^{p+1}.$$

Then $\mathcal{E}$ is a Lyapunov functional for (9.7); more precisely,

$$\frac{d}{ds}\mathcal{E}\big(w(s)\big) = -\beta\big(w'(s)\big)^2 \leq 0. \tag{9.8}$$

Denoting $x := w$ and $y := w'$, problem (9.7) can be written in the form

$$\begin{pmatrix} x \\ y \end{pmatrix}' = \begin{pmatrix} y \\ -\beta y - x^p + \gamma x \end{pmatrix} =: F(x, y) \tag{9.9}$$

where $x > 0$ and $(x, y) \to (0, 0)$ as $s \to -\infty$. Problem (9.9) possesses two equilibria, $(0,0)$ and $(c_p, 0)$ lying in the half-space $\{(x, y) : x \geq 0\}$. Denote

$$A_1 := \nabla F(0,0) = \begin{pmatrix} 0 & 1 \\ \gamma & -\beta \end{pmatrix}, \qquad A_2 := \nabla F(c_p, 0) = \begin{pmatrix} 0 & 1 \\ -\gamma(p-1) & -\beta \end{pmatrix}.$$

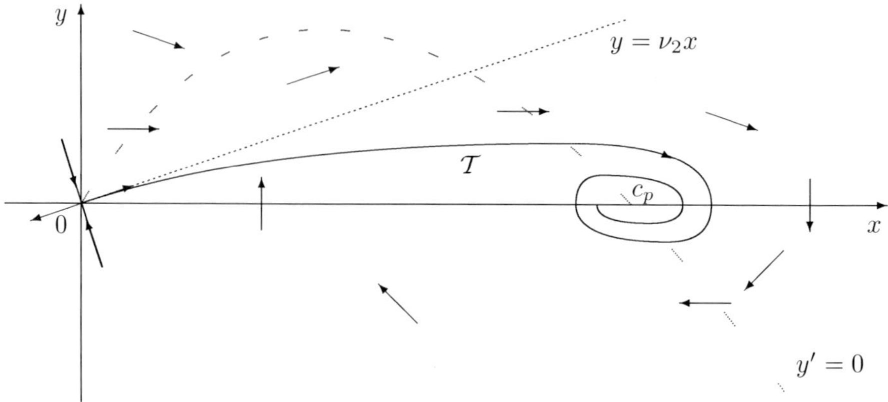

Figure 5: The flow generated by (9.9) for $p_S < p < p_{JL}$.

First consider the case $p > p_S$. Then $\beta > 0$ and the matrix A_1 has two real eigenvalues $\nu_{1,2} := -\frac{1}{2}\big(\beta \pm \sqrt{\beta^2 + 4\gamma}\big)$ with $\nu_1 < 0 < \nu_2 = 2/(p-1)$. The corresponding eigenvectors (x_i, y_i) satisfy $y_i = \nu_i x_i$, $i = 1, 2$. The eigenvalues $\tilde{\nu}_{1,2} := -\frac{1}{2}\big(\beta \pm \sqrt{\beta^2 - 4\gamma(p-1)}\big)$ of A_2 are real iff $\beta^2 \geq 4\gamma(p-1)$, that is iff $p \geq p_{JL}$.

Assume $p_S < p < p_{JL}$. In this case, the eigenvalues $\tilde{\nu}_1, \tilde{\nu}_2$ are complex and their real parts are negative so that the critical point $(c_p, 0)$ is a stable spiral. The flow for the planar system (9.9) is illustrated in Figure 5.

We are interested in the trajectory $\mathcal{T}$ emanating from the origin to the right half-space, since it represents the graph of any positive solution of (9.7) in the w-w' plane. This trajectory cannot hit the axis $x = 0$ again since the energy functional $\mathcal{E}$ is nonnegative on this axis, $\mathcal{E}(0,0) = 0$, $\beta > 0$ and (9.8) is true. Moreover, the corresponding solutions w exists for all $s \in \mathbb{R}$ and w, w' remain bounded for all $s \in \mathbb{R}$ due to (9.8). Consequently, $\mathcal{T}$ has to converge to the critical point $(c_p, 0)$ which corresponds to the singular solution $w_*(s) = r^{2/(p-1)}U_*(r) \equiv c_p$. Thus, if U_α is the unique local solution of (9.2) such that $U_\alpha(0) = \alpha > 0$, then its transform $w_\alpha(s) = r^{2/(p-1)}U_\alpha(r)$ exists globally and satisfies $w_\alpha(s) \to c_p$ as $s \to \infty$. Consequently, U_α exists globally and $r^{2/(p-1)}U_\alpha(r) \to c_p$ as $r \to \infty$. It is easily verified that the function $\tilde{U}_\alpha(r) := \alpha U_1(\alpha^{(p-1)/2}r)$ is a solution of (9.2) satisfying $\tilde{U}_\alpha(0) = \alpha$, hence $\tilde{U}_\alpha = U_\alpha$ by uniqueness. The graphs of w_α and w_1 in the w-w' plane are identical, so that there exists $s_\alpha \in \mathbb{R}$ such that $U_\alpha(e^s) = w_\alpha(s) = w_1(s - s_\alpha)$ for all $s \in \mathbb{R}$. Hence, given $\alpha_1 > \alpha_2 > 0$, $U_{\alpha_1}(r) = U_{\alpha_2}(r)$ for some $r > 0$ iff $w_1(s - s_{\alpha_1}) = w_1(s - s_{\alpha_2})$ for some $s \in \mathbb{R}$. This happens for infinitely many s since $\mathcal{T}$ spirals around the point $(c_p, 0)$. Similarly, $w_{\alpha_1}(s) = c_p$

for infinitely many s, hence U_{α_1} and U_* intersect infinitely many times.

Next consider the case $p \geq p_{JL}$. On the halfline $y = -\frac{\beta}{2}(x - c_p)$, $x < c_p$, we have for suitable $x_\theta \in (x, c_p)$:

$$\frac{y'}{x'} = -\beta - \frac{x}{y}(x^{p-1} - \gamma) = -\beta + \frac{2x(x^{p-1} - c_p^{p-1})}{\beta(x - c_p)}$$

$$= -\beta + \frac{2}{\beta}x(p-1)x_\theta^{p-2} < -\beta + \frac{2}{\beta}(p-1)\gamma \leq -\frac{\beta}{2}.$$

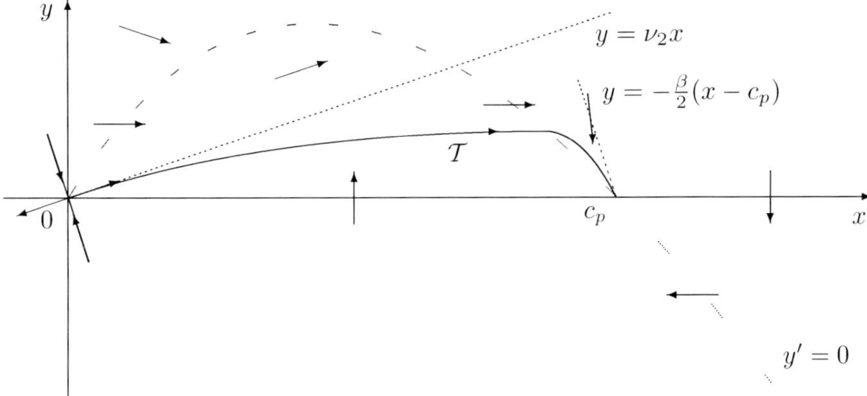

Figure 6: The flow generated by (9.9) for $p \geq p_{JL}$.

Consequently, the trajectory $\mathcal{T}$ ends up at $(c_p, 0)$ again but the x-coordinate is increasing along $\mathcal{T}$ (see Figure 6). Hence, the solutions U of (9.2) are ordered according to their values at $r = 0$, $U_* > U_{\alpha_1} > U_{\alpha_2}$ if $\alpha_1 > \alpha_2$.

Finally consider the case $p = p_S$. Then $\beta = 0$ and the energy functional $\mathcal{E}$ is constant along any solution. Since $\mathcal{E}(c_p, 0) < 0$ and $\mathcal{E}(0, y) > 0$ for $y \neq 0$, the trajectory $\mathcal{T}$ is a homoclinic orbit (see Figure 7).

Let w_α, s_α have the same meaning as above. Given $\alpha_1 \neq \alpha_2$, there exists a unique $s \in \mathbb{R}$ such that $w_1(s - s_{\alpha_1}) = w_1(s - s_{\alpha_2})$. Hence, the corresponding solutions $U_{\alpha_1}, U_{\alpha_2}$ of (9.2) intersect exactly once. Similarly, given $\alpha > 0$, we have $w_\alpha(s) = c_p$ for two values of s, so that U_α and U_* intersect twice. One can easily check that the function U_1 defined by (9.5) is a solution of (9.2) satisfying the initial condition $U_1(0) = 1$. $\square$

Remarks 9.2. (i) The exponent p_{JL} appeared for the first time in [293] where the authors studied mainly problems with the nonlinearities $f(u) = \lambda(1 + au)^p$ and $f(u) = \lambda e^u$, $\lambda, a > 0$.

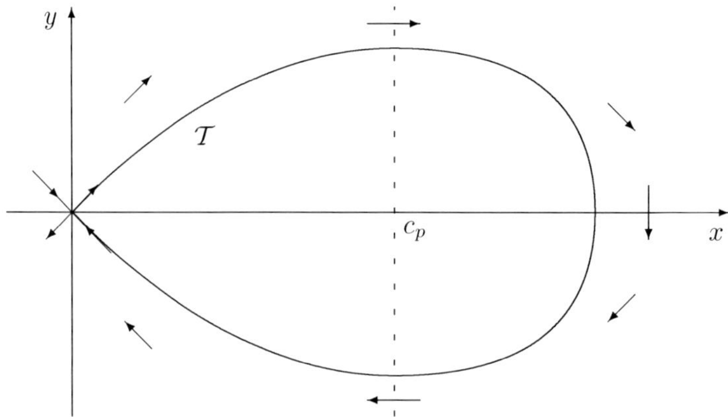

Figure 7: The flow generated by (9.9) for $p = p_S$.

(ii) The intersection properties of the solutions U in Theorem 9.1 play an important role in the study of stability and asymptotic behavior of solutions of the corresponding parabolic problem, see Sections 22, 23. □

Remark 9.3. Let $p = p_S$ and $a > 0$. For all $\alpha \geq M_0(a)$ with $M_0(a) > 0$ large enough, if V is a positive classical solution of

$$V'' + \frac{n-1}{r}V' + V^p = 0, \qquad 0 < r < a,$$

such that $V(a) = U_\alpha(a)$ and $\lim_{r \to 0} V(r) = \infty$, then V has to intersect U_α in $(0, a)$.

In fact, denoting $w_\alpha(s) := r^{2/(p-1)}U_\alpha(r)$, $s = \log r$, the rescaled function from the last proof, it suffices to chose $M_0(a)$ such that

$$w'_{M_0(a)}(\log a) < 0 \qquad\qquad (9.10)$$

(hence $w'_\alpha(\log a) < 0$ for all $\alpha \geq M_0(a)$). Indeed the trajectory of $W(s) := r^{2/(p-1)}V(r)$, $s \in (-\infty, \log a)$, has to be a subset of a periodic orbit lying inside the trajectory $\mathcal{T}$ (see Figure 7). Due to (9.10) there exists $s_0 \in (-\infty, \log a)$ such that $w_\alpha(s_0) = W(s_0)$, hence $U_\alpha(e^{s_0}) = V(e^{s_0})$.

Note also that there exist infinitely many periodic orbits of (9.7) for $p = p_S$, corresponding to positive singular solutions of $u'' + \frac{n-1}{r}u' + u^p = 0$ for $r > 0$. □

Remark 9.4. Let $p > p_{JL}$. Since the trajectory $\mathcal{T}$ approaches the limit point $(c_p, 0)$ below the dotted line with slope $-\beta/2$ and $\tilde{\nu}_2 < -\beta/2 < \tilde{\nu}_1 < 0$, it has to converge along the eigenvector $(1, \tilde{\nu}_1)$ corresponding to the eigenvalue $\tilde{\nu}_1$, hence

$$\frac{y(s)}{x(s) - c_p} \to \tilde{\nu}_1 \quad \text{as} \quad s \to \infty.$$

Returning to the original variables and denoting $V(r) := U(r) - U_*(r)$ we obtain

$$\lim_{r \to \infty} \frac{rV'(r)}{V(r)} = \tilde{\nu}_1 - m, \tag{9.11}$$

where $m := 2/(p-1)$. Assuming that $V(r) = cr^{-\alpha} + h.o.t.$ for some $c \neq 0$ and $\alpha > m$, (9.11) guarantees $c < 0$ and $\alpha = m + \lambda_-$, where

$$\lambda_- := -\tilde{\nu}_1 = \frac{1}{2}\left(\beta - \sqrt{\beta^2 - 4\gamma(p-1)}\right)$$
$$= \frac{1}{2}\left(n - 2 - 2m - \sqrt{(n-2-2m)^2 - 8(n-2-m)}\right).$$

This expansion is indeed true: In fact, a more precise asymptotic expansion of V was established in [260] and [334]. $\square$

10. A priori bounds via the method of Hardy-Sobolev inequalities

A priori estimates of solutions can be used for the proof of existence and multiplicity results. Unlike the variational methods in sections 6 and 7, this approach does not require any variational structure of the problem and enables one to prove the existence of continuous branches of solutions.

Due to Theorem 7.8(ii) one cannot hope for a priori estimates of all solutions. The bifurcation diagrams in Figure 2 suggest that there is some hope for such estimates if we restrict ourselves to positive solutions and to the subcritical case.[3]

In the present and the following three sections we introduce four different methods which are often used in the proofs of a priori bounds for positive solutions of superlinear elliptic problems. We will study mainly the scalar problem

$$\left.\begin{aligned} -\Delta u &= f(x, u, \nabla u), & x \in \Omega, \\ u &= 0, & x \in \partial\Omega, \end{aligned}\right\} \tag{10.1}$$

where Ω is bounded and f is a sufficiently smooth function with superlinear growth in the u-variable. Some of the possible generalizations and modifications will be mentioned as remarks, others can be found in the subsequent chapters.

This section is devoted to the method of [99], which is based on a Hardy-type inequality and enables one to treat rather general nonlinearities f. On the

---

[3]In fact, in the subcritical case one can get a priori estimates of all solutions with bounded Morse indices (without the positivity assumption), see [49], [539], [32].

other hand, it requires an upper growth restriction corresponding to the limiting exponent

$$p_{BT} := \begin{cases} \infty & \text{if } n = 1, \\ (n+1)/(n-1) & \text{if } n > 1, \end{cases}$$

which is stronger than what is imposed by the methods in Sections 12 and 13 (for instance, in the particular case $f(x, u, \nabla u) = u^p$, we have to assume $p < p_{BT}$). However, the exponent p_{BT} is not technical and its role will be clarified in the next section.

Theorem 10.1. *Let $\Omega \subset \mathbb{R}^n$ be bounded, $n \geq 3$, $\beta := p_{BT}$. Let $f : \overline{\Omega} \times \mathbb{R}_+ \times \mathbb{R}^n \to \mathbb{R}_+$ be continuous and bounded on $\overline{\Omega} \times M \times \mathbb{R}^n$ for $M \subset \mathbb{R}_+$ bounded. Let*

$$\liminf_{u \to \infty} \frac{f(x, u, s)}{u} > \lambda_1, \quad \lim_{u \to \infty} \frac{f(x, u, s)}{u^\beta} = 0, \quad \text{uniformly for } (x, s) \in \overline{\Omega} \times \mathbb{R}^n. \tag{10.2}$$

Then there exists $C > 0$ with the following property: If $t \geq 0$ and $u \in H_0^1 \cap L^\infty(\Omega)$ is a positive variational solution of

$$\left. \begin{array}{ll} -\Delta u = f(x, u, \nabla u) + t\varphi_1, & x \in \Omega, \\ u = 0, & x \in \partial\Omega, \end{array} \right\} \tag{10.3}$$

then

$$\|u\|_\infty + t \leq C. \tag{10.4}$$

Proof. We shall denote by C various positive constants which may vary from step to step but which are independent of u and t. Let $t \geq 0$ and u be a positive solution of (10.3). The proof of (10.4) will consist of the following three steps:

1. $\int_\Omega u\delta \, dx \leq C$, $t \leq C$ and $\int_\Omega f(x, u, \nabla u)\delta \, dx \leq C$,
2. $\|\nabla u\|_2 \leq C$,
3. $\|u\|_\infty \leq C$.

Step 1. Due to (10.2) there exist $C_1 > \lambda_1$ and $C_2 > 0$ such that $f(x, u, s) \geq C_1 u - C_2$ for all (x, u, s). Multiplying the equation in (10.3) by φ_1 yields

$$\lambda_1 \int_\Omega u\varphi_1 \, dx = \int_\Omega u(-\Delta\varphi_1) \, dx = \int_\Omega (-\Delta u)\varphi_1 \, dx = \int_\Omega (f\varphi_1 + t\varphi_1^2) \, dx$$
$$\geq C_1 \int_\Omega u\varphi_1 \, dx - C_2 \int_\Omega \varphi_1 \, dx + t \int_\Omega \varphi_1^2 \, dx, \tag{10.5}$$

where $f = f(x, u(x), \nabla u(x))$. This estimate can be written in the form

$$(C_1 - \lambda_1) \int_\Omega u\varphi_1 \, dx + t \int_\Omega \varphi_1^2 \, dx \leq C,$$

hence

$$\int_\Omega u\varphi_1 \, dx \le C \quad \text{and} \quad t \le C. \tag{10.6}$$

Now (10.5) and $\delta \le C\varphi_1$ guarantee

$$\int_\Omega f\delta \, dx \le C\int_\Omega f\varphi_1 \, dx = C\lambda_1 \int_\Omega u\varphi_1 \, dx - Ct \int_\Omega \varphi_1^2 \, dx \le C. \tag{10.7}$$

Step 2. Multiplying the equation in (10.3) by u yields

$$\|\nabla u\|_2^2 = \int_\Omega |\nabla u|^2 \, dx = \int_\Omega fu \, dx + t\int_\Omega \varphi_1 u \, dx \le \int_\Omega fu \, dx + C. \tag{10.8}$$

Denoting $\alpha := 2/(n+1) \in (0,1)$ we have $\beta + 1/(1-\alpha) = 2/(1-\alpha)$. Given $\varepsilon > 0$ there exists $C_\varepsilon > 1$ such that

$$f(x,u,s) \le \varepsilon u^\beta + C_\varepsilon. \tag{10.9}$$

Using Hölder's inequality, Step 1, (10.9) and Lemma 50.4 we obtain

$$\int_\Omega fu \, dx = \int_\Omega (f^\alpha \delta^\alpha)\left(f^{1-\alpha}\frac{u}{\delta^\alpha}\right) dx \le \left(\int_\Omega f\delta \, dx\right)^\alpha \left(\int_\Omega f\frac{u^{1/(1-\alpha)}}{\delta^{\alpha/(1-\alpha)}} \, dx\right)^{1-\alpha}$$

$$\le \varepsilon^{1-\alpha}\left(\int_\Omega \frac{u^{\beta+1/(1-\alpha)}}{\delta^{\alpha/(1-\alpha)}} \, dx\right)^{1-\alpha} + C_\varepsilon\left(\int_\Omega \frac{u^{1/(1-\alpha)}}{\delta^{\alpha/(1-\alpha)}} \, dx\right)^{1-\alpha}$$

$$= \varepsilon^{1-\alpha}\left\|\frac{u}{\delta^{\alpha/2}}\right\|_{2/(1-\alpha)}^2 + C_\varepsilon\left\|\frac{u}{\delta^\alpha}\right\|_{1/(1-\alpha)} \le \varepsilon^{1-\alpha}C\|\nabla u\|_2^2 + CC_\varepsilon\|\nabla u\|_2.$$

This estimate and (10.8) guarantee

$$\|\nabla u\|_2 \le C. \tag{10.10}$$

Step 3. Choose $p \in (n/2, n)$. Then

$$W^{2,p}(\Omega) \hookrightarrow L^\infty(\Omega) \quad \text{and} \quad W^{1,2}(\Omega) \hookrightarrow L^{p(\beta-1)}(\Omega)$$

due to $n(\beta - 1) < 2^*$. These embeddings, L^p-estimates (see Appendix A), (10.9), Step 1 and (10.10) imply

$$\|u\|_\infty \le C\|u\|_{2,p} \le C\|f + t\varphi_1\|_p \le \varepsilon\|u^\beta\|_p + C(C_\varepsilon + 1)$$
$$\le \varepsilon\|u\|_{p(\beta-1)}^{\beta-1}\|u\|_\infty + \tilde{C}_\varepsilon \le \varepsilon\|\nabla u\|_2^{\beta-1}\|u\|_\infty + \tilde{C}_\varepsilon \le \varepsilon C\|u\|_\infty + \tilde{C}_\varepsilon.$$

Now choosing $\varepsilon > 0$ small enough yields $\|u\|_\infty < C$. $\square$

Remarks 10.2. (i) The proof of Theorem 10.1 can be easily modified for more general second-order elliptic differential operators. In the case of a nonsymmetric operator one has to work with the first eigenfunction of the adjoint operator, of course. One could also allow more general nonlinearities (nonlocal, for example). The boundedness assumption on f could be relaxed as well.

(ii) The term $t\varphi_1$ in (10.3) is needed for the proof of existence of a positive solution of (10.3) with $t = 0$ (see Corollary 10.3 below). This lower order term does not play any significant role in a priori estimates in the following sections provided $t \le C$. Since this bound for t was proved in Step 1 of the proof of Theorem 10.1 by using only the lower bound for f in (10.2), in the following sections we shall restrict ourselves to the case $t = 0$ only.

(iii) A priori estimates of solutions of problems like (10.3) appeared first in [400] and [517]. The assumptions on the growth of f or the dimension n in these articles are more restrictive than those in Theorem 10.1 which is due to [99]. $\quad\square$

Corollary 10.3. *Let Ω and f be as in Theorem 10.1 and let*

$$\limsup_{u \to 0+} \frac{f(x, u, s)}{u} < \lambda_1 \qquad \text{uniformly for } (x, s) \in \overline{\Omega} \times \mathbb{R}^n. \tag{10.11}$$

Then problem (10.3) with $t = 0$ possesses at least one positive solution u, with $u \in W^{2,q} \cap C_0(\Omega)$ for all finite q.

Proof. Set $X := C^1(\overline{\Omega})$. Given $u \in X$ and $t \ge 0$, let $F_t(u) = w$ be the unique solution of the linear problem

$$\left.\begin{aligned} -\Delta w &= f(x, u, \nabla u) + t\varphi_1, &\quad x \in \Omega, \\ w &= 0, &\quad x \in \partial\Omega \end{aligned}\right\} \tag{10.12}$$

(cf. Theorem 47.3(i)). Note that, since $f(\cdot, u, \nabla u) \in L^\infty(\Omega)$, we have $u \in W^{2,q} \cap C_0(\Omega)$ for all finite q. Then $F_t : X \to X$ is compact and we are looking for a positive fixed point of F_0.

Let $\|u\|_X = r \ll 1$, $\tau \in [0, 1]$ and assume $\tau F_0(u) = u$. Multiplying the equation in (10.12) by u and applying (10.11) yield

$$\int_\Omega |\nabla u|^2 \, dx = \tau \int_\Omega fu \, dx \le (\lambda_1 - \varepsilon) \int_\Omega u^2 \, dx,$$

which contradicts (1.3). Hence $\tau F_0(u) \ne u$ and the homotopy invariance of the topological degree implies

$$\deg\left(I - F_0, 0, B_r\right) = \deg\left(I, 0, B_r\right) = 1, \tag{10.13}$$

where I denotes the identity and $B_r := \{u \in X : \|u\|_X < r\}$.

Let $\|u\|_X = R$. If R is large enough, then Theorem 10.1 and L^p-estimates (see Appendix A) imply $F_t(u) \neq u$ for any $t \geq 0$. The same theorem implies also $F_T(u) \neq u$ provided T is large enough. Consequently,

$$\deg\left(I - F_0, 0, B_R\right) = \deg\left(I - F_T, 0, B_R\right) = 0. \tag{10.14}$$

Now (10.13) and (10.14) guarantee $\deg\left(I - F_0, 0, B_R \setminus \bar{B}_r\right) = -1$, hence there exists $u \in B_R \setminus \bar{B}_r$ such that $F_0(u) = u$. The positivity of u is a consequence of the maximum principle. $\quad\square$

In what follows we present an alternative proof of Theorem 10.1 in the special case $f(x, u, s) = |u|^{p-1}u$, $1 < p < p_{BT}$, $n \geq 1$. Instead of Hardy's inequality we shall use the following lemma (see [89], [450], and cf. also [143] and the references in [450, Remark 4.1]). It provides a useful singular test-function and will also be used later in Section 26.

Lemma 10.4. *Assume Ω bounded and $0 < \alpha < 1$. Then the problem*

$$\left.\begin{array}{ll} -\Delta\xi = \varphi_1^{-\alpha}, & x \in \Omega, \\ \xi = 0, & x \in \partial\Omega \end{array}\right\} \tag{10.15}$$

admits a unique classical solution $\xi \in C(\overline{\Omega}) \cap C^2(\Omega)$. Moreover, we have $\varphi_1^{-\alpha} \in L^1(\Omega)$, $\xi \in H_0^1(\Omega)$, and

$$\xi(x) \leq C(\Omega, \alpha)\delta(x), \quad x \in \Omega. \tag{10.16}$$

Proof. Define $h(s) = 3s - s^{2-\alpha}$, $s \geq 0$. The function $h \in C^1([0, \infty)) \cap C^2((0, \infty))$ satisfies

$$h' = 3 - (2 - \alpha)s^{1-\alpha}, \quad -h'' = (2 - \alpha)(1 - \alpha)s^{-\alpha}, \quad s > 0$$

and

$$h(s) \leq 3s, \quad h'(s) \geq 1, \quad \text{for all } s \in [0, 1].$$

Let $\varphi = \|\varphi_1\|_\infty^{-1}\varphi_1$, and set $v(x) = h(\varphi(x))$. Simple computation yields

$$\begin{aligned} -\Delta v &= -h''(\varphi)|\nabla\varphi|^2 - h'(\varphi)\Delta\varphi \\ &= C_1\varphi^{-\alpha}|\nabla\varphi|^2 + \lambda_1 h'(\varphi)\varphi \\ &\geq C_1\varphi^{-\alpha}|\nabla\varphi|^2 + \lambda_1\varphi. \end{aligned}$$

Now, for $\delta(x) \leq \varepsilon$ small enough, we have $|\nabla\varphi|^2 \geq \eta > 0$, hence $-\Delta v \geq C_1\eta\varphi^{-\alpha}$. On the other hand, for $\delta(x) \geq \varepsilon$, we have $\varphi \geq c > 0$, hence $-\Delta v \geq \lambda_1 c \geq C_2\varphi^{-\alpha}$. We conclude that for some $c > 0$, $w := cv$ satisfies

$$-\Delta w \geq \varphi_1^{-\alpha} \quad \text{and} \quad w(x) \leq C_3\delta(x), \quad \text{for all } x \in \Omega. \tag{10.17}$$

Next, for all $\varepsilon > 0$, let ξ_ε be the (classical) solution of $-\Delta\xi_\varepsilon = (\varphi_1 + \varepsilon)^{-\alpha}$ in Ω, with $\xi_\varepsilon = 0$ on $\partial\Omega$. By (10.17) and the maximum principle, we have

$$\xi_\varepsilon(x) \le w(x) \le C_3\delta(x) \le C_4, \quad x \in \Omega \tag{10.18}$$

and ξ_ε is increasing as ε decreases to 0. Denote by ξ the (pointwise) limit of ξ_ε. Elliptic estimates along with (10.18) imply that $\xi \in C(\overline\Omega) \cap C^2(\Omega)$, that ξ satisfies (10.16) and is a classical solution of (10.15). The uniqueness follows immediately from the maximum principle.

The fact that $\varphi_1^{-\alpha} \in L^1(\Omega)$ can be easily deduced from the inequality $\varphi_1 \ge c\delta$, by flattening the boundary and using a partition of unity (see e.g. [485] for details). Finally, to show that $\xi \in H_0^1(\Omega)$, it suffices to note that, since $\alpha < 1$,

$$\int_\Omega |\nabla\xi_\varepsilon|^2 = -\int_\Omega \xi_\varepsilon\Delta\xi_\varepsilon = \int_\Omega \xi_\varepsilon(\varphi_1 + \varepsilon)^{-\alpha} \le C_4 \int_\Omega \varphi_1^{-\alpha} < \infty. \quad \square$$

Alternative proof of Theorem 10.1 for $f = u^p$, $t = 0$. Let $\varepsilon > 0$ be small and $\alpha := r'/r$, where r is defined by $1/r = 1/2 - \varepsilon/(p-1)$. Let ξ be the solution of (10.15). As in Step 1 of the proof of Theorem 10.1 we obtain $\int_\Omega u^p\delta\, dx \le C$. Testing the equation with ξ, we obtain

$$\int_\Omega u\varphi_1^{-\alpha}\, dx = \int_\Omega \nabla u \cdot \nabla\xi\, dx = \int_\Omega (-\Delta u)\xi\, dx = \int_\Omega u^p\xi\, dx \le C$$

(where we used $\varphi_1^{-\alpha} \in L^1(\Omega)$ and $\xi \in H_0^1(\Omega)$). Denoting $p_\varepsilon := (p+1)/2 - \varepsilon$, we get

$$\int_\Omega u^{p_\varepsilon}\, dx = \int_\Omega \left(u^{p/r}\varphi_1^{1/r}\right)\left(u^{1/r'}\varphi_1^{-1/r}\right) dx$$

$$\le \left(\int_\Omega u^p\varphi_1\, dx\right)^{1/r}\left(\int_\Omega u\varphi_1^{-\alpha}\, dx\right)^{1/r'} \le C.$$

Define $\theta \in (0, p+1)$ by $\theta/p_\varepsilon + (p+1-\theta)/2^* = 1$. Then $p + 1 - \theta < 2$ provided ε is small enough and the interpolation inequality yields

$$\int_\Omega |\nabla u|^2\, dx = \int_\Omega u^{p+1}\, dx = \|u\|_{p+1}^{p+1} \le \|u\|_{p_\varepsilon}^\theta \|u\|_{2^*}^{p+1-\theta} \le C\|\nabla u\|_2^{p+1-\theta},$$

which guarantees a bound for u in $W^{1,2}(\Omega)$. The rest of the proof is the same as in the proof of Theorem 10.1 (Step 3). $\quad\square$

11. A priori bounds via bootstrap in L^p_δ-spaces

This section is devoted to the L^p_δ bootstrap method, which, in the scalar case, was developed independently in [85], [449]. It applies to problem (10.1) under essentially the same assumptions on the nonlinearities f as in the method of the previous section, with a growth restriction still given by the exponent p_{BT} of Section 10. However, unlike that method (and those in the next two sections), it applies to very weak solutions. The optimality of the L^p_δ bootstrap method was studied in [489] and it turns out that the exponent p_{BT} is optimal for the regularity of very weak solutions, thus showing the critical role played by this exponent for problems of the form (10.1).

Let us point out that in the case of systems, studied in [449], the growth restrictions of the L^p_δ bootstrap method become much weaker than those imposed by the (generalization of the) method of Hardy-Sobolev inequalities (see Section 31).

In this section, by a solution u of (10.1), we understand a very weak (or L^1_δ-) solution, cf. Definition 3.1. Namely, if f does not depend on ∇u, this means that

$$u \in L^1(\Omega), \quad f(\cdot, u) \in L^1_\delta(\Omega), \tag{11.1}$$

and

$$-\int_\Omega u \Delta\varphi = \int_\Omega f(\cdot, u)\varphi, \quad \text{for all } \varphi \in C^2(\overline{\Omega}), \ \varphi_{|\partial\Omega} = 0. \tag{11.2}$$

If f depends on ∇u, we assume in addition that ∇u is a function, i.e. $\nabla u \in L^1_{\mathrm{loc}}(\Omega)$ and we replace $f(\cdot, u)$ by $f(\cdot, u, \nabla u)$ in (11.1)–(11.2).

Remark 11.1. If $u \in L^1(\Omega)$ and $\Delta u \in L^1_\delta(\Omega)$ (where Δu is understood in the distribution sense), we say that $u = 0$ on $\partial\Omega$ in the weak sense if

$$\int_\Omega u \, \Delta\varphi = \int_\Omega \varphi \, \Delta u \quad \text{for all } \varphi \in C^2(\overline{\Omega}), \ \varphi_{|\partial\Omega} = 0.$$

If (11.1) is satisfied (and $\nabla u \in L^1_{\mathrm{loc}}(\Omega)$ in case f depends on ∇u), then u is a very weak solution of (10.1) if and only if it solves the differential equations in (10.1) in the distribution sense and the boundary conditions in the weak sense. $\square$

Theorem 11.2. *Assume Ω bounded and $1 < p < p_{BT}$. Let $f : \overline{\Omega} \times \mathbb{R}_+ \times \mathbb{R}^n \to \mathbb{R}_+$ be continuous. Assume*

$$f(x, u, s) \le C_1(1 + u^p), \quad x \in \Omega, \quad u \ge 0, \quad s \in \mathbb{R}^n \tag{11.3}$$

and

$$f(x, u, s) \ge \lambda u - C_1, \quad x \in \Omega, \quad u \ge 0, \quad s \in \mathbb{R}^n \qquad \text{for some } \lambda > \lambda_1. \tag{11.4}$$

There exists $C > 0$ such that if u is a nonnegative very weak solution of (10.1), then $u \in L^\infty(\Omega)$ and

$$\|u\|_\infty \leq C.$$

Condition (11.4) can be weakened or replaced by other conditions of different form. For instance, by applying the same method, we obtain regularity and a priori estimates for the following simple equation:

$$\left. \begin{aligned} -\Delta u &= a(x)u^p, & x \in \Omega, \\ u &= 0, & x \in \partial\Omega. \end{aligned} \right\} \tag{11.5}$$

Theorem 11.3. *Assume Ω bounded and $a \in L^\infty(\Omega)$, $a \geq 0$, $a \not\equiv 0$ and $1 < p < p_{BT}$. Then the conclusions of Theorem 11.2 remain valid for problem (11.5).*

Remarks 11.4. (i) The growth condition (11.3) in Theorem 11.2 is slightly stronger than that in Theorem 10.1 (where (10.2) allows some "almost critical" f's).

(ii) Under the assumptions of Theorems 11.2 and 11.3, as a consequence of standard regularity results for linear elliptic equations, we moreover obtain $u \in C_0 \cap W^{2,q}(\Omega)$ for all finite q (argue similarly as in the proof of Corollary 3.4, using the uniqueness part of Theorem 49.1 instead of Proposition 52.3). $\quad\square$

The optimality of the exponent p_{BT} in Theorems 11.2 and 11.3 is shown by the following result from [489].

Theorem 11.5. *Assume Ω bounded and $p > p_{BT}$. Then there exists a function $a \in L^\infty(\Omega)$, $a \geq 0$, $a \not\equiv 0$, such that problem (11.5) admits a positive very weak solution u such that*

$$u \notin L^\infty(\Omega).$$

The method of proof of Theorems 11.2–11.3 is based on bootstrap and uses the L^p_δ regularity theory of the Laplacian (cf. Theorem 49.2 and Proposition 49.5 in Appendix C).

Proof of Theorem 11.2. *Step 1.* Initialization. By (10.6), (10.7) in the proof of Theorem 10.1, we know that

$$\|u\|_{1,\delta} \leq C, \qquad \|f(\cdot, u, \nabla u)\|_{1,\delta} \leq C. \tag{11.6}$$

Since $p < p_{BT}$, we may fix $\rho > 1$ and k_0 such that

$$\max\left(p, \frac{n+1}{2}\left(p - \frac{1}{\rho}\right)\right) < k_0 < \frac{n+1}{n-1}.$$

By (11.6) and Proposition 49.5, it follows that $\|u\|_{k_0,\delta} \leq C$.

Step 2. Bootstrap. Put $k_i = k_0 \rho^i$, $i = 1, 2, \ldots$. Assume that there holds

$$\|u\|_{k_i,\delta} \leq C(i) \tag{11.7}$$

for some $i \geq 0$ (this is true for $i = 0$ by Step 1). Since

$$\frac{p}{k_i} - \frac{1}{k_{i+1}} = \frac{1}{k_0 \rho^i}\left(p - \frac{1}{\rho}\right) < \frac{2}{n+1},$$

by using Theorem 49.2(i) and (11.3), we obtain

$$\|u\|_{k_{i+1},\delta} \leq C\|\Delta u\|_{k_i/p,\delta} = C\|f\|_{k_i/p,\delta}$$
$$\leq C(1 + \|v^p\|_{k_i/p,\delta}) = C(1 + \|v\|^p_{k_i,\delta}) \leq C.$$

By induction, it follows that (11.7) is true for all integers i. Taking i large enough, we thus have (11.7) for some $k_i > (n+1)p/2$. Applying Theorem 49.2(i) and (11.3) once more, and Remark 1.1, we obtain $\|u\|_\infty \leq C$. $\quad\square$

Proof of Theorem 11.3. We only need to modify Step 1, the bootstrap step being then unchanged.

Assume that u is a nonnegative (very weak) solution of (11.5). It follows from the quantitative version of Hopf's lemma (see Remark 49.12(i) in Appendix C) that

$$u \geq c\left(\int_\Omega au^p \delta\, dy\right)\delta \geq c_1\left(\int_\Omega au^p \varphi_1\, dy\right)\varphi_1,$$

for some constant $c_1 > 0$ depending only on Ω. We deduce that

$$\int_\Omega au^p \varphi_1\, dx \geq c_1^p\left(\int_\Omega au^p \varphi_1\, dx\right)^p \int_\Omega a\varphi_1^{p+1}\, dx \geq 2\int_\Omega au^p \varphi_1\, dx - C,$$

hence

$$\lambda_1 \int_\Omega u\varphi_1\, dx = \int_\Omega au^p \varphi_1\, dx \leq C. \quad\square$$

We now turn to the proof of Theorem 11.5. It is based on Lemma 49.13 from Appendix C, where a singular solution of the linear Laplace equation with an appropriate right-hand side belonging to L^1_δ is constructed. The right-hand side has to possess suitable boundary singularities, supported in a conical subdomain of Ω. In order to re-construct a posteriori the coefficient $a(x)$, the key point is the lower estimate (11.8) for the solution in the same cone.

Proof of Theorem 11.5. Assume that $0 \in \partial\Omega$ without loss of generality. Let $\alpha = 2/(p-1)$. By assumption, we have $\alpha < n - 1$. By Lemma 49.13, there exist $R > 0$ and a revolution cone Σ_1 of vertex 0, with $\Sigma := \Sigma_1 \cap B_{2R} \subset \Omega$, such that the function

$$\phi := |x|^{-(\alpha+2)}\chi_\Sigma$$

belongs to L^1_δ and such that the (very weak) solution $u > 0$ of

$$\left.\begin{aligned} -\Delta u &= \phi, & x \in \Omega, \\ u &= 0, & x \in \partial\Omega \end{aligned}\right\}$$

satisfies

$$u \geq C|x|^{-\alpha}\chi_\Sigma. \tag{11.8}$$

Therefore, we have $u \notin L^\infty$ and

$$u^p \geq C|x|^{-\alpha p}\chi_\Sigma = C|x|^{-(\alpha+2)}\chi_\Sigma = C\phi.$$

Setting $a(x) = \phi/u^p \geq 0$, we get $-\Delta u = \phi = a(x)u^p$ and $a(x) \leq 1/C$, hence $a \in L^\infty$. The proof is complete. $\square$

Remarks 11.6. Localization of singularities. (a) In Theorem 11.5, it is to be noted that, in spite of the imposed homogeneous Dirichlet boundary condition, the singularity of the solution occurs at a boundary point, actually a single point. The boundary conditions continue to be satisfied not only in the weak sense but also in the sense of traces (see Remark 49.4(c) in Appendix C).

(b) If we assume that $p < p_{sg}$ and that a given weak solution of (11.5) is bounded near the boundary, then one can use usual Lebesgue spaces instead of L^p_δ-spaces in the proof of Theorem 11.2, to show that the solution is bounded in Ω. Therefore, the occurrence of boundary singularities is necessary if $p_{BT} < p < p_{sg}$. On the other hand, when $p > p_{sg}$, the situation is different and much easier, since it is then not difficult to construct examples of similar equations with only an interior singularity (see Remarks 3.6).

(c) The support of a in Theorem 11.5 can be localized in an arbitrarily small neighborhood of a boundary point. However, it is also possible to construct an example where the function a is positive in Ω, uniformly away from $\partial\Omega$ (see [489] for details). $\square$

Remarks 11.7. (a) **The cases** $f(u) = u^p$ **and** $p = p_{BT}$. Similar counterexamples as in Theorem 11.5 have been constructed recently in [155] for the model problem (3.10) $(a(x) \equiv 1)$ when $p > p_{BT}$ is close to p_{BT}. Moreover the critical case $p = p_{BT}$ was shown to belong to the singular case. Related results have also been announced in [81].

(b) **Variable critical exponents in nonsmooth domains.** The notion of very weak solution has been recently extended in [362] to the case of some nonsmooth domains, namely Lipschitz domains, and generalizations of Theorems 11.2 and 11.5 have been obtained. For suitable cone-shaped domains, the analogue of the exponent p_{BT} was computed. Interestingly, it was found to depend on the domain and to be smaller than $(n+1)/(n-1)$. $\square$

12. A priori bounds via the rescaling method

In this section we present a priori estimates of solutions of (10.3) based on rescaling and Liouville-type theorems. In this context, this method was first used in [241]. In comparison to the method of Section 10, it requires a rather precise asymptotic behavior for f as $u \to \infty$ (f has to behave like u^p for u large) but the growth condition on f is optimal ($p < p_S$). The method also works for general second-order elliptic operators but for simplicity we restrict ourselves to the Laplace operator. As explained in Remark 10.2(ii) we consider the case $t = 0$ only.

Theorem 12.1. *Assume Ω bounded, $1 < p < p_S$, $a \in C(\overline{\Omega})$, $a(x) \geq a_0 > 0$ for all $x \in \overline{\Omega}$, $g \in C(\overline{\Omega} \times \mathbb{R} \times \mathbb{R}^n)$, and*

$$|g(x, u, s)| \leq C\big(1 + |u|^q + |s|^r\big), \qquad where \ \ q < p, \ r < \frac{2p}{p+1}. \tag{12.1}$$

Then there exists $C > 0$ such that any positive strong solution $u \in C^1(\overline{\Omega})$ of

$$\left.\begin{aligned} -\Delta u &= a(x)u^p + g(x, u, \nabla u), & x &\in \Omega, \\ u &= 0, & x &\in \partial\Omega \end{aligned}\right\} \tag{12.2}$$

satisfies $\|u\|_\infty \leq C$.

Remark 12.2. Here, u being a strong solution means that $u \in W^{2,1}_{loc}(\Omega)$ and u satisfies the differential equation a.e. in Ω. Since we also assume $u \in C^1(\overline{\Omega})$, Remarks 47.4(i) and (iii), actually imply $u \in W^{2,q}(\Omega)$ for all finite q. $\square$

Proof of Theorem 12.1. Assume the contrary. Then there exist positive solutions u_j of (12.2) such that $\|u_j\|_\infty \to \infty$ as $j \to \infty$. Let $x_j \in \Omega$ be such that

$$u_j(x_j) + |\nabla u_j(x_j)|^{2/(p+1)} = \sup_\Omega\big(u_j + |\nabla u_j|^{2/(p+1)}\big) =: M_j$$

and let $d_j := \mathrm{dist}\,(x_j, \partial\Omega)$. Since $\overline{\Omega}$ is compact, we may assume $x_j \to x_0$ for some $x_0 \in \overline{\Omega}$. Set $\kappa_j := M_j^{-(p-1)/2}$. The sequence d_j/κ_j is either unbounded or bounded. In the former case we may assume $d_j/\kappa_j \to \infty$, in the latter $d_j/\kappa_j \to c \geq 0$.

Case 1. Let $d_j/\kappa_j \to \infty$. Set

$$v_j(y) := \frac{1}{M_j} u_j(x), \qquad y := \frac{x - x_j}{\kappa_j},$$

and $\Omega_j := \{y \in \mathbb{R}^n : |y| < d_j/\kappa_j\}$. Then

$$v_j + |\nabla v_j|^{2/(p+1)} \leq v_j(0) + |\nabla v_j(0)|^{2/(p+1)} = 1 \tag{12.3}$$

and

$$-\Delta v_j(y) = a(\kappa_j y + x_j)v_j^p(y) + g_j(y), \qquad y \in \Omega_j, \tag{12.4}$$

where

$$g_j(y) := \kappa_j^{2p/(p-1)} g\big(\kappa_j y + x_j, \kappa_j^{-2/(p-1)} v_j(y), \kappa_j^{(p+1)/(p-1)} \nabla v_j(y)\big)$$

satisfies

$$|g_j| \le C\kappa_j^\varepsilon, \qquad \varepsilon := \min\big(2(p-q), 2p - (p+1)r\big)/(p-1). \tag{12.5}$$

Interior elliptic L^p-estimates (see Appendix A) guarantee that v_j are locally bounded in $W^{2,z}$ for any $z > 1$ (uniformly with respect to j). Let $\alpha \in (0,1)$, $R > 0$ and $B_R := \{y \in \mathbb{R}^n : |y| < R\}$. There exists $z = z(\alpha) > 1$ such that $W^{2,z}(B_R)$ is compactly embedded into $BUC^{1+\alpha}(B_R)$. Consequently, we may assume $v_j \to v$ in $C^{1+\alpha}$. Passing to the limit in (12.4) and (12.3) we see that v is a positive (classical) solution of

$$-\Delta v = a(x_0)v^p \quad \text{in } \mathbb{R}^n,$$

which contradicts Theorem 8.1.

Case 2. Let $d_j/\kappa_j \to c \ge 0$. Let $\tilde{x}_j \in \partial\Omega$ be such that $d_j = |x_j - \tilde{x}_j|$. For any j we can choose a local coordinate $z = z_{(j)} = (z^1, z^2, \ldots, z^n)$ in an ε-neighborhood U_j of $\tilde{x}_j$ such that the image of the boundary $\partial\Omega$ will be contained in the hyperplane $z^1 = 0$, $\tilde{x}_j$ becomes 0, x_j becomes $z_j := (d_j, 0, 0, \ldots, 0)$, and the image of U_j will contain the set $\{z : |z| < \varepsilon'\}$ for some $\varepsilon' > 0$. We may assume that $\varepsilon, \varepsilon'$ are independent of j and the local charts are uniformly bounded in C^2. In these new coordinates, the equation for $w = w_j(z) = u_j(x)$ becomes

$$\left. \begin{aligned} -\sum_{i,k} a^{ik}(z)\frac{\partial^2 w}{\partial z^i \partial z^k} - \sum_i b^i(z)\frac{\partial w}{\partial z^i} = a(x(z))w^p + \tilde{g}(z), \quad |z| < \varepsilon, \; z^1 > 0, \\ w = 0, \qquad\qquad\qquad\qquad\qquad\qquad |z| < \varepsilon, \; z^1 = 0, \end{aligned} \right\} \tag{12.6}$$

where $\tilde{g}(z) := g\big(x(z), w(z), D(z)\nabla_z w(z)\big)$, $D = D_{(j)} = (\partial z^i/\partial x^k)_{i,k}$, $b^i = b^i_{(j)} = \Delta z^i$, $a^{ik} = a^{ik}_{(j)} = \sum_\ell \frac{\partial z^i}{\partial x^\ell} \frac{\partial z^k}{\partial x^\ell}$, hence $A = A_{(j)} := (a^{ik}_{(j)})_{i,k} = D \cdot {}^tD$, and the $A_{(j)}$ are uniformly elliptic. Also, since $\partial\Omega$ is uniformly C^2, it follows that the $a^{ik}_{(j)}$ are uniformly bounded in C^1 and the $b^i_{(j)}$ in L^∞. Moreover, since $D(0)$ is a Euclidean transformation, it follows that $A_{(j)}(0) = D(0) \cdot {}^tD(0) = \mathrm{Id}$. Set

$$v_j(y, s) := \frac{1}{M_j} w_j(\kappa_j y + z_j),$$

where

$$y \in \Omega_j := \left\{ y : \left| y - \frac{z_j}{\kappa_j} \right| < \frac{\varepsilon'}{\kappa_j}, \; y^1 > -\frac{d_j}{\kappa_j} \right\}.$$

Then v_j is a solution of

$$-\sum_{i,k} a^{ik}(\kappa_j y + z_j)\frac{\partial^2 v}{\partial y^i \partial y^k} - \kappa_j \sum_i b^i(\kappa_j y + z_j)\frac{\partial v}{\partial y^i}$$

$$= a(x(\kappa_j y + z_j))v^p + g_j \qquad \text{in } \Omega_j,$$

$$v = 0 \qquad \text{on } \{y \in \partial\Omega_j : y^1 = -d_j/\kappa_j\},$$

where

$$g_j(y) := \kappa_j^{2p/(p-1)} g\big(x(\kappa_j y + z_j), \kappa_j^{-2/(p-1)} v(y), \kappa_j^{-(p+1)/(p-1)} D(\kappa_j y + z_j)\nabla v(y)\big)$$

satisfies (12.5). Interior-boundary L^p-estimates (see Appendix A) and the bounds on the coefficients $a^{ik}_{(j)}$, $b^i_{(j)}$ again yield a subsequence of $\{v_j\}$ converging to a positive (classical) solution v of

$$\Delta v = a(x_0)v^p, \qquad y_1 > -c,$$

$$v = 0, \qquad y_1 = -c,$$

which contradicts Theorem 8.2. $\quad\square$

Remarks 12.3. (i) If g is independent of the gradient variable, then it is sufficient to choose $M_k := \sup u_k$ in the proof of Theorem 12.1.

(ii) **Indefinite coefficients.** Assume that the function a in problem (12.2) changes sign. Under suitable assumptions on a, g and p one can still use the method of [241] in order to get a priori bounds for positive solutions (see [74], [19] and [167], for example). In addition to the limiting problems in the proof of Theorem 12.1 one has to deal with problems of the form

$$-\Delta u = h(y)u^p, \qquad y \in \mathbb{R}^n,$$

where typically $h(y) = |y_1|^\alpha y_1$ for some $\alpha \geq 0$. In some cases, a combination of the above approach with other arguments (moving planes, energy, ...) yields the a priori bounds, see [124], [453], [237] and the references therein. Of course, if the problem has variational structure, then the existence of nontrivial solutions can often be proved by variational or dynamical methods, see [8], [75], [7], [257], [121], [3] and the references therein.

(iii) The rescaling method is sometimes referred to as the "blow-up method", because one performs a zoom of the microscopic scales of the solution. Here we shall not use this terminology, in order to avoid confusion with the blow-up phenomenon in the parabolic problem. $\quad\square$

13. A priori bounds via moving planes and Pohozaev's identity

In this section we describe the method of a priori estimates of solutions of (2.1) due to [183]. Similarly as in the preceding section, the growth condition for function f will be optimal. The advantage of this method consists in the fact that it does neither require precise asymptotic behavior of f for u large nor Liouville-type theorems. On the other hand, the symmetry of the Laplace operator plays an important role, f cannot depend on ∇u in a general way and we also have to assume that either Ω is convex or f satisfies a restrictive monotonicity condition, see (13.3) below. The assumptions for a general function $f = f(x, u)$ are rather complicated (see [183, Remark 1.5]) and therefore we restrict ourselves to the case $f = f(u)$. Hence, we shall study positive solutions of the problem

$$\left.\begin{aligned} -\Delta u &= f(u), & x \in \Omega, \\ u &= 0, & x \in \partial\Omega. \end{aligned}\right\} \tag{13.1}$$

Theorem 13.1. *Assume* $n \geq 2$ *and* Ω *bounded. Let* $f : \mathbb{R}_+ \to \mathbb{R}$ *be locally Lipschitz continuous and assume*

$$\liminf_{u\to\infty} \frac{f(u)}{u} > \lambda_1, \qquad \lim_{u\to\infty} \frac{f(u)}{u^\sigma} = 0,$$

where $\sigma = p_S$ *if* $n \geq 3$, $\sigma < \infty$ *is arbitrary if* $n = 2$. *Let one of the following assumptions be satisfied:*

   (i) Ω *is convex and*

$$\limsup_{u\to\infty} \frac{uf(u) - \theta F(u)}{u^2 f(u)^\kappa} \leq 0, \qquad \theta \in [0, 2^*), \tag{13.2}$$

     where $\kappa = 2/n$.

  (ii) *Condition* (13.2) *is satisfied with* $\kappa = 2/n$ *and, in the case* $n \geq 3$,

$$\text{the function } u \mapsto f(u)u^{-p_S} \text{ is nonincreasing on } (0, \infty). \tag{13.3}$$

  (iii) *Condition* (13.2) *is satisfied with* $\kappa = 2/(n+1)$, $n \geq 3$, $\partial\Omega = \Gamma_1 \cup \Gamma_2$, *where* Γ_1, Γ_2 *are closed and satisfy*

     (1) *at every point of* Γ_1, *all sectional curvatures of* Γ_1 *are bounded away from 0 by a positive constant* a;

     (2) *there exists* $x_0 \in \mathbb{R}^n$ *such that* $(x - x_0, \nu(x)) \leq 0$ *for all* $x \in \Gamma_2$.

Then there exists $C > 0$ *such that* $\|u\|_\infty < C$ *for any positive classical solution* u *of* (13.1).

In view of the proof we set some notation. For each $\varepsilon > 0$, let

$$\Omega_\varepsilon := \{z \in \Omega : \delta(z) < \varepsilon\}.$$

For $y \in \partial\Omega$ and $\lambda > 0$, we define

$$T(y,\lambda) := \{x \in \mathbb{R}^n : (y - x, \nu(y)) = \lambda\},$$
$$\Sigma(y,\lambda) := \{x \in \Omega : (y - x, \nu(y)) \leq \lambda\},$$

we denote by $R(y,\lambda)$ the reflection with respect to the hyperplane $T(y,\lambda)$ and we set $\Sigma'(y,\lambda) := R(y,\lambda)\Sigma(y,\lambda)$. We need the following lemma.

Lemma 13.2. *Assume Ω bounded and convex, $\lambda_0 > 0$, and $0 \leq u \in C(\overline{\Omega}) \cap C^1(\Omega)$. Assume that*

$$(\nabla u(x), \nu(y)) \leq 0, \qquad y \in \partial\Omega, \ x \in \Sigma(y,\lambda_0). \tag{13.4}$$

Then

$$\sup_{\Omega_\varepsilon} u \leq C \int_\Omega u\varphi_1 \, dx,$$

where $\varepsilon, C > 0$ depend only on Ω and λ_0.

Proof. Let us first recall that

$$\nu(\partial\Omega) = S^{n-1}. \tag{13.5}$$

This follows from a standard degree argument. We give the proof for completeness. Assume without loss of generality that $0 \in \Omega$ and select $\tilde{\nu}$, a continuous extension of ν to $\overline{\Omega}$. The homotopy $H_1(t,x) := t\tilde{\nu}(x) + (1-t)x$ has no zero on $\partial\Omega$, since $(x, \nu(x)) \geq 0$ on $\partial\Omega$ due to the convexity of Ω. Therefore $d(\tilde{\nu}, 0, \Omega) = d(id, 0, \Omega) = 1$, where d denotes the Brouwer degree. Assume for contradiction that $\eta \notin \nu(\partial\Omega)$ for some $\eta \in S^{n-1}$. Then the homotopy $H_2(t,x) = t\tilde{\nu}(x) - (1-t)\eta$ has no zero on $\partial\Omega$. Consequently $d(\tilde{\nu}, 0, \Omega) = d(-\eta, 0, \Omega) = 0$, a contradiction which proves (13.5).

Next, by decreasing λ_0 if necessary, we may assume that

$$\{y - \lambda\nu(y) \in \mathbb{R}^n : \lambda \in (0, \lambda_0]\} \subset \Omega, \qquad y \in \partial\Omega. \tag{13.6}$$

Let $\varepsilon \in (0, \lambda_0/4]$, $x \in \Omega_\varepsilon$, and let $\tilde{x} \in \partial\Omega$ satisfy $|x - \tilde{x}| = \delta(x)$. Notice that $\tilde{x}$ is uniquely determined and $(\tilde{x} - x)/|\tilde{x} - x| = \nu(\tilde{x})$ if ε is small. Let $\alpha \in (0,1)$ and let $\eta \in S^{n-1}$ be such that $(\eta, \nu(\tilde{x})) \geq \alpha$. Using the fact that Ω is contained in the half-space $\{z \in \mathbb{R}^n : (z - x, \nu(\tilde{x})) \leq |\tilde{x} - x|\}$ (due to the convexity of Ω), we obtain

$$(y(\eta)-x,\eta) \leq (y(\eta)-x,\nu(\tilde{x}))+|y(\eta)-x||\eta-\nu(\tilde{x})| \leq \varepsilon+\mathrm{diam}(\Omega)\sqrt{2(1-\alpha)} \leq \lambda_0/2,$$

provided α is close to 1 and ε is small enough, say $1 - \alpha + \varepsilon < \varepsilon_0 = \varepsilon_0(\Omega, \lambda_0)$. This along with (13.6) implies

$$\{x - \lambda\eta \in \mathbb{R}^n : \lambda \in [0, \lambda_0]\} \subset \Sigma(y(\eta), \lambda_0).$$

It then follows from (13.4) that $[0, \varepsilon] \ni \lambda \mapsto u(x - \lambda\eta)$ is nondecreasing for any $\eta \in S^{n-1}$ satisfying $(\eta, \nu(\tilde{x})) \geq \alpha$. This property guarantees the existence of $\gamma = \gamma(\Omega, \lambda_0) > 0$ such that

$$\left. \begin{array}{l} \text{for all } x \in \Omega_\varepsilon \text{ there exists a measurable set } I_x \subset \Omega \setminus \Omega_\varepsilon \\[2mm] \text{satisfying meas } I_x \geq \gamma \ \text{ and } \ u(\xi) \geq u(x) \text{ for all } \xi \in I_x. \end{array} \right\} \quad (13.7)$$

Indeed (decreasing the value of ε if necessary), it is sufficient to take a conical piece
$$I_x = \Omega_\varepsilon^c \cap \{x - \lambda\eta : \eta \in S^{n-1}, \ (\eta, \nu(\tilde{x})) \geq \alpha, \ \lambda \in [0, \lambda_0]\}.$$

Since $\varphi_1 \geq C_\varepsilon$ on $\Omega \setminus \Omega_\varepsilon$ for some $C_\varepsilon > 0$, we deduce from (13.7) that

$$C_\varepsilon \gamma u(x) \leq C_\varepsilon \int_{I_x} u(\xi)\, d\xi \leq \int_{I_x} u(\xi)\varphi_1(\xi)\, d\xi \leq \int_\Omega u(\xi)\varphi_1(\xi)\, d\xi$$

and the lemma is proved. $\quad \square$

Proof of Theorem 13.1. First assume (i). The proof will consist of the following four steps:

1. $\int_\Omega u\delta\, dx \leq C$, $\int_\Omega |f(u)|\delta\, dx \leq C$, where $\delta(x) = \text{dist}\,(x, \partial\Omega)$,
2. $u + |\nabla u| \leq C$ in a neighborhood of $\partial\Omega$,
3. $\|\nabla u\|_2 \leq C$,
4. $\|u\|_\infty \leq C$.

Step 1. This step is almost the same as Step 1 in the proof of Theorem 10.1 and we leave the detailed proof to the reader.

Step 2. Since Ω is convex and smooth, we can find $\lambda_0, c_0 > 0$ such that

$$\Sigma'(y, \lambda) \subset \Omega, \quad \lambda \leq \lambda_0 \quad \text{and} \quad (\nu(x), \nu(y)) > c_0, \quad x \in \partial\Sigma(y, \lambda_0) \cap \partial\Omega.$$

We shall now apply the moving planes method (cf. [239], [183]) to show that

$$u(R(y, \lambda)x) \geq u(x), \qquad y \in \partial\Omega, \ x \in \Sigma(y, \lambda), \ \lambda \leq \lambda_0. \qquad (13.8)$$

Without loss of generality, we may assume that $y = 0$ and that $\nu(0) = -e_1$ (in particular, Ω lies entirely in the upper half-space $\{x_1 > 0\}$). For each $x = (x_1, x')$, we denote $x^\lambda := R(0, \lambda)x = (2\lambda - x_1, x')$, $\Sigma_\lambda := \Sigma(0, \lambda) = \Omega \cap \{x_1 < \lambda\}$, and $\Sigma'_\lambda := \Sigma'(0, \lambda)$. Define

$$w^\lambda(x) = u(x^\lambda) - u(x), \qquad \text{for } x \in \Sigma_\lambda, \ 0 < \lambda \leq \lambda_0,$$

and set

$$E := \{\mu \in (0, \lambda_0] : w^\lambda(x) \geq 0 \text{ for all } x \in \Sigma_\lambda \text{ and } \lambda \in (0, \mu)\}.$$

Since $\frac{\partial u}{\partial x_1}(0) > 0$ by Hopf's lemma, we have $\lambda \in E$ for $\lambda > 0$ small. Assume for contradiction that $\bar{\lambda} := \sup E < \lambda_0$. We have

$$w^\lambda \geq 0, \qquad \text{for all } x \in \Sigma_\lambda \text{ and } \lambda \in (0, \bar{\lambda}], \tag{13.9}$$

and there exists a sequence $\lambda_i \to \bar{\lambda}$, with $\bar{\lambda} < \lambda_i < \lambda_0$, such that $\min_{\overline{\Sigma_{\lambda_i}}} w^{\lambda_i} < 0$.

Since $w^\lambda = 0$ on $\{x_1 = \lambda\} \cap \overline{\Omega}$ and

$$w^\lambda > 0 \quad \text{on } \{x_1 < \lambda\} \cap \partial\Omega, \quad \text{for all } \lambda < \lambda_0, \tag{13.10}$$

it follows that this minimum is attained at a point $q_i \in \Sigma_{\lambda_i}$. Therefore $\nabla w^{\lambda_i}(q_i) = 0$. On the other hand, since $\frac{\partial u}{\partial x_1} = (e_1 \cdot \nu) \frac{\partial u}{\partial \nu} \geq c > 0$ on $\{x_1 \leq \lambda_0\} \cap \partial\Omega$ and

$$w^\lambda(x) = u(2\lambda - x_1, x') - u(x_1, x') = 2(\lambda - x_1)\frac{\partial u}{\partial x_1}(\xi(x)),$$

with $|\xi(x) - x| \leq 2(\lambda - x_1)$, we see that $w^\lambda(x) \geq 0$ for x in an ε-neighborhood of $\{x_1 = \lambda\} \cap \partial\Omega$, with $\varepsilon > 0$ independent of $\lambda \in (0, \lambda_0]$. Therefore, we may assume that $q_i \to \bar{q} \in \overline{\Sigma_{\bar{\lambda}}}$, $\bar{q} \notin \{x_1 = \bar{\lambda}\} \cap \partial\Omega$, and by continuity we get

$$w^{\bar{\lambda}}(\bar{q}) = 0 \quad \text{and} \quad \nabla w^{\bar{\lambda}}(\bar{q}) = 0. \tag{13.11}$$

But (13.9) implies

$$-\Delta w^{\bar{\lambda}}(x) = f\big(u(x^{\bar{\lambda}})\big) - f(u(x)) \geq -cw^{\bar{\lambda}}(x) \quad \text{and} \quad w^{\bar{\lambda}}(x) \geq 0, \qquad x \in \Sigma_{\bar{\lambda}},$$

for some constant $c > 0$ (depending on u). By Hopf's lemma (cf. Proposition 52.1 and Remark 52.2), this along with (13.11) implies $w^{\bar{\lambda}} = 0$ in $\Sigma_{\bar{\lambda}}$, contradicting (13.10). Consequently, $\bar{\lambda} = \lambda_0$, which proves (13.8). This guarantees that u satisfies (13.4). By Lemma 13.2 and Step 1, we deduce that $u \leq C$ on Ω_ε for some $\varepsilon, C > 0$ depending only on Ω. Now the bound for ∇u in $\Omega_{\varepsilon/2}$ follows from interior-boundary elliptic L^p-estimates (see Appendix A) and the embedding $W^{2,p} \hookrightarrow C^1$ for $p > n$. In particular, we have shown that

$$\left|\frac{\partial u}{\partial \nu}\right| \leq C, \quad x \in \partial\Omega. \tag{13.12}$$

Step 3. Notice that Steps 1 and 2 imply

$$\|f(u)\|_1 \leq C. \tag{13.13}$$

First consider the case $n \geq 3$. The Hölder and Sobolev inequalities and (13.13) guarantee

$$\int_\Omega u^2 |f(u)|^{2/n}\, dx \leq \|u\|_{2^*}^2 \|f(u)\|_1^{2/n} \leq C\|\nabla u\|_2^2.$$

Pohozaev's identity (5.1) and (13.12) yield

$$\left| \int_\Omega |\nabla u|^2 \, dx - 2^* \int_\Omega F(u) \, dx \right| \le C.$$

Since $\int_\Omega |\nabla u|^2 \, dx = \int_\Omega u f(u) \, dx$, the last two estimates and (13.2) imply

$$2^* \int_\Omega F(u) \, dx \le \int_\Omega u f(u) \, dx + C \le \theta \int_\Omega F(u) \, dx + \varepsilon \int_\Omega u^2 |f(u)|^{2/n} \, dx + C_\varepsilon$$

$$\le (\theta + \varepsilon C) \int_\Omega F(u) \, dx + \tilde{C}_\varepsilon.$$

Choosing $\varepsilon < (2^* - \theta)/C$ we obtain $\int_\Omega F(u) \, dx \le C$, hence $\|\nabla u\|_2 \le C$.

Next let $n = 2$. Set $\gamma := 1 - 1/\sigma$. Given $\varepsilon > 0$, the assumption $\lim_{u \to \infty} f(u)/u^\sigma = 0$ guarantees the existence of $C_\varepsilon > 0$ such that

$$u f(u) \le \varepsilon u^2 f(u)^\gamma + C_\varepsilon.$$

Similarly as above we obtain

$$\|\nabla u\|_2^2 = \int_\Omega u f(u) \, dx \le \varepsilon \int_\Omega u^2 |f(u)|^\gamma \, dx + C_\varepsilon$$

$$\le \varepsilon \|u\|_{2/(1-\gamma)}^2 \|f(u)\|_1^\gamma \le \varepsilon C \|\nabla u\|_2^2 + C_\varepsilon,$$

which proves the assertion.

Step 4. If

$$f(u) \le C(1 + u^p) \quad \text{for some } p < p_S \tag{13.14}$$

(which is always true if $n = 2$), then one can use standard bootstrap estimates based on L^q-estimates (see Appendix A) to show that the $W^{1,2}$-bound from Step 3 guarantees an L^∞-bound. If $n \ge 3$ and (13.14) is not true, then we use the following estimates (see [96] and cf. the proof of Proposition 3.3).

Let $p > 1$, $a_p := (p+1)^2/4$ and $q := (p+1)n/(n-2)$. Then

$$\left(\int_\Omega u^q \, dx \right)^{(n-2)/n} = \|u^{(p+1)/2}\|_{2^*}^2 \le C \int_\Omega |\nabla u^{(p+1)/2}|^2 \, dx = C a_p \int_\Omega |\nabla u|^2 u^{p-1} \, dx$$

$$= C \frac{a_p}{p} \int_\Omega f(u) u^p \, dx \le \varepsilon \int_\Omega u^{p+\sigma} \, dx + C_\varepsilon,$$

where $\sigma = (n+2)/(n-2)$. Next Hölder's inequality and Step 3 yield

$$\int_\Omega u^{p+\sigma} \, dx = \int_\Omega u^{q(n-2)/n + 4/(n-2)} \, dx \le \left(\int_\Omega u^q \, dx \right)^{(n-2)/n} \left(\int_\Omega u^{2^*} \, dx \right)^{2/n}$$

$$\le C \left(\int_\Omega u^q \, dx \right)^{(n-2)/n}.$$

These estimates imply $\|u\|_q \le C$, hence $\|f(u)\|_{q/\sigma} \le C$. Since q can be made arbitrarily large, the L^p-estimates (see Appendix A) conclude the proof in case (i).

Next consider assumption (ii). Instead of Ω being convex we now assume (13.3). Since the convexity assumption was used only in the proof of Step 2, it is sufficient to modify the proof of this step. Choose $x_0 \in \partial\Omega$. Then there exists a ball $B_r \subset \mathbb{R}^n \setminus \Omega$ of radius r such that $x_0 \in \partial B_r$. The radius r can be chosen independent of x_0 and, without loss of generality, we may assume $r = 1$. Choose a coordinate system such that B_r is centered at the origin and $x_0 = (1, 0, \ldots, 0)$. Set $y = J(x) := x/|x|^2$ and $w(y) = |x|^{n-2}u(x)$. Then

$$-\Delta w(y) = g(y, w) \quad \text{in } \mathcal{O} := J(\Omega),$$

where $g(y, w) := f(|y|^{n-2}w)/|y|^{n+2}$ is nonincreasing in y due to (13.3). Since $\mathcal{O} \subset B_r$ is smooth and $x_0 \in \partial\mathcal{O} \cap \partial B_r$ we can use the moving planes method in order to get the existence of $\varepsilon_{x_0}, \gamma_{x_0} > 0$ with the following property: for any $y \in \mathcal{O}$, $|y - x_0| < \varepsilon_{x_0}$, there exists a set $K_y \subset \{z \in \mathcal{O} : \operatorname{dist}(z, \partial\mathcal{O}) > \varepsilon_{x_0}\}$ satisfying meas $K_y \ge \gamma_{x_0}$ and $w(\xi) \ge w(y)$ for all $\xi \in K_y$. Going back to the original variables and using the compactness of $\partial\Omega$ we get the existence of $\varepsilon, \gamma, c > 0$ such that (13.7) is true, with $u(\xi) \ge u(x)$ replaced by $u(\xi) \ge cu(x)$. The rest of the proof of Step 2 is the same as in case (i).

Finally consider case (iii). Then Steps 1 and 4 can be proved in the same way as in case (i). Repeating the arguments in the proof of Step 2 of case (i) we obtain a uniform bound for u and $|\nabla u|$ in a neighborhood of Γ_1. Without loss of generality we may assume $x_0 = 0$, hence $x \cdot \nu(x) \le 0$ for all $x \in \Gamma_2$. These facts and Pohozaev's identity (5.1) imply

$$2^* \int_\Omega F(u)\, dx - \int_\Omega u f(u)\, dx \le C. \tag{13.15}$$

Next using Lemma 50.4 with $\tau := 1/(n+1)$ and $q := 2(n+1)/(n-1)$, Step 1 and Hölder's inequality, we obtain

$$\int_\Omega u f(u)\, dx = \|\nabla u\|_2^2 \ge c_1 \left\|\frac{u}{\delta^\tau}\right\|_q^2 \ge c_2 \left\|\frac{u}{\delta^\tau}\right\|_q^2 \|f(u)\delta\|_1^{1-2/q}$$

$$\ge c_2 \int_\Omega \frac{u^2}{\delta^{2\tau}} \left(|f(u)|\delta\right)^{1-2/q} dx = c_2 \int_\Omega u^2 |f(u)|^{2/(n+1)}\, dx.$$

Now (13.2) with $\kappa = 2/(n+1)$, (13.15) and the last estimate imply

$$\int_\Omega u f(u)\, dx \le \theta \int_\Omega F(u)\, dx + \varepsilon \int_\Omega u^2 |f(u)|^{2/(n+1)}\, dx + C_\varepsilon$$

$$\le (\theta/2^* + \varepsilon C) \int_\Omega u f(u)\, dx + C_\varepsilon$$

and the choice of ε small enough concludes the proof. $\square$

The following corollary can be proved in the same way as Corollary 10.3.

Corollary 13.3. *Let* $f : \mathbb{R}_+ \to \mathbb{R}_+$ *satisfy the assumptions in Theorem* 13.1 *and* $\limsup_{u \to 0+} f(u)/u < \lambda_1$. *Then problem* (2.1) *possesses at least one positive classical solution.*

Remark 13.4. If one is interested only in the existence of positive solutions of (2.1) without knowing their a priori bounds, then the technical assumption (13.2) can be omitted, see [183]. The proof is based on an approximation of the function f, on the mountain pass theorem (including uniform bounds for the energy of approximating solutions) and Pohozaev's identity. $\square$

Chapter II

Model Parabolic Problems

14. Introduction

In Chapter II, we mainly consider semilinear parabolic problems of the form

$$\left.\begin{aligned}
u_t - \Delta u &= f(u), & x \in \Omega,\ t > 0, \\
u &= 0, & x \in \partial\Omega,\ t > 0, \\
u(x,0) &= u_0(x), & x \in \Omega,
\end{aligned}\right\} \tag{14.1}$$

where f is a C^1-function with a superlinear growth. For simplicity, we formulate most of our assertions for the model case $f(u) = |u|^{p-1}u$ with $p > 1$, but the methods of our proofs can be applied to more general parabolic problems (not necessarily of the form (14.1)). Some of possible generalizations and modifications will be mentioned as remarks, other can be found in the subsequent chapters.

15. Well-posedness in Lebesgue spaces

Definition 15.1. Given a Banach space X of functions defined in Ω, $u_0 \in X$ and $T \in (0, \infty]$, we say that the function $u \in C([0, T), X)$ is a **solution** (more precisely, a **classical X-solution**) of (14.1) in $[0, T)$ if $u \in C^{2,1}(\Omega \times (0, T)) \cap C(\overline{\Omega} \times (0, T))$, $u(0) = u_0$ and u is a classical solution of (14.1) for $t \in (0, T)$. If Ω is unbounded, then we also require $u \in L^\infty_{loc}((0, T), L^\infty(\Omega))$.

If $X = L^\infty(\Omega)$, then, instead of the condition $u \in C([0, T), X)$, we require $u \in C((0, T), X)$ and $\|u(t) - e^{-tA}u_0\|_\infty \to 0$ as $t \to 0$, where e^{-tA} is the Dirichlet heat semigroup in Ω (cf. Appendix B).

We say that (14.1) is **well-posed** in X if, given $u_0 \in X$, there exist $T > 0$ and a unique classical X-solution of (14.1) in $[0, T]$. $\square$

It is well known that (14.1) is well-posed in $X = W_0^{1,q}(\Omega)$ for any $q > n$ if Ω is bounded, or in $X = L^\infty(\Omega)$ for any Ω (see Example 51.9 and Remark 51.11). In this section we study the well-posedness of the model problem

$$\left.\begin{aligned}
u_t - \Delta u &= |u|^{p-1}u, & x \in \Omega,\ t > 0, \\
u &= 0, & x \in \partial\Omega,\ t > 0, \\
u(x,0) &= u_0(x), & x \in \Omega,
\end{aligned}\right\} \tag{15.1}$$

in the Lebesgue spaces $L^q(\Omega)$, $1 \le q < \infty$, since this well-posedness will play a crucial role in many subsequent sections. The following two results show that the exponent

$$q_c := n(p-1)/2$$

is critical for the well-posedness. The existence/nonexistence part of these results is due to [528], [529] (where uniqueness was proved in a more restrictive class of solutions). The uniqueness and nonuniqueness parts were proved in [93] and [53], [425], respectively. An alternative proof of the existence-uniqueness part of Theorem 15.2 based on interpolation and extrapolation spaces can be found in Appendix E (see Theorem 51.25 and Example 51.27).

In what follows we write shortly L^q-solution instead of $L^q(\Omega)$-solution.

Theorem 15.2. *Let $p > 1$, $u_0 \in L^q(\Omega)$, $1 \le q < \infty$, $q > q_c$. Then there exists $T = T(\|u_0\|_q) > 0$ such that problem (15.1) possesses a unique classical L^q-solution in $[0, T)$ and the following smoothing estimate is true:*

$$\|u(t)\|_r \le C\|u_0\|_q t^{-\alpha_r}, \qquad \alpha_r := \frac{n}{2}\left(\frac{1}{q} - \frac{1}{r}\right), \tag{15.2}$$

for all $t \in (0, T)$ and $r \in [q, \infty]$, with $C = C(n, p, q) > 0$. In addition, $u \ge 0$ provided $u_0 \ge 0$.

Theorem 15.3. *Let $p > 1 + 2/n$ and $1 \le q < q_c$.*

(i) *There exists a nonnegative function $u_0 \in L^q(\Omega)$, such that (15.1) does not admit any nonnegative classical L^q-solution in $[0, T)$ for any $T > 0$.*

(ii) *Assume $p < p_S$, $\Omega = B_R$, and let $u_0 \in L^\infty(\Omega)$, $u_0 \ge 0$, be radial nonincreasing. Then there exists a time $T > 0$ such that (15.1) possesses infinitely many positive radial nonincreasing classical L^q-solutions in $[0, T)$.*

Remarks 15.4. (i) **The critical case.** It was also proved in [529], [93] that when $u_0 \in L^q(\Omega)$ and $q = q_c > 1$, then there exists $T = T(u_0) > 0$ such that (15.1) possesses a unique classical L^q-solution in $[0, T)$ (see Example 51.27 and cf. Remark 20.24(i)). The same arguments as in Remark 51.26(vi) guarantee that this solution satisfies (15.2) in $(0, T)$. In addition, it is nonnegative if $u_0 \ge 0$.

Unlike in the case $q > q_c$, T cannot be chosen uniform for all u_0 lying in a bounded subset of $L^q(\Omega)$. Indeed assume without loss of generality that $\Omega \supset B(0, 1)$ and choose $0 \le u_0 \in L^q \cap L^\infty(\Omega)$ such that $T_0 := T_{\max}(u_0) < \infty$ (see Section 16 for the definition of the maximal existence time $T_{\max}(u_0)$ and Section 17 for the existence of such solution). For each $j \ge 1$, set $\omega_j = B(0, 1/j)$ and define

$$u_{0,j}(x) := \begin{cases} j^{2/(p-1)} u_0(jx), & x \in \omega_j, \\ 0, & x \in \Omega \setminus \omega_j. \end{cases} \tag{15.3}$$

By direct computation, we see that $\tilde{u}_j(x,t) := j^{2/(p-1)}u(jx, j^2 t)$ solves (15.1) in $\omega_j \times (0, j^{-2}T_0)$ with initial data $u_{0,j}|_{\omega_j}$. Let u_j be the solution of problem (15.1) (in Ω) with initial data $u_{0,j}$. Since $u_j \geq 0$ on $\partial \omega_j$, it follows from the comparison principle that $u_j \geq \tilde{u}_j$ in ω_j as long as u_j exists. Consequently, $T_{\max}(u_{0,j}) \leq j^{-2}T_0 \to 0$, as $j \to \infty$, while $\|u_{0,j}\|_{q_c} = \|u_0\|_{q_c}$ due to $q_c = n(p-1)/2$. See Remark 22.10(iii), Remark 27.8(g) and [36] for further results in that direction.

If $q = q_c = 1$ (i.e. $q = 1$, $p = 1 + 2/n$), then there exists a positive function $u_0 \in L^1(\Omega)$ for which (15.1) does not possess any nonnegative classical L^1-solution in $[0, T)$ for any $T > 0$; see [93, Theorem 11]) and see also [116] for a similar example with the weaker notion of integral solution.

(ii) **Nonuniqueness in $\mathbb{R}^n$.** Assume $\Omega = \mathbb{R}^n$, $u_0 = 0$, $1 + 2/n < p < p_S$. Then there exists a function u which is positive for $t > 0$ and which is a global classical L^q-solution of (15.1) for any $q < q_c$ (and a $W^{1,q}$-solution for any $q < n(p-1)/(p+1)$, see [269]). Moreover, there exist infinitely many nontrivial functions which are global classical L^q-solutions of (15.1) for any $q < q_c$ (see [532]). All these solutions are (forward) **self-similar**, that is

$$u(x,t) = \lambda^{2/(p-1)}u(\lambda x, \lambda^2 t), \quad \lambda > 0.$$

Such solutions can be found in the form $u(x,t) = t^{-1/(p-1)}w(x/\sqrt{t})$, where $w = w(y)$ solves the problem

$$\Delta w + \frac{y}{2} \cdot \nabla w + \frac{1}{p-1}w + |w|^{p-1}w = 0, \qquad y \in \mathbb{R}^n. \tag{15.4}$$

In [269] and [532], radial positive and infinitely many radial nontrivial solutions of (15.4) (with a rapid decay at infinity) were found by ODE techniques (see also [533]). Variational methods for solving (15.4) were used in [174].

(iii) **Uniqueness and nonuniqueness in the class of <u>mild</u> solutions.** If u is a classical L^q-solution of (14.1) in $[0, T)$, then it satisfies the **variation-of-constants formula**

$$u(t) = e^{-(t-\tau)A}u(\tau) + \int_\tau^t e^{-(t-s)A}f(u(s))\,ds, \qquad 0 < \tau < t < T. \tag{15.5}$$

Indeed, applying the operator $e^{-(t-s)A}$ to the equation $u_t(s) + Au(s) = f(u(s))$, integrating in $s \in (\tau, t)$ and using $\frac{d}{ds}(e^{-(t-s)A}u(s)) = e^{-(t-s)A}(u_t(s) + Au(s))$ we obtain (15.5).

Any function $u \in C([0, T), L^q(\Omega))$ satisfying $f(u) \in L^1_{loc}((0, T), L^1 + L^\infty(\Omega))$, $u(0) = u_0$ and (15.5) is called a **mild L^q-solution** of (15.1). (If $q = \infty$, then we modify this definition in the same way as in the case of classical solutions.)

Now assume $q \geq p$ and let u be a mild L^q-solution of (15.1). Then we can pass to the limit in (15.5) as $\tau \to 0$ to get

$$u(t) = e^{-tA}u_0 + \int_0^t e^{-(t-s)A}|u(s)|^{p-1}u(s)\,ds. \tag{15.6}$$

On the other hand, any solution of (15.6) in $C\big([0,T),L^q(\Omega)\big)$ is obviously a mild L^q-solution. If, in addition, $q \geq q_c$ (and $q > p$ if $q = q_c$), then each mild L^q-solution is a classical L^q-solution so that the uniqueness in Theorem 15.2 and (i) holds in the class of mild L^q-solutions (see [93], [536]). This is not true for the limiting case $q = q_c = p = n/(n-2)$. In fact, if Ω is the unit ball and $q = p = n/(n-2)$, then there exists a singular stationary solution $u_s \in L^q(\Omega) \setminus C(\Omega)$ of (15.1) (see [394] and cf. Remark 3.6(ii)). The function $u(t) := u_s$ is a mild L^q-solution of (15.1) with $u_0 := u_s$ which is not classical for $t > 0$. On the other hand, (i) guarantees the existence of a classical L^q-solution. A similar example for $\Omega = \mathbb{R}^n$ was constructed in [511].

(iv) **Integral solutions.** Consider problem (14.1) with f nonnegative and $u_0 \geq 0$. We say that u is an **integral solution** of (14.1) in $[0,T)$ if $u : \Omega \times [0,T) \to [0,\infty]$ is measurable, finite a.e. and

$$u(x,t) = \int_\Omega G(x,y,t)u_0(y)\,dy + \int_0^t \int_\Omega G(x,y,t-s)f(u(y,s))\,dy\,ds \qquad (15.7)$$

for a.e. $(x,t) \in Q_T$, where G is the Dirichlet heat kernel in Ω (cf. Appendix B). If $u_0 \in L^q(\Omega)$ is nonnegative and u is a mild L^q-solution of (14.1), then u is also an integral solution of (14.1). In fact, since

$$e^{-tA}w(x) = \int_\Omega G(x,y,t)w(y)\,dy,$$

the functions u, f are nonnegative and $u : [0,T) \to L^q(\Omega)$ is continuous, it is easy to pass to the limit in (15.5) as $\tau \to 0$ in order to obtain (15.7). Let us mention that the nonexistence statement in Theorem 15.3 is true in the class of integral solutions.

(v) **Weak solutions.** Assume that Ω is bounded and $u_0 \in L^1_\delta(\Omega)$. A function $u \in C([0,T),L^1_\delta(\Omega))$ is called a **weak solution** (more precisely **weak L^1_δ-solution**) of (14.1) in $[0,T)$ if the functions $u, \delta f(u)$ belong to $L^1_{loc}((0,T),L^1(\Omega))$, $u(0) = u_0$ and

$$\int_\tau^t \int_\Omega f(u)\varphi = -\int_\tau^t \int_\Omega u(\varphi_t + \Delta\varphi) - \int_\Omega u(\tau)\varphi(\tau)$$

for any $0 < \tau < t < T$ and any $\varphi \in C^2(\overline{\Omega} \times [\tau,t])$ such that $\varphi = 0$ on $\partial\Omega \times [\tau,t]$ and $\varphi(t) = 0$. One can prove that any mild L^q-solution (hence any classical L^q-solution) is a weak L^1_δ-solution for any $q \geq 1$ (see Corollary 48.11) and that the linear problem

$$
\begin{aligned}
u_t - \Delta u &= f, & & x \in \Omega,\ t \in [0,T), \\
u &= 0, & & x \in \partial\Omega,\ t \in [0,T), \\
u(x,0) &= u_0(x), & & x \in \Omega
\end{aligned}
\qquad (15.8)
$$

possesses a unique weak L^1_δ-solution in $[0, T)$ for any $f \in L^1_{loc}([0, T), L^1_\delta(\Omega))$ and $u_0 \in L^1_\delta(\Omega)$ (see Proposition 48.9). It is also easy to see that the notions of integral solution and weak L^1_δ-solution coincide if Ω is bounded, f is nonnegative and we consider nonnegative, locally integrable solutions only (see Corollary 48.10). Notice that uniqueness of weak solutions need not be true for the nonlinear problem (see (iii)).

(vi) **Initial traces.** In view of Theorems 15.2 and 15.3, it is a natural question to ask what should be the most general admissible initial data for local existence in problem (15.1). This question can be formulated as a problem of initial traces and has been studied in [56], [28]. In the case $\Omega = \mathbb{R}^n$, it is known that for any nonnegative classical solution u of

$$u_t - \Delta u = u^p, \qquad x \in \mathbb{R}^n, \ 0 < t < T \tag{15.9}$$

with $p > 1$, there exists a unique nonnegative Radon measure μ such that

$$u(t) \to \mu \ \text{ in the sense of measures,} \qquad \text{as } t \to 0. \tag{15.10}$$

The measure μ is called the **initial trace** of u. If, moreover, $p < 1 + 2/n$, then μ is uniformly locally finite, i.e.:

$$\sup_{x \in \mathbb{R}^n} \int_{B(x,1)} d\mu < \infty, \tag{15.11}$$

and the result is optimal. Namely, if $p < 1 + 2/n$ and μ is any nonnegative Radon measure verifying (15.11), then there exists a nonnegative classical solution of (15.9) which satisfies (15.10), and it is unique in a suitable class. Actually these results remain valid for properly defined weak solutions. On the other hand, in the range $p \geq 1 + 2/n$, the initial trace of a given solution u has to satisfy conditions stronger than (15.11) (in particular the Dirac measure $\mu = \delta_0$ is not admissible), but the necessary and sufficient condition on initial traces seems to be unknown. See also Theorem 15.11 and Remark 15.12 below for related results. $\square$

Remark 15.5. Independence of the local solution with respect to q. If $u_0 \in L^{q_1} \cap L^{q_2}(\Omega)$ for some $1 \leq q_1, q_2 \leq \infty$, with $q_1, q_2 > q_c$ or $q_1, q_2 \geq q_c > 1$, then the corresponding solutions u^i on $[0, T^i)$, given by Theorem 15.2 (or Remark 15.4(i)), coincide for $t < \min(T^1, T^2)$. This is a consequence of the following general argument.

By decreasing one of the T_i's, we may assume $T_1 = T_2$. The solution u_i is obtained as the unique fixed point of a contraction $\Phi^i_{u_0} : X^i \to X^i$, where X^i is a complete metric space (of functions of $t \in [0, T_1)$). For u_0 as above, $\Phi^1_{u_0}$ coincides with $\Phi^2_{u_0}$ on $X := X^1 \cap X^2$, and it is a contraction on the complete metric space X (with norm $\|\cdot\|_X = \max(\|\cdot\|_{X_1}, \|\cdot\|_{X_2})$). It thus has a unique fixed point u. By uniqueness in each X^i, we immediately deduce that $u_1 = u = u_2$. $\square$

Proof of Theorem 15.2. It is divided into several steps.

Step 1. Fixed-point argument. To handle the singularity of the initial data, the idea is to introduce a Banach space of functions with a temporal weight which has a suitable decay as $t \to 0$. We may assume that $\|u_0\|_q > 0$. Let $T > 0$ be small and consider the Banach space

$$Y_T := \{u \in L_{loc}^\infty((0,T), L^{pq}(\Omega)) \colon \|u\|_{Y_T} < \infty\}, \qquad \|u\|_{Y_T} := \sup_{0<t<T} t^\alpha \|u(t)\|_{pq},$$

where $\alpha := n(p-1)/2pq < 1/p < 1$. Choose $M > \|u_0\|_q$ and let $B_M = B_{M,T}$ denote the closed ball in Y_T with center 0 and radius M. We will use the Banach fixed point theorem for the mapping $\Phi_{u_0} : B_M \to B_M$, where

$$\Phi_{u_0}(u)(t) := e^{-tA} u_0 + \int_0^t e^{-(t-s)A} |u(s)|^{p-1} u(s) \, ds. \tag{15.12}$$

Using the L^p-L^q-estimates (see Proposition 48.4) we obtain for any $u, v \in B_M$ and $v_0 \in L^q(\Omega)$,

$$t^\alpha \|\Phi_{u_0}(u)(t) - \Phi_{v_0}(v)(t)\|_{pq}$$

$$\leq t^\alpha \|e^{-tA}(u_0 - v_0)\|_{pq} + t^\alpha \int_0^t \|e^{-(t-s)A}(|u(s)|^{p-1}u(s) - |v(s)|^{p-1}v(s))\|_{pq} \, ds$$

$$\leq (4\pi)^{-\alpha} \|u_0 - v_0\|_q + t^\alpha \int_0^t [4\pi(t-s)]^{-\alpha} \|(|u(s)|^{p-1}u(s) - |v(s)|^{p-1}v(s))\|_q \, ds$$

$$\leq (4\pi)^{-\alpha} \|u_0 - v_0\|_q$$

$$\qquad + C'(p)t^\alpha \int_0^t (t-s)^{-\alpha} (\|u(s)\|_{pq}^{p-1} + \|v(s)\|_{pq}^{p-1}) \|u(s) - v(s)\|_{pq} \, ds$$

$$\leq (4\pi)^{-\alpha} \|u_0 - v_0\|_q + C(p)M^{p-1}t^\alpha \int_0^t (t-s)^{-\alpha} s^{-(p-1)\alpha} \|u(s) - v(s)\|_{pq} \, ds. \tag{15.13}$$

In particular, choosing $v_0 = 0$ and $v = 0$ in (15.13) we have

$$\|\Phi_{u_0}(u)\|_{Y_T} \leq (4\pi)^{-\alpha} \|u_0\|_q + \sup_{0<t<T} C(p)M^{p-1}t^\alpha \int_0^t (t-s)^{-\alpha} s^{-p\alpha} \, ds \, \|u\|_{Y_T}$$

$$\leq (4\pi)^{-\alpha} \|u_0\|_q + C(p,\alpha)M^{p-1}T^{1-p\alpha} \|u\|_{Y_T}.$$

Let $T_0 = T_0(M,n,p,q) > 0$ be such that

$$C(p,\alpha)M^{p-1}T_0^{1-p\alpha} < \min(1 - (4\pi)^{-\alpha}, 1/2) \quad \text{and} \quad C(p)M^{p-1}T_0^{1-\alpha} < 1/2. \tag{15.14}$$

Then the above estimate implies

$$\|\Phi_{u_0}(u)\|_{Y_T} < (4\pi)^{-\alpha}M + (1 - (4\pi)^{-\alpha})M = M \qquad \text{for any } T \leq T_0, \tag{15.15}$$

hence Φ_{u_0} maps B_M into B_M for $T \leq T_0$. Choosing $v_0 = u_0$ in (15.13) we obtain

$$\|\Phi_{u_0}(u) - \Phi_{u_0}(v)\|_{Y_T} \leq C(p,\alpha) M^{p-1} T^{1-p\alpha} \|u - v\|_{Y_T} \leq \frac{1}{2}\|u - v\|_{Y_T}$$

for any $T \leq T_0$. Consequently, Φ_{u_0} is a contraction in B_M and it possesses a unique fixed point u in this set.

Note for further reference that in fact, for any $T \leq T_0$,

$$u \text{ is the only solution of } \Phi_{u_0}(u) = u \text{ in } Y_T. \tag{15.16}$$

Indeed, given any two solutions, both belong to $B_{M',T_0'}$ for some large M' and small T_0' satisfying (15.14). Therefore they coincide for small $t > 0$, hence on $(0,T)$ by an obvious continuation argument.

Step 2. Regularity. The function u satisfies $|u|^{p-1}u \in L^1\big((0,T), L^q(\Omega)\big)$ hence $u = \Phi_{u_0}(u) \in C\big([0,T], L^q(\Omega)\big)$. Choose $\varepsilon > 0$ small and set $\kappa_1 := pq$. Then

$$u \in L^\infty\big([\varepsilon,T], L^{\kappa_1}(\Omega)\big)$$

and

$$u(t+\varepsilon) = e^{-tA}u(\varepsilon) + \int_0^t e^{-(t-s)A}|u(s+\varepsilon)|^{p-1}u(s+\varepsilon)\,ds. \tag{15.17}$$

Choose $\kappa_2 > \kappa_1$ such that $\beta_1 := \frac{n}{2}\big(\frac{p}{\kappa_1} - \frac{1}{\kappa_2}\big) < 1$ and set $\beta_2 := \frac{n}{2}\big(\frac{1}{\kappa_1} - \frac{1}{\kappa_2}\big)$. Using (15.17) and the L^p-L^q-estimates we get

$$\|u(t+\varepsilon)\|_{\kappa_2} \leq t^{-\beta_2}\|u(\varepsilon)\|_{\kappa_1} + \int_0^t (t-s)^{-\beta_1}\|u(s+\varepsilon)\|_{\kappa_1}^p\,ds \leq C(\varepsilon)$$

for $t \in [\varepsilon, T-\varepsilon]$. Hence $u \in L^\infty\big([2\varepsilon,T], L^{\kappa_2}(\Omega)\big)$ and an obvious bootstrap argument shows $u \in L^\infty_{loc}\big((0,T], L^\infty(\Omega)\big)$. Now standard existence and regularity results for linear parabolic equations (see Appendix B) guarantee that u is a classical solution for $t > 0$, hence a classical L^q-solution. Let us explain this in more detail in the case of bounded domains; in the general case one can use smooth cut-off functions and use localized versions of the regularity statements in Appendix B.

Fix $\delta > 0$ small and let $\psi : \mathbb{R} \to [0,1]$ be a smooth function satisfying $\psi(t) = 0$ for $t \leq \delta$ and $\psi(t) = 1$ for $t \geq 2\delta$. Since u is a mild solution, it is also a weak (L^1_δ-) solution (see Corollary 48.11). Consequently, ψu is a weak solution of the linear problem (15.8) with $f := \psi_t u + \psi|u|^{p-1}u \in L^\infty(Q)$, where $Q := Q_T$. Now Theorem 48.1(iii) guarantees that this linear problem has a strong solution $v \in W^{2,1;q}(Q)$ for any $q \in (1,\infty)$. This strong solution is obviously a weak solution and the uniqueness of weak solutions (see Proposition 48.9) guarantees $\psi u = v$, consequently $u \in W^{2,1;q}(\Omega \times (2\delta, T))$. Now fixing $q > n + 2$ we see that $f(u)$ is

Hölder continuous in $\Omega \times (2\delta, T)$. Next consider the function $\psi(t - 2\delta)u(t)$ and use Theorem 48.2(ii) to see that u is a classical solution for $t > 4\delta$.

Step 3. Continuous dependence. Let us denote by $U(t)u_0$ the solution $u(t)$ constructed above. The existence proof shows that $U(\cdot)v_0$ is defined and belongs to $B_{M,T}$ for any v_0 satisfying $\|v_0\|_q < M$ and any $T \leq T_0$. In addition, (15.13) guarantees

$$\|U(\cdot)u_0 - U(\cdot)v_0\|_{Y_T} \leq \|u_0 - v_0\|_q + C(p, \alpha)M^{p-1}T^{1-p\alpha}\|U(\cdot)u_0 - U(\cdot)v_0\|_{Y_T},$$

hence the choice of T_0 implies

$$\|U(\cdot)u_0 - U(\cdot)v_0\|_{Y_T} \leq 2\|u_0 - v_0\|_q. \tag{15.18}$$

It follows that

$$\|U(t)u_0 - U(t)v_0\|_q$$
$$\leq \|u_0 - v_0\|_q + \int_0^t \||U(s)u_0|^{p-1}U(s)u_0 - |U(s)v_0|^{p-1}U(s)v_0\|_q \, ds$$
$$\leq \|u_0 - v_0\|_q + C(p)M^{p-1}T_0^{1-\alpha}\|U(\cdot)u_0 - V(\cdot)v_0\|_{Y_T}$$
$$\leq 2\|u_0 - v_0\|_q$$
$$\tag{15.19}$$

whenever $t \leq T_0$. Consequently, the map $L^q(\Omega) \to L^q(\Omega) : v_0 \to U(t)v_0$ is Lipschitz continuous in a neighborhood of u_0.

Step 4. Uniqueness. Let v be a classical L^q-solution of (15.1) in an interval $[0, T_1)$, that is $v \in C\big([0, T_1), L^q(\Omega)\big) \cap L^\infty_{loc}((0, T_1), L^\infty(\Omega))$, $v(0) = u_0$ and v is a classical solution of (15.1) for $t \in (0, T_1)$. Due to the uniqueness property (15.16), it is sufficient to show that $v(t) = U(t)u_0$ for small t. Decreasing T_1 if necessary we may thus assume that $T_1 \leq T_0$ and $\|v(s)\|_q < M$ for all $s \in [0, T_1)$. Let $T = T_1/2$. For each $\tau \in (0, T)$, since $v_\tau := v(\cdot + \tau) \in Y_T$ and v_τ satisfies (15.6), property (15.16) implies

$$v(t + \tau) = U(t)v(\tau) \quad \text{for all } t \in (0, T).$$

Passing to the limit as $\tau \to 0$ and using (15.19), we obtain $v(t) = U(t)u_0$ for all $t \in (0, T)$, hence the solution u is unique.

Step 5. Smoothing estimate. Fix $M = 2\|u_0\|_q$ and notice that $T_0 = T_0(\|u_0\|_q)$ (provided we suppress the dependence of T_0 on n, p, q). Choose $r \geq q$. If $r = q$ or $r = pq$, then (15.2) follows from (15.19) (with $v_0 = 0$) or (15.15), respectively. Assume that

$$\|u(t)\|_m \leq C\|u_0\|_q t^{-\alpha_m} \tag{15.20}$$

for some $m \geq \max(p, q)$, where α_m is defined in (15.2). We shall prove that we may increase the value of m in this estimate (by enlarging C if necessary) in such a way that we can reach the value $m = \infty$ in a finite number of iterations. Then (15.2) follows for any $r \in [q, \infty]$ from the interpolation inequality

$$\|u(t)\|_r \leq \|u(t)\|_q^{q/r} \|u(t)\|_\infty^{1-q/r}.$$

Similarly as above we obtain

$$\|u(t)\|_r \leq \|e^{-tA/2} u(t/2)\|_r + \int_{t/2}^t (t-s)^{-(n/2)(p/m-1/r)} \|u(s)\|_m^p \, ds$$

$$\leq t^{-(n/2)(1/m-1/r)} \|u(t/2)\|_m$$

$$+ C^p \|u_0\|_q^p \int_{t/2}^t (t-s)^{-(n/2)(p/m-1/r)} s^{-p(n/2)(1/q-1/m)} \, ds$$

$$\leq C \|u_0\|_q t^{-\alpha_r}$$

$$\times \left(1 + t^{1-n(p-1)/(2q)} \int_{1/2}^1 (1-s)^{-(n/2)(p/m-1/r)} s^{-p(n/2)(1/q-1/m)} \, ds \right)$$

$$\leq C \|u_0\|_q t^{-\alpha_r}$$

provided $p/m - 1/r < 2/n$. Since $p/m - 1/m < 2/n$ due to $m \geq q$, the conclusion follows.

Step 6. Positivity. The positivity statement follows from the nonnegativity of the semigroup e^{-tA} and the construction of the solution as a limit of nonnegative iterations $u_{k+1} = \Phi_{u_0}(u_k)$, $u_1(t) \equiv 0$. $\square$

In view of the proof of Theorem 15.3, we prepare the following lemma from [535] (see also [529]). It implies in particular a (weighted) a priori estimate for any local nonnegative (integral) solution of (15.1) (see Corollary 15.8), which will be used in the proof of Theorem 18.3. Given a measurable function $\Phi : \Omega \to [0, \infty]$, we set

$$(e^{-tA}\Phi)(x) := \int_\Omega G(x, y, t) \Phi(y) \, dy,$$

where $G = G_\Omega$ is the Dirichlet heat kernel in Ω (see Appendix B).

Lemma 15.6. *Let $u_0 : \Omega \to [0, \infty]$ and $u : \Omega \times [0, T] \to [0, \infty]$ be measurable and satisfy*

$$u(t) \geq e^{-tA} u_0 + \int_0^t e^{-(t-s)A} u^p(s) \, ds \qquad a.e. \ in \ Q_T. \qquad (15.21)$$

Assume that $u(x, t) < \infty$ for a.a. $(x, t) \in Q_T$. Then there holds

$$t^{1/(p-1)} \|e^{-tA} u_0\|_\infty \leq k_p := (p-1)^{-1/(p-1)} \qquad for \ all \ t \in (0, T]. \qquad (15.22)$$

Proof. In this proof, operations such as interchange of integrals and moving of e^{-tA} inside integrals are justified by Fubini's theorem for nonnegative measurable functions. First notice that

$$e^{-tA}\Phi = e^{-(t-s)A}e^{-sA}\Phi \quad \text{for all } 0 < s < t$$
$$\text{and any measurable } \Phi : \Omega \to [0, \infty]. \tag{15.23}$$

Also, we deduce from Jensen's inequality and $\int_\Omega G(x, y, t)\, dy \le 1$ that

$$e^{-tA}\Phi^p \ge (e^{-tA}\Phi)^p \text{ for all measurable } \Phi : \Omega \to [0, \infty]. \tag{15.24}$$

Now, by redefining u on a null set, we may assume that (15.21) actually holds everywhere in $\Omega \times (0, T)$. By assumption, for a.a. $\tau \in (0, T)$, we have $u(\cdot, \tau) < \infty$ a.e. in Ω. Fix such τ and let $\Omega_\tau := \{x \in \Omega : u(x, \tau) < \infty\}$. For $t \in [0, \tau]$, it follows from (15.21), (15.23) and (15.24) that

$$e^{-(\tau-t)A}u(t) \ge e^{-\tau A}u_0 + \int_0^t e^{-(\tau-s)A}u^p(s)\, ds$$
$$\ge e^{-\tau A}u_0 + \int_0^t \left(e^{-(\tau-s)A}u(s)\right)^p ds =: h(\cdot, t). \tag{15.25}$$

By the second inequality in (15.25), we see that

$$h(\cdot, t) \le e^{-\tau A}u_0 + \int_0^\tau e^{-(\tau-s)A}u^p(s)\, ds \le u(\cdot, \tau); \tag{15.26}$$

and so $h(x, t) < \infty$ for all $(x, t) \in \Omega_\tau \times [0, \tau]$. Fix $x \in \Omega_\tau$. Then the function $\phi(t) := h(x, t)$ is absolutely continuous on $[0, \tau]$ and (15.25) yields

$$\phi'(t) = \left(e^{-(\tau-t)A}u(t)\right)^p(x) \ge \phi^p(t) \quad \text{for a.a. } t \in [0, \tau]. \tag{15.27}$$

Also $\phi(t) \ge \left(e^{-\tau A}u_0\right)(x) > 0$, and so (15.27) can be rewritten as $[\phi^{1-p}]' \le -(p-1)$. Integrating this inequality over $[0, \tau]$, we get

$$\left[\left(e^{-\tau A}u_0\right)(x)\right]^{1-p} = \phi^{1-p}(0) \ge \phi^{1-p}(\tau) + (p-1)\tau \ge (p-1)\tau. \tag{15.28}$$

It follows that $\tau^{1/(p-1)}\|e^{-\tau A}u_0\|_\infty \le k$. This guarantees in particular that $e^{-tA}u_0 \in L^\infty(\Omega)$ for a.a. $t \in (0, T)$, Since $t \mapsto \|e^{-tA}v\|_\infty$ is continuous for $v \in L^\infty(\Omega)$ and $t > 0$, we deduce from (15.23) that the function

$$t \mapsto t^{1/(p-1)}\|e^{-tA}u_0\|_\infty$$

is continuous in $(0, T)$, hence (15.22). $\quad\square$

Remark 15.7. If $0 \leq u_0 \in L^\infty(\Omega)$, and u is a (sufficiently regular) supersolution of (14.1) on $[0,T]$, then estimate (15.22) can be alternatively obtained as follows (cf. [363]). Let

$$\underline{u}(x,t) := \left[\left(e^{-tA}u_0\right)^{1-p}(x) - (p-1)t\right]_+^{-1/(p-1)},$$

which is finite in Q_{T_1}, where $T_1 := \inf\{t \in [0,T] : t^{1/(p-1)}\|e^{-tA}u_0\|_\infty \geq k_p\} \in (0,T]$. A direct computation reveals that $\underline{u}_t - \Delta\underline{u} \leq \underline{u}^p$ in Q_{T_1}. In view of the comparison principle, since $\underline{u} = 0$ on S_{T_1} and $\underline{u}(\cdot,0) = u_0$, we obtain the lower estimate

$$u \geq \underline{u} \quad \text{in } Q_{T_1}. \tag{15.29}$$

In particular, we have $T_1 = T$, hence (15.22).

On the other hand, let us observe that estimate (15.29) also follows from (15.26) and (15.28). $\square$

Corollary 15.8. *Assume that* (15.21) *is true with the inequality sign replaced by the equality sign. Then*

$$\|t^{1/(p-1)}e^{-tA}u(\tau)\|_\infty \leq k_p \qquad \text{for all } t \in (0, T - \tau] \text{ and a.a. } \tau \in (0,T).$$

Proof. Set $v(t) := u(t + \tau)$. Then (15.23) and Fubini's theorem guarantee, for a.a. $\tau \in (0,T)$ and a.a. $t \in (\tau, T)$,

$$v(t) = e^{-(t+\tau)A}u_0 + \int_0^{t+\tau} e^{-(t+\tau-s)A}u^p(s)\,ds$$

$$= e^{-tA}e^{-\tau A}u_0 + \int_0^\tau e^{-tA}e^{-(\tau-s)A}u^p(s)\,ds + \int_\tau^{t+\tau} e^{-(t+\tau-s)A}u^p(s)\,ds$$

$$= e^{-tA}\left(e^{-\tau A}u_0 + \int_0^\tau e^{-(\tau-s)A}u^p(s)\,ds\right) + \int_0^t e^{-(t-s)A}v^p(s)\,ds$$

$$= e^{-tA}u(\tau) + \int_0^t e^{-(t-s)A}v^p(s)\,ds.$$

Hence, we may use Lemma 15.6 with u_0 replaced by $u(\tau)$ and T replaced by $T - \tau$ for a.a. $\tau \in (0,T)$. $\square$

Proof of Theorem 15.3. (i) Fix $\alpha \in (0, n/q)$, assume (without loss of generality) that $B(0,2\rho) \subset \Omega$, $\rho > 0$, and define

$$u_0(y) = |y|^{-\alpha}\chi_{B(0,\rho)}(y).$$

Clearly, we have $0 \leq u_0 \in L^q(\Omega)$. Using the heat kernel estimate in Proposition 49.10, we obtain, for $t > 0$ small,

$$\left(e^{-tA}u_0\right)(0) = \int_{|y|<\rho} G(0,y,t)|y|^{-\alpha}\,dy$$
$$\geq c_1 t^{-n/2} \int_{\{\sqrt{t}/2<|y|<\sqrt{t}\}} |y|^{-\alpha}\,dy \geq ct^{-\alpha/2}. \qquad (15.30)$$

Taking α close enough to n/q, we have $\alpha/2 > 1/(p-1)$. Combining Lemma 15.6 and (15.30), it follows that (15.1) cannot have any integral solution on $[0,T]$ (cf. Remark 15.4(iv)) for any $T > 0$.

(ii) The assertion is a consequence of Proposition 28.1 below. $\square$

For certain applications (see Section 26), it is useful to study well-posedness and regularization properties in different types of Lebesgue spaces. We first consider bounded domains and the spaces $L^q_\delta(\Omega)$, the Lebesgue spaces weighted by the function distance to the boundary. Based on the linear theory in these spaces (see Theorem 49.7 in Appendix C), we obtain the following results [200], in a completely similar manner as in Theorems 15.2 and 15.3(i). They show that the critical exponent for local well-posedness is now $q = (n+1)(p-1)/2$.

Theorem 15.9. *Assume Ω bounded and $p > 1$. Let $u_0 \in L^q_\delta(\Omega)$, $1 \leq q < \infty$, $q > (n+1)(p-1)/2$. Then there exists $T - T(\|u_0\|_{q,\delta}) > 0$ such that problem (15.1) possesses a unique classical L^q_δ-solution in $[0,T)$ and the following smoothing estimate is true:*

$$\|u(t)\|_{r,\delta} \leq C\|u_0\|_{q,\delta}t^{-\beta_r}, \qquad \beta_r := \frac{n+1}{2}\left(\frac{1}{q}-\frac{1}{r}\right), \qquad (15.31)$$

for all $t \in (0,T]$ and $r \in [q,\infty]$, with $C = C(n,p,q,\Omega) > 0$. In addition, $u \geq 0$ provided $u_0 \geq 0$.

Theorem 15.10. *Assume Ω bounded,*

$$p > 1 + \frac{2}{n+1} \qquad and \qquad 1 \leq q < \frac{(n+1)(p-1)}{2}.$$

Then there exists $u_0 \in L^q_\delta(\Omega)$, such that (15.1) does not admit any nonnegative L^q_δ-solution in $[0,T)$ for any $T > 0$.

In the case $\Omega = \mathbb{R}^n$, let us finally consider the uniformly local Lebesgue spaces $L^q_{ul}(\mathbb{R}^n)$. Using the linear smoothing effect in these spaces (see Proposition 49.15 in Appendix C or [37]), we obtain the following smoothing estimate (cf. [253]) by similar arguments as in the proof of Theorem 15.2. Here e^{-tA} denotes the heat semigroup in $\mathbb{R}^n$.

Theorem 15.11. *Let $p > 1$, $q > q_c$ and $1 \leq q < \infty$. Let $u_0 \in L^\infty(\mathbb{R}^n)$, $T > 0$ and assume that $u \in L^\infty((0,T), L^\infty(\mathbb{R}^n))$ is a solution of*

$$u(t) = e^{-tA}u_0 + \int_0^t e^{-(t-s)A}|u(s)|^{p-1}u(s)\,ds, \quad 0 < t < T.$$

Then there exist $C = C(n,p,q) > 0$, $T_0 = T_0(\|u_0\|_{q,ul}) > 0$ such that

$$\|u(t)\|_{r,ul} \leq C\|u_0\|_{q,ul}t^{-\alpha_r}, \qquad \alpha_r := \frac{n}{2}\left(\frac{1}{q} - \frac{1}{r}\right),$$

for all $t \in (0, \min(T, T_0)]$ and $r \in [q, \infty]$.

Remark 15.12. A local well-posedness result similar to Theorem 15.2 can also be proved in $L^q_{ul}(\mathbb{R}^n)$ (see [253], and cf. also [28]). $\quad\square$

16. Maximal existence time. Uniform bounds from L^q-estimates

In this section we are interested in sufficient conditions guaranteeing global existence. More precisely, we want to show that any solution satisfying suitable bounds in the Lebesgue space $L^q(\Omega)$ is global.

Let us start with a simple proposition which defines the maximal solution and existence time. We formulate the statement only for the model problem (14.1) but it will be clear from the proof that the same statement is true for a much more general class of equations and systems.

Proposition 16.1. *Let X be a Banach space of functions defined in Ω. Assume that problem (14.1) possesses for each $u_0 \in X$ a unique (classical X-) solution u on the interval $[0,T]$, where $T = T(u_0)$. Then there exists $T_{\max} = T_{\max}(u_0) \in (T, \infty]$ with the following properties.*

(i) The solution u can be continued (in a unique way) to a classical X-solution on the interval $[0, T_{\max})$.

(ii) If $T_{\max} < \infty$, then u cannot be continued to a classical X-solution on $[0, \tau)$ for any $\tau > T_{\max}$. We call u the maximal (classical X-) solution starting from u_0 and $T_{\max}$ its maximal existence time.

(iii) Assume further that $T = T(\|u_0\|_X)$. Then

$$\text{either } T_{\max} = \infty \quad \text{or} \quad \lim_{t \to T_{\max}} \|u(t)\|_X = \infty. \tag{16.1}$$

Proof. Let $u_0 \in X$ be fixed. If u_1 and u_2 are solutions of (14.1) on $[0, T_1)$ and $[0, T_2)$, respectively, then $u_1 = u_2$ on $[0, \min(T_1, T_2))$ due to the uniqueness. Let

$\{u_\alpha : [0, T_\alpha) \to X\}$ be the set of all solutions of (14.1) and $\tilde{T} := \sup T_\alpha$. Define $u : [0, \tilde{T}) \to X$ by $u(t) := u_\alpha(t)$, where α is any index such that $T_\alpha > t$. Then u is obviously a solution of (14.1) on $[0, \tilde{T})$, and properties (i)(ii) are verified.

Under the assumption in property (iii), suppose that

$$\tilde{T} < \infty \qquad \text{and} \qquad \liminf_{t \to \tilde{T}} \|u(t)\|_X < \infty.$$

Choose $C > 0$ and $t_k \to \tilde{T}$ such that $\|u(t_k)\|_X < C$ for all $k = 1, 2, \ldots$. Due to our assumptions there exists $T > 0$ independent of k such that the problem (14.1) with initial data $u(t_k)$ possesses a unique solution $u_k : [0, T] \to X$, $k = 1, 2, \ldots$. By uniqueness, $u_k(t) = u(t + t_k)$ for t small. Fix k such that $t_k \in (\tilde{T} - T, \tilde{T})$ and set

$$\tilde{u}(t) := \begin{cases} u(t), & t \in [0, t_k], \\ u_k(t - t_k), & t \in [t_k, t_k + T]. \end{cases}$$

Then $\tilde{u}$ is a solution of (14.1) on $[0, t_k + T]$ and $t_k + T > \tilde{T}$ which contradicts the definition of $\tilde{T}$. $\square$

Remarks 16.2. (i) **Maximal L^q-solution.** Consider problem (15.1) and set $X = L^q(\Omega)$, where $1 \le q \le \infty$ satisfies $q > q_c = n(p-1)/2$ or $q = q_c > 1$. If $u_0 \in X$, then Theorem 15.2 and Proposition 16.1 (or Remark 51.11 if $q = \infty$) guarantee the existence of a maximal (classical L^q-) solution u, up to a maximal existence time $T_{\max}(u_0)$. Moreover, property (16.1) is true if $q > q_c$. Similarly as in Example 51.9, u in fact satisfies

$$u \in BC^{2,1}(\overline{\Omega} \times [t_1, t_2]), \qquad 0 < t_1 < t_2 < T_{\max}(u_0). \tag{16.2}$$

If $u_0 \ge 0$, then $u \ge 0$. If u_0 is radial (resp. nonnegative and radial nonincreasing), then u enjoys the same property, as a consequence of Proposition 52.17.

(ii) **Independence of the maximal solution with respect to q.** Let q, u_0 and u be as in remark (i). We show that, if u_0 belongs to several L^q-spaces, then u and $T_{\max}(u_0)$ do not depend on q.

Thus assume that $u_0 \in L^{q_1} \cap L^{q_2}(\Omega)$ for some q_1, q_2 as above, and denote by u^i, $i = 1, 2$, the corresponding maximal, classical L^{q_i}-solution, of existence time T^i. We know that $u^1 = u^2$ for $t > 0$ small (cf. Remark 15.5). Using $u^i \in C([0, T^i), L^{q_i}(\Omega))$, we deduce easily that $u^1 = u^2$ on $[0, \min(T^1, T^2))$. Assume for contradiction that $T^1 < T^2$ (hence $T^1 < \infty$). Since, by the definition of a maximal classical L^q-solution, $u^2 \in L^\infty_{loc}((0, T^2), L^\infty(\Omega))$, it follows that $\||u^1|^{p-1}u^1| \le C|u^1|$ on $(T^1/2, T^1)$, which readily implies $\sup_{[T^1/2, T^1]} \|u^1(t)\|_{q_1} < \infty$. If $q_1 > q_c$, then this contradicts (16.1). If $q = q_c$, then the variation-of-constants formula implies $u^1 \in C([0, T^1], L^{q_1}(\Omega))$. The local existence theorem can then be used to extend u^1 after T^1 and we again reach a contradiction.

(iii) **Lower bounds on supercritical L^q-norms.** By using the local theory of problem (14.1), developed in Section 15, it is actually possible to obtain lower estimates of the supercritical L^q-norms, in case $T_{\max}(u_0) < \infty$. Namely, let $q \geq 1$ satisfy $q_c < q < \infty$ and assume $u_0 \in L^q(\Omega)$. Then the proof of Theorem 15.2 (see in particular formula (15.14)) shows that $(T_{\max}(u_0))^{1-n(p-1)/2q}\|u_0\|_q^{p-1} \geq C(n,p,q) > 0$. After a time shift, this yields

$$\|u(t)\|_q \geq C(n,p,q)(T_{\max}(u_0) - t)^{n/(2q)-1/(p-1)}, \quad 0 \leq t < T. \tag{16.3}$$

(iv) **Critical L^q-space.** If $u_0 \in L^q(\Omega)$ with $q = q_c > 1$, then it is not known in general whether $T_{\max}(u_0) < \infty$ implies $\lim_{t \to T_{\max}(u_0)} \|u(t)\|_q = \infty$. A positive result in that direction (cf. [91], [535]) is given by the following proposition. The proof, which relies on simple energy arguments, is postponed to the next section. For further positive results, see [530], [535], [246], [357]. $\square$

Proposition 16.3. *Consider problem* (15.1) *with* $p = 1 + 4/n$ *(so that* $q_c = n(p-1)/2 = 2$*). Let* $u_0 \in L^2(\Omega)$ *and assume* $T := T_{\max}(u_0) < \infty$. *Then*

$$\|u(t)\|_2 \geq C(n,p)|\log(T-t)|^{1/2}, \quad t \to T. \tag{16.4}$$

We say that (14.1) possesses a **global solution** if $T_{\max} = \infty$. Proposition 16.1 provides a simple criterion for global existence: If $\|u(t)\|_X$ remains bounded, then $T_{\max} = \infty$. Since the assumptions of Proposition 16.1 are satisfied with $X = L^\infty(\Omega)$ if $f \in C^1$ (see Remark 51.11) we see that the boundedness of the solution in $L^\infty(\Omega)$ is sufficient for its global existence. Note that the same statement is true for a much more general class of equations and systems.

Unfortunately, it is not easy to obtain the L^∞-estimate for solutions of (14.1). As we shall see, standard methods usually yield only an L^q-estimate for some $q < \infty$. Therefore, it is important to find q as small as possible and such that the L^q-estimate guarantees the L^∞-estimate, hence global existence. We will call this property of L^q the **continuation property.**

Theorem 15.2, Proposition 16.1 and Remarks 16.2 guarantee the global existence of a solution of the model problem (15.1) provided the solution is bounded in $L^q(\Omega)$ for some $q > q_c$.

As we shall see in Corollary 24.2, this condition is optimal (up to the equality sign): If $1 < q < q_c$, then there exists a radial positive solution of (15.1) in a ball such that $T_{\max} < \infty$ but the solution stays bounded in $L^q(\Omega)$. Therefore, the exponent q_c for problem (15.1) is "critical" both for well-posedness and the global existence. This is due to the simple structure of the nonlinearity. We will see in Chapter III that for more complicated problems, the critical exponents for the well-posedness and the continuation property may differ. Therefore it is important to find methods guaranteeing the global existence of a solution under the assumption of its boundedness in L^q and not using any well-posedness result.

In this section we present a method due to [11], [12] (cf. also [459]) which is based on Moser-type iterations and can be efficiently used for a very general class of problems (including degenerate problems, problems on nonsmooth domains etc). In order to make it as clear as possible, we again restrict ourselves to the model problem (15.1).

Another method for obtaining L^∞-bounds from L^q-bounds is presented in Appendix E (see Proposition 51.34). That method is based on the variation-of-constants formula and interpolation-extrapolation spaces and is due to [14].

Hence, our aim is to prove the following theorem (which is a consequence of Theorem 15.2 and Proposition 16.1), without using the well-posedness results.

Theorem 16.4. *Let $p > 1$ and let u be a classical solution of (15.1) defined on $[0, T)$. Assume $q > 1$ and $U_q := \sup_{t<T} \|u(t)\|_q < \infty$. If $p < 1 + 2q/n$, then $U_\infty < \infty$.*

For simplicity we will assume that Ω is bounded and $n \geq 3$.

Lemma 16.5. *Let u be a classical solution of (15.1) on $[0, T)$, $r \geq q \geq 1$, $p < 1 + 2q/n$ and $\tilde{U}_r := \max\{1, \|u_0\|_\infty, \sup_{t<T} \|u(t)\|_r\} < \infty$. Set*

$$\sigma(r) := \frac{n+2}{2n}\left(\frac{2r}{n} + 1 - p\right)^{-1} \quad \text{and} \quad \rho(r) := 1 + (p-1)\sigma(r).$$

Then there exists a constant $C_1 = C_1(p, q, n, \Omega) > 0$ such that

$$\tilde{U}_{2r} \leq C_1^{1/r} r^{\sigma(r)} \tilde{U}_r^{\rho(r)}.$$

Proof. Multiplying the equation in (15.1) by $|u|^{2r-2}u$ we obtain

$$\frac{d}{dt}\frac{1}{2r}\int_\Omega |u|^{2r}\,dx + \frac{2r-1}{r^2}\int_\Omega |\nabla|u|^r|^2\,dx = \int_\Omega |u|^{p+2r-1}\,dx.$$

Denote $w := |u|^r$, $\alpha = \alpha(r) := (p + 2r - 1)/(2r)$ and let β be defined by $1/(2\alpha) = \beta + (1-\beta)/2^*$. Then the assumption $1 < p < 1 + 2r/n$ guarantees $\beta \in (0,1)$ and $\alpha(1-\beta) < 1$. The above identity, interpolation, the Sobolev embedding theorem and Young's inequality imply

$$\frac{d}{dt}\frac{1}{2r}\|w\|_2^2 + \frac{2r-1}{r^2}\|\nabla w\|_2^2 = \|w\|_{2\alpha}^{2\alpha} \leq \left(\|w\|_1^\beta \|w\|_{2^*}^{1-\beta}\right)^{2\alpha}$$

$$\leq C\left(\|w\|_1^\beta \|\nabla w\|_2^{1-\beta}\right)^{2\alpha} = \left(\frac{1}{2r}\|\nabla w\|_2^2\right)^{\alpha(1-\beta)}\left(Cr^{1-\beta}\|w\|_1^{2\beta}\right)^\alpha$$

$$\leq \frac{1}{2r}\|\nabla w\|_2^2 + Cr^{\alpha(1-\beta)/\delta}\|w\|_1^{2\alpha\beta/\delta} = \frac{1}{2r}\|\nabla w\|_2^2 + Cr^{2r\sigma(r)-1}\|w\|_1^{2\rho(r)},$$

where $\delta := 1 - \alpha(1-\beta)$. Consequently, there exist $C, c > 0$ such that

$$e^{-ct}\frac{d}{dt}\left(e^{ct}\|w\|_2^2\right) = \frac{d}{dt}\|w\|_2^2 + c\|w\|_2^2 \leq Cr^{2r\sigma(r)}\|w\|_1^{2\rho(r)} \leq Cr^{2r\sigma(r)}\tilde{U}_r^{2r\rho(r)}.$$

Since $\|w(0)\|_2^2 \leq C\|u_0\|_\infty^{2r} \leq C\tilde{U}_r^{2r}$ and $\|w\|_2^2 = \|u\|_{2r}^{2r}$, integration of the above estimate implies the assertion. $\quad\square$

Proof of Theorem 16.4. We shall use the notation from Lemma 16.5. Notice that $\gamma := q\sigma(q) \geq r\sigma(r)$ for any $r \geq q$. Using repeatedly Lemma 16.5 with $r := q$, $r := 2q$, $r := 4q$ etc, one can easily verify that, given $\nu \in \{0, 1, 2, \dots\}$,

$$\tilde{U}_{2^{\nu+1}q} \leq (C_1 q^\gamma)^{k_1} 2^{k_2} \tilde{U}_q^{k_3},$$

where

$$k_1 = k_1(\nu) = \frac{1}{2^\nu q} + \frac{\rho(2^\nu q)}{2^{\nu-1}q} + \cdots + \frac{\rho(2^\nu q) \cdot \cdots \cdot \rho(2q)}{q},$$

$$k_2 = k_2(\nu) = \frac{\gamma}{q}\left(\frac{\nu}{2^\nu} + \frac{\nu-1}{2^{\nu-1}}\rho(2^\nu q) + \cdots + \frac{1}{2}\rho(2^\nu q) \cdot \cdots \cdot \rho(2^2 q)\right),$$

$$k_3 = k_3(\nu) = \rho(2^\nu q) \cdot \cdots \cdot \rho(q).$$

Since $\sigma(2r) \leq \sigma(r)/2$ we obtain $\rho(2^i q) \leq 1 + (p-1)\sigma(q)2^{-i}$. Now using the inequality $\log(1+x) \leq x$ for $x \geq 0$ we get

$$\log k_3 \leq \sum_{i=0}^{\nu}(p-1)\frac{\sigma(q)}{2^i} \leq 2(p-1)\sigma(q) =: C_3 < \infty.$$

Finally,

$$k_1 \leq \frac{k_3}{q}\sum_{i=0}^{\nu}\frac{1}{2^i} \leq \frac{2}{q}e^{C_3} < \infty \qquad \text{and} \qquad k_2 \leq \frac{\gamma}{q}k_3\sum_{i=1}^{\nu}\frac{i}{2^i} \leq \frac{\gamma}{q}e^{C_3}\sum_{i=1}^{\infty}\frac{i}{2^i} < \infty,$$

which concludes the proof. $\quad\square$

17. Blow-up

In this section we mainly consider the model problem

$$\left.\begin{aligned} u_t - \Delta u &= \lambda u + |u|^{p-1}u, & x \in \Omega,\ t > 0, \\ u &= 0, & x \in \partial\Omega,\ t > 0, \\ u(x,0) &= u_0(x), & x \in \Omega, \end{aligned}\right\} \tag{17.1}$$

with $p > 1$ and $\lambda \in \mathbb{R}$, and we derive some criteria for u_0 which guarantee blow-up of the solution of (17.1) in finite time. More general nonlinearities $f(u)$ will be briefly considered.

We will always assume that u_0 belongs to a function space X where (17.1) is well-posed and we denote by $T_{\max}(u_0)$ the maximal existence time of the solution of (17.1). We start with a simple criterion. In the bounded domain case, it is based on the **eigenfunction method** due to [296].

Theorem 17.1. *Consider problem (17.1) with $p > 1$ and $\lambda \in \mathbb{R}$.*

(i) *Assume Ω bounded. Let $u_0 \in L^\infty(\Omega)$ satisfy*

$$u_0 \geq 0, \qquad \int_\Omega u_0 \varphi_1 \, dx > c := \left(\max(0, \lambda_1 - \lambda)\right)^{1/(p-1)}. \tag{17.2}$$

Then $T_{\max}(u_0) < \infty$.

(ii) *Assume $\Omega = \mathbb{R}^n$. Then assertion (i) remains valid if we replace λ_1 by $2n$ and φ_1 by the function $\varphi(x) = \pi^{-n/2} e^{-|x|^2}$.*

Proof. (i) Recall that $u \geq 0$ and denote $y = y(t) := \int_\Omega u(t)\varphi_1 \, dx$. Multiplying the equation in (17.1) with φ_1, integrating by parts, and using $\Delta\varphi_1 = -\lambda_1\varphi_1$ and Jensen's inequality yields

$$y' = \int_\Omega u_t\varphi_1 \, dx = \int_\Omega u\Delta\varphi_1 \, dx + \lambda \int_\Omega u\varphi_1 \, dx + \int_\Omega u^p\varphi_1 \, dx \geq y^p - c^{p-1}y. \tag{17.3}$$

Since $y(0) > c$, we infer from (17.3) that

$$y' \geq \varepsilon y^p, \qquad 0 < t < T_{\max}(u_0),$$

with $\varepsilon = 1 - (c/y(0))^{p-1} > 0$. This differential inequality guarantees that u cannot exist globally.

(ii) The proof is the same except that we now use $\Delta\varphi \geq -2n\varphi$. The calculation in (17.3) can be easily justified by integrating by parts over B_R and letting $R \to \infty$, using property (16.2) and the exponential decay of φ and $\nabla\varphi$. $\square$

Remarks 17.2. (i) **Estimation of the blow-up time.** The proof of Theorem 17.1 shows that if, for instance, $\lambda = 0$, Ω is bounded and $\int_\Omega u_0\varphi_1 \, dx \geq (2\lambda_1)^{1/(p-1)}$, then

$$T_{\max}(u_0) \leq \frac{2}{p-1}\left(\int_\Omega u_0\varphi_1 \, dx\right)^{1-p}.$$

We refer to [324], [262] for more precise results concerning upper estimates of the blow-up time.

(ii) **Neumann boundary conditions.** If we replace the homogeneous Dirichlet boundary conditions in problem (17.1) with the homogeneous Neumann boundary conditions $\partial_\nu u = 0$, then all positive solutions blow up in finite time when $\lambda \geq 0$. Indeed, by integrating the equation over Ω, we see that the function $y(t) := \int_\Omega u(t) \, dx$ satisfies $y'(t) \geq \int_\Omega u^p \, dx \geq |\Omega|^{1-p} y^p$ with $y(0) > 0$. Alternatively, for $t_0 > 0$ small, the strong maximum principle guarantees that $u(x, t_0) \geq \varepsilon > 0$ in $\overline{\Omega}$, and it suffices to use the solution of the ODE $z' = z^p$, $z(t_0) = \varepsilon$, as subsolution. $\square$

The previous result can be easily extended to problem (14.1) under suitable convexity and superlinearity conditions.

Theorem 17.3. *Consider problem (14.1) where $f : \mathbb{R} \to \mathbb{R}$ is a convex C^1-function and Ω is bounded. Assume that, for some $a > 0$, we have $f(s) > 0$ for all $s \geq a$ and*

$$\int^{\infty} \frac{ds}{f(s)} < \infty. \tag{17.4}$$

Then Theorem 17.1(i) remains true provided the constant c in (17.2) is replaced by $C = C(\Omega, f) > 0$ large enough.

Proof. Denote again $y = y(t) := \int_{\Omega} u(t)\varphi_1 \, dx$. Arguing as in the previous proof, we obtain

$$y' = \int_{\Omega} u_t \varphi_1 \, dx = \int_{\Omega} u \Delta \varphi_1 \, dx + \int_{\Omega} f(u)\varphi_1 \, dx \geq -\lambda_1 y + f(y). \tag{17.5}$$

Since f is convex, the function $g(s) := \frac{f(s) - f(a)}{s - a}$ is nondecreasing for $s > a$ and $g(s) \to \infty$ as $s \to \infty$, due to (17.4). Therefore, there exists $C \geq a$ such that $f(s) \geq 2\lambda_1 s$ for all $s \geq C$. If $y(0) \geq C$, it follows from (17.5) that, as long as u exists, $y(t) \geq C$ and

$$y' \geq f(y) - \lambda_1 y \geq \frac{1}{2} f(y),$$

hence

$$\lambda_1 t/2 \leq \int_0^t \frac{y'(\tau)}{f(y(\tau))} \, d\tau = \int_{y(0)}^{y(t)} \frac{ds}{f(s)} \leq \int_{y(0)}^{\infty} \frac{ds}{f(s)} < \infty.$$

Therefore u cannot exist globally. $\square$

Remark 17.4. It is well known that condition (17.4) is necessary and sufficient for the existence of blow-up solutions of the ODE $u' = f(u)$, $t \geq 0$. The convexity condition in Theorem 17.3 can be replaced by the assumption that $f \geq \tilde{f}$ for s large, where $\tilde{f}$ satisfies the assumptions of the theorem. As a typical "weakly superlinear" f satisfying (17.4), one may take a function f such that $f(s) = (1 + s)\log^p(1 + s)$ for $s \geq 0$, with $p > 1$. $\square$

The next criterion is based on the fact that the energy functional

$$E(u) = \frac{1}{2} \int_{\Omega} \left(|\nabla u|^2 - \lambda u^2 \right) dx - \frac{1}{p+1} \int_{\Omega} |u|^{p+1} \, dx \tag{17.6}$$

is nonincreasing along any solution of (17.1). More precisely we have:

Lemma 17.5. *Consider problem (17.1) with $p > 1$, $\lambda \in \mathbb{R}$, $u_0 \in L^{\infty} \cap H_0^1(\Omega)$, and let $T = T_{\max}(u_0)$. Then $E(u(\cdot)) \in C([0,T)) \cap C^1((0,T))$ and*

$$\frac{d}{dt} E(u(t)) = - \int_{\Omega} u_t^2(t) \, dx. \tag{17.7}$$

Proof. Example 51.28 guarantees $u \in C\big([0,T), H_0^1 \cap L^{p+1}(\Omega)\big)$, hence $E(u(\cdot)) \in C([0,T))$, and

$$u \in C\big((0,T), H^2(\Omega)\big) \cap C^1\big((0,T), L^2 \cap L^{p+1}(\Omega)\big). \tag{17.8}$$

Denote $E_1(t) = \int_\Omega |\nabla u(t)|^2 \, dx$. For $t, s \in (0,T)$, $s \neq t$, using integration by parts, we obtain

$$\begin{aligned}
\frac{E_1(t) - E_1(s)}{t - s} &= \frac{1}{t-s} \int_\Omega \nabla(u(t) - u(s)) \cdot \nabla(u(t) + u(s)) \, dx \\
&= -\int_\Omega \left(\frac{u(t) - u(s)}{t-s}\right) \Delta(u(t) + u(s)) \, dx \to -2 \int_\Omega u_t(t) \Delta u(t) \, dx
\end{aligned}$$

as $s \to t$, due to (17.8). Consequently, $E(u(\cdot)) \in C^1((0,T))$ and

$$\frac{d}{dt} E\big(u(t)\big) = \int_\Omega (-\Delta u - \lambda u - |u|^{p-1} u) u_t \, dx = -\int_\Omega u_t^2(t) \, dx. \quad \square$$

The following result is due to [327].

The simpler proof in the case Ω bounded, $\lambda = 0$, is from [515], [51].

Theorem 17.6. *Consider problem (17.1) with $p > 1$, $\lambda \in \mathbb{R}$ and $u_0 \in L^\infty \cap H_0^1(\Omega)$. Assume either Ω bounded or $\lambda \leq 0$. If $E(u_0) < 0$, then $T_{\max}(u_0) < \infty$.*

Proof. (i) First assume that Ω is bounded.

Set $\psi(t) := \|u(t)\|_2^2$. Multiplying the equation in (17.1) by u and using Hölder's inequality we obtain

$$\begin{aligned}
\frac{1}{2} \psi'(t) &= \int_\Omega u u_t(t) \, dx = -\int_\Omega |\nabla u(t)|^2 \, dx + \lambda \int_\Omega u^2 \, dx + \int_\Omega |u(t)|^{p+1} \, dx \\
&= -2E\big(u(t)\big) + \frac{p-1}{p+1} \int_\Omega |u(t)|^{p+1} \, dx \geq -2E(u_0) + c\psi(t)^{(p+1)/2},
\end{aligned} \tag{17.9}$$

where $c := (p-1)/[(p+1)|\Omega|^{(p-1)/2}]$. This inequality implies $T_{\max}(u_0) < \infty$ provided $E(u_0) < 0$ (or $\psi(0)^{(p+1)/2} > 2E(u_0)/c$).

(ii) Next consider the case Ω unbounded, $\lambda \leq 0$. (The following argument works also if $\lambda \leq \lambda_1$ and Ω is bounded). We will use the **concavity method** due to [327].

Assume $T_{\max}(u_0) = \infty$ and denote $M(t) := \frac{1}{2} \int_0^t \|u(s)\|_2^2 \, ds$. Then we have $M'(t) = \frac{1}{2} \|u(t)\|_2^2$ and

$$\begin{aligned}
M''(t) &= \int_\Omega u u_t(t) \, dx = -\int_\Omega |\nabla u(t)|^2 \, dx + \lambda \int_\Omega u^2(t) \, dx + \int_\Omega |u(t)|^{p+1} \, dx \\
&= -(p+1) E\big(u(t)\big) + \frac{p-1}{2} \int_\Omega \big(|\nabla u(t)|^2 - \lambda u^2(t)\big) \, dx \\
&\geq -(p+1) E(u_0) > 0,
\end{aligned}$$

which implies $M'(t) \to \infty$ and $M(t) \to \infty$ as $t \to \infty$. Moreover, this estimate and

$$\int_0^t \|u_t(s)\|_2^2 \, ds = E(u_0) - E\big(u(t)\big) < -E\big(u(t)\big) \tag{17.10}$$

(cf. (17.7)) imply

$$M''(t) \geq -(p+1)E\big(u(t)\big) \geq (p+1)\int_0^t \|u_t(s)\|_2^2 \, ds,$$

hence

$$\begin{aligned}
M(t)M''(t) &\geq \frac{p+1}{2}\left(\int_0^t \|u_t(s)\|_2^2 \, ds\right)\left(\int_0^t \|u(s)\|_2^2 \, ds\right) \\
&\geq \frac{p+1}{2}\left(\int_0^t \int_\Omega u(x,s)u_t(x,s) \, dx \, ds\right)^2 \\
&= \frac{p+1}{2}\big(M'(t) - M'(0)\big)^2.
\end{aligned}$$

Since $M'(t) \to \infty$ as $t \to \infty$, the last estimate implies existence of $\alpha, t_0 > 0$ such that

$$M(t)M''(t) \geq (1+\alpha)\big(M'(t)\big)^2, \quad t \geq t_0.$$

This inequality guarantees that the nonincreasing function $t \mapsto M^{-\alpha}(t)$ is concave on $[t_0, \infty)$ which contradicts the fact $M^{-\alpha}(t) \to 0$ as $t \to \infty$. $\quad\square$

Proof of Proposition 16.3. By Example 51.27 in Appendix E, after a time-shift, we may assume $u_0 \in H_0^1(\Omega)$. Similarly as in (17.9) (but without assuming Ω bounded), we have

$$\frac{1}{2}\frac{d}{dt}\int_\Omega u^2(t) \, dx \geq -2E(u_0) + \frac{p-1}{p+1}\int_\Omega |u(t)|^{p+1} \, dx.$$

Integrating and using (16.3) with $q = p+1 > q_c$, it follows that

$$\begin{aligned}
\int_\Omega u^2(t) \, dx &\geq -2E(u_0)t + \frac{p-1}{p+1}\int_0^t \int_\Omega |u(s)|^{p+1} \, dx \, ds \\
&\geq -C + C(n,p)\int_0^t (T-s)^{(n/2)-(p+1)/(p-1)} \, ds.
\end{aligned}$$

Since $(n/2) - (p+1)/(p-1) = -1$, (16.4) follows. $\quad\square$

Remarks 17.7. (i) The proof of Theorem 17.6 does *not* imply blow-up of the L^2-norm of u. Indeed, as was observed in [51], the solution u might cease to exist *before* the time obtained by integrating the differential inequality in (17.9). Examples where the L^2-norm of u remains bounded will be given in Corollary 24.2. A similar remark holds concerning the quantity $y(t)$ in the proof of Theorem 17.1.

(ii) The first part of the proof of Theorem 17.6 shows that

$$\|u(t)\|_2 \le \big(2E(u_0)/c\big)^{1/(p+1)}$$

for any global solution u of (17.1) provided Ω is bounded. Now the results of the preceding section guarantee that $\|u(t)\|_\infty \le C\big(\|u_0\|_\infty, E(u_0)\big)$ if $p < 1 + 4/n$. As we shall see later in Section 22, this assertion is true for any $p < p_S$.

(iii) If u is a global solution of (17.1) and Ω is bounded or $\lambda \le 0$, then Theorem 17.6 guarantees $0 \le E\big(u(t)\big) \le E(u_0)$ for all $t > 0$.

(iv) Inequality (17.9) also shows the following: Given $\delta > 0$ there exists $C_\delta > 0$ such that $T_{\max}(u_0) < \delta$ whenever $E(u_0) < -C_\delta$.

(v) Let $\varphi \in L^\infty \cap H_0^1(\Omega)$ be a fixed function, $\varphi \not\equiv 0$. Then $T_{\max}(\alpha\varphi) < \infty$ for $\alpha > 0$ large enough. This follows from Theorem 17.6 and the fact that

$$E(\alpha\varphi) = \alpha^2 \int_\Omega \frac{|\nabla\varphi|^2 - \lambda\varphi^2}{2}\, dx - \alpha^{p+1} \int_\Omega \frac{\varphi^{p+1}}{p+1}\, dx.$$

Note that if we assume $0 \le \varphi \in L^\infty(\Omega)$ instead, then the same conclusion follows from Theorem 17.1. □

Further blow-up conditions involving the energy will be given in Theorem 19.5. We now give a third criterion (cf. [342]), which guarantees blow-up if one starts above a positive equilibrium.

Theorem 17.8. *Assume Ω bounded, $p > 1$ and $\lambda \in \mathbb{R}$. Assume that problem* (17.1) *has a (classical) equilibrium v, with $v > 0$ in Ω. If $u_0 \in L^\infty(\Omega)$ satisfies $u_0 \ge v$, $u_0 \not\equiv v$, then $T_{\max}(u_0) < \infty$.*

For the proof, we prepare the following separation lemma, which will be used again later.

Lemma 17.9. *Assume Ω bounded and consider problem* (14.1) *where $f : \mathbb{R} \to \mathbb{R}$ is a convex C^1-function with $f(0) = 0$. Let $u_0, \underline{u}_0 \in L^\infty(\Omega)$ be such that $u_0 \ge \underline{u}_0$, $u_0 \not\equiv \underline{u}_0$. Let $u, \underline{u}$ be the corresponding solutions of* (14.1), *and fix $\tau \in (0, T_{\max}(u_0))$. Then $T_{\max}(\underline{u}_0) \ge T_{\max}(u_0)$ and there exists $\alpha > 1$ such that*

$$u \ge \alpha\underline{u}, \qquad \tau \le t < T_{\max}(u_0). \tag{17.11}$$

Proof. Since $\underline{u} \leq u$ by the comparison principle and $f(s) \geq f'(0)s$, $s \in \mathbb{R}$, by the convexity of f and $f(0) = 0$, we have $T_{\max}(\underline{u}_0) \geq T_{\max}(u_0)$. By the strong and the Hopf maximum principles, we have

$$u(x,\tau) > \underline{u}(x,\tau) \ \text{ in } \Omega \quad \text{and} \quad \frac{\partial u}{\partial \nu}(x,\tau) < \frac{\partial \underline{u}}{\partial \nu}(x,\tau) \ \text{ on } \partial\Omega.$$

Therefore, there exists $\alpha > 1$ such that $u(x,\tau) \geq \alpha\underline{u}(x,\tau)$ in Ω. Since $f(\alpha\underline{u}) \geq \alpha f(\underline{u})$, due to f convex and $f(0) = 0$, we infer that

$$(\alpha\underline{u})_t - \Delta(\alpha\underline{u}) - f(\alpha\underline{u}) \leq \alpha(\underline{u}_t - \Delta\underline{u} - f(\underline{u})) = 0,$$

and the lemma follows from the comparison principle. $\square$

Proof of Theorem 17.8. By Lemma 17.9, applied with $\underline{u}_0 = v$, there exist $\alpha > 1$ and $\tau \in (0, T_{\max}(u_0))$, such that

$$u \geq \alpha v, \quad t \in [\tau, T_{\max}(u_0)). \tag{17.12}$$

Denote $z = z(t) := \int_\Omega u(t)v\,dx$. Multiplying the equation in (15.1) with v, integrating by parts, and using (17.12) and Hölder's (or Jensen's) inequality, we obtain

$$z' = \int_\Omega u_t v\,dx = \int_\Omega u\Delta v\,dx + \int_\Omega (u^p + \lambda u)v\,dx$$

$$= \int_\Omega (u^p v - v^p u)\,dx = \int_\Omega \left(1 - (v/u)^{p-1}\right)u^p v\,dx$$

$$\geq (1 - \alpha^{1-p})\int_\Omega u^p v\,dx \geq (1 - \alpha^{1-p})\left(\int_\Omega v\,dx\right)^{1-p} z^p,$$

for $t \in [\tau, T_{\max}(u_0))$. It follows that u cannot exist globally. $\square$

By using an alternative linearization argument based on an idea from [311], one can extend Theorem 17.10 to more general convex nonlinearities.

Theorem 17.10. *Consider problem (14.1) with f and Ω as in Theorem 17.3. Assume in addition that $f(0) = 0$, f' is nonconstant near 0, and that problem (14.1) has a (classical) equilibrium v, with $v > 0$ in Ω. If $u_0 \in L^\infty(\Omega)$ satisfies $u_0 \geq v$, $u_0 \not\equiv v$, then $T_{\max}(u_0) < \infty$.*

Proof. Let μ and $\psi > 0$ denote the first eigenvalue and the corresponding eigenfunction of the problem

$$\Delta\psi + f'(v)\psi = \mu\psi \quad \text{in } \Omega, \qquad \psi = 0 \quad \text{on } \partial\Omega,$$

$\int_\Omega \psi\,dx = 1$. Multiplying the above equation by v, the equation $\Delta v + f(v) = 0$ by ψ, integrating and subtracting the resulting identities, we obtain

$$\mu \int_\Omega v\psi\,dx = \int_\Omega (vf'(v) - f(v))\psi\,dx.$$

Due to $v, \psi > 0$, $f(0) = 0$, f convex and f' nonconstant near 0, the last integral is positive, hence $\mu > 0$ (the solution v is linearly unstable).

Since $u \geq v$ by the comparison principle, we have $y(t) := \int_\Omega (u(t) - v)\psi \, dx \geq 0$. In addition,

$$y'(t) = \int_\Omega (\Delta u + f(u))\psi \, dx = \int_\Omega \left((u - v)\Delta\psi + (f(u) - f(v))\psi\right) dx$$
$$= \int_\Omega \mu(u - v)\psi + \left(f(u) - f(v) - f'(v)(u - v)\right)\psi \, dx.$$

Since $f(u) - f(v) - f'(v)(u - v) \geq 0$ by convexity, we have $y'(t) \geq \mu y(t)$. Assume for contradiction that $T_{\max}(u_0) = \infty$. Then $\lim_{t\to\infty} y(t) = \infty$. Since $\psi \leq c\varphi_1$ due to (1.4), it follows that $\lim_{t\to\infty} \int_\Omega u(t)\varphi_1 \, dx = \infty$. But this contradicts Theorem 17.3. $\square$

Remark 17.11. The proofs of Theorems 17.8 and 17.10 were based on the convexity of the nonlinearity. However, if v is a maximal, unstable equilibrium, then blow-up of solutions starting above v can be shown for general superlinear f with subcritical growth.

Assume first
$$f(cu) \geq cf(u) \qquad \text{for } c > 1 \tag{17.13}$$

and let $u_0 \geq v$, $u_0 \not\equiv v$. Fix $\tau > 0$. Due to the maximum principle, there exists $\varepsilon > 0$ such that $u(\tau) \geq (1 + \varepsilon)v =: \tilde{u}_0$ (cf. the proof of Lemma 17.9). Let $\tilde{u}$ denote the solution with initial data $\tilde{u}_0$. Since $u(t + \tau) \geq \tilde{u}(t)$ by the maximum principle, it suffices to prove $T_{\max}(\tilde{u}_0) < \infty$. Assume on the contrary that $\tilde{u}$ exists globally. Since $\Delta\tilde{u}_0 + f(\tilde{u}_0) \geq 0$, we have $\tilde{u}_t \geq 0$. Lemma 53.10 and the maximality of v guarantee that $\tilde{u}$ cannot stay bounded, hence $\|u(t)\|_\infty \to \infty$ as $t \to \infty$. Since the growth of f is subcritical, we have also $\|u(t)\|_{1,2} \to \infty$ as $t \to \infty$. Now a simple modification of the concavity method (cf. the proof of Theorem 17.6(ii) and see [186] for details) yields a contradiction.

If f is a general function (not necessarily satisfying (17.13)), then [354] guarantees the existence of a time increasing solution w defined for $t \in (-\infty, 0]$ and satisfying $w(t) \to v$ in $C^1(\overline{\Omega})$ as $t \to -\infty$. Fix $\tau > 0$. Since $u(\tau) \geq w(t)$ for suitable $t \leq 0$ we can proceed as above. This approach can be used for more general problems provided one can show boundedness of global increasing solutions (see [185], for example). $\square$

Our last criterion [55], [324] concerns the Cauchy problem and asserts that finite-time blow-up occurs whenever the nonnegative initial data has a sufficiently *slow decay at infinity*.

Theorem 17.12. *Let $p > 1$ and consider problem (15.1) with $\Omega = \mathbb{R}^n$. Let $-\mu < 0$ be the first eigenvalue of the Dirichlet Laplacian in the unit ball of $\mathbb{R}^n$. If*

$0 \leq u_0 \in L^\infty(\mathbb{R}^n)$ *satisfies*

$$\liminf_{|x|\to\infty} |x|^{2/(p-1)} u_0(x) > \mu^{1/(p-1)}, \tag{17.14}$$

then $T_{\max}(u_0) < \infty$.

Remarks 17.13. (i) **Slow decay in more general domains.** A similar result holds (with a different constant on the RHS of (17.14)) if the inferior limit is taken on a cone Σ, instead of the whole space (see [497]). The proof of [497], is different, based on scaling and comparison arguments. Similar blow-up conditions still hold for more general domains Σ, typically a paraboloid of the form $\Sigma = \{x = (x', x_n) \in \mathbb{R}^n : x_n > 0, |x'_n| < x_n^\beta\}$ for some $0 < \beta < 1$, the power $2/(p-1)$ in (17.14) being then replaced by the smaller number $2\beta/(p-1)$ (see [386], [461]). Also, similar results can be proved when Ω itself is replaced by such a domain.

(ii) **Sign-changing initial data with slow decay.** An extension of Theorem 17.12 to sign-changing solutions has been obtained in [386]. $\square$

Proof of Theorem 17.12. Assume that $T_{\max}(u_0) = \infty$. For $R > 0$, denote by $\lambda_{1,R}$ the first Dirichlet eigenvalue of $-\Delta$ in the ball B_R. Let $\varphi_{1,R}$ be the corresponding eigenfunction satisfying $\int_{B_R} \varphi_{1,R}\, dx = 1$. We know that $T_{\max}(u_0) = \infty$ implies

$$\int_{B_R} u_0\, \varphi_{1,R}\, dx \leq \lambda_{1,R}^{1/(p-1)}. \tag{17.15}$$

Indeed, this follows from the proof of Theorem 17.1, using the fact that

$$\int_{B_R} \varphi_{1,R}\, \Delta u\, dx = \int_{B_R} u\, \Delta\varphi_{1,R}\, dx - \int_{\partial B_R} u\, \partial_\nu\varphi_{1,R}\, d\sigma \geq -\lambda_{1,R} \int_{B_R} u\, \varphi_{1,R}\, dx.$$

Set $\psi := \varphi_{1,1}$. By standard scaling properties of eigenfunctions and eigenvalues, we have $\varphi_{1,R}(x) = R^{-n}\psi(R^{-1}x)$, $x \in B_R$, and $\lambda_{1,R} = R^{-2}\mu$. For each $\varepsilon \in (0,1)$, (17.15) implies

$$\mu^{\frac{1}{p-1}} R^{-\frac{2}{p-1}} \geq \int_{\varepsilon R < |x| < R} u_0(x)\varphi_{1,R}(x)\, dx$$

$$\geq \left(\inf_{\varepsilon R < |x| < R} u_0(x)\right) \int_{\varepsilon R < |x| < R} R^{-n}\psi(R^{-1}x)\, dx,$$

hence

$$\mu^{\frac{1}{p-1}} \geq \left(\inf_{\varepsilon R < |x| < R} |x|^{\frac{2}{p-1}} u_0(x)\right) \int_{\varepsilon < |y| < 1} \psi(y)\, dy.$$

Setting $\ell = \liminf_{|x|\to\infty} |x|^{2/(p-1)} u_0(x)$ and letting $R \to \infty$, we get

$$\mu^{\frac{1}{p-1}} \geq \ell \int_{\varepsilon < |y| < 1} \psi(y)\, dy,$$

hence $\mu^{\frac{1}{p-1}} \geq \ell$ upon letting $\varepsilon \to 0$. The result follows. $\square$

Remark 17.14. Comparison of domains. Assume that $\dot{\Omega}_1 \subset \Omega_2$ are (possibly unbounded) smooth domains. Let $0 \leq u_0 \in L^\infty(\Omega_1)$ and extend u_0 by 0 outside Ω_1. Denote by u^i, $i \in \{1,2\}$, the solution of problem (17.1) in $\Omega = \Omega_i$ with initial data u_0. If u^1 is nonglobal, then so is u^2 (this follows from the comparison principle applied in Ω_1). This simple fact illustrates the heuristic principle that "larger domains are more instable" (cf. [328]). From this and, e.g., Theorem 17.1, one can derive blow-up criteria in general unbounded domains. $\quad\square$

Remark 17.15. Quenching. The so-called **quenching** phenomenon is closely related to blow-up. Instead of $f = |u|^{p-1}u$ or f satisfying condition (17.4), consider a "singular" nonlinearity $f : [0, a) \to [0, \infty)$ for some $a \in (0, \infty)$, of class C^1, and satisfying

$$\lim_{s \to a-} f(s) = \infty;$$

typically

$$f(u) = \lambda(a - u)^{-p} \quad \text{for some } \lambda, p > 0. \tag{17.16}$$

Assume that $0 \leq u_0 \in L^\infty(\Omega)$ is such that $\operatorname{ess\,sup}_\Omega u_0 < a$. Then problem (14.1) still admits a unique, maximal, classical solution $u \geq 0$, and it is easy to show that either

$$T := T_{\max}(u_0) = \infty, \quad \text{or } T < \infty \text{ and } \lim_{t \to T} \|u(t)\|_\infty = a.$$

The latter case is called (finite time) quenching: The solution itself remains bounded, but a singularity appears in the RHS. In fact, it can be shown that under suitable assumptions, quenching implies blow-up of u_t, namely $\lim_{t \to T} \|u_t(t)\|_\infty = \infty$. Different, but related, is the phenomenon of gradient blow-up; see Sections 40 and 41.

If, for instance, f is given by (17.16) with λ large enough, then quenching occurs for all u_0 as above. The quenching problem, first considered in [303], has been investigated in numerous articles. We refer to, e.g., [329], [120] for surveys on this subject. $\quad\square$

18. Fujita-type results

Consider problem (15.1) with $\Omega = \mathbb{R}^n$:

$$\left.\begin{aligned}
u_t - \Delta u &= |u|^{p-1}u, & x \in \mathbb{R}^n,\ t > 0, \\
u(x,0) &= u_0(x), & x \in \mathbb{R}^n.
\end{aligned}\right\} \tag{18.1}$$

Assume $p > 1$ and $u_0 \geq 0$. The results of the preceding section guarantee that the solution of (18.1) blows up in finite time if u_0 is sufficiently large. In this section we show that *all* solutions of (18.1) with $u_0 \geq 0$, $u_0 \not\equiv 0$, blow up in finite time if and only if $p \leq p_F$, where

$$p_F := 1 + \frac{2}{n}.$$

The number p_F thus plays the role of a *critical exponent* for the Cauchy problem (18.1). This result was proved in [220] for $p \neq p_F$ and in [271], [307] (see also [35], [530]) for $p = p_F$.

Numerous generalizations and modifications can be found in (the references of) the survey articles [328], [159] and in [372]. Some of them are described in the remarks at the end of this section, in Theorem 32.7, and in Sections 37 and 45. On the other hand, an application to a model arising in population genetics will be given at the end of this section.

Theorem 18.1. (i) *Let $1 < p \leq p_F$. Then the equation*

$$u_t - \Delta u = u^p \tag{18.2}$$

does not admit any nontrivial distributional solution $u \geq 0$ in $Q := \mathbb{R}^n \times (0, \infty)$.

(ii) *Let $p > p_F$. Then problem (18.1) has a global, classical solution for some positive $u_0 \in L^\infty(\Omega)$.*

Remark 18.2. (i) By a distributional solution, we here mean a function $u \in L^p_{loc}(Q)$ which satisfies (18.2) in $\mathcal{D}'(Q)$. The proof of assertion (i) given below shows that this remains true for distributional solutions of the inequality $u_t - \Delta u \geq u^p$ in Q. We use a modification of arguments in [372], based on rescalings of a simple, compactly supported test-function, depending on x and t.

A related proof can be found in [56], where the test-functions are obtained by solving an adjoint problem. The original proof of [220] involved Gaussian test-functions depending on x only (given by the heat kernel with t as a parameter), hence requiring more regularity of the solutions in time.

(ii) In the result of Theorem 18.1(i), the roles played by the behaviors of the nonlinearity as $u \to 0$ and as $u \to \infty$ are different; see Remark 18.8(iii) for details. $\square$

Proof of Theorem 18.1(i). Let $u \geq 0$ be a distributional solution of (18.2) in Q and fix $t_0 > 0$.

Step 1. We claim that, for each $\xi \in \mathcal{D}(\mathbb{R}^n)$, $\psi \in C^\infty([t_0, \infty))$, with $\xi, \psi \geq 0$, $\psi(t) = 1$ near $t = t_0$ and $\psi(t) = 0$ for t large, there holds

$$\int_{t_0}^\infty \int_{\mathbb{R}^n} u^p \xi \psi \, dx \, dt \leq -\int_{t_0}^\infty \int_{\mathbb{R}^n} u(\xi \partial_t \psi + \psi \Delta \xi) \, dx \, dt. \tag{18.3}$$

To show (18.3), observe that there exists a sequence of functions $\psi_j \in \mathcal{D}((0, \infty))$ such that $\psi_j = 0$ on $(0, t_0 - 1/j]$, $\partial_t \psi_j \geq 0$ on $[t_0 - 1/j, t_0]$, and $\psi_j = \psi$ on $[t_0, \infty)$.

Taking $\varphi(x,t) := \xi(x)\psi_j(t)$ as test-function, it follows that

$$\int_{t_0}^{\infty}\int_{\mathbb{R}^n} u^p \xi\psi \, dx \, dt \leq \int_0^{\infty}\int_{\mathbb{R}^n} u^p \xi\psi_j \, dx \, dt = -\int_0^{\infty}\int_{\mathbb{R}^n} u(\xi\partial_t\psi_j + \psi_j\Delta\xi)\,dx\,dt$$

$$\leq -\int_{t_0}^{\infty}\int_{\mathbb{R}^n} u(\xi\partial_t\psi + \psi\Delta\xi)\,dx\,dt - \int_{t_0-1/j}^{t_0}\int_{\mathbb{R}^n} u\psi_j\Delta\xi\,dx\,dt.$$

Since the last integral goes to 0 by dominated convergence, this yields (18.3).

Step 2. Now we take $\zeta \in \mathcal{D}(B_1)$ and $\phi \in \mathcal{D}((-1,1))$, such that $\zeta = 1$ in $B_{1/2}$, $\phi = 1$ in $[0,1/2)$, and $0 \leq \zeta, \phi \leq 1$. Let $m = 2p/(p-1)$ and define

$$\xi_R(x) = \zeta^m\left(\frac{x}{R}\right), \quad x \in \mathbb{R}^n, \qquad \psi_R(t) = \phi^m\left(\frac{t-t_0}{R^2}\right), \quad t \geq t_0.$$

We observe that

$$\Delta\xi_R(x) = mR^{-2}\big[\zeta^{m-1}\Delta\zeta + (m-1)\zeta^{m-2}|\nabla\zeta|^2\big]\left(\frac{x}{R}\right)$$

and

$$\partial_t\psi_R(t) = mR^{-2}\big[\phi^{m-1}\phi_t\big]\left(\frac{t-t_0}{R^2}\right),$$

hence

$$\big|\xi_R\partial_t\psi_R + \psi_R\Delta\xi_R\big| \leq CR^{-2}(\xi_R\psi_R)^{1/p}\chi_{\{R/2<|x|<R\}}\chi_{\{R^2/2<t-t_0<R^2\}}.$$

Using (18.3) with $\xi = \xi_R$, $\psi = \psi_R$, and applying Hölder's inequality, we obtain

$$\int_{t_0}^{\infty}\int_{\mathbb{R}^n} u^p \xi_R\psi_R \, dx \, dt \leq CR^{-2}\int_{t_0+R^2/2}^{t_0+R^2}\int_{R/2<|x|<R} u(\xi_R\psi_R)^{1/p}\,dx\,dt$$

$$\leq CR^{-2+(n+2)(p-1)/p}\left(\int_{t_0+R^2/2}^{t_0+R^2}\int_{R/2<|x|<R} u^p\xi_R\psi_R\,dx\,dt\right)^{1/p}. \tag{18.4}$$

In particular, it follows that

$$\int_{t_0}^{\infty}\int_{\mathbb{R}^n} u^p\xi_R\psi_R\,dx\,dt \leq CR^{n+2-2p/(p-1)}. \tag{18.5}$$

If $p < p_F$, i.e. $n+2-2p/(p-1) < 0$, this implies $u \equiv 0$ upon letting $R \to \infty$ and then $t_0 \to 0$. If $p = p_F$, then (18.5) implies $\int_{t_0}^{\infty}\int_{\mathbb{R}^n} u^p < \infty$. Therefore, the RHS of (18.4) goes to 0 as $R \to \infty$ and we again conclude that $u \equiv 0$. $\quad\square$

The proof of assertion (ii) is postponed to Section 20, where more detailed global existence results will be given. Below we present two other proofs of (different formulations of) the nonexistence part of Theorem 18.1. We shall start with the proof which is due to [529], [530]. Recall that G_t denotes the Gaussian heat kernel, defined in (48.5).

Theorem 18.3. *Let $1 < p \leq p_F$, $u_0 : \mathbb{R}^n \to [0, \infty]$ be measurable, $u_0 > 0$ in a set of positive measure. Then there is no nonnegative measurable global solution $u : \mathbb{R}^n \times [0, \infty) \to [0, \infty]$ to the integral equation*

$$u(x,t) = \int_{\mathbb{R}^n} G_t(x - y)u_0(y)\, dy + \int_0^t \int_{\mathbb{R}^n} G_{t-s}(x - y)u^p(y, s)\, dy\, ds \tag{18.6}$$

$$and \quad u(x,t) < \infty \ for \ a.e. \ (x,t) \in \mathbb{R}^n \times (0, \infty).$$

Proof. Assume that there exists a global solution of (18.6). Lemma 15.6 implies

$$t^{1/(p-1)} G_t * u_0 \leq C. \tag{18.7}$$

Given a measurable function $v : \mathbb{R}^n \to [0, \infty]$, we have

$$\lim_{t \to \infty} (4\pi t)^{n/2} G_t * v = \|v\|_1 \quad \text{pointwise in } \mathbb{R}^n, \tag{18.8}$$

where $\|v\|_1 := \infty$ if $v \notin L^1(\mathbb{R}^n)$. If $p < p_F$, then (18.7) implies $t^{n/2}\|G_t * u_0\|_\infty \to 0$ as $t \to \infty$ which contradicts (18.8) with $v = u_0$.

Hence we may assume $p = p_F$. By redefining u on a null set, we may assume that (18.6) actually holds everywhere in $\mathbb{R}^n \times (0, \infty)$ and it is easy to check that

$$u(t + t_0) = G_t * u(t_0) + \int_0^t G_{t-s} * u^p(s + t_0)\, ds, \qquad \text{for all } t, t_0 > 0. \tag{18.9}$$

We first note that Corollary 15.8 and (18.8) imply the existence of $C_1 > 0$ such that

$$\|u(\tau)\|_1 \leq C_1, \qquad \text{for a.e. } \tau > 0. \tag{18.10}$$

On the other hand, since $|x - z|^2/(4t) \leq (|x|^2 + |z|^2)/(2t)$, we obtain

$$u(x,t) \geq (G_t * u_0)(x) \geq (4\pi t)^{-n/2} e^{-|x|^2/(2t)} \int_{\mathbb{R}^n} e^{-|z|^2/(2t)} u_0(z)\, dz.$$

In particular, we have

$$u(x, 2) \geq k G_1(x), \qquad x \in \mathbb{R}^n, \tag{18.11}$$

for some $k > 0$. Using (18.9) and (48.6), we deduce that

$$u(s + 2) \geq G_s * u(2) \geq k G_s * G_1 = k G_{s+1}, \qquad s > 0. \tag{18.12}$$

Now, Proposition 48.4(a) and $(p - 1)n/2 = 1$ imply

$$\|G_{s+1}^p\|_1 = (4\pi(s + 1))^{-(p-1)n/2} p^{-n/2} \|G_{(s+1)/p}\|_1 = C_2(s + 1)^{-1}$$

for some $C_2 > 0$. This calculation, (18.9) with $t_0 = 2$, (18.12) and Proposition 48.4(b) guarantee

$$\|u(t+2)\|_1 \geq \int_0^t \|G_{t-s} * u^p(s+2)\|_1 \, ds \geq \int_0^t \|G_{t-s} * (kG_{s+1})^p\|_1 \, ds$$

$$= k^p \int_0^t \|G_{s+1}^p\|_1 \, ds = k^p C_2 \int_0^t (s+1)^{-1} \, ds \to \infty$$

as $t \to \infty$, which contradicts (18.10). $\quad\square$

The following method is based on the approach of [299] (in Lemma 18.4 below we also use some ideas from [384]). We introduce the **forward similarity variables**

$$y = \frac{x}{\sqrt{t+1}}, \qquad s = \log(1+t)$$

and define the rescaled function

$$v(y,s) = e^{\beta s} u(e^{s/2}y, e^s - 1), \qquad \beta = \frac{1}{p-1} \tag{18.13}$$

(in other words, $v(y,s) = t^\beta u(x,t)$). Problem (18.1) can then be written in the form

$$\begin{aligned}
v_s + Lv &= |v|^{p-1}v + \beta v, & y \in \mathbb{R}^n, s > 0, \\
v(y,0) &= u_0(y), & y \in \mathbb{R}^n,
\end{aligned} \tag{18.14}$$

where

$$Lv := -\Delta v - \frac{y \cdot \nabla v}{2} = -g^{-1}\nabla \cdot (g\nabla v), \qquad g(y) := e^{|y|^2/4}. \tag{18.15}$$

Set

$$L_g^q := \{f \in L^q(\mathbb{R}^n) : \int_{\mathbb{R}^n} |f(y)|^q g(y) \, dy < \infty\},$$

$$H_g^1 := \{f \in L_g^2 : \nabla f \in L_g^2\} \tag{18.16}$$

and $H_g^2 := \{f \in H_g^1 : \nabla f \in H_g^1\}$. Then

$$(Lv, w)_g = -\int_{\mathbb{R}^n} \nabla \cdot (g\nabla v) w \, dy = \int_{\mathbb{R}^n} (\nabla v \cdot \nabla w) g \, dy, \qquad v, w \in H_g^2,$$

where $(u,v)_g := \int_{\mathbb{R}^n} uvg \, dy$ denotes the scalar product in L_g^2. Lemmas 47.9, 47.10 and 47.13 show that L is a positive self-adjoint operator in L_g^2 with compact inverse, domain of definition H_g^2 and eigenvalues $\lambda_k^L = (n+k-1)/2$, $k = 1, 2, \ldots$. In addition, $\phi_1(y) := e^{-|y|^2/4}$ is the eigenfunction corresponding to the first eigenvalue λ_1^L.

Denote

$$E(v) := \int_{\mathbb{R}^n} \left[\frac{|\nabla v|^2}{2} - \frac{\beta}{2} v^2 - \frac{1}{p+1} |v|^{p+1} \right] g(y)\, dy.$$

Notice that E is well defined in H_g^1 if $p \le p_S$ since $H_g^1 \hookrightarrow L_g^{p+1}$ due to Lemma 47.11. Let $T \in (0, \infty]$ and assume that $v \in C([0, T), H_g^1)$ is a solution of (18.14). Example 51.24 shows that $v \in C((0, T), H_g^2) \cap C^1((0, T), L_g^2)$, hence the mapping $s \mapsto E(v(s))$ belongs to $C([0, T), \mathbb{R}) \cap C^1((0, T), \mathbb{R})$.

Lemma 18.4. *Let $1 < p < p_S$.*

(i) *The function $s \mapsto E(v(s))$ is nonincreasing.*

(ii) *If $\beta \le \lambda_1^L$ and $E(v(s_0)) < 0$ for some s_0, then $T < \infty$.*

(iii) *If $T = \infty$, then there exist positive constants $C_0 = C_0(n, p)$, $C_1 = C_1(\|u_0\|_{H_g^1})$ and $C_2 = C_2(u_0)$ such that*

$$-C_0 \le E(v(s)) \le C_1, \tag{18.17}$$

$$\|v(s)\|_{L_g^2} \le C_1, \tag{18.18}$$

$$\|v(s)\|_{H_g^1} \le C_2 \tag{18.19}$$

for all $s \ge 0$.

Proof. (i) The assertion follows from

$$\frac{d}{ds} E(v(s)) = - \int_{\mathbb{R}^n} v_s^2(s) g\, dy \le 0.$$

(ii) Without loss of generality we may assume $s_0 = 0$. Then the proof follows by repeating word-by-word part (ii) of the proof of Theorem 17.6.

(iii) Set $Av := Lv - \beta v$, choose $\varepsilon \in (0, (p-1)/2)$ and denote $c_0 := 1 - (2 + 2\varepsilon)/(p+1) > 0$. Multiplying the equation $v_s + Av = |v|^{p-1} v$ by vg we obtain

$$\begin{aligned}
\frac{1}{2} \frac{d}{ds} \|v(s)\|_{L_g^2}^2 &= -(Av(s), v(s))_g + \|v(s)\|_{L_g^{p+1}}^{p+1} \\
&= -(2 + 2\varepsilon) E(v(s)) + \varepsilon (Av(s), v(s))_g + c_0 \|v(s)\|_{L_g^{p+1}}^{p+1}.
\end{aligned} \tag{18.20}$$

Recall that A is a self-adjoint operator with compact resolvent and its eigenvalues are $\lambda_k^L - \beta$, $k = 1, 2, \ldots$. Choose k_0 such that $\lambda_{k_0}^L > \beta$, let P be the spectral projection in L_g^2 corresponding to the spectral set $\{\lambda_{k_0}^L - \beta, \lambda_{k_0+1}^L - \beta, \ldots\}$ and $Q = I - P$. Notice that

$$\dim Q L_g^2 < \infty, \quad P L_g^2 \perp Q L_g^2 \tag{18.21}$$

and there exist $c_1, c_2 > 0$ such that, for all $w \in H_g^2$,

$$(Aw, w)_g = (APw, Pw)_g + (AQw, Qw)_g$$
$$\geq c_1 \|Pw\|_{H_g^1}^2 - c_2 \|Qw\|_{H_g^1}^2 = c_1 \|w\|_{H_g^1}^2 - (c_1 + c_2)\|Qw\|_{H_g^1}^2. \tag{18.22}$$

Fix $s > 0$. If

$$c_0 \|v(s)\|_{L_g^{p+1}}^{p+1} \leq \varepsilon(c_1 + c_2)\|Qv(s)\|_{H_g^1}^2, \tag{18.23}$$

then Hölder's inequality and (18.21) guarantee the existence of $c_3, c_4 > 0$ such that

$$\frac{1}{c_3} \|Qv(s)\|_{H_g^1}^2 \leq \|Qv(s)\|_{L_g^2}^2 = (Qv(s), Qv(s))_g = (v(s), Qv(s))_g$$

$$\leq \|v(s)\|_{L_g^{p+1}} \|Qv(s)\|_{L_g^{(p+1)'}} \leq c_4 \|Qv(s)\|_{H_g^1}^{2/(p+1)} \|Qv(s)\|_{H_g^1},$$

hence $\|Qv(s)\|_{H_g^1} \leq c_5$ and (18.20), (18.22) guarantee

$$\frac{1}{2}\frac{d}{ds}\|v(s)\|_{L_g^2}^2 \geq -(2 + 2\varepsilon)E(v(s)) + \tilde{c}_1 \|v(s)\|_{H_g^1}^2 - \tilde{c}_5 \tag{18.24}$$

for some $\tilde{c}_1, \tilde{c}_5 > 0$. If (18.23) fails, then (18.20), (18.22) guarantee (18.24) with $\tilde{c}_5 = 0$.

Assume $s_0 \geq 0$ and

$$-(2 + 2\varepsilon)E(v(s_0)) > \tilde{c}_5 + 1 \quad \text{or} \quad \tilde{c}_1 \|v(s_0)\|_{L_g^2}^2 > \tilde{c}_5 + 1 + (2 + 2\varepsilon)E(u_0). \tag{18.25}$$

Then (18.24), the inequality $E(v(s_0)) \leq E(u_0)$ and the identity

$$E(v(s)) - E(v(s_0)) - \int_{s_0}^s \|v_s(t)\|_{L_g^2}^2 \, dt, \qquad s \geq s_0, \tag{18.26}$$

imply

$$\frac{1}{2}\frac{d}{ds}\|v(s)\|_{L_g^2}^2 \geq (2 + 2\varepsilon) \int_{s_0}^s \|v_s(t)\|_{L_g^2}^2 \, dt + 1, \qquad s \geq s_0.$$

Set $f(s) := \frac{1}{2}\int_{s_0}^s \|v(t)\|_{L_g^2}^2 \, dt$. Then the same arguments as in the proof of Theorem 17.6 show that the function $s \mapsto f(s)^{-\varepsilon}$ is concave for s large which contradicts the assumption $T = \infty$. Consequently, (18.25) fails and (18.17), (18.18) are true.

Notice that (18.26) and (18.17) imply

$$\int_0^\infty \|v_s(s)\|_{L_g^2}^2 \, ds \leq C_1 + C_0 \tag{18.27}$$

and (18.24), (18.17), (18.18) and Cauchy's inequality guarantee the existence of $c_6, c_7 > 0$ such that

$$\|v_s(s)\|_{L_g^2}^2 \geq c_6 \|v(s)\|_{H_g^1}^4 - c_7. \tag{18.28}$$

Set $\Lambda_t := \{s \geq t : c_6\|v(s)\|_{H_g^1}^4 > c_7 + 1\}$ and let $|\Lambda_t|$ denote the measure of Λ_t. Then $|\Lambda_t| \to 0$ as $t \to \infty$ due to (18.27) and (18.28). The well-posedness of (18.14) in H_g^1 (see Example 51.24) guarantees the existence of $\eta, c_8 > 0$ such that $\|v(s + \tau)\|_{H_g^1} \leq c_8$ whenever $\tau \in [0, \eta]$ and $s \notin \Lambda_0$. Fix $t > 0$ such that $|\Lambda_t| < \eta$. Then $\|v(s)\|_{H_g^1} \leq c_8$ for all $s \geq t + \eta$, which proves (18.19). $\quad \square$

Remark 18.5. The constant C_2 in (18.19) depends on $\|u_0\|_{H_g^1}$ only. In fact, let Λ_t be the set in the proof of Lemma 18.4(iii). Since $|\Lambda_0| < C_1 + C_0$ due to (18.27) and (18.28), in any interval of the form $[s, s + C_1 + C_0]$, $s > 0$ we can find s_0 such that $\|v(s_0)\|_{H_g^1} \leq C_3$ for some $C_3 = C_3(\|u_0\|_{H_g^1})$ (and the same is true for all $s_0 > 0$ close to zero). Due to the smoothing estimates in Example 51.24 we may also assume $\|v(s_0)\|_\infty \leq C_3$. Now estimate (5.26) in [440, Theorem 5.3] (used with $\tilde{u}(x,t) = (t+1)^{-1/(p-1)}v(x/\sqrt{t+1}, \log(t+1) + s_0)$) guarantees $\|v(s)\|_\infty \leq C_4$ for some $C_4 = C_4(\|u_0\|_{H_g^1})$ and all $s \in [s_0, s_0 + 2(C_0 + C_1)]$. Consequently, $||v|^{p-1}v| \leq C_4^{p-1}|v|$ and an easy estimate based on the variation-of-constants formula guarantees $\|v(s)\|_{H_g^1} \leq C_5$ for some $C_5 = C_5(\|u_0\|_{H_g^1})$ and all $s \in [s_0, s_0 + 2(C_0 + C_1)]$. $\quad\square$

Another proof of Theorem 18.1(i) for classical solutions. Let $p \leq p_F$, $0 \leq u_0 \in L^\infty(\mathbb{R}^n)$, $u_0 \not\equiv 0$, and assume that the corresponding maximal classical solution u of (18.1) is global. Similarly as in the proof of Theorem 18.3, we have (18.11), hence $u(\cdot, 2) \geq c_0\phi_1$ for some $c_0 > 0$. Due to the maximum principle, the solution v of (18.14) starting at $v_0 := c_0\phi_1$ exists globally.

First assume $p < p_F$. Since the solution v is global, Lemma 18.4(iii) guarantees that it is bounded in H_g^1. On the other hand,

$$v(t) \geq e^{-t(L-\beta)}(c_0\phi_1) = c_0 e^{t(\beta-\lambda_1^L)}\phi_1$$

and $\beta - \lambda_1^L > 0$, which yields a contradiction.

Now assume $p = p_F$. Using $(L\phi_1, \phi_1)_g = \lambda_1^L(\phi_1, \phi_1)_g$ and $\beta = \lambda_1^L$ we obtain

$$E(c_0\phi_1) = -\frac{c_0^{p+1}}{p+1} \int_{\mathbb{R}^n} \phi_1^{p+1}(y)g(y)\,dy < 0,$$

which contradicts Lemma 18.4(ii). $\quad\square$

Remarks 18.6. (i) **Alternative proof.** In [299], another contradiction argument was used in the case $p < p_F$: Let v be the global solution starting at $v_0 := c_0\phi_1$. Set $\psi := b_\varepsilon \phi_1^{1+\varepsilon}$ where $\varepsilon > 0$ and $b_\varepsilon > 0$ is such that $\int_{\mathbb{R}^n} \psi g\,dy = 1$. Notice that

$$L\psi = (1+\varepsilon)\lambda_1^L\psi - \varepsilon(1+\varepsilon)b_\varepsilon \phi_1^{\varepsilon-1}|\nabla\phi_1|^2 \leq (1+\varepsilon)\lambda_1^L\psi$$

and set $f(s) := (v(s), \psi)_g$. Then Jensen's inequality implies

$$\frac{d}{ds}f(s) = \int_{\mathbb{R}^n} v(y,s)^p \psi(y)g(y)\,dy + \beta(v(s), \psi)_g - (v(s), L\psi)_g$$

$$\geq \left(\int_{\mathbb{R}^n} v(y,s)\psi(y)g(y)\,dy\right)^p + \left[\beta - (1+\varepsilon)\lambda_1^L\right](v(s), \psi)_g$$

$$= f(s)^p + \left[\beta - (1+\varepsilon)\lambda_1^L\right]f(s).$$

Due to $p < p_F$ there exists $\varepsilon > 0$ such that $\beta = 1/(p-1) \geq (1+\varepsilon)n/2 = (1+\varepsilon)\lambda_1^L$, hence $f' \geq f^p$, $f(0) > 0$, which contradicts the global existence of f.

(ii) **Other domains.** Consider problem (15.1) in the half-space $\Omega = \mathbb{R}_+^n :=$ $\{x \in \mathbb{R}^n : x_n > 0\}$. Repeating the last proof of Theorem 18.3 we obtain a self-adjoint operator $\tilde{L}$ with the first eigenvalue $\lambda_1^{\tilde{L}} = (n+1)/2$ and the corresponding eigenfunction $\tilde{\phi}_1(y) = y_n e^{-|y|^2/4}$. Consequently, the problem does not possess nontrivial nonnegative global solutions if $p \leq 1 + 2/(n+1)$. Of course, instead of G_t one has to work with the kernel $\tilde{G}_t(x,z) = G_t(x-z)\big(1 - e^{-x_n z_n/t}\big)$. If $\Omega = (0,\infty)^n$, then analogous arguments show nonexistence of global positive solutions for $p \leq 1 + 1/n$ (the first eigenfunction is $y_1 y_2 \ldots y_n e^{-|y|^2/4}$).

(iii) **A characterization of the critical exponent.** It has been observed in [363] that, for any domain Ω, the critical Fujita exponent $p_F = p_F(\Omega)$ can be characterized in terms of maximal decay rate of the heat semigroup. Namely, denoting

$$a^* := \sup\Big\{a > 0 : \sup_{t \in (0,\infty)} t^a \|e^{-tA} u_0\|_\infty < \infty \ \text{ for some } 0 \leq u_0 \in L^\infty(\Omega), \ u_0 \not\equiv 0\Big\},$$

$$(18.29)$$

there holds

$$p_F = 1 + \frac{1}{a^*}.$$

Indeed, if $1 < p < 1 + (1/a^*)$, then for any $0 \leq u_0 \in L^\infty(\Omega)$, $u_0 \not\equiv 0$, we have

$$\sup_{t \in (0,\infty)} t^{1/(p-1)} \|e^{-tA} u_0\|_\infty = \infty,$$

hence $T_{\max}(u_0) < \infty$ by Lemma 15.6 or Remark 15.7. If $p > 1 + (1/a^*)$, by taking u_0 such that a in (18.29) satisfies $1/(p-1) < a < a^*$, we deduce from the proof of Theorem 20.2 below (with $\mathbb{R}^n$ replaced by Ω) that $T_{\max}(u_0) = \infty$.

(iv) **Sign-changing solutions.** Consider problem (18.1) with $n = 1$ and set

$$\Lambda_k = \{u : u \text{ has exactly } k \text{ sign changes}\}.$$

Then there exists a global solution of (18.1) with $u_0 \in \Lambda_k$ if and only if $p > 1 + 2/(k+1)$. In addition, if $p > 1 + 2/(k+1)$, then there exists a global solution of (18.1) with $u_0 \in \Lambda_k \cap H_g^1$ (see [384] and [385]). Notice that $1 + 2/(k+1) = 1 + 1/\lambda_{k+1}^L$. $\square$

We close this section with an application of Theorem 18.1 to a model arising in population genetics [211], [35]. In that model, a biological species possesses a gene existing in two allelic forms A and a, leading to the three genotypes AA, Aa and aa. It is assumed that the death rate of the individuals is determined by this particular gene, and the death rates corresponding to the genotypes AA, Aa, aa are respectively denoted by k_1, k_2, k_3. Moreover, it is assumed that $k_2 = k_3$

and that the genotype AA is advantageous in the sense that $k_1 < k_2$. Denote by $u : \mathbb{R}^n \times [0, \infty) \to [0, 1]$ the relative density of the gene A at point x and time t, and set $k = k_2 - k_1 > 0$. Under suitable physical assumptions, the equation for u is then given by

$$u_t - \Delta u = ku^2(1 - u), \qquad x \in \mathbb{R}^n,\ t > 0. \tag{18.30}$$

This equation is supplemented with the initial condition

$$u(x, 0) = u_0(x), \qquad x \in \mathbb{R}^n, \tag{18.31}$$

where $u_0 \in X := \{\phi \in C(\mathbb{R}^n) : 0 \le \phi(x) \le 1,\ x \in \mathbb{R}^n\}$. It follows from Remark 51.11 and the comparison principle that problem (18.30)–(18.31) admits a unique, global, classical solution u and that $0 \le u(x, t) \le 1$ in $\mathbb{R}^n \times (0, \infty)$.

We have the following result [35] concerning the asymptotic behavior of solutions (our proof is a simplification of arguments in [35]).

Theorem 18.7. *Consider problem* (18.30)–(18.31).

(i) *If $n = 1$ or 2, then $u = 1$ is globally stable in the following sense: For any $u_0 \in X$, $u_0 \not\equiv 0$, there holds*

$$\lim_{t \to \infty} u(x, t) = 1,$$

uniformly on compact subsets.

(ii) *If $n \ge 3$, then there exist positive $u_0 \in X$ such that*

$$\lim_{t \to \infty} \|u(t)\|_\infty = 0.$$

Remarks 18.8. (i) The phenomenon displayed in Theorem 18.7(i) is called the "hair-trigger effect": Any small perturbation from the rest-state $u \equiv 0$ drives the solution to the equilibrium $u \equiv 1$, leading to the eventual extinction of the gene a.

(ii) Equation (18.30) is a special case of a more general class of equations of the form $u_t - \Delta u = f(u)$, where the nonlinearity satisfies $f(0) = f(1) = 0$, which arise in various biological models and also in flame propagation models from combustion theory. An important case is the so-called Fisher-KPP equation, corresponding to $f(u) = u(1 - u)$. Starting with the pioneering works [211], [308], a very large amount of literature has been devoted to these problems, in particular to the existence of traveling wave solutions and to their analysis. These are solutions of the form $u(x, t) = w(x_1 - ct)$, connecting the equilibria $u \equiv 0$ and $u \equiv 1$ (i.e. $w(-\infty) = 1$, $w(+\infty) = 0$). See [35], [179], and e.g. [266] and the references therein for more recent results.

(iii) Simple modifications of the proof of Theorem 18.7 show the following. Assume that the nonlinearity in (18.1) is replaced by any C^1 function $f : [0, \infty) \to [0, \infty)$ such that

$$f(u) \ge ku^p, \quad u \in [0, b], \qquad \text{for some } k, b > 0 \text{ and } 1 \le p \le p_F. \tag{18.32}$$

Then any positive solution is either nonglobal or satisfies $\liminf_{t\to\infty} u(x,t) \geq b$, uniformly on compact subsets. In that sense, the Fujita-type result can be seen as an instability property of $u = 0$ for small perturbations, which essentially depends on the behavior of f for small u. This explains the fact that the instable range corresponds to small p's. On the other hand, if in addition to (18.32) we assume that f is convex and satisfies the blow-up condition (17.4), then it is easy to show that any positive solution is nonglobal. $\square$

For the proof of assertion (i), we need the following lemma.

Lemma 18.9. *For all $\varepsilon > 0$, there exist $R_\varepsilon > 0$ and a function $\phi_\varepsilon \in C^2(\overline{B}_{R_\varepsilon})$ such that*

$$0 \leq \phi_\varepsilon(x) \leq 1 - \varepsilon, \qquad x \in B_{R_\varepsilon}$$

and

$$v_\varepsilon(0,t) \geq 1 - \varepsilon, \qquad t \geq 0, \tag{18.33}$$

where v_ε is the solution of the problem

$$\left.\begin{aligned}
v_t - \Delta v &= v^2(1-v), & x &\in B_{R_\varepsilon},\ t > 0,\\
v &= 0, & x &\in \partial B_{R_\varepsilon},\ t > 0,\\
v(x,0) &= \phi_\varepsilon(x), & x &\in B_{R_\varepsilon}.
\end{aligned}\right\}$$

Proof. Assume $\varepsilon \in (0, 1/2)$ without loss of generality. Fix a nontrivial nonnegative radial function $h \in \mathcal{D}(\mathbb{R}^n)$ such that $h(x) = 0$ for $|x| \geq 1/2$. Let φ be the classical solution of

$$\left.\begin{aligned}
-\Delta\varphi &= h, & |x| &< 1,\\
\varphi &= 0, & |x| &= 1,
\end{aligned}\right\}$$

and observe that φ is positive, radial nonincreasing. Let

$$\phi(x) = \phi_\varepsilon(x) := (1-\varepsilon)\frac{\varphi(x/R)}{\varphi(0)} \leq 1 - \varepsilon, \qquad |x| \leq R,$$

where $R > 0$ is to be fixed. For $|x| \leq R/2$, we have $\phi \geq c := (2\varphi(0))^{-1}\varphi(1/2) > 0$, hence

$$\Delta\phi + \phi^2(1-\phi) \geq \Delta\phi + \varepsilon c^2 \geq -(\varphi(0))^{-1}R^{-2}\|\Delta\varphi\|_\infty + \varepsilon c^2 > 0$$

provided we take $R = R_\varepsilon > 0$ large enough. Since $\Delta\phi = 0$ for $|x| \geq R/2$, we obtain

$$\Delta\phi + \phi^2(1-\phi) \geq 0, \qquad x \in B_{R_\varepsilon}.$$

It follows from Proposition 52.19 that $\partial_t v_\varepsilon \geq 0$, hence in particular

$$v_\varepsilon(0, t) \geq \phi(0) = 1 - \varepsilon, \qquad t \geq 0. \qquad \square$$

Proof of Theorem 18.7(i). We may assume $k = 1$ without loss of generality.

Step 1. Let $v_0 \in X$, $v_0 \not\equiv 0$, be such that $v_0(x_0 + \cdot)$ is radial nonincreasing for some $x_0 \in \mathbb{R}^n$, and let v be the solution of (18.30) with initial data v_0. Then $v(x + x_0, t)$ is also radial nonincreasing (cf. Proposition 52.17). We claim that

$$\limsup_{t \to \infty} v(x_0, t) = 1.$$

Assume the contrary. Then there exist $\varepsilon \in (0, 1)$ and $T > 0$ such that $v(x, t) \leq 1 - \varepsilon$ in $\mathbb{R}^n \times [T, \infty)$. Consequently, $w := \varepsilon v$ satisfies $w_t - \Delta w \geq w^2$ in $\mathbb{R}^n \times [T, \infty)$. Since $2 \leq p_F$ due to $n \leq 2$, it follows from Theorem 18.1(i) and Remark 18.2(i) that w is nonglobal: a contradiction.

Step 2. Let $u_0 \in X$, $u_0 \not\equiv 0$. We claim that for all $\varepsilon, R > 0$, there exists $t_0 > 0$ such that

$$u(x, t_0) \geq 1 - \varepsilon, \qquad |x| \leq R. \tag{18.34}$$

By a time shift, we may assume without loss of generality that $u_0 > 0$ in $\mathbb{R}^n$. Therefore, for any $x_0 \in \mathbb{R}^n$, u_0 dominates some nontrivial $v_0 \in X$ such that $v_0(x_0 + \cdot)$ is radial nonincreasing. If follows from Step 1 and the comparison principle that for all $x_0 \in \mathbb{R}^n$,

$$\limsup_{t \to \infty} u(x_0, t) = 1.$$

If u_0 is radial nonincreasing, then this readily implies (18.34). The general case follows from the fact that u_0 dominates some nontrivial, radial nonincreasing $v_0 \in X$.

Step 3. Let $u_0 \in X$, $u_0 \not\equiv 0$. Fix $\varepsilon \in (0, 1)$ and $M > 0$. Let $R_\varepsilon, \phi_\varepsilon$ be given by Lemma 18.9. By Step 2, applied with $R = R_\varepsilon + M$, there exists $t_0 > 0$ such that

$$u(x_0 + x, t_0) \geq 1 - \varepsilon \geq \phi_\varepsilon(x), \qquad |x| \leq R_\varepsilon, \ |x_0| \leq M.$$

By the comparison principle and (18.33), we conclude that

$$u(x_0, t) \geq v_\varepsilon(0, t) \geq 1 - \varepsilon, \qquad |x_0| \leq M, \ t \geq t_0.$$

The assertion is proved.

(ii) Since $2 > p_F$ due to $n \geq 3$, this is an immediate consequence of Theorem 18.1(ii) and of the comparison principle. $\qquad \square$

19. Global existence for the Dirichlet problem

19.1. Small data global solutions

We start with a basic result of global existence for small initial data for the problem

$$\left.\begin{aligned}
u_t - \Delta u &= f(u), & x \in \Omega,\ t > 0,\\
u &= 0, & x \in \partial\Omega,\ t > 0,\\
u(x,0) &= u_0(x), & x \in \Omega.
\end{aligned}\right\} \tag{19.1}$$

Definition 19.1. Assume that $f(0) = 0$ (so that $u \equiv 0$ is a solution to (19.1)) and that (19.1) is locally well-posed in a space X. We say that the zero solution is **asymptotically stable** in X if there exists a constant $\eta > 0$ such that, for all $u_0 \in X$ with $\|u_0\|_X \le \eta$, there holds $T_{\max}(u_0) = \infty$ and

$$\lim_{t \to \infty} \|u(t)\|_X = 0.$$

We say that the zero solution is **exponentially asymptotically stable** in X if there exist constants $\eta, \mu > 0$ and $K \ge 1$ such that, for all $u_0 \in X$ with $\|u_0\|_X \le \eta$, there holds $T_{\max}(u_0) = \infty$ and

$$\|u(t)\|_X \le K\|u_0\|_X e^{-\mu t}, \quad t > 0. \quad \square$$

Theorem 19.2. *Consider problem (19.1), where Ω is bounded and $f : \mathbb{R} \to \mathbb{R}$ is a C^1-function such that $f(0) = 0$ and $f'(0) < \lambda_1$. Then the zero solution is exponentially asymptotically stable in $L^\infty(\Omega)$.*

Theorem 19.2 can be given a simple proof based on the comparison principle (see [296] for similar arguments).

Proof. By assumption, there exist $\eta > 0$, $\varepsilon \in (0, \lambda_1/2)$ such that

$$|f(s)| \le (\lambda_1 - 2\varepsilon)|s|, \qquad |s| \le \eta. \tag{19.2}$$

We claim that there exists a function $\varphi \in C^2(\overline{\Omega})$ such that

$$-\Delta\varphi = (\lambda_1 - \varepsilon)\varphi \quad \text{and} \quad \varphi \ge 1, \qquad x \in \overline{\Omega}. \tag{19.3}$$

Indeed, it suffices to consider $\varphi = 1 + \psi$, where ψ is the solution of

$$\left.\begin{aligned}
-\Delta\psi &= (\lambda_1 - \varepsilon)\psi + (\lambda_1 - \varepsilon), & x \in \Omega,\\
\psi &= 0, & x \in \partial\Omega,
\end{aligned}\right\}$$

and to note that $\psi \geq 0$ by the maximum principle. Next set

$$\overline{u}(x,t) = \overline{\eta}\, e^{-\varepsilon t}\varphi(x), \qquad \text{where } \overline{\eta} = \left(\max_{\overline{\Omega}} \varphi\right)^{-1}\eta.$$

An obvious computation and (19.3), (19.2) yield

$$\overline{u}_t - \Delta \overline{u} = (\lambda_1 - 2\varepsilon)\overline{u} \geq f(\overline{u}), \quad x \in \Omega, \ t > 0.$$

Assume that $\|u_0\|_\infty \leq \eta$, hence $|u_0| \leq \overline{u}(\cdot,0)$. By the comparison principle, we deduce that $u \leq \overline{u}$ in $\Omega \times (0, T_{\max}(u_0))$, and we get $u \geq -\overline{u}$ by arguing on $-u$. The conclusion follows. $\square$

Let us now consider in more detail the case of the model problem (15.1). We would like to extend Theorem 19.2 in two directions:

- unbounded domains;

- L^q- instead of L^∞-stability. Note that this is a legitimate question for $q > q_c :=$ $n(p-1)/2$ or $q = q_c > 1$, since we know (cf. Theorems 15.2 and 15.3 and Remark 15.4) that problem (15.1) is locally well-posed in $L^q(\Omega)$ for (and only for) such q.

Domains that admit such extension can be characterized geometrically through the notion of **inradius.** Recall (see Appendix D) that the inradius of Ω is defined by:

$$\rho(\Omega) = \sup\{r > 0 : \Omega \text{ contains a ball of radius } r\} = \sup_{x \in \Omega} \operatorname{dist}(x, \partial\Omega)$$

and that, for any $q \in [1, \infty]$, the condition $\rho(\Omega) < \infty$ is equivalent to the Poincaré inequality

$$\|\phi\|_q \leq C(\Omega, q)\|\nabla\phi\|_q, \quad \phi \in W_0^{1,q}(\Omega) \tag{19.4}$$

(provided Ω is uniformly smooth).

The following result of [481], [483] asserts in particular that for any $q_c < q \leq \infty$, the zero solution is asymptotically stable in $L^q(\Omega)$ *if and only if* Ω has finite inradius.

Theorem 19.3. *Consider problem* (15.1) *with $p > 1$ and let $1 \leq q \leq \infty$.*

(i) *Assume $q > q_c$ or $q = q_c > 1$. If $\rho(\Omega) < \infty$, then the zero solution is exponentially asymptotically stable in $L^q(\Omega)$.*

(ii) *Assume $q > q_c$. If $\rho(\Omega) = \infty$, then the zero solution is not asymptotically stable in $L^q(\Omega)$. More precisely, there exist initial data $u_0 \in L^q(\Omega)$ of arbitrarily small L^q-norm such that $T_{\max}(u_0) < \infty$.*

Remarks 19.4. (a) **Critical case.** The result of Theorem 19.3(ii) is no longer true in the critical case $q = q_c$: We shall see later that for any domain Ω (including the whole space), the zero solution is asymptotically stable in $L^{q_c}(\Omega)$ — see Corollary 20.20 and Remark 20.21. However, the stability is exponential only if $\rho(\Omega) < \infty$ (see [481]).

(b) **Different methods of proof.** Theorem 19.3(i) for $1 < q < \infty$ can be proved by a multiplier argument, using multiplication by powers of u and the Poincaré inequality [481]. We shall employ this method here, but for simplicity we shall prove the result only in the range $2 \leq q < \infty$ (the idea for $1 < q < 2$ is the same, but some additional technical difficulties arise).

An alternative proof, covering the extremal cases $q = 1$ and $q = \infty$ as well, can be carried out by using the variation-of-constants formula and the exponential decay of the heat semigroup for $\rho(\Omega) < \infty$ (see [483]). Such arguments can be used to prove more general results of linearized stability; see Theorems 51.17, 51.19 and 51.33 in Appendix E.

As an advantage, the energy proof might also apply to certain quasilinear problems. $\square$

Proof of Theorem 19.3(i) for $2 \leq q < \infty$. To simplify notation, if k is any positive number, we write u^k for $\mathrm{sign}(u)\,|u|^k$. Since $u_0 \in L^q(\Omega)$, it follows from Example 51.27 in Appendix E that $u \in C([0,\infty), L^q(\Omega)) \cap C((0,\infty), W^{2,q} \cap W_0^{1,q}(\Omega)) \cap C^1((0,\infty), L^q(\Omega))$. Multiplying the equation by u^{q-1} and integrating by parts, we obtain

$$\frac{1}{q}\frac{d}{dt}\|u(t)\|_q^q = \langle u^{q-1}, \Delta u \rangle + \|u\|_{q+p-1}^{q+p-1} = -\frac{4(q-1)}{q^2}\|\nabla(u^{q/2})\|_2^2 + \|u\|_{q+p-1}^{q+p-1},$$

$$(19.5)$$

for all $t \in (0, T_{\max}(u_0))$. For the last term of inequality (19.5), we next establish the estimate

$$\|u\|_{q+p-1} \leq C\|u\|_q^{\theta}\|\nabla(u^{q/2})\|_2^{2(1-\theta)/q}, \tag{19.6}$$

with

$$\theta = 1 - \frac{n(p-1)}{2(q+p-1)} \in (0,1). \tag{19.7}$$

To do so, let us consider separately the cases $n \geq 3$ and $n \leq 2$. If $n \geq 3$, since $q \geq n(p-1)/2$, we have $q+p-1 \leq nq/(n-2)$ hence, by Hölder's inequality,

$$\|u\|_{q+p-1} \leq \|u\|_q^{\theta}\,\|u\|_{nq/(n-2)}^{1-\theta},$$

with

$$\theta = \left(\frac{1}{q+p-1} - \frac{1}{nq/(n-2)}\right)\left(\frac{1}{q} - \frac{1}{nq/(n-2)}\right)^{-1} = 1 - \frac{n(p-1)}{2(q+p-1)},$$

and Sobolev's inequality then yields (19.6).

If $n \leq 2$, we use the Gagliardo-Nirenberg inequalities

$$\|v\|_a \leq C_a \|v\|_2^{(a+2)/2a} \|v'\|_2^{(a-2)/2a}, \quad a \geq 2, \quad v \in H_0^1(\Omega) \quad (n = 1)$$

and

$$\|v\|_a \leq C_a \|v\|_2^{2/a} \|\nabla v\|_2^{1-(2/a)}, \quad a \geq 2, \quad v \in H_0^1(\Omega) \quad (n = 2).$$

Applying this with $v = u^{q/2}(t)$ and $a = 2(q + p - 1)/q > 2$ yields (19.6) with $\theta = q/(q+p-1)$ if $n = 2$ and $\theta = (2q+p-1)/2(q+p-1)$ if $n = 1$, that is (19.7).

The next step is to use the Poincaré inequality (50.2) in $W_0^{1,q}(\Omega)$ (valid due to $\rho(\Omega) < \infty$; see Proposition 50.1 in Appendix D) to obtain a lower estimate of the first term in the right-hand side of (19.5). It follows from (50.2) that, for all $\alpha \in [0, 1]$,

$$\begin{aligned}
\|\nabla(u^{q/2})\|_2^2 &\geq C\|u\|_q^q + C\|\nabla(u^{q/2})\|_2^{2\alpha} \|\nabla(u^{q/2})\|_2^{2(1-\alpha)} \\
&\geq C\|u\|_q^q + C\|u\|_q^{q\alpha} \|\nabla(u^{q/2})\|_2^{2(1-\alpha)}.
\end{aligned} \tag{19.8}$$

On the other hand, one has $(1 - \theta)(q + p - 1)/q = n(p - 1)/2q \leq 1$. Therefore, we may choose

$$\alpha = 1 - (1 - \theta)(q + p - 1)/q,$$

and by combining (19.5), (19.6) and (19.8), it follows that

$$\frac{1}{q}\frac{d}{dt}\|u(t)\|_q^q \leq -C\|u\|_q^q + C\|\nabla(u^{q/2})\|_2^{2(1-\alpha)} \|u\|_q^{q\alpha} \big(\|u\|_q^{p-1} - C'\big).$$

It follows from this differential inequality that if $\|u_0\|_q$ is sufficiently small, then for all $t > 0$,

$$\frac{d}{dt}\|u(t)\|_q^q \leq -C\|u\|_q^q,$$

hence

$$\|u(t)\|_q \leq e^{-C't}\|u_0\|_q, \tag{19.9}$$

as long as the solution exists. If $q > q_c$, we know from Theorem 15.2 that the L^q-norm must blow up if $T_{\max}(u_0)$ is finite. The estimate (19.9) thus ensures global existence. If $q = q_c$, global existence when $\|u_0\|_q$ is small follows from Corollary 20.20 below. $\quad\square$

Proof of Theorem 19.3(ii). Fix a test-function $\varphi \in \mathcal{D}(\mathbb{R}^n)$, $\varphi \geq 0$, $\varphi \not\equiv 0$ with $\mathrm{supp}(\varphi) \subset B := B(0, 1)$, and let w be the solution of problem (15.1) with Ω replaced by B and u_0 replaced by φ. Due to e.g. Theorem 17.1, we can assume that w blows up in a finite time T (replacing φ by a sufficiently large multiple).

Now, since $\rho(\Omega) = \infty$, Ω contains some ball $B_k = B(x_k, k)$ for any integer $k \geq 1$. Let us set

$$u_k(x,t) := k^{-2/(p-1)} w(k^{-1}(x - x_k), k^{-2}t), \quad u_{0,k}(x) := k^{-2/(p-1)} \varphi(k^{-1}(x - x_k)).$$

Due to the invariance of the equation under this scaling, it is easily verified that u_k solves the problem

$$\left.\begin{aligned}
\partial_t u_k - \Delta u_k &= |u_k|^{p-1} u_k, & x \in B_k,\ 0 < t < k^2 T, \\
u_k &= 0, & x \in \partial B_k,\ 0 < t < k^2 T, \\
u_k(x,0) &= u_{0,k}(x), & x \in B_k.
\end{aligned}\right\}$$

Let $\tilde{u}_k$ be the solution of problem (15.1) with $u_0 = u_{0,k}$. Since each B_k is included in Ω and $\tilde{u}_k \geq 0$ on ∂B_k, it follows from the comparison principle that $\tilde{u}_k \geq u_k$, hence $\tilde{u}_k$ blows up in finite time.

Last, an easy calculation yields

$$\|u_{0,k}\|_q = k^{-2/(p-1)+n/q} \|\varphi\|_q \to 0, \quad k \to \infty,$$

which concludes the proof. $\quad\square$

We shall now describe the **potential well method**. It will enable us to obtain alternative sufficient conditions for global existence (and nonexistence) for the model problem (15.1).

In the rest of this subsection we assume Ω bounded and $1 < p \leq p_S$. Recall that the energy functional E is given by

$$E(u) = \frac{1}{2} \int_\Omega |\nabla u|^2\, dx - \frac{1}{p+1} \int_\Omega |u|^{p+1}\, dx, \qquad u \in H_0^1(\Omega). \tag{19.10}$$

We define the **Nehari functional** I by

$$I(u) = \int_\Omega |\nabla u|^2\, dx - \int_\Omega |u|^{p+1}\, dx, \qquad u \in H_0^1(\Omega).$$

The **potential well** associated with problem (15.1) is the set

$$W := \big\{ u \in H_0^1(\Omega) : E(u) < d,\ I(u) > 0 \big\} \cup \{0\},$$

where d, the depth of the potential well, is defined by

$$d := \inf\big\{ E(u) : u \in H_0^1(\Omega) \setminus \{0\},\ I(u) = 0 \big\}. \tag{19.11}$$

We shall show in Lemma 19.7(i) below that

$$d = \frac{p-1}{2(p+1)} \Lambda^{2(p+1)/(p-1)}, \tag{19.12}$$

where $\Lambda = \Lambda_{p+1}(\Omega)$ denotes the best constant in the Sobolev embedding $H_0^1(\Omega) \hookrightarrow L^{p+1}(\Omega)$, i.e.,

$$\Lambda := \inf\left\{ \frac{\|\nabla u\|_2}{\|u\|_{p+1}} : u \in H_0^1(\Omega),\ u \neq 0 \right\}. \tag{19.13}$$

The exterior of the potential well is the set

$$Z := \left\{ u \in H_0^1(\Omega) : E(u) < d,\ I(u) < 0 \right\}.$$

In what follows, for $u_0 \in H_0^1(\Omega)$, u denotes the maximal L^{p+1}-classical solution of problem (15.1) (recall from Section 15 that (15.1) is well-posed in $L^{p+1}(\Omega)$, since $p + 1 \geq q_c$ due to $p \leq p_S$).

Theorem 19.5. *Consider problem (15.1) with Ω bounded.*

(i) *Assume $1 < p < p_S$. If $u_0 \in W$, then $T_{\max}(u_0) = \infty$,*

$$u(t) \in W \quad \text{for all } t > 0,$$

and

$$\|u(t)\|_\infty \to 0, \quad t \to \infty. \tag{19.14}$$

(ii) *Assume $1 < p \leq p_S$. If $u_0 \in Z$, then $T_{\max}(u_0) < \infty$.*

The potential well method was introduced in [467] to obtain global existence results for nonlinear hyperbolic equations. As for parabolic problems, the global existence part in Theorem 19.5(i) is due to [515] and the decay property is essentially from [290]. Theorem 19.5(ii) is due to [409] (see also [290]), where the potential well method was extended to obtain nonexistence results for hyperbolic and parabolic problems.

Remarks 19.6. (a) Theorem 19.5(i) provides in particular a sufficient smallness condition on u_0 for global existence when $p < p_S$. Indeed, we have $u_0 \in W$ whenever $u_0 \in H_0^1(\Omega)$ satisfies $\|\nabla u_0\| < \sqrt{2d}$ (cf. Lemma 19.7(iii)).

(b) The quantity d can be interpreted as a *mountain-pass energy* (cf. Section 7). Indeed, for $p \leq p_S$, it is easy to show that

$$d = \inf_{u \in H_0^1(\Omega) \setminus \{0\}} \ \max_{s \geq 0} \ E(su).$$

Note that for $p < p_S$, there exist least-energy stationary solutions v, i.e.: such that $E(v) = d$ (this follows from Theorem 7.2, applied with $u_0 = 0$ and u_1 such that $E(u_1) < 0$, and from the easy fact that $d = \beta$, where β is defined in (7.1)).

(c) The sets W and Z are invariant under the semiflow associated with problem (15.1) for $p \leq p_S$. This follows from the proof of Theorem 19.5.

(d) Theorem 19.5 admits a converse (cf. [287]). Namely, if $p \leq p_S$ and u is a global solution satisfying (19.14), then $u(t) \in W$ for large t. If $p < p_S$ and u is a blowing-up solution, then $u(t) \in Z$ for t close to $T_{\max}(u_0)$. These facts respectively follow from smoothing effects and Theorem 19.5(ii), and from property (22.28) in Proposition 22.11 and $I(u) \leq 2E(u)$.

(e) Theorems 17.6 and 19.5 give an essentially complete characterization of global existence/nonexistence in the subcritical range for initial data with energy less than d. See [236] and the references therein for additional information, including some partial results for higher energy data. It seems an open question whether or not $I(u_0) < 0$ is a sufficient condition for blow-up. $\square$

In view of the proof of Theorem 19.5, we need the following properties of the potential well.

Lemma 19.7. *Let Ω be bounded and let $1 < p \leq p_S$.*

(i) *Then property (19.12) is true. If moreover $p < p_S$, then the infimum in (19.11) is attained.*

(ii) *For any $\varepsilon > 0$, there holds*

$$d_\varepsilon := \inf\{E(u) : u \in H_0^1(\Omega), \ I(u) = -\varepsilon\} \geq d - \frac{\varepsilon}{p+1}. \tag{19.15}$$

(iii) *For all $u \in H_0^1(\Omega)$, we have*

$$\|\nabla u\|_2 < \sqrt{2d} \implies u \subset W \implies \|\nabla u\|_2 < \sqrt{\tfrac{2(p+1)}{p-1}d}. \tag{19.16}$$

Proof. Denote $D = \frac{p-1}{2(p+1)}\Lambda^{2(p+1)/(p-1)}$ and fix $\varepsilon \geq 0$. Let $u \in H_0^1(\Omega)$ satisfy $I(u) = -\varepsilon$, and assume in addition that $u \neq 0$ if $\varepsilon = 0$. Then

$$E(u) = \frac{p-1}{2(p+1)} \int_\Omega |\nabla u|^2 \, dx - \frac{\varepsilon}{p+1}. \tag{19.17}$$

Since, by (19.13),

$$\int_\Omega |\nabla u|^2 \, dx \leq \int_\Omega |u|^{p+1} \, dx \leq \Lambda^{-(p+1)} \left(\int_\Omega |\nabla u|^2 \, dx \right)^{(p+1)/2}$$

and $u \neq 0$, we get $\int_\Omega |\nabla u|^2 \, dx \geq \Lambda^{2(p+1)/(p-1)}$. This combined with (19.17) implies $d \geq D$ and

$$d_\varepsilon \geq D - (p+1)^{-1}\varepsilon, \qquad \varepsilon > 0. \tag{19.18}$$

Let now u_j be a minimizing sequence for (19.13). By multiplying u_j with suitable $\mu_j > 0$, we may assume that $I(u_j) = 0$. Therefore

$$\int_\Omega |\nabla u_j|^2 \, dx = \int_\Omega |u_j|^{p+1} \, dx = (\Lambda + \eta_j)^{-(p+1)} \left(\int_\Omega |\nabla u_j|^2 \, dx \right)^{(p+1)/2},$$

where $\eta_j \to 0+$. Combining this with (19.17) for $\varepsilon = 0$, we obtain

$$E(u_j) = \frac{p-1}{2(p+1)}(\Lambda + \eta_j)^{2(p+1)/(p-1)} \to D,$$

hence $d = D$, i.e. (19.12). If $p < p_S$, then the infimum in (19.13) is attained for some $v \in H_0^1(\Omega)$, due to the compactness of the embedding $H_0^1(\Omega) \hookrightarrow L^{p+1}(\Omega)$. Arguing similarly as above, with u_j replaced by v, we see that the infimum in (19.11) is also attained.

Assertion (ii) follows from (19.18).

Finally, let us prove assertion (iii). Assume $0 < \|\nabla u\|_2 < \sqrt{2d}$. Then $E(u) < d$. Next, using (19.13) and $\|\nabla u\|_2 < \sqrt{2d} < \Lambda^{(p+1)/(p-1)}$, we obtain

$$\int_\Omega |u|^{p+1}\, dx \leq \Lambda^{-(p+1)}\left(\int_\Omega |\nabla u|^2\, dx\right)^{(p+1)/2} < \int_\Omega |\nabla u|^2\, dx.$$

Consequently $I(u) > 0$, hence $u \in W$.

On the other hand, for any $u \in W$, the conditions $E(u) < d$ and $I(u) \geq 0$ imply

$$\frac{p-1}{2(p+1)}\int_\Omega |\nabla u|^2\, dx \leq E(u) < d,$$

hence the last inequality in (19.16). $\square$

Proof of Theorem 19.5. Set $T := T_{\max}(u_0)$. By (17.7), we have

$$E(t) \leq E(u_0) < d, \qquad t \in [0, T). \tag{19.19}$$

(i) If $u(t) = 0$ for some $t \geq 0$, then by uniqueness, $u(s) = 0$ for all $s \geq t$ and the conclusion is true. Hence we may assume that $u(t) \neq 0$ for all $t \in [0, T)$. Since $I(u_0) > 0$, using (19.11) and (19.19), it follows by continuity that, for all $t \in [0, T)$, $I(u(t)) > 0$, hence $u(t) \in W$. By Lemma 19.7(iii), we deduce that $u(t)$ is bounded in $H_0^1(\Omega)$, hence in $L^{p+1}(\Omega)$. Remarks 16.2 then guarantee that $T = \infty$. On the other hand, by Example 53.7 (and in particular the existence of a strict Lyapunov functional given by (19.10)), the ω-limit set $\omega(u_0)$ in the $H_0^1(\Omega)$-topology is nonempty and consists of (classical) equilibria. But for any nontrivial equilibrium v, we have $I(v) = 0$, hence $E(v) \geq d$ by (19.11). Consequently $v \notin \omega(u_0)$ in view of (19.19). In other words, $\lim_{t\to\infty}\|u(t)\|_{1,2} = 0$, hence $\lim_{t\to\infty}\|u(t)\|_{p+1} = 0$. By the smoothing estimate (15.2), this guarantees (19.14).

(ii) Fix $\varepsilon > 0$ such that

$$\varepsilon < \min\big(-I(u_0), d - E(u_0)\big).$$

By (19.15) and (19.19), we have $E(t) \leq E(u_0) < d_\varepsilon$ for $t \in [0, T)$. Since $I(u_0) < -\varepsilon$, using the definition of d_ε in (19.15), it follows by continuity that $I(u(t)) < -\varepsilon$ for all $t \in [0, T)$, hence

$$\frac{1}{2}\frac{d}{dt}\int_\Omega u^2\, dx = -I(u(t)) > \varepsilon \qquad (19.20)$$

(cf. (17.9)). But on the other hand, we know from Remark 17.7(ii) that $T = \infty$ implies $\sup_{t\geq 0}\|u(t)\|_2 < \infty$. In view of (19.20), we conclude that $T < \infty$. $\square$

19.2. Structure of global solutions in bounded domains

In this subsection we study some properties of the set of initial data giving rise to global solutions of problem (19.1). Throughout this subsection we assume that the domain Ω is bounded. We define the sets

$$\mathcal{G} = \{u_0 \in L^\infty(\Omega) : T_{\max}(u_0) = \infty\},$$

$$\mathcal{B} = \{u_0 \in L^\infty(\Omega) : T_{\max}(u_0) = \infty \ \text{ and } \ \sup_{t\geq 0}\|u(t)\|_\infty < \infty\},$$

and

$$\mathcal{D} = \{u_0 \in L^\infty(\Omega) : T_{\max}(u_0) = \infty \ \text{ and } \ \|u(t)\|_\infty \to 0,\ t \to \infty\}$$

($\mathcal{D}$ is the **domain of attraction** of 0). When (19.1) admits both global and nonglobal solutions, these sets are natural and interesting objects. Clearly, $\mathcal{D} \subset \mathcal{B} \subset \mathcal{G}$.

In order to describe the properties of these sets, we first need some properties of the steady states of (19.1), i.e. (classical) solutions of

$$\left.\begin{aligned} -\Delta u &= f(u), & x \in \Omega, \\ u &= 0, & x \in \partial\Omega. \end{aligned}\right\} \qquad (19.21)$$

The following result implies in particular that two ordered positive steady states cannot exist when the nonlinearity is strictly convex. Note that the result fails in general if $\Omega = \mathbb{R}^n$ (with $u, v \to 0$ at infinity), as shown by Theorem 9.1 with $p \geq p_{JL}$.

Proposition 19.8. *Assume Ω bounded and let $f : \mathbb{R} \to \mathbb{R}$ be a strictly convex C^1-function, with $f(0) = 0$. Assume that $u, v \in C^2(\Omega) \cap C^1(\overline{\Omega})$ are respectively sub- and supersolutions to (19.21), in the sense that*

$$\left.\begin{aligned} -\Delta u &\leq f(u), & x \in \Omega, \\ -\Delta v &\geq f(v), & x \in \Omega, \\ u &= v = 0, & x \in \partial\Omega. \end{aligned}\right\} \qquad (19.22)$$

If $v \geq u > 0$ in Ω, then $u \equiv v$.

Proof. Multiplying the inequalities in (19.22) by v and u respectively, we obtain

$$\int_\Omega f(u)v\,dx \geq \int_\Omega (-\Delta u)v\,dx = \int_\Omega \nabla u \cdot \nabla v\,dx = \int_\Omega (-\Delta v)u\,dx \geq \int_\Omega f(v)u\,dx,$$

hence

$$\int_\Omega \left(\frac{f(u)}{u} - \frac{f(v)}{v} \right) uv\,dx \geq 0.$$

But in view of the strict convexity of f, the integrand is nonpositive in Ω, and (strictly) negative at each x such that $v(x) > u(x)$. The conclusion follows. $\quad\square$

The next result describes some basic geometrical and topological properties of the sets $\mathcal{D}, \mathcal{B}, \mathcal{G}$. Here we refer to the L^∞-topology (but other choices are possible). Also, for a given convex subset $\mathcal{K}$ of a vector space, we recall that $x \in \mathcal{K}$ is called an **extremal point** if it cannot be written under the form $x = \theta y + (1 - \theta)z$ with $y, z \in \mathcal{K}$ and $\theta \in (0, 1)$. Theorem 19.9 is due to [342]. However the present proof of assertion (iv) is different and simpler than that in [342].

Theorem 19.9. *Consider problem (19.1) where Ω is bounded and $f : \mathbb{R} \to \mathbb{R}$ is a C^1-function, with $f(0) = f'(0) = 0$.*

(i) *Then $\mathcal{D}$ is an open neighborhood of 0.*

(ii) *Assume that f is convex. Then the sets $\mathcal{G}, \mathcal{B}$ and $\mathcal{D}$ are convex.*

Now assume that f is strictly convex.

(iii) *If u_0 is not an extremal point of $\mathcal{G}$ (resp., of $\mathcal{B}$), then u_0 is an interior point. This is true in particular if $0 \leq u_0 \leq v_0$, with $v_0 \in \mathcal{G}$ (resp., $\mathcal{B}$), $v_0 \not\equiv u_0$.*

(iv) *There holds $\mathrm{int}(\mathcal{B}) = \mathcal{D}$.*

Proof. (i) This is a consequence of Theorem 19.2 and of the continuous dependence of solutions on initial values.

(ii) Let $\theta \in (0, 1)$, $u_0, v_0 \in L^\infty(\Omega)$, $u_0 \not\equiv v_0$, $w_0 = \theta u_0 + (1 - \theta)v_0$, and denote by u, v, w the solutions of (19.1) with initial data u_0, v_0, w_0, respectively. Set $\overline{w} = \theta u + (1 - \theta)v$. By the convexity of f, for all $x \in \Omega$ and $t \in (0, \min(T_{\max}(u_0), T_{\max}(v_0)))$, we have

$$\overline{w}_t - \Delta\overline{w} = \theta f(u) + (1 - \theta)f(v) \geq f(\theta u + (1 - \theta)v) = f(\overline{w}), \qquad (19.23)$$

hence

$$w \leq \theta u + (1 - \theta)v, \qquad (19.24)$$

in view of the comparison principle. On the other hand, the assumptions on f imply $f(s) \geq 0$, $s \in \mathbb{R}$. Denoting by e^{-tA} the heat semigroup in Ω with homogeneous

Dirichlet boundary conditions, the maximum principle and Proposition 48.5 in Appendix B imply

$$w \geq e^{-tA}w_0 \geq -Ce^{-\lambda_1 t}. \tag{19.25}$$

It then follows from (19.24) and (19.25) that $w_0 \in \mathcal{G}$ (resp., $\mathcal{B}$, $\mathcal{D}$) whenever $u_0, v_0 \in \mathcal{G}$ (resp., $\mathcal{B}$, $\mathcal{D}$) and the convexity assertion is proved.

Assume now that f is strictly convex.

(iii) Let w_0 be a nonextremal point of $\mathcal{B}$, i.e. $w_0 = \theta u_0 + (1-\theta)v_0$, with $\theta \in (0,1)$, $u_0, v_0 \in \mathcal{B}$, $u_0 \not\equiv v_0$. Then, by continuity and the strict convexity of f, we have $\theta f(u(\cdot,t)) + (1-\theta)f(v(\cdot,t)) \not\equiv f(\overline{w}(\cdot,t))$ for $t \in [0,\tau]$, with $\tau > 0$ small. By (19.23) and the strong maximum principle, we deduce that

$$w(x,\tau) < \overline{w}(x,\tau) \quad \text{in } \Omega \quad \text{and} \quad \frac{\partial w}{\partial \nu}(x,\tau) > \frac{\partial \overline{w}}{\partial \nu}(x,\tau) \text{ on } \partial\Omega.$$

Due to (51.28), we know that for small $\tau > 0$, the map $L^\infty(\Omega) \ni \tilde{u}_0 \mapsto u(\cdot,\tau;\tilde{u}_0) \in C^1(\overline{\Omega})$ is well-defined and continuous on a neighborhood of u_0. Therefore, if $\|w_0 - \tilde{u}_0\|_\infty$ is small enough, then $T_{\max}(\tilde{u}_0) > \tau$ and $\tilde{u}(\tau) \leq \overline{w}(\tau)$, where $\tilde{u} := u(\cdot;\tilde{u}_0)$. This, along with $\tilde{u}(t) \geq e^{-tA}\tilde{u}_0$ guarantees that $\tilde{u}_0 \in \mathcal{B}$ and w_0 is an interior point. The same argument applies for $\mathcal{G}$.

To justify the last part of assertion (iii), write $u_0 = \theta v_0 + (1-\theta)\tilde{v}_0$, with $\tilde{v}_0 := (1-\theta)^{-1}(u_0 - \theta v_0) \not\equiv v_0$. For $\theta > 0$ small, we have $v_0 \geq \tilde{v}_0 \geq -(1-\theta)^{-1}\theta v_0 \in \mathcal{D}$ due to Theorem 19.2, hence $\tilde{v}_0 \in \mathcal{B}$ (resp., $\mathcal{G}$) by comparison.

(iv) Let $u_0 \subset \mathrm{int}(\mathcal{B})$. In particular, there exists $v_0 \in \mathcal{B}$, with $u_0 \leq v_0$, $u_0 \not\equiv v_0$. Denote by u, v the solutions of (19.1) with initial data u_0, v_0, respectively. Due to Example 53.7, we know that uniformly bounded solutions are relatively compact in $X := H^1 \cap C_0(\Omega)$ for $t \geq 1$ and that the ω-limit set $\omega(u_0)$ (in the X-topology) is nonempty and consists of equilibria. Let $z \in \omega(u_0)$. By definition, there exists a sequence $t_k \to \infty$, such that $u(t_k) \to z$ in X. Since $\{v(t) : t \geq 1\}$ is precompact in X, there exist $\tilde{z} \in \omega(v_0)$ and a subsequence t_{k_j} such that $v(t_{k_j}) \to \tilde{z}$ in X. By Lemma 17.9, there exist $\tau > 0$ and $\alpha > 1$, such that $v \geq \alpha u$ for $t \geq \tau > 0$, hence

$$\tilde{z} \geq \alpha z. \tag{19.26}$$

Due to $f \geq 0$, we have $z \geq 0$ by the maximum principle. Since $z, \tilde{z}$ are steady states of (19.1), we then deduce that $z \equiv 0$, because otherwise (19.26) would contradict Proposition 19.8. Consequently, $u_0 \in \mathcal{D}$. In particular, $\mathrm{int}(\mathcal{B}) \subset \mathcal{D}$. Conversely, it is clear that $\mathcal{D} \subset \mathrm{int}(\mathcal{B})$ since $\mathcal{D}$ is open. $\quad\square$

Remark 19.10. Instability of positive equilibria. Theorem 19.9 shows that the only way for the ω-limit set of a global bounded solution of (19.1) to contain positive equilibria is to have $u(t)$ be an extremal (or boundary) point of $\mathcal{B}$ for all $t \geq 0$. By the same token, if v is a positive steady state and $u_0 \geq v$, $u_0 \not\equiv v$, then $u_0 \notin \mathcal{B}$. In the special case $f(u) = |u|^p$, we have the following stronger property (which generalizes Theorem 17.8). $\quad\square$

Proposition 19.11. *Consider problem* (19.1) *with* Ω *bounded and* $f(u) = |u|^p$, $p > 1$. *Assume that* $v_0 \in \mathcal{B} \setminus \mathcal{D}$ *and that* $u_0 \geq v_0$, $u_0 \not\equiv v_0$. *Then* $T_{\max}(u_0) < \infty$.

Proof. Let u, v be the corresponding solutions and assume for contradiction that u is global. Then, by Lemma 17.9, we have $u(t) \geq \alpha v(t)$, for all $t \geq 1$ and some $\alpha > 1$. On the other hand, by Example 53.7, $\omega(v_0)$ contains a nonzero steady state z and there exists a sequence $t_k \to \infty$ such that $v(t_k) \to z$ in $C^1(\overline{\Omega})$. Moreover, since $v(t_k) \geq e^{-tA}v_0$, we have $z \geq 0$ in Ω, hence $\partial z / \partial \nu < 0$ on $\partial\Omega$ by the Hopf maximum principle. It follows that for large k, $v(t_k) > \alpha^{-1}z$ in Ω. Consequently, $u(t_k) > z$, contradicting Theorem 17.8. $\square$

Remark 19.12. Further properties of $\mathcal{D}, \mathcal{B}, \mathcal{G}$**.** Consider problem (19.1) with $f(u) = |u|^p$ and $p > 1$, and let us restrict ourselves to nonnegative initial data. Let us define $\mathcal{G}^+ := \{u \in \mathcal{G} : u \geq 0\}$ and $\mathcal{B}^+, \mathcal{D}^+$ similarly. We first note that the set $\mathcal{D}^+$ is unbounded, due to the existence of nonnegative global classical solutions such that $\|u(t)\|_\infty \to \infty$ as $t \to 0+$. Indeed, since 0 is an asymptotically stable solution of problem (15.1) in L^q for $q \geq q_c$ by Theorem 19.3, this occurs for any $0 \leq u_0 \in L^q(\Omega) \setminus L^\infty(\Omega)$ with $\|u_0\|_q$ small enough. On the other hand, for $p < p_S$, it follows from Theorem 6.2 that $\mathcal{B}^+ \neq \mathcal{D}^+$. Further related results will be obtained later, in particular in Subsection 28.4, where we study the transition between global existence and blow-up along each ray of nonnegative initial data starting from 0. Among the consequences of these results, let us mention the following properties:

(a) if $p < p_S$, then $\mathcal{G}^+ = \mathcal{B}^+ \neq \mathcal{D}^+$ and $\mathcal{G}^+$ is a closed subset of $L^\infty(\Omega)$ (cf. Theorem 22.1);

(b) if $p \geq p_S$ and Ω is starshaped, then $\mathcal{B}^+ = \mathcal{D}^+$ (cf. Corollary 5.2 and Theorem 28.7(iv));

(c) if $p = p_S$ and Ω is a ball, then $\mathcal{G}^+ \neq \mathcal{B}^+ = \mathcal{D}^+$ (cf. Theorem 28.7);

(d) if $p > p_S$, Ω is a ball and we consider radial solutions only, then $\mathcal{G}^+ = \mathcal{B}^+ = \mathcal{D}^+$ (see Theorem 22.4). $\square$

Remark 19.13. Stabilization towards an equilibrium. In the proofs of Theorem 19.9 and Proposition 19.11 we used the fact that the ω-limit set of any global bounded solution of (19.1) in a bounded domain Ω is a nonempty compact connected set consisting of equilibria. If all equilibria (at a given energy level) are isolated, then this fact guarantees that each global bounded solution converges to a single equilibrium. If $n = 1$ or if we consider radial solutions in a ball, then this convergence is true without any information on the set of equilibria, see [543], [353], [265], [102], [268], [127]. Similar stabilization result is true for general bounded domains in $\mathbb{R}^n$, $n \geq 1$, provided the nonlinearity is analytic, see [474], [291]. Other sufficient conditions can be found in the survey article [421]. Nonconvergent global bounded solutions were constructed in [426] and [427] for spatially inhomogeneous nonlinearities of the form $f = f(x, u)$. $\square$

Remark 19.14. Global solutions and very weak stationary solutions.
Consider problem (14.1), where $f : [0, \infty) \to (0, \infty)$ is a C^1 nondecreasing convex function satisfying the blow-up condition (17.4). It was shown in [94] that strong relations exist between the existence of global (classical) solutions of (14.1) and the existence of very weak solutions of the stationary problem (13.1) (throughout this remark, "solution" implicitly means "nonnegative solution"):

(i) if $T_{\max}(u_0) = \infty$ for some $0 \le u_0 \in L^\infty(\Omega)$, then (13.1) admits a very weak solution;

(ii) conversely, if (13.1) admits a very weak solution v, then for any $u_0 \in L^\infty(\Omega)$ with $0 \le u_0 \le v$, we have $T_{\max}(u_0) = \infty$.

Note that (i) provides a further blow-up criterion: if (13.1) has no very weak solution, then all solutions of (14.1) have to blow up in finite time. As for (ii), it gives a new sufficient condition for global existence (see Theorem 20.5 for a related result concerning the Cauchy problem).

Assertion (i) is not immediate since no bound is assumed on u. As for assertion (ii), it would be a direct consequence of the comparison principle if we were assuming $v \in L^\infty(\Omega)$, but it is far from obvious in general since the inequality $u \le v$ in itself does not a priori prevent $\|u(t)\|_\infty$ from blowing up in finite time.

On the other hand, the existence of a global solution of (14.1) does not in general imply the existence of a *classical* steady state. In fact, there are situations where (13.1) has a singular (very weak) solution but no classical solution (see Remark 3.7) and where (14.1) admits global unbounded solutions which stabilize to a singular solution as $t \to \infty$ (see Remark 22.6(b)).

The idea of the proof of assertion (i) is as follows. Assume $u_0 = 0$ without loss of generality (u is then also global by the comparison principle). Since $u_t \ge 0$ by Proposition 52.19, we may let $v(x) := \lim_{t \to \infty} u(x, t) \le \infty$. Theorem 17.3 implies $\int_\Omega u(t)\varphi_1 \, dx \le C$ for $t \ge 0$. Integrating (17.5) in time between t and $t + 1$ and using $u_t \ge 0$, it follows that

$$\int_\Omega f(u(t))\varphi_1 \, dx \le \int_t^{t+1} \int_\Omega f(u)\varphi_1 \, dx \, ds$$

$$= \lambda_1 \int_t^{t+1} \int_\Omega u\varphi_1 \, dx \, ds + \left[\int_\Omega u(s)\varphi_1 \, dx \right]_t^{t+1} \le (1 + \lambda_1)C.$$

Let now $\Theta \in C^2(\overline{\Omega})$, $\Theta \ge 0$, be the classical solution of the problem

$$\left. \begin{array}{ll} -\Delta\Theta = 1 & \text{in } \Omega, \\ \Theta = 0 & \text{on } \partial\Omega. \end{array} \right\} \tag{19.27}$$

Multiplying the equation in (14.1) by the function Θ defined in (19.27), integrating over $\Omega \times (t, t + 1)$, and using $u_t \ge 0$, we obtain

$$\int_\Omega u(t) \, dx \le \int_t^{t+1} \int_\Omega u \, dx \, ds = \int_t^{t+1} \int_\Omega f(u)\Theta \, dx \, ds - \left[\int_\Omega u(s)\Theta \, dx \right]_t^{t+1} \le C'.$$

In particular, we get $f(v) \in L_\delta^1(\Omega)$ and $v \in L^1(\Omega)$. By arguing similarly as in the (alternative) proof of Lemma 53.10, using again $u_t \geq 0$, we then easily conclude that v is a very weak solution of (13.1).

The proof of assertion (ii) is more delicate and will not be given here. It is based on a perturbation argument which relies on a variant of Lemma 27.4 and on Lemma 27.5 below (used in the study of complete blow-up). $\square$

19.3. Diffusion eliminating blow-up

In Section 17, we used the convexity of the function $f(u) = \lambda u + u^p$, $u > 0$, in order to prove blow-up of solutions of (17.1) for suitable initial data. On the other hand, it follows from Theorem 19.2 that any solution of (17.1) with Ω bounded, $\lambda < \lambda_1$ and u_0 small does exist globally and tends to zero as $t \to \infty$. A similar assertion is true for $\Omega = \mathbb{R}^n$ if, for example, $\lambda = 0$ and $p > p_F$ (see Theorem 18.1). Since all positive solutions of the ODE $U' = U^p$ blow up in finite time, we see that diffusion and the Dirichlet boundary conditions (or just the diffusion if $\Omega = \mathbb{R}^n$) can prevent blow-up for some initial data. Next we show that for some particular nonlinearities f, diffusion with the Dirichlet boundary condition can completely eliminate blow-up. This result (and its modification for unbounded domains) is due to [197].

Hence, let $f : [0, \infty) \to [0, \infty)$ be smooth, $f(u) > 0$ for $u > 0$ and consider the ODE

$$U_t = f(U), \qquad t > 0,$$
$$U(0) = U_0, \tag{19.28}$$

where $U_0 > 0$, and the related Cauchy-Dirichlet problem

$$\left. \begin{array}{ll} u_t - d\Delta u = f(u), & x \in \Omega,\ t > 0, \\ u = 0, & x \in \partial\Omega,\ t > 0, \\ u(x, 0) = u_0(x), & x \in \Omega, \end{array} \right\} \tag{19.29}$$

where $\Omega \subset \mathbb{R}^n$ is a bounded domain, $d > 0$ and $u_0 \geq 0$ is a an L^∞-function. It is well known that condition (17.4) is sufficient and necessary for blow-up of the solution of (19.28), and we have seen in Theorem 17.3 that if f satisfies (17.4) and is *convex*, then the solution of (19.29) blows up for large initial data. We will prove that there exist (nonconvex) f satisfying (17.4) such that (19.29) possesses a global and bounded solution for any u_0 and any $d > 0$.

Theorem 19.15. *There is a C^∞-function $f : [0, \infty) \to [0, \infty)$, $f(u) > 0$ for $u > 0$, such that the following holds:*

(i) *All solutions of (19.28) with $U(0) > 0$ blow up in finite time.*

(ii) *If Ω is bounded and $d > 0$, then all solutions of (19.29) with $0 \le u_0 \in L^\infty(\Omega)$ exist and remain bounded for all $t \ge 0$.*

Of course Theorem 19.15 cannot be true in the case of Neumann boundary conditions $u_\nu = 0$, since solutions of (19.28) also solve the PDE. On the other hand, Theorem 19.15 remains true in the case of Robin boundary conditions $\kappa u_\nu + (1 - \kappa)u = 0$, $\kappa \in (0, 1)$, see [196].

The idea of the construction of the function f for Theorem 19.15 is to start with a typical blow-up function satisfying (17.4), like $f(u) = cu^p$, with $c > 0$, $p > 1$, and then to modify it in an infinite number of intervals $I_k = (a_k, b_k)$ with $a_k < b_k < a_{k+1}$, $a_k \to \infty$. The modified function will be small enough in subintervals of I_k in order to provide us with suitable supersolutions of (19.29) but it will still satisfy condition (17.4).

Lemma 19.16. *Let $\{a_k\}$ be an increasing sequence, $a_1 \ge 1$, $\lim_{k \to \infty} a_k = \infty$. Then there are a C^∞ function $f : [0, \infty) \to [0, \infty)$ with $f(u) > 0$ for $u > 0$, and a sequence $\{b_k\}$ such that*

$$a_k < b_k < a_{k+1}, \tag{19.30}$$

$$\int_1^\infty \frac{du}{f(u)} < \infty, \tag{19.31}$$

$$\int_{a_k}^{b_k} \frac{du}{\sqrt{F(b_k) - F(u)}} \ge k, \tag{19.32}$$

for $k = 1, 2, \ldots$, where $F' = f$.

Proof. Take any C^1-function $g : [0, \infty) \to [0, \infty)$, with $g(u) > 0$ for $u > 0$, such that

$$\int_1^\infty \frac{ds}{g(s)} < \infty, \qquad g(s) \ge 1 \quad \text{for } s \ge 1.$$

Choose also a positive sequence $\{\beta_k\}$ such that

$$\sum_k \beta_k < \infty, \quad \beta_k < k^2, \quad 2\beta_k^2 g(a_k)k^{-2} < a_{k+1} - a_k,$$

and define

$$\gamma_k := 1 - \beta_k k^{-2} > 0, \quad b_k := a_k + \beta_k^2 g(a_k)k^{-2} < a_{k+1}.$$

We will also choose sequences $\{c_k\}$ and $\{d_k\}$ ($a_k < b_k < c_k < a_{k+1}$, $d_k > 0$) specified later. Then, we construct an auxiliary function $\tilde{g}$ by modifying the function g on the intervals on $[a_k, b_k]$ and $[b_k, c_k]$ in the following way (see Figure 8)

$$\tilde{g}(u) = \begin{cases} d_k + \dfrac{g(a_k) - d_k}{(b_k - a_k)^{\gamma_k}}(b_k - u)^{\gamma_k} & \text{for } a_k \le u \le b_k, \\[4mm] d_k + \dfrac{g(c_k) - d_k}{c_k - b_k}(u - b_k) & \text{for } b_k \le u \le c_k. \end{cases}$$

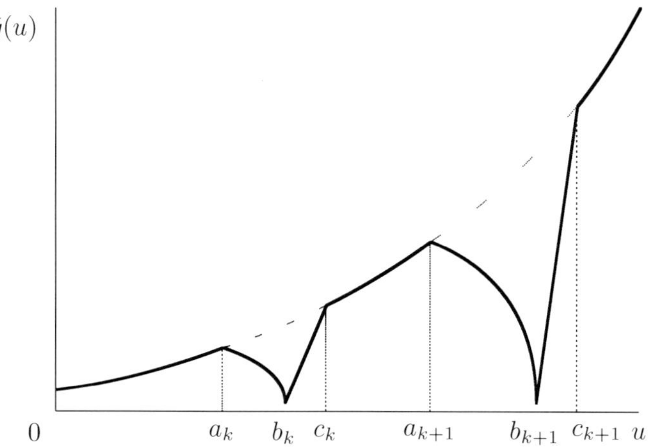

Figure 8: Graph of $\tilde{g}$.

Set also

$$\widetilde{G}(u) = d_k(u - b_k) - \frac{g(a_k) - d_k}{(\gamma_k + 1)(b_k - a_k)^{\gamma_k}}(b_k - u)^{\gamma_k + 1}$$

on the interval $[a_k, b_k]$. Then $\left(\widetilde{G}\right)' = \tilde{g}$ and

$$\int_{a_k}^{b_k} \frac{du}{\sqrt{\widetilde{G}(b_k) - \widetilde{G}(u)}}$$

$$= \int_{a_k}^{b_k} \left(d_k(b_k - u) + \frac{g(a_k) - d_k}{(\gamma_k + 1)(b_k - a_k)^{\gamma_k}}(b_k - u)^{\gamma_k + 1} \right)^{-1/2} du$$

$$\to \int_{a_k}^{b_k} \left(\frac{g(a_k)}{(\gamma_k + 1)(b_k - a_k)^{\gamma_k}}(b_k - u)^{\gamma_k + 1} \right)^{-1/2} du$$

as $d_k \to 0$. Thus we obtain

$$\lim_{d_k \to 0} \int_{a_k}^{b_k} \frac{du}{\sqrt{\widetilde{G}(b_k) - \widetilde{G}(u)}} = \frac{2}{1 - \gamma_k} \sqrt{\frac{(\gamma_k + 1)(b_k - a_k)}{g(a_k)}}.$$

We choose $d_k \in (0, 1/2)$ small enough so that

$$\int_{a_k}^{b_k} \frac{du}{\sqrt{\widetilde{G}(b_k) - \widetilde{G}(u)}} \geq \frac{1}{1 - \gamma_k} \sqrt{\frac{b_k - a_k}{g(a_k)}} = k. \tag{19.33}$$

Using $g(s) \geq 1 > 2d_k > 0$ for $s \geq 1$ we obtain

$$\int_{a_k}^{b_k} \frac{du}{\widetilde{g}(u)} = \int_{a_k}^{b_k} \left(d_k + \frac{g(a_k) - d_k}{(b_k - a_k)^{\gamma_k}} (b_k - u)^{\gamma_k} \right)^{-1} du$$

$$\leq \int_{a_k}^{b_k} \left(\frac{g(a_k) - d_k}{(b_k - a_k)^{\gamma_k}} (b_k - u)^{\gamma_k} \right)^{-1} du$$

$$= \frac{b_k - a_k}{(1 - \gamma_k)(g(a_k) - d_k)} = \frac{g(a_k)\beta_k}{g(a_k) - d_k} \leq 2\beta_k,$$

and

$$\int_{b_k}^{c_k} \frac{du}{\widetilde{g}(u)} = \int_{b_k}^{c_k} \left(d_k + \frac{g(c_k) - d_k}{c_k - b_k} (u - b_k) \right)^{-1} du$$

$$= \frac{c_k - b_k}{g(c_k) - d_k} \log \frac{g(c_k)}{d_k} \leq 2(c_k - b_k) \log \frac{g(c_k)}{d_k} \leq \beta_k$$

provided $c_k \in (b_k, a_{k+1})$ is sufficiently close to b_k. The above estimates imply

$$\int_{a_k}^{c_k} \frac{du}{\widetilde{g}(u)} \leq 3\beta_k.$$

This inequality and (19.33) guarantee that (19.31) and (19.32) are satisfied for $\widetilde{g}$. Take a C^∞-function f such that

$$\frac{1}{2}\widetilde{g}(u) \leq f(u) \leq \widetilde{g}(u). \tag{19.34}$$

We can easily check (19.31). Since

$$\int_u^{b_k} f(s)ds = F(b_k) - F(u),$$

we have

$$F(b_k) - F(u) \leq \widetilde{G}(b_k) - \widetilde{G}(u),$$

by integrating the second inequality in (19.34) over $[b_k, u]$. This guarantees that f also satisfies (19.32).　□

The existence of supersolutions immediately follows from the previous lemma.

Lemma 19.17. *Let f be as in Lemma 19.16 and $d, L > 0$. Then for sufficiently large k there is a solution u_k of*

$$d(u_k)_{xx} + f(u_k) = 0 \quad for \ -L < x < L, \tag{19.35}$$

$$(u_k)_x(0) = 0, \qquad u_k(x) \geq a_k \quad for \ -L < x < L. \tag{19.36}$$

Proof. Since the solution of the initial value problem

$$d(u_k)_{xx} + f(u_k) = 0,$$
$$(u_k)_x(0) = 0, \quad u_k(0) = b_k,$$

is given by

$$\int_{u_k(x)}^{b_k} \frac{du}{\sqrt{F(b_k) - F(u)}} = \sqrt{\frac{2}{d}} |x|,$$

the assertion follows from (19.32). $\square$

Proof of Theorem 19.15. Since Ω is bounded, we may choose $L > 0$ such that

$$\{x_1 \mid x = (x_1, \hat{x}) \in \Omega\} \subset [-L, L].$$

Let $u_0 \in L^\infty(\Omega)$, $u_0 \geq 0$, and let u be the solution of (19.29). For large enough positive integer k, the function u_k defined in Lemma 19.17 becomes a supersolution of (19.29) and we have

$$u_0(x) < a_k \leq u_k(x_1), \qquad x \in \Omega.$$

Since there is no problem in comparing the data on the lateral boundary, the comparison principle thus implies $u(x, t) \leq u_k(x_1)$ for $t \in (0, T_{\max}(u_0))$, hence $T_{\max}(u_0) = \infty$. $\square$

20. Global existence for the Cauchy problem

20.1. Small data global solutions

As announced in Section 18 (cf. Theorem 18.1(ii)), we show that, when $p > p_F$, small positive initial data yield global solutions. A simple example is provided by data dominated by a small multiple of a Gaussian, in which case the solution remains controlled by the heat kernel. In all this section we use the notation $(G_t)_{t>0}$ set in (48.5).

Theorem 20.1. *Consider problem* (18.1) *with* $p > p_F$, $u_0 \in L^\infty(\mathbb{R}^n)$, *and let* $\gamma > 0$. *There exists* $\varepsilon = \varepsilon(\gamma) > 0$ *such that, if*

$$0 \leq u_0(x) \leq \varepsilon G_\gamma(x), \quad x \in \mathbb{R}^n, \tag{20.1}$$

then $T_{\max}(u_0) = \infty$ *and* u *satisfies*

$$u(x, t) \leq C G_{t+\gamma}(x), \quad x \in \mathbb{R}^n, \ t > 0 \tag{20.2}$$

for some $C = C(\gamma) > 0$.

Theorem 20.1 is due to [220], where it was obtained by a contraction mapping argument. Here we shall derive Theorem 20.1 as a consequence of a more general criterion on u_0 for global existence, due to [530].

Theorem 20.2. *Consider problem (18.1) with $p > p_F$. Assume that $0 \le u_0 \in L^\infty(\mathbb{R}^n)$ satisfies*

$$\int_0^\infty \|e^{-sA}u_0\|_\infty^{p-1}\, ds < 1/(p-1). \tag{20.3}$$

Then $T_{\max}(u_0) = \infty$ and u behaves like the solution of the linear part of the equation, up to multiplicative constants, i.e.:

$$\big(e^{-tA}u_0\big)(x) \le u(x,t) \le C\big(e^{-tA}u_0\big)(x), \quad x \in \mathbb{R}^n,\ t > 0, \tag{20.4}$$

for some $C > 1$ (depending on u_0).

Remarks 20.3. (a) Inequality (20.2) corresponds in a sense to the minimal growth in time and space for positive solutions. Indeed, any positive solution u of (18.1) satisfies

$$u(x,t+\tau) \ge cG_{t+\alpha}(x), \quad x \in \mathbb{R}^n,\ t > 0$$

for some $\tau, \alpha, c > 0$ (this follows from the argument preceding formula (18.12)).

(b) Since

$$\|e^{-tA}u_0\|_\infty \ge ct^{-n/2}, \quad t \to \infty \tag{20.5}$$

for any nontrivial $u_0 \ge 0$, it follows that condition (20.3) cannot be satisfied for $p \le p_F$.

(c) A different smallness condition on u_0 ensuring global existence appears in Corollary 20.20 below.

(d) The constant C in (20.4) can be explicitly computed from the proof below. In particular C converges to 1 as the LHS of (20.3) goes to 0. $\square$

Proof of Theorem 20.2. We look for a supersolution of the form

$$\overline{u}(x,t) = h(t)\big(e^{-tA}u_0\big)(x), \quad x \in \mathbb{R}^n,\ t > 0,$$

where

$$h(t) = \left(1 - (p-1)\int_0^t \|e^{-sA}u_0\|_\infty^{p-1}\, ds\right)^{-1/(p-1)}.$$

Since

$$h'(t) = \|e^{-tA}u_0\|_\infty^{p-1}\left(1 - (p-1)\int_0^t \|e^{-sA}u_0\|_\infty^{p-1}\, ds\right)^{-1/(p-1)-1}$$

$$= \|e^{-tA}u_0\|_\infty^{p-1}h^p(t),$$

it follows that

$$\overline{u}_t = h(t)\big(e^{-tA}u_0\big)_t + h'(t)e^{-tA}u_0 = \Delta\overline{u} + \|e^{-tA}u_0\|_\infty^{p-1}h^p(t)e^{-tA}u_0 \ge \Delta\overline{u} + \overline{u}^p.$$

Since $\overline{u}(x,0) = u_0(x)$, we infer from the comparison principle that

$$0 \leq u(x,t) \leq \overline{u}(x,t), \quad x \in \mathbb{R}^n,\ t < T_{\max}(u_0).$$

The conclusion follows. $\square$

Proof of Theorem 20.1. By (48.6) we have $e^{-tA}G_\gamma = G_t * G_\gamma = G_{t+\gamma}$. Since $\|G_{t+\gamma}\|_\infty^{p-1} = (4\pi(t+\gamma))^{-n(p-1)/2}$ and $n(p-1)/2 > 1$, we deduce that (20.3) is satisfied with strict inequality for $\varepsilon > 0$ small. The conclusion then follows from Theorem 20.2. $\square$

Remarks 20.4. (i) Global existence under assumption (20.1) can be shown by a simpler comparison argument, by looking for a supersolution of the form $v(x,t) = \eta t^\alpha G(x,t)$, where $\alpha, \eta > 0$. Using $\partial_t G - \Delta G = 0$, we obtain

$$v_t - \Delta v - v^p = \eta\alpha t^{\alpha-1}G - \eta^p t^{\alpha p}G^p$$
$$= \eta t^{\alpha-1}\left[\alpha - \eta^{p-1}t^{1+\alpha(p-1)-n(p-1)/2}e^{-(p-1)|x|^2/(4t)}\right]G \geq 0,$$

provided we choose $\alpha = (n/2) - 1/(p-1) > 0$ and $\eta = \alpha^{1/(p-1)}$. It then suffices to compare u with $v(x,t+\gamma)$. However, this argument does not yield estimate (20.2) nor the sharp decay rate in $t^{-n/2}$.

(ii) Let $\Omega \subset \mathbb{R}^n$ be an arbitrary domain and e^{-tA} denote the Dirichlet heat semigroup in Ω. Then, for any $p > 1$ and u_0 satisfying condition (20.3) (with $0 \leq u_0 \in L^\infty(\Omega)$, say), problem (15.1) has a unique global nonnegative (mild) solution and estimate (20.4) is true for $x \in \Omega$ and $t > 0$.

Indeed, the local in time solution u is constructed by the Banach fixed point theorem as a limit of iterations $u_{k+1} = \Phi_{u_0}(u_k)$, $u_1(t) \equiv 0$ (cf. (15.12)). But one easily shows that the function $\overline{u}(t)$ in the proof of Theorem 20.2 satisfies $\overline{u} \geq \Phi_{u_0}(\overline{u})$. By induction, it follows that $\overline{u} \geq u_k$.

(iii) If $p \geq p_S$, then problem (18.1) possesses positive stationary solutions (see Theorem 9.1). If $p_F < p$ and $p(n-4) < n$, then the existence of global positive solutions of (18.1) with exponentially decaying initial data also follows from Example 51.24 (the zero solution of (18.14) is exponentially stable), cf. Proposition 20.13 and Remark 20.14(ii) below. $\square$

When $n \geq 3$ and $p > p_{sg}$, we have a simple global existence criterion, for solutions starting below the singular steady state (cf. [232, Theorem 10.4], where a more general result is proved).

Theorem 20.5. *Consider problem* (18.1) *with* $n \geq 3$, $p > p_{sg}$, *and* $u_0 \in L^\infty(\mathbb{R}^n)$. *Assume that* $|u_0| \leq u_*$ *in* $\mathbb{R}^n \setminus \{0\}$, *where* $u_*(x) := U_*(|x|)$ *is defined in* (3.9). *Then* $T_{\max}(u_0) = \infty$.

Proof. The proof is based on the strong maximum principle, along with a space-shift argument. Assume for contradiction that $T := T_{\max}(u_0) < \infty$.

We first claim that

$$u(x,t) \leq u_*(x), \qquad x \neq 0, \ 0 < t < T. \tag{20.6}$$

Fix $0 < \tau < T$. Since u is bounded in $\mathbb{R}^n \times (0, \tau)$, there exists $\varepsilon > 0$ such that $u \leq u_*$ in $\{0 < |x| \leq \varepsilon\} \times (0, \tau)$. By applying the comparison principle in the domain $\{|x| > \varepsilon\}$, it follows that $u \leq u_*$ in $(\mathbb{R}^n \setminus \{0\}) \times (0, \tau)$, hence (20.6). In particular, by parabolic estimates, u extends to a continuous function in $(\mathbb{R}^n \setminus \{0\}) \times (0, T]$.

Now fix $0 < t_0 < T$. There exist $a, \varepsilon > 0$ such that

$$u(x,t_0) \leq u_*(x) - \varepsilon, \qquad 0 < |x| \leq 3a. \tag{20.7}$$

As a consequence of (20.6), (20.7) and of the strong maximum principle, applied in the domain $\{a < |x| < 3a\}$, we deduce that

$$u(x,t) \leq u_*(x) - \eta, \qquad |x| = 2a, \ t_0 \leq t \leq T \tag{20.8}$$

for some $\eta > 0$. By (20.7), (20.8) and continuity, one can find $b \in \mathbb{R}^n$, $0 < |b| < a$, such that $v(x,t) := u(x+b,t)$ satisfies

$$v(x,t_0) < u_*(x), \qquad 0 < |x| \leq 3a$$

and

$$v(x,t) < u_*(x), \qquad |x| = 2a, \ t_0 \leq t \leq T.$$

Since v is a solution, arguing as for (20.6), we deduce that

$$v(x,t) \leq u_*(x), \qquad 0 < |x| \leq 2a, \ t_0 \leq t < T,$$

hence

$$u(x,t) \leq u_*(x-b), \qquad 0 < |x| \leq a, \ t_0 \leq t < T. \tag{20.9}$$

Finally, (20.6) together with (20.9) imply $\sup_{\mathbb{R}^n \times (t_0,T)} u < \infty$. Applying the above argument to $-u$ we obtain $\sup_{\mathbb{R}^n \times (t_0,T)} |u| < \infty$ which contradicts $T < \infty$. $\square$

We have seen in Theorem 17.12 that if the nonnegative initial data decays slower than $|x|^{-2/(p-1)}$, then the solution of (18.1) blows up in finite time. We shall now show that if the initial data decays faster (and satisfies a global smallness condition), then the solution exists for all times. Moreover, we shall show that these global solutions exhibit a typical parabolic feature: they have a temporal decay whose exponent is precisely half that of the spatial decay of the initial data, with an upper limit of $n/2$. The following result is due to [324].

Theorem 20.6. *Consider problem* (18.1) *with* $p > p_F$, *and let* $k \geq 2/(p-1)$. *There exists* $c = c(n, p, k) > 0$ *such that, if* $u_0 \in L^\infty(\mathbb{R}^n)$ *satisfies*

$$0 \leq u_0(x) \leq c(1 + |x|)^{-k}, \quad x \in \mathbb{R}^n, \tag{20.10}$$

then $T_{\max}(u_0) = \infty$ *and we have, for all* $t \geq 1$,

$$\|u(t)\|_\infty \leq \begin{cases} t^{-n/2} & \text{if } k > n, \\ t^{-n/2} \log t & \text{if } k = n, \\ t^{-k/2} & \text{if } 2/(p-1) \leq k < n. \end{cases}$$

If moreover $k > 2/(p-1)$, *then* (20.4) *is satisfied.*

Remarks 20.7. (i) The decay rates in Theorem 20.6 are sharp for the choice $u_0(x) = c(1 + |x|)^{-k}$. This follows from $u(x, t) \geq (e^{-tA} u_0)(x)$ and the lower estimates in Lemma 20.8.

(ii) For any $p > 1$, problem (18.1) admits some nontrivial global classical solutions. Of course, they have to change sign if $p \leq p_F$. For instance, for any $p > 1$, there exist self-similar solutions of the form $u(x, t) = (t + 1)^{-1/(p-1)} w(x/\sqrt{t+1})$, with $w \in L^\infty(\mathbb{R}^n)$ (see [269, Theorem 5]).

(iii) All the solutions constructed in Theorem 20.6 decay at least like $t^{-1/(p-1)}$. We shall see in Section 26 (see Theorem 26.9) that if p is less than a suitable exponent, then this is actually true for *any* nonnegative global classical solution of (18.1), cf. also Theorem 28.10.

On the other hand, if $p \geq p_{JL}$ and $u_*(x) = c_p |x|^{-2/(p-1)}$ denotes the singular steady state (see (3.9)), then there are bounded positive initial data u_0 satisfying $u_0 < u_*$ such that the corresponding solutions exist globally, decay to zero, but

$$\lim_{t \to \infty} t^{1/(p-1)} \|u(t)\|_\infty = \infty,$$

see [261]. In addition, if $p > p_{JL}$ and $k \in (0, 2/(p-1))$, then one can find $\ell > 2/(p-1)$ such that for any bounded nonnegative continuous radial u_0 satisfying $u_0 \leq u_*$ and $u_0(x) - u_*(x) \sim |x|^{-\ell}$ for large $|x|$, there exist $C_1, C_2 > 0$ such that

$$C_1 t^{-k/2} \leq \|u(t)\|_\infty \leq C_2 t^{-k/2}, \qquad t \geq 1, \tag{20.11}$$

see Section 29. Recent results (see [306]) indicate that similar behavior of suitable positive solutions can also be expected for $p = p_S$ provided $n < 6$ (if $n = 4$, then the decay rate $t^{-k/2}$ should be replaced by $t^{-k/2} \log t$).

(iv) If $p < p_S$, $u_0 \geq 0$ has exponential decay (more precisely, $u_0(x)e^{|x|^2/8} \in L^2(\mathbb{R}^n)$) and the corresponding solution u exists globally, then (20.11) is true with either $k = n$ or $k = 2/(p-1)$ (and both possibilities occur). This follows from Theorem 28.9 below. $\square$

For the proof of Theorem 20.6 we need the following lemma concerning the linear heat equation. Here and in the rest of this subsection, $f(t) \sim g(t)$ means that

$$C_1 g(t) \leq f(t) \leq C_2 g(t) \quad \text{for some constants } C_1, C_2 > 0.$$

Lemma 20.8. *Let $\phi(x) = (1 + |x|)^{-k}$ with $k > 0$. There holds*

$$\|e^{-tA}\phi\|_\infty \sim \begin{cases} t^{-n/2} & \text{if } k > n, \\ t^{-n/2}\log t & \text{if } k = n, \\ t^{-k/2} & \text{if } k < n, \end{cases}$$

for $t \geq 1$.

Proof. Due $(1 + |x|)^{-k} \leq (1 + |x|^2)^{-k/2} \leq C(1 + |x|)^{-k}$, we may replace ϕ by $\phi(x) := (1 + |x|^2)^{-k/2}$. For each $t > 0$, the function $e^{-tA}\phi(x)$ is radially symmetric in x and nonincreasing in $r = |x|$ (see Proposition 52.17). Consequently, we have

$$\|e^{-tA}\phi\|_\infty = (e^{-tA}\phi)(0) = \int_{|y|\leq 1} (4\pi t)^{-n/2} e^{-|y|^2/4t}\phi(y)\, dy$$

$$+ \int_{|y|>1} (4\pi t)^{-n/2} e^{-|y|^2/4t}\phi(y)\, dy =: I_1(t) + I_2(t).$$

First it is clear that $I_1(t) \sim t^{-n/2}$ for $t \geq 1$. If $k > n$, the conclusion follows from $I_2(t) \leq t^{-n/2}\int_{|y|>1} |y|^{-k}\, dy = Ct^{-n/2}$. Now assume $k \leq n$ and observe that

$$I_2(t) = \pi^{-n/2}\int_{|z|>1/(2\sqrt{t})} e^{-|z|^2}(1 + 4|z|^2 t)^{-k/2}\, dz$$

$$\sim t^{-k/2}\int_{|z|>1/(2\sqrt{t})} e^{-|z|^2}|z|^{-k}\, dz,$$

for $t \geq 1$. If $k < n$, we simply use $\int_{\mathbb{R}^n} e^{-|z|^2}|z|^{-k}\, dz < \infty$, hence $I_2(t) \sim t^{-k/2}$ for $t \geq 1$. If $k = n$, we use $\int_{|z|>R} e^{-|z|^2}|z|^{-n}\, dz = \int_R^\infty e^{-r^2}r^{-1}\, dr \sim \log(1/R)$ for $R \in (0, 1/2]$, hence $I_2(t) \sim t^{-n/2}\log t$ for $t \geq 1$. The lemma follows. $\square$

Proof of Theorem 20.6. In view of the comparison principle, it is sufficient to prove the theorem when $u_0(x) = c(1 + |x|)^{-k}$ for some $c > 0$.

Let us first consider the case $k > 2/(p - 1)$. Since $\min(k, n)(p - 1)/2 > 1$, it follows from Lemma 20.8 that $\int_0^\infty \|e^{-sA}u_0\|_\infty^{p-1}\, ds < 1/(p-1)$ for $c = c(n, p, k) > 0$ small enough. The result is then a consequence of Theorem 20.2.

Let us turn to the case $k = 2/(p - 1)$. If $n \geq 3$ and $p > n/(n - 2)$, the result follows from the observation that the function $\bar{u}(x, t) = \varepsilon(1 + |x|^2 + \varepsilon t)^{-1/(p-1)}$ is a (self-similar) supersolution for $\varepsilon > 0$ sufficiently small (which can be checked by a simple computation). If $(n - 2)p \leq n$ (or in the general case $p > p_F$), the result is a consequence of Theorem 20.19(i) and (ii) below (applied with $u_0 = c|x|^{-2/(p-1)}$ and $v_0 = c\min(1, |x|^{-2/(p-1)})$) and of the comparison principle. $\square$

Similar results as in Theorem 20.6 can also be obtained for sign-changing solutions. Let us first prove two auxiliary lemmas concerning the linear heat equation on a half-line and in a cone in $\mathbb{R}^2$.

Lemma 20.9. *Let $k \in [1,2)$ and $\phi(x) = (1 - e^{-x})(1+x)^{-k}$ for $x \geq 0$. Let e^{-tA} denote the Dirichlet heat semigroup in $(0, \infty)$. Then*

$$\|e^{-tA}\phi\|_\infty \sim t^{-k/2}, \qquad t \geq 1.$$

Proof. We will use the formula

$$e^{-tA}\phi(x,t) = \frac{1}{\sqrt{4\pi t}} \int_0^\infty \left(1 - e^{-xy/t}\right) e^{-|x-y|^2/4t} \phi(y)\, dy$$

and estimates

$$\left(1 - e^{-xy/t}\right)\phi(y) \leq \min\left(1, \frac{xy}{t}\right) \frac{1}{(1+y)^k}, \qquad x, y, t > 0,$$

$$\left(1 - e^{-xy/t}\right) e^{-|x-y|^2/4t} \phi(y) \geq \frac{c}{(1+y)^k} \geq ct^{-k/2}, \qquad \text{for } y \in [x, 2x],\ x^2 = t \geq 1,$$

where $c > 0$ denotes a generic constant which may change from step to step.

If $x^2 = t \geq 1$, then the above estimates imply

$$e^{-tA}\phi(x,t) \geq \frac{1}{\sqrt{4\pi t}} \int_x^{2x} ct^{-k/2}\, dy = ct^{-k/2}.$$

On the other hand, if $t \geq x^2$, $t \geq 1$, then

$$e^{-tA}\phi(x,t) \leq \frac{1}{\sqrt{4\pi t}} \left(\int_0^x \frac{xy}{t} \frac{1}{(1+y)^k}\, dy + \int_x^\infty e^{-|x-y|^2/4t} \frac{x}{t} \frac{1}{y^{k-1}}\, dy \right)$$

$$\leq \frac{c}{t} \int_0^x \frac{1}{(1+y)^{k-1}}\, dy + c\frac{x^{2-k}}{t} \int_x^\infty \frac{1}{\sqrt{t}} e^{-|x-y|^2/4t}\, dy \leq ct^{-k/2}.$$

Finally, if $1 \leq t < x^2$, then

$$e^{-tA}\phi(x,t) \leq \frac{1}{\sqrt{4\pi t}} \left(\int_0^{x/2} e^{-x^2/16t} \frac{x}{t} \frac{1}{(1+y)^{k-1}}\, dy \right.$$

$$\left. + \int_{x/2}^\infty e^{-|x-y|^2/4t} \frac{c}{(1+x/2)^k}\, dy \right)$$

$$\leq c\left(\frac{x^2}{16t}\right)^{(3-k)/2} e^{-x^2/16t} t^{-k/2} + \frac{c}{x^k} \leq ct^{-k/2},$$

which concludes the proof. $\square$

In the following lemma we will use polar coordinates (r, φ) in $\mathbb{R}^2$.

Lemma 20.10. *Let $k \in \mathbb{N} \setminus \{0\}$, $\Omega = \Omega_k := \{(r, \varphi) : r > 0, \ \varphi \in (-\pi/2k, \pi/2k)\}$. Then there exists $u_0 \in L^\infty(\Omega)$, $u_0 \geq 0$, such that $\|e^{-tA}u_0\|_\infty \sim t^{-1-k/2}$ for $t \geq 1$.*

Sketch of proof (see [363] for details). Let $G(r, \rho; \varphi, \psi; t)$ denote the Dirichlet heat kernel in Ω and $w(r, \varphi, t) := G(r, 1; \varphi, 0; t + t_0)$, where $t_0 > 0$ is fixed. Then

$$w(r, \varphi, t)$$

$$= \frac{1}{4\pi(t + t_0)} \sum_{j=0}^{2k-1} (-1)^j \exp\left[-\frac{(r\cos\varphi - \cos j\pi/k)^2 + (r\sin\varphi - \sin j\pi/k)^2}{4(t + t_0)}\right].$$

Set $s(r, t) := w(r, 0, t)$. Then one can show that $\|w(\cdot, \cdot, t)\|_\infty = \sup_{r>0} s(r, t)$ and

$$s(r, t) = C_0 \frac{1}{t + t_0} \exp\left(-\frac{r^2 + 1}{4(t + t_0)}\right)\left(\frac{r}{t + t_0}\right)^k \left[1 + \frac{r}{t + t_0} R\left(\frac{r}{t + t_0}\right)\right],$$

where C_0 is a positive constant and R is bounded on bounded sets. Obviously,

$$s(\sqrt{t + t_0}, t) \geq ct^{-1-k/2} \quad \text{for} \quad t \geq 1.$$

On the other hand, one can also show that $\sup_r s(r, t)$ is attained at some $r_M(t)$ which satisfies $r_M(t) \leq C\sqrt{t + t_0}$, hence $s(r_M(t), t) \leq Ct^{-1-k/2}$. $\quad\square$

Theorem 20.11. *Let $n \geq 3$, $p > 1$ and $\alpha > 1/(p-1)$. Then there exists $u_0 \in L^\infty(\mathbb{R}^n)$ such that the solution u of (18.1) is global and $\|u(t)\|_\infty \sim t^{-\alpha}$ for $t \geq 1$.*

Proof. If $\alpha \leq n/2$, then $p > p_F$ and the assertion follows from Theorem 20.6.

Let $n/2 < \alpha < 2$. Then $n = 3$. Set $\gamma := \alpha/3$ and $\phi(x) := \prod_{i=1}^3 \psi(x_i)$, where

$$\psi(r) := \text{sign}(r)(1 - e^{-|r|})(1 + |r|)^{-2\gamma}. \tag{20.12}$$

Let $-A_m$ denote the Laplacian in $\mathbb{R}^m$. Then $e^{-tA_3}\phi(x, t) = \prod_{i=1}^3 e^{-tA_1}\psi(x_i, t)$, hence $\|e^{-tA_3}\phi\|_\infty \sim t^{-\alpha}$ for $t \geq 1$ due to Lemma 20.9 and the oddness of $e^{-tA_1}\psi(\cdot, t)$. Now choosing $u_0 = \varepsilon\phi$, $\varepsilon > 0$ small, we obtain the result from Remark 20.4(ii) used with $\Omega = (0, \infty)^3$.

Finally, let $\alpha \geq 2$. Fix $k \in \mathbb{N}$ such that $\gamma := \alpha - 1 - k/2 \in [1/2, 1)$ and consider the cone Ω_k and the function $w(t) := e^{-tA}u_0$ from Lemma 20.10. Extend the function w to $\mathbb{R}^2 \times [0, \infty)$ by $w(r, \varphi, t) = -w(r, \pi/k - \varphi, t)$ for $\varphi \in (\pi/2k, 3\pi/2k)$ and $w(r, \varphi + 2j\pi/k, t) = w(r, \varphi, t)$, $j = 1, 2, \ldots, k - 1$. Then $w = w(x_1, x_2, t)$ is a solution of the heat equation in $\mathbb{R}^2$ and $\|w(t)\|_\infty \sim t^{-1-k/2}$ for $t \geq 1$. Set $\phi(x) = w(x_1, x_2, 0)\psi(x_3)$, where ψ is defined by (20.12). Then, similarly as above, $e^{-tA_n}\phi(x, t) = w(x_1, x_2, t)e^{-tA_1}\psi(x_3, t)$, hence $\|e^{-tA_3}\phi\|_\infty \sim t^{-\alpha}$ for $t \geq 1$. Now choosing $u_0 = \varepsilon\phi$, $\varepsilon > 0$ small, we obtain the result from Remark 20.3(d) used with $\Omega = \Omega_k \times (0, \infty) \times \mathbb{R}^{n-3}$. $\quad\square$

Remark 20.12. (i) **Solutions with exponential time decay.** In addition to the solutions with polynomial time decay in Theorem 20.11 one can also easily construct sign-changing global solutions with exponential time decay. In fact, let $1 < p < p_S$, $\lambda > 0$, $A = \pi/2\sqrt{\lambda}$ and let w be the positive solution of the problem $w'' + w^p = 0$ in $(-A, A)$, $w(-A) = w(A) = 0$. Choose $\alpha \in (0, 1)$ and set

$$u_0(x) := \begin{cases} \alpha w(x - 4kA) & \text{if } x \in [(4k - 1)A, (4k + 1)A), \\ -\alpha w(x - (4k + 2)A) & \text{if } x \in [(4k + 1)A, (4k + 3)A), \end{cases}$$

where $k \in \mathbb{Z}$. Then the solution u of (18.1) with $n = 1$ satisfies $\|u(t)\|_\infty \sim e^{-\lambda t}$. This follows from Theorem 51.19 and Theorem 19.9(iv).

(ii) **Decay of global solutions.** Let $1 < p < p_S$. Assume $u_0 \geq 0$ and either u_0 is radial or

$$p < p_B := \begin{cases} \infty & \text{if } n = 1, \\ n(n + 2)/(n - 1)^2 & \text{if } n > 1. \end{cases}$$

If the solution u of (18.1) is global, then $\|u(t)\|_\infty \to 0$ as $t \to \infty$ (see Theorem 26.9). This result is also true for all nonnegative data lying in the energy space $\mathcal{E}$ (see [478] and Example 51.27), but it fails for sign-changing radial solutions (consider the choice $\alpha = 1$ in (i)).

(iii) Notice that the radially symmetric initial data of the form $c(1 + |x|)^{-k}$, $k \geq 2/(p - 1)$, appearing in Theorem 20.6 belong to the energy space $\mathcal{E} := \{u \in L^{p+1}(\mathbb{R}^n) : \nabla u \in L^2(\mathbb{R}^n)\}$ if $p < p_S$ (see Example 51.27 for the well-posedness in this space). $\square$

20.2. Global solutions with exponential spatial decay

We have seen in Theorem 20.11 and Remark 20.12 that there is a wide range of possibilities for the decay of global solutions of the Cauchy problem (18.1). In this subsection we show that the situation is much simpler if we restrict ourselves to the initial data with exponential spatial decay. More precisely, we will consider initial data in the space H_g^1 (see (18.16)) and exponents $p \in (1, p_S)$. We will use the rescaled solutions v (see (18.13)) and operator L (see (18.15)). As above, let $\lambda_k^L = (n + k - 1)/2$ denote the eigenvalues of L. In addition, we denote $\lambda_0^L := 1/(p - 1)$.

Proposition 20.13. (i) *Let $1 < p < p_S$, $p \notin \{1 + 1/\lambda_k^L : k = 1, 2, \dots\}$ and let k_0 be the minimal $k \in \{1, 2, \dots\}$ with the property $p > 1 + 1/\lambda_{k_0}^L$. If u is a global solution of (18.1) with $u_0 \in H_g^1 \setminus \{0\}$ and $t_0 > 0$, then there exist $C_1, C_2 > 0$ and $k \in \{0\} \cup \{k_0, k_0 + 1, k_0 + 2, \dots\}$ such that*

$$C_1 t^{-\lambda_k^L} \leq \|u(t)\|_\infty \leq C_2 t^{-\lambda_k^L}, \quad t \geq t_0. \tag{20.13}$$

Conversely, if $k = 0$ or $k \geq k_0$, then there exists $u_0 \in H_g^1$ such that the corresponding solution of (18.1) is global and satisfies (20.13).

(ii) *Let $p_F < p < p_S$ and u be a global solution of (18.1) with $u_0 \in H_g^1$ satisfying (20.13) with $k = 1$. Set $q := \min((n+1)/2, (pn-2)/2) > n/2$. Then there exists $M \neq 0$ such that*

$$\| u(t) - M(t+1)^{-n/2} e^{-|x|^2/4(t+1)} \|_\infty \leq C(t+1)^{-q}, \qquad t \geq 0.$$

Remark 20.14. (i) If $p = 1 + 1/\lambda_{k_0}^L$ for some $k_0 \geq 1$, then the proof of Proposition 20.13(i) guarantees the following: Let u be a global solution of (18.1) with $u_0 \in H_g^1 \setminus \{0\}$, $t_0 > 0$. Then $\|u(t)\|_\infty \leq Ct^{-\lambda_0^L}$ for $t \geq t_0$. If there exist $C > 0$ and $\lambda > \lambda_0^L$ such that $\|u(t)\|_\infty \leq Ct^{-\lambda}$ for $t \geq t_0$, then there exist $C_1, C_2 > 0$ and $k > k_0$ such that (20.13) is true. Conversely, if $k = 0$ or $k > k_0$, then there exists $u_0 \in H_g^1$ such that the corresponding solution of (18.1) is global and satisfies (20.13).

(ii) Some of the results in Proposition 20.13(i) concerning the decay (20.13) with $k > 0$ can also be obtained for supercritical p, $p(n-4) < n$ (cf. Example 51.24).

(iii) Sufficient conditions for the initial data u_0 to satisfy the assumptions of Proposition 20.13(ii) can be found in Theorem 28.9. For related results see also the following subsection and [288], for example. $\square$

Proof of Proposition 20.13. (i) Let u be a global solution of (18.1) with $u_0 \in H_g^1 \setminus \{0\}$. Then the rescaled solution v (see (18.13)) is a global solution of (18.14). If $\|v(s)\|_{H_g^1} \to 0$ as $s \to \infty$, then (20.13) is true with some $k \geq k_0$ due to Example 51.24 (see (51.72)). If $\|v(s)\|_{H_g^1} \nrightarrow 0$, then Lemma 18.4(iii) and Example 51.24 show that $\|v(s)\|_\infty \leq C_2$ for all $s \geq s_0$ and $s_0 > 0$.

Assume $\liminf_{s \to \infty} \|v(s)\|_\infty = 0$. Then the same estimates as at the end of Example 51.24 guarantee $\liminf_{s \to \infty} \|v(s)\|_{H_g^1} = 0$. Consequently, choosing $\delta > 0$ small, there exist $s_j \to \infty$ such that $\|v(s_j)\|_{H_g^1} = \delta$. Using the compactness of the semiflow we may assume $v(s_j) \to w$ in H_g^1, where w belongs to the ω-limit set of v, hence w is an equilibrium of (18.14), $\|w\|_{H_g^1} = \delta$. However, the zero equilibrium is isolated due to $p \notin \{1 + 1/\lambda_k^L : k = 1, 2, \dots\}$ which yields a contradiction. Hence $C_1 \leq \|v(s)\|_\infty \leq C_2$ for $s \geq s_0$ which implies (20.13) with $k = 0$.

To prove the converse statement, assume first $k = 0$. By [174] there exist a sequence of nontrivial stationary solutions v_j, $j = 1, 2, \dots$, of problem (18.14). The corresponding rescaled solutions u_j satisfy (20.13) with $k = 0$.

If $k \geq k_0$, then the existence of $u_0 \in H_g^1$ such that the solution u satisfies (20.13) follows from Example 51.24.

(ii) The proof is a direct consequence of assertion (ii) in Example 51.24. $\square$

20.3. Asymptotic profiles for small data solutions

More information on the asymptotic behavior of positive solutions than in Subsection 20.1 can be gained if one considers suitably small initial data in L^1.

Theorem 20.15. *Consider problem* (18.1) *with* $p > p_F$. *Assume that* $0 \leq u_0 \in L^\infty \cap L^1(\mathbb{R}^n)$ *satisfies*

$$\|u_0\|_1 \|u_0\|_\infty^{(n(p-1)/2)-1} \leq c(n,p), \tag{20.14}$$

with $c(n,p) > 0$ *sufficiently small. Then* $T_{\max}(u_0) = \infty$ *and* (20.4) *is satisfied. Moreover,* $u(t)$ *behaves like a multiple of the heat kernel. Namely, the limit*

$$I_\infty := \lim_{t \to \infty} \|u(t)\|_1 \quad \text{exists and is finite,} \tag{20.15}$$

and there holds

$$\|u(t) - I_\infty G_t\|_1 \to 0, \quad t \to \infty. \tag{20.16}$$

Theorem 20.15 is a variant of a result of [302] (see also [145], [322]). We prove (20.15), as a consequence of Theorem 20.2. This proof is simpler than those in [302] (based on energy estimates) or in [145], [322] (based on the variation-of-constants formula). As for property (20.16), it will be a consequence of the following lemma from [86] (see also [322]) concerning the inhomogeneous heat equation.

Lemma 20.16. *Let* $u_0 \in L^1(\mathbb{R}^n)$, $f \in L^1(\mathbb{R}^n \times (0,\infty))$ *with* $u_0, f \geq 0$, $u_0 \not\equiv 0$ *and let* u *be given by*

$$u(t) = e^{-tA}u_0 + \int_0^t e^{-(t-s)A} f(s)\,ds, \quad t > 0.$$

Then $M := \lim_{t \to \infty} \|u(t)\|_1$ *exists in* $(0,\infty)$ *and we have*

$$\|u(t) - M\,G_t\|_1 \to 0, \quad t \to \infty.$$

Proof. By the variation-of-constants formula, we have

$$\|u(t) - e^{-(t-t_0)A}u(t_0)\|_1 = \int_{t_0}^t \|f(s)\|_1\,ds, \quad t \geq t_0 \geq 0.$$

Since $\|e^{-(t-t_0)A}u(t_0)\|_1 = \|u(t_0)\|_1$ we see that $\lim_{t \to \infty} \|u(t)\|_1$ exists and is finite. Since $u(t) \geq e^{-tA}u_0$ this limit is positive. Denoting $M(t) = \|u(t)\|_1$, it follows that

$$\|u(t) - M\,G_t\|_1 \leq \int_{t_0}^\infty \|f(s)\|_1\,ds + \left\|e^{-(t-t_0)A}u(t_0) - M(t_0)\,G_t\right\|_1 + |M(t_0) - M|.$$

Using

$$\left\| e^{-sA}\phi - \|\phi\|_1\, G_s \right\|_1 \to 0, \ s \to \infty, \quad 0 \le \phi \in L^1(\mathbb{R}^n)$$

(see Proposition 48.6 in Appendix B) and letting $t \to \infty$, we obtain

$$\limsup_{t\to\infty} \|u(t) - M\,G_t\|_1 \le \int_{t_0}^{\infty} \|f(s)\|_1\, ds + |M(t_0) - M|.$$

The lemma follows by letting $t_0 \to \infty$. $\quad\square$

Proof of Theorem 20.15. Assume (20.14) with $c = c(n,p) > 0$ small. Using the L^p-L^q-estimate (cf. Proposition 48.4(d)) and choosing $\tau = (\|u_0\|_1/\|u_0\|_\infty)^{2/n}$, we obtain

$$
\begin{aligned}
\int_0^\infty \|e^{-sA}u_0\|_\infty^{p-1}\, ds &\le \int_0^\tau \|u_0\|_\infty^{p-1}\, ds + \int_\tau^\infty \|u_0\|_1^{p-1} s^{-n(p-1)/2}\, ds \\
&= \tau\|u_0\|_\infty^{p-1} + C(n,p)\|u_0\|_1^{p-1}\tau^{1-n(p-1)/2} \\
&\le C(n,p)\|u_0\|_1^{2/n}\|u_0\|_\infty^{p-1-2/n} \le 1/2(p-1).
\end{aligned}
$$

By Theorem 20.2, we deduce that $T_{\max}(u_0) = \infty$ and that

$$u(t) \le C e^{-tA}u_0. \tag{20.17}$$

Applying the L^p-L^q-estimate again, we obtain

$$\|u(t)\|_p^p \le C\|e^{-tA}u_0\|_p^p \le C\min(1, t^{-n(p-1)/2}). \tag{20.18}$$

On the other hand, by Proposition 48.4(b) and the variation-of-constants formula, we have

$$\int_{\mathbb{R}^n} u(t)\, dx = \int_{\mathbb{R}^n} u_0\, dx + \int_0^t \|u(s)\|_p^p\, ds. \tag{20.19}$$

We deduce from (20.18), (20.19) and $n(p-1)/2 > 1$ that $\|u(t)\|_1$ is nondecreasing and bounded. The conclusion then follows from Lemma 20.16. $\quad\square$

Remark 20.17. Estimates similar to (20.16) are also true for other L^q-norms. In particular there holds

$$t^{n/2}\|u(t) - I_\infty G_t\|_\infty \to 0, \quad t \to \infty. \tag{20.20}$$

This is a consequence of [70, Theorem 4.1] and inequality (20.17). Estimates for all L^q-norms follow immediately by interpolating between (20.16) and (20.20). $\quad\square$

It follows from Theorems 17.12 and 20.6 that initial data which decay at the rate $|x|^{-2/(p-1)}$ constitute a borderline between blow-up and global existence. Our next results concern some particular classes of initial data with this asymptotic behavior which are especially interesting.

First it turns out that initial data which are *homogeneous of degree* $-2/(p-1)$ (and suitably small) give rise to (forward) self-similar solutions, cf. Remark 15.4(ii). Moreover, these solutions enjoy stability properties. For instance, if a (small) initial data coincides for large x with a homogeneous function of degree $-2/(p-1)$, then the solution is asymptotically self-similar. These results will be proved by semigroup techniques and suitable fixed point arguments, that are refinements of the methods introduced in Section 15.

We will also describe the global properties of the equation in a space which naturally arises in this connection, namely the critical L^q-space. Its special role as an *invariant space* can be explained as follows (cf. e.g. [112], and see also [115] for earlier references and for similar considerations concerning the Navier-Stokes and nonlinear Schrödinger equations). Consider the scaling transformations

$$\mathcal{S}_\lambda : u \mapsto u_\lambda(x,t) := \lambda^{2/(p-1)} u(\lambda x, \lambda^2 t)$$

for $\lambda > 0$. Observe that the equation in (18.1) is invariant under these transformations. On the other hand, for spatial functions $\phi = \phi(x)$, we have

$$\|\phi_\lambda\|_q = \lambda^{2/(p-1)-(n/q)} \|\phi\|_q, \quad 1 \le q \le \infty, \tag{20.21}$$

so that the only L^q-norm left invariant by the transformations $\mathcal{S}_\lambda$ is the critical norm, i.e. $q = q_c = n(p-1)/2$. Now assume that there exists q with the property that the solution of (18.1) is global whenever the initial data u_0 is small in L^q. If $q \ne q_c$, then, by (20.21) applied to $\phi = u_0$, global existence will hold for any $u_0 \in L^q$. But this is a contradiction to Theorem 17.1; hence $q = q_c$ is the only possible value with that property.

In accordance with these observations, we will indeed prove **global existence for small initial data in** L^{q_c}, provided that $q_c > 1$. Note that the critical exponent $p = p_F$ corresponds to the case when $q_c = 1$, and the requirement that $q_c > 1$ is thus consistent with the Fujita-type result Theorem 18.1. Furthermore, still using the techniques mentioned in the previous paragraph, we will establish the asymptotic stability of the zero solution in the space L^{q_c}. More generally, we will show that the above mentioned self-similar solutions are in a sense stable with respect to critical L^q-perturbations. Note, in turn, that the transformations $\mathcal{S}_\lambda$ also leave invariant the homogeneous functions of degree $-2/(p-1)$ (from which the self-similar solutions arise).

We shall use the following definition of mild solution of problem (18.1).

Definition 20.18. Let $u_0 \in L^1(\mathbb{R}^n) + L^\infty(\mathbb{R}^n)$. We say that u is (global) mild solution of (18.1) if $u \in L^\infty_{loc}((0,\infty), L^r(\mathbb{R}^n))$ for some $r \ge p$ and satisfies

$$u(t) = e^{-tA} u_0 + \int_0^t e^{-(t-s)A} |u|^{p-1} u(s)\, ds, \quad t > 0,$$

where for each $t > 0$ the integral is absolutely convergent in $L^r(\mathbb{R}^n)$. In particular, there holds $u(t) - e^{-tA}u_0 \to 0$ in $L^r(\mathbb{R}^n)$ as $t \to 0$. $\quad\square$

This definition is slightly different from that in Remark 15.4(iii). Note that the definition makes sense since $e^{-(t-s)A}|u|^{p-1}u(s) \in L^\infty_{loc}((0,t), L^r(\mathbb{R}^n))$, due to $r \geq p$ and the L^p-L^q-estimates. The following result is due to [457], [115] for assertion (i), [115], [475] for assertion (ii). Assertion (iii) for $u_0 = 0$ (i.e. Corollary 20.20) is from [481], improving on earlier results of [530], whereas the case $u_0 \neq 0$ seems new.

Theorem 20.19. *Let $p > p_F$, $\omega \in L^\infty(S^{n-1})$ and set*

$$u_0(x) := |x|^{-2/(p-1)}\omega(x/|x|), \quad x \in \mathbb{R}^n \setminus \{0\}. \tag{20.22}$$

There exists $\mu_0 = \mu_0(n,p) > 0$ such that, if $\|\omega\|_\infty \leq \mu_0$, then the following properties hold.

(i) Problem (18.1) admits a global mild solution u (in the sense of Definition 20.18). Moreover, u is self-similar, i.e. is of the form

$$u(x,t) = t^{-1/(p-1)}f(x/\sqrt{t}), \quad x \in \mathbb{R}^n, \ t > 0,$$

with $f(y) = u(y,1) \in L^\infty(\mathbb{R}^n)$, and u is a classical solution for $t > 0$. Furthermore, the solution u is stable in the sense indicated in parts (ii) and (iii) hereafter.

(ii) Let $v_0 = \varphi u_0$, where $\varphi \in L^\infty(\mathbb{R}^n)$ satisfies $\varphi = 1$ for $|x|$ large. Assume that $\|\omega(\cdot/|\cdot|)\varphi\|_\infty \leq \mu_0$. Then problem (18.1) with initial data v_0 admits a global solution v with $v(t) \in L^\infty(\mathbb{R}^n)$ for each $t > 0$, and v is a classical solution for $t > 0$. Furthermore, v is asymptotically self-similar, with profile f, i.e.:

$$t^{1/(p-1)}\|u(t) - v(t)\|_\infty = \sup_{y \in \mathbb{R}^n} \left|t^{1/(p-1)}v(y\sqrt{t},t) - f(y)\right| \to 0, \quad t \to \infty. \tag{20.23}$$

(iii) Let $q := q_c$. Assume that $v_0 \in L^1(\mathbb{R}^n) + L^\infty(\mathbb{R}^n)$ satisfies $u_0 - v_0 \in L^q(\mathbb{R}^n)$ and

$$\|u_0 - v_0\|_q < \mu_0.$$

Then problem (18.1) with initial data v_0 admits a global solution v which satisfies (20.23), together with

$$\sup_{t>0} \|u(t) - v(t)\|_q \leq 2\|u_0 - v_0\|_q \tag{20.24}$$

and

$$\|u(t) - v(t)\|_q \to 0, \quad t \to \infty. \tag{20.25}$$

Corollary 20.20. *Let $p > p_F$ and $q := q_c$. Then the zero solution is asymptotically stable in L^q. More precisely, if $v_0 \in L^q(\mathbb{R}^n)$ satisfies*

$$\|v_0\|_q \le \mu_0$$

with $\mu_0 = \mu_0(n, p) > 0$ sufficiently small, then (18.1) admits a global mild solution v which satisfies

$$\sup_{t>0} \|v(t)\|_q \le 2\|v_0\|_q \quad and \quad \|v(t)\|_q \to 0, \quad t \to \infty.$$

Furthermore $v(t) \in L^\infty(\mathbb{R}^n)$ for each $t > 0$, v is a classical solution for $t > 0$, and there holds

$$t^{1/(p-1)}\|v(t)\|_\infty \to 0, \quad t \to \infty.$$

Remarks 20.21. (i) **Nonuniqueness.** The solutions u and v constructed in Theorem 20.19 are unique in a suitable class of functions (see Lemma 20.22 and cf. also Remark 20.24(iii) below). When ω is a suitably small positive constant, nonuniqueness in a larger class of functions has been proved in [499]. Another nonuniqueness result can be found in [389].

(ii) **Decay rates.** The convergence statement in Theorem 20.19(ii) says that $u(t) - v(t)$ decays in L^∞ faster that $u(t)$ or $v(t)$ separately. More precise estimates on the decay of $u(t) - v(t)$ when u is radial can be found in [205]. Observe that the asymptotic behaviors in Theorem 20.19 and in Corollary 20.20 are different (note that u_0 in Theorem 20.19 just fails to be in L^q for $q = q_c$ if $u_0 \not\equiv 0$). In particular, in Theorem 20.19 with $u_0 \not\equiv 0$, $\|v(t)\|_\infty$ decays like $t^{-1/(p-1)}$ as $t \to \infty$, whereas in Corollary 20.20 it decays faster.

(iii) **Radial self-similar solutions.** The self-similar solutions constructed in Theorem 20.19 are not radial unless u_0 is radial. Radial self-similar solutions of (18.1) have been constructed by ODE or variational techniques (see Remarks 15.4 and the references there). In the radial case, the decay of the profile $f(y)$ as $y \to \infty$ has also been studied. The profile can decay either like $|y|^{-2/(p-1)}$ or exponentially.

(iv) **Other domains.** Consider problem (15.1) in a (possibly unbounded) domain Ω. By the comparison principle and Corollary 20.20, it follows that the zero solution is asymptotically stable in L^q for $q = q_c$. This is in contrast with the situation for $q > q_c$ (cf. Theorem 19.3). $\square$

In view of the proof, we introduce the following notation. For p, q as above, we fix r such that

$$1 \le r/p < q < r. \tag{20.26}$$

Although r is not uniquely determined, we assume that it is fixed once and for all (see also Remark 20.24(ii) below). We let $\beta = \frac{n}{2}(\frac{1}{q} - \frac{1}{r}) = \frac{1}{p-1} - \frac{n}{2r}$ and we define the following function spaces:

$$X = \big\{u \in L^\infty_{loc}((0, \infty), L^r(\mathbb{R}^n)) : \|u\|_X < \infty\big\}, \quad \text{where } \|u\|_X = \sup_{t>0} t^\beta \|u(t)\|_r,$$

$$Y = \left\{ u \in L^\infty_{loc}((0,\infty), L^\infty(\mathbb{R}^n)) : \|u\|_Y < \infty \right\}, \qquad \text{where } \|u\|_Y = \sup_{t>0} t^{\frac{1}{p-1}} \|u(t)\|_\infty,$$

and $Z = X \cap Y$, with norm

$$\|u\|_Z = \|u\|_X + \|u\|_Y.$$

For $\delta \geq 0$, we also define

$$E_\delta = \left\{ u_0 \in L^1(\mathbb{R}^n) + L^\infty(\mathbb{R}^n) : \mathcal{N}_\delta(u_0) < \infty \right\},$$
$$\text{where} \quad \mathcal{N}_\delta(u_0) = \sup_{t>0} t^{\beta+\delta} \|e^{-tA} u_0\|_r, \tag{20.27}$$

and for $0 < T < \infty$ we set

$$\|u\|_{X,\delta,T} = \sup_{0<t<T} t^{\beta+\delta} \|u(t)\|_r < \infty, \quad u \in X,$$

$$\|u\|_{Y,\delta,T} = \sup_{0<t<T} t^{\frac{1}{p-1}+\delta} \|u(t)\|_\infty < \infty, \quad u \in Y,$$

and

$$\|u\|_{Z,\delta,T} = \|u\|_{X,\delta,T} + \|u\|_{Y,\delta,T}, \quad u \in Z.$$

We note right away that for all $1 \leq m \leq q$, due to the L^p-L^q-estimates (see Proposition 48.4), we have $L^m(\mathbb{R}^n) \subset E_\delta$ and

$$\mathcal{N}_\delta(u_0) \leq \|u_0\|_m, \quad \text{with } \delta = \delta(m) = \frac{n}{2m} - \frac{1}{p-1}. \tag{20.28}$$

Moreover, we set $E := E_0$, $\mathcal{N} := \mathcal{N}_0$ and for $M > 0$, we denote by $B_X(M)$ (resp., $B_Y(M), B_Z(M)$) the closed ball of radius M in X (resp., Y, Z). The main ingredient of the proof of Theorem 20.19 is the following lemma.

Lemma 20.22. *(i) There exists $\varepsilon_0 = \varepsilon_0(n, p, r) > 0$ such that if $u_0 \in E$ satisfies $\mathcal{N}(u_0) \leq \varepsilon_0$, then (18.1) admits a unique global mild solution $u \in B_Z(M)$ with $M = C(p)\mathcal{N}(u_0)$. Moreover u is a classical solution of (18.1) for $t > 0$.*

(ii) Let $0 \leq \delta < \bar{\delta} := np/2r - 1/(p-1)$. There exists $\varepsilon_1 = \varepsilon_1(n, p, r, \delta) \in (0, \varepsilon_0]$ such that if $u_0, v_0 \in E$ satisfy $\mathcal{N}(u_0), \mathcal{N}(v_0) \leq \varepsilon_1$ and $u_0 - v_0 \in E_\delta$, then the corresponding solutions u, v of (18.1) given by part (i) satisfy

$$\sup_{t>0} \left[t^{\frac{1}{p-1}+\delta} \|u(t) - v(t)\|_\infty + t^{\beta+\delta} \|u(t) - v(t)\|_r \right] \leq C(p)\mathcal{N}_\delta(u_0 - v_0). \tag{20.29}$$

(iii) Let $m \in (r/p, q]$ and set $\delta = n/2m - 1/(p-1)$. There exists $\varepsilon_2 = \varepsilon_2(n, p, r, m) \in (0, \varepsilon_0]$ such that if $u_0, v_0 \in E$ satisfy $\mathcal{N}(u_0), \mathcal{N}(v_0) \leq \varepsilon_2$ and $u_0 - v_0 \in L^m(\mathbb{R}^n)$, then the corresponding solutions u, v of (18.1) given by part (i) satisfy

$$\sup_{t>0} t^\delta \|u(t) - v(t)\|_q \leq 2\|u_0 - v_0\|_m. \tag{20.30}$$

Proof. For $u_0 \in L^1(\mathbb{R}^n) + L^\infty(\mathbb{R}^n)$ and $u \in X$, we define the mapping

$$\mathcal{T}_{u_0} u(t) = e^{-tA} u_0 + \int_0^t e^{-(t-s)A} |u|^{p-1} u(s)\, ds.$$

Let $M > 0$. We fix $0 \le \delta < \bar{\delta} \, (\le 1)$ and $u_0, v_0 \in E$, with $u_0 - v_0 \in E_\delta$.

Step 1. Estimates of the mapping $\mathcal{T}$ in X. For all $u, v \in X$ and $0 < s < t < T < \infty$, we have

$$\left\| e^{-(t-s)A} \big(|u|^{p-1} u(s) - |v|^{p-1} v(s) \big) \right\|_r$$
$$\le (t-s)^{-n(p-1)/2r} \left\| |u|^{p-1} u(s) - |v|^{p-1} v(s) \right\|_{r/p}$$
$$\le (t-s)^{-q/r} \big(\|u(s)\|_r^{p-1} + \|v(s)\|_r^{p-1} \big) \|u(s) - v(s)\|_r$$
$$\le (t-s)^{-q/r} s^{-(\beta p + \delta)} \big(\|u\|_X^{p-1} + \|v\|_X^{p-1} \big) \|u - v\|_{X,\delta,T}$$

On the other hand, using $1 - \beta(p-1) - q/r = 0$, we have

$$t^{\beta + \delta} \int_0^t (t-s)^{-q/r} s^{-(\beta p + \delta)}\, ds = t^{1 - \beta(p-1) - q/r} \int_0^1 (1-\sigma)^{-q/r} \sigma^{-(\beta p + \delta)}\, d\sigma$$
$$= C(n, p, r, \delta),$$

where the integrals are finite, due to $q/r < 1$ and $\beta p + \delta < \beta p + \bar{\delta} = 1$. It follows that

$$t^{\beta + \delta} \| \mathcal{T}_{u_0} u(t) - \mathcal{T}_{v_0} v(t) \|_r \le t^{\beta + \delta} \| e^{-tA} (u_0 - v_0) \|_r$$
$$+ t^{\beta + \delta} \int_0^t \left\| e^{-(t-s)A} \big(|u|^{p-1} u(s) - |v|^{p-1} v(s) \big) \right\|_r ds$$
$$\le \mathcal{N}_\delta(u_0 - v_0) + C \big(\|u\|_X^{p-1} + \|v\|_X^{p-1} \big) \|u - v\|_{X,\delta,T},$$

hence

$$\| \mathcal{T}_{u_0} u - \mathcal{T}_{v_0} v \|_{X,\delta,T} \le \mathcal{N}_\delta(u_0 - v_0) + C_1 M^{p-1} \|u - v\|_{X,\delta,T}, \qquad u, v \in B_X(M),$$
$$(20.31)$$

with $C_1 = C_1(n, p, r, \delta) > 0$.

Step 2. Estimates of the mapping $\mathcal{T}$ in Z and fixed-point. For all $u, v \in Z$, $0 < t < T < \infty$ and $t/2 < s < t$, we have

$$\left\| e^{-(t-s)A} \big(|u|^{p-1} u(s) - |v|^{p-1} v(s) \big) \right\|_\infty$$
$$\le \left\| |u|^{p-1} u(s) - |v|^{p-1} v(s) \right\|_\infty$$
$$\le p \big[\|u(s)\|_\infty^{p-1} + \|v(s)\|_\infty^{p-1} \big] \|u(s) - v(s)\|_\infty$$
$$\le p(t/2)^{-\frac{p}{p-1} - \delta} \big(\|u\|_Y^{p-1} + \|v\|_Y^{p-1} \big) \|u - v\|_{Y,\delta,T}.$$

Using the fact that

$$\mathcal{T}_{u_0} u(t) = e^{-(t/2)A}\big(\mathcal{T}_{u_0} u(t/2)\big) + \int_{t/2}^{t} e^{-(t-s)A} |u|^{p-1} u(s)\, ds,$$

it follows that, for all $u, v \in B_Z(M)$,

$$
\begin{aligned}
t^{\frac{1}{p-1}+\delta}&\|\mathcal{T}_{u_0} u(t) - \mathcal{T}_{v_0} v(t)\|_\infty \\
&\leq t^{\frac{1}{p-1}+\delta}\big\| e^{-(t/2)A}\big(\mathcal{T}_{u_0} u(\tfrac{t}{2}) - \mathcal{T}_{u_0} v(\tfrac{t}{2})\big)\big\|_\infty \\
&\quad + t^{\frac{1}{p-1}+\delta}\int_{t/2}^{t}\big\| e^{-(t-s)A}\big(|u|^{p-1} u(s) - |v|^{p-1} v(s)\big)\big\|_\infty\, ds \\
&\leq 2^{\frac{1}{p-1}+\delta}(4\pi)^{-\frac{n}{2r}}\big(\tfrac{t}{2}\big)^{\frac{1}{p-1}-\frac{n}{2r}+\delta}\|\mathcal{T}_{u_0} u(\tfrac{t}{2}) - \mathcal{T}_{u_0} v(\tfrac{t}{2})\|_r \\
&\quad + p2^{\frac{1}{p-1}+\delta}\big(\|u\|_Y^{p-1} + \|v\|_Y^{p-1}\big)\|u - v\|_{Y,\delta,T} \\
&\leq C(p)\|\mathcal{T}_{u_0} u - \mathcal{T}_{u_0} v\|_{X,\delta,T} + C(p)M^{p-1}\|u - v\|_{Y,\delta,T}.
\end{aligned}
$$

Taking supremum for $t \in (0, T)$ and combining this with (20.31), we obtain

$$\|\mathcal{T}_{u_0} u - \mathcal{T}_{u_0} v\|_{Z,\delta,T} \leq C_2 \mathcal{N}_\delta(u_0 - v_0) + C_3 M^{p-1}\|u - v\|_{Z,\delta,T}, \quad u, v \in B_Z(M), \tag{20.32}$$

with $C_2 = C_2(p) \geq 1$ and $C_3 = C_3(n, p, r, \delta) > 0$. In particular, letting $T \to \infty$ in (20.32) with $\delta = 0$, we get

$$\|\mathcal{T}_{u_0} u - \mathcal{T}_{v_0} v\|_Z \leq C_2 \mathcal{N}(u_0 - v_0) + C_3 M^{p-1}\|u - v\|_Z, \quad u, v \in B_Z(M). \tag{20.33}$$

Choose $\varepsilon_0 = \varepsilon_0(n, p, r) > 0$ such that $2^p C_3(n, p, r, 0)(C_2 \varepsilon_0)^{p-1} \leq 1$ and assume that $\mathcal{N}(u_0) \leq \varepsilon_0$. Taking $M = 2C_2 \mathcal{N}(u_0)$, we have $C_3 M^{p-1} \leq 1/2$ and $C_2 \mathcal{N}(u_0) + C_3 M^p \leq M$. It follows from (20.33) (with the choices $v_0 = 0$, $v = 0$ and $u_0 = v_0$) that $\mathcal{T}_{u_0}$ is a strict contraction on the complete metric space $B_Z(M)$, endowed with the distance induced by the norm $\|\cdot\|_Z$. Therefore it possesses a unique fixed point, that we denote by $u(t) = W_t u_0$. In particular $u(t) \in L^\infty(\mathbb{R}^n)$ for $t > 0$ and u is a classical solution of (18.1) for $t > 0$. This proves the existence-uniqueness statement of assertion (i).

Next, assume in addition that $\mathcal{N}(v_0) \leq \varepsilon_0$ and put $v(t) = W_t v_0$. Replacing ε_0 by $\varepsilon_1 > 0$ possibly smaller and depending also on δ, we have $C_3(n, p, r, \delta)M^{p-1} \leq 1/2$. It then follows from (20.32) that

$$\|u - v\|_{Z,\delta,T} \leq 2C_2 \mathcal{N}_\delta(u_0 - v_0).$$

Assertion (ii) follows by letting $T \to \infty$.

Step 3. L^q-estimates. Fix $m \in (r/p, q]$ and put $\delta = \delta(m) = \frac{n}{2m} - \frac{1}{p-1}$. Note that $\delta \in [0, \bar{\delta})$. Assume that $u_0, v_0 \in E$ satisfy $\mathcal{N}(u_0), \mathcal{N}(v_0) \leq \varepsilon_2(n, p, r, m) :=$

$\varepsilon_1(n, p, r, \delta)$ and $u_0 - v_0 \in L^m(\mathbb{R}^n)$. Let u, v be the corresponding solutions of (18.1) given by Steps 1 and 2. Similarly as in the beginning of Step 1, we obtain for $0 < s < t$:

$$\left\| e^{-(t-s)A}\left(|u|^{p-1}u(s) - |v|^{p-1}v(s)\right) \right\|_q$$
$$\leq (t-s)^{-\frac{n}{2}\left(\frac{p}{r} - \frac{1}{q}\right)}s^{-(\beta p + \delta)}\left(\|u\|_X^{p-1} + \|v\|_X^{p-1}\right)\sup_{\sigma > 0}\sigma^{\beta+\delta}\|u(\sigma) - v(\sigma)\|_r.$$
$$(20.34)$$

On the other hand, using $1 - \frac{n}{2}\left(\frac{p}{r} - \frac{1}{q}\right) - \beta p = 0$, we have

$$\int_0^t (t-s)^{-\frac{n}{2}\left(\frac{p}{r} - \frac{1}{q}\right)}s^{-(\beta p + \delta)}\,ds = t^{-\delta}\int_0^1 (1-\sigma)^{-\frac{n}{2}\left(\frac{p}{r} - \frac{1}{q}\right)}\sigma^{-(\beta p + \delta)}\,d\sigma$$
$$(20.35)$$
$$= C(n, p, r, \delta)t^{-\delta},$$

where the integrals are finite, due to $\frac{n}{2}\left(\frac{p}{r} - \frac{1}{q}\right) < n(p-1)/2q = 1$ and $\beta p + \delta < \beta p + \bar\delta = 1$. Combining (20.34), (20.35) and (20.28) (and taking $\varepsilon_2(n, p, r, m)$ possibly smaller), we obtain

$$t^\delta\|u(t) - v(t)\|_q$$
$$\leq t^\delta\|e^{-tA}(u_0 - v_0)\|_q + t^\delta\int_0^t \left\| e^{-(t-s)A}\left(|u|^{p-1}u(s) - |v|^{p-1}v(s)\right) \right\|_q\,ds$$
$$\leq \|u_0 - v_0\|_m + C(n, p, r, \delta)\left(\|u\|_X^{p-1} + \|v\|_X^{p-1}\right)\sup_{\sigma > 0}\sigma^{\beta+\delta}\|u(\sigma) - v(\sigma)\|_r$$
$$\leq \|u_0 - v_0\|_m + C(n, p, r, \delta)M^{p-1}\mathcal{N}_\delta(u_0 - v_0) \leq 2\|u_0 - v_0\|_m.$$

This proves assertion (iii). $\square$

The next lemma shows that the homogeneous initial data u_0 belong to the class E used in Lemma 20.22.

Lemma 20.23. *Let $0 < k < n$, $L > 0$, and let $u_0 \in L^1(\mathbb{R}^n) + L^\infty(\mathbb{R}^n)$ satisfy*

$$|u_0(x)| \leq L|x|^{-k}.$$

Then, for $n/k < s \leq \infty$, there holds

$$\sup_{t > 0} t^{k/2 - n/(2s)}\|e^{-tA}u_0\|_s \leq cL$$

where $c = c(n, k, s) = \|e^{-A}|x|^{-k}\|_s < \infty$.

Proof. Set $\phi(x) = |x|^{-k}$ and decompose $\phi = \phi_1 + \phi_2$, where $\phi_1 = \chi_{\{|x| < 1\}}\phi$, $\phi_2 = \chi_{\{|x| \geq 1\}}\phi$. Then $\phi_1 \in L^m(\mathbb{R}^n)$, $m < n/k$ and $\phi_2 \in L^s(\mathbb{R}^n)$, $s > n/k$. Consequently,

$$e^{-A}\phi = e^{-A}\phi_1 + e^{-A}\phi_2 \in L^s(\mathbb{R}^n), \quad s > n/k.$$

Now using $\phi(\lambda x) = \lambda^{-k}\phi(x)$, we obtain

$$
\begin{aligned}
\left|(e^{-tA}u_0)(x)\right| &= \left|\int_{\mathbb{R}^n} (4\pi t)^{-n/2} e^{-|y|^2/4t} u_0(x-y)\,dy\right| \\
&\leq (4\pi)^{-n/2} L \int_{\mathbb{R}^n} e^{-|z|^2/4} \phi(x - zt^{1/2})\,dz \\
&= Lt^{-k/2}(4\pi)^{-n/2} \int_{\mathbb{R}^n} e^{-|z|^2/4} \phi(xt^{-1/2} - z)\,dz \\
&= Lt^{-k/2}(e^{-A}\phi)(xt^{-1/2}).
\end{aligned}
$$

In particular,

$$
\|e^{-tA}u_0\|_s \leq Lt^{(n/2s)-(k/2)}\|e^{-A}\phi\|_s, \qquad s > n/k,
$$

and the lemma follows. $\square$

Proof of Theorem 20.19. In this proof we shall take μ_0 as small as necessary to apply Lemma 20.22, but μ_0 will depend only on n, p, r.

(i) Since $\mathcal{N}(u_0) \leq c(n,p,r)\|w\|_\infty$ by Lemma 20.23, the existence of u follows from Lemma 20.22(i).

Let us show that u is self-similar. This is equivalent to showing that, for each $\lambda > 0$, $u_\lambda(x,t) := \lambda^{2/(p-1)}u(\lambda x, \lambda^2 t)$ satisfies $u_\lambda \equiv u$ (indeed, consider $\lambda = t^{-1/2}$). Since $\|u_\lambda\|_X = \|u\|_X$, it is thus sufficient, in view of the uniqueness part of Lemma 20.22(i), to check that u_λ is also a mild solution of (18.1). To this end, we define the dilation operator $(d_\lambda f)(x) := f(\lambda x)$ and note that $u_\lambda(t) = \lambda^{2/(p-1)}d_\lambda u(\lambda^2 t)$. A direct computation involving the heat kernel yields

$$
e^{-tA}(d_\lambda f) = d_\lambda(e^{\lambda^2 - tA}f). \tag{20.36}
$$

Applying (20.36) with $f = u^p$, we see that the function

$$
(Su)(t) := \int_0^t e^{-(t-s)A}u^p(s)\,ds
$$

satisfies

$$
\begin{aligned}
S(u_\lambda)(t) &= \lambda^{2p/(p-1)} \int_0^t e^{-(t-s)A}d_\lambda u^p(\lambda^2 s)\,ds \\
&= \lambda^{2p/(p-1)} d_\lambda \int_0^t e^{\lambda^2 - (t-s)A}u^p(\lambda^2 s)\,ds \\
&= \lambda^{2p/(p-1)} d_\lambda \int_0^{\lambda^2 t} e^{-(\lambda^2 t - \sigma)A}u^p(\sigma)\,\lambda^{-2}\,d\sigma \\
&= \lambda^{2/(p-1)} d_\lambda(Su)(\lambda^2 t) =: (Su)_\lambda(t).
\end{aligned} \tag{20.37}
$$

Now, since u_0 satisfies (20.22), we have $d_\lambda u_0 = \lambda^{-2/(p-1)} u_0$, hence

$$(e^{-tA} u_0)_\lambda := \lambda^{2/(p-1)} d_\lambda (e^{\lambda^2 - tA} u_0) = \lambda^{2/(p-1)} e^{-tA} (d_\lambda u_0) = e^{-tA} u_0. \qquad (20.38)$$

Combining (20.37) and (20.38), it follows that

$$e^{-tA} u_0 + S(u_\lambda)(t) = (e^{-tA} u_0)_\lambda + (Su)_\lambda(t) = u_\lambda(t).$$

We have thus shown that u is self-similar.

(ii) Since $\mathcal{N}(v_0) \le c(n, p, r) \|\omega(\cdot/|\cdot|)\varphi\|_\infty$ by Lemma 20.23, the existence of v follows from Lemma 20.22(i). Next, since $|v_0 - u_0| \le C|x|^{-2/(p-1)} \chi_{\{|x|<R\}}$ for some $C, R \in (0, \infty)$ and $2/(p-1) < n$, we have $v_0 - u_0 \in L^m(\mathbb{R}^n)$ for all $m \in [1, q)$, hence $v_0 - u_0 \in E_\delta$ for all $\delta \in (0, \bar\delta)$, by (20.28). The assertion then follows from Lemma 20.22(ii).

(iii) Using (20.28), we get

$$\mathcal{N}(v_0) \le \mathcal{N}(v_0 - u_0) + \mathcal{N}(u_0) \le \|v_0 - u_0\|_q + c(n, p, r)\|\omega\|_\infty \le (1 + c(n, p, r))\mu_0.$$

The existence of a global solution v with initial data v_0 is a consequence of Lemma 20.22(i). Property (20.24) is just (20.30) with $m = q$ (hence $\delta = 0$).

Let us show (20.25) and (20.23). To this end, we fix $m \in (r/p, q)$, we let $\eta := u_0 - v_0$ and we introduce the sequence $\eta_i := \eta \chi_{\{|x|<i\}} \in L^m \cap L^q(\mathbb{R}^n)$, which satisfies $\eta_i \to \eta$ in $L^q(\mathbb{R}^n)$. Let $\psi_i := u_0 + \eta_i$. We have $\|\psi_i - u_0\|_q = \|\eta_i\|_q \le \|v_0 - u_0\|_q$. Consequently, the existence of a global solution v_i of (18.1) with initial data ψ_i is ensured. By (20.30) and (20.28) we have

$$\|v_i(t) - u(t)\|_q \le 2\|\eta_i\|_m t^{-\delta(m)} \quad \text{and} \quad \|v_i(t) - v(t)\|_q \le 2\|\eta_i - \eta\|_q.$$

For each i, it follows that

$$\limsup_{t \to \infty} \|u(t) - v(t)\|_q \le \limsup_{t \to \infty} \|u(t) - v_i(t)\|_q + \limsup_{t \to \infty} \|v_i(t) - v(t)\|_q \le 2\|\eta_i - \eta\|_q,$$

and property (20.25) follows by letting $i \to \infty$. On the other hand, (20.29) and (20.28) imply

$$t^{1/(p-1)} \|v_i(t) - u(t)\|_\infty \le C(p)\mathcal{N}_{\delta(m)}(\eta_i) t^{-\delta(m)} \le C(p)\|\eta_i\|_m t^{-\delta(m)}$$

and

$$t^{1/(p-1)} \|v_i(t) - v(t)\|_\infty \le C(p)\mathcal{N}(\eta_i - \eta) \le C(p)\|\eta_i - \eta\|_q.$$

For each i, it follows that

$$\limsup_{t \to \infty} t^{1/(p-1)} \|u(t) - v(t)\|_\infty$$
$$\le \limsup_{t \to \infty} t^{1/(p-1)} \|u(t) - v_i(t)\|_\infty + \limsup_{t \to \infty} t^{1/(p-1)} \|v_i(t) - v(t)\|_\infty$$
$$\le C(p)\|\eta_i - \eta\|_q,$$

and property (20.23) follows by letting $i \to \infty$. $\qquad\square$

Remarks 20.24. (i) When $u_0 \in L^{q_c}$ is not small, (18.1) still admits a local in time solution (cf. Remark 15.4(i)). The existence can be proved by arguments similar to those in the proof of Lemma 20.22(i).

(ii) It can be shown that the solution u constructed in Lemma 20.22 does not depend on the choice of r if u_0 is suitably small. More precisely, given another $\hat{r}$ satisfying (20.26) and denoting by $\hat{\mathcal{N}} = \hat{\mathcal{N}}_0$ the corresponding norm in (20.27), there exists $\tilde{\varepsilon}_0 \in (0, \min(\varepsilon_0(n,p,q,r), \varepsilon_0(n,p,q,\hat{r}))]$ such that the two solutions coincide if $\hat{\mathcal{N}}(u_0), \mathcal{N}(u_0) \leq \tilde{\varepsilon}_0$.

(iii) It follows from the proof of Lemma 20.22 that (18.1) admits a mild solution which is unique in the larger class $B_X(K)$ with $K = C(p)\mathcal{N}(u_0)$. $\square$

21. Parabolic Liouville-type results

In Section 18 on Fujita-type results, we have seen that the equation $u_t - \Delta u = u^p$ with $p > 1$ has no global positive (classical) solution in $\mathbb{R}^n \times (0, \infty)$ if (and only if) $p \leq p_F$. In view of the Liouville-type results proved in Section 8 for the elliptic equation $-\Delta u = u^p$, it is natural to look also for parabolic Liouville-type theorems. More precisely, if one now considers positive solutions that are global for both positive *and* negative time, i.e. solutions on the whole space $\mathbb{R}^{n+1} = \mathbb{R}^n \times \mathbb{R}$, can one prove nonexistence for a larger range of p's than in the Fujita problem ? We will also study the same question on a half-space.

As it will turn out, we shall see in Section 26 that such results have many applications in the study of a priori estimates and (blow-up) singularities.

Let us first consider the case of radial solutions, for which we have the following optimal result from [422].

Theorem 21.1. *Let $1 < p < p_S$. Then the equation*

$$u_t - \Delta u = u^p, \qquad x \in \mathbb{R}^n, \quad t \in \mathbb{R} \tag{21.1}$$

has no positive, radial, bounded classical solution.

Theorem 21.1 is optimal in view of the existence of bounded positive radial stationary solutions for $n \geq 3$ and $p \geq p_S$ (see Section 9). It is very likely that Theorem 21.1 should hold without the radial symmetry assumption, but this has not been proved so far. However, under the stronger restriction $p < p_B$, where

$$p_B := \begin{cases} \infty & \text{if } n = 1, \\ \dfrac{n(n+2)}{(n-1)^2} & \text{if } n > 1, \end{cases}$$

we have the following Liouville-type theorem in the general (nonradial) case. It is a consequence of [79, Theorem 0.1].

Theorem 21.2. *Let $1 < p < p_B$. Then equation (21.1) has no positive classical solution.*

Remark 21.3. Theorem 21.1 remains true for nontrivial nonnegative radial classical solutions, bounded or not, whereas Theorem 21.2 remains true for nontrivial nonnegative classical solutions (see Remark 26.10(i) and cf. [425]). $\square$

The proofs of Theorems 21.1 and 21.2 are completely different, based on intersection-comparison and integral estimates, respectively.

For the proof of Theorem 21.1, we need some simple preliminary observations concerning radial steady states. Let ψ_1 be the solution of the equation

$$\psi'' + \frac{n-1}{r}\psi' + |\psi|^{p-1}\psi = 0, \quad r > 0, \tag{21.2}$$

satisfying $\psi(0) = 1$, $\psi'(0) = 0$. Obviously $\psi_1''(0) < 0$. It is known that the solution is defined on some interval and it changes sign due to $p < p_S$ (this follows for instance from Theorem 8.1). We denote by $r_1 > 0$ its first zero. By uniqueness for the initial-value problem, $\psi_1'(r_1) < 0$. We thus have

$$\psi_1(r) > 0 \text{ in } [0, r_1) \text{ and } \psi_1(r_1) = 0 > \psi_1'(r_1).$$

Clearly, $\psi_\alpha(r) := \alpha\psi_1(\alpha^{\frac{p-1}{2}}r)$ is the solution of (21.2) with $\psi(0) = \alpha$, $\psi'(0) = 0$, and with the first positive zero $r_\alpha = \alpha^{-\frac{p-1}{2}}r_1$. As an elementary consequence of the properties of ψ_1 we obtain the following

Lemma 21.4. *Given any $m > 0$, we have*

$$\lim_{\alpha \to \infty} \left(\sup\{\psi_\alpha'(r) : r \in [0, r_\alpha] \text{ is such that } \psi_\alpha(r) \leq m\}\right) = -\infty.$$

Proof of Theorem 21.1. The proof is by contradiction. Assume that u is a positive, bounded classical solution of (21.1), $u(x, t) = U(r, t)$, where $r = |x|$. By the boundedness assumption and parabolic estimates, U and U_r are bounded on $[0, \infty) \times \mathbb{R}$. It follows from Lemma 21.4 that if α is sufficiently large, then $U(\cdot, t) - \psi_\alpha$ has exactly one zero in $[0, r_\alpha]$ for any t and the zero is simple.

We next claim that

$$z_{[0,r_\alpha]}(U(\cdot, t) - \psi_\alpha) \geq 1 \quad t \leq 0, \ \alpha > 0, \tag{21.3}$$

where $z_{[0,r_\alpha]}(w)$ denotes the zero number of the function w in the interval $[0, r_\alpha]$ (see Appendix F). Indeed, if not, then $U(\cdot, t_0) > \psi_\alpha$ in $[0, r_\alpha]$ for some t_0. By Theorem 17.8 we know that each solution of the Dirichlet problem

$$\left.\begin{array}{ll} \overline{u}_t - \Delta\overline{u} = \overline{u}^p, & |x| < r_\alpha, \ t > 0, \\ \overline{u} = 0, & |x| = r_\alpha, \ t > 0, \\ \overline{u}(x, t_0) = \overline{U}_0(|x|), & |x| < r_\alpha \end{array}\right\}$$

blows up in finite time provided $\overline{U}_0 > \psi_\alpha$ in $[0, r_\alpha)$. Choosing the initial function $\overline{U}_0$ between ψ_α and $U(\cdot, t_0)$ we conclude, by comparison, that $\overline{u}$ and u both blow up in finite time, in contradiction to the global existence assumption on u. This proves the claim.

Set

$$\alpha_0 := \inf\{\beta > 0 : z_{[0, r_\alpha]}(U(\cdot, t) - \psi_\alpha) = 1 \text{ for all } t \le 0 \text{ and } \alpha \ge \beta\}.$$

In view of the above remark on large α, we have $\alpha_0 < \infty$. Also $\alpha_0 > 0$. Indeed, for small $\alpha > 0$ we have $\psi_\alpha(0) < U(0, t)$ for $t = 0$ and for $t > 0$ small. By the properties of the zero number (see Theorem 52.28), we can choose $t < 0$ small such that $\psi_\alpha(0) - U(\cdot, t)$ has only simple zeros and then, by (21.3), $z_{[0, r_\alpha]}(U(\cdot, t) - \psi_\alpha) \ge 2$.

By definition of α_0 (and (21.3)), there are sequences $\alpha_k \nearrow \alpha_0$ and $t_k \le 0$ such that

$$z_{[0, r_{\alpha_k}]}(U(\cdot, t_k) - \psi_{\alpha_k}) \ge 2, \quad k = 1, 2, \ldots.$$

Using Theorem 52.28 again, we get

$$z_{[0, r_{\alpha_k}]}(U(\cdot, t_k + t) - \psi_{\alpha_k}) \ge 2, \quad t \le 0, \ k = 1, 2, \ldots. \tag{21.4}$$

This in particular allows us to assume, choosing different t_k if necessary, that $t_k \to -\infty$. By the boundedness assumption and parabolic estimates, passing to a subsequence, we may further assume that

$$u(x, t_k + t) \to v(x, t), \quad x \in \mathbb{R}^n, \ t \in \mathbb{R},$$

with convergence in $C^{2,1}(\mathbb{R}^n \times \mathbb{R})$. Clearly then, there is $\delta > 0$ such that for each fixed t,

$$U(\cdot, t_k + t) - \psi_{\alpha_k} \to V(\cdot, t) - \psi_{\alpha_0}$$

in $C^1[0, r_{\alpha_0} + \delta]$, where $v(x, t) = V(|x|, t)$. This and (21.4) guarantee that for each $t \le 0$, $V(\cdot, t) - \psi_{\alpha_0}$ has at least two zeros or a multiple zero in $[0, r_{\alpha_0})$. By the properties of the zero number (see Theorem 52.28), we may choose $t < 0$ so that $V(\cdot, t) - \psi_{\alpha_0}$ has only simple zeros (and, hence at least two of them). Since $U(\cdot, t_k + t) - \psi_{\alpha_0}$ is close to $V(\cdot, t) - \psi_{\alpha_0}$ in $C^1[0, r_{\alpha_0}]$, if k is large, it has at least two simple zeros in $[0, r_{\alpha_0})$ as well. But then, for $\alpha > \alpha_0$, α close to α_0, the function $u(\cdot, t_k + t) - \psi_\alpha$ has at least two zeros in $[0, r_\alpha)$, contradicting the definition of α_0.

We have thus shown that the assumption $u \not\equiv 0$ leads to a contradiction, which proves the theorem. $\square$

We now turn to the proof of Theorem 21.2. It will be a direct consequence of the following space-time integral estimates [79] for (local) solutions of (21.1).

Proposition 21.5. *Let $1 < p < p_B$ and let B_1 be the unit ball in $\mathbb{R}^n$. There exists $r = r(n, p) > (n+2)(p-1)/2$ such that if $0 < u \in C^{2,1}(B_1 \times (-1, 1))$ is a solution of*

$$u_t - \Delta u = u^p, \qquad |x| < 1, \quad -1 < t < 1,$$

then

$$\int_{-1/2}^{1/2} \int_{|x|<1/2} u^r \, dx \, dt \leq C(n, p).$$

Let us first prove Theorem 21.2 assuming Proposition 21.5. It suffices to apply a simple homogeneity argument.

Proof of Theorem 21.2. Let $R > 0$. Let u be a solution of (21.1). Then, for each $R > 0$, $v(x, t) := R^{2/(p-1)} u(Rx, R^2 t)$ solves (21.1) in $B_1 \times (-1, 1)$. It follows from Proposition 21.5 that

$$\int_{-R^2/2}^{R^2/2} \int_{|y|<R/2} u^r(y, s) \, dy \, ds = R^{n+2} \int_{-1/2}^{1/2} \int_{|x|<1/2} u^r(Rx, R^2 t) \, dx \, dt$$

$$= R^{n+2-2r/(p-1)} \int_{-1/2}^{1/2} \int_{|x|<1/2} v^r(x, t) \, dx \, dt \leq C(n, p) R^{n+2-2r/(p-1)}.$$

Since $r > (n+2)(p-1)/2$, by letting $R \to \infty$, we conclude that $\int_{-\infty}^{\infty} \int_{\mathbb{R}^n} u^r \, dy \, ds = 0$, hence $u \equiv 0$. $\square$

The proof of Proposition 21.5 uses the following key gradient estimate, which is the analogue of the one used in Section 8 to prove the Liouville-type theorem and the local estimates for the elliptic equation $-\Delta u = u^p$. In the rest of this section, we use the notation $\int\int = \int_{-T}^{T} \int_\Omega$ for simplicity.

Lemma 21.6. *(i) Let Ω be an arbitrary domain in $\mathbb{R}^n$, $T > 0$, and $0 \leq \varphi \in \mathcal{D}(\Omega \times (-T, T))$. Let $0 < u \in C^{2,1}(\Omega \times (-T, T))$, be a solution of (21.1) in $\Omega \times (-T, T)$. Fix $k \in \mathbb{R}$ with $k \neq -1$ and denote*

$$I = \int\int \varphi \, u^{-2} |\nabla u|^4, \qquad L = \int\int \varphi \, u^{2p},$$

where, here and below, integrals are over $\Omega \times (-T, T)$. Then there holds

$$\alpha I + \delta L \leq C(n, p, k) \int\int \varphi \left[(u_t)^2 + |u_t| \, u^{-1} |\nabla u|^2 \right] + |\nabla u|^2 |\Delta \varphi|$$

$$+ C(n, p, k) \int\int (u^p + |u_t| + u^{-1}|\nabla u|^2)|\nabla u \cdot \nabla \varphi| + u^{p+1}|\varphi_t|, \tag{21.5}$$

where

$$\alpha = -((n-1)k + n)\frac{k}{n}, \qquad \delta = -\frac{n-1+(n+2)k/p}{n}. \tag{21.6}$$

Assume $1 < p < p_B$. Then there exist $k = k(n,p) \in \mathbb{R}$, $k \neq -1$, such that the constants α, δ defined in (21.6) satisfy

$$\alpha, \delta > 0. \tag{21.7}$$

The main ingredient in the proof of Lemma 21.6 is Lemma 8.9, proved in Section 8, which provides a family of integral estimates relating any C^2-function with its gradient and its Laplacian.

Proof. (i) We apply Lemma 8.9 with $q = 0$. Denoting

$$J = \int\int \varphi\, u^{-1}|\nabla u|^2 \Delta u, \quad K = \int\int \varphi\,(\Delta u)^2,$$

this gives us

$$-\left(\frac{n-1}{n}k+1\right)kI + \frac{n+2}{n}kJ - \frac{n-1}{n}K$$
$$\leq \frac{1}{2}\int\int |\nabla u|^2 \Delta\varphi + \int\int\left[\Delta u - ku^{-1}|\nabla u|^2\right]\nabla u \cdot \nabla\varphi. \tag{21.8}$$

Now, since $\Delta u = u_t - u^p$, integrating by parts in t and/or in x, we obtain

$$K = \int\int \varphi\,(u_t)^2 + \int\int \varphi\, u^{2p} - 2\int\int \varphi\, u^p u_t$$
$$= \int\int \varphi\,(u_t)^2 + L + \frac{2}{p+1}\int\int u^{p+1}\varphi_t$$

and

$$pJ = -\int\int \varphi \nabla u \cdot \nabla(u^p) + p\int\int \varphi\, u_t\, u^{-1}|\nabla u|^2$$
$$= \int\int \varphi\,(\Delta u)u^p + \int\int (\nabla\varphi \cdot \nabla u)u^p + p\int\int \varphi\, u_t\, u^{-1}|\nabla u|^2$$
$$= -L - \frac{1}{p+1}\int\int u^{p+1}\varphi_t + \int\int (\nabla\varphi \cdot \nabla u)u^p + p\int\int \varphi\, u_t\, u^{-1}|\nabla u|^2.$$

Substituting in (21.8), we obtain (21.5).

(ii) For $k < 0$, the condition $\alpha, \delta > 0$ is equivalent to

$$(n-1)p/(n+2) < -k < n/(n-1).$$

Such choice of $k < 0$ is clearly possible if $p < p_B$. $\square$

Proof of Proposition 21.5. Taking k as in Lemma 21.6(ii), we shall estimate the terms on the RHS of (21.5). Let us first prepare a suitable test-function. We take $\xi \in \mathcal{D}(B_1 \times (-1,1))$, such that $\xi = 1$ in $B_{1/2} \times (-1/2, 1/2)$ and $0 \leq \xi \leq 1$. By taking $\varphi = \xi^{4p/(p-1)}$, we have

$$|\nabla \varphi| \leq C\varphi^{(3p+1)/4p}, \quad |\Delta \varphi| \leq C\varphi^{(p+1)/2p}, \quad |\varphi_t| \leq C\varphi^{(3p+1)/4p} \leq C\varphi^{(p+1)/2p}. \tag{21.9}$$

We first observe that

$$\int\!\!\int |\nabla u|^2 \big(|\Delta \varphi| + \varphi^{-1}|\nabla \varphi|^2 + |\varphi_t|\big) \leq \eta(I + L) + C(\eta), \qquad \eta > 0. \tag{21.10}$$

Indeed, this follows from Young's inequality and (21.9), by writing

$$|\nabla u|^2 \big(|\Delta \varphi| + \varphi^{-1}|\nabla \varphi|^2 + |\varphi_t|\big)$$
$$\leq \eta\varphi\, u^{-2}|\nabla u|^4 + C(\eta)\varphi^{-1}u^2\big(|\Delta \varphi| + \varphi^{-1}|\nabla \varphi|^2 + |\varphi_t|\big)^2$$
$$\leq \eta\varphi\, u^{-2}|\nabla u|^4 + C(\eta)\varphi^{1/p}u^2$$
$$\leq \eta\varphi\, u^{-2}|\nabla u|^4 + \eta\varphi\, u^{2p} + C(\eta).$$

Now fix $\varepsilon > 0$. Using Young's inequality, (21.9) and (21.10), we estimate the RHS of (21.5) as follows:

$$\int\!\!\int \varphi\left[(u_t)^2 + |u_t|\, u^{-1}|\nabla u|^2\right] + |\nabla u|^2|\Delta \varphi|$$
$$+ \int\!\!\int (u^p + |u_t| + u^{-1}|\nabla u|^2)|\nabla u \cdot \nabla \varphi| + u^{p+1}|\varphi_t|$$
$$\leq \varepsilon \int\!\!\int \varphi\big[u^{2p} + u^{-2}|\nabla u|^4\big]$$
$$+ C(\varepsilon)\int\!\!\int \left[\varphi(u_t)^2 + |\nabla u|^2(\varphi^{-1}|\nabla \varphi|^2 + |\Delta \varphi|) + (\varphi^{-(p+1)}|\varphi_t|^{2p})^{1/(p-1)}\right]$$
$$\leq 2\varepsilon(I + L) + C(\varepsilon)\Big(1 + \int\!\!\int \varphi(u_t)^2\Big). \tag{21.11}$$

Let us handle the last term in the above inequality. Multiplying equation (21.1) by $\varphi\, u_t$, integrating by parts in x and t, and using Young's inequality and (21.9), we get, for each $\eta > 0$,

$$\int\!\!\int \varphi\,(u_t)^2 = \int\!\!\int \varphi\,\partial_t\Big(\frac{u^{p+1}}{p+1} - \frac{|\nabla u|^2}{2}\Big) - (\nabla \varphi \cdot \nabla u)u_t$$
$$= \int\!\!\int \Big(\frac{|\nabla u|^2}{2} - \frac{u^{p+1}}{p+1}\Big)\varphi_t - (\nabla \varphi \cdot \nabla u)u_t$$
$$\leq \frac{1}{2}\int\!\!\int |\nabla u|^2\big(|\varphi_t| + |\nabla \varphi|^2\varphi^{-1}\big) + \frac{1}{2}\int\!\!\int \varphi\,(u_t)^2 + \frac{1}{p+1}\int\!\!\int u^{p+1}|\varphi_t|.$$

By (21.10) and (21.9), for all $\eta > 0$, it follows that

$$\int\int \varphi\,(u_t)^2 \leq \int\int |\nabla u|^2 \big(|\varphi_t| + |\nabla\varphi|^2\varphi^{-1}\big) + \frac{2}{p+1}\int\int u^{p+1}|\varphi_t|$$

$$\leq \eta(I+L) + C(\eta) + \eta\int\int \varphi\,u^{2p} + C(\eta)\int\int \varphi^{-(p+1)/(p-1)}|\varphi_t|^{2p/(p-1)}$$

$$\leq 2\eta(I+L) + C(\eta).$$

$$\tag{21.12}$$

Combining (21.12), applied with $\eta = \varepsilon(2C(\varepsilon))^{-1}$, (21.11) and (21.5), we obtain

$$\alpha I + \delta L \leq C(n,p)\varepsilon(I+L) + C(\varepsilon).$$

Since $\alpha, \delta > 0$, by choosing $\varepsilon = \varepsilon(n,p)$ sufficiently small, we conclude that $I, L \leq C$. $\quad\square$

Remark 21.7. In the above proof, it is a priori possible to use Lemma 8.9 with values other than $q = 0$ (at the expense of additional complications in the estimate of the terms on the RHS of (21.5)). However, this does not seem to enable one to go beyond the condition $p < p_B$. $\quad\square$

We now consider the case of a half-space. The following result was proved in [425].

Theorem 21.8. *Let $p > 1$. Assume $n \leq 2$, or $p < (n-1)(n+1)/(n-2)^2$ and $n \geq 3$. Then the problem*

$$\left.\begin{array}{ll} u_t - \Delta u = u^p, & x \in \mathbb{R}^n_+,\ t \in \mathbb{R}, \\[2mm] u = 0, & x \in \partial\mathbb{R}^n_+,\ t \in \mathbb{R} \end{array}\right\} \tag{21.13}$$

has no positive bounded classical solution.

Remarks 21.9. (a) We note that $(n-1)(n+1)/(n-2)^2$ is the exponent p_B of Theorem 21.2 in dimension $n - 1$, and that this number is greater than p_S.

(b) Any nontrivial bounded classical solution $u \geq 0$ of (21.13) is positive (this follows from the argument after (26.44) in the proof of Theorem 26.8). On the other hand, it can be shown [425] that if $p < p_B = n(n+2)/(n-1)^2$, then problem (21.13) has no nontrivial nonnegative classical solution, bounded or not. $\quad\square$

Theorem 21.8 is a consequence of Theorem 21.2 and the following monotonicity result [425] concerning the more general problem

$$\left.\begin{array}{ll} u_t - \Delta u = f(u), & x \in \mathbb{R}^n_+,\ t \in \mathbb{R}, \\[2mm] u = 0, & x \in \partial\mathbb{R}^n_+,\ t \in \mathbb{R}, \end{array}\right\} \tag{21.14}$$

where f is a C^1-function.

Theorem 21.10. *Assume $f : [0, \infty) \to \mathbb{R}$ is a C^1-function satisfying $f(0) = 0$ and $f'(0) \leq 0$. Then the following statements hold true.*

(i) *Each positive bounded solution u of (21.14) is increasing in x_1:*

$$\partial_{x_1} u(x, t) > 0, \qquad x \in \mathbb{R}^n_+, \ t \in \mathbb{R}.$$

(ii) *If there is a positive bounded solution of (21.14), then there exists a positive bounded solution of*

$$u_t - \Delta u = f(u), \quad x \in \mathbb{R}^{n-1}, \ t \in \mathbb{R}. \tag{21.15}$$

For $n = 1$, equation (21.15) should be understood as the ordinary differential equation $u_t = f(u)$.

The proofs of both statements (i) and (ii) use extensions of moving plane arguments of [150] to parabolic equations. A straightforward modification of the proof below shows that (i), (ii) hold for positive bounded solutions defined on $(-\infty, T)$ for some $T > 0$.

Proof. First we prove (i). We use the following notation. For $\lambda > 0$ let

$$\mathbb{T}_\lambda = \{x \in \mathbb{R}^n : 0 < x_1 < \lambda\}.$$

For a function z defined on $\mathbb{R}^n_+$ let z^λ and $V_\lambda z$ be functions on $\mathbb{T}_\lambda$ defined by

$$\left. \begin{aligned} z^\lambda(x) &= z(2\lambda - x_1, x'), \\ V_\lambda z(x) &= z^\lambda(x) - z(x), \end{aligned} \right\} \tag{21.16}$$

where $x' = (x_2, x_3, \ldots, x_n)$.

Let u be a positive bounded solution of (21.14). Observe that for each $\lambda > 0$, $v = V_\lambda u$ satisfies

$$\left. \begin{aligned} v_t - \Delta v &= c^\lambda(x, t)v, && x \in \mathbb{T}_\lambda, \ t \in \mathbb{R}, \\ v &= 0, && x_1 = \lambda, \ x' \in \mathbb{R}^{n-1}, \ t \in \mathbb{R}, \\ v &> 0, && x_1 = 0, \ x' \in \mathbb{R}^{n-1}, \ t \in \mathbb{R}, \end{aligned} \right\} \tag{21.17}$$

where

$$c^\lambda(x, t) = \int_0^1 f'(u(x, t) + s(u^\lambda(x, t) - u(x, t))) \, ds. \tag{21.18}$$

Our goal is to prove that the statement

$$V_\lambda u(x, t) \geq 0 \quad (x \in \mathbb{T}_\lambda, \ t \in \mathbb{R}) \tag{S$_\lambda$}$$

holds for each $\lambda > 0$. Once this is done, the maximum principle applied to the above linear problem guarantees that we have in fact the strict inequality in (S)$_\lambda$ and the Hopf boundary principle then gives

$$2\partial_{x_1} u(x, t)\big|_{x_1 = \lambda} = -\partial_{x_1} V_\lambda u(x, t)\big|_{x_1 = \lambda} > 0$$

for each $\lambda > 0$, proving (i).

We shall use the following lemma [150].

Lemma 21.11. *Given any positive constants q, λ satisfying $\lambda^{-2}\pi^2 > q$, there exists a smooth function h on $\bar{\mathbb{T}}_\lambda$ such that*

$$\left.\begin{aligned}
\Delta h + qh = 0, && x \in \mathbb{T}_\lambda, \\
h(x) > 0, && x \in \bar{\mathbb{T}}_\lambda, \\
h(x) \to \infty, && |x| \to \infty, \ x \in \bar{\mathbb{T}}_\lambda.
\end{aligned}\right\} \tag{21.19}$$

Moreover, h satisfies $h \geq \eta$ for some constant $\eta > 0$.

Proof. A straightforward computation shows that

$$h = h(x_1, x_2, \ldots, x_n) = \cos\left[\frac{\pi(2x_1 - \lambda)}{2(\lambda + \varepsilon)}\right] \prod_{i=2}^{n} \cosh(\varepsilon x_i)$$

satisfies the required properties for $\varepsilon > 0$ small. $\quad\square$

We first prove that $(S)_\lambda$ holds for λ small. Fix a positive constant γ and set

$$q := \sup_{t \in \mathbb{R},\, x \in \mathbb{R}^n_+} f'(u(x,t)) + \gamma. \tag{21.20}$$

If $\lambda > 0$ is sufficiently small, so that $\lambda^{-2}\pi^2 > q$, we can apply Lemma 21.11. With the resulting function h, we consider the problem satisfied by $w := e^{\gamma t}v/h$, where $v = V_\lambda u$. A simple computation using (21.17), (21.19) shows that

$$\left.\begin{aligned}
w_t - \Delta w - \frac{2\nabla h}{h} \cdot \nabla w - (\gamma + c^\lambda(x,t) - q)w = 0, && x \in \mathbb{T}_\lambda,\ t \in \mathbb{R}, \\
w \geq 0, && x \in \partial\mathbb{T}_\lambda,\ t \in \mathbb{R}, \\
w(x,t) \to 0, && |x| \to \infty,\ x \in \bar{\mathbb{T}}_\lambda,\ t \in \mathbb{R}.
\end{aligned}\right\} \tag{21.21}$$

The choice of q implies $\gamma + c^\lambda - q \leq 0$ in $\mathbb{T}_\lambda \times \mathbb{R}$. Applying the maximum principle on $\mathbb{T}_\lambda \times (t_0, t)$, for each $t_0 < t$, we obtain

$$\sup_{x \in \mathbb{T}_\lambda} w_-(x,t) \leq \sup_{x \in \mathbb{T}_\lambda} w_-(x,t_0). \tag{21.22}$$

For v the above inequality means

$$\sup_{x \in \mathbb{T}_\lambda} \frac{v_-(x,t)}{h(x)} \leq e^{-\gamma(t-t_0)} \sup_{x \in \mathbb{T}_\lambda} \frac{v_-(x,t_0)}{h(x)}. \tag{21.23}$$

In view of boundedness of $v = V_\lambda u$, letting $t_0 \to -\infty$ we obtain that $v \geq 0$ everywhere. Using the maximum principle again we conclude that v is positive in $\mathbb{T}_\lambda \times \mathbb{R}$, hence $(S)_\lambda$ holds.

In the next step we denote

$$\lambda_0 = \sup\{\mu > 0 : (S)_\lambda \text{ holds for all } \lambda \in (0,\mu)\}. \tag{21.24}$$

As proved above, $\lambda_0 > 0$. We now show by contradiction that $\lambda_0 = \infty$. Assume $\lambda_0 < \infty$. Then there is a sequence $\lambda_k \geq \lambda_0$ such that $\lambda_k \to \lambda_0$ and the set

$$Z_k := \{(x,t) \in \mathbb{T}_{\lambda_k} \times \mathbb{R} : V_{\lambda_k} u(x,t) < 0\}$$

is nonempty. Set

$$m_k := \sup\{u(y_1, x', t) : y_1 \in (0, \lambda_k),\ x' \in \mathbb{R}^{n-1},\ t \in \mathbb{R},\ \text{and}$$
$$\text{there exists } x_1 \in (0, \lambda_k) \text{ such that } (x_1, x', t) \in Z_k\}.$$

We consider the following two possibilities.

(a) $m_k \to 0$,

(b) passing to a subsequence we have $m_k \geq \varepsilon_0$ for some $\varepsilon_0 > 0$.

First assume that (b) holds. Then there are sequences $x_1^k, y_1^k \in (0, \lambda_k)$, $z^k \in \mathbb{R}^{n-1}$, $t^k \in \mathbb{R}$ such that $V_{\lambda_k} u(x_1^k, z^k, t^k) < 0$ and $u(y_1^k, z^k, t^k) \geq \varepsilon_0$. We may assume that $x_1^k \to a$ and $y_1^k \to b$ for some $a, b \in [0, \lambda_0]$. Consider the functions

$$u_k(x,t) := u(x_1, x' + z^k, t + t^k), \qquad x = (x_1, x') \in \mathbb{R}^n,\ t \in \mathbb{R}.$$

Each of them is a positive solution of (21.14) satisfying $V_{\lambda_k} u_k(x_1^k, 0, 0) < 0$, $u_k(y_1^k, 0, 0) \geq \varepsilon_0$ and $V_{\lambda_0} u_k \geq 0$ in $\mathbb{T}_{\lambda_0} \times \mathbb{R}$ (the last inequality follows from the definition of λ_0 and continuity). Moreover, the sequence u_k is uniformly bounded. Using standard parabolic estimates, one shows that if u_k is replaced by a subsequence, then it converges in $C^{2,1}(\mathbb{R}^{n+1})$ to a nonnegative solution $\tilde{u}$ of (21.14). The above properties of u_k imply $V_{\lambda_0}\tilde{u}(a,0,0) \leq 0$, $\tilde{u}(b,0,0) \geq \varepsilon_0$, and $V_{\lambda_0}\tilde{u} \geq 0$ in $\mathbb{T}_{\lambda_0} \times \mathbb{R}$. Since $\tilde{u}$ is nontrivial and $f(0) = 0$ the maximum principle guarantees that $\tilde{u}$ is positive everywhere. Consequently, $\tilde{v} := V_{\lambda_0}\tilde{u}$ solves the corresponding problem (21.17) with $\lambda = \lambda_0$ and therefore $\tilde{v} > 0$ in $\mathbb{T}_{\lambda_0} \times \mathbb{R}$. It follows in particular that necessarily $a = \lambda_0$. By the Hopf principle,

$$2\tilde{u}_{x_1}(\lambda_0, 0, 0) = -\partial_{x_1} V_{\lambda_0}\tilde{u}(x_1, 0, 0)\big|_{x_1 = \lambda_0} > 0.$$

Consequently, $\tilde{u}_{x_1}(x_1, 0, 0)$ is bounded below by a positive constant on an interval around λ_0 and this remains valid if $\tilde{u}$ is replaced by u_k for k large. That is, there is $\delta > 0$ such that

$$\partial_{x_1} u(x_1, z^k, t^k) = \partial_{x_1} u_k(x_1, 0, 0) > 0, \qquad x_1 \in [\lambda_0 - \delta, \lambda_0 + \delta], \tag{21.25}$$

for all sufficiently large k. However, since $2\lambda_k - x_1^k > x_1^k$ both belong to $[\lambda_0 - \delta, \lambda_0 + \delta]$ for large k, (21.25) contradicts the assumption that $V_{\lambda_k} u(x_1^k, z^k, t^k) < 0$.

We have shown that (b) leads to a contradiction. Assume now that (a) holds. Consider problem (21.17) with $\lambda = \lambda_k$ and k sufficiently large. We are going to apply the maximum principle on the set Z_k (assumed to be nonempty). The boundary conditions in (21.17) imply $v = V_{\lambda_k} u = 0$ on ∂Z_k. Next observe that property (a), in conjunction with (21.18) and the definition of m_k, guarantees that for

$$\tilde{q}_k := \sup_{(x,t) \in Z_k} c^{\lambda_k}(x,t)$$

we have

$$\limsup_{k \to \infty} \tilde{q}_k \leq 0.$$

Fix k so large that $q := \tilde{q}_k + \gamma < \lambda_k^{-2}\pi^2$, where γ is some positive constant, and set $\lambda = \lambda_k$. Apply Lemma 21.11 and let h be the resulting function. As in our arguments above, $w := e^{\gamma t}v/h$ satisfies problem (21.21). This time we know that $\gamma + c^\lambda - q \leq 0$ on Z_k only. However, since v vanishes on ∂Z_k, we can still apply the maximum principle on Z_k to conclude that (21.22) holds and, consequently, that $v \geq 0$ in Z_k. This of course contradicts the definition of Z_k. Thus possibility (a) leads to a contradiction, too, which proves that $\lambda_0 = \infty$.

We have completed the proof of assertion (i).

To prove assertion (ii), let u be a positive bounded solution of (21.14). For $k = 1, 2, \dots$ consider the functions

$$u_k(x_1, x', t) := u(x_1 + k, x', t), \qquad (x_1, x', t) \in (-k, \infty) \times \mathbb{R}^{n-1} \times \mathbb{R}.$$

Each of them solves the equation $u_t - \Delta u = f(u)$ on its domain. Since the sequence is uniformly bounded, using parabolic estimates one shows that a subsequence of u_k converges uniformly on each compact to a bounded nonnegative solution $\tilde{u}$ of $u_t - \Delta u = f(u)$ on $\mathbb{R}^n \times \mathbb{R}$. From the monotonicity of u proved in (c1), we further conclude that $\tilde{u}$ is positive and independent of x_1. This proves assertion (ii). $\square$

Remark 21.12. Liouville-type result under a decay assumption at $-\infty$. A different parabolic Liouville-type theorem was proved in [367] for $1 < p < p_S$. Namely, if u is a classical solution of

$$u_t - \Delta u = |u|^{p-1}u \tag{21.26}$$

on $\mathbb{R}^n \times (-\infty, 0)$ and is such that

$$\sup_{t<0} |t|^{1/(p-1)}\|u(t)\|_\infty < \infty, \tag{21.27}$$

then u depends only on t. This result was used in [367] to obtain refined blow-up estimates for problem (18.1) near the blow-up time. It implies in particular that (21.1) has no positive bounded classical solution satisfying (21.27) for $1 < p < p_S$. However, it does not seem possible to use this form of Liouville-type theorem to establish universal blow-up estimates on the whole existence interval $(0, T)$, like those which will be derived from Theorems 21.1 and 21.2 in Section 21. $\square$

22. A priori bounds

Consider the model problem

$$\left.\begin{aligned}
u_t - \Delta u &= |u|^{p-1}u, & x \in \Omega,\ t > 0, \\
u &= 0, & x \in \partial\Omega,\ t > 0, \\
u(x,0) &= u_0(x), & x \in \Omega,
\end{aligned}\right\} \qquad (22.1)$$

where Ω is bounded and $p > 1$. We have seen that (22.1) admits both:

- finite-time blow-up solutions — cf. Section 17; and

- global bounded solutions (in particular small data solutions decaying to 0 as $t \to \infty$, and stationary solutions if $p < p_S$) — cf. Sections 19 and 6.

In order to understand the structure of solutions of problem (22.1), it is natural to investigate whether or not it admits other kinds of solutions (namely global unbounded classical solutions). In the case when all global solutions are bounded, one can further look for an **a priori estimate** of global solutions, that is, an estimate of the form

$$\sup_{t \geq 0} \|u(t)\|_\infty \leq C(\|u_0\|_\infty), \qquad \text{with } C \text{ bounded on bounded sets.} \qquad (22.2)$$

This estimate means that, given $K > 0$, there exists $C = C(K) > 0$ such that all global solutions with $\|u_0\|_\infty \leq K$ satisfy $\|u(t)\|_\infty \leq C$ for all $t \geq 0$. The existence of stronger *universal bounds* (independent of initial data) will be studied in Section 26.

We shall see that the answers to these questions (boundedness of global solutions vs. existence of unbounded global solutions, existence vs. nonexistence of a priori estimates) strongly depend on the value of p. Besides the intrinsic interest of such questions, let us emphasize that the results and techniques of proofs have many applications (see Section 28 and cf. also, for instance, Theorem 22.13, the proof of Theorem 23.7, Remark 23.14, and the proof of Theorem 27.2).

22.1. A priori bounds in the subcritical case

In this subsection we establish a priori estimates of global solutions in the subcritical case $p < p_S$. As we shall see below, the assumption $p < p_S$ is necessary for the bound (22.2) (at least if Ω is starshaped).

Theorem 22.1. *Assume Ω bounded and $1 < p < p_S$. Then the bound (22.2) is true for all global solutions of (22.1).*

This result was proved in [243] for $u_0 \geq 0$ and in [437] in the general case. Earlier partial results in that direction can be found in [403], [396], [114], [186].

We shall first prove the above theorem under the additional assumption $u_0 \geq 0$. This proof is due to [243] and it is based on rescaling arguments (similar to those used in the proof of Theorem 12.1) and on the energy functional E.

Proof of Theorem 22.1 for nonnegative solutions. Assume that the bound (22.2) does not hold for global nonnegative solutions. Then there exist $t_k > 0$ and $u_{0,k} \geq 0$ such that $\|u_{0,k}\|_\infty \leq C_0$ and the solutions $u_k := u(\cdot\,; u_{0,k})$ satisfy

$$M_k := u_k(x_k, t_k) = \sup\{u_k(x,t) : x \in \Omega,\ t \in [0, t_k]\} \to \infty \quad \text{as } k \to \infty. \tag{22.3}$$

Let ψ be the solution of $\psi(0) = C_0$, $\psi'(t) = \psi^p(t)$ for $t > 0$, and let $\delta = \delta(C_0, p) > 0$ be such that $\psi(\delta) = 2C_0$. Then the comparison principle shows $u_k(x,t) \leq \psi(t) \leq 2C_0$ for all $x \in \Omega$ and $t \in [0, \delta]$, hence $t_k \geq \delta$ for k large enough. Now the variation-of-constants formula (15.5) and the estimate

$$\|e^{-tA} w\|_{1,2} \leq C_1 t^{-1/2} \|w\|_2 \leq C_2 t^{-1/2} \|w\|_\infty$$

easily imply $\|u_k(\delta/2)\|_{1,2} \leq C$, where by C we denote a positive constant which does not depend on k. This estimate and Theorem 17.6 guarantee

$$0 \leq E\big(u_k(\delta/2)\big) < C. \tag{22.4}$$

Denote $\nu_k := M_k^{-(p-1)/2}$ and set

$$v_k(y, s) := \frac{1}{M_k} u_k(x_k + \nu_k y, t_k + \nu_k^2 s), \qquad (y, s) \in \overline{Q_k},$$

where $Q_k := \{(y, s) : (x_k + \nu_k y, t_k + \nu_k^2 s) \in \Omega \times (0, t_k)\}$. Then $0 \leq v_k(y,s) \leq 1 = v_k(0,0)$ and v_k solves the problem

$$\partial_s v_k - \Delta_y v_k = v_k^p \qquad \text{in } Q_k,$$

$$v_k = 0 \qquad \text{for } (y, s) \in \partial Q_k, \quad -\frac{t_k}{\nu_k^2} < s < 0.$$

Denote $d_k := \mathrm{dist}\,(x_k, \partial\Omega)$. Passing to a subsequence we may assume that one of the following cases occurs: (i) $d_k/\nu_k \to \infty$, (ii) $d_k/\nu_k \to c \geq 0$.

Case (i). Set

$$\tilde{Q}_k := \{(y, s) : |y| < \frac{d_k}{\nu_k},\ -\frac{t_k}{2\nu_k^2} < s < 0\}.$$

Then $\tilde{Q}_k \subset Q_k$ and the parabolic L^p-estimates (see Appendix B) together with standard embedding theorems guarantee the boundedness of v_k in the space $C^{\alpha,\alpha/2}(\mathbb{R}^n \times (-\infty, 0))$ for some $\alpha > 0$. Consequently, given $\beta \in (0, \alpha)$, we may assume $v_k \to v$ in $C^{\beta,\beta/2}(\mathbb{R}^n \times (-\infty, 0))$, where v is a classical solution of

$$v_s - \Delta v = v^p \qquad \text{in } \mathbb{R}^n \times (-\infty, 0) \tag{22.5}$$

satisfying $0 \le v \le v(0,0) = 1$. Now setting $\sigma := 4/(p-1) - (n-2) > 0$ and using (22.4) we obtain

$$\iint_{\tilde{Q}_k} |\partial_s v_k|^2 \, dy \, ds = \nu_k^\sigma \int_{t_k/2}^{t_k} \int_{|x-x_k|<d_k} |\partial_t u_k|^2 \, dx \, dt \le \nu_k^\sigma \int_{\delta/2}^{\infty} \int_{\Omega} |\partial_t u_k|^2 \, dx \, dt$$

$$\le \nu_k^\sigma \Big[E\big(u_k(\delta/2)\big) - \lim_{t\to\infty} E\big(u_k(t)\big) \Big] \to 0.$$

Since $\partial_s v_k \to v_s$ in $\mathcal{D}'(\mathbb{R}^n \times (-\infty, 0))$, it follows that $v_s \equiv 0$. Now (22.5) contradicts Theorem 8.1.

Case (ii). In this case we obtain, similarly as in Case (i), a function v solving the problem

$$\begin{aligned} v_s - \Delta v &= v^p && \text{in } H_c^n \times (-\infty, 0), \\ v &= 0 && \text{on } \partial H_c^n \times (-\infty, 0), \end{aligned} \tag{22.6}$$

and satisfying $0 \le v \le v(0,0) = 1$, where $H_c^n := \{y \in \mathbb{R}^n : y_1 > -c\}$ (see [243] for details and cf. also the proof of Theorem 12.1). As in Case (i) we obtain $v_s \equiv 0$, hence (22.6) contradicts Theorem 8.2. $\square$

Now we are going to prove Theorem 22.1 in the general case. The proof is based on energy estimates, interpolation, maximal regularity, and a bootstrap argument. The first two ingredients were first used in [114], where the authors had to assume $p(3n-4) < (3n+8)$. The bootstrap argument (which enables one to get rid of this additional assumption on p) appeared for the first time in [437].

Proof of Theorem 22.1. Let $M > 0$ and let u be a global solution of (22.1) with $\|u_0\|_\infty \le M$. We shall denote by C, C_1, C_2 various positive constants which depend on u_0 through M only and which may vary from step to step. Also, by the word "bounded", we mean that the bound depends on u_0 through M only.

As in the proof of Theorem 22.1 for nonnegative solutions, there exists $\delta = \delta(M) > 0$ such that $\|u(t)\|_\infty \le C$ for $t \in [0, \delta]$ and $\|u(\delta)\|_{1,2} \le C$. Hence we may assume $\|u_0\|_{1,2} \le C$. Since u is global, Theorem 17.6 and Remark 17.7 guarantee

$$0 \le E\big(u(t)\big) \le C, \qquad t \ge 0, \tag{22.7}$$

and

$$\|u(t)\|_2 \le C, \qquad t \ge 0. \tag{22.8}$$

Consequently,

$$\int_0^\infty \int_\Omega u_t^2 \, dx \, dt = E(u_0) - \lim_{t\to\infty} E\big(u(t)\big) \le C. \tag{22.9}$$

This estimate and (22.8) guarantee that

$$u \text{ is bounded in } W^{1,2}\big([t, t+1], L^2(\Omega)\big) \text{ uniformly for } t \ge 0. \tag{22.10}$$

Multiplying the equation in (22.1) by u we get

$$\int_\Omega u u_t \, dx = -\int_\Omega |\nabla u(t)|^2 \, dx + \int_\Omega |u(t)|^{p+1} \, dx = -2E\big(u(t)\big) + C \int_\Omega |u(t)|^{p+1} \, dx,$$

so that, for each $r \geq 1$, (22.7) implies

$$\int_t^{t+1} \Big(\int_\Omega |u|^{p+1} \, dx\Big)^r \, ds \leq C\Big[1 + \int_t^{t+1} \Big(\int_\Omega |u u_t| \, dx\Big)^r \, ds\Big], \qquad t \geq 0. \quad (22.11)$$

Notice that Cauchy's inequality, (22.8) and (22.9) imply

$$\int_t^{t+1} \Big(\int_\Omega |u u_t| \, dx\Big)^2 \, ds \leq \int_t^{t+1} \Big(\int_\Omega u^2 \, dx\Big)\Big(\int_\Omega u_t^2 \, dx\Big) \, ds \leq C,$$

hence we infer from (22.11) that

$$u \text{ is bounded in } L^{(p+1)r}\big([t, t+1], L^{p+1}(\Omega)\big) \text{ uniformly for } t \geq 0, \quad (22.12)$$

if $r = 2$. Now (22.10), (22.12) and (51.6) guarantee

$$\|u(t)\|_q \leq C_q \qquad \text{for all } t \geq 0 \text{ and } q < q_r := p + 1 - \frac{p-1}{r+1}, \quad (22.13)$$

where $r = 2$. Theorem 15.2 or Remark 51.37(iii) (see also Theorem 16.4) imply our assertion provided $\sup_{t \geq 0} \|u(t)\|_q \leq C$ for some $q > n(p-1)/2$. This estimate follows from (22.13) if

$$\frac{n}{2}(p-1) < p + 1 - \frac{p-1}{r+1}. \quad (22.14)$$

If $r = 2$, then (22.14) is equivalent to $p(3n - 4) < 3n + 8$ (which is the condition of [114]). In what follows we shall use a bootstrap argument to show that (22.13) is true for any $r \geq 2$. Since (22.14) reduces to $p < p_S$ if $r \to \infty$ we shall be done.

We already know (see the beginning of the proof) that there exists $\delta = \delta(M) > 0$ such that $\|u(t)\|_\infty \leq C$ for $t \in [0, \delta]$. We claim that for any interval $I \subset [0, \infty)$ of length δ there exists $\tau \in I$ such that $\|u(\tau)\|_{BC^2} \leq C$. In fact, let $I = (t, t+\delta)$ and set $J := (t, t + \delta/2)$. Then (22.7) and (22.12) with $r = 2$ imply

$$\int_J \Big(\int_\Omega |\nabla u|^2 \, dx\Big)^2 \, ds \leq C\Big[1 + \int_J \Big(\int_\Omega |u|^{p+1} \, dx\Big)^2 \, ds\Big] \leq C,$$

hence there exist $C_1 > 0$ and $\tau_J \in J$ such that $\|u(\tau_J)\|_{1,2} \leq C_1$. The well-posedness of (22.1) in $W_0^{1,2}(\Omega)$ (see Example 51.10 and Theorem 51.7) guarantees the existence of $\eta = \eta(C_1) > 0$ and $C_2 = C_2(C_1) > 0$ such that $\eta < \delta/2$ and $\|u(s)\|_{1,2} \leq C_2$ for all $s \in [\tau_J, \tau_J + \eta]$. Now standard regularity results (see Example 51.27 and

Appendix B) guarantee $\|u(\tau_J + \eta)\|_{BC^2} \leq C$, where $C = C(\eta, C_2)$. Hence it is sufficient to put $\tau := \tau_J + \eta$.

Next assume that $r \geq 2$ and

$$\int_t^{t+1} \left(\int_\Omega |u|^{p+1} \, dx \right)^r ds \leq C \quad \text{for all } t \geq \delta. \tag{22.15}$$

We shall show that the same estimate is true with r replaced by $\tilde{r}$ for any $\tilde{r} \in (r, r+2)$. Since (22.15) is true for $r = 2$, an obvious bootstrap argument will guarantee (22.15) for any $r \geq 2$. Since (22.15) implies (22.13), the conclusion will follow.

Hence let $\tilde{r} \in (r, r+2)$, and consider $q < q_r$ (q close to q_r). Set

$$\hat{p} := (p+1)/p, \qquad \theta := \frac{p+1}{p-1} \frac{q-2}{q} \in (0,1), \qquad \beta := 2/(\tilde{r}(1-\theta)) > 1.$$

Choose $t \geq \delta$ and $\tau \in (t - \delta, t)$ such that $\|u(\tau)\|_{BC^2} \leq C$. Using successively (22.11), Hölder's inequality and (22.13), interpolation, Hölder's inequality, (22.9), the maximal regularity property (51.8), and $\|u(\tau)\|_{BC^2} \leq C$, we obtain

$$\int_\tau^{t+1} \left(\int_\Omega |u|^{p+1} \, dx \right)^{\tilde{r}} ds \leq C \left[1 + \int_\tau^{t+1} \|uu_t\|_1^{\tilde{r}} \, ds \right]$$

$$\leq C \left[1 + \int_\tau^{t+1} \|u_t\|_{q'}^{\tilde{r}} \, ds \right]$$

$$\leq C \left[1 + \int_\tau^{t+1} \|u_t\|_{\hat{p}}^{\tilde{r}\theta} \|u_t\|_2^{\tilde{r}(1-\theta)} \, ds \right]$$

$$\leq C \left[1 + \left(\int_\tau^{t+1} \|u_t\|_{\hat{p}}^{\tilde{r}\theta\beta'} \, ds \right)^{1/\beta'} \left(\int_\tau^{t+1} \|u_t\|_2^2 \, ds \right)^{1/\beta} \right]$$

$$\leq C \left[1 + \|u(\tau)\|_{BC^2} + \left(\int_\tau^{t+1} \||u|^{p-1} u\|_{\hat{p}}^{\tilde{r}\theta\beta'} \, ds \right)^{1/\beta'} \right]$$

$$\leq C \left[1 + \left(\int_\tau^{t+1} \left(\int_\Omega |u|^{p+1} \, dx \right)^{\tilde{r}\theta\beta' p/(p+1)} ds \right)^{1/\beta'} \right].$$

Since $\tilde{r} < r+2$ we can choose q close to q_r so that $\tilde{r}\theta\beta' p/(p+1) < \tilde{r}$. Consequently, (22.15) and the last estimate guarantee (22.15) with r replaced by $\tilde{r}$. $\square$

Remarks 22.2. Uniform bound in terms of the energy. Let Ω be bounded, $p < p_S$, u be a solution of (22.1) on the time interval $[0, T)$, $T < \infty$ and $M > 0$.

(i) If

$$\|u_0\|_\infty \leq M, \qquad E(u(t)) \geq -M \quad \text{and} \quad \|u(t)\|_2 \leq M, \quad t \in [0, T),$$

then
$$\|u(t)\|_\infty \le C(M), \qquad t \in [0, T).$$

This follows from the above proof of Theorem 22.1 by replacing the interval $[0, \infty)$ with $[0, T)$.

(ii) If
$$\|u_0\|_\infty \le M \quad \text{and} \quad E(u(t)) \ge -M, \quad t \in [0, T), \tag{22.16}$$

then
$$\|u(t)\|_\infty \le C(M, T), \qquad t \in [0, T).$$

In fact, (22.16), (17.10) and Gronwall's inequality guarantee
$$\|u(t)\|_2 \le C(K, M, T), \qquad t < T,$$

where K stands for a bound on $\|u_0\|_2$. Therefore the assertion follows from (i). $\square$

Remark 22.3. Cauchy problem. Let $\Omega = \mathbb{R}^n$ and $1 < p < p_S$. Then (22.2) is still true for positive radial solutions (and for all positive solutions provided $p < p_B$), see Theorem 26.9 below. Using the same approach as in the proof of Theorem 22.1 for nonnegative solutions one can also show weaker estimate
$$\|u(t)\|_\infty \le C(\|u_0\|_\infty, E(u_0)) \quad \text{for all } t \ge 0$$

for nonnegative initial data $u_0 \in H^1(\mathbb{R}^n)$.

If we consider problem (17.1) with $\lambda < 0$, $1 < p < p_S$, $\Omega = \mathbb{R}^n$ and initial data in $X := L^\infty \cap L^{(p+1)/p} \cap H^1(\mathbb{R}^n)$, then any global (not necessarily positive) solution satisfies the estimate
$$\|u(t)\|_X \le C(\|u_0\|_X) \quad \text{for all } t \ge 0,$$

see [440]. The same result remains true for $X := H^1(\mathbb{R}^n)$ or $X := L^\infty \cap H^1(\mathbb{R}^n)$ due to a recent result in [537]. $\square$

22.2. Boundedness of global solutions in the supercritical case

Consider problem (22.1), where Ω is a ball and $u_0 \in L^\infty(\Omega)$ is a nonnegative radial function. If the solution u is global and $p < p_S$, then Theorem 22.1 guarantees the boundedness of u, i.e.:
$$\sup_{t \ge 0} \|u(t)\|_\infty < \infty. \tag{22.17}$$

In this subsection we show that this property remains true for $p > p_S$.

Let us emphasize that the bound (22.17) does *not* imply the stronger a priori estimate (22.2). In fact, we will see in Theorem 28.7(iv) that estimate (22.2) fails whenever $p \ge p_S$.

Theorem 22.4. *Assume $p > p_S$, $\Omega = B_R$, and let $u_0 \in L^\infty(\Omega)$, $u_0 \geq 0$, be radial. If the solution u of (22.1) is global, then property (22.17) is true.*

We will prove Theorem 22.4 only under the additional assumption $p < p_L$, where

$$p_L := \begin{cases} \infty & \text{if } n \leq 10, \\ 1 + \dfrac{6}{n-10} & \text{if } n > 10. \end{cases} \tag{22.18}$$

Notice that $p_L > p_{JL}$ if $n > 10$, where p_{JL} is defined in (9.3). If $n > 10$ and $p > p_{JL}$, then the statement of Theorem 22.4 follows from [378]. See also Remark 23.13 for an alternative proof, due to [125], in the case $p < p_{JL}$.

In the proof of the above theorem we will need the following result.

Proposition 22.5. *Let $p_S < p < p_L$ and let c_p be the constant defined in (3.9). Then there exists a positive bounded solution of the problem*

$$\varphi'' + \left(\frac{n-1}{y} - \frac{y}{2} \right)\varphi' + \varphi^p - \frac{1}{p-1}\varphi = 0, \qquad y > 0,$$
$$\varphi'(0) = 0,$$

satisfying $\lim_{y\to\infty} \varphi(y)y^{2/(p-1)} = B \in (0, c_p)$.

Given $T \in \mathbb{R}$, set

$$w(r,t) := (T-t)^{-1/(p-1)}\varphi\big(r/\sqrt{T-t}\big) \qquad \textit{for } r \geq 0, \ t < T,$$
$$w(r,T) := \lim_{t\to T+} w(r,t) \qquad\qquad\qquad \textit{for } r > 0.$$

Then

$$w_t - w_{rr} - \frac{n-1}{r}w_r = w^p, \qquad\qquad r > 0, \ t < T,$$
$$w(r,T) = Br^{-2/(p-1)}, \qquad r > 0.$$

The function w in the preceding proposition is a **backward self-similar solution** of problem (22.1). Proposition 22.5 follows from [105], [325] (if $p < p_{JL}$) and [326] (if $p \geq p_{JL}$). Since the corresponding proofs are quite long, we will prove it just in the case $p = 2$ when one can find an explicit formula for φ (due to [225]). Let us note that in the case $p < p_{JL}$ there exist infinitely many functions φ with the required properties, and that both numerical and analytic results indicate that such solutions do not exist if $p > p_L$, see [418], [376].

Proof of Proposition 22.5 for $p = 2$. Let $p = 2$ and $6 < n < 16$ (this corresponds to $p_S < 2 < p_L$). Set

$$\varphi(y) := \frac{A}{(a + y^2)^2} + \frac{B}{a + y^2},$$

where

$$A := 48(10D - (n + 14)), \quad B := 24(D - 2), \quad D := \sqrt{1 + n/2}.$$

It is easy to see that φ possesses the required properties. In particular, $B < c_2 = 2(n - 4)$. $\quad\square$

The following proof is due to [232].

Proof of Theorem 22.4 for $p < p_L$**.** Let $U_*(r) = c_p r^{-2/(p-1)}$ be the singular solution defined in (3.9). Assume on the contrary that u is a global unbounded classical solution. Since u is radial (see Remark 16.2(i)), we have $u(x, t) = U(|x|, t)$ for some $U : [0, R] \times (0, \infty) \to \mathbb{R}$.

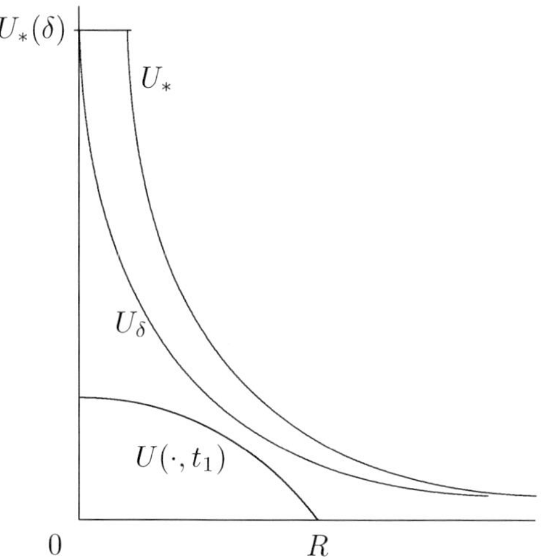

Figure 9: Graphs of $U_*, U_\delta, U(\cdot, t_1)$ if $z(U(\cdot, t_0) - U_*) = 0$.

Assume $z(U(\cdot, t_0) - U_*) \leq 1$ for some $t_0 > 0$, where $z(\psi)$ denotes the zero number of the function ψ in the interval $(0, R)$ (see Appendix F). Since $U(0, t_0) < U_*(0) = \infty$ and $0 = U(R, t_0) < U_*(R)$ we have $z(U(\cdot, t_0) - U_*) = 0$. Consequently $U(\cdot, t_0) \leq U_*$. Fix $t_1 > t_0$. Then by the maximum principle there exists $\varepsilon > 0$ such that $U(\cdot, t_1) \leq U_* - \varepsilon$ and we may find $\delta > 0$ such that the function $U_\delta(r) := U_*(r + \delta)$ lies above $U(\cdot, t_1)$. Since

$$-U_\delta'' - \frac{n - 1}{r} U_\delta' \geq U_\delta^p, \qquad 0 < r < R,$$

with $-U_\delta'(0) > 0$ and $U_\delta(R) > 0$, it follows from the maximum principle that $U(r,t) \le U_\delta(r) \le U_*(\delta)$ for all $r \in [0,R]$ and $t \ge t_1$, see Figure 9. However, this contradicts our assumptions. Consequently,

$$z(U(\cdot,t) - U_*) \ge 2 \quad \text{for all } t > 0. \tag{22.19}$$

Fix $\tau > 0$ small. Since $U(r,\tau) > 0$ for $r \in [0,R)$ and $U_r(R,\tau) < 0$ by the maximum principle, we can find T large enough such that the backward self-similar solution w from Proposition 22.5 satisfies $z(U(\cdot,\tau) - w(\cdot,\tau)) = 1$, see Figure 10.

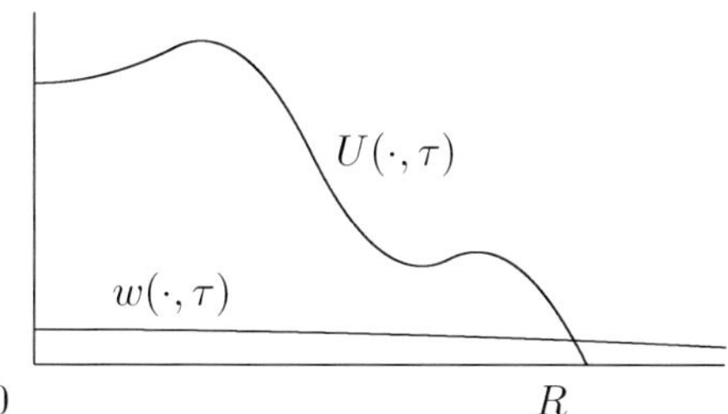

Figure 10: Graphs of $U(\cdot,\tau), w(\cdot,\tau)$.

Consequently, Theorem 52.28 implies

$$z(U(\cdot,t) - w(\cdot,t)) \le 1 \quad \text{for all } t \in [\tau, T). \tag{22.20}$$

However, $w(\cdot,T) < U_*$ so that (22.19) implies (see Figure 11)

$$z(U(\cdot,t) - w(\cdot,t)) \ge 2 \quad \text{for } t < T,\ t \text{ close to } T,$$

which contradicts (22.20). $\square$

Remarks 22.6. (a) **Cauchy problem.** Let $\Omega = \mathbb{R}^n$, $u_0 \in C(\mathbb{R}^n)$ be nonnegative, bounded and radially symmetric and let the solution u of (22.1) be global. Then the boundedness of u is known in each of the following (supercritical) cases:

(i) $p_S < p < p_L$, $u_0 \in C^1$ has compact support and its local minima are bounded away from zero (see [374]);

(ii) $p_S < p < p_{JL}$ and u_0 has just finitely many intersections with the singular stationary solution (see [356]);

(iii) $p > p_{JL}$, $u_0(x) \le U_*(|x|) - c_0|x|^{-|\alpha|}$, where $c_0 > 0$,

$$\alpha = \frac{1}{2}\left[-(n-2) + \sqrt{\beta^2 - 4(p-1)c_p^{p-1}}\right],$$

$\beta = n - 2 - 4/(p-1)$ and c_p, U_* are defined in (3.9) (see [378]).

On the other hand, if $p \geq p_{JL}$ it was shown in [428] that there exists a continuous radial function u_0 satisfying $0 < u_0(x) \leq U_*(|x|)$ such that the corresponding solution u is global and unbounded. See also Section 29 for more precise information on the asymptotic behavior of such solutions.

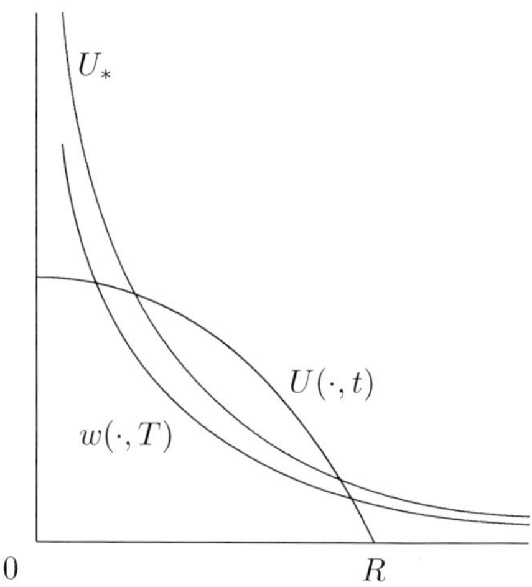

Figure 11: Graphs of $U_*, w(\cdot, T), U(\cdot, t)$ if t is close to T.

(b) **Inhomogeneous boundary conditions.** The result in Theorem 22.4 is sensitive to the boundary conditions. Indeed consider problem (22.1) in $\Omega = B_1$ with the boundary conditions replaced by $u = a > 0$ on $\partial\Omega \times (0, \infty)$. Note that this is equivalent to problem (14.1) with

$$f(v) := (v + a)^p \qquad (\text{resp.,}\ f(w) := \lambda(w + a)^p,\ \ \lambda = a^{1-p})$$

via the transformation $v = u - a$ (resp., $w = a^{-1}u - 1$). If $p_S < p < p_{JL}$, then global radial solutions are still bounded [125]. But if $p > p_{JL}$ and $a = c_p$, where c_p is given by (3.9), then there exist unbounded global solutions [316]. More precisely, any initial data $u_0 \in L^\infty(\Omega)$ satisfying $0 \leq u_0(x) \leq u_*(x) = U_*(|x|)$ gives rise to an unbounded global classical solution, which stabilizes to u_* as $t \to \infty$. The rate of approach has been studied in [164]. $\square$

Remark 22.7. Eventual radial monotonicity of global radial solutions. The following property was shown in [395] (actually for more general nonlinearities). Let $p > 1$, $\Omega = B_R$, and assume that $u \geq 0$ is a radial, global classical solution of (22.1) (not necessarily bounded). Then there exists $t_0 > 0$ such that u becomes radial nonincreasing for $t \geq t_0$. $\square$

Remark 22.8. Exponential nonlinearity. Consider the problem

$$\left. \begin{aligned} u_t - \Delta u &= \lambda e^u, & x \in \Omega,\ t > 0, \\ u &= 0, & x \in \partial\Omega,\ t > 0, \\ u(x,0) &= u_0(x), & x \in \Omega, \end{aligned} \right\} \tag{22.21}$$

where $\Omega \subset \mathbb{R}^n$, $\lambda > 0$ and $u_0 \in L^\infty(\Omega)$. Embedding theorems, intersection properties of stationary solutions of (22.21) for $\Omega = \mathbb{R}^n$ (see [510]), bifurcation diagrams for stationary solutions of (22.21) for Ω being a ball (see Remark 6.10(ii)), and several results for time dependent solutions of (22.21) indicate that the cases $n \leq 2$, $3 \leq n \leq 9$ and $n \geq 10$ correspond to the cases $p < p_S$, $p_S < p < p_{JL}$ and $p > p_{JL}$ for problem (22.1), respectively. In fact, many (but not all) proofs in this chapter can be adapted to the case of exponential nonlinearity. Unfortunately, similarly as in the case of the power nonlinearity, a lot of basic questions for (22.21) remain open. For example, a priori bounds (22.2) are known if $n = 1$ (see [440]) but not for $n = 2$. On the other hand, the boundedness of global solutions of (22.21) with general bounded Ω is true if $n \leq 2$ (see [186], [109]) and the boundedness of radial global solutions in a ball for $3 \leq n \leq 9$ is known as well (see [198]). We refer to [188] for a survey on problem (22.21). $\square$

22.3. Global unbounded solutions in the critical case

The following result due to [232] shows that the situation in the Sobolev critical case is very different from both the subcritical and the supercritical cases.

To formulate it, we introduce the notion of threshold solution. Let $\varphi \in L^\infty(\Omega)$ be a fixed nonnegative function, $\varphi \not\equiv 0$, $\alpha > 0$, and set $u_0 = \alpha\varphi$. If α is small enough, then the solution $u = u(t; \alpha\varphi)$ of (22.1) exists globally. (Moreover $u(t) \to 0$ in $L^\infty(\Omega)$, as $t \to \infty$.) This follows from Theorem 19.2. We may thus define

$$\alpha^* = \alpha^*(\varphi) := \sup\{\alpha > 0 : T_{\max}(\alpha\varphi) = \infty\}. \tag{22.22}$$

Note that $\alpha^* \in (0, \infty)$ due to Remark 17.7(v). The function $u^* = u(t; \alpha^*\varphi)$ is called the **threshold solution** (associated with φ), due to the fact that u^* lies on the borderline between blow-up and global existence. Further properties of threshold and non-threshold solutions will be studied in Sections 27 and 28.

Theorem 22.9. *Consider problem (22.1) with $p = p_S$ and $\Omega = B_R$. Let $u_0 = \alpha^*\varphi$, where $\varphi(x) = \Phi(|x|)$, with $0 \leq \Phi \in L^\infty(0, R)$, Φ nonincreasing, and α^* defined by (22.22). Then the solution u^* is global and unbounded. More precisely,*

$$\lim_{t \to \infty} \|u^*(t)\|_q = \infty \qquad \text{for any } q > p_S + 1,$$

$$\liminf_{t \to \infty} \|u^*(t)\|_{p_S+1} < \infty. \tag{22.23}$$

Proof. First assume that u^* blows up in finite time T. Let $\alpha_k \nearrow \alpha^*$, $\alpha_1 > 0$ and let v_k, $k = 1, 2, \ldots$, denote the (global) solution with the initial data $\alpha_k \varphi$. The solutions u^*, v_k are radial and radially decreasing, $u^*(x, t) = U^*(|x|, t)$, $v_k(x, t) = V_k(|x|, t)$. Let $t_1 \in (0, T)$ be fixed. Since V_1 is positive on $Q_1 := [0, R/2] \times [t_1, T+1]$, there exists $c_1 > 0$ such that $V_k \geq V_1 > c_1$ on Q_1 for any k. In addition, $U^* \geq V_k$ on $[0, R/2] \times [t_1, T)$. The functions $U^*(\cdot, t_1)$ and $V_k(\cdot, t_1)$, $k = 1, 2, \ldots$, are uniformly bounded in $C^1([0, R])$. In particular, there exists $c_2 > 0$ such that $V_k(\cdot, t_1) \leq U^*(\cdot, t_1) < c_2$.

Let U_M be the unique positive solution of (9.2) satisfying $U_M(0) = M$, see Theorem 9.1. Since $U_M(R/2) \to 0$ as $M \to \infty$, there exists $M_1 > 0$ such that $U_M(R/2) < c_1$ for all $M \geq M_1$. Enlarging M_1 if necessary we may also assume that the function $M \mapsto U_M(R/2)$ is decreasing for $M \geq M_1$, and that $M_1 \geq M_0(R/2)$, where the function M_0 is defined in Remark 9.3. Finally, since $U'_M(r) \to -\infty$ as $M \to \infty$ uniformly on $\{r : U_M(r) \in [c_1, c_2]\}$, we may assume that U_M intersects any of the functions $V_k(\cdot, t_1)$, $k = 1, 2, \ldots$, exactly once in $[0, R/2]$, for all $M \geq M_1$, see Figure 12.

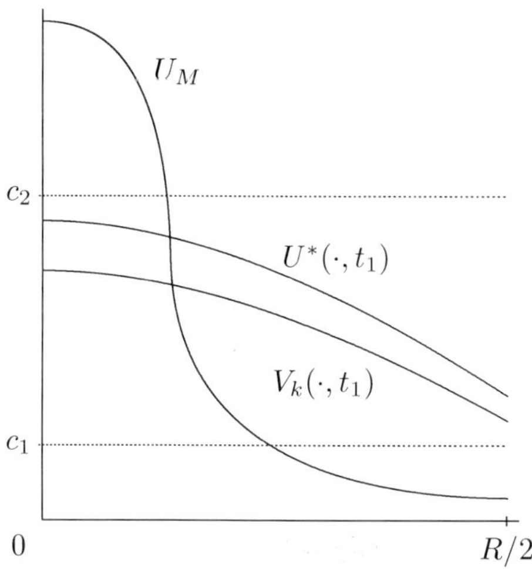

Figure 12: Graphs of $U^*(\cdot, t_1), V_k(\cdot, t_1), U_M$.

Consequently, denoting by $z(\psi)$ the zero number of the function ψ in the interval $[0, R/2]$ (see Appendix F), we have

$$z(U_M - V_k(\cdot, t_1)) = 1, \qquad k = 1, 2, \ldots, \ M \geq M_1. \tag{22.24}$$

Fix $M_2 > M_1$ (see Figure 13) and let $\tilde{U}$ be the solution of the problem

$$\tilde{U}_t - \tilde{U}_{rr} - \frac{n-1}{r}\tilde{U}_r = \tilde{U}^p, \qquad\qquad r \in (0, R/2),\ t > 0,$$

$$\tilde{U}_r(0,t) = 0,\ \tilde{U}(R/2, t) = U_{M_1}(R/2), \qquad t > 0,$$

$$\tilde{U}(r,0) = \max(U_{M_2}(r), U_{M_1}(r)), \qquad\qquad r \in (0, R/2).$$

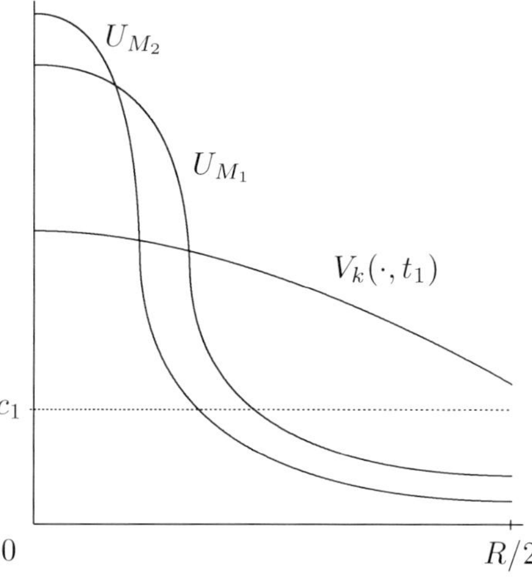

Figure 13: Graphs of $V_k(\cdot, t_1), U_{M_1}, U_{M_2}$.

We have $\tilde{U}_r \le 0$ for $t > 0$ by Proposition 52.17. Moreover, the function $\tilde{U}(r,0)$ is a subsolution for this problem, hence

$$\tilde{U}_t \ge 0, \qquad \text{for } t > 0 \tag{22.25}$$

(in fact this follows from a simple modification of the proof of Proposition 52.19). We claim that:

$$\tilde{U} \text{ blows up in a finite time } \tilde{T}. \tag{22.26}$$

Assume for contradiction that $\tilde{U}$ exists globally and let $\tilde{V}(r) := \lim_{t\to\infty} \tilde{U}(r,t)$. First we have $\tilde{V}(r) < \infty$ for $0 < r \le R/2$ (otherwise we would have

$$\lim_{t\to\infty} \tilde{U}(r,t) = \infty \quad \text{uniformly on } [0, r_0) \text{ for some } r_0 > 0,$$

which would imply finite-time blow-up by an eigenfunction argument — cf. the proof of Theorem 17.1). It follows from (22.25) and Lemma 53.10 that $\tilde{V} \in C^2((0, R/2])$ is a solution of $\tilde{V}_{rr} + \frac{n-1}{r}\tilde{V}_r + \tilde{V}^p = 0$ on $0 < r \leq R/2$. Moreover we have $\tilde{V} > \max(U_{M_2}, U_{M_1})$ and $\tilde{V}_r \leq 0$ on $(0, R/2)$, and $\tilde{V}(R/2) = U_{M_1}(R/2)$. But since $M_1 \geq M_0(R/2)$, Remark 9.3 implies $\tilde{V}(r_0) = U_{M_1}(r_0)$ for some $r_0 \in (0, R/2)$: a contradiction. Consequently, (22.26) is true.

Fix $\beta \geq 1$ such that $T_\beta := \tilde{T}\beta^{1-p} < 1$, set $R_\beta := \beta^{-(p-1)/2}R/2$ and notice that $W(r,t) := \beta\tilde{U}(r\beta^{(p-1)/2}, t\beta^{p-1})$ is a solution of the problem

$$W_t - W_{rr} - \frac{n-1}{r}W_r = W^p, \qquad\qquad r \in (0, R_\beta),\ t > 0,$$
$$W_r(0,t) = 0,\ W(R_\beta, t) = U_{\beta M_1}(R_\beta), \qquad t > 0,$$
$$W(r,0) = \max(U_{\beta M_2}(r), U_{\beta M_1}(r)), \qquad r \in (0, R_\beta),$$

which blows up at time $T_\beta < 1$. Since U^* blows up at time T and is decreasing in r, and since $V_k(0,t) \to U^*(0,t)$ as $k \to \infty$ for any $t < T$, there exist k and $t_0 \in (t_1, T)$ such that $V_k(0, t_0) > U_{\beta M_2}(0) = \beta M_2 > U_{\beta M_1}(0)$. Notice also that

$$V_k(R/2, t) > c_1 > U_{\beta M_1}(R/2) > U_{\beta M_2}(R/2) \quad \text{for all } t \in [t_1, T+1].$$

Now (22.24) and the monotonicity of the zero number (see Theorem 52.28) imply $V_k(\cdot, t) > U_{\beta M_i}$ on $[0, R/2]$ for all $t \in [t_0, T+1]$ and $i = 1, 2$, hence $V_k(\cdot, t_0) > W(\cdot, 0)$ on $[0, R_\beta]$. Since $V_k(R_\beta, t) > U_{\beta M_1}(R_\beta) = W(R_\beta, t)$ for $t \in [t_0, T+1]$, we have $V_k(\cdot, t+t_0) > W(\cdot, t)$ whenever $t > 0$, $t + t_0 \leq T+1$. However W blows up at $T_\beta < 1$ which yields a contradiction. Consequently, u^* is global.

Next assume that $\liminf_{t\to\infty} \|u^*(t)\|_q < \infty$ for some $q > p_S + 1$. Then there exist $C > 0$ and $t_k \to \infty$ such that $\|u^*(t_k)\|_q < C$. Fix $\gamma \in (1/2, 1)$. Since $q > n(p_S - 1)/2$, Theorem 51.25, Remark 51.26(vi) and Example 51.27 (with $z = q$ and $\alpha = 1$) show the existence of $\delta > 0$ such that the sequence $\{u^*(t_k + \delta)\}$ is bounded in $W^{2\gamma,q} \cap W_0^{1,q}(\Omega)$, hence relatively compact in $X := H_0^1 \cap L^q(\Omega)$. Next Example 51.28 and Proposition 53.6 guarantee that a subsequence of $\{u^*(t_k + \delta)\}$ converges in X to an equilibrium v. The maximum principle implies $v \geq 0$. Assume $v = 0$. Then $\alpha^*\varphi$ belongs to the domain of attraction of the zero solution (which is an open set) hence the same is true for $\alpha\varphi$ with some $\alpha > \alpha^*$. But this contradicts the definition of α^*. Consequently, $v > 0$. However, this contradicts Corollary 5.2.

Finally assume $\lim_{t\to\infty} \|u^*(t)\|_{p_S+1} = \infty$. Then estimate (17.9) shows that the L^2-norm of $u^*(t)$ has to blow up in finite time which is absurd. In fact, Theorem 17.6 also shows that the energy of $u^*(t)$ remains bounded and the proof of Theorem 22.1 guarantees that the norm of u^* in $L^4((t, t+1), H^1(\Omega))$ is bounded uniformly with respect to $t \geq t_0 > 0$. $\square$

Remarks 22.10. (i) **Grow-up rates in a ball.** The assertion in Theorem 22.9 remains true even if the function Φ is not monotone (cf. the proof of Theorem 28.7

below). In addition, if $R = 1$, then all such global unbounded radial positive solutions exhibit the following asymptotic behavior as $t \to \infty$ (see [223]):

$$\log \|u^*(\cdot, t)\|_\infty = \frac{\pi^2}{4} t(1 + o(1)) \quad \text{if } n = 3,$$

$$\log \|u^*(\cdot, t)\|_\infty = 2\sqrt{t}(1 + o(1)) \quad \text{if } n = 4,$$

$$\|u^*(\cdot, t)\|_\infty = \gamma_0 t^{(n-2)/2(n-4)}(1 + o(1)) \quad \text{if } n \geq 5,$$

where the constant $\gamma_0 > 0$ depends only on the spatial dimension n.

(ii) **Grow-up rates for the Cauchy problem.** Let $\Omega = \mathbb{R}^n$, $p = p_S$, $\gamma > 2/(p-1)$, $\Phi : [0, \infty) \to (0, \infty)$ satisfy $\Phi(r) \sim Cr^{-\gamma}$ for r large and consider initial data $u_0(x) = \alpha^* \Phi(|x|)$, where α^* has the same meaning as in Theorem 22.9. If $n = 3$, then formal matched asymptotics expansions ([306]) suggest that for t large, $\|u(t)\|_\infty$ behaves like $t^{(\gamma-1)/2}$ or $t^{1/2}$ provided $\gamma \in (1/2, 2)$ or $\gamma > 2$, respectively. On the other hand, the same arguments indicate that this solution remains bounded if $n > 3$.

(iii) **Nonuniformity of the smoothing time in the critical L^q-space.** Let u^* be the global unbounded solution from Theorem 22.9. Fix $C_1 > 0$ and $t_k \to \infty$ such that $\|u^*(t_k)\|_{p_S+1} < C_1$. Since $p_S + 1 = q_c = n(p_S - 1)/2$, Remark 15.4(i) guarantees that problem (22.1) is well-posed in $L^{p_S+1}(\Omega)$ and, in particular, there exist $C_2 > 0$ and $T_k > 0$ such that

$$\|u^*(t_k + t)\|_\infty \leq C_2 \|u^*(t_k)\|_{p_S+1} t^{-\alpha} \leq C_1 C_2 t^{-\alpha}, \qquad t \in (0, T_k),$$

where $\alpha = (n/2)(p_S + 1) = (n-2)/4$, cf. (15.2). Since $\|u^*(t_k + t)\|_\infty \to \infty$ for any $t \geq 0$, we see that $T_k \to 0$ (in spite of the fact that $T_{\max}(u^*(t_k)) = \infty$ and $\|u^*(t_k)\|_{p_S+1} < C_1$). $\square$

22.4. Estimates for nonglobal solutions

The estimates in Theorem 22.1 can be extended to nonglobal solutions in the following way.

Proposition 22.11. *Assume Ω bounded, $1 < p < p_S$, $K, \delta > 0$ and $\|u_0\|_\infty \leq K$. If u is the solution of (22.1), then*

$$\|u(t)\|_\infty \leq C(\delta, K) \quad \text{for all } t \in [0, T_{\max}(u_0) - \delta), \tag{22.27}$$

(where $T_{\max}(u_0) - \delta := \infty$ if $T_{\max}(u_0) = \infty$) and

$$E\big(u(t)\big) \to -\infty \quad \text{as } t \to T_{\max}(u_0), \quad \text{whenever } T_{\max}(u_0) < \infty. \tag{22.28}$$

Remark 22.12. Related blow-up rate estimates of the form

$$\|u(t)\|_\infty \le M(T-t)^{-1/(p-1)}, \qquad 0 < t < T := T_{\max}(u_0),$$

will be proved in Section 23. In some cases the constant M will be known to depend on u_0 through a bound on $\|u_0\|_\infty$ only (see e.g. Remark 23.9). However, up to now, such a priori estimates are not available under the general assumptions of Proposition 22.11 (one has to assume either Ω convex, or $u \ge 0$ and a stronger restriction on p) $\quad\square$

Proof of Proposition 22.11. If $T_{\max}(u_0) = \infty$, then estimate (22.27) follows from Theorem 22.1.

Assume $T_{\max}(u_0) < \infty$, and set $T := T_{\max}(u_0) - \delta$. As in the proof of Theorem 22.1 we may assume that $\|u_0\|_{1,2} \le C$, hence $E(u(t)) \le C$ for $t \ge 0$. Denoting $\psi(t) = \|u(t)\|_2^2$ we have (cf. (17.9))

$$\frac{1}{2}\psi'(t) \ge -2E(u(t)) + c_1\psi^{(p+1)/2}(t),$$

where $c_1 = c_1(p,\Omega) > 0$. Set $M := ((p-1)c_1\delta/2)^{-2/(p-1)}/\delta$ and assume $E(u(t_0)) \le -M$ for some $t_0 \in [0, T]$. Then $\psi'(t) \ge 4M$ for $t \ge t_0$, hence $\psi(t_0 + \delta/2) \ge 2\delta M$. Since $\psi' \ge 2c_1\psi^{(p+1)/2}$ and the solution of the problem

$$y(0) = 2\delta M, \qquad y' = 2c_1 y^{(p+1)/2}$$

blows up at $t < \delta/2$, ψ cannot exist on the whole interval $[t_0 + \delta/2, t_0 + \delta)$ which yields a contradiction. Consequently, $E(u(t)) \ge -M$ for all $t \in [0, T]$ and similar arguments show $\|u(t)\|_2 = \sqrt{\psi(t)} \le \tilde{M}$ for all $t \in [0, T]$ and suitable $\tilde{M} = \tilde{M}(K, \delta)$. Now (22.27) follows from Remark 22.2(i).

Assertion (22.28) follows from Remark 22.2(ii). $\quad\square$

Estimates (22.27) and (22.28) can be proved for a fairly general class of superlinear subcritical parabolic problems in bounded domains, including problems with nonlocal nonlinearities (see [440]).

Property (22.28) plays an important role in the proof of complete blow-up (see Remark 27.8(b) below).

As an easy application of estimate (22.27) we obtain the following important theorem concerning the *continuity of the existence time*.

Theorem 22.13. *Assume Ω bounded, $1 < p < p_S$, and let $T_{\max}(u_0)$ denote the maximal existence time of the solution of (22.1). Then the function*

$$T_{\max} : L^\infty(\Omega) \to (0, \infty] : u_0 \mapsto T_{\max}(u_0)$$

is continuous.

Proof. If $0 < T < T_{\max}(u_0)$, then the continuous dependence of solutions of (22.1) on initial data (see (51.28)) guarantees the existence of $\varepsilon > 0$ such that $T_{\max}(v_0) > T$ for any v_0 satisfying $\|u_0 - v_0\|_\infty < \varepsilon$. Hence $T_{\max}$ is lower semicontinuous. Next assume $u_{0,k} \to u_0$ in $L^\infty(\Omega)$ and $T_{\max}(u_{0,k}) > T + \delta > 0$ for some $\delta > 0$ and all k. Then (22.27) guarantees that the corresponding solutions u_k satisfy $\|u_k(t)\|_\infty \leq C$ for all $t \in [0, T]$ and $k = 1, 2, \dots$. Passing to the limit we obtain $T_{\max}(u_0) \geq T$ and $\|u(t)\|_\infty \leq C$. Consequently, $T_{\max}$ is upper semicontinuous. $\square$

The function $T_{\max}$ need *not* be continuous in the supercritical case even in the model case (22.1) (consider the threshold trajectory u^* from Theorem 28.7 below: if u^* blows up in finite time, then $T_{\max}$ is not continuous at $u^*(0) = \alpha^* \phi$).

23. Blow-up rate

In this section we consider the model problem (22.1) and assume that $T_{\max}(u_0) < \infty$. The solution of the ODE

$$y' = y^p, \quad t > 0, \qquad y(0) = y_0 > 0, \tag{23.1}$$

is given by

$$y(t) = k(T - t)^{-1/(p-1)}, \ 0 < t < T, \quad \text{where } k = (p-1)^{-1/(p-1)}, \tag{23.2}$$

with $T = (p-1)^{-1} y_0^{1-p}$. It is natural to ask whether the blow-up rate for (22.1) will be of the same order. More precisely, do there exist positive constants $C_1, C_2 > 0$ such that

$$C_1(T - t)^{-1/(p-1)} \leq \|u(t)\|_\infty \leq C_2(T - t)^{-1/(p-1)}, \tag{23.3}$$

where $T := T_{\max}(u_0)$? It is not difficult to show that the lower bound in (23.3) is always satisfied, in fact with the same constant as for the ODE.

Proposition 23.1. *Consider problem (22.1) with $p > 1$. Let $u_0 \in L^\infty(\Omega)$ and assume that $T := T_{\max}(u_0) < \infty$. Then*

$$\|u(t)\|_\infty \geq k(T - t)^{-1/(p-1)}, \quad 0 < t < T.$$

Proof. Assume for contradiction that there exists $t_0 \in [0, T)$ such that $\|u(t_0)\|_\infty < y(t_0)$, where y is given by (23.2). Therefore $\|u(t_0)\|_\infty \leq y(t_0 - \varepsilon)$ for some $\varepsilon > 0$. Since $y' = y^p$, we deduce from the comparison principle that $\pm u(x, t) \leq y(t - \varepsilon)$ for $(x, t) \in \Omega \times (t_0, T)$. If follows that u is bounded in $\Omega \times (t_0, T)$, a contradiction. $\square$

We present an alternative proof from [219]. It is slightly less simple but the argument may be useful for other problems (see e.g. the proofs of Theorems 44.2(i), 44.17(ii) and 46.4(i)).

Alternative proof for Ω bounded and $u_0 \geq 0$. We may assume

$$M(t) := \max_{x \in \overline{\Omega}} u(x,t) > 0$$

for all $t \in (0,T)$ and pick $x_0(t) \in \Omega$ such that $M(t) = u(x_0(t),t)$. For $0 < s < t < T$, we have

$$M(t) - M(s) \leq u(x_0(t),t) - u(x_0(t),s) = (t-s)u_t\big(x_0(t), s + \theta(t-s)\big) \quad (23.4)$$

and

$$M(t) - M(s) \geq u(x_0(s),t) - u(x_0(s),s) = (t-s)u_t\big(x_0(s), s + \tilde{\theta}(t-s)\big)$$

for some $\theta, \tilde{\theta} \in (0,1)$. Since u_t is locally bounded in $\overline{\Omega} \times (0,T)$, it follows that the function M is locally Lipschitz. In particular, M is a.e. differentiable.[4] Dividing (23.4) by $t-s$, passing to the limit $s \to t$, and using $\Delta u(x_0(t),t) \leq 0$, we obtain

$$M'(t) \leq u_t(x_0(t),t) \leq u^p(x_0(t),t) = M^p(t), \quad \text{a.e. in } (0,T).$$

Integrating between t and $s \in (t,T)$ we get $M^{1-p}(t) \leq M^{1-p}(s) + (p-1)(s-t)$ and the conclusion follows by letting $s \to T$ and using $\lim_{s \to T} M(s) = \infty$. $\quad \square$

Remarks 23.2. (i) **Radial case.** In the case when $\Omega = B_R$ and $u \geq 0$ is radial decreasing in r, then the above proof is just reduced to the obvious observation that $x_0(t) = 0$ and $M'(t) = u_t(0,t) \leq u^p(0,t) = M^p(t)$.

(ii) **Alternative proof.** By simple arguments based on the variation-of-constants formula, one obtains still another proof (cf. [530]) of the lower bound in (23.3) (without the sharp constant). Indeed, by (15.5), we have

$$\|u(s)\|_\infty \leq \|u(t)\|_\infty + \int_t^s \|u(\tau)\|_\infty^p \, d\tau, \quad 0 < t < s < T$$

and, by choosing $s = \min\{\tau \in (t,T) : \|u(\tau)\|_\infty = 2\|u(t)\|_\infty\}$, we obtain $\|u(t)\|_\infty = \|u(s)\|_\infty - \|u(t)\|_\infty \leq 2^p(T-t)\|u(t)\|_\infty^p$, hence the lower bound in (23.3). For similar estimates concerning L^q-norms (also based on the variation-of-constants formula), see Remark 16.2(iii).

(iii) **Estimation of the blow-up time.** An upper estimate of the blow-up time was given in Remark 17.2(i). Proposition 23.1 provides the lower estimate

$$T_{\max}(u_0) \geq \frac{1}{p-1}\|u_0\|_\infty^{1-p}. \quad \square$$

---

[4] Alternatively, one could avoid employing this fact and use an argument involving the derivative of M in the sense of distributions.

The upper blow-up rate estimate

$$\|u(t)\|_\infty \le M(T - t)^{-1/(p-1)}, \quad 0 \le t < T \tag{23.5}$$

(for some constant $M > 0$ possibly depending on u) is much less trivial and need not be always true. It was first obtained in [531], [534] for special classes of solutions. Estimate (23.5) is sometimes referred to as type I blow-up, whereas blow-up is said to be of type II if (23.5) fails (cf. [355]). In this section we prove this upper estimate in three cases:

(i) for all $p > 1$ when $u \ge 0$ is increasing in time, with Ω bounded — cf. Theorem 23.5. This result is due to [219] if Ω is convex; similar ideas were used before in [502] to estimate blow-up times.

(ii) for $1 < p < p_S$ when $u \ge 0$ and $\Omega = \mathbb{R}^n$ (cf. Theorem 23.7, a result due to [245]);

(iii) for $p_S \le p < p_{JL}$ when $\Omega = B_R$ and $u \ge 0$ is radial nonincreasing (cf. Theorem 23.10, a result due to [355], see also [206] for a related result in the case $p = p_S$).

On the contrary, type II blow-up may occur if $\Omega = \mathbb{R}^n$, $n \ge 11$ and $p > p_{JL}$: There exist radial nonincreasing solutions $u \ge 0$ such that (23.5) fails (see [277], [278] and [377]). The proof of this important result is quite long and delicate and will not be given here. Formal arguments indicate that this upper estimate should also fail for some radial (sign-changing) solutions if $\Omega = \mathbb{R}^n$, $3 \le n \le 6$ and $p = p_S$ (see [206]). If $p \ge p_S$, nothing seems to be known for solutions which are neither radial nor increasing in time.

Remarks 23.3. (a) **Extensions.** The result of case (i) above remains true for $\Omega = \mathbb{R}^n$ if we assume in addition that u_0 is radial nonincreasing (see [358]). The result of case (ii) is true also for Ω bounded convex (see [245]) and without the assumption $u \ge 0$ (see [247], [248]). If $u \ge 0$ and $p < p_B$, the convexity assumption can be removed (see Theorem 26.8 below). As for the result of case (iii), it remains true for all radial solutions if $p_S < p < p_{JL}$ and for all positive radial solutions if $p = p_S$ (see [355]). In the case $\Omega = \mathbb{R}^n$, it is true under an additional assumption on u_0.

(b) **Different methods of proof.** The three proofs corresponding to cases (i), (ii) and (iii) above are quite different. They are based respectively on the maximum principle (applied to a suitable auxiliary function), on similarity variables, rescaling and energy, and on rescaling and intersection-comparison. In particular cases, different rescaling (resp., intersection-comparison) arguments were used before in [534] (resp., [229]).

(c) **Neumann problem.** For problem (22.1) with Neumann instead of Dirichlet boundary conditions, results on (type I) blow-up rate can be found in [219], [375], for example. $\square$

Remark 23.4. Refined blow-up rate estimates. Assume $p < p_S$ and $u \geq 0$. When $\Omega = \mathbb{R}^n$ or Ω is a bounded convex domain, the refined asymptotic behavior $\lim_{t \to T} (T - t)^{1/(p-1)} \|u(t)\|_\infty = k$ was proved in [367] (see also [368] for a higher order asymptotic expansion). On the other hand, estimates similar to (23.5), but with M independent of u, will be studied in Section 26 on universal bounds. $\quad\square$

Theorem 23.5. *Consider problem* (22.1) *with* $p > 1$, Ω *bounded and* $0 \leq u_0 \in L^\infty(\Omega)$. *Assume that* u *is nondecreasing in time and nonstationary. Then* $T := T_{\max}(u_0) < \infty$ *and blow-up is of type I, i.e.* (23.5) *is true.*

Remark 23.6. The assumption $u_t \geq 0$ is guaranteed if, for instance, $0 \leq u_0 \in C_0 \cap C^2(\Omega)$ and $\Delta u_0 + u_0^p \geq 0$ (see Proposition 52.19, and also Proposition 52.20 for weaker regularity conditions on u_0). $\quad\square$

Proof of Theorem 23.5. It is a modification of the corresponding proof in [219]. The idea is to apply the maximum principle to the **auxiliary function** J defined in (23.6) below. By Example 51.10, we have $u_t \in C^{2,1}(\overline{\Omega} \times (0, T))$. Since $v := u_t \geq 0$ is a nontrivial solution of $v_t - \Delta v = f'(u)v$ in Q_T vanishing on S_T, it follows from the Hopf maximum principle (cf. Proposition 52.7) that $u_t > 0$ in Q_T and $\partial_\nu u_t < 0$ on S_T. Choosing $\eta \in (0, T)$ we can thus find $\delta > 0$ such that $u_t(x, \eta) \geq \delta u^p(x, \eta)$ for all $x \in \Omega$. Set

$$J := u_t - \delta f, \quad \text{where } f = f(u) := u^p, \tag{23.6}$$

and note that $J \in C^{2,1}(Q_T) \cap C(\overline{\Omega} \times (0, T))$ (due to $u > 0$ in Q_T). We compute

$$J_t - \Delta J = f' u_t - \delta f' f + \delta f'' |\nabla u|^2,$$

hence

$$J_t - \Delta J - f' J = \delta f'' |\nabla u|^2 \geq 0 \quad \text{in } Q^\eta := \Omega \times (\eta, T).$$

Since $J \geq 0$ on the parabolic boundary of Q^η, it follows from the maximum principle (cf. Proposition 52.4) that $J \geq 0$ in Q^η. Consequently, $u_t \geq \delta u^p$ in Q^η. For each $x \in \Omega$, by integrating this inequality between t and $s \in (t, T)$, and then letting $s \to T$, we obtain

$$u^{1-p}(x, t) \geq (p-1)\delta(T - t), \quad \eta < t < T.$$

This gives $T < \infty$ and (23.5). $\quad\square$

Theorem 23.7. *Consider problem* (22.1) *with* $\Omega = \mathbb{R}^n$, $1 < p < p_S$, $0 \leq u_0 \in L^\infty(\mathbb{R}^n)$, *and assume that* $T := T_{\max}(u_0) < \infty$. *Then blow-up is of type I, i.e.* (23.5) *is true.*

In view of the proof of Theorem 23.7 we introduce the notion of **backward similarity variables** (cf. [244], [228]). This is a fundamental tool in the study of the asymptotic behavior of blow-up solutions to problem (22.1), and it will be used again in Section 25. Namely, let $0 < T < \infty$ and let u be a solution of (22.1) with $\Omega = \mathbb{R}^n$, such that u exists on $\mathbb{R}^n \times (0, T)$. For each fixed $a \in \mathbb{R}^n$, we set

$$y := \frac{x - a}{\sqrt{T - t}}, \qquad s := -\log(T - t), \tag{23.7}$$

and we define the rescaled function

$$w(y, s) = w_a(y, s) := e^{-\beta s} u(a + e^{-s/2} y, T - e^{-s}), \qquad \beta := \frac{1}{p - 1} \tag{23.8}$$

(in other words, $w(y, s) = (T - t)^\beta u(x, t)$). Let $s_0 := -\log T$. Then w is a *global* solution of

$$w_s - \Delta w + \frac{1}{2} y \cdot \nabla w = |w|^{p-1} w - \beta w, \qquad y \in \mathbb{R}^n, \ s \in (s_0, \infty), \tag{23.9}$$

with

$$w(y, s_0) = T^\beta u_0(a + y\sqrt{T}), \qquad y \in \mathbb{R}^n. \tag{23.10}$$

Observe that (23.5) is equivalent to the uniform estimate $|w_a(y, s)| \leq M$, where M does not depend on y, s. To prove (23.5) we shall use this fact and the rescaling arguments from the proof of Theorem 22.1 for nonnegative solutions.

Note that equation (23.9) can be rewritten as

$$\rho w_s - \nabla \cdot (\rho \nabla w) = \rho |w|^{p-1} w - \beta \rho w, \tag{23.11}$$

where the Gaussian weight ρ is defined by

$$\rho(y) := e^{-|y|^2/4}.$$

An important property of the rescaled equation (23.9) is the existence of a **weighted energy functional**, defined by

$$E(w) := \int_{\mathbb{R}^n} \left(\frac{1}{2} |\nabla w|^2 + \frac{\beta}{2} w^2 - \frac{1}{p + 1} |w|^{p+1} \right) \rho \, dy. \tag{23.12}$$

We shall first establish some auxiliary results involving this energy (these results will also be used in Section 25).

Proposition 23.8. *Let $p > 1$ and let w be a global solution of (23.11) with $w(\cdot, s_0) \in BC^1(\mathbb{R}^n)$. Then, for all $s > s_0$, we have*

$$\frac{1}{2} \frac{d}{ds} \int_{\mathbb{R}^n} w^2 \rho \, dy = -2E\big(w(s)\big) + \frac{p - 1}{p + 1} \int_{\mathbb{R}^n} |w|^{p+1} \rho \, dy, \tag{23.13}$$

$$\frac{d}{ds}E\big(w(s)\big) = -\int_{\mathbb{R}^n} w_s^2 \rho\, dy, \tag{23.14}$$

$$E\big(w(s)\big) \geq 0, \tag{23.15}$$

and

$$\int_{\mathbb{R}^n} w^2 \rho\, dy \leq C(n,p)\big[E\big(w(s_0)\big)\big]^{2/(p+1)}. \tag{23.16}$$

Moreover,

$$\int_{s_0}^{\infty} \int_{\mathbb{R}^n} w_s^2 \rho\, dy\, ds \leq E\big(w(s_0)\big) \tag{23.17}$$

and

$$a \mapsto E\big(w_a(s_0)\big) \quad \textit{is smooth and bounded.} \tag{23.18}$$

Proof. Problems (18.1) and (23.9)-(23.10) are equivalent via the transformation (23.7)–(23.8). Let $0 < t_1 < t_2 < T$. By Proposition 48.7 and a simple use of the variation-of-constants formula, we see that $u, \nabla u \in L^\infty(\mathbb{R}^n \times (0, t_2))$ (see also (51.29) in Remark 51.11). On the other hand, applying interior parabolic L^q- and Schauder estimates, we obtain that $D^2 u, u_t \in L^\infty(\mathbb{R}^n \times (t_1, t_2))$. Next applying Remark 48.3(i) and, again, interior Schauder estimates, we get $\nabla u \in C^{2,1}(\mathbb{R}^n \times (0, T))$ and $\nabla u_t \in L^\infty(\mathbb{R}^n \times (t_1, t_2))$. Consequently, given $s_2 \in (s_0, \infty)$, the rescaled function $w = w(s)$ satisfies

$$\sup_{\mathbb{R}^n \times (s_0, s_2)} (|w| + |\nabla w|) < \infty \tag{23.19}$$

and

$$\sup_{\mathbb{R}^n \times (s_1, s_2)} \big(|D^2 w| + (1 + |y|)^{-1}(|w_s| + |\nabla w_s|)\big) < \infty, \qquad s_0 < s_1 < s_2. \tag{23.20}$$

We shall write shortly $\int f$ instead of $\int_{\mathbb{R}^n} f(y)\, dy$. We compute

$$\frac{1}{2}\frac{d}{ds}\int w^2 \rho = \int w w_s \rho = \int w\big[\nabla \cdot (\rho \nabla w) + \rho|w|^{p-1} w - \beta \rho w\big],$$

and

$$\frac{1}{2}\frac{d}{ds}\int |\nabla w|^2 \rho = \int \rho(\nabla w_s \cdot \nabla w),$$

for $s > s_0$. Note that the differentiability of the integrals is guaranteed by (23.19), (23.20), and the exponential decay of ρ. By using integration by parts, we deduce that

$$\frac{1}{2}\frac{d}{ds}\int w^2 \rho = \int \big[-|\nabla w|^2 - \beta w^2 + |w|^{p+1}\big]\rho = -2E(w) + \frac{p-1}{p+1}\int |w|^{p+1}\rho$$

i.e., (23.13), and

$$\frac{1}{2}\frac{d}{ds}\int |\nabla w|^2 \rho = -\int w_s(\nabla \cdot \rho \nabla w).$$

This procedure can be easily justified by using again (23.19), (23.20), and the exponential decay of ρ: It suffices to integrate by parts on B_R and then let $R \to \infty$. On the other hand, we have

$$\frac{d}{ds}\int \Big(\frac{\beta}{2}w^2 - \frac{1}{p+1}|w|^{p+1}\Big)\rho = \int (\beta w - |w|^{p-1}w)w_s\rho.$$

Summing the last two identities and using equation (23.11), we obtain (23.14).

Denote $\psi(s) := \int w^2(s)\rho$. Then (23.13) and Jensen's inequality imply

$$\frac{1}{2}\frac{d\psi}{ds} \geq -2E\big(w(s)\big)+C(n,p)\psi^{(p+1)/2}(s).$$

Since $E\big(w(s)\big)$ is nonincreasing due to (23.14), this guarantees (23.15) and (23.16) (otherwise ψ has to blow up in finite time). Next, (23.15) and (23.14) imply (23.17).

Finally, to check (23.18), we note that

$$\int w_a^2(s_0)\rho\, dy = T^{2\beta}\int_{\mathbb{R}^n} u_0^2(x)\rho\Big(\frac{x-a}{\sqrt{T}}\Big)\frac{dx}{T^{n/2}} \leq T^{2\beta}(4\pi)^{n/2}\sup u_0^2,$$

which shows the smoothness and boundedness of the second term appearing in the definition of $E\big(w_a(s_0)\big)$, see (23.12). The proof for the remaining terms is similar. $\square$

Now we are ready to repeat the idea of the proof of Theorem 22.1 for nonnegative solutions.

Proof of Theorem 23.7. By a time shift we may assume $u_0 \in BC^1(\mathbb{R}^n)$, see (51.28). Assume, on the contrary, that there exist t_k such that

$$M_k := \sup_{\mathbb{R}^n \times [0,t_k]} (T-t)^{\beta}u(x,t) = \sup_{\mathbb{R}^n}(T-t_k)^{\beta}u(x,t_k) \to \infty.$$

We may assume $t_k \geq \tilde{t}$ for some $\tilde{t} > 0$. Choose $x_k \in \mathbb{R}^n$ such that

$$(T-t_k)^{\beta}u(x_k,t_k) \geq M_k/2.$$

Rewriting u in similarity variables around $a = x_k$ (cf. (23.7)–(23.8)), we denote $w_k := w_{x_k}$, $s_k := -\log(T-t_k)$. Then $s_k - s_0 \geq \delta^2$ for some $\delta > 0$, $0 \leq w_k(y,s) \leq M_k$ for $s \leq s_k$ and $w_k(0,s_k) \in [M_k/2, M_k]$. Denote

$$v_k(z,\tau) := \frac{1}{M_k}w_k(\nu_k z, \nu_k^2\tau + s_k), \qquad \nu_k := M_k^{-(p-1)/2}.$$

Then $0 \leq v_k(z,\tau) \leq 1$ for $(z,\tau) \in Q_k := \mathbb{R}^n \times (-(\delta/\nu_k)^2, 0]$, $v_k(0,0) \in [1/2, 1]$ and

$$\partial_\tau v_k - \Delta v_k = v_k^p - \nu_k^2\left(\frac{1}{2}z \cdot \nabla v_k + \beta v_k\right) \qquad \text{in } Q_k.$$

Since $Q(r) := \{(z,\tau) : |z| < r, \ -r^2 < \tau \leq 0\} \subset Q_k$ for k large enough, uniform parabolic L^p-estimates used for the operators $A_k v := -\Delta v + \frac{1}{2}\nu_k^2 z \cdot \nabla v$ (see Appendix B) imply the boundedness of v_k in C^α. Consequently, we may pass to the limit to get a solution v of the problem

$$v_\tau - \Delta v = v^p \qquad \text{in } \mathbb{R}^n \times (-\infty, 0),$$

satisfying $0 \leq v \leq 1$ and $v(0,0) \geq 1/2$. Finally, setting $\sigma := -n + 2 + 4/(p-1) > 0$ and using (23.17) and (23.18) we obtain

$$\iint_{Q(\delta/\nu_k)} |\partial_\tau v_k|^2 \, dz \, d\tau = \nu_k^\sigma \int_{s_k - \delta^2}^{s_k} \int_{|y| < \delta} |\partial_s w_k|^2 \, dy \, ds$$

$$\leq \nu_k^\sigma C(\delta) \int_{s_0}^\infty \int_{\mathbb{R}^n} |\partial_s w_k|^2 \rho(y) \, dy \, ds \to 0 \quad \text{as } k \to \infty,$$

hence $v_\tau \equiv 0$ and we get a contradiction with Theorem 8.1. $\square$

Remark 23.9. A priori estimate of the blow-up rate. A simple modification of the proof shows that in Theorem 23.7, the constant M in (23.5) depends on u_0 through a bound on $\|u_0\|_\infty$ only. The same property is true in the case of bounded convex domains (this follows from the arguments in [245], see also [368]). $\square$

Theorem 23.10. *Consider problem* (22.1) *with* $p_S \leq p < p_{JL}$ *and* $\Omega = B_R$. *Let* $u_0 \in L^\infty(\Omega)$, $u_0 \geq 0$, *be radial nonincreasing. If* $T := T_{\max}(u_0) < \infty$, *then blow-up is of type I, i.e.* (23.5) *is true.*

As a preparation to the proof, we first derive the following result, valid for all $p \geq p_S$ and of independent interest. It shows that in case of type II blow-up (i.e. if (23.5) is violated) or of unbounded global solutions, suitably rescaled solutions should converge along some sequence to the positive radial steady state U_1, solution of

$$\left.\begin{aligned}
U'' + \frac{n-1}{r}U' + U^p &= 0, \qquad r \in (0, \infty), \\
U(0) = 1, \qquad U'(0) &= 0
\end{aligned}\right\} \qquad (23.21)$$

(which is known to be unique, cf. Theorem 9.1).

Proposition 23.11. *Consider problem* (22.1) *with* $p \geq p_S$ *and* $\Omega = B_R$. *Let* $u_0 \in L^\infty(\Omega)$, $u_0 \geq 0$, *be radial nonincreasing, and set* $T := T_{\max}(u_0)$. *Assume that either*

$$T < \infty \quad and \quad \limsup_{t \to T} (T - t)^{1/(p-1)} \|u(t)\|_\infty = \infty, \tag{23.22}$$

or

$$T = \infty \quad and \quad \limsup_{t \to T} \|u(t)\|_\infty = \infty. \tag{23.23}$$

Then there exists a sequence $t_j \to T$ *such that*

$$\frac{1}{m(t_j)} \, u\left(\frac{r}{m^{\frac{p-1}{2}}(t_j)}, t_j \right) \to U_1(r), \qquad j \to \infty, \tag{23.24}$$

uniformly for bounded $r \geq 0$, *where* $m(t) := u(0, t)$ *and* $U_1 = U_1(r)$ *is the unique solution of* (23.21).

In the case $T < \infty$, Proposition 23.11 was actually established in [355] for general radial solutions without assuming $u \geq 0$ nor $u_r \leq 0$ (replacing $m(t)$ by $\|u(t)\|_\infty$ and U_1 by $\pm U_1$). Here in the radial decreasing case, we give a simpler proof, which is due to [125] (and which applies to $T \leq \infty$). Theorem 23.10 will then be deduced as a consequence of intersection-comparison arguments involving U_1 and the singular steady state U_*.

In the proof of Proposition 23.11, we shall need the following general monotonicity property of unbounded, positive radial nonincreasing solutions, valid for all $p > 1$ (see [395], [229]).

Lemma 23.12. *Consider problem* (22.1) *with* $p > 1$ *and* $\Omega = B_R$. *Let* $u_0 \in L^\infty(\Omega)$, $u_0 \geq 0$, *be radial nonincreasing, and set* $T := T_{\max}(u_0)$. *Assume that either* $T < \infty$ *or* (23.23) *holds. Denote by* $N(t) := z_{[0,R]}(u_t(\cdot, t))$ *the zero number of the function* $u_t(\cdot, t)$ *in the interval* $[0, R]$ *(see Appendix F). Then there exists* $t_0 \in (0, T)$ *such that*

$$u_t(0, t) > 0 \quad and \quad N(t) = \text{Const.}, \qquad t_0 < t < T. \tag{23.25}$$

Proof. We note that the function u_t is a radial classical solution of

$$\left. \begin{array}{ll} V_t - \Delta V = p u^{p-1} V, & x \in B_R, \ 0 < t < T, \\ V = 0, & x \in \partial B_R, \ 0 < t < T. \end{array} \right\} \tag{23.26}$$

By Theorem 52.28, $N(t)$ is finite and nonincreasing, hence constant on (t_0, T) for some $t_0 \in (0, T)$. Moreover, by the symmetry of the solution, if $u_t(0, t) = 0$, then $u_t(\cdot, t)$ has a degenerate zero at $r = 0$, so that the function N drops at time t. Consequently, $u_t(0, t)$ does not change sign on (t_0, T). Since $\limsup_{t \to T} u(0, t) = \infty$, the claim follows. $\square$

Proof of Proposition 23.11. By a time shift, we may assume that $t_0 = 0$ in Lemma 23.12. Our assumptions imply the existence of a sequence $t_j \to T$ such that

$$\frac{u_t(0, t_j)}{u^p(0, t_j)} \to 0 \tag{23.27}$$

(otherwise we would have $u_t(0, t) \geq cu^p(0, t)$ as $t \to T$ for some $c > 0$, due to (23.25); but this contradicts each of (23.22) and (23.23)). Set $M_j = u(0, t_j)$. By comparing with the solution of the ODE $\psi' = \psi^p$, $\psi(t_j) = M_j$, we easily obtain the existence of $s^* = s^*(p) > 0$ such that

$$\tilde{t}_j := t_j + s^* M_j^{1-p} < T \quad \text{and} \quad u(0, t) \leq 2M_j, \quad t_j \leq t \leq \tilde{t}_j. \tag{23.28}$$

Let $\lambda_j = M_j^{-(p-1)/2}$ and define the rescaled solutions

$$v_j(y, s) = \frac{1}{M_j} u(\lambda_j y, t_j + \lambda_j^2 s), \qquad (y, s) \in D_j := B_{R\lambda_j^{-1}} \times \left(-t_j \lambda_j^{-2}, (T - t_j)\lambda_j^{-2} \right).$$

Then

$$\partial_s v_j - \Delta v_j = v_j^p, \qquad (y, s) \in D_j$$

and, by (23.28) and (23.25), we have $0 \leq v_j \leq 2$ in $B_{R\lambda_j^{-1}} \times (-t_j \lambda_j^{-2}, s^*)$. Moreover, $v_j(0, 0) = 1$ and $\partial_s v_j(0, 0) \to 0$, due to (23.27). Let $D := \mathbb{R}^n \times (-\infty, s^*)$. By interior parabolic estimates, it follows that (some subsequence of) v_j converges in $C^{2+\alpha, 1+\alpha/2}(D)$ to a radial, nonnegative solution of

$$\partial_s v - \Delta v = v^p, \qquad (y, s) \in D,$$

such that $v(0, 0) = 1$ and $\partial_s v(0, 0) = 0$. By using equation (23.26), we see also that

$$\partial_s v_j \to \partial_s v \quad \text{in } C^{1,0}(D). \tag{23.29}$$

We shall now show that $\partial_s v(\cdot, 0) \equiv 0$. Suppose not. Then there exist $A > 0$ and $\varepsilon \in (0, s^*)$ such that

$$\partial_s v(A, s) \neq 0, \quad |s| \leq \varepsilon. \tag{23.30}$$

Since $\partial_s v(\cdot, 0)$ has a degenerate zero at $r = 0$, it follows from Theorem 52.28 that the zero number of $\partial_s v$ on $[0, A]$ drops at $s = 0$. Namely, we can fix $-\varepsilon < s_1 < 0 < s_2 < \varepsilon$ such that $\partial_s v(\cdot, s_i)$ has only simple zeroes on $[0, A]$ and such that

$$z_{[0,A]}\big(\partial_s v(\cdot, s_1)\big) \geq z_{[0,A]}\big(\partial_s v(\cdot, s_2)\big) + 1.$$

Owing to (23.29), we deduce that for j large enough,

$$z_{[0,A]}\big(\partial_s v_j(\cdot, s_1)\big) \geq z_{[0,A]}\big(\partial_s v_j(\cdot, s_2)\big) + 1,$$

hence

$$z_{[0,A\lambda_j]}\big(u_t(\cdot,t_j+\lambda_j^2 s_1)\big)\geq z_{[0,A\lambda_j]}\big(u_t(\cdot,t_j+\lambda_j^2 s_2)\big)+1. \qquad (23.31)$$

Since, on the other hand, (23.30) implies $u_t(A\lambda_j,t_j+\lambda_j^2 s)\neq 0$ for $|s|\leq\varepsilon$, Remark 52.29(ii) implies

$$z_{[A\lambda_j,R]}\big(u_t(\cdot,t_j+\lambda_j^2 s_1)\big)\geq z_{[A\lambda_j,R]}\big(u_t(\cdot,t_j+\lambda_j^2 s_2)\big). \qquad (23.32)$$

By (23.31) and (23.32), we deduce that $N(t_j+\lambda_j^2 s_1)\geq N(t_j+\lambda_j^2 s_2)+1$, which contradicts (23.25). It follows that $v_s(\cdot,0)\equiv 0$, hence $v(\cdot,0)\equiv U_1$ due to $v(0,0)=1$, and the proposition follows. $\square$

Proof of Theorem 23.10. Assume that (23.5) is false and let the sequence $t_j\to T<\infty$ be given by Proposition 23.11. We treat the supercritical and critical cases separately.

Case 1: $p_S<p<p_{JL}$. By Theorem 52.28, there exists an integer K such that

$$z_{[0,R]}(u(\cdot,t_j)-U_*)\leq K,\quad j=1,2,\ldots. \qquad (23.33)$$

On the other hand, by Theorem 9.1, U_1 and U_* intersect infinitely many times. Moreover, these intersections are transversal by local uniqueness for the ODE $U''+\frac{n-1}{r}U'+U^p=0$. Pick $A>0$ such that

$$z_{[0,A]}(U_1-U_*)\geq K+1. \qquad (23.34)$$

Also it is clear that

$$z_{[0,R]}\big(u(r,t_j)-U_*(r)\big)$$

$$=z_{\left[0,R\,m^{\frac{p-1}{2}}(t_j)\right]}\left(\frac{1}{m(t_j)}\,u\left(\frac{r}{m^{\frac{p-1}{2}}(t_j)},t_j\right)-\frac{1}{m(t_j)}\,U_*\left(\frac{r}{m^{\frac{p-1}{2}}(t_j)}\right)\right)$$

$$=z_{\left[0,R\,m^{\frac{p-1}{2}}(t_j)\right]}\left(\frac{1}{m(t_j)}\,u\left(\frac{r}{m^{\frac{p-1}{2}}(t_j)},t_j\right)-U_*(r)\right).$$

By (23.34) and (23.24), it follows that $z_{[0,R]}\big(u(\cdot,t_j)-U_*\big)\geq K+1$ for j large: a contradiction with (23.33).

Case 2: $p=p_S$. Fix $t_0\in(0,T)$ and take $c_2>c_1>0$ such that

$$u(r,t)\geq (e^{-tA}u_0)(r)\geq c_1,\quad 0\leq r\leq R/2,\ t_0\leq t<T$$

and

$$u(r,t_0)\leq c_2,\quad 0\leq r\leq R.$$

Let U_M be the unique positive solution of (9.2) satisfying $U_M(0)=M$, see Theorem 9.1. Since $U_M(R/2)\to 0$ as $M\to\infty$ and $U_M'(r)\to-\infty$ as $M\to\infty$

uniformly on $\{r : U_M(r) \in [c_1, c_2]\}$, there exists $M_0 > c_2$ such that, for all $M \geq M_0$, $u(R/2, t) > U_M(R/2)$ for $t \in [t_0, T)$ and $z_{[0,R/2]}(u(\cdot, t_0) - U_M) = 1$. By the nonincreasing property of the zero-number (see Theorem 52.28), we deduce that

$$z_{[0,R/2]}\big(u(\cdot, t) - U_M\big) \leq 1, \qquad t_0 \leq t < T, \ M \geq M_0.$$

Since $\lim_{t \to T} u(0, t) = \infty$, for each $M \geq M_0$, there exists a first $\tau(M) \in (t_0, T)$ such that $u(0, \tau(M)) = U_M(0) = M$. By symmetry of the solutions, $u(\cdot, \tau(M))$ has a double zero at the origin. Therefore, by Theorem 52.28(iii), $z_{[0,R/2]}(u(\cdot, t) - U_M)$ must drop at $t = \tau(M)$, hence

$$u(r, t) > U_M(r), \qquad 0 \leq r \leq R/2, \ \tau(M) < t < T. \tag{23.35}$$

Now, for large j, (23.24) implies

$$u(0, t_j) > \frac{m(t_j)}{2} U_1(0) = U_{m(t_j)/2}(0).$$

Consequently, $t_j > \tau(m(t_j)/2)$. Using (23.35) and (9.4), it follows that

$$u(r, t_j) > U_{m(t_j)/2}(r) = m(t_j) U_{1/2}(m^{\frac{p-1}{2}}(t_j)r), \qquad 0 \leq r \leq R/2, \ t_j \leq t < T.$$

Therefore,

$$\frac{1}{m(t_j)}\, u\left(\frac{\rho}{m^{\frac{p-1}{2}}(t_j)}, t_j\right) > U_{1/2}(\rho), \qquad 0 \leq \rho \leq (R/2)\, m^{\frac{p-1}{2}}(t_j), \ t_j \leq t < T.$$

Using (23.24) again and letting $j \to \infty$, we obtain

$$U_1(\rho) \geq U_{1/2}(\rho), \quad 0 \leq \rho < \infty,$$

contradicting Theorem 9.1. $\quad\square$

Remark 23.13. By combining Proposition 23.11 for $T = \infty$ and Case 1 of the proof of Theorem 23.10, we obtain an alternative proof [125] of Theorem 22.4 on boundedness of global solutions in the case $p_S < p < p_{JL}$ and $u_r \leq 0$. Moreover, as a consequence of Remark 22.7, this proof can be used without assuming $u_r \leq 0$. $\quad\square$

Remarks 23.14. (i) **Sign-changing solutions.** The proof of Theorem 23.7 can be considered as an analogue to the proof of Theorem 22.1 for nonnegative solutions. In [247] the authors prove Theorem 23.7 without the positivity assumption on u_0 and the proof is again an analogue of the (interpolation) proof of Theorem 22.1 in the general, sign-changing case. However, the localization of the arguments of this interpolation proof is nontrivial: The authors of [247] have to use two kinds of localized version of weighted energies,

$$E_\varphi(w) := \frac{1}{2} \int_{\mathbb{R}^n} \big(|\nabla(\varphi w)|^2 + (\beta \varphi^2 - |\nabla \varphi|^2)w^2\big)\rho\, dy - \frac{1}{p+1} \int_{\mathbb{R}^n} \varphi^2 |w|^{p+1} \rho\, dy,$$

$$\mathcal{E}_\varphi(w) := \frac{1}{2} \int_{\mathbb{R}^n} \varphi^2 \big(|\nabla w|^2 + \beta w^2\big)\rho\, dy - \frac{1}{p+1} \int_{\mathbb{R}^n} \varphi^2 |w|^{p+1} \rho\, dy,$$

and the corresponding bounds $E_\varphi(w) \geq 0$, $\mathcal{E}_\varphi(w) \leq C$ and $|E_\varphi(w) - \mathcal{E}_\varphi(w)| \leq C$.

(ii) **Applications of blow-up rate estimates.** The knowledge of the blow-up rate has important consequences in the study of the blow-up behavior. In particular (23.5) is the first step in the description of asymptotically self-similar blow-up (see Section 25). On the other hand, it can be used for the proof of the Hölder continuity of the maximal existence time $T_{\max} : L^\infty(\Omega) \to (0, \infty]$ (see [254], [255] and cf. Theorem 22.13). $\square$

We close this section with a simple result which shows that the upper blow-up estimate (23.5) implies a similar estimate for the gradient. This property will be useful in Section 25.

Proposition 23.15. *Consider problem* (22.1) *with* $p > 1$ *and* $u_0 \in L^\infty(\Omega)$. *Assume that* $T := T_{\max}(u_0) < \infty$ *and that* (23.5) *is satisfied for some* $M > 0$. *Then*

$$\|\nabla u(t)\|_\infty \leq M_1(T - t)^{-1/(p-1)-1/2}, \quad T/2 \leq t < T$$

for some $M_1 = M_1(M, p, \Omega, T) > 0$.

Proof. Fix $T/2 \leq t < T$ and put $s = 2t - T \in [0, t)$. By the variation-of-constants formula, the gradient estimate in Proposition 48.7 and (23.5), we have

$$\|\nabla u(t)\|_\infty \leq \|\nabla e^{-(t-s)A} u(s)\|_\infty + \int_s^t \|\nabla e^{-(t-\tau)A} |u|^{p-1} u(\tau)\|_\infty \, d\tau$$

$$\leq C(t-s)^{-1/2} \|u(s)\|_\infty + C \int_s^t (t-\tau)^{-1/2} \|u(\tau)\|_\infty^p \, d\tau$$

$$\leq CM(t-s)^{-1/2}(T-s)^{-1/(p-1)} + CM^p \int_s^t (t-\tau)^{-1/2}(T-\tau)^{-p/(p-1)} \, d\tau.$$

Since $T - t = t - s = (T - s)/2$, we have

$$\|\nabla u(t)\|_\infty$$

$$\leq 2^{-1/(p-1)} CM(T-t)^{-1/(p-1)-1/2} + CM^p(T-t)^{-p/(p-1)} \int_s^t (t-\tau)^{-1/2} \, d\tau$$

$$\leq C[2^{-1/(p-1)} M + 2M^p] (T-t)^{-1/(p-1)-1/2}$$

and the proposition is proved. $\square$

24. Blow-up set and space profile

This and the subsequent section are devoted to the space and space-time description of singularities of blowing-up solutions of the model problem (22.1). Assume that the solution u blows up in finite time $T := T_{\max}(u_0)$ and denote by $B(u_0)$ its blow-up set:

$$B(u_0) := \{x \in \overline{\Omega} : \exists (x_k, t_k) \in \Omega \times (0,T) \text{ such that } t_k \to T \text{ and } |u(x_k, t_k)| \to \infty\}. \tag{24.1}$$

The following theorem, due to [219], guarantees single-point blow-up for radial decreasing solutions and provides an upper estimate for the blow-up profile.

Theorem 24.1. *Consider problem* (22.1) *with* $p > 1$ *and* $\Omega = B_R$. *Let* $u_0 \in L^\infty(\Omega)$, $u_0 \geq 0$, *be radial nonincreasing, and assume* $T := T_{\max}(u_0) < \infty$. *Then* $B(u_0) = \{0\}$ *and, for any* $\alpha > 2/(p-1)$, *there exists* $C_\alpha > 0$ *such that*

$$u(x,t) \leq C_\alpha |x|^{-\alpha}, \qquad 0 < |x| < R, \ 0 < t < T.$$

Corollary 24.2. *Under the assumptions of Theorem* 24.1, *we have*

$$\limsup_{t \to T} \|u(t)\|_q < \infty, \qquad 1 \leq q < q_c = n(p-1)/2.$$

Proof of Theorem 24.1. As in the proof of Theorem 23.5, the idea is to apply the maximum principle to a (different) auxiliary function J, defined in (24.3) below. Note that our assumptions imply

$$u_r(r,t) < 0 \quad \text{for all } r \in (0,R], \ t \in (0,T), \tag{24.2}$$

due to Proposition 52.17. We split the proof in two steps.

Step 1. Let $\gamma \in (1,p)$, $\eta \in (0,T)$ and $\delta > 0$. We will show that there exists $\varepsilon > 0$ such that $J(r,t) \leq 0$ in $\Omega \times (\eta, T)$, where

$$\left. \begin{aligned} J &= J(u) := w + c(r)F(u), \\ w(r) &:= r^{n-1} u_r(r,t), \quad c(r) := \varepsilon r^{n+\delta}, \quad F(u) := u^\gamma. \end{aligned} \right\} \tag{24.3}$$

Denote $f(u) := u^p$, $\Omega_1 := \Omega \cap \{x : x_1 > 0\}$ and notice that $v := u_{x_1}$ satisfies $v_t - \Delta v = f'(u)v$ in Ω_1. Since $v = 0$ for $x \in \partial\Omega_1$, $x_1 = 0$, and $v < 0$ for $x \in \partial\Omega_1$, $x_1 > 0$, the maximum principle implies $v < 0$ for $x \in \Omega_1$ and $v_{x_1}(0,t) = u_{x_1 x_1}(0,t) < 0$ for $t > 0$. Hence $u_{rr}(0,t) < 0$. Since $J = r^{n-1}(u_r + \varepsilon r^{1+\delta}u^\gamma)$, this inequality and (24.2) imply $J(r,\eta) \leq 0$ for all r provided ε is small enough. We have also $J \in C^{2,1}((0,R) \times (0,T)) \cap C([0,R] \times (0,T))$ (due to $u > 0$), with

$J(R,t) \le 0$ and $J(0,t) = 0$ for $t > 0$. Now the claim follows from the maximum principle in Proposition 52.4, provided we show

$$J_t + \frac{n-1}{r}J_r - J_{rr} - bJ \le 0 \qquad \text{in } (0,R) \times (\eta,T), \tag{24.4}$$

where b is bounded on $(0,R) \times (\eta, T - \tau)$ for all $\tau > 0$. Summing the identities

$$J_t = w_t + cF'u_t,$$
$$\frac{n-1}{r}J_r = \frac{n-1}{r}\left(w_r + cF'u_r + c'F\right),$$
$$-J_{rr} = -\left(w_{rr} + cF'u_{rr} + cF''u_r^2 + 2c'F'u_r + c''F\right),$$

and using the equations

$$u_t - u_{rr} - \frac{n-1}{r}u_r = f(u), \qquad w_t + \frac{n-1}{r}w_r - w_{rr} = f'(u)w,$$

and $F'' \ge 0$ we get

$$J_t + \frac{n-1}{r}J_r - J_{rr} \le f'w + cF'f + \frac{2(n-1)}{r}cF'u_r + \frac{n-1}{r}c'F - 2c'F'u_r - c''F.$$

Using $w = -cF + J$ and $u_r = w/r^{n-1} = (-cF + J)/r^{n-1}$ we see that the RHS of the last inequality can be written in the form $bJ - cH$, where the function $b = f' + 2(n-1)cF'r^{-n} - 2c'F'r^{1-n}$ is bounded on $(0,R) \times (\eta, T-\tau)$ for all $\tau > 0$ and

$$H = Ff' - F'f - \frac{2}{r^{n-1}}F'F\left(c' - \frac{n-1}{r}c\right) + \frac{F}{c}\left(c'' - \frac{n-1}{r}c'\right).$$

Now $H \ge 0$ is equivalent to

$$(p - \gamma)u^{p-1} + (n + \delta)\delta r^{-2} \ge 2\varepsilon\gamma(1+\delta)u^{\gamma-1}r^\delta \tag{24.5}$$

which is obviously satisfied if ε is small enough. Consequently, (24.4) is true.

Step 2. Let $t \in (\eta, T)$ and $\alpha := (2 + \delta)/(\gamma - 1)$. Notice that $J(u) \le 0$ implies $u_r/u^\gamma \le -\varepsilon r^{1+\delta}$. Integrating this inequality we arrive at $u(r,t) \le Cr^{-\alpha}$, where $C := (\alpha/\varepsilon)^{1/(\gamma-1)}$. This estimate guarantees the assertion. $\square$

Under an additional assumption of monotonicity in time, the corresponding lower estimate on the blow-up profile can be established by relatively simple arguments (cf. [490]). For more precise information on the blow-up profiles, see Remark 25.8 below.

Theorem 24.3. *Consider problem* (22.1) *with* $p > 1$ *and* $\Omega = B_R$ *or* $\Omega = \mathbb{R}^n$. *Let* $u_0 \in L^\infty(\Omega)$, $u_0 \geq 0$, *be radial nonincreasing and such that* $T := T_{\max}(u_0) < \infty$. *Assume in addition that* $u_t \geq 0$ *in* Q_T. *Then there holds*

$$u(x, T) \geq C|x|^{-2/(p-1)}, \qquad 0 < |x| < \eta, \tag{24.6}$$

for some $C = C(p) > 0$ *and* $\eta = \eta(u_0) > 0$.

Proof. We assume $\Omega = B_R$ (the case $\Omega = \mathbb{R}^n$ can be treated by straightforward modifications). Since $u_t \geq 0$ and $u_r \leq 0$, we have

$$\frac{\partial}{\partial r}\left(\frac{1}{2}u_r^2 + \frac{1}{p+1}u^{p+1}\right) = (u_{rr} + u^p)u_r = \left(u_t - \frac{n-1}{r}u_r\right)u_r \leq 0,$$

hence

$$\left(\frac{1}{2}u_r^2 + \frac{1}{p+1}u^{p+1}\right)(r, t) \leq \frac{1}{p+1}u^{p+1}(0, t).$$

Therefore, we get

$$\|u_r(t)\|_\infty \leq C_1 u^{(p+1)/2}(0, t).$$

For $0 < t < T$, let $r_0(t)$ be such that $u(r_0(t), t) = \frac{1}{2}u(0, t)$. Note that, in view of (24.2), the implicit function theorem guarantees that $r_0(t)$ is unique and is a continuous function of t. We may assume that 0 is the only blow-up point, since otherwise the result is trivial. Using $u(0, t) = \|u(t)\|_\infty \to \infty$ as $t \to T$, this implies $r_0(t) \to 0$ as $t \to T$. Now we have

$$-u_r \leq C_2 u^{(p+1)/2}, \qquad 0 \leq r \leq r_0(t).$$

Integrating, we get

$$u^{-(p-1)/2}(r_0(t), t) \leq u^{-(p-1)/2}(0, t) + C_3 r_0(t)$$
$$= 2^{-(p-1)/2}u^{-(p-1)/2}(r_0(t), t) + C_3 r_0(t)$$

hence $u(r_0(t), t) \geq C_4(r_0(t))^{-2/(p-1)}$. Using $u_t \geq 0$, it follows that

$$u(r_0(t), T) \geq C_4(r_0(t))^{-2/(p-1)}, \qquad 0 < t < T.$$

Since r_0 is continuous and $r_0(t) \to 0$ as $t \to T$, we deduce that the range $r_0((0, T))$ contains an interval of the form $(0, \eta)$ and the conclusion follows. $\square$

Remark 24.4. The lower bound (24.6) has been proved in [232] (see also [229]) for radial solutions without assuming $u_t \geq 0$ nor $u_r \leq 0$, but under the condition $p \leq p_S$. The method therein is different, based on intersection-comparison with stationary solutions. $\square$

We get back to the question of the blow-up set. In the case of $\mathbb{R}^n$, the following result, due to [246], gives a necessary condition, involving the weighted energy, for a given point to be a blow-up point and a sufficient condition for the blow-up set to be compact. The proof is postponed to the next section.

Theorem 24.5. *Consider problem* (22.1) *with* $1 < p \leq p_S$ *and* $u_0 \in BC^1(\mathbb{R}^n)$. *Assume that* $T := T_{\max}(u_0) < \infty$ *and that the upper blow-up rate estimate* (23.5) *is satisfied.*

(i) *Let* $E_a(u_0) := E\big(w_a(s_0)\big)$ *be defined in* (23.12). *There exists* $\eta = \eta(n,p) > 0$ *such that, if* $E_a(u_0) < \eta$, *then* a *is not a blow-up point.*

(ii) *Assume in addition that* $u_0(x), |\nabla u_0(x)| \to 0$ *as* $|x| \to \infty$, *then the blow-up set is compact.*

Remarks 24.6. (i) **Single-point blow-up.** The first example of a single-point blow-up for problem (22.1) was found in [531] with $n = 1$ and $u_0 = k\psi$, where ψ is a positive stationary solution of (22.1) and $k, p \gg 1$. On the other hand, under the assumptions of Theorem 24.1 with $\Omega = \mathbb{R}^n$ instead of a ball, we still have a single-point blow-up (see [388]; the proof is different from that of Theorem 24.1).

(ii) **Blow-up at infinity.** By a careful reading of the proof of Theorem 24.5(ii) one obtains the stronger conclusion that $\sup\{|u(x,t)| : |x| > R, \ 0 < t < T\} < \infty$ for some large $R > 0$. On the other hand, the result may fail if no decay is assumed on u_0, as shown by the simple example $u = k(T - t)^{-1/(p-1)}$. Furthermore, it has been shown in [249] that, if $\lim_{|x|\to\infty} u_0(x) = L > 0$ and $0 \leq u_0 \leq L$ in $\mathbb{R}^n$, $u_0 \not\equiv L$, then u remains bounded on compact subsets of $\mathbb{R}^n$ up to $t = T_{\max}(u_0) < \infty$ and blows up only at space infinity (see also [312] for related results). Under the same assumption, denoting by y the solution of the ODE (23.1) with $y(0) = L$, it was also proved in [249] that u and y share the same blow-up time T and that $\lim_{|x|\to\infty} u(x,t) = y(t)$, uniformly for t bounded away from T.

(iii) **One-dimensional case.** Consider problem (22.1) with $n = 1$ and Ω bounded.

Assume first $u_0 \geq 0$ and $T = T_{\max}(u_0) < \infty$. Then the results of [126] guarantee that $B(u_0)$ is finite and its cardinality is bounded above by the number of local maxima of u_0. Moreover, given $x \notin B(u_0)$, there exists $\varphi(x) := \lim_{t\to T} u(x,t)$ and $\varphi \in C^2\big(\Omega \setminus B(u_0)\big)$.

On the other hand, given $x_1, x_2, \ldots x_k \in \Omega$, there exists $u_0 \geq 0$ such that $B(u_0) = \{x_1, \ldots, x_k\}$, see [364].

The arguments in [209] and the universal bounds in Section 26 show that there exists $T^* < \infty$ with the following property: If $u_0 \geq 0$ and $T_{\max}(u_0) > T^*$, then $B(u_0)$ consists of a single point.

(iv) **Blow-up in the interior.** Consider problem (22.1) with Ω bounded and convex. If $u_0 \geq 0$ and $T = T_{\max}(u_0) < \infty$, then $B(u_0)$ is always a compact subset of Ω (see [219]). The idea of the proof is the following: Choose $y \in \partial\Omega$. Without loss of generality we may assume that $y = 0$ and that the hyperplane $\{x = (x_1, \ldots, x_n) : x_1 = 0\}$ is tangential to $\partial\Omega$ at the origin, with $x_1 < 0$ for all $x \in \Omega$. The method of moving planes guarantees $u_{x_1} < 0$ for all $t \geq \eta > 0$ and $x \in \Sigma_\lambda := \{z \in \Omega : z_1 > -\lambda\}$ provided $\lambda > 0$ is small enough. Now, similarly as in the proof of Theorem 24.1 one obtains $J := u_{x_1} + \varepsilon(x_1 - \lambda)^{1+\delta} u^\gamma \leq 0$ in

$\Sigma_\lambda \times (\eta, T)$ which implies $u(x) \leq C(\lambda - \tilde{\lambda})^{-\alpha}$ for any $\tilde{\lambda} \in (0, \lambda)$ and $x \in \Sigma_{\tilde{\lambda}}$ (where $\alpha, \gamma, \delta, \eta, \varepsilon$ have the same meaning as in the proof of Theorem 24.1). This estimate guarantees $B(u_0) \cap \partial\Omega = \emptyset$.

(v) **Global and regional blow-up.** Consider problem (14.1) with

$$f(u) = (1 + u)\big(\log(1 + u)\big)^b, \qquad b > 1.$$

Assume first that Ω is bounded, $u_0 \geq 0$ and $T_{\max}(u_0) < \infty$ (such functions do exist). If $b < 2$, then the blow-up is global, that is $B(u_0) = \overline{\Omega}$. If $b = 2$, then the blow-up is either global or regional (that is $B(u_0)$ contains a nonempty open set, but $B(u_0) \neq \overline{\Omega}$), depending on the size of Ω. These results were proved in [313]. Similarly, if $\Omega = \mathbb{R}$, $b = 2$ and $u_0 \geq 0$ is symmetric and radially nonincreasing, $u_0 \not\equiv 0$, then $T_{\max}(u_0) < \infty$, the measure of $B(u_0)$ is at least 2π (and $B(u_0) = [-\pi, \pi]$ under some additional assumptions on u_0), see [230]. On the other hand, if Ω is a ball and $b > 2$, then there are positive initial data such that the corresponding solutions blow up at a single point (this follows from the proof of Theorem 24.1 with the choice $F(u) = (1 + u)\big(\log(1 + u)\big)^{b-1}$).

Regional or global blow-up cannot happen for positive solutions of (18.1) if $p < p_S$ and u_0 is continuous, bounded and nonconstant. In this case, the $(n - 1)$-dimensional Hausdorff measure of $B(u_0) \cap M$ is finite for any bounded measurable set $M \subset \mathbb{R}^n$, see [520]. This is optimal, in view of examples from [246] of solutions blowing up on a sphere. Moreover, results on the regularity of $B(u_0)$ near a nonisolated blow-up point have been obtained in [541], [542].

(vi) **Small and large diffusion limits.** Consider positive solutions of problem (22.1) with a diffusion coefficient $D > 0$ in front of the Laplacian, and with either the Dirichlet or the Neumann boundary conditions. Then, under suitable additional assumptions, the blow-up set of u concentrates near the maxima of u_0 as $D \to 0$. In the limit $D \to \infty$, for the Neumann case, it concentrates near $\mathcal{M}$, where $\mathcal{M}$ is the set of maxima of the L^2-projection of u_0 onto the second Neumann eigenspace (see [289] and the references therein). $\square$

Remark 24.7. Limitations concerning comparison arguments. If Ω is a bounded domain, then two ordered sub-/supersolutions cannot share the same existence time unless both are global. For instance, if u is the solution of (22.1) with $T_{\max}(u_0) < \infty$ and if $v \not\equiv u$ is a supersolution of (22.1) on $(0, T)$ such that $u_0 \leq v(\cdot, 0) \in L^\infty(\Omega)$, then $T < T_{\max}(u_0)$. This follows from Proposition 27.3 below. In particular the knowledge of the blow-up rate or set of v does not provide direct information on that of u (but the situation can be different in unbounded domains, cf. the end of Remark 24.6(ii)). Nevertheless, in bounded domains, one can sometimes use indirect comparison arguments (see Proposition 23.1 for a simple case) or more sophisticated intersection-comparison arguments. $\square$

25. Self-similar blow-up behavior

In this section we apply the method of similarity variables, introduced in the proof of Theorem 23.7, to study the space-time behavior of solutions of the model problem (22.1) near blow-up points as t approaches the blow-up time.

The following theorem is due to [244], [246]. A similar result for $n = 1$ was obtained independently in [229]. Theorem 25.1 can be extended to bounded convex domains (see [246]), but here we restrict ourselves to the case of the whole space for simplicity.

Theorem 25.1. *Consider problem* (22.1) *with* $\Omega = \mathbb{R}^n$, $1 < p \le p_S$, $u_0 \in L^\infty(\mathbb{R}^n)$, *and let* $k = (p-1)^{-1/(p-1)}$. *Assume that* $T := T_{\max}(u_0) < \infty$ *and that the upper blow-up rate estimate* (23.5) *is satisfied. If* a *is a blow-up point of* u, *then we have*

$$\lim_{t \to T} (T-t)^{1/(p-1)} u(a + y\sqrt{T-t}, t) = \pm k, \qquad (25.1)$$

uniformly on compact sets $|y| \le C$.

Remark 25.2. Let $1 < p < p_S$, $\Omega = \mathbb{R}^n$, $u_0 \in L^\infty(\mathbb{R}^n)$ and let a be a blow-up point of u (in the sense of the definition in (24.1)). Theorem 23.7 and Remark 23.3(a) guarantee that the upper blow-up rate estimate (23.5) is satisfied. Consequently, it follows in particular from Theorem 25.1 that u <u>does</u> blow-up *at* $x = a$ and the blow-up rate of $u(a, t)$ is exactly that given by the ODE (cf. (23.1)–(23.2)). $\square$

In the proof, among other things, we shall use another result from [246], which is of independent interest since it is valid for any $p > 1$. If a is a blow-up point, then this result provides a lower bound on the blow-up rate. Note that no boundary conditions are assumed and that this is a purely local result. The proof in [246] was based on a cut-off, the variation-of-constants formula, parabolic estimates and bootstrap. We here give a simpler proof, based on a comparison argument using a quadratic change of unknown and a cut-off (cf. (25.6)).

Theorem 25.3. *Let* $p > 1$, $T > 0$, $\rho > 0$, $a \in \mathbb{R}^n$ *and denote* $Q = B(a, \rho) \times (T - \rho^2, T)$. *There exists* $\varepsilon = \varepsilon(n, p) > 0$ *such that if* u *is a classical solution of*

$$u_t - \Delta u = |u|^{p-1} u, \qquad (x, t) \in Q,$$

and u *satisfies*

$$|u(x, t)| \le \varepsilon (T - t)^{-1/(p-1)}, \qquad (x, t) \in Q, \qquad (25.2)$$

then u *is uniformly bounded in a neighborhood of* (a, T).

Proof. By a space-time translation, we may assume $a = 0$ and $T = \rho^2$. By scaling, we may also assume $\rho = 1$. Indeed, $\tilde{u}(x, t) := \rho^{2/(p-1)} u(\rho x, \rho^2 t)$ solves the same equation in $B_1 \times (0,1)$, and (25.2) is equivalent to $|\tilde{u}(x,t)| \le \varepsilon(1-t)^{-1/(p-1)}$ for $|x| < 1$ and $t \in (0,1)$. Set

$$\alpha = \min(1/2, (p-1)/4). \tag{25.3}$$

For each $R > 0$, we may find $\phi \in C^2(\mathbb{R}^n)$ such that

$$\phi(x) = 0 \ \text{ for } |x| \ge R/\sqrt{2}, \qquad \phi(x) \ge 1 \ \text{ for } |x| \le R/2, \tag{25.4}$$

and

$$|\nabla\phi|^2 + |\Delta\phi^2| \le C(R,n)\phi^{2(1-\alpha)} \tag{25.5}$$

(it suffices to consider $\phi(x) = (2 - 4R^{-2}|x|^2)_+^m$ for $m > 2$ large enough). Choose $R = 1$ and put

$$v = u^2\phi^2. \tag{25.6}$$

For $(x,t) \in B_1 \times (0,T)$, we have

$$v_t - \Delta v = 2uu_t\phi^2 - 2\phi^2(u\Delta u + |\nabla u|^2) - 8u\phi\nabla u \cdot \nabla\phi - u^2\Delta\phi^2.$$

Since $4|u\phi\nabla u \cdot \nabla\phi| \le \phi^2|\nabla u|^2 + 4u^2|\nabla\phi|^2$, we deduce that

$$v_t - \Delta v \le 2\phi^2|u|^{p+1} + u^2(8|\nabla\phi|^2 + |\Delta\phi^2|). \tag{25.7}$$

Using (25.5) and assumption (25.2), it follows that

$$\begin{aligned}
v_t - \Delta v &\le 2|u|^{p-1}v + u^{2\alpha}v^{1-\alpha}\phi^{-2(1-\alpha)}(8|\nabla\phi|^2 + |\Delta\phi^2|) \\
&\le 2|u|^{p-1}v + Cu^{2\alpha}(1 + v) \\
&\le 2\varepsilon^{p-1}(T-t)^{-1}v + C\varepsilon^{2\alpha}(T-t)^{-2\alpha/(p-1)}(1 + v).
\end{aligned}$$

Here and below, C is a generic positive constant depending only on n, p, whereas K will denote a generic positive constant depending on n, p, u. Assuming $0 < \varepsilon < 1$ and recalling (25.3), we obtain

$$v_t - \Delta v \le C\varepsilon^{2\alpha}(T-t)^{-1}v + C(T-t)^{-2\alpha/(p-1)}. \tag{25.8}$$

Let $\bar{v} = (T-t)^{-C\varepsilon^{2\alpha}} + K(T-t)^{1-2\alpha/(p-1)}$ for $0 \le t < T$. We have

$$\bar{v}_t = C\varepsilon^{2\alpha}(T-t)^{-1}\bar{v} + K(1 - 2\alpha/(p-1) - C\varepsilon^{2\alpha})(T-t)^{-2\alpha/(p-1)}.$$

For $\varepsilon = \varepsilon(n,p) > 0$ small and $K > 0$ large, it follows that $\bar{v}$ is a supersolution to (25.8) and we also have $\bar{v}(1/2) \ge \|v(\cdot, 1/2)\|_\infty$. Since $v = 0$ on $\partial B_1 \times (0,T)$, we deduce from the comparison principle that $v \le \bar{v}$ in $B_1 \times [1/2, T)$, hence

$$u \le K(T-t)^{-C\varepsilon^{2\alpha}}, \quad |x| < 1/2, \ 1/2 \le t < T. \tag{25.9}$$

Now considering $v = u^2\phi^2$ with $R = 1/2$ instead of $R = 1$ in (25.4), and taking a smaller $\varepsilon(n,p) > 0$, inequalities (25.7) and (25.9) imply $v_t - \Delta v \le K(T-t)^{-1/2}$ in $B_{1/2} \times (1/2, T)$. Using a supersolution of the form $\tilde{K} - K(T-t)^{1/2}$, we conclude that u is bounded in a neighborhood of $(x = 0, t = T)$. $\square$

Before going into the proof of Theorem 25.1, let us first observe that, considering the rescaled solution in similarity variables (cf. (23.7)–(23.8)), the conclusion can be restated as:

$$\lim_{s\to\infty} w_a(y,s) = \pm k, \text{ uniformly on compact sets } |y| \le C.$$

The basic idea of the proof is to apply dynamical systems arguments to show that the global bounded solution w_a is attracted by the set of equilibria, i.e. solutions of

$$\Delta z - \frac{1}{2}y \cdot \nabla z + |z|^{p-1}z - \beta z = 0, \qquad z \in \mathbb{R}^n \tag{25.10}$$

(cf. Lemma 25.6(i)). On the other hand, under the assumption $p \le p_S$, we shall show that the only (bounded) equilibria are the constant solutions $z = k, -k$ and 0 (Proposition 25.4). The last task will then be to show the nondegeneracy, i.e. to exclude the possibility of w_a approaching 0, which will be achieved by combining Theorem 25.3 with suitable energy arguments (cf. Lemma 25.6(ii)(iii)).

Thus, let us define

$$\mathcal{S} = \big\{z \in C^2 \cap L^\infty(\mathbb{R}^n) : z \text{ is a solution of } (25.10)\big\}.$$

For given $a \in \mathbb{R}^n$, we denote

$$\omega(w_a) = \big\{z \in \mathcal{S} : \exists s_n \to \infty, \ w_a(y, s_n) \to z(y) \text{ in } C^1(\mathbb{R}^n).\big\}$$

Proposition 25.4. *If* $1 < p \le p_S$, *then* $\mathcal{S} = \{0, k, -k\}$.

Remarks 25.5. Supercritical case. (i) Proposition 25.4, and consequently Theorem 25.1, are no longer true for $p > p_S$ (provided $p < p_L$ defined in (22.18)). Indeed, in that range, Proposition 22.5 shows the existence of backward self-similar solutions with (positive) bounded nonconstant profile.

(ii) However, if $\Omega = B_R$ and $u \ge 0$ is radial and satisfies $u_t \ge 0$ and $u_r \le 0$, then assertion (25.1) with $a = 0$ remains true for all $p > p_S$ (see [62], and also [358], [359] for further results). $\square$

Proof of Proposition 25.4. Let $w \in \mathcal{S}$. We first claim that $|\nabla w|$ is bounded. Indeed, by setting $u(x,t) = (1-t)^{-\beta}w(x/\sqrt{1-t})$, we define a (self-similar) solution of (18.1) in $\mathbb{R}^n \times (0,1)$, with $u_0 = w \in L^\infty(\mathbb{R}^n)$. Since $\nabla u(x, 1/2) = 2^{\beta+1/2}w(\sqrt{2}x)$ and $\nabla u(\cdot, 1/2) \in L^\infty(\mathbb{R}^n)$ by smoothing effect, the claim follows.

Let us show that w satisfies the Pohozaev-type identity

$$\left(\frac{n}{p+1} - \frac{n-2}{2}\right) \int_{\mathbb{R}^n} |\nabla w|^2 \rho \, dy + \frac{1}{2}\left(\frac{1}{2} - \frac{1}{p+1}\right) \int_{\mathbb{R}^n} |y|^2 |\nabla w|^2 \rho \, dy = 0. \quad (25.11)$$

We shall obtain (25.11) as a linear combination of three other identities. The first one is

$$\int |\nabla w|^2 \rho \, dy + \beta \int w^2 \rho \, dy - \int |w|^{p+1} \rho \, dy = 0. \quad (25.12)$$

(Here and in what follows all integrals are taken over $\mathbb{R}^n$.) Rewriting (25.10) as

$$\nabla \cdot (\rho \nabla w) - \beta \rho w + \rho |w|^{p-1} w = 0, \quad (25.13)$$

(25.12) is obtained by multiplying (25.13) by $-w$ and using integration by parts. This procedure can be easily justified since w and $|\nabla w|$ are bounded and ρ decays exponentially: It suffices to integrate by parts on B_R and then let $R \to \infty$. This argument will be used in the rest of the proof without further mention.

The second identity is

$$\int |y|^2 |\nabla w|^2 \rho \, dy + \int \left[\left(\beta + \frac{1}{2}\right)|y|^2 - n\right] w^2 \rho \, dy - \int |y|^2 |w|^{p+1} \rho \, dy = 0. \quad (25.14)$$

It is obtained by multiplying (25.13) by $-|y|^2 w$ and using integration by parts, since

$$-\int |y|^2 w \nabla \cdot (\rho \nabla w) \, dy = \int |y|^2 |\nabla w|^2 \rho \, dy + \int (y \cdot \nabla w^2) \rho \, dy$$

$$= \int |y|^2 |\nabla w|^2 \rho \, dy - n \int w^2 \rho \, dy + \frac{1}{2} \int |y|^2 w^2 \rho \, dy.$$

The third identity is

$$\int \left(\frac{|y|^2}{4} - \frac{n-2}{2}\right) |\nabla w|^2 \rho \, dy + \int \left(\frac{\beta |y|^2}{4} - \frac{n\beta}{2}\right) w^2 \rho \, dy$$
$$- \int \left(\frac{|y|^2}{2(p+1)} - \frac{n}{p+1}\right) |w|^{p+1} \rho \, dy = 0. \quad (25.15)$$

To get (25.15), we multiply (25.13) by $-(y \cdot \nabla w)$ and we use

$$\int (y \cdot \nabla w) \rho (\beta w - |w|^{p-1} w) \, dy = \int \rho y \cdot \nabla \left(\frac{\beta w^2}{2} - \frac{|w|^{p+1}}{p+1}\right) dy$$

$$= \int \left(\frac{|y|^2}{2} - n\right) \rho \left(\frac{\beta w^2}{2} - \frac{|w|^{p+1}}{p+1}\right) dy$$

and

$$- \int (y \cdot \nabla w) \nabla \cdot (\rho \nabla w) \, dy = \int (\rho \nabla w) \cdot \nabla (y \cdot \nabla w) \, dy$$

$$= \int \rho |\nabla w|^2 \, dy + \frac{1}{2} \int (\rho y) \cdot \nabla (|\nabla w|^2) \, dy$$

$$= \int \rho |\nabla w|^2 \, dy + \frac{1}{2} \int \left(\frac{|y|^2}{2} - n \right) \rho |\nabla w|^2 \, dy.$$

Now, to complete the proof of (25.11), we eliminate the terms involving w^2 and $|w|^{p+1}$ by taking the linear combination $\frac{n}{p+1} \cdot (25.12) - \frac{1}{2(p+1)} \cdot (25.14) + (25.15)$.

Finally, (25.11) and our assumption $p \leq p_S$ imply $\nabla w \equiv 0$, hence $w \equiv 0$, $w \equiv k$ or $w \equiv -k$. $\quad \square$

Lemma 25.6. *Consider problem (18.1) with $p > 1$ and $u_0 \in L^\infty(\mathbb{R}^n)$. Assume that the upper blow-up rate estimate (23.5) is satisfied. Then we have:*

(i) For any sequence $s_j \to \infty$, there exists a subsequence, still denoted s_j, and a function $z \in \mathcal{S}$ such that $w_a(\cdot, s_j) \to z$ in $C^1(\mathbb{R}^n)$.

(ii) Assume that $\omega(w_a) \ni 0$ (resp., $\pm k$). Then $E\big(w_a(s)\big) \to 0$ (resp., $E\big(w_a(s)\big) \to \eta(n, p) > 0$).

(iii) If $\omega(w_a) \ni 0$, then a is not a blow-up point.

(iv) If $p \leq p_S$, then $\omega(w_a)$ is one of the sets $\{0\}$, $\{k\}$, $\{-k\}$.

Proof. (i) Assumption (23.5) implies

$$|w_a| \leq M, \qquad y \in \mathbb{R}^n, \ s \geq s_0. \tag{25.16}$$

By Proposition 23.15, since $\nabla w_a(y, s) = (T - t)^{\beta + 1/2} \nabla u(x, t)$, this implies

$$|\nabla w_a| \leq M_1, \qquad y \in \mathbb{R}^n, \ s \geq \bar{s}_0 := s_0 + \log 2. \tag{25.17}$$

Let $z_j(y, s) = w_a(y, s + s_j)$. By (23.9), (25.16), (25.17) and parabolic estimates, the sequence $\{z_j\}$ is precompact in $C^{2,1}(\mathbb{R}^n \times [0, 1])$. Consequently, there exists a subsequence of s_j (still denoted s_j) and a solution z of

$$z_s - \Delta z + \frac{1}{2} z \cdot \nabla z = |z|^{p-1} z - \beta z, \qquad y \in \mathbb{R}^n, \ s \in [0, 1],$$

such that $w_a(\cdot, \cdot + s_j) \to z$ in $C^{2,1}(\mathbb{R}^n \times [0, 1])$. Moreover $z, \nabla z$ are bounded in $\mathbb{R}^n \times [0, 1]$. On the other hand, using (23.17), we have

$$\int_0^1 \int_{\mathbb{R}^n} (\partial_s z_j)^2 \rho \, dy \leq \int_{s_j}^\infty \int_{\mathbb{R}^n} (\partial_s w_a)^2 \rho \, dy \to 0$$

as $j \to \infty$. By Fatou's lemma we deduce that $\partial_s z = 0$ and the assertion follows.

(ii) Assume that $w_a(\cdot, s_j) \to 0$ (resp., $\pm k$) and $\nabla w_a(\cdot, s_j) \to 0$, uniformly for $|y|$ bounded. Using (25.16), (25.17) and dominated convergence, we infer $E(w_a(s_j)) \to 0$, resp.

$$E(w_a(s_j)) \to \left(\int_{\mathbb{R}^n} \rho \, dy \right) \left(\frac{\beta}{2} - \frac{k^{p-1}}{p+1} \right) k^2 = \frac{(4\pi)^{n/2} k^2}{2(p+1)} =: \eta(n,p) > 0.$$

The assertion then follows from the monotonicity of $E(w_a(s))$.

(iii) Let $b \in \mathbb{R}^n$. Similar to (25.16) and (25.17), we have

$$|w_b| \le M, \quad |\nabla w_b| \le M_1, \qquad y \in \mathbb{R}^n, \ s \ge \bar{s}_0, \tag{25.18}$$

with M, M_1 independent of b. We shall use the interpolation inequality

$$|v(0)| \le C(n, \theta) \left[\|v\|_{L^2(B_1)}^{\theta} \|\nabla v\|_{L^\infty(B_1)}^{1-\theta} + \|v\|_{L^2(B_1)} \right], \quad v \in C^1(\overline{B_1}), \tag{25.19}$$

where $0 < \theta < 2/(n+2)$ if $n \ge 2$ and $\theta = 1/2$ if $n = 1$. (Inequality (25.19) can be shown by applying the mean-value theorem to the difference $|v(0)|^{1/(1-\theta)} - |v(x)|^{1/(1-\theta)}$, integrating over $x \in B_1$ and using Hölder's inequality.) Fix θ as above. By (23.16), since w_b exists globally, we have, for any $s_1 \ge \bar{s}_0$,

$$\|w_b(0, s)\|_{L^2(B_1)}^2 \le C(n, p) \int_{\mathbb{R}^n} w_b^2 \rho \, dy \le C(n, p) E^{2/(p+1)}(w_b(s_1)), \quad s \ge s_1. \tag{25.20}$$

Using (25.18), (25.19) and (25.20), it follows that

$$|w_b(0, s)| \le C(n, p) \left[M_1^{1-\theta} E^{\theta/(p+1)}(w_b(s_1)) + E^{1/(p+1)}(w_b(s_1)) \right], \quad s \ge s_1.$$

Consequently, there exists $\gamma(M_1, \varepsilon) > 0$ such that

$$E(w_b(s_1)) < \gamma(M_1, \varepsilon) \quad \text{implies} \quad |w_b(0, s)| \le \varepsilon, \quad s \ge s_1.$$

Assume that $\omega(w_a) \ni 0$. Assertion (ii) implies $E(w_a(s_1)) < \gamma(M_1, \varepsilon)$ for s_1 large. But since, for given s, $E(w_b(s))$ depends continuously on b (cf. Proposition 23.8), we infer that $E(w_b(s_1)) < \gamma(M_1, \varepsilon)$ for $|b - a|$ small. It follows that $|w_b(y, s)| \le \varepsilon$, $s \ge s_1$, hence $(T - t)^{1/(p-1)} |u(b, t)| \le \varepsilon$, for (b, t) close to (a, T). By Theorem 25.3, we conclude that a is not a blow-up point.

(iv) In view of Proposition 25.4, this follows from an obvious connectedness argument. $\quad \square$

Proof of Theorem 25.1. This is an immediate consequence of assertions (iii) and (iv) of Lemma 25.6. $\quad \square$

As a consequence of the above arguments, we are also now in a position to prove the result stated in the previous section concerning the blow-up set.

Proof of Theorem 24.5. (i) If $E(u_0) < \eta$, where η is given by Lemma 25.6(ii), then $\lim_{s\to\infty} E(w_a(s)) < \eta$. Consequently, w_a cannot converge to $\pm k$, due to Lemma 25.6(ii). So it converges to 0 and a is not a blow-up point in view of Theorem 25.1.

(ii) By dominated convergence, under the current assumption on u_0, we have $\lim_{a\to\infty} E(w_a(s_0)) = 0$. The conclusion thus follows from assertion (i). $\square$

Remark 25.7. Radial nonincreasing case. In Theorem 25.1, assume in addition that $u_0 \geq 0$ is radial nonincreasing. Since

$$u(0,t) = \|u(t)\|_\infty \geq k(T-t)^{-1/(p-1)}$$

by Proposition 23.1, the conclusion (with the $+$ sign) follows directly from Lemma 25.6(i) and Proposition 25.4, and the nondegeneracy result is not needed. $\square$

Remark 25.8. Blow-up profile. The results in Theorems 24.1 and 24.3 concerning the blow-up profile in the original variables can be strongly improved, at the expense of significant technical difficulty, though. In order to do this one has to linearize problem (23.9) around the nontrivial equilibrium $w_\infty \equiv k$ and study the convergence to w_∞ in more detail. Assume $\Omega = \mathbb{R}$, $u_0 \in BC(\mathbb{R})$, $u_0 \geq 0$ and $0 \in B(u_0)$. Then one obtains (see [275], [276] and cf. also [207], [208] for earlier results in that direction) that one of the following alternatives must hold:

1. $(T - t)^\beta u(x,t) \equiv k$,

2. $(T - t)^\beta u\big(y\big((T-t)|\log(T-t)|\big)^{1/2}, t\big) \to k\big(1 + (p-1)y^2/(4p)\big)^{-\beta}$ as $t \to T$,

3. there exists $C > 0$ and an even integer $m \geq 4$ such that

$$(T - t)^\beta u\big(y(T-t)^{1/m}, t\big) \to k(1 + Cy^m)^{-\beta} \quad \text{as } t \to T,$$

where the convergence is uniform for y lying in a bounded set. (Note also that the second alternative is always true if u_0 is symmetric and has a unique local maximum at $x = 0$.) As a consequence of this assertion, the following general result was obtained in [519]. Assume that $u(x,t)$ is a positive solution of $u_t - u_{xx} = u^p$ for $x \in (-R, R)$ and $t \in (0, T)$ which blows up at $t = T$. Assume also that its blow-up set B is contained in $[-\delta, \delta]$ for some $\delta < R$. Then B is isolated and for any $\bar{x} \in B$ one of the following holds

$$\lim_{x\to\bar{x}} \left(\frac{|x - \bar{x}|^2}{|\log|x - \bar{x}||}\right)^\beta u(x,T) = \left(\frac{8p}{(p-1)^2}\right)^\beta,$$

$$\lim_{x\to\bar{x}} |x - \bar{x}|^{m\beta} u(x,T) = kC^{-\beta},$$

where β, k, C, m are as above. It is also known that any of the possibilities mentioned above may happen (see [100], [366]). For results in higher dimensions we refer to [61], [521], [367], [360]. $\square$

26. Universal bounds and initial blow-up rates

The a priori estimate (22.2) with a universal constant C cannot be true for all global solutions of (22.1) for the following reasons. First, such an estimate would imply an a priori bound for stationary solutions and we know from Theorem 7.8(ii) that such bound is not true for sign-changing solutions in the subcritical case. Second, we know from Remark 19.12 that there exist nonnegative global classical solutions such that $\|u(t)\|_\infty \to \infty$ as $t \to 0+$. Anyhow, in the subcritical case, we can still hope for a universal bound of global nonnegative solutions of (22.1) on the interval (τ, ∞), where $\tau > 0$. In other words, we are interested in the estimate

$$\sup_{t \geq \tau} \|u(t)\|_\infty \leq C(\tau) \qquad \text{for all } \tau > 0. \tag{26.1}$$

(Note that (26.1) cannot be true in the critical or supercritical case — at least in starshaped domains — due to Theorem 28.7.) It will be natural at the same time to ask about the dependence of the constant $C(\tau)$, as $\tau \to 0$.

In fact, this question can be also considered from a different point of view, which gives rise to interesting connections and unifications with questions studied in Sections 23 and 21. Consider *local nonnegative* classical solutions of

$$\left.\begin{array}{ll} u_t - \Delta u = u^p, & x \in \Omega, \ 0 < t < T, \\ u = 0, & x \in \partial\Omega, \ 0 < t < T \end{array}\right\} \tag{26.2}$$

(without any prescribed initial conditions). Do there exist estimates of the form

$$\|u(t)\|_\infty \leq Ct^{-\alpha}, \quad 0 < t \leq T/2, \tag{26.3}$$

and

$$\|u(t)\|_\infty \leq C(T-t)^{-\beta}, \quad T/2 \leq t < T, \tag{26.4}$$

where $C = C(p, \Omega, T) > 0$ is a universal constant, independent of u? If (26.4) were true with $\beta = 1/(p-1)$, one would in particular recover the (final) blow-up estimates of Section 23, now with a universal constant. Analogously, estimate (26.3) would provide (universal) **initial blow-up rates**. An interesting question is what should be the optimal value of α. Of course, (26.3) or (26.4) implies in particular the universal bound (26.1) for global nonnegative solutions. Furthermore, we will see that these estimates are strongly connected with parabolic Liouville-type theorems and decay of global solutions of the Cauchy problem (see Remark 26.10(i)).

The bound (26.1) for all global nonnegative solutions of (22.1) in bounded domains was first proved in [200] for $p < p_{BT}$ (note that this exponent has already appeared in an elliptic context in Section 10). As for the initial and final blow-up rate estimates (26.3) and (26.4), they have first been established in [28] for the Cauchy problem with $p < p_F$. Those results have been improved and extended in a number of subsequent works, using various techniques. We shall present some of these results and techniques. Some of the proofs rely on rescaling arguments and apply essentially only to the model problem (26.2), while some others allow to treat nonlinearities $f(u)$ without precise power behavior (see Remarks 26.5 and 26.12).

We start with a result whose proof is relatively simple. Better results will be given later for the model problem (see Theorems 26.6 and 26.8), but the present approach, besides its simplicity, has the advantage to be applicable to more general nonlinearities (see Remark 26.5). It is based on integral bounds obtained by test-function arguments (in particular using the first eigenfunction) and on smoothing properties in L^q- or L^q_δ-spaces (see Theorem 26.14 below for further results obtained by using L^q_δ-spaces).

Theorem 26.1. *Assume Ω bounded and $1 < p < p_{BT}$. For all $\tau > 0$, there exists $C(\Omega, p, \tau) > 0$ such that any global nonnegative classical solution of (26.2) satisfies*

$$\sup_{t \geq \tau} \|u(t)\|_\infty \leq C(\Omega, p, \tau). \tag{26.5}$$

Remarks 26.2. (i) **Instantaneous attractors.** In other words, Theorem 26.1 (and similar subsequent results) shows the existence of "instantaneous attractors" for global nonnegative trajectories of (26.2). Note that, by standard smoothing effects, (26.5) guarantees that for each $\tau > 0$, there is a compact (absorbing) subset K_τ of $C^2(\overline{\Omega}) \cap C_0(\Omega)$, such that any global nonnegative solution of (26.2) remains in K_τ for $t \geq \tau$ (otherwise u has to blow up in finite time).

In terms of the set $\mathcal{G}^+$ introduced in Remark 19.12, Theorem 26.1 says that, although $\mathcal{G}^+$ itself is unbounded, for each $\tau > 0$, $S(\tau)\mathcal{G}^+$ is a bounded subset of $L^\infty(\Omega)$ (where $S(t)u_0$ denotes the solution $u(t)$ of problem (15.1)).

(ii) **Differences from equations with absorption.** We emphasize that such localization results are of a quite different nature from what occurs in equations with absorption, such as $u_t - \Delta u + |u|^{p-1}u = 0$ with $p > 1$. Indeed, for this equation, it is straightforward that *all* solutions of the Dirichlet or Cauchy problem (with bounded initial data) satisfy the universal estimate $\|u(t)\|_\infty \leq C(p)t^{-1/(p-1)}$ for all $t > 0$. This immediately follows by comparing with the solution $y(t) \equiv C(p)t^{-1/(p-1)}$ of the ODE $y' + y^p = 0$.

In the case of problem (26.2), this is of course not true, due to the existence of blowing-up solutions. The universal bound (26.5) is verified by a solution u, *under the assumption* that u exists globally (or on some time interval $(0, T)$ in the case of estimates (26.3) and (26.4)).　□

We give a first proof of Theorem 26.1 based on L^q_δ-spaces, due to [200]. We first derive some basic estimates for positive solutions of (26.2).

Lemma 26.3. *Assume Ω bounded, $p > 1$, and $0 < T < \infty$. Let u be a nonnegative classical solution of (26.2) on $(0, T)$. Then for all $t \in (0, T/2]$, there holds*

$$\int_\Omega u(t)\varphi_1 \, dx \le C(p, \Omega)(1 + T^{-1/(p-1)}), \tag{26.6}$$

and

$$\int_0^t \int_\Omega u^p \varphi_1 \, dx \, ds \le C(p, \Omega)\big(1 + t\big)\big(1 + T^{-1/(p-1)}\big). \tag{26.7}$$

Proof. As in the proof of Theorem 17.1, denote $y = y(t) := \int_\Omega u(t)\varphi_1 \, dx$, multiply the equation in (26.2) by φ_1 and integrate by parts. We obtain

$$\frac{d}{dt} \int_\Omega u(t)\varphi_1 \, dx + \lambda_1 \int_\Omega u(t)\varphi_1 \, dx = \int_\Omega u^p(t)\varphi_1 \, dx. \tag{26.8}$$

By Jensen's inequality, we infer that

$$\frac{d}{dt} \int_\Omega u(t)\varphi_1 \, dx \ge \left(\int_\Omega u(t)\varphi_1 \, dx \right)^p - \lambda_1 \int_\Omega u(t)\varphi_1 \, dx.$$

Since u exists on $(0, T)$, we deduce easily that

$$\int_\Omega u(t)\varphi_1 \, dx \le C(p, \Omega)(1 + (T - t)^{-1/(p-1)}), \quad 0 < t < T,$$

hence (26.6). Integrating (26.8) in time over (τ, t) $(0 < \tau < t \le T/2)$ and using (26.6), we obtain

$$\int_\tau^t \int_\Omega u^p \varphi_1 \, dx \, ds = \lambda_1 \int_\tau^t \int_\Omega u\varphi_1 \, dx \, ds + \int_\Omega u(t)\varphi_1 \, dx - \int_\Omega u(\tau)\varphi_1 \, dx$$
$$\le C(p, \Omega)\big(1 + t\big)\big(1 + T^{-1/(p-1)}\big)$$

and (26.7) follows by letting $\tau \to 0$.　□

Proof of Theorem 26.1. By Theorem 22.1, we know that global solutions of (22.1) satisfy the a priori estimate

$$\|u(t)\|_\infty \le C(\Omega, p, \|u(t_0)\|_\infty), \quad t \ge t_0 \ge 0,$$

where C remains bounded for $\|u(t_0)\|_\infty$ bounded. Therefore, it is sufficient to show the existence of $C(p,\Omega,\tau) > 0$ such that any global classical solution u of (26.2) satisfies

$$\inf_{t\in(0,\tau)} \|u(t)\|_\infty \le C(p,\Omega,\tau). \tag{26.9}$$

Moreover, by the L^q_δ-smoothing estimate in Theorem 15.9, (26.9) will follow if we can show that, for some $q > (n+1)(p-1)/2$,

$$\inf_{t\in(0,\tau/2)} \|u(t)\|_{q,\delta} \le C(p,\Omega,q,\tau). \tag{26.10}$$

But (26.7) guarantees that (26.10) is true for $q = p$ and, since $p < p_{BT}$, we have $p > (n+1)(p-1)/2$. $\square$

We now give a second proof (see [200, Section 6]), which does not use L^q_δ-spaces. Instead it requires the following estimate, whose proof uses the special test-function constructed in Lemma 10.4 by considering a singular elliptic problem.

Lemma 26.4. *Assume Ω bounded, $p > 1$, $0 < T < \infty$, and $\varepsilon \in (0,(p+1)/2]$. Let u be a nonnegative classical solution of (26.2) on $(0,T)$. Then for all $t \in (0,T/2]$, there holds*

$$\int_0^t \int_\Omega u^{\frac{p+1}{2}-\varepsilon}\, dx\, ds \le C(p,\Omega,\varepsilon)\bigl(1+t\bigr)\bigl(1+T^{-1/(p-1)}\bigr).$$

Proof. For given $0 < \alpha < 1$, Lemma 10.4 ensures the existence of a function $\xi \in C(\overline{\Omega}) \cap C^2(\Omega) \cap H_0^1(\Omega)$ such that $-\Delta\xi = \varphi_1^{-\alpha}$ in Ω. Moreover, ξ satisfies

$$\xi(x) \le C(\Omega,\alpha)\delta(x), \quad x \in \Omega. \tag{26.11}$$

Here we choose $\alpha = 1 - \frac{4\varepsilon}{p-1+2\varepsilon}$. Taking ξ as a test-function in (26.2) (which is possible due to $\xi \in H_0^1(\Omega)$) and integrating in time over (τ,t), we obtain

$$\int_\tau^t \int_\Omega u\varphi_1^{-\alpha}\, dx\, ds = \int_\tau^t \int_\Omega u^p\xi\, dx\, ds + \int_\Omega u(\tau)\xi\, dx - \int_\Omega u(t)\xi\, dx.$$

Due to (26.11), (26.6) and (26.7) readily imply

$$\int_0^t \int_\Omega u\varphi_1^{-\alpha}\, dx\, ds \le C(p,\Omega,\varepsilon)\bigl(1+t\bigr)\bigl(1+T^{-1/(p-1)}\bigr).$$

Using Hölder's inequality, the last estimate and (26.7) imply the lemma. $\square$

Second proof of Theorem 26.1. As in the first proof, it is sufficient to show the existence of $C(p,\Omega,\tau) > 0$ such that any global classical solution u of (26.2) satisfies (26.9). Moreover, by the smoothing estimate in Theorem 15.2, (26.9) will follow if we can show that, for some $q > n(p-1)/2$,

$$\inf_{t\in(0,\tau/2)} \|u(t)\|_q \le C(p,\Omega,q,\tau). \tag{26.12}$$

But Lemma 26.4 guarantees that (26.12) is true for all $q \in [1,(p+1)/2)$ and, since $p < p_{BT}$, we have $q > n(p-1)/2$ for $q < (p+1)/2$ close to $(p+1)/2$. $\square$

Remark 26.5. The assumption $p < p_{BT}$ in Theorem 26.1 is not optimal for the model problem (26.2), see Theorems 26.6 and 26.8 below. However, unlike the proofs of those theorems, the proof of Theorem 26.1 does not rely on rescaling and can be applied to more general nonlinearities $f(x, u)$ satisfying $C_1 u^q - C \le f(x, u) \le C_2 u^p + C$ with $p < p_{BT}$, under suitable assumption on $q \in (1, p)$ (see [450]). Note that the proof uses a priori estimates of global solutions obtained in Theorem 22.1. However, the proof of Theorem 22.1 based on interpolation can be also extended to such nonlinearities. $\square$

Now we give an optimal result [438] in dimensions $n \le 3$ concerning universal bounds of global nonnegative solutions of the Dirichlet problem. The method is completely different. It is based on energy, measure arguments, rescaling and elliptic Liouville-type theorems.

Theorem 26.6. *Let $n \le 3$ and $1 < p < p_S$. Assume Ω bounded. Then the conclusion of Theorem 26.1 is true.*

Proof. As in the (first) proof of Theorem 26.1, it is sufficient to show the existence of $C(p, \Omega, \tau) > 0$ such that any global classical solution u of (26.2) satisfies (26.9). Moreover, since $p + 1 > n(p - 1)/2$ due to $p < p_S$, by the smoothing property in Theorem 15.2, (26.9) will follow if we can show that

$$\inf_{t \in (0, \tau/2)} \|u(t)\|_{p+1} \le C(p, \Omega, \tau).$$

We argue by contradiction and assume that for each $k = 1, 2, \ldots$, there exists a global solution $u_k \ge 0$ of (26.2) such that

$$\|u_k(t)\|_{p+1}^{p+1} > k \qquad \text{for all } t \in (0, \tau/2). \tag{26.13}$$

Denote

$$E_k(t) = E\big(u_k(t)\big) = \frac{1}{2} \int_\Omega |\nabla u_k(t)|^2 \, dx - \frac{1}{p+1} \int_\Omega u_k^{p+1}(t) \, dx.$$

Recall that $E_k'(t) = -\|\partial_t u_k(t)\|_2^2 \le 0$ and that u_k satisfies the identity

$$\frac{1}{2} \frac{d}{dt} \int_\Omega u_k^2(t) \, dx = \int_\Omega u_k^{p+1}(t) \, dx - \int_\Omega |\nabla u_k(t)|^2 \, dx$$

$$= -2E_k(t) + \frac{p-1}{p+1} \int_\Omega u_k^{p+1}(t) \, dx. \tag{26.14}$$

We now proceed in several steps. From now on, C will denote a positive constant and k_0 a positive integer, both depending only on p, Ω, τ (and also on q in Steps 4 and 5).

Step 1. We claim that

$$E_k(\tau/4) \geq k^{1/2}, \tag{26.15}$$

for all $k \geq k_0$ large enough.

Assume (26.15) is false. Using (26.14), $E_k' \leq 0$ and Hölder's inequality, we obtain, for all $t \geq \tau/4$,

$$\frac{1}{2}\frac{d}{dt}\int_\Omega u_k^2(t)\,dx \geq -2k^{1/2} + \frac{p-1}{p+1}\int_\Omega u_k^{p+1}(t)\,dx, \tag{26.16}$$

hence

$$\frac{1}{2}\frac{d}{dt}\int_\Omega u_k^2(t)\,dx \geq -2k^{1/2} + C\left(\int_\Omega u_k^2(t)\,dx\right)^{(p+1)/2}.$$

This implies

$$\int_\Omega u_k^2(t)\,dx \leq Ck^{\frac{1}{p+1}}, \quad t \geq \tau/4, \tag{26.17}$$

since otherwise $\int_\Omega u_k^2(t)\,dx$ has to blow up in finite time. Integrating (26.16) over $(\tau/4, \tau/2)$ and using (26.13), (26.17), we obtain

$$\frac{1}{4}k\tau \leq \int_{\tau/4}^{\tau/2}\int_\Omega u_k^{p+1}\,dx\,dt \leq C(k^{\frac{1}{p+1}} + k^{1/2}\tau),$$

a contradiction for $k \geq k_0$ large.

Step 2. Let $a > 0$ to be fixed later and set $F_k = \{t \in (0,\tau/4] \ : \ -E_k'(t) \geq E_k^{1+1/a}(t)\}$. We claim that $|F_k| < \tau/8$ for all $k \geq k_0$ large enough.

Note that $E_k > 0$ on $(0,\tau/4]$ for $k \geq k_0$ by (26.15), since $E_k' \leq 0$. By definition of F_k, it follows that

$$(aE_k^{-1/a})' = -E_k'E_k^{-1-1/a} \geq \chi_{F_k} \quad \text{on } (0,\tau/4].$$

By integration, we deduce that $aE_k^{-1/a}(\tau/4) \geq |F_k|$. The claim then follows from (26.15).

Step 3. Choose

$$a \geq (p+1)/(p-1). \tag{26.18}$$

We claim that for all $k \geq k_0$ large,

$$\|\partial_t u_k(t)\|_2^2 \leq C\left(\int_\Omega u_k^{p+1}(t)\,dx\right)^{(a+1)/a} \quad \text{for all } t \in (0,\tau/4] \setminus F_k. \tag{26.19}$$

For all $t \in (0,\tau/4] \setminus F_k$, we have

$$\|\partial_t u_k(t)\|_2^2 = -E_k'(t) \leq E_k^{1+1/a}(t) \leq \|\nabla u_k(t)\|_2^{2(1+1/a)}. \tag{26.20}$$

Hence, by (26.14) as well as Hölder's and Young's inequalities,

$$
\begin{aligned}
\|\nabla u_k(t)\|_2^2 &\leq \int_\Omega u_k^{p+1}(t)\,dx + \|u_k(t)\|_2 \|\partial_t u_k(t)\|_2 \\
&\leq \int_\Omega u_k^{p+1}(t)\,dx + \|u_k(t)\|_2 \|\nabla u_k(t)\|_2^{1+1/a} \\
&\leq \int_\Omega u_k^{p+1}(t)\,dx + C\|u_k(t)\|_{p+1} \|\nabla u_k(t)\|_2^{1+1/a} \\
&\leq \int_\Omega u_k^{p+1}(t)\,dx + C\|u_k(t)\|_{p+1}^{2a/(a-1)} + \frac{1}{2}\|\nabla u_k(t)\|_2^2 \\
&\leq C \int_\Omega u_k^{p+1}(t)\,dx + \frac{1}{2}\|\nabla u_k(t)\|_2^2,
\end{aligned}
$$

where we have used (26.18) and (26.13). Consequently,

$$
\|\nabla u_k(t)\|_2^2 \leq C \int_\Omega u_k^{p+1}(t)\,dx.
$$

This along with (26.20) implies (26.19).

Step 4. Let $0 < q < (p+1)/2$, $b = (p+1-q)(a+1)/a$ and

$$
G_k = \{t \in (0,\tau/4] \,:\, \|\partial_t u_k(t)\|_2^2 \leq C\|u_k(t)\|_\infty^b\}.
$$

We claim that $|G_k| > 0$.

Due to Lemma 26.4, for $A = A(p,q,\Omega,\tau) > 0$ large enough, the set

$$
\widetilde{G}_k := \{t \in (0,\tau/4] \,:\, \int_\Omega u_k^q(t)\,dx \geq A\}
$$

satisfies

$$
|\widetilde{G}_k| < \tau/8. \tag{26.21}
$$

We deduce from (26.13) that, for all $t \in (0,\tau/4] \setminus \widetilde{G}_k$,

$$
\int_\Omega u_k^{p+1}(t)\,dx \leq C\|u_k(t)\|_\infty^{p+1-q} \int_\Omega u_k^q(t)\,dx \leq C\|u_k(t)\|_\infty^{p+1-q}.
$$

Therefore, $G_k \supset (0,\tau/4] \setminus (F_k \cup \widetilde{G}_k)$ by Step 3. The claim then follows from Step 2 and (26.21).

Step 5. We will now obtain a contradiction by using a rescaling argument.

By Step 4, for each large k, we may pick $t_k \in G_k$. By (26.13), we have $M_k := \|u_k(t_k)\|_\infty \to \infty$. Choose $x_k \in \overline{\Omega}$ such that $u_k(x_k, t_k) = M_k$, denote $\nu_k = M_k^{-(p-1)/2}$ and put

$$
\begin{aligned}
w_k(y) &= M_k^{-1} u_k(x_k + \nu_k y, t_k), \\
\widetilde{w}_k(y) &= M_k^{-p} \partial_t u_k(x_k + \nu_k y, t_k).
\end{aligned}
$$

Then the functions $w_k, \widetilde{w}_k$ satisfy

$$\left.\begin{array}{rcll} \Delta w_k + w_k^p &=& \widetilde{w}_k & \text{in } \Omega_k, \\[4pt] w_k &=& 0 & \text{on } \partial\Omega_k, \end{array}\right\} \tag{26.22}$$

where $\Omega_k = \nu_k^{-1}(\Omega - x_k)$. Moreover, $0 \le w_k \le 1 = w_k(0)$. Now passing to the limit we will obtain a contradiction in the same way as in [241]; we only have to show that the functions w_k are (locally) uniformly Hölder continuous and $\widetilde{w}_k \to 0$ in an appropriate way.

Hence let $R > 0$, $B_R(x_0) = \{x \in \Omega : |x - x_0| < R\}$ and $B_R^k = \{y \in \Omega_k : |y| < R\}$. Since $t_k \in G_k$, we have

$$\int_{B_R^k} |\widetilde{w}_k(y)|^2 \, dy = M_k^{-2p} \int_{B_R^k} |\partial_t u_k(x_k + \nu_k y, t_k)|^2 \, dy$$

$$= M_k^{-2p} \nu_k^{-n} \int_{B_{R\nu_k}(x_k)} |\partial_t u_k(x, t_k)|^2 \, dx$$

$$\le C M_k^{-2p} M_k^{n(p-1)/2} M_k^b = C M_k^\gamma$$

for $k \ge k_0$, where

$$\gamma = -2p + \frac{a+1}{a}(p+1-q) + \frac{n(p-1)}{2}.$$

By taking q close to $(p+1)/2$ and a sufficiently large, γ will be negative provided $(n-3)p < n-1$. (In particular this is true due to $p < p_S$ if $n \le 4$.)

Consequently,

$$\int_{B_R^k} |\widetilde{w}_k(y)|^2 \, dy \to 0$$

for any $R > 0$. Since $0 \le w_k \le 1$ and w_k solves (26.22), standard regularity theory guarantees that w_k is uniformly bounded in $W^{2,2}(B_R^k)$. Since $W^{2,2}$ is embedded in the space of Hölder continuous functions due to $n \le 3$, we may pass to the limit in (26.22), similarly as in the proof of Theorem 12.1, in order to get a limiting solution $w \ge 0$ satisfying the equation $\Delta w + w^p = 0$ either in $\mathbb{R}^n$ or in a half-space (and satisfying the homogeneous Dirichlet boundary conditions in the latter case). Moreover $w \le 1$ and $w(0) = 1$, which contradicts the Liouville-type Theorems 8.1 and 8.2. $\square$

Remark 26.7. By a (nontrivial) modification of the proof of Theorem 26.6, one can show that the result remains true for $n = 4$, and for $n \ge 5$ under the stronger restriction $p < (n-1)/(n-3) < p_S$ (see [450]). $\square$

We now turn to universal initial and final blow-up rates. Recall that the exponent p_B in (26.23) has appeared in Section 21.

Theorem 26.8. *Let $p > 1$, $T > 0$ and u be a nonnegative classical solution of (26.2) on Q_T. Assume that either*

$$p < p_B, \qquad or \ \ p < p_S, \ \Omega = \mathbb{R}^n \ or \ \Omega = B_R, \ and \ u \ radial. \qquad (26.23)$$

Then there holds

$$u(x,t) \leq C(n,p)\big(t^{-1/(p-1)} + (T-t)^{-1/(p-1)}\big), \quad x \in \mathbb{R}^n, \quad 0 < t < T \quad (26.24)$$

if $\Omega = \mathbb{R}^n$, and

$$u(x,t) \leq C(p,\Omega)\big(1 + t^{-1/(p-1)} + (T-t)^{-1/(p-1)}\big), \quad x \in \Omega, \quad 0 < t < T \quad (26.25)$$

otherwise.

Estimates (26.24), (26.25) provide a universal localization in $L^\infty(\Omega)$ throughout the time interval $(0,T)$ for positive trajectories of (26.2). Note that Theorem 26.8 partially improves the above results on universal bounds of global solutions to the Dirichlet problem. Theorem 26.8 in the case $\Omega = \mathbb{R}^n$ with $p < p_B$ is due to [79], where it follows from integral estimates for local solutions (cf. Proposition 21.5). The other cases are due to [425] and the proof is based on a doubling lemma, a rescaling argument, and the parabolic Liouville-type theorems established in Section 21. The methods are thus different from those in Theorems 26.1 and 26.6.

As an interesting consequence of Theorem 26.8, one obtains the decay of all nonnegative global solutions to the Cauchy problem.

Theorem 26.9. *Let $p > 1$ and u be a global nonnegative classical solution of*

$$u_t - \Delta u = u^p, \quad x \in \mathbb{R}^n, \quad t > 0. \qquad (26.26)$$

Assume

$$p < p_B, \qquad or \ \ p < p_S \ and \ u \ radial.$$

Then there holds

$$u(x,t) \leq C(n,p)\, t^{-\frac{1}{p-1}}, \quad x \in \mathbb{R}^n, \quad t > 0. \qquad (26.27)$$

Remarks 26.10. (i) If the parabolic Liouville-type Theorem 21.2 were known for all $p < p_S$, then this would imply Theorems 26.8 and 26.9 for all $p < p_S$ as well. Conversely, it is clear that estimate (26.24) or (26.27) implies nonexistence of positive solutions of (21.1). We see that Liouville-type theorems and these universal estimates are thus equivalent. On the other hand, Theorem 26.8 guarantees that Theorem 21.1 remains true for nontrivial nonnegative radial classical solutions, bounded or not, and that Theorem 21.2 remains true for nontrivial nonnegative classical solutions.

(ii) For all $p < p_S$ and without radial symmetry assumption, it is however known that the solution of the Cauchy problem (18.1) satisfies $\|u(t)\|_\infty \to 0$ as $t \to \infty$ (without a universal estimate), provided u is global and $0 \le u_0 \in L^\infty \cap L^2(\mathbb{R}^n)$; see [478] and cf. also [299].

(iii) In Theorems 26.8 and 26.9, no conditions at space infinity are assumed on the solution u.

(iv) Consider problem (26.2) with the nonlinearity u^p replaced by $f(u)$. Assume that $f : [0, \infty) \to \mathbb{R}$ is continuous and is such that $\lim_{s \to \infty} s^{-p} f(s)$ exists in $(0, \infty)$. Then Theorem 26.8 remains valid (with C in (26.24)-(26.25) depending also on f and with an additive constant 1 in estimate (26.24) as well). If we assume in addition that f is C^1 and verifies $|f'(s)| \le C(1 + s^{p-1})$, $s \ge 0$, then Theorem 26.6 remains valid.

(v) When Ω is a convex bounded domain, estimate (26.4) with $\beta = 1/(p-1)$ and $C = C(p, \Omega, T)$ is known also for $p < p_S$, $n \le 4$ [450]. This follows by combining Theorem 26.6 (cf. also Remark 26.7) with the a priori estimate of the blow-up rate (cf. Remark 23.9). Let us point out that the method of proof of Theorem 26.6 can be modified to establish initial blow-up rate estimates as well [450], but the values of $\alpha = \alpha(n, p)$ obtained in this way are not optimal. $\quad\square$

We will use the following key doubling lemma [424].

Lemma 26.11. *Let (X, d) be a complete metric space and let $\emptyset \ne D \subset \Sigma \subset X$, with Σ closed. Set $\Gamma = \Sigma \setminus D$. Finally let $M : D \to (0, \infty)$ be bounded on compact subsets of D and fix a real $k > 0$. If there exists $y \in D$ such that*

$$M(y) \operatorname{dist}(y, \Gamma) > 2k, \tag{26.28}$$

then there exists $x \in D$ such that

$$M(x) \operatorname{dist}(x, \Gamma) > 2k, \qquad M(x) \ge M(y), \tag{26.29}$$

and

$$M(z) \le 2M(x) \quad \text{for all } z \in D \cap \overline{B}_X\big(x, k\, M^{-1}(x)\big).$$

Proof. Assume that the lemma is not true. Then we claim that there exists a sequence (x_j) in D such that

$$M(x_j) \operatorname{dist}(x_j, \Gamma) > 2k, \tag{26.30}$$

$$M(x_{j+1}) > 2M(x_j), \tag{26.31}$$

and

$$d(x_j, x_{j+1}) \le kM^{-1}(x_j) \tag{26.32}$$

for all $j \in \mathbb{N}$. We choose $x_0 = y$. By our contradiction assumption, there exists $x_1 \in D$ such that

$$M(x_1) > 2M(x_0)$$

and

$$d(x_0, x_1) \leq k\, M^{-1}(x_0).$$

Fix some $i \geq 1$ and assume that we have already constructed $x_0, \ldots, x_i$ such that (26.30)–(26.32) hold for $j = 0, \ldots, i - 1$. We have

$$\operatorname{dist}(x_i, \Gamma) \geq \operatorname{dist}(x_{i-1}, \Gamma) - d(x_{i-1}, x_i) > (2k - k)\, M^{-1}(x_{i-1}) > 2k\, M^{-1}(x_i),$$

hence

$$M(x_i)\operatorname{dist}(x_i, \Gamma) > 2k.$$

By our contradiction assumption, it follows that there exists $x_{i+1} \in D$ such that

$$M(x_{i+1}) > 2M(x_i)$$

and

$$d(x_i, x_{i+1}) \leq k\, M^{-1}(x_i).$$

We have thus proved the claim by induction.

Now, we have

$$M(x_i) \geq 2^i M(x_0) \quad \text{and} \quad d(x_i, x_{i+1}) \leq k\, 2^{-i} M^{-1}(x_0), \qquad i \in \mathbb{N}. \qquad (26.33)$$

In particular, (x_i) is a Cauchy sequence, hence it converges to some $a \in \overline{D} \subset \Sigma$. Moreover,

$$d(x_0, x_i) \leq \sum_{j=0}^{i-1} d(x_j, x_{j+1}) \leq k\, M^{-1}(x_0) \sum_{j=0}^{i-1} 2^{-j} \leq 2k\, M^{-1}(x_0),$$

hence

$$\operatorname{dist}(x_i, \Gamma) \geq \operatorname{dist}(x_0, \Gamma) - 2k\, M^{-1}(x_0) =: \delta > 0.$$

Therefore, $K := \{x_i : i \in \mathbb{N}\} \cup \{a\}$ is a compact subset of $\Sigma \setminus \Gamma = D$. Since $M(x_i) \to \infty$ as $i \to \infty$ by (26.33), this contradicts the assumption that M is bounded on compact subsets of D. The lemma is proved. $\square$

Proof of Theorem 26.8. We first consider the nonradial case and assume $p < p_B$. We will show (26.25). Note that if $\Omega = \mathbb{R}^n$, by a simple scaling argument (replacing $u(x,t)$ by $\tilde{u}(y,s) := T^{1/(p-1)}u(\sqrt{T}y, Ts)$), (26.25) with $T = 1$, implies (26.24) for any $T > 0$.

Assume that estimate (26.25) fails. Then, there exist sequences $T_k \in (0, \infty)$, u_k, $y_k \in \Omega$, $s_k \in (0, T_k)$, such that u_k solves (26.2) (with T replaced by T_k) and the functions

$$M_k := u_k^{\frac{p-1}{2}}, \quad k = 1, 2, \ldots, \tag{26.34}$$

satisfy

$$M_k(y_k, s_k) > 2k\,(1 + d_k^{-1}(s_k)), \tag{26.35}$$

where $d_k(t) := (\min(t, T_k - t))^{1/2}$. We will use Lemma 26.11 with $X = \mathbb{R}^{n+1}$, equipped with the parabolic distance

$$d_P\big((x, t), (y, s)\big) = |x - y| + |t - s|^{1/2},$$

$\Sigma = \Sigma_k = \overline{\Omega} \times [0, T_k]$, $D = D_k = \overline{\Omega} \times (0, T_k)$, and $\Gamma = \Gamma_k = \overline{\Omega} \times \{0, T_k\}$. Notice that

$$d_k(t) = \mathrm{dist}_P\big((x, t), \Gamma_k\big), \qquad (x, t) \in \Sigma_k.$$

By Lemma 26.11, it follows that there exists $x_k \in \Omega$, $t_k \in (0, T_k)$ such that

$$M_k(x_k, t_k) > 2k\, d_k^{-1}(t_k), \tag{26.36}$$

$$M_k(x_k, t_k) \geq M_k(y_k, s_k) > 2k,$$

and

$$M_k(x, t) \leq 2M_k(x_k, t_k), \qquad (x, t) \in D_k \cap \tilde{B}_k, \tag{26.37}$$

where

$$\tilde{B}_k := \big\{ (x, t) \in \mathbb{R}^{n+1} : |x - x_k| + |t - t_k|^{1/2} \leq k\,\lambda_k \big\},$$

and

$$\lambda_k := M_k^{-1}(x_k, t_k) \to 0. \tag{26.38}$$

Observe that for all $(x, t) \in \tilde{B}_k$, we have $|t - t_k| \leq k^2 \lambda_k^2 < d_k^2(t_k) = \min(t_k, T_k - t_k)$ by (26.36), hence $t \in (0, T_k)$. It follows that

$$\Big(\Omega \cap \{|x - x_k| < \tfrac{k\lambda_k}{2}\}\Big) \times \big(t_k - \tfrac{k^2 \lambda_k^2}{4}, t_k + \tfrac{k^2 \lambda_k^2}{4}\big) \subset D_k \cap \tilde{B}_k.$$

Now we rescale u_k by setting

$$v_k(y, s) := \lambda_k^{2/(p-1)} u_k(x_k + \lambda_k y, t_k + \lambda_k^2 s), \qquad (y, s) \in \tilde{D}_k, \tag{26.39}$$

where

$$\tilde{D}_k := \big(\lambda_k^{-1}(\Omega - x_k) \cap \{|y| < k/2\}\big) \times (-k^2/4, k^2/4).$$

The function v_k solves

$$\left.\begin{array}{ll} \partial_s v_k - \Delta_y v_k = v_k^p, & (y, s) \in \tilde{D}_k, \\[2mm] v_k = 0, & y \in \lambda_k^{-1}(\partial\Omega - x_k),\ |y| < k/2,\ |s| < k^2/4. \end{array}\right\} \tag{26.40}$$

Moreover we have $v_k(0,0) = 1$ and (26.37) implies

$$v_k \leq C := 2^{\frac{2}{p-1}}, \quad (y,s) \in \tilde{D}_k. \tag{26.41}$$

Let $\rho_k := \text{dist}(x_k, \partial\Omega)$. By passing to a subsequence, we may assume that either

$$\rho_k/\lambda_k \to \infty, \tag{26.42}$$

or

$$\rho_k/\lambda_k \to c \geq 0. \tag{26.43}$$

In case (26.42) holds, by using (26.40), (26.41), (26.38), interior parabolic estimates and the embedding (1.2), we deduce that some subsequence of v_k converges in $C^\alpha(\mathbb{R}^{n+1})$, $0 < \alpha < 1$, to a bounded classical solution $u \geq 0$ of (21.1) with $u(0,0) = 1$. Moreover, as a consequence of the strong maximum principle, we have either $u > 0$ in $\mathbb{R}^{n+1}$, or

$$u = 0 \text{ in } \mathbb{R}^n \times (-\infty, t_0] \quad \text{and} \quad u > 0 \text{ in } Q := \mathbb{R}^n \times (t_0, \infty), \tag{26.44}$$

for some $t_0 < 0$. But, in the latter case, since $u \leq C$, we have $u_t - \Delta u \leq C^{p-1} u$ in Q and we infer from the maximum principle in Proposition 52.4 that $u = 0$ in Q, a contradiction. Therefore $u > 0$, which contradicts Theorem 21.1.

In case (26.43) holds, denote $H_c := \{y \in \mathbb{R}^n : y_1 > -c\}$. By performing a suitable orthogonal change of coordinates, similarly as in the proof of Theorem 12.1, using (26.38), (26.40), (26.41), interior-boundary parabolic estimates and the embedding (1.2), we obtain a subsequence of v_k which converges in $C^\alpha(\overline{H}_c)$, $0 < \alpha < 1$, to a bounded classical solution $v \geq 0$ of

$$\left. \begin{aligned} \partial_s v - \Delta_y v &= \ell\, v^p, & y \in H_c,\ s \in \mathbb{R}, \\ v &= 0, & y \in \partial H_c,\ s \in \mathbb{R}, \end{aligned} \right\} \tag{26.45}$$

with $v(0,0) = 1$ (hence $c > 0$). Similarly as in the previous case, we obtain $v > 0$, which contradicts Theorem 21.8.

In the radial case, let us assume in addition that $u(|x|, t)$ is nonincreasing as a function of $|x|$. Then we may take $x_k = 0$ in the above proof and the rescaling procedure yields a positive, bounded, radial, classical solution of (21.1), contradicting the radial Liouville-type Theorem 21.2. For the general (nonmonotone) radial case, which is slightly more delicate, we refer to [425]. $\square$

Remark 26.12. Lemma 26.11 and the method of proof of Theorem 26.8 are a generalization of an idea in [283] (see also, e.g., [130], [448], [361], [339]). In those works, blow-up estimates and a priori bounds of global solutions, with nonuniversal constants, were derived for various types of superlinear parabolic problems. By using a property similar to Lemma 26.11 (but concerning functions of the time variable only), it was shown that if a solution u were violating a suitable estimate, then the function $M(t) := \|u(t)\|_\infty$ would satisfy $M(s) \le 2M(t_k)$ for all $s \in [t_k, t_k + kM^{1-p}(t_k)]$ and some sequence of times t_k. Then, by a rescaling argument similar to that used in the proof of Theorem 26.8, one was led to a contradiction with a Fujita-type theorem. Note that these approaches do not use any variational structure of the problem, unlike the methods in the proofs of Theorems 22.1 and 23.7 for instance. This advantage will be exploited in Sections 38 and 44. $\square$

A natural question is whether the exponent $1/(p-1)$ in Theorem 26.8 is optimal. As for the (final) blow-up rates, this is indeed the case, due to Proposition 23.1. Interestingly, the situation is different for the initial blow-up rate, as it appears from the following results, which show that for p close to 1, the optimal initial blow-up rate exponents are in fact less than $1/(p-1)$. Moreover, they are different for the Cauchy and for the Dirichlet problems.

Theorem 26.13. *Let $p > 1$, $T > 0$, and $\Omega = \mathbb{R}^n$.*

(i) *Assume $p < p_F$. Then any nonnegative classical solution of* (26.2) *on $\mathbb{R}^n \times (0, T)$ satisfies*

$$u(x,t) \le C(n,p,T)\, t^{-n/2}, \quad x \in \mathbb{R}^n, \quad 0 < t < T/2.$$

(ii) *Let*

$$\alpha_0 := \min\left(\frac{n}{2}, \frac{1}{p-1}\right).$$

For all $\varepsilon > 0$, there exist $T > 0$, a positive classical solution u of (26.2)*, and $C > 0$ such that*

$$\|u(t)\|_\infty \ge Ct^{-\alpha_0+\varepsilon}, \quad \text{for } t > 0 \text{ small.}$$

Theorem 26.14. *Let $p > 1$, $T > 0$, and assume Ω bounded.*

(i) *Assume $p < 1 + 2/(n+1)$. Then any nonnegative classical solution of* (26.2) *on Q_T satisfies*

$$u(x,t) \le C(p,\Omega,T)t^{-(n+1)/2}, \quad x \in \Omega, \quad 0 < t < T/2.$$

(ii) *Let*

$$\alpha_1 := \min\left(\frac{n+1}{2}, \frac{1}{p-1}\right).$$

For all $\varepsilon > 0$, there exist $T > 0$, a positive classical solution u of (26.2)*, and $C > 0$ such that*

$$\|u(t)\|_\infty \ge Ct^{-\alpha_1+\varepsilon}, \quad \text{for } t > 0 \text{ small.}$$

Remarks 26.15. (i) The proof of Theorem 26.13 yields

$$C(n, p, T) = C(n, p)T^{n/2 - 1/(p-1)}.$$

(ii) As already mentioned in Remark 15.4(ii), it is known [269] that if $p_F < p < p_S$, then (26.26) possesses global, positive self-similar solutions of the form $u(x, t) = t^{-1/(p-1)}w(|x|/\sqrt{t})$, with $w \in C^2([0, \infty))$, radial and decreasing. In particular we have $\|u(t)\|_\infty = w(0)t^{-1/(p-1)}$, $t > 0$ (compare with Theorem 26.9).

(iii) By minor modifications of the proof, one can show that Theorem 26.14(i) remains valid for more general nonlinearities $f(u)$ instead of u^p, see [450]. Namely one may assume that f, of class C^1, satisfies $C_1 s^q - C_2 \le f(s) \le C_2(1 + s^p)$, $s \ge 0$, for some $1 < q < p < 1 + 2/(n+1)$ and $C, C_1, C_2 > 0$. A similar generalization is true for Theorem 26.13(i). $\quad\square$

Theorem 26.13(i) was proved in [79] by using Harnack inequality for the linear parabolic equation $u_t - \Delta u = V(x, t)u$, which holds under suitable integrability conditions on the potential V. An alternative proof relying on local regularity estimates from [29] (based on Moser's iteration arguments) was also given in [79]. Here we provide a more elementary proof (based on a modification of ideas from [361]), which relies on smoothing in uniformly local Lebesgue spaces (cf. Section 15). The introduction of these spaces in our problem is natural. Indeed, a simple application of the eigenfunction method (cf. Section 17) provides the following uniformly local L^1 a priori estimate.

Lemma 26.16. *Let u be a nonnegative classical solution of $u_t - \Delta u = u^p$ in $\mathbb{R}^n \times (0, T)$. Then there holds*

$$\|u(t)\|_{1,ul} = \sup_{a \in \mathbb{R}^n} \int_{|y-a|<1} |u(y, t)|\, dy \le C(n, p)(1 + T^{-1/(p-1)}), \quad 0 < t < T/2.$$

$$(26.46)$$

Proof. Let φ_1 be the first positive eigenfunction of $-\Delta$ in the ball $B_2 \subset \mathbb{R}^n$, with zero Dirichlet conditions. As usual, we normalize φ_1 by $\int_{B_2} \varphi_1 = 1$ and denote by λ_1 the corresponding eigenvalue. Multiplying the equation by φ_1, integrating by parts over B_2, using $\partial_\nu \varphi_1 \le 0$ on ∂B_2 and Jensen's inequality, we obtain

$$\frac{d}{dt} \int_{B_2} u(t)\varphi_1\, dx \ge \int_{B_2} u^p(t)\varphi_1\, dx - \lambda_1 \int_{B_2} u(t)\varphi_1\, dx$$

$$\ge \left(\int_{B_2} u(t)\varphi_1\, dx \right)^p - \lambda_1 \int_{B_2} u(t)\varphi_1\, dx$$

for all $0 < t < T$. By a standard differential inequality argument, it follows that

$$\int_{B_1} u(t)\, dx \le C(n) \int_{B_2} u(t)\varphi_1\, dx \le C(n, p)(1 + T^{-1/(p-1)}), \quad 0 < t < T/2.$$

The estimate then follows by applying this to $u(x - a, t)$ and taking the supremum over $a \in \mathbb{R}^n$. $\square$

Proof of Theorem 26.13. (i) By a simple scaling argument (replacing $u(x, t)$ by $\tilde{u}(y, s) := T^{1/(p-1)} u(\sqrt{T} y, T s)$), it is enough to show the estimate for $T = 1$. By Lemma 26.16, we have

$$\|u(t)\|_{1,ul} \le C(n, p), \quad 0 < t < 1/2.$$

We may then apply uniformly local L^q-smoothing results as follows. Fix $t \in (0, 1/2]$ and let $t_1, t_2 > 0$ be such that $t = t_1 + t_2$. Since $p < p_F$, we have $1 > n(p-1)/2$. Moreover, due to Theorem 26.8, we have $u \in L^\infty_{loc}((0, 1), L^\infty(\mathbb{R}^n))$. It then follows from Theorem 15.11 with $q = 1$ and (26.46) that

$$\|u(t)\|_\infty \le C(n, p)\|u(t_1)\|_{1,ul} \, t_2^{-n/2} \le C(n, p) \, t_2^{-n/2},$$

provided $t_2 \le \tau = \tau(n, p)$.

If $t \le \tau$, we take $t_1 = t_2 = t/2$. If $\tau < t \le 1/2$, we take $t_2 = \tau$, $t_1 = t - \tau$. In both cases, we thus obtain

$$\|u(t)\|_\infty \le C(n, p)t^{-n/2}.$$

(ii) Let $q \in (q_0, \infty)$ with $q_0 := \max(1, n(p-1)/2)$. By Theorem 15.2, we know that problem (22.1) is locally well-posed in $L^q(\mathbb{R}^n)$. Let $u_0(x) = |x|^{-2k} \chi_{\{|x|<1\}}$, with $k < n/2q_0$. Then $u_0 \in L^q(\mathbb{R}^n)$ for $q > q_0$ close to q_0 and, by estimate (15.30), we have $\|e^{-tA}u_0\|_\infty \ge (e^{-tA}u_0)(0) = Ct^{-k}$ for $t > 0$ small. It follows that the local solution u of (22.1) with initial data u_0 satisfies $\|u(t)\|_\infty \ge Ct^{-k}$ for small $t > 0$. Since $k \to \alpha_0$ as $k \to n/2q_0$, the conclusion follows. $\square$

The proof of Theorem 26.14(i) is similar to that of Theorem 26.13(i), except that we now use smoothing in L^q_δ-spaces (cf. Section 15).

Proof of Theorem 26.14(i). Due to (1.4), estimate (26.6) can be restated as an L^1_δ-estimate:

$$\|u(t)\|_{1,\delta} \le M := C(p, \Omega)(1 + T^{-1/(p-1)}), \quad 0 < t \le T/2. \tag{26.47}$$

We can now apply L^q_δ-smoothing results as follows. Fix $t \in (0, T/2]$ and let $t_1, t_2 > 0$ be such that $t = t_1 + t_2$. Since $1 > (n+1)(p-1)/2$, due to $p < 1 + 2/(n+1)$, we may apply Theorem 15.9 with $q = 1$, and we deduce from (26.47) that

$$\|u(t)\|_\infty \le C(p, \Omega, M) \, t_2^{-(n+1)/2},$$

provided $t_2 \le \tau_M := \tau_M(p, \Omega, M)$.

If $t \leq \tau_M$, we take $t_1 = t_2 = t/2$. If $\tau_M < t \leq T/2$, we take $t_2 = \tau_M$, $t_1 = t - \tau_M$, and we note that $t_2^{-(n+1)/2} \leq (\frac{T}{\tau_M})^{(n+1)/2} t^{-(n+1)/2}$. In both cases we thus obtain

$$\|u(t)\|_\infty \leq C(p, \Omega, T) t^{-(n+1)/2}, \quad 0 < t \leq T/2.$$

(ii) Let $q \in (q_1, \infty)$ with $q_1 := \max(1, (n+1)(p-1)/2)$. By Theorem 15.9, we know that problem (22.1) is locally well-posed in $L_\delta^q(\Omega)$. By Theorem 49.7(ii) in Appendix C, for any $k < (n+1)/2q$, there exists $u_0 \in L_\delta^q(\Omega)$ such that

$$\|e^{-tA} u_0\|_\infty \geq Ct^{-k}$$

for small $t > 0$. It follows that the local solution u of (22.1) with initial data u_0 satisfies $\|u(t)\|_\infty \geq Ct^{-k}$ for small $t > 0$. Since $k \to \alpha_1$ as $k \to (n+1)/2q_1$, the conclusion follows. $\square$

27. Complete blow-up

In this section we consider the question whether or not nonglobal classical solutions of problem (22.1) can be continued in some weak sense after the blow-up time $T_{\max}(u_0)$.

A natural way to look at this question is via monotone approximation. Let u be the solution of problem (22.1) and assume $u_0 \geq 0$ and $T_{\max}(u_0) < \infty$. Set

$$f_k(v) := \min(v^p, k), \quad v \geq 0, \; k = 1, 2, \ldots$$

and let u_k be the solution of the problem

$$\left. \begin{array}{rll} v_t - \Delta v = f_k(v), & x \in \Omega, \; t > 0, \\ v = 0, & x \in \partial\Omega, \; t > 0, \\ v(x, 0) = u_0(x), & x \in \Omega. \end{array} \right\} \tag{27.1}$$

The function u_k is globally defined and $u_{k+1} \geq u_k$. Define

$$\bar{u}(x, t) := \lim_{k \to \infty} u_k(x, t).$$

Notice that u_k solves the integral equation

$$u_k(x, t) = \int_\Omega G(x, y, t) u_0(y) \, dy + \int_0^t \int_\Omega G(x, y, t - s) f_k(u_k(y, s)) \, dy \, ds, \tag{27.2}$$

$$x \in \Omega, \; t > 0,$$

where G is the Dirichlet heat kernel in Ω. Since $G > 0$ and $u_{k+1} \geq u_k$, we may pass to the limit in (27.2) in order to get

$$\bar{u}(x,t) = \int_\Omega G(x,y,t)u_0(y)\,dy + \int_0^t \int_\Omega G(x,y,t-s)\bar{u}^p(y,s)\,dy\,ds, \qquad x \in \Omega,\ t > 0,$$

$$(27.3)$$

where the double integral may be infinite. Obviously $\bar{u}(\cdot,t) = u(\cdot,t)$ for $t < T_{\max}(u_0)$. Set

$$T^c = T^c(u_0) := \inf\{t \geq T_{\max}(u_0) : \bar{u}(x,t) = \infty \text{ for all } x \in \Omega\}$$

and notice that $T^c(u_0) \geq T_{\max}(u_0)$. Moreover, due to (27.3) and

$$u_k(x,t) \geq \int_\Omega G(x,y,t-s)u_k(y,s)\,dy, \qquad x \in \Omega,\ t > s > 0,$$

we have $\bar{u} < \infty$ a.e. in $\Omega \times (0,T^c)$, $\bar{u}(\cdot,t) < \infty$ a.e. in Ω for all $t \in (0,T^c)$, and $\bar{u} = \infty$ in $\Omega \times (T^c,\infty)$.

Definition 27.1. We say that u blows up at $t = T_{\max}(u_0)$ **completely** if $T_{\max}(u_0) = T^c(u_0)$. $\square$

As we shall see below the notion of complete blow-up is different from the notion of global blow-up in Remark 24.6(v) and Section 43. In fact, the following theorem guarantees that the solution u from Theorem 24.1 (satisfying $u(x,T) \leq C_\alpha |x|^{-\alpha}$) blows up completely.

Theorem 27.2. *Consider problem (22.1) with $p > 1$, Ω bounded, $0 \leq u_0 \in L^\infty(\Omega)$, and $T_{\max}(u_0) < \infty$. Assume either*

(i)
$$p < p_S$$

or

(ii)
$$u_t \geq 0 \quad in\ (0,T_{\max}(u_0)).$$

Then u blows up completely at $t = T_{\max}(u_0)$.

See Remark 23.6 for conditions ensuring that $u_t \geq 0$. Theorem 27.2 is due to [54]. In Proposition 27.7 below, we shall see that the result may fail for $p > p_S$. Before presenting the full proof of Theorem 27.2, we shall first give a proof of a special, one-dimensional case. This alternative approach is simpler than that in the general case and, as an advantage, it can be used for problems with nonconvex nonlinearities. However, although the argument can be extended to dimensions $n > 1$, the nonlinearity then has to satisfy severe growth restrictions and it requires the solution u to be increasing in time; see [54] for details.

Proof of Theorem 27.2 in a special case. We shall prove the assertion in case (ii), under the additional assumptions that $n = 1$, $\Omega = (-1, 1)$, and u_0 is radial nonincreasing. These assumptions and Proposition 52.17 guarantee that $u_x \leq 0$ for $x \in [0, 1)$ and $t \in (0, T)$. Denote $T := T_{\max}(u_0)$.

Step 1. Denote $f(u) = u^p$. We shall prove $\|f(u(t))\|_1 \to \infty$ as $t \to T-$.

Since $u \geq 0$ and $u_t \geq 0$ we see that the function $\psi : t \mapsto \|f(u(t))\|_1$ is nondecreasing. Assume, by contrary, that ψ is bounded. Then the L^p-L^q-estimates and the variation-of-constants formula guarantee

$$\|u(t)\|_q \leq \|u_0\|_q + \int_0^t (t - s)^{-\alpha} \|f(u(s))\|_1 \, ds, \qquad \alpha := \frac{n}{2}\left(1 - \frac{1}{q}\right).$$

Since $n = 1$, in the particular case $q = \infty$ we obtain

$$\|u(t)\|_\infty \leq \|u_0\|_\infty + C \int_0^t (t - s)^{-1/2} \, ds < C(T),$$

which contradicts $T < \infty$.

Step 2. Denote $\varphi(x) := \lim_{t \to T-} u(x, t)$ and let $\varepsilon \in (0, 1)$. Then

$$\int_{-1+\varepsilon}^{1-\varepsilon} f(\varphi(x)) \, dx = \lim_{t \nearrow T} \int_{-1+\varepsilon}^{1-\varepsilon} f(u(x, t)) \, dx$$

$$\geq \liminf_{t \to T-} (1 - \varepsilon) \int_{-1}^{1} f(u(x, t)) \, dx = \infty,$$

where we have used successively the monotone convergence of u to φ, $u_x \leq 0$ for $x \geq 0$ and Step 1.

Step 3. Choose $x \in (-1, 1)$, $t > T$. Then there exists $\varepsilon > 0$ such that $t - T \geq 2\varepsilon$ and $|x| < 1 - \varepsilon$. We have

$$(e^{-sA} w)(x) = \int_{-1}^{1} G(x, y, s) w(y) \, dy \geq \tilde{C}_\varepsilon \int_{-1+\varepsilon}^{1-\varepsilon} w(y) \, dy, \qquad s \in (\varepsilon, 2\varepsilon),$$

and

$$\tilde{C}_\varepsilon := \inf\{G(x, y, s) : |x|, |y| < 1 - \varepsilon, \ s \in (\varepsilon, 2\varepsilon)\} > 0.$$

Since $e^{-tA} u_0 \geq 0$ and $f_k(u_k(y, s)) \geq f_k(u_k(y, T))$ for $s \geq T$, we obtain

$$u_k(x, t) \geq \int_0^t \left[e^{-(t-s)A} f_k(u_k(s))\right](x) \, ds$$

$$\geq \tilde{C}_\varepsilon \int_{t-2\varepsilon}^{t-\varepsilon} \int_{-1+\varepsilon}^{1-\varepsilon} f_k(u_k(y, s)) \, dy \, ds \geq C_\varepsilon \int_{-1+\varepsilon}^{1-\varepsilon} f_k(u_k(y, T)) \, dy.$$

$$(27.4)$$

Step 4. Using (27.4) and Step 2 we see that, as $k \to \infty$,

$$u_k(x,t) \geq C_\varepsilon \int_{-1+\varepsilon}^{1-\varepsilon} f_k(u_k(y,T))\, dy \to C_\varepsilon \int_{-1+\varepsilon}^{1-\varepsilon} f(\varphi(y))\, dy = \infty,$$

which proves the assertion. $\square$

An essential ingredient in the proof of Theorem 27.2 in the general case is the following result (cf. [54, Lemma 2.1]), of independent interest, which is true for all $p > 1$ and without monotonicity assumption on u.

Proposition 27.3. *Consider problem (22.1) with $p > 1$ and Ω bounded. Assume that $u_0, v_0 \in L^\infty(\Omega)$ satisfy $0 \leq v_0 \leq u_0$ and $u_0 \not\equiv v_0$. Then either $T_{\max}(v_0) = T^c(u_0) = \infty$ or $T_{\max}(v_0) > T^c(u_0)$.*

In [345], by employing arguments different from those in [54], complete blow-up was proved for problem (19.1) for a rather general class of (convex nondecreasing) nonlinearities $f(u)$, but only for monotone solutions in time (cf. case (ii) of Theorem 27.2). Our proofs of Theorem 27.2 and Proposition 27.3 are based on a modification of ideas of [345] which enable one to cover case (i) as well. Let us also mention that the proof of Theorem 27.2(ii) in [54] can be used for more general functions f only if either f satisfies serious growth restrictions or f is convex and the function $f(u)/u^\gamma$ is nondecreasing for large u, where $\gamma > 1$, see [54, Theorem 1].

In the proof of Proposition 27.3, we shall use the following two lemmas from [94]. The first one is an approximation lemma which will enable us to construct a suitable perturbation of the equation in (22.1).

Lemma 27.4. *Let $p > 1$ and set $\varepsilon_0 := 1/(p+1)$. For each $\varepsilon \in (0, \varepsilon_0)$, there exists a concave function $\phi_\varepsilon \in C^2([0,\infty))$ with the following properties:*

$$\phi_\varepsilon(0) = 0, \qquad 0 < \phi_\varepsilon(s) \leq s \ \text{ for all } s > 0, \tag{27.5}$$

$$1 \geq \phi_\varepsilon'(s) \geq s^{-p}\big(\phi_\varepsilon^p(s) - (p+1)\varepsilon\big)_+, \quad s > 0, \tag{27.6}$$

$$\lim_{\varepsilon \to 0+} \phi_\varepsilon'(s) = 1 \quad \text{uniformly on } [0, M], \text{ for every } M > 0, \tag{27.7}$$

$$\sup_{s \geq 0} \phi_\varepsilon(s) < \infty. \tag{27.8}$$

Proof. Let $z = z_\varepsilon$ be the solution of the ODE

$$z'(s) = s^{-p}(z^p(s) - \varepsilon), \ s \geq 1, \quad \text{with } z(1) = 1 - \varepsilon. \tag{27.9}$$

We claim that z exists and satisfies

$$0 < z'(s) < 1, \ z(s) < s \quad \text{for all } s \geq 1. \tag{27.10}$$

First observe that $\varepsilon^{1/p} < z(1) < 1$, due to $(1-\varepsilon)^p > 1 - p\varepsilon > \varepsilon$. The claim thus easily follows from the fact that $z(s) - s < 0$ implies $z'(s) - 1 \leq (z(s)/s)^p - \varepsilon s^{-p} - 1 < 0$, and that $z(s) > \varepsilon^{1/p}$ implies $z'(s) > 0$. Differentiating (27.9) and using (27.10), we get

$$z''(s) = s^{-p} p z^{p-1} z' - p s^{-p-1}(z^p - \varepsilon) = p s^{-p}(z^{p-1} - s^{p-1})z' \leq 0, \quad s \geq 1.$$

Now extend $z(s)$ to a concave function $\phi_\varepsilon \in C^2([0,\infty))$ verifying (27.5) and $0 \leq \phi'_\varepsilon \leq 1$. (This is clearly possible since $z'(1) < z(1) < 1$.) We see that

$$\phi'_\varepsilon(s) \geq \phi'_\varepsilon(1) = (1-\varepsilon)^p - \varepsilon > 1 - (p+1)\varepsilon \geq s^{-p}(\phi^p_\varepsilon(s) - (p+1)\varepsilon)_+, \quad 0 < s \leq 1. \tag{27.11}$$

This along with (27.9), (27.10) and $0 \leq \phi'_\varepsilon \leq 1$ proves (27.6). Next, by (27.9), we have

$$\int_{z(1)}^{z(s)} \frac{d\tau}{\tau^p - \varepsilon} = \int_1^s \frac{d\tau}{\tau^p} < C := \int_1^\infty \frac{d\tau}{\tau^p} < \int_{z(1)}^\infty \frac{d\tau}{\tau^p - \varepsilon}, \quad s \geq 1,$$

which yields (27.8). Finally, (27.7) is a consequence of the first inequality in (27.11), together with the continuous dependence of the solution z_ε of (27.9) on ε (observe that $z(s) = s$ is solution of (27.9) for $\varepsilon = 0$). $\quad\square$

Lemma 27.5. *Assume Ω bounded and let $0 < T_0 < \infty$. There exists $K = K(T_0) > 0$ such that the solution of the problem*

$$\left.\begin{array}{ll} Z_t - \Delta Z = 1, & x \in \Omega, \ 0 < t \leq T_0, \\[4pt] Z = 0, & x \in \partial\Omega, \ 0 < t \leq T_0, \\[4pt] Z(x,0) = -K\delta(x), & x \in \Omega, \end{array}\right\} \tag{27.12}$$

satisfies $Z \leq 0$ in $\overline{\Omega} \times [0, T_0]$.

Proof. Decompose Z as $Z = Z_1 - KZ_2$ where Z_1 solves (27.12) with $K = 0$ and $Z_2 = e^{-tA}\delta$. In view of (1.4), we have $Z_2(x,t) \geq c_2^{-1} e^{-\lambda_1 t}\varphi_1 \geq c_1 c_2^{-1} e^{-\lambda_1 t}\delta$. Combining this with $Z_1 \leq c_3(T_0)\delta$ (due to $Z_1 \in C^{1,0}(\overline{\Omega} \times [0, T_0])$) we obtain $Z \leq (c_3 - Kc_1 c_2^{-1} e^{-\lambda_1 t})\delta$, and the lemma follows by choosing $K = c_1^{-1} c_2 c_3 e^{\lambda_1 T_0}$. $\quad\square$

Proof of Proposition 27.3. Assume for contradiction that $T_{\max}(v_0) \leq T^c(u_0) < \infty$ or $T_{\max}(v_0) < \infty = T^c(u_0)$. Let v be the solution of (22.1) with initial data v_0. Notice that $T_{\max}(u_0) \leq T_{\max}(v_0)$ and fix $\tau \in (0, T_{\max}(u_0))$. By the assumptions on v_0 and Proposition 52.7, there exists $\eta > 0$ such that

$$v(x,\tau) + 2\eta\delta(x) \leq u(x,\tau), \quad x \in \Omega. \tag{27.13}$$

Fix $T \in (\tau, T_{\max}(v_0))$.

    Step 1. L^1- and L^p_δ-estimates. We claim that

$$\overline{u}, \ \overline{u}^p \delta \in L^1(Q_T). \tag{27.14}$$

Let u_k be the solution of (27.1). Using Proposition 49.11 in Appendix C and $T < T^c(u_0)$, we deduce that, for all small $t > 0$,

$$c(x,t) \int_\Omega u_k(y,T)\delta(y)\,dy \leq \int_\Omega G(x,y,t)u_k(y,T)\,dy$$

$$\leq u_k(x,T+t) \leq \overline{u}(x,T+t) < \infty$$

for a.e. $x \in \Omega$ and some constant $c(x,t) > 0$. It follows that

$$\sup_k \int_\Omega u_k(y,T)\delta(y)\,dy < \infty. \tag{27.15}$$

Now, for any $0 \leq \varphi \in C^{2,1}(\overline{\Omega} \times [0,T])$ such that $\varphi = 0$ on $\partial\Omega \times [0,T]$, by testing (27.1) with φ we obtain

$$\int_\Omega u_k(y,T)\varphi(y,T)\,dy = \int_\Omega u_0(y)\varphi(y,0)\,dy + \int_0^T \int_\Omega \{u_k(\varphi_t + \Delta\varphi) + f_k(u_k)\varphi\}\,dy\,ds.$$

First taking $\varphi(x,t) = e^{\lambda_1 t}\varphi_1(x)$ and using (1.4) and (27.15), we obtain

$$\sup_k \int_0^T \int_\Omega f_k(u_k)\delta\,dy\,ds \leq C \sup_k \int_\Omega u_k(y,T)\delta(y)\,dy < \infty, \tag{27.16}$$

hence $\overline{u}^p\delta \in L^1(Q_T)$ by monotone convergence. Finally taking $\varphi(x,t) = \Theta(x)$, where Θ is defined by (19.27), and using (27.16), we similarly obtain $\overline{u} \in L^1(Q_T)$.

Step 2. Derivation of a penalized weak inequality for $\phi_\varepsilon(\overline{u})$. Now fix $\varepsilon \in (0,\varepsilon_0)$ to be determined later and let ϕ_ε be given by Lemma 27.4. For each k, a direct computation yields

$$\partial_t(\phi_\varepsilon(u_k)) - \Delta(\phi_\varepsilon(u_k)) = \phi_\varepsilon'(u_k)(\partial_t u_k - \Delta u_k) - \phi_\varepsilon''(u_k)|\nabla u_k|^2 \geq \phi_\varepsilon'(u_k)f_k(u_k)$$

in $\Omega \times (\tau,T)$. For any $0 \leq \varphi \in C^{2,1}(\overline{\Omega} \times [\tau,T])$ such that $\varphi = 0$ on $\partial\Omega \times [\tau,T]$ and $\varphi(T) = 0$, multiplying by φ and integrating by parts, it follows that

$$\int_\Omega (\phi_\varepsilon(u_k)\varphi)(y,\tau)\,dy + \int_\tau^T \int_\Omega \{\phi_\varepsilon(u_k)(\varphi_t + \Delta\varphi) + \phi_\varepsilon'(u_k)f_k(u_k)\varphi\}\,dy\,ds \leq 0. \tag{27.17}$$

Set $w := \phi_\varepsilon(\overline{u})$ and observe that $w \in L^\infty(\Omega \times (0,\infty))$ by (27.8). Using (27.14), (27.5), (27.6) and passing to the limit in (27.17) via dominated convergence, we obtain

$$\int_\Omega (\phi_\varepsilon(u)\varphi)(y,\tau)\,dy + \int_\tau^T \int_\Omega \{w(\varphi_t + \Delta\varphi) + \phi_\varepsilon'(\overline{u})\overline{u}^p\varphi\}\,dy\,ds \leq 0.$$

For sufficiently small $\varepsilon_1 \in (0, \varepsilon_0)$ and all $\varepsilon \in (0, \varepsilon_1]$, owing to (27.13) and (27.7), we have $\phi_\varepsilon(u(\cdot, \tau)) \geq v(\cdot, \tau) + \eta\delta(\cdot)$. Using (27.6), it follows that

$$\int_\Omega ((v+\eta\delta)\varphi)(y, \tau)\, dy + \int_\tau^T \int_\Omega \{w(\varphi_t + \Delta\varphi) + (w^p - (p+1)\varepsilon)_+ \varphi\}\, dy\, ds \leq 0. \quad (27.18)$$

Step 3. Construction of a supersolution to the original problem and conclusion. Now let K and Z be given by Lemma 27.5 for $T_0 := T_{\max}(v_0)$, select $\varepsilon = \min\{\varepsilon_1, \eta(K(p+1))^{-1}\}$, and set

$$z(\cdot, t) := w(\cdot, t) + (p+1)\varepsilon Z(\cdot, t - \tau) \leq w(\cdot, t), \quad \tau < t < T_0.$$

By (27.8), we have

$$\sup_{\tau < s < T_0} \|z(t)\|_\infty \leq M := \sup_{s \geq 0} \phi_\varepsilon(s) + (p+1)\varepsilon \sup_{0 < s < T_0} \|Z(s)\|_\infty < \infty.$$

Combining (the weak formulation of) (27.12) with (27.18), it follows that

$$\int_\Omega (v\varphi)(y, \tau)\, dy + \int_\tau^T \int_\Omega \{z(\varphi_t + \Delta\varphi) + z^p \varphi\}\, dy\, ds \leq 0$$

for all φ as in Step 2. In other words, z is a bounded, weak supersolution to problem (22.1) on $[\tau, T_{\max}(v_0))$, with initial data $v(\cdot, \tau)$. By the weak comparison principle (cf. Appendix F), it follows that $v \leq z \leq M$ in $\Omega \times (\tau, T_{\max}(v_0))$: a contradiction. $\quad\square$

Proof of Theorem 27.2 in case (i). By Theorem 22.13 and $p < p_S$ we know that the function

$$T_{\max} : L^\infty(\Omega) \to (0, \infty] : u_0 \mapsto T_{\max}(u_0) \quad (27.19)$$

is continuous. Fix $u_0 \geq 0$ with $T_{\max}(u_0) < \infty$. Proposition 27.3 then implies

$$T_{\max}(u_0) = \lim_{\alpha \to 1-} T_{\max}(\alpha u_0) \geq T^c(u_0). \quad\square$$

In view of the proof in case (ii), we first make a simple observation.

Lemma 27.6. *Let $u_0 \in L^\infty(\Omega)$, $u_0 \geq 0$, and $0 < \tau < T_{\max}(u_0)$. Then $T^c(u(\tau)) = T^c(u_0) - \tau$.*

Proof. Let v_k be the solution of (27.1) with u_0 replaced by $u(\tau)$. For k large, we have $f_k(u(\cdot, t)) = u^p(\cdot, t)$ on $[0, \tau]$, hence $u_k = u$ on $[0, \tau]$ by uniqueness. In particular $v_k(0) = u(\tau) = u_k(\tau)$, hence $v_k(t) = u_k(t + \tau)$ for all $t \geq 0$, and the conclusion follows from the definition of T^c. $\quad\square$

Proof of Theorem 27.2 in case (ii). Fix $\tau \in (0, T_{\max}(u_0))$. Then $u(\tau) \geq u_0$, and $u(\tau) \not\equiv u_0$ (since otherwise u would be a stationary solution, hence $T_{\max}(u_0) = \infty$). Proposition 27.3 and Lemma 27.6 then guarantee that

$$T_{\max}(u_0) \geq T^c(u(\tau)) = T^c(u_0) - \tau.$$

Consequently, $T_{\max}(u_0) \geq T^c(u_0)$. $\quad\square$

We now present a different proof of Proposition 27.3, which is based on the original ideas of [54] (cf. [54, Lemma 2.1]). This proof can be easily adapted to problems with nonlinear boundary conditions (see [445]) and also to problems on unbounded domains. In the unbounded domain case, the modification of the proof below guarantees $T_{\max}(\alpha u_0) > T^c(u_0)$ provided $\alpha < 1$ and $T^c(u_0) < \infty$. This information is still sufficient for the proof of complete blow-up if we know that the function $\alpha \mapsto T_{\max}(\alpha u_0)$ is continuous (see the proof of Theorem 27.2 in case (i)).

Alternative proof of Proposition 27.3. Owing to Lemma 27.6, we may assume that $u_0 \in C^1(\overline{\Omega})$, $u_0 > 0$ in Ω, $u_0 = 0$ and $\partial u_0 / \partial n < 0$ on $\partial\Omega$, and $v_0 := \alpha u_0$ for some $\alpha \in (0,1)$ (otherwise, just replace u_0 by $u(\tau)$ for some small $\tau > 0$ and observe that $v(\tau) \leq \alpha u(\tau)$ for some $\alpha \in (0,1)$). Let $T \in (0, \infty)$, $T \leq T^c(u_0)$, hence $\bar{u}(x,t) < \infty$ a.e. in Q_T. We shall prove that there exists a constant $C_\alpha < \infty$ such that $u(x,t; \alpha u_0) \leq C_\alpha$ for all $x \in \Omega$ and $t < T$, which implies the conclusion (since then, $T_{\max}(\alpha u_0) > T$, hence $T_{\max}(\alpha u_0) > T^c(u_0)$ if $T^c(u_0) < \infty$, $T_{\max}(\alpha u_0) = \infty$ otherwise).

Let $V := e^{-tA} v_0$, and let u_λ^k, $\lambda \in \{\alpha, 1\}$, $k = 1, 2, \ldots$, be given by

$$\partial_t u_\lambda^k - \Delta u_\lambda^k = (u_\lambda^{k-1})^p \qquad \text{in } Q_T,$$
$$u_\lambda^k = 0 \qquad \text{on } S_T,$$
$$u_\lambda^k(x, 0) = \lambda u_0(x), \qquad x \in \Omega,$$

where $u_\lambda^0 :\equiv 0$. Notice that $u_\lambda^k \in C^{2,1}(\overline{\Omega} \times (0, T))$ and that the maximum principle implies

$$\left.\begin{array}{c} 0 \leq u_\lambda^k \leq u_\lambda^{k+1} \leq \bar{u} \\[2mm] u_\alpha^k \leq \alpha u_1^k \end{array}\right\} \qquad \text{in } Q_T. \qquad (27.20)$$

For $m \in \mathbb{N}$, $\mu > 1$, set

$$E_\mu^m := \{(x,t) \in Q_T : u_\alpha^m(x,t) > \mu V(x,t)\},$$

$$g_k^m(\mu) := \inf_{(x,t) \in E_\mu^m} \frac{u_1^k(x,t)}{u_\alpha^m(x,t)},$$

$$w(x,t) := u_1^{k+1}(x,t) - g_k^m(\mu)^p u_\alpha^m(x,t) + \mu\big(g_k^m(\mu)^p - g_{k+1}^m(\mu)\big) V(x,t).$$

(Here and below, we write $g_k^m(\mu)^p$ in place of $(g_k^m(\mu))^p$ for simplicity.) Observe that $E_{\mu'}^m \subset E_\mu^m$ for $\mu' > \mu$, hence the functions g_k^m are nondecreasing in μ. Set

$$M := \sup\{\mu > 1 : E_\mu^m \ne \emptyset\} = \inf\{\mu > 1 : E_\mu^m = \emptyset\}$$

and assume $1 < \mu < M$. Then $w \in C^{2,1}(\overline{\Omega} \times (0,T))$ and there exists $\delta = \delta(m,\mu) > 0$ such that $t > \delta$ for all $(x,t) \in E_\mu^m$. For $k \ge m > 1$ we have

$$w_t - \Delta w = (u_1^k)^p - \left(g_k^m(\mu)u_\alpha^{m-1}\right)^p,$$
$$w \ge g_{k+1}^m(\mu)u_\alpha^m - g_k^m(\mu)^p u_\alpha^m + \mu\left(g_k^m(\mu)^p - g_{k+1}^m(\mu)\right)V \quad \text{in } E_\mu^m,$$

and, by (27.20),
$$g_k^m(\mu) \ge 1/\alpha > 1, \tag{27.21}$$
$$(u_1^k)^p \ge \left(g_k^m(\mu)u_\alpha^m\right)^p \ge \left(g_k^m(\mu)u_\alpha^{m-1}\right)^p \quad \text{in } E_\mu^m,$$

hence
$$w_t - \Delta w \ge 0 \quad \text{in } E_\mu^m.$$

Since $u_\alpha^m = \mu V$ on $\partial E_\mu^m \setminus (\Omega \times \{T\})$, we also have

$$w \ge 0 \quad \text{on } \partial E_\mu^m \setminus (\Omega \times \{T\}),$$

and we deduce from the maximum principle[5] that $w \ge 0$ in E_μ^m.

Assume that $M > \mu' > \mu > 1$. We claim that

$$g_{k+1}^m(\mu') \ge g_k^m(\mu)^p - \left(g_k^m(\mu)^p - g_{k+1}^m(\mu)\right)\frac{\mu}{\mu'}. \tag{27.22}$$

If $g_k^m(\mu)^p - g_{k+1}^m(\mu) \ge 0$, then (27.22) follows by combining $w \ge 0$ on E_μ^m,

$$V(x,t) < \frac{1}{\mu'}u_\alpha^m(x,t) \quad \text{for all } (x,t) \in E_{\mu'}^m$$

and $E_{\mu'}^m \subset E_\mu^m$. If $g_k^m(\mu)^p - g_{k+1}^m(\mu) < 0$, then, using $g_{k+1}^m(\mu') \ge g_{k+1}^m(\mu)$, inequality (27.22) reduces to $\mu' \ge \mu$.

Now, the sequence $\{g_k^m(\mu)\}_{k\in\mathbb{N}}$ is nondecreasing and bounded by $\inf_{E_\mu^m} \bar{u}/u_\alpha^m < \infty$. Its limit $g^m(\mu)$ satisfies

$$g^m(\mu') \ge g^m(\mu)^p - \left(g^m(\mu)^p - g^m(\mu)\right)\frac{\mu}{\mu'},$$

---

[5]The set E_μ^m need not be connected nor cylindrical, but the corresponding maximum principle can be proved by using similar arguments as in the proof of Proposition 52.4.

hence,

$$\frac{g^m(\mu') - g^m(\mu)}{\mu' - \mu} \geq \frac{g^m(\mu)^p - g^m(\mu)}{\mu'}. \tag{27.23}$$

Fix $\mu_0 \in (1, M)$, set

$$f(\mu) := g^m(\mu_0) + \int_{\mu_0}^{\mu} \frac{g^m(s)^p - g^m(s)}{s}\, ds, \qquad \mu \in [\mu_0, M),$$

and note that

$$f'(\mu) = \frac{g^m(\mu)^p - g^m(\mu)}{\mu} \qquad \text{a.e. in } [\mu_0, M). \tag{27.24}$$

As the function g^m is nondecreasing, we know that its derivative exists a.e. and that

$$g^m(\mu) \geq g^m(\mu_0) + \int_{\mu_0}^{\mu} (g^m)'(\xi)\, d\xi \qquad \text{in } [\mu_0, M). \tag{27.25}$$

Since $(g^m)' \geq (g^m(\mu)^p - g^m(\mu))/\mu$ a.e. due to (27.23), it follows from (27.25) that

$$g^m \geq f \qquad \text{in } [\mu_0, M). \tag{27.26}$$

Combining (27.24), (27.26) and (27.21), we infer $f'(\mu) \geq (f(\mu)^p - f(\mu))/\mu$ a.e. Integrating this inequality and using (27.21) again we obtain

$$\log(\mu/\mu_0) \leq \int_{g^m(\mu_0)}^{\infty} \frac{d\sigma}{\sigma^p - \sigma} \leq \int_{1/\alpha}^{\infty} \left(\frac{\sigma^{p-2}}{\sigma^{p-1} - 1} - \frac{1}{\sigma}\right) d\sigma = \log\left((1 - \alpha^{p-1})^{-1/(p-1)}\right)$$

for all $1 < \mu_0 < \mu < M$. Consequently,

$$M \leq c_\alpha := (1 - \alpha^{p-1})^{-1/(p-1)}.$$

Since $E_\mu^m = \emptyset$ for $\mu > M$, we have

$$u_\alpha^m(x, t) \leq c_\alpha V(x, t) \qquad \text{in } Q_T.$$

Since the limit $U_\alpha(x, t) := \lim_{m \to \infty} u_\alpha^m(x, t)$ is a bounded integral solution of (22.1) with u_0 replaced by αu_0, it coincides with $u(x, t; \alpha u_0)$ for $t < T$. This concludes the proof. $\square$

The following result shows that *incomplete blow-up* may occur when $p > p_S$. Parts (i) and (ii) are respectively due to [396] and [232].

Proposition 27.7. *Consider problem (22.1) with Ω bounded and $p > 1$. Let $0 \leq \varphi \in L^\infty(\Omega)$, $\varphi \not\equiv 0$, let α^* be defined by (22.22) and set $u_0 = \alpha^* \varphi$.*

(i) *Then $T^c(u_0) = \infty$.*

(ii) *Assume in addition $\Omega = B_R$, φ radial and $p > p_S$. Then $T_{\max}(u_0) < \infty$. Consequently, u blows up incompletely as $t = T_{\max}$.*

Proof. (i) Let $0 \leq \alpha < \alpha^*$. As a consequence of the definition of α^* and of the comparison principle, we have $T_{\max}(\alpha\varphi) = \infty$. Let v_α be the solution of (22.1) with initial data $\alpha\varphi$, and let $v_{\alpha,k}$ and u_k be the (global) solutions of (27.1), with initial data $\alpha\varphi$ and u_0 respectively. Since $T_{\max}(\alpha\varphi) = \infty$, Theorem 17.1 implies

$$\int_\Omega v_\alpha(t)\varphi_1 \, dx \leq C = C(\Omega, p), \qquad \text{for all } t > 0.$$

Since $v_{\alpha,k} \leq v_\alpha$, it follows that $\int_\Omega v_{\alpha,k}(t)\varphi_1 \, dx \leq C$, hence

$$\int_\Omega u_k(t)\varphi_1 \, dx \leq C, \qquad \text{for all } t > 0,$$

by continuous dependence. Letting $k \to \infty$ and using monotone convergence, we deduce that, for each $t > 0$, $\int_\Omega \overline{u}(t)\varphi_1 \, dx \leq C$, hence $\overline{u}(x,t) < \infty$ for a.e. $x \in \Omega$. We conclude that $T^c(u_0) = \infty$.

(ii) This assertion is a consequence of Theorem 28.7 below. $\square$

Remarks 27.8. (a) If Ω is a ball and u_0 is radial, then the assumption $p < p_S$ in Theorem 27.2 can be weakened to $p \leq p_S$ (or can be removed if we know $B(u_0) \neq \{0\}$), see [232, the proof of Theorem 5.1].

(b) For some class of problems (including (22.1)), property (22.28) is sufficient for complete blow-up, see [54, Corollary 3.1].

(c) **Genericity of complete blow-up.** Consider the situation in Proposition 27.7. If $u_0 = \alpha\varphi$, $\alpha > \alpha^*$, then $T^c(u_0) < \infty$. More precisely, $T_{\max}(\alpha\varphi) \leq T^c(\alpha\varphi) < T_{\max}(\beta\varphi)$ for any $\alpha > \beta \geq \alpha^*$. This follows from Proposition 27.3 (see also [317, Theorem 2] and [232, Theorem 14.1]). Since the function

$$[\alpha^*, \infty) \to (0, \infty) : \alpha \mapsto T_{\max}(\alpha\varphi) \tag{27.27}$$

is decreasing, it has to be (left) continuous a.e., hence $T^c(\alpha\varphi) = T_{\max}(\alpha\varphi)$ for a.e. $\alpha > \alpha^*$. Note that if we were able to prove the blow-up rate (23.5) for $u_0 = \alpha\varphi$, $\alpha \geq \alpha^*$, with $M = M(\alpha)$ being locally bounded, then the proof of [254, Theorem 1.2] would guarantee the continuity of (27.27) everywhere. Recall also that (23.5) is true if $p_S \leq p < p_{JL}$, cf. Theorem 23.10.

(d) **Peaking solutions.** The facts mentioned in (c) show that solutions which blow up incompletely are rather exceptional. Many interesting results on the behavior of such solutions in the interval $(T_{\max}(u_0), T^c(u_0))$ can be found in [194], [379], [380], [382], [195].

In [232] the authors considered problem (18.1) with $p_S < p < p_L$ and constructed global weak positive radial solutions of the self-similar form

$$u(r,t) = \begin{cases} (T-t)^{-1/(p-1)} f(r/\sqrt{T-t}), & t < T, \\ (t-T)^{-1/(p-1)} g(r/\sqrt{t-T}), & t > T, \end{cases}$$

where $f = f(\rho)$ and $g = g(\rho)$ are suitable bounded positive solutions of the ODE's

$$f'' + \frac{n-1}{\rho} f' - \frac{1}{2} f' \rho - \frac{1}{p-1} f + f^p = 0, \quad \rho > 0, \qquad f'(0) = 0,$$

and

$$g'' + \frac{n-1}{\rho} g' + \frac{1}{2} g' \rho + \frac{1}{p-1} g + g^p = 0, \quad \rho > 0, \qquad g'(0) = 0,$$

respectively. These solutions have singularity (peak) only at the point $(0, T)$ and their blow-up profile is $\lim_{t \to T} u(r,t) = Cr^{-2/(p-1)}$ for some $C < c_p$. Similar solutions for problem (22.21) had been previously constructed in [318].

(e) **Incomplete blow-up in the subcritical case.** Another explicit example of incomplete blow-up, for the nonautonomous equation

$$u_t - \Delta u = a(|x|, t)u^2, \qquad x \in \mathbb{R}^n, \ t \in \mathbb{R}, \tag{27.28}$$

(with $a > 0$ being bounded above and bounded away from zero), is due to [414]. Let $\varphi \in BC^1(\mathbb{R})$ be nonnegative. Set

$$u(x,t) := \frac{1}{\varphi(t) + r^2}, \qquad \text{where } r = |x|.$$

Then a straightforward computation yields

$$u_t - \Delta u = a(|x|, t)u^2 \quad \text{if } x \neq 0 \text{ or } \varphi(t) \neq 0,$$

where

$$a(r,t) := 2n - \varphi'(t) - \frac{8r^2}{\varphi(t) + r^2}$$

is bounded above and

$$a \geq 2(n-4) - \sup \varphi' > 0$$

provided $n > 4$ and $\sup \varphi'$ is small enough. In addition, it is easily verified that u is a weak solution of (27.28) if $n > 4$. Notice that $p = 2$ is subcritical if $n = 5$.

In particular, if $n > 4$ and $\varphi(t) = t^2$ for $|t| < n - 4$, then $\infty > C_2 \geq a(|x|,t) \geq C_1 > 0$ in $\mathbb{R}^n \times (-1/2, 1/2)$ and the solution u exhibits incomplete blow-up at $t = 0$. Similarly, the choices $\varphi(t) = [t(1-t)]^2$ or $\varphi(t) = t^3(\sin\frac{1}{t})^2$ yield examples of functions u which blow up incompletely multiple or infinitely many times (cf. [380], [382] in the case $a \equiv 1$).

Using the example above one can easily construct explicit examples of incomplete blow-up for the problem

$$u_t - \Delta u = a(x,t)u^2 + b(x,t), \qquad x \in \Omega, \ t > 0,$$
$$u = 0, \qquad x \in \partial\Omega, \ t > 0,$$
$$u(x,0) = u_0(x), \qquad x \in \Omega,$$

where $\Omega \subset \mathbb{R}^n$ is bounded, $n > 4$, a, b are bounded and $a > 0$ is bounded away from zero, see [442]. On the other hand, it was shown in [442] that incomplete blow-up cannot occur in such problems in the subcritical range $p < p_S$ if $a, b \in BUC^1$.

(f) **Analytic continuation.** A different possible way of continuing the solution after $T_{\max}$ was studied in [351]. It is based on a suitable notion of analytic continuation where the time variable t is extended to a sector in the complex plane. The existence of such continuation was proved there for the equation $u_t - \Delta u = u^2$, under Neumann boundary conditions and suitable assumptions on the initial data.

(g) **Critical L^q-space.** Like the classical existence time, the complete blow-up time $T^c(u_0)$ is not uniformly positive for bounded sets of initial data in $L^q(\Omega)$ when $q = q_c = n(p-1)/2$ (cf. [36]). Indeed, assume for instance Ω bounded, $\Omega \supset B(0,1)$, and let $\tilde{u}_{0,j} = 2u_{0,j}$, where $u_{0,j}$ is given by (15.3). Then Proposition 27.3 and Remark 15.4(i) imply that $T^c(\tilde{u}_{0,j}) < T_{\max}(u_{0,j}) \to 0$ as $j \to \infty$, while $\|\tilde{u}_{0,j}\|_{q_c} = \text{Const.}$ $\square$

28. Applications of a priori bounds

We have seen in previous sections that a priori and universal estimates of solutions play a key role in the proofs of several important statements. In this section we provide further applications of such estimates. Other applications (concerning existence of nodal equilibria and connecting orbits) can be found in [128] and [3], for example. These articles are devoted to superlinear problems with nonlinear boundary conditions and indefinite nonlinearities, respectively.

28.1. A nonuniqueness result

In this subsection we use universal bounds from Section 26 and arguments of [53] in order to prove Theorem 15.3(ii). More precisely, we prove the following proposition.

Proposition 28.1. *Let $\Omega = B_R$ and $p_F < p < p_S$. Fix $r > q_c = n(p-1)/2$ and assume that $u_0 \in L^r(\Omega)$, $u_0 \geq 0$, is radial nonincreasing. Let T_m denote the maximal existence time of the corresponding classical $L^r(\Omega)$-solution u_m of (15.1) (cf. Theorem 15.2 and Proposition 16.1) and let $T \in (0, T_m)$. Then there exists a* *function $u \geq u_m$, $u \neq u_m$, such that u is a classical $L^q(\Omega)$-solution of (15.1) for any $q \in [1, q_c)$, $u(\cdot, t)$ is radial nonincreasing,*

$$\lim_{t \to 0} \|u(\cdot, t)\|_q = \infty \qquad \textit{for any } q > q_c, \tag{28.1}$$

$$\lim_{t \to T} \|u(\cdot, t)\|_q = \infty \qquad \textit{for any } q > q_c. \tag{28.2}$$

In the proof of Proposition 28.1 we will also need the following lemma.

Lemma 28.2. *Let $\Omega = B_R$, $p > p_F$, $0 < T < \infty$ and let u be a positive, radial nonincreasing classical solution of (15.1) in the time interval $(0, T)$. Let $\delta > 0$ and $T' \in (0, T)$. Then there exist constants c depending only on R, p, n and the indicated quantities, such that*

$$u(x, t) \leq c(\delta)|x|^{-2/(p-1)}, \qquad |x| \leq R, \ t \in (0, T - \delta], \tag{28.3}$$

$$\|u(\cdot, t)\|_q \leq c(q, \delta), \qquad t \in (0, T - \delta], \ 1 \leq q < q_c, \tag{28.4}$$

$$\int_0^{T'} \|u(\cdot, s)\|_p^p \, ds < c(T', T - T'). \tag{28.5}$$

Proof. Let $t \in (0, T - \delta]$. Denote $\beta = 1/(p-1)$ and $v := u(\cdot, t)$. Then (15.22) guarantees

$$\|s^\beta e^{-sA} v\|_\infty \leq C_p, \text{ for all } s \in (0, \delta]. \tag{28.6}$$

Let us show the existence of constants $C, k > 0$ such that

$$v(x) \leq C e^{k\delta/R^2} (R^2/\delta)^\beta |x|^{-2\beta}, \qquad |x| \leq R. \tag{28.7}$$

If $k > 0$ is sufficiently large, then there exists $\eta \in \mathcal{D}(B_1)$ radial, radially decreasing, $\eta \neq 0$, such that $-\Delta \eta \leq k\eta$ (one can take $\eta(x) = \exp[-1/(1 - 2|x|^2)_+]$, for example). Set $\eta_\lambda(x) := \eta(\lambda x)$, $\lambda \geq 1/R$. Then the support of η_λ is a subset of Ω and

$$-\Delta \eta_\lambda \leq k\lambda^2 \eta_\lambda,$$

hence

$$e^{-sA} \eta_\lambda \geq e^{-k\lambda^2 s} \eta_\lambda$$

by the maximum principle. Consequently, (28.6) guarantees

$$C_p \int_\Omega \eta_\lambda \, dx \geq \int_\Omega s^\beta (e^{-sA} v) \eta_\lambda \, dx = s^\beta \int_\Omega v(e^{-sA} \eta_\lambda) \, dx \geq s^\beta e^{-k\lambda^2 s} \int_\Omega v\eta_\lambda \, dx.$$

Since $v(x) \geq v(1/\lambda)$ on the support of η_λ, we obtain

$$v\left(\frac{1}{\lambda}\right) \leq C_p s^{-\beta} e^{k\lambda^2 s}, \qquad \lambda \geq \frac{1}{R}, \ s \in [0, \delta].$$

Choosing $\lambda = 1/|x|$ and $s = \delta|x|^2/R^2$ we obtain (28.7).

Notice that (28.7) guarantees (28.3) and (28.4) is a consequence of (28.3). For $\tau \in (0,t)$, multiplying the variation-of-constants formula between τ and t by η_λ we obtain

$$\int_\Omega (e^{-(t-\tau)A}u(\tau))\eta_\lambda \, dx + \int_\tau^t \int_\Omega u^p(s)(e^{-(t-s)A}\eta_\lambda) \, dx \, ds = \int_\Omega u(t)\eta_\lambda \, dx.$$

It follows that

$$\int_0^t \int_\Omega u^p(s)e^{-k\lambda^2(t-s)}\eta_\lambda \, dx \, ds \leq \int_\Omega u(t)\eta_\lambda \, dx.$$

Fixing $\lambda_0 \geq 1/R$ and $r_0 > 0$ such that $\eta_{\lambda_0}(r_0) > 0$ and using (28.3) we obtain

$$\eta_{\lambda_0}(r_0)e^{-k\lambda_0^2 T'} \int_0^t \int_{B_{r_0}} u^p(s) \, dx \, ds \leq \int_\Omega u(t)\eta_{\lambda_0} \, dx \leq C(T - T'), \qquad t \leq T' < T.$$

Since $u(\cdot,t)$ is radial decreasing, the last estimate guarantees (28.5). $\qquad\square$

Proof of Proposition 28.1. Fix $T \in (0, T_m)$. Let $\eta_k \in \mathcal{D}(B_{1/k})$, $k = 1, 2, \ldots$, be nonnegative, radial, radially decreasing and $\eta_k \not\equiv 0$. Fix k. Due to the continuous dependence on initial data (see Remark 51.8(iii)) we have $T_{\max}(u_0 + \alpha\eta_k) > T$ for $\alpha > 0$ small. On the other hand, as a consequence of Remark 17.2(i) (see also Remark 17.7(iv)), we have $T_{\max}(u_0 + \alpha\eta_k) < T$ for $\alpha > 0$ large. Since the mapping $\alpha \mapsto T_{\max}(u_0 + \alpha\eta_k)$ is continuous (see Theorem 22.13 and (51.92)) there exists $\alpha_k > 0$ such that $T_{\max}(u_0 + \alpha_k\eta_k) = T$.

Let u_k denote the $L^r(\Omega)$-solution of (15.1) with initial data $u_0 + \alpha_k\eta_k$. Due to Theorem 26.8 the sequence $\{u_k\}$ is uniformly bounded on $\Omega \times [\delta, T - \delta]$ for any $\delta > 0$. Now parabolic regularity estimates (see Theorems 48.1 and 48.2) imply a uniform bound in $BUC^{2+\alpha,1+\alpha/2}(\Omega \times [\delta, T - \delta])$ for some $\alpha > 0$, hence we may assume $u_k \to u$ in $C^{2,1}(\overline{\Omega} \times [\delta, T - \delta])$ for all $\delta > 0$, where u is a classical solution of (26.2). Passing to the limit in the variation-of-constants formula for u_k we obtain

$$u(t) = e^{-(t-s)A}u(s) + \int_s^t e^{-(t-\sigma)A}u^p(\sigma) \, d\sigma. \qquad 0 < s < t < T. \qquad (28.8)$$

Moreover, applying Lemma 28.2 to the u_k's, and then Fatou's lemma, we deduce that u satisfies (28.3)–(28.5). Next fix $q \in [1, q_c)$ and let $t \in (0,T)$. Inequality (28.4) and the compactness of e^{-tA} show that there exist $s_m \to 0$ and $w \in L^q(\Omega)$ such that

$$u(s_m) \to w \text{ weakly in } L^q(\Omega) \text{ and}$$

$$e^{-(t-s_m)A}u(s_m) \to e^{-tA}w \text{ in } L^q(\Omega).$$

Since (28.5) guarantees $u^p \in L^1(Q_{T'})$ for all $T' < T$, using (28.8) with $s = s_m$ and passing to the limit we obtain

$$u(t) = e^{-tA}w + \int_0^t e^{-(t-\sigma)A}u^p(\sigma)\,d\sigma, \qquad 0 < t < T. \tag{28.9}$$

Next we show $w = u_0$ a.e. Let $\phi \in \mathcal{D}(\Omega)$. Multiplying the equation for u_k with ϕ and integrating we obtain

$$\int_\Omega u_k(t)\phi\,dx - \int_\Omega (u_0 + \alpha_k \eta_k)\phi\,dx + \int_0^t \int_\Omega u_k(s)(-\Delta\phi)\,dx\,ds = \int_0^t \int_\Omega u_k(s)^p \phi\,dx\,ds.$$

Assume $\phi \equiv 0$ on B_ε for some $\varepsilon > 0$. Since $u_k \le C(\varepsilon)$ on $(\Omega \setminus B_\varepsilon) \times (0, T)$, we may pass to the limit via dominated convergence in the above identity, and we arrive at

$$\int_\Omega u(t)\phi\,dx - \int_\Omega u_0\phi\,dx + \int_0^t \int_\Omega u(s)(-\Delta\phi)\,dx\,ds = \int_0^t \int_\Omega u^p(s)\phi\,dx\,ds. \tag{28.10}$$

On the other hand, (26.2) shows that

$$\int_\Omega u(t)\phi\,dx - \int_\Omega u(s_m)\phi\,dx + \int_{s_m}^t \int_\Omega u(s)(-\Delta\phi)\,dx\,ds = \int_{s_m}^t \int_\Omega u^p(s)\phi\,dx\,ds. \tag{28.11}$$

Passing to the limit in (28.11) and comparing the resulting identity with (28.10) yields $\int_\Omega u_0\phi\,dx = \int_\Omega w\phi\,dx$ for all $\phi \in \mathcal{D}(\Omega)$ which vanish in a neighborhood of the origin, hence $u_0 = w$ a.e.

Now (28.9) guarantees $\|u_0 - u(t)\|_1 \to 0$ as $t \to 0$. This convergence, (28.4) and interpolation yield $\|u_0 - u(t)\|_q \to 0$ as $t \to 0$ for any $q < q_c$, hence u is an $L^q(\Omega)$-solution of (15.1) for any $q < q_c$. It remains to prove (28.1) and (28.2). Fix $q > q_c$. We know that $u_k(T/2) \to u(T/2)$ in $L^q(\Omega)$. Due to the continuity of $T_{\max}$ (cf. above) we have

$$T_{\max}(u(T/2)) = \lim_{k \to \infty} T_{\max}(u_k(T/2)) = T/2,$$

hence $\|u(t)\|_q \to \infty$ as $t \to T$ due to Remarks 16.2. Next assume that there exist $C > 0$ and $t_k \to 0$ such that $\|u(t_k)\|_q < C$. Choose $\tilde{q} \in (q_c, q)$. Then interpolation yields $u(t_k) \to u_0$ in $L^{\tilde{q}}(\Omega)$ and the continuity of $T_{\max}$ in $L^{\tilde{q}}(\Omega)$ shows

$$T = \lim_{k \to \infty} T_{\max}(u(t_k)) = T_{\max}(u_0) = T_m > T,$$

a contradiction. This shows (28.2) and concludes the proof. $\square$

Remark 28.3. Entire solutions. Let us mention another, simple application of the universal bounds in Section 26. Let $\Omega = B_R$, $1 < p < p_S$, and denote by ϕ the unique positive solution of (6.1) with $\lambda = 0$ (cf. Remark 6.9(ii)); we know that ϕ is radial. One can show that any entire, radial, positive classical solution u of (15.1) (i.e. defined for all $t \in \mathbb{R}$) is either ϕ or a connection from ϕ to 0. Moreover, this remains true without the assumption that u be radial if we assume $1 < p < p_B$.

Indeed, due to Theorem 26.8, any such solution satisfies $\sup_{t \in \mathbb{R}} \|u(t)\|_\infty < \infty$, hence $\sup_{t \in \mathbb{R}} \|u(t)\|_{BUC^{1+\alpha}(\Omega)} < \infty$ by smoothing effects. Owing to the (strict) Lyapunov functional given by the energy E (cf. (17.6)), we know from Proposition 53.5 that the ω-limit set of u (in the BUC^1-topology) is nonempty and consists of nonnegative equilibria. By the same token, this is also true for the α-limit set (obtained by taking $t_k \to -\infty$ instead of $+\infty$ in formula (53.1)). Now, using the fact that ϕ and 0 are the only nonnegative steady states, that $E'(t) \le 0$, and that $E(\phi) > E(0) = 0$, one easily obtains the conclusion. $\square$

28.2. Existence of periodic solutions

Analogously as in Corollary 10.3, a priori estimates for positive periodic solutions of (suitable) parabolic problems with periodic superlinear nonlinearities guarantee their existence. For example, consider the problem

$$\left.\begin{aligned} u_t - \Delta u &= f(x, t, u), & x \in \Omega, \ t > 0, \\ u &= 0, & x \in \partial\Omega, \ t > 0. \end{aligned}\right\} \qquad (28.12)$$

If $\Omega = B_R$, $f = f(|x|, t, u)$ is continuous, T-periodic in t, $1 < p < p_S$,

$$-C_1 \le f(x, t, u) \le C_1(1 + u^p), \quad x \in \Omega, \ t > 0, \ u \ge 0,$$

and, for all (x, t) in the closure of $Q := Q_T$,

$$\lim_{u \to \infty, \ Q \ni (z, \tau) \to (x, t)} u^{-p} f(z, \tau, u) = m(x, t) \in (0, \infty),$$

then a straightforward generalization of Theorem 26.8 shows that any positive T-periodic solution of (28.12) is bounded by a universal constant $C = C(f, \Omega)$ (see [425] for more general statements). Consequently, if f satisfies additional assumptions guaranteeing the well-posedness of (28.12) in a suitable function space, then a topological degree argument shows the existence of a positive T-periodic solution of (28.12) (see [176] for details concerning the use of the topological degree).

Of course, instead of the radial symmetry assumption we could have assumed $p < p_B$. Let us sketch another proof of universal estimates of positive T-periodic solutions of (28.12) in the general nonradial case and the full subcritical range $1 < p < p_S$. Unfortunately, this alternative proof requires quite restrictive assumptions concerning the nonlinearity f.

Proposition 28.4. *Assume Ω bounded and $f(x,t,u) = m(t)|u|^{p-1}u$, where $1 < p < p_S$ and $m \in W^{1,\infty}(\mathbb{R}_+)$ is positive and T-periodic. Assume also*

$$\operatorname*{ess\,sup}_{t>0} \frac{m'(t)_-}{m(t)} < \frac{2n - (n-2)(p+1)}{r^2(\Omega)}, \tag{28.13}$$

where $r(\Omega)$ denotes the radius of the smallest ball containing Ω. Then there exists a constant $C > 0$ such that any positive T-periodic solution of (28.12) satisfies

$$\|u(t)\|_\infty \le C \qquad \text{for all } t > 0. \tag{28.14}$$

Consequently, there exists at least one positive T-periodic solution of (28.12).

Sketch of proof. Let u be a positive T-periodic solution of (28.12). Multiplying by φ_1 one easily gets

$$\psi' \ge -\lambda_1 \psi + \Big[\inf_{t>0} m(t)\Big]\psi^p, \qquad \psi(t) := \int_\Omega u(t)\varphi_1\, dx,$$

hence $\int_\Omega u(t)\varphi_1\, dx \le C$. If Ω is convex, then the method of moving planes guarantees $u(x,t), |\nabla u(x,t)| \le C$ for all x in a neighborhood of $\partial\Omega$. If Ω is not convex, then the same estimate can be obtained by using the Kelvin transform (cf. the proof of Theorem 13.1). Now, the Pohozaev-type identity

$$\int_0^T \int_\Omega \Big[\frac{n}{p+1} - \frac{n-2}{2}\Big] u^{p+1} m(t)\, dx\, dt = \int_0^T \int_\Omega \Big[\frac{m'(t)}{p+1} u^{p+1} + u_t^2\Big] \frac{|x|^2}{2}\, dx\, dt$$
$$+ \frac{1}{2}\int_0^T \int_{\partial\Omega} |\nabla u|^2 \big(x \cdot \nu(x)\big)\, d\sigma\, dt,$$

the identity

$$\int_0^T \int_\Omega u_t^2\, dx\, dt = -\frac{1}{p+1}\int_0^T \int_\Omega m' u^{p+1}\, dx\, dt$$

and the assumption (28.13) guarantee an a priori bound for u in $W^{1,2}(Q_T)$. Finally it is sufficient to use the bootstrap procedure from the proof of Theorem 22.1. $\square$

Remarks 28.5. (i) The estimates in Proposition 28.4 were first proved in [176] and [177] under the additional assumptions $p(3n-4) < 3n+8$ and $p(n-2) < n$, respectively. The general case was proved in [441], cf. also [286]. Analogous results for $f(x,t,u) = |u|^{p-1}u + h(x,t)$, h "small", can be found in [280].

(ii) If Ω, f are as in Proposition 28.4 and we consider problem (28.12) complemented with the initial condition $u(x,0) = u_0(x)$, then the a priori bound (22.27) is true for all solutions of (28.12) (not necessarily positive or periodic) and even without assuming (28.13), see [441, Theorem 5.1(i)]. $\square$

28.3. Existence of optimal controls

The a priori estimate (22.27) also plays an important role in the proof of existence of optimal controls for problems with final observation. Let $\Omega \subset \mathbb{R}^n$ be bounded, $T > 0$, $1 < p < p_S$, $q \geq 2$, $u_d \in L^q(\Omega)$, $u_0 \in C^2(\overline{\Omega}) \cap C_0(\Omega)$, and let us consider the model optimal control problem

$$\text{Minimize } J(u(w), w) \text{ over } w \in L^2(\Omega), \tag{28.15}$$

where

$$J(u, w) = \int_\Omega |u(x, T) - u_d(x)|^q \, dx + \int_\Omega w^2 \, dx,$$

and $u(w)$ is the solution of the governing equation

$$\left.\begin{aligned}
u_t - \Delta u &= |u|^{p-1}u + w, & x \in \Omega,\ t \in (0, T], \\
u &= 0, & x \in \partial\Omega,\ t \in (0, T], \\
u(x, 0) &= u_0(x), & x \in \Omega,
\end{aligned}\right\} \tag{28.16}$$

(we set $J(u(w), w) := \infty$ if (28.16) does not possess global solution up to time T). Then we have the following

Proposition 28.6. *Under the above assumptions, let*

$$q \in \left(\frac{n}{2}(p - 1), \frac{2n}{(n - 4)_+} \right).$$

Assume that problem (28.16) possesses a global solution at least for one $w \in L^2(\Omega)$. Then the optimal control problem (28.15) has a solution.

The statement of Proposition 28.6 remains true for more general time-dependent controls $w \in L^r([0, T], L^2(\Omega))$ (where r is large enough) and more general cost functionals J, see [22]. In addition, one can also derive optimality conditions for optimal controls and show that the assumption $p < p_S$ is essentially optimal (see [22]).

Sketch of proof of Proposition 28.6. Let $\{w_k\} \subset L^2(\Omega)$ be a minimizing sequence for J and $u_k := u(w_k)$. Then $\{w_k\}$ is bounded in $L^2(\Omega)$ (and we may assume $w_k \to w$ weakly in $L^2(\Omega)$) and $\{u_k(T)\}$ is bounded in $L^q(\Omega)$, due to the boundedness of $J(u_k, w_k)$. Since the problem (28.16) is well-posed in $L^q(\Omega)$ we may find $\delta > 0$ such that the solutions u_k can be continued on the interval $[T, T + \delta]$. A straightforward modification of the proof of estimate (22.27) shows that the solutions u_k are uniformly bounded in $L^\infty((0, T), L^{2p}(\Omega))$. The Sobolev maximal regularity (see Theorem 51.1(vi)) guarantees that u_k are uniformly bounded in $W^{1,r}([0, T], L^2(\Omega)) \cap L^r([0, T], W^{2,2} \cap W_0^{1,2}(\Omega))$ for any $r > 1$. Since this space is compactly embedded in $X := C([0, T], W_0^{1,2} \cap L^q(\Omega))$ for r sufficiently large (see Proposition 51.3), we may assume $u_k \to u$ in X. Now it is easy to pass to the limit to show $u = u(w)$ and $J(u, w) \leq \lim_{k\to\infty} J(u_k, w_k)$. $\square$

28.4. Transition from global existence to blow-up and stationary solutions

Let us consider problem (22.1) with either Ω bounded and $p > 1$, or $\Omega = \mathbb{R}^n$ and $p > p_F$, and let us go back to the situation introduced in Subsection 22.3. Namely, fix a function $\phi \geq 0$, $\phi \not\equiv 0$, with $\phi \in L^\infty(\Omega)$ if Ω is bounded and, for instance, $\phi \in \mathcal{D}(\Omega)$ if $\Omega = \mathbb{R}^n$. Let α^* be again defined by

$$\alpha^* = \alpha^*(\phi) := \sup\{\alpha > 0 : T_{\max}(\alpha\phi) = \infty\}, \tag{28.17}$$

and note that $\alpha^* \in (0, \infty)$ (cf. Subsection 22.3 for Ω bounded; when $\Omega = \mathbb{R}^n$ this follows from similar arguments by using Theorem 20.1). By definition of α^*, we have

$$T(\alpha\phi) < \infty \quad \text{for } \alpha > \alpha^*$$

and, as a consequence of the comparison principle,

$$T(\alpha\phi) = \infty \quad \text{for } 0 \leq \alpha < \alpha^*.$$

Now if we consider the threshold solution $u^* := u(t; \alpha^*\phi)$ of (22.1) starting at $u_0 = \alpha^*\phi$, we have the following three possibilities for u^*:

   (a) u^* is global and bounded in $L^\infty(\Omega)$,
   (b) u^* is global but unbounded,
   (c) u^* blows up in finite time.

It turns out that any of these three possibilities may occur.

Theorem 28.7. *Consider the situation described above.*

(i) Assume either $1 < p < p_S$ and Ω bounded, or $p_F < p < p_S$, $\Omega = \mathbb{R}^n$ and ϕ radial. Then case (a) occurs.

(ii) Let $p = p_S$. Assume $\Omega = B_R$ and ϕ radial. Then case (b) occurs.

(iii) Let $p > p_S$. Assume either $\Omega = B_R$ and ϕ radial, or $\Omega = \mathbb{R}^n$ and ϕ radial nonincreasing. Then case (c) occurs.

(iv) If Ω is bounded and case (a) occurs, then the ω-limit set of the solution u^* is a nonempty compact connected set consisting of positive equilibria. As a consequence, if Ω is a bounded starshaped domain and $p \geq p_S$, then (b) or (c) occurs, and the a priori bound (22.2) fails.

Proof. First let us show that the bound (22.2) guarantees alternative (a). For any $\alpha \in (0, \alpha^*)$, the solution $u_\alpha(t) := u(t; \alpha\phi)$ exist globally. If (22.2) is true, then $\|u_\alpha(t)\|_\infty \leq C^*$ for some C^* independent of α and the continuous dependence of the solutions on the initial data shows $\|u^*(t)\|_\infty \leq C^*$, hence case (a) occurs. Since (22.2) is true if $p < p_S$ and either Ω is bounded or $\Omega = \mathbb{R}^n$ and u_0 is radial (see Theorem 22.1 or Theorem 26.9, respectively), we have (a) in these cases.

238 II. Model Parabolic Problems

If (a) is true and Ω is bounded, then Example 53.7 guarantees that the ω-limit set $\omega(\alpha^*\phi)$ consists of positive equilibria. Since (22.1) does not possess positive equilibria if Ω is starshaped and $p \geq p_S$ (see Corollary 5.2), the alternative (a) and hence the estimate (22.2) cannot be true in this case.

Now assume $p = p_S$, $\Omega = B_R$ and ϕ radial. If (c) occurred, then u would blow up completely at $t = T_{\max}(u_0)$, due to [232, the proof of Theorem 5.1]. But this would contradict Proposition 27.7(i). Consequently, (b) is the only remaining possibility and assertion (ii) is proved. In the case of radial nonincreasing functions the assertion follows from Theorem 22.9.

Finally, let $p > p_S$. If Ω is a ball and ϕ is radially symmetric, then again (a) cannot happen, and (b) is ruled out by Theorem 22.4. Consequently, (c) is true. If $\Omega = \mathbb{R}^n$ and ϕ is radial nonincreasing, then the result follows from [374, Theorem 1.1]) provided $p < p_L$. If $p \geq p_L$, then one can use [374, Lemma 3.2] and [378]. $\square$

Remarks 28.8. (i) **Non-threshold solutions.** Assume that either $p < p_S$ and Ω is bounded, or $p > p_S$, $\Omega = B_R$ and ϕ is radial. Then $\lim_{t\to\infty} \|u(t; \alpha\phi)\|_\infty = 0$ for all $0 \leq \alpha < \alpha^*$. This follows from Proposition 19.11 and the boundedness of global solutions (cf. Theorems 22.1 and 22.4).

(ii) **Dynamical proofs of existence of equilibria.** Let Ω be bounded and $p < p_S$. Then similarly as above, $\omega(\alpha^*\phi)$ consists of nontrivial equilibria for any (possibly sign-changing) $\phi \in L^q(\Omega) \setminus \{0\}$, $q > q_c$, and this fact (together with a topological degree argument) can be used for the proof of existence of positive and sign-changing stationary solutions of (22.1) and related problems (see [114], [436], [439], [441], [3]).

(iii) **Threshold solutions in the nonradial supercritical case.** Assume that Ω is bounded and convex, $p > p_S$, and $\phi \in L^\infty(\Omega)$ is nonnegative, $\phi \not\equiv 0$. Let $\alpha^* = \alpha^*(\phi)$ and u^* have the same meaning as above. Fix $\alpha_k \nearrow \alpha^*$ and denote

$$\bar{u}(t) := \lim_{k\to\infty} u(t; \alpha_k\phi).$$

Then estimates in [396] show that $\bar{u}$ is a global weak solution of (22.1) and $\bar{u}(t) = u^*(t)$ for $t \in [0, T_{\max}(\alpha^*\phi))$ (cf. also Section 27). Recent results in [132] guarantee that $T_{\max}(\alpha^*\phi) < \infty$ and there exists a compact set $\mathcal{S} \subset \Omega \times [T_{\max}(\alpha^*\phi), \infty)$ such that the Hausdorff measure $\mathcal{H}^{n-4/(p-1)}(\mathcal{S})$ is zero, $\bar{u}$ is continuous in $\overline{\Omega} \times (0, \infty) \setminus \mathcal{S}$, $\lim_{t\to\infty} \|\bar{u}(t)\|_\infty = 0$. In particular, Theorem 28.7(iii) remains true in the nonradial case if Ω is bounded and convex.

(iv) Further results and references on threshold solutions can be found in the following subsection, Remarks 27.8 and Section 29. $\square$

28.5. Decay of the threshold solution of the Cauchy problem

In this subsection we denote

$$\beta := \frac{1}{p-1},$$

and the notation $f(t) \sim g(t)$ for $t \geq t_0$ means that

$$C_1 g(t) \leq f(t) \leq C_2 g(t) \quad \text{for all } t \geq t_0 \text{ and some constants } C_1, C_2 > 0.$$

Consider the Cauchy problem (18.1) with $p > p_F$. We continue to study the situation described at the beginning of the previous subsection. In what follows, by **non-threshold solutions** we more specifically mean solutions corresponding to $\alpha \in (0, \alpha^*)$.

Let us first consider the case of initial data with exponential spatial decay, more precisely $\phi \in H_g^1$, and assume also $p < p_S$. Recall from Proposition 20.13 that if u is global and $t_0 > 0$, then there exists $k \geq 0$ such that

$$\|u(t)\|_\infty \sim t^{-\lambda_k^L}, \quad t \geq t_0, \tag{28.18}$$

where $\lambda_k^L = (n + k - 1)/2$ for $k \geq 1$ and $\lambda_0^L = \beta$. The following theorem is due to [301].

Theorem 28.9. *Let* $p_F < p < p_S$, $\phi \in H_g^1$, $\phi \geq 0$, $\phi \not\equiv 0$. *For* $\alpha > 0$, *denote by* u_α *the solution of* (18.1) *with initial data* $u_0 = \alpha\phi$ *and let* α^* *be defined by* (28.17). *Then* $\alpha^* \in (0, \infty)$. *Moreover:*

(a) u_α *is global and* $\|u_\alpha(t)\|_\infty \sim t^{-n/2}$ *for* $t \geq 1$ *if* $0 < \alpha < \alpha^*$;

(b) u_{α^*} *is global and* $\|u_{\alpha^*}(t)\|_\infty \sim t^{-\beta}$ *for* $t \geq 1$;

(c) u_α *blows up in finite time if* $\alpha > \alpha^*$.

Proof. Assertion (c) follows from the definition of α^*.

Let v_α denote the rescaled solution (see (18.13)). The asymptotic stability of the zero equilibrium of (18.14) (see Example 51.24) shows that v_α is global, $v_\alpha(s) \to 0$ in H_g^1 (and L^∞) if α is small and $v_{\alpha^*}(s) \not\to 0$ in H_g^1 as $s \to \infty$. In particular, $\alpha^* > 0$.

If ϕ is radial, then Theorem 26.9 guarantees that u_{α^*} (hence v_{α^*}) are global. In the general case one can use the estimates in [301] or [478, Theorem 1] (see also [440, Theorem 1.2] in the case of sign-changing solutions). The arguments in the proof of Proposition 20.13 show $C_1 \leq \|v_{\alpha^*}(s)\|_\infty \leq C_2$, hence (28.18) is true with $k = 0$. In addition, the compactness of the semiflow for problem (18.14), the existence of the Lyapunov functional and the stability of the zero equilibrium guarantee that the ω-limit set $\omega(v_{\alpha^*})$ of v_{α^*} in H_g^1 is nonempty and consists of positive equilibria (cf. Theorem 28.7). For further reference fix $w^* \in \omega(v_{\alpha^*})$ and a sequence $s_j \to \infty$ such that $v_{\alpha^*}(s_j) \to w^*$.

Fix $\alpha < \alpha^*$ and assume that $v_\alpha(s) \not\to 0$. Then the arguments above show that there exists a subsequence of $v_\alpha(s_j)$ which converges to a positive equilibrium w. Now the proof of Theorem 19.9(ii) guarantees that $v_\alpha(s) \leq (\alpha/\alpha^*)v_{\alpha^*}(s)$, hence $w < w^*$. However, the proof of Proposition 19.8 shows that (18.14) does not possess ordered positive equilibria. Consequently, $v_\alpha(s) \to 0$ as $s \to \infty$. Now the upper bound in (28.18) with $k = 1$ follows from Example 51.24 and the lower bound from the comparison with the solution of the linear problem (cf. (20.5)). $\square$

Theorem 28.9 shows that for positive $\phi \in H_g^1$ and $p < p_S$, the threshold solution decays with the self-similar rate $t^{-\beta}$ while the non-threshold solutions decay with the same rate as the corresponding solutions of the linear heat equation. The next theorem [443] and subsequent remarks show that the same behavior of non-threshold solutions can be expected in a more general case, while the behavior of the threshold solution strongly depends on the exponent p.

Theorem 28.10. *Assume $p > p_F$. Let $\phi \in C(\mathbb{R}_+)$ be nonnegative, $\phi \not\equiv 0$, and*

$$\lim_{r \to \infty} \phi(r)r^{2\beta} = 0. \tag{28.19}$$

Denote by u_α the solution of (18.1) with $u_0(x) = \alpha\phi(|x|)$, $\alpha > 0$, and let α^* be defined by (28.17). Then $\alpha^* \in (0, \infty)$ and the following assertions are true.
(i) Let $p < p_S$. Then u_{α^*} is global and

$$\|u_{\alpha^*}(t)\|_\infty \sim t^{-\beta}, \qquad t \geq 1. \tag{28.20}$$

If $\alpha \in (0, \alpha^*)$, then
$$\lim_{t \to \infty} \|u_\alpha(t)\|_\infty t^\beta = 0. \tag{28.21}$$

(ii) Let $p \geq p_S$. If u_{α^*} is global, then

$$\limsup_{t \to \infty} \|u_{\alpha^*}(t)\|_\infty t^\beta = \infty. \tag{28.22}$$

If $\alpha \in (0, \alpha^*)$ and $\|u_\alpha(t)\|_\infty \leq ct^{-\beta}$ for all $t > 0$, then (28.21) is true.

Remarks 28.11. (i) If $\lim_{r \to \infty} \phi(r)r^{2\beta} = \infty$, then u_α blows up in finite time for any $\alpha > 0$ due to Theorem 17.12. If

$$0 < \liminf_{r \to \infty} \phi(r)r^{2\beta} \leq \limsup_{r \to \infty} \phi(r)r^{2\beta} < \infty \tag{28.23}$$

and $p < p_S$, then (28.20) remains true. In fact, the proof of Theorem 28.10 shows that the threshold solution u_{α^*} satisfies the upper bound in (28.20). The lower bound $\|u_{\alpha^*}(t)\| \geq ct^{-\beta}$ follows from the comparison with the solution of the linear problem and Lemma 20.8. Similarly, if $p \geq p_{JL}$, then one can replace condition (28.19) in the proof of (28.22) with the condition

$$\limsup_{r \to \infty} \alpha^* \phi(r) r^{2\beta} < c_p, \tag{28.24}$$

where c_p is the constant from (3.9), see [443].

(ii) Assume that

$$\limsup_{r \to \infty} \phi(r) r^{2\eta} < \infty \qquad \text{for some } \eta > \beta.$$

Then estimate (28.21) guarantees that the solution behaves like the solution of the linear problem. In fact, set $h(t) := \|u_\alpha(t)\|_\infty$ and notice that the function $w(t) := \exp[-\int_0^t h(s)^{p-1}\, ds] u_\alpha(t)$ is a subsolution of the linear heat equation (cf. [507, Proposition 2.6]). Assuming $\eta \in (\beta, n)$ without loss of generality, Lemma 20.8 thus implies

$$h(t) = \exp\left[\int_0^t h(s)^{p-1}\, ds\right] \|w(t)\|_\infty \leq Ct^{-\eta} \exp\left[\int_0^t h(s)^{p-1}\, ds\right], \quad t > 1,$$

and (28.21) guarantees $h(t)t^\beta \to 0$ as $t \to \infty$. Choose $\varepsilon > 0$ such that $\kappa := \eta - \varepsilon^{p-1} > \beta$ and fix $t_0 > 1$ such that $h(t) \leq \varepsilon t^{-\beta}$ for $t \geq t_0$. Let $t \geq t_0$. Then

$$\int_0^t h(s)^{p-1}\, ds \leq \int_0^{t_0} h(s)^{p-1}\, ds + \varepsilon^{p-1} \log\left(\frac{t}{t_0}\right) =: I_0 + \varepsilon^{p-1} \log\left(\frac{t}{t_0}\right),$$

hence $h(t) \leq Ct^{-\eta} e^{I_0} (t/t_0)^{\varepsilon^{p-1}} = C_0 t^{-\kappa}$, thus $H := \int_0^\infty h^{p-1}(t)\, dt < \infty$. Now we see that

$$e^{-tA}(\alpha\phi) \leq u_\alpha(t) \leq e^H w(t) \leq e^H e^{-tA}(\alpha\phi), \qquad t > 1.$$

In particular, Lemma 20.8 implies $\|u_\alpha(t)\|_\infty \leq Ct^{-\eta}$ for all $t > 0$ provided $\eta < n/2$. The proof of $H < \infty$ above is based on [191, Lemma 2.3].

(iii) Let $p = p_S$ and ϕ be as in Theorem 28.10. Then u_{α^*} exists globally (see [232]) so that either its time decay is slower than the self-similar one or the solution does not decay at all. Remark 22.10(ii) suggests that both possibilities can occur. Assume in addition that ϕ has only finitely many local minima and belongs to the energy space $\{u \in L^{p+1}(\mathbb{R}^n) : |\nabla u| \in L^2(\mathbb{R}^n)\}$. Then [423] guarantees that $\lim_{t \to \infty} \|u_{\alpha^*}(t)\|_\infty t^\beta = \infty$ and all non-threshold global solutions satisfy (28.21).

(iv) Let $p > p_S$ and let $\phi \in L^\infty \cap H^1(\mathbb{R}^n)$ satisfy the assumptions in Theorem 28.10. If $T_{\max}(\alpha\phi) = \infty$, then [356] implies $\|u_\alpha(t)\|_\infty \leq Ct^{-n/4}$ for $t > 1$. Since $n/4 > \beta$, the threshold solution u_{α^*} has to blow up in finite time. $\square$

In the proof of Theorem 28.10 we will need the following result on stationary solutions of the rescaled equation (see [269], [412], [538], [163] and [390]).

Proposition 28.12. *Let $p > 1$, $\lambda \geq 0$ and let $w_\lambda = w_\lambda(\rho)$ be the solution of the problem*

$$w'' + \frac{n-1}{\rho}w' + \frac{\rho}{2}w' + \beta w + |w|^{p-1}w = 0 \quad \text{for } \rho > 0, \qquad w(0) = \lambda, \quad w'(0) = 0.$$

Then w_λ is defined for all $\rho > 0$ and there exists finite $\lim_{\rho \to \infty} w_\lambda(\rho)\rho^{2\beta} =: A(\lambda)$. Given $\lambda > 0$, set $\rho_\lambda := \sup\{\rho > 0 : w_\lambda > 0 \text{ on } [0, \rho)\}$. Then the following is true:

(i) If $p \leq p_F$ and $\lambda > 0$, then $\rho_\lambda < \infty$.

(ii) If $p_F < p < p_S$, then there exists $\lambda_0 \in (0, \infty)$ such that $\rho_\lambda < \infty$ if and only if $\lambda > \lambda_0$. In addition, $A(\lambda) > 0$ for $\lambda \in (0, \lambda_0)$ and $A(\lambda_0) = 0$.

(iii) If $p \geq p_S$, then $\rho_\lambda = \infty$ and $A(\lambda) > 0$ for all $\lambda > 0$.

(iv) If $p \geq p_{JL}$, then the mapping $\lambda \mapsto w_\lambda(\rho)$ is strictly increasing for each fixed $\rho > 0$ and $\sup_\lambda A(\lambda) = c_p$, where c_p is the constant from (3.9).

Proof of Theorem 28.10. We have $\alpha^* > 0$ due to Theorem 20.6 and $\alpha^* < \infty$ due to Theorem 17.1.

Since the solutions u_α are radial we will consider them as functions $u_\alpha(t) = u_\alpha(r, t)$, where $r = |x|$. Set $v(\rho, s) = e^{\beta s}u(e^{s/2}\rho, e^s - 1)$, $\rho, s \geq 0$. Then v solves the equation

$$v_s - v_{\rho\rho} - \frac{n-1}{\rho}v_\rho = \frac{\rho}{2}v_\rho + \beta v + v^p, \qquad (28.25)$$

cf. (18.13), (18.14).

(i) Assume $p < p_S$. Theorem 26.9 guarantees that any global positive radial solution $u = u(r, t)$ satisfies

$$\|u(t)\|_\infty \leq C_0 t^{-\beta}, \qquad \text{where } C_0 = C_0(n, p).$$

This estimate, continuous dependence on initial data and the definition of α^* show that the solution u_{α^*} is global and satisfies the upper bound in (28.20).

If u is a solution (18.1), then the rescaled solution v of (28.25) satisfies

$$\|v(s)\|_\infty = (t + 1)^\beta \|u(t)\|_\infty, \qquad t = e^s - 1, \qquad (28.26)$$

hence

$$\|v(s)\|_\infty \leq C_0\left(1 + \frac{1}{t}\right)^\beta \qquad \text{for all } s \geq 0, \qquad (28.27)$$

whenever u is global positive and radial. Since $\alpha^*\phi \in L^\infty(\mathbb{R}^n)$, the solution u_{α^*} remains bounded in $L^\infty(\mathbb{R}^n)$ on a small time interval. Now using (28.26) and (28.27) we can find $C_1 > 0$ such that

$$\|v_{\alpha^*}(s)\|_\infty < C_1 \qquad \text{for all } s \geq 0. \qquad (28.28)$$

Let λ_0 be from Proposition 28.12 and fix $\lambda \in (0, \lambda_0)$. Then $A := A(\lambda) > 0$. Fix $a \in (0, A)$ and set $W_a(\rho) := a\rho^{-2\beta}$. Choose $\delta > 0$ such that

$$W_a(\delta) > C_1 + 1. \tag{28.29}$$

An easy computation shows that the function $V(\rho) = V_a(\rho) := W_a(\rho - R_1)$ is a supersolution of (28.25) for $\rho \geq R_1 + \delta$ provided $R_1 > 0$ is large enough. In fact,

$$V_{\rho\rho} + \frac{n-1}{\rho}V_\rho + \frac{\rho}{2}V_\rho + \beta V + V^p \leq V_{\rho\rho} + \frac{\rho}{2}V_\rho + \beta V + V^p$$
$$= a(\rho - R_1)^{-2\beta-2}\left[2\beta(2\beta+1) + a^{p-1} - \beta R_1(\rho - R_1)\right] < 0,$$

provided $\rho \geq R_1 + \delta$ and $R_1 > (2\beta(2\beta+1) + a^{p-1})/\beta\delta$. Increasing R_1 if necessary we may also assume

$$V(\rho) > (\alpha^* + 1)\phi(\rho) \qquad \text{for all } \rho \geq R_1, \tag{28.30}$$

due to (28.19). Fix $R_2 > R_1 + \delta$ such that

$$w_\lambda(\rho) > V(\rho) \qquad \text{for } \rho \geq R_2$$

where w_λ is the solution from Proposition 28.12. We will show that $v_{\alpha^*}(\cdot, s)$ and w_λ intersect in $[0, R_2]$ for any $s \geq 0$, cf. Figure 14. This intersection guarantees the lower estimate in (28.20).

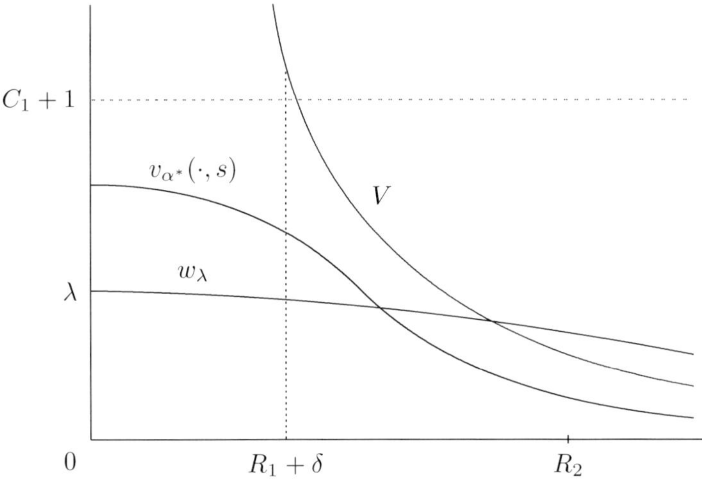

Figure 14: Intersection of $v_{\alpha^*}(\cdot, s)$ and w_λ in $[0, R_2]$.

Assume on the contrary that $v_{\alpha^*}(\rho, s_0) < w_\lambda(\rho)$ for some $s_0 \geq 0$ and all $\rho \in [0, R_2]$ and set

$$\varepsilon := \inf_{[0,R_2]} \left(w_\lambda - v_{\alpha^*}(\cdot, s_0)\right) > 0.$$

We have $\|v_\alpha(s) - v_{\alpha^*}(s)\|_\infty < \min(\varepsilon, 1)$ for all $s \leq s_0$ and α close to α^*, due to the continuous dependence of solutions u of (18.1) on initial data. Fix such $\alpha \in (\alpha^*, \alpha^* + 1)$. Then

$$v_\alpha(\rho, s_0) < w_\lambda(\rho), \qquad \rho \in [0, R_2], \tag{28.31}$$

and

$$v_\alpha(\rho, s) < C_1 + 1, \qquad \rho \in [0, R_2], \ s \leq s_0. \tag{28.32}$$

Since $v_\alpha(\rho, 0) = \alpha\phi(\rho) < V(\rho)$ for $\rho \geq R_1$ due to (28.30) and $v_\alpha(R_1 + \delta, s) < C_1 + 1 < V(R_1 + \delta)$ for $s \leq s_0$ due to (28.32) and (28.29), the comparison principle (see Proposition 52.6) implies

$$v_\alpha(\rho, s) \leq V(\rho) \qquad \text{for } \rho \geq R_1 + \delta, \ s \leq s_0. \tag{28.33}$$

Since $V(\rho) < w_\lambda(\rho)$ for $\rho \geq R_2$, estimates (28.33) and (28.31) imply $v_\alpha(s_0) < w_\lambda$, hence v_α exists globally due to the comparison principle. But this contradicts the choice of α^* and concludes the proof of (28.20).

Next choose $\alpha \subset (0, \alpha^*]$. Since v_α is uniformly bounded due to $v_\alpha \leq v_\alpha^*$ and (28.28), the ω-limit set of $\{v_\alpha(0, s)\}_{s \geq 0}$ is a compact interval $J \subset [0, C_1]$. Assume that J is not a singleton and fix $\lambda \in (\inf J, \sup J) \setminus \{\lambda_0\}$. Then there exist an infinite sequence $s_1 < s_2 < s_3 < \ldots$ such that $v_\alpha(0, s_k) = \lambda$ for $k = 1, 2, \ldots$. If $\lambda > \lambda_0$, then $w_\lambda(\rho_\lambda) = 0$ and the zero number $z_{[0,\rho_\lambda]}(v_\alpha(s) - w_\lambda)$ is finite for $s > 0$. However, this number has to drop at each s_k, which yields a contradiction. Consequently, $\lambda \in (0, \lambda_0)$. Let $A := \lim_{\rho \to \infty} w_\lambda(\rho)\rho^{2\beta}$, $a \in (0, A)$, and let $V = V_a$, δ and R_1 be as above. Then $v_\alpha(\rho, s) < V(\rho)$ for $\rho \geq R_1 + \delta$ and any s. Fix $R_2 > R_1 + \delta$ such that $w_\lambda(R_2) > V(R_2)$. Then we obtain the same contradiction as above by considering the zero number $z_{[0,R_2]}(v_\alpha(s) - w_\lambda)$. Consequently, there exists $\lambda = \lambda(\alpha) \geq 0$ such that

$$v_\alpha(0, s) \to \lambda \qquad \text{as } s \to \infty. \tag{28.34}$$

Due to the parabolic estimates the trajectory $\{v_\alpha(s)\}_{s \geq 0}$ is relatively compact in $C(\mathbb{R}_+)$ (considered with the locally uniform convergence) and its ω-limit set ω_α is a nonempty compact connected set, invariant under the semiflow generated by (28.25). In addition, (28.34) implies $\psi(0) = \lambda$ for any $\psi \in \omega_\alpha$. Assume that there exists $\psi \in \omega_\alpha \setminus \{w_\lambda\}$ and consider the solution $v = v_\psi$ of (28.25) with initial data ψ. Fix $\rho_0 > 0$ and $s_0 > 0$ such that $v_\psi(\rho_0, s) \neq w_\lambda(\rho_0)$ for all $s \in [0, s_0]$. Then the zero number $z_{[0,\rho_0]}(v_\psi(s) - w_\lambda)$ is finite for $s > 0$ and has to drop at each $s \in (0, s_0)$ (due to $v_\psi(0, s) = \lambda = w_\lambda(0)$) which yields a contradiction.

Consequently, $\omega_\alpha = \{w_\lambda\}$. Since $v_\alpha \geq 0$ we have $\lambda \leq \lambda_0$. Similarly, estimates of the form $v_\alpha(s) \leq V$ for $\rho \geq R_1 + \delta$ show $\lambda \notin (0, \lambda_0)$. Hence, $\lambda(\alpha) \in \{0, \lambda_0\}$ for any $\alpha \in (0, \alpha^*]$.

Given $0 < \alpha_1 < \alpha_2 \leq \alpha^*$, the function $\tilde{v} = (\alpha_2/\alpha_1)v_{\alpha_1}$ is a subsolution of (28.25) and $\tilde{v}(\cdot, 0) = v_{\alpha_2}(\cdot, 0)$, hence $v_{\alpha_2} \geq \tilde{v}$. Consequently, $\lambda(\alpha_2) \geq (\alpha_2/\alpha_1)\lambda(\alpha_1)$. This inequality guarantees $\lambda(\alpha) = 0$ for all $\alpha < \alpha^*$ (and $\lambda(\alpha^*) = \lambda_0$). Hence, given $\alpha < \alpha^*$, we have $v_\alpha(s) \to w_0 = 0$ locally uniformly in $[0, \infty)$ as $s \to \infty$ and the estimate $v_\alpha(s) \leq V$ on $[R_1 + \delta, \infty)$ concludes the proof of $\|v_\alpha(s)\|_\infty \to 0$. Consequently, (28.21) is true.

(ii) Assume $p \geq p_S$. If $\alpha \in (0, \alpha^*)$, then our assumptions imply the existence of $C_1 > 0$ such that the rescaled solution v_α satisfies $\|v_\alpha(s)\|_\infty < C_1$ for all $s \geq 0$. Now the same arguments as in the proof of (i) show the existence of $\lambda \in [0, C_1]$ such that $\|v_\alpha(s) - w_\lambda\|_\infty \to 0$ as $s \to \infty$, where w_λ is the solution from Proposition 28.12. However, for any $a \in (0, 1)$ we have an estimate of the form

$$v_\alpha(\rho, s) \leq a(\rho - R_1)^{-2\beta}, \qquad \rho > R_1 + \delta,$$

for some $R_1 = R_1(a) > 0$ (cf. (28.33)). In particular, assuming $\lambda > 0$, the choice $a < A(\lambda)$ leads to a contradiction. Hence $\lambda = 0$ and (28.21) is true.

Finally consider the threshold solution u_{α^*} and assume on the contrary that $\|v_{\alpha^*}(s)\|_\infty \leq C_1$ for all $s \geq 0$. Then the arguments above guarantee

$$\|v_{\alpha^*}(s)\|_\infty \to 0 \quad \text{as } s \to \infty. \tag{28.35}$$

Fix $\lambda > 0$, $a \in (0, A(\lambda))$ and choose δ, R_1 and R_2 as in the proof of (i). Then the same arguments as in that proof show that $v_\alpha^*(s)$ and w_λ intersect in $[0, R_2]$ for all $s \geq 0$, which contradicts (28.35). $\square$

29. Decay and grow-up of threshold solutions in the super-supercritical case

In this section we consider positive solutions of the Cauchy problem

$$\left.\begin{array}{ll} u_t - \Delta u = u^p, & x \in \mathbb{R}^n, \ t > 0, \\ u(x, 0) = u_0(x), & x \in \mathbb{R}^n, \end{array}\right\} \tag{29.1}$$

where $n \geq 11$ and $p > p_{JL}$. Set

$$m := 2/(p - 1)$$

and let $U_*(r) = c_p r^{-m}$ be the singular stationary solution defined in (3.9). We will use matched asymptotics to study the asymptotic behavior of solutions of (29.1) with initial data $u_0 \in L^\infty(\mathbb{R}^n)$ satisfying

$$0 \leq u_0(x) \leq U_*(|x|) \qquad \text{for } x \neq 0 \tag{29.2}$$

and

$$U_*(|x|) - c_1|x|^{-\ell} \leq u_0(x) \leq U_*(|x|) - c_2|x|^{-\ell} \qquad \text{for } |x| > c_3, \qquad (29.3)$$

for some $c_1, c_2, c_3 > 0$ and $\ell > m$. Note that solutions u with such initial data are global (due to Theorem 20.5) and they are also threshold solutions in the sense of Subsections 22.3, 28.4 since $T_{\max}(\lambda u_0) < \infty$ for $\lambda > 1$ (due to [260]). Set

$$\lambda_\pm := \frac{1}{2}\left(n - 2 - 2m \pm \sqrt{(n - 2 - 2m)^2 - 8(n - 2 - m)}\right).$$

Due to Remark 9.4 and (9.4) there exists $a > 0$ such that the positive radial steady state $U_\alpha = U_\alpha(r)$ of (29.1) satisfying $U_\alpha(0) = \alpha > 0$ has the asymptotic expansion

$$U(r) = U_*(r) - a_\alpha r^{-m-\lambda_-} + o(r^{-m-\lambda_-}) \quad \text{as} \quad r \to \infty, \qquad (29.4)$$

where $a_\alpha := \alpha^{-\lambda_-/m} a$. We will sketch the proof of the following theorem due to [203], [189], [204].

Theorem 29.1. *Let $p > p_{JL}$, $\ell \in (m, m + \lambda_+ + 2)$. Suppose that $u_0 \in L^\infty(\mathbb{R}^n)$ satisfies (29.2) and (29.3). Then there exist positive constants C_1, C_2 such that the solution of (29.1) satisfies*

$$C_1(t + 1)^\alpha \leq \|u(\cdot, t)\|_\infty \leq C_2(t + 1)^\alpha \qquad \text{for all } t \geq 0, \qquad (29.5)$$

where $\alpha := m(\ell - m - \lambda_-)/(2\lambda_-)$.

Remarks 29.2. (i) The above theorem shows that threshold solutions can decay to zero with an arbitrarily slow decay rate (if $\ell \in (m, m + \lambda_-)$) and also can grow up with any rate of the form t^α, $\alpha \in (0, m(2 + \lambda_+ - \lambda_-)/(2\lambda_-))$. The upper bound for α is known to be optimal. More precisely, if $p > p_{JL}$ and $u_0 \in L^\infty(\mathbb{R}^n)$ satisfies (29.2) (but not necessarily (29.3)), then u is global and $\|u(t)\|_\infty \leq C(t + 1)^{\alpha^*}$, where $\alpha^* = m(2 + \lambda_+ - \lambda_-)/(2\lambda_-)$. In addition, there exists $u_0 \in L^\infty(\mathbb{R}^n)$ satisfying (29.2) such that $\|u(t)\|_\infty \geq c(t + 1)^{\alpha^*}$ (see [381]).

(ii) Let $p = p_{JL}$. Then $\lambda_- = \lambda_+ =: \lambda$. If $u_0 \in L^\infty(\mathbb{R}^n)$ satisfies (29.2) and (29.3) with some $\ell \in (m + \lambda, m + \lambda + 2)$, then (29.5) remains true with $\|u(\cdot, t)\|_\infty$ replaced by $\|u(\cdot, t)\|_\infty (\log(t + 2))^{-m/\lambda}$, see [190].

(iii) Assume $p_{sg} < p < p_{JL}$. If $u_0 \in L^\infty(\mathbb{R}^n)$ satisfies (29.2), then u is global and satisfies

$$\|u(t)\|_\infty \leq C(n, p)t^{-1/(p-1)}, \quad t > 0.$$

This is a consequence of [232, Theorem 10.1(i)] (see also [499]). Therefore the condition $p \geq p_{JL}$, for grow-up or slow decay below the singular steady-state, is optimal. $\square$

The idea of matched asymptotics is to find a suitable asymptotic expansion for the solution in an inner region (for "small" $|x|$) and an outer region (for "large" $|x|$).

Matching these expansions on the boundary of the inner and outer regions (that is, comparing the coefficients of the leading terms of the expansions) determines the quantity that we are looking for. This formal approach not only provides a guess for the behavior of solutions but often also suggests the form of sub- and supersolutions that enable one to prove the result rigorously. It should be mentioned that in many cases the approach is more complicated: For example, in addition to the inner and outer regions one also has to consider an intermediate region.

We will only consider the case $\ell < m + \lambda_-$ in Theorem 29.1 since the case $\ell > m + \lambda_-$ can be treated by similar arguments and the proof in the case $\ell = m + \lambda_-$ follows from the fact that the solution remains between two positive stationary solutions of (29.1) for $t \geq t_0 > 0$ due to the comparison principle and (29.4). In addition, we will only describe in detail the formal part of the proof; the rigorous part will be sketched. Although the detailed rigorous proof in [204] represents one of the simplest applications of matched asymptotics, it is still quite long and technical and lies beyond the scope of this book. Another relatively simple example of matched asymptotics is mentioned in Remark 40.9(c).

Throughout the rest of this section we will write $f \sim g$ if $\tilde{C}_1 g \leq f \leq \tilde{C}_2 g$ for some constants $\tilde{C}_1, \tilde{C}_2 > 0$ and $f \approx g$ (or $f = g + h.o.t.$) if $f - g = o(f)$.

Sketch of proof of Theorem 29.1 for $\ell < m + \lambda_-$.

Part 1: Formal matched asymptotics. We will consider radial solutions $u = u(r,t)$, $r = |x|$ of (29.1). Such solutions satisfy

$$\left.\begin{aligned}
u_t &= u_{rr} + \frac{n-1}{r} u_r + u^p, && r > 0, \ t > 0, \\
u(r,0) &= u_0(r), && r > 0.
\end{aligned}\right\} \tag{29.6}$$

Assume that u_0 is continuous and radial nonincreasing and that

$$\eta(t) := u(0,t) \ \text{ behaves like } \ (t+1)^\alpha \ \text{ for some } \ \alpha \in (-m/2, 0) \ \text{ and } t \gg 1. \tag{29.7}$$

Notice that introducing a new variable $\zeta = \zeta(t,r) := \eta^{1/m}(t)r$ and assuming that u can be written in the form $u = \eta(t)\varphi(\zeta)$, (29.6) is transformed to

$$\eta_t \eta^{-p}\left(\varphi + \frac{1}{m}\zeta\varphi_\zeta\right) = \varphi_{\zeta\zeta} + \frac{n-1}{\zeta}\varphi_\zeta + \varphi^p,$$

where $\eta_t \eta^{-p} \to 0$ as $t \to \infty$. Consequently,

$$\text{the solution } u \text{ should asymptotically behave like } \eta(t)\varphi(\eta(t)^{1/m}r), \tag{29.8}$$

where φ is a solution of

$$\varphi_{\zeta\zeta} + \frac{n-1}{\zeta}\varphi_\zeta + \varphi^p = 0, \quad \zeta > 0, \qquad \varphi(0) = 1, \ \varphi_\zeta(0) = 0. \tag{29.9}$$

It turns out that before making the transformation mentioned above it is useful to apply the self-similar change of variables

$$v(\rho, s) = (t+1)^{m/2} u(r, t), \qquad \rho = \frac{r}{\sqrt{t+1}}, \qquad s = \log(t+1),$$

which transforms (29.6) into

$$\left.\begin{aligned}
v_s &= v_{\rho\rho} + \frac{n-1}{\rho} v_\rho + v^p + \frac{\rho}{2} v_\rho + \frac{m}{2} v, \qquad &&\rho > 0,\ s > 0, \\
v(\rho, 0) &= v_0(\rho) := u_0(\rho), \qquad &&\rho > 0.
\end{aligned}\right\} \tag{29.10}$$

Notice that $v(0, s) = (t+1)^{m/2} u(0, t) \to \infty$ as $s \to \infty$ due to (29.7).

Let us first consider the inner region (where ρ is small). The equation in (29.10) is "similar" to that in (29.6) for small ρ: The additional two terms at the end of the RHS are expected to be small in comparison to the remaining ones if v is large and ρ small. Therefore, taking into account (29.8) and (29.9), for small ρ we will look for solution v in the form

$$v(\rho, s) = \sigma(s)\big(\psi(\xi) - R(s, \xi)\big) \tag{29.11}$$

where $\sigma(s) := v(0, s)$, $\xi := \sigma^{1/m}\rho$, ψ is the solution of

$$\psi_{\xi\xi} + \frac{n-1}{\xi}\psi_\xi + \psi^p = 0, \quad \xi > 0, \qquad \psi(0) = 1,\ \psi_\zeta(0) = 0, \tag{29.12}$$

and R represents the higher order terms (remainder). Plugging the ansatz (29.11) into (29.10) we obtain $R \approx \sigma_s \sigma^{-p}\Psi(\xi)$ for ρ small and s large, where

$$\left.\begin{aligned}
\Psi_{\xi\xi} + \frac{n-1}{\xi}\Psi_\xi + p\psi^{p-1}\Psi &= \left(\frac{m\sigma}{2\sigma_s} - 1\right)\left(\psi + \frac{1}{m}\xi\psi_\xi\right), \quad &&\xi > 0, \\
\Psi(0) &= \Psi_\xi(0) = 0.
\end{aligned}\right\} \tag{29.13}$$

Since we expect $\sigma(s)$ to behave like $e^{(m/2+\alpha)s}$ for some $\alpha \in (-m/2, 0)$ due to (29.7), the coefficient $\left(\frac{m\sigma}{2\sigma_s} - 1\right)$ in (29.13) behaves like a positive constant and [204, Lemma 3.1], [203, Lemma 4.2] guarantee that there exists $K > 0$ such that $\Psi(\xi) \approx K\xi^{2-m-\lambda_-}$ as $\xi \to \infty$. Fixing $\rho > 0$, we have $\xi = \sigma^{1/m}(s)\rho \to \infty$ as $s \to \infty$, hence

$$R(s, \xi) \approx \frac{\sigma_s}{\sigma^p}\Psi(\xi) \approx K_1\frac{1}{\sigma^{p-1}}\Psi(\xi) \approx K_2\xi^{-2}\Psi(\xi) \approx K_3\xi^{-m-\lambda_-},$$

where K_1, K_2, K_3 are positive constants. Due to (29.4), the solution ψ of (29.12) satisfies

$$\psi(\xi) = c_p\xi^{-m} - a\xi^{-m-\lambda_-} + o(\xi^{-m-\lambda_-}), \qquad \text{as } \xi \to \infty,$$

where $a > 0$. Consequently, we obtain the two-term inner expansion

$$v \approx \sigma(c_p \xi^{-m} - \tilde{a}\xi^{-m-\lambda_-}) = c_p \rho^{-m} - \tilde{a}\sigma^{-\lambda_-/m}\rho^{-m-\lambda_-}, \qquad (29.14)$$

where $\tilde{a} = a + K_3 > 0$.

Next we consider the formal expansion in the outer region (where $\rho \gg 1$) as $s \to \infty$. Setting

$$v = c_p \rho^{-m} - w$$

and assuming $w \ll \rho^{-m}$ for $\rho \gg 1$, we have

$$w_s = w_{\rho\rho} + \frac{n-1}{\rho}w_\rho + \frac{pc_p^{p-1}}{\rho^2}w + \frac{\rho}{2}w_\rho + \frac{m}{2}w + h.o.t., \qquad \rho \gg 1.$$

If we look for a solution w in the form

$$w(\rho, s) = e^{-\beta s}\tilde{W}(\rho) + h.o.t.,$$

then $\tilde{W}$ has to solve the equation

$$\tilde{W}_{\rho\rho} + \frac{n-1}{\rho}\tilde{W}_\rho + \frac{pc_p^{p-1}}{\rho^2}\tilde{W} + \frac{\rho}{2}\tilde{W}_\rho + \left(\beta + \frac{m}{2}\right)\tilde{W} = 0. \qquad (29.15)$$

In addition, due to our assumption (29.3), $\tilde{W}$ is required to satisfy the condition

$$0 < \liminf_{\rho \to \infty} \rho^\ell \tilde{W}(\rho) \le \limsup_{\rho \to \infty} \rho^\ell \tilde{W}(\rho) < \infty. \qquad (29.16)$$

If $\rho \gg 1$, then the last two terms in (29.15) are much greater than the remaining ones so that we have to guarantee $\frac{\rho}{2}\tilde{W}_\rho \approx -(\beta + \frac{m}{2})\tilde{W}$. Due to (29.16) we have to set $\beta := (\ell - m)/2$. In order that the outer expansion matches with the inner expansion (29.14), $\tilde{W}$ should also satisfy

$$0 < \liminf_{\rho \to 0} \rho^{m+\lambda_-}\tilde{W}(\rho) \le \limsup_{\rho \to 0} \rho^{m+\lambda_-}\tilde{W}(\rho) < \infty. \qquad (29.17)$$

It is known (see [1] or [204]) that the problem (29.15), (29.16), (29.17) with $\beta = (\ell - m)/2$ possesses a positive solution $\tilde{W}$ provided $\ell \in (m, m + \lambda_+ + 2)$ (this solution can be expressed explicitly in terms of Kummer's functions). Hence, we obtain the two-term outer expansion

$$v \approx c_p \rho^{-m} - e^{-(\ell-m)s/2}\tilde{W}(\rho). \qquad (29.18)$$

If we now match the inner expansion (29.14) with the outer expansion (29.18) at $\rho = \rho_0 > 0$, then we obtain

$$\sigma(s) \sim e^{m(\ell-m)s/(2\lambda_-)}, \qquad (29.19)$$

hence

$$u(0,t) \sim t^\alpha, \qquad \text{where } \alpha = \frac{m(\ell - m - \lambda_-)}{2\lambda_-}.$$

This gives a formal proof of Theorem 29.1 for $\ell < m + \lambda_-$.

Part 2: Sketch of the rigorous proof. We will find a subsolution $\underline{v}$ and a supersolution $\overline{v}$ for the solution v of (29.10) such that the estimates $\underline{v} \leq v \leq \overline{v}$ will guarantee (29.5).

It is relatively easy to check that the subsolution $\underline{v}$ can be chosen as

$$\underline{v}(\rho, s) := \max\big(0, c_p \rho^{-m} - be^{-(\ell-m)s/2}\tilde{W}(\rho)\big),$$

where $\tilde{W}$ is a fixed solution of (29.15), (29.16), (29.17) with $\beta = (\ell - m)/2$, and $b > 0$ is large enough.

The supersolution $\overline{v}$ is defined by

$$\overline{v}(\rho, s) := \begin{cases} \overline{v}_1(\rho, s), & s \geq 0, \ \rho \leq \rho_M(s), \\ \overline{v}_2(\rho, s), & s \geq 0, \ \rho > \rho_M(s), \end{cases}$$

where $\rho_M(s) := \inf\{\rho > 0 : \overline{v}_2(\rho, s) < \overline{v}_1(\rho, s)\}$ and $\overline{v}_1, \overline{v}_2$ are supersolutions in the corresponding domains.

It is again relatively easy to check that the supersolution $\overline{v}_2$ can be chosen in the form

$$\overline{v}_2(\rho, s) := c_p \rho^{-m} - be^{-(\ell-m)s/2}W(\rho),$$

where W is the solution of

$$W_{\rho\rho} + \frac{n-1}{\rho}W_\rho + \frac{\rho}{2}W_\rho + \frac{\ell}{2}W = 0, \quad \rho > 0, \qquad W(0) = 1, \quad W_\rho(0) = 0,$$

(which can be again expressed in terms of Kummer's functions) and b is small enough.

The most difficult part is the choice of the supersolution $\overline{v}_1$. Recall that in the inner region, we expect

$$v(\rho, s) \approx \sigma(s)\Big(\psi(\xi) - \frac{\sigma_s}{\sigma^p}\Psi(\xi)\Big),$$

where Ψ solves (29.13). Plugging (29.19) into (29.13) we see that Ψ solves the problem

$$\left.\begin{array}{c} \Psi_{\xi\xi} + \dfrac{n-1}{\xi}\Psi_\xi + p\psi^{p-1}\Psi = \dfrac{m + \lambda_- - \ell}{\ell - m}\Big(\psi + \dfrac{1}{m}\xi\psi_\xi\Big) + R_\Psi, \quad \xi > 0, \\[2ex] \Psi(0) = \Psi_\xi(0) = 0, \end{array}\right\} \quad (29.20)$$

where R_Ψ represents higher order terms. Now it turns out that one can set

$$\overline{v}_1(\rho, s) := \sigma(s)\Big(\psi(\xi) - \frac{\sigma_s}{\sigma^p}\Psi(\xi)\Big),$$

where Ψ is the solution of (29.20) with $R_\Psi := A/(1 + \xi^{m+\lambda_-})$ and A is a suitable positive constant. (The term R_Ψ is purely technical.) $\qquad \square$

Chapter III

Systems

30. Introduction

Chapter III is devoted to systems of elliptic and parabolic types. In Section 31, we study the questions of a priori estimates and existence for weakly coupled elliptic systems which are natural extensions of the scalar equations considered in Chapter I. In Section 32, we study a simple parabolic system which is the analogue of the scalar model problem (15.1) studied in Chapter II. For this system, we treat the questions of well-posedness, global existence and blow-up. In Section 33, we discuss the different possible effects of adding linear diffusion (and some boundary conditions) to a system of ODE's. It will turn out that quite opposite effects can be observed. This will lead us to consider some systems arising in biological or physical contexts, such as mass-preserving and Gierer-Meinhardt systems.

31. Elliptic systems

In Sections 10–13, we have studied several methods to derive a priori estimates of positive solutions of scalar elliptic equations. The aim of this section is to present analogous results and methods in the case of elliptic systems. The three methods that we shall describe are extensions of the methods of Sections 11–13 from the scalar case, but they require substantial additional work and several new ideas. As mentioned before, a priori estimates can be used for the proof of existence, and they do not require any variational structure of the problem. Therefore they are well-suited for elliptic systems, which do not possess such structure in general.

We will devote our attention to the Dirichlet problem for superlinear systems, especially of cooperative type, of the form:

$$\left.\begin{aligned} -\Delta u &= f(x, u, v), & x &\in \Omega, \\ -\Delta v &= g(x, u, v), & x &\in \Omega, \\ u = v &= 0, & x &\in \partial\Omega. \end{aligned}\right\} \tag{31.1}$$

A simple model case of such systems, and the analogue of the scalar problem (3.10), is the **Lane-Emden system**:

$$\left.\begin{aligned} -\Delta u &= v^p, & x &\in \Omega, \\ -\Delta v &= u^q, & x &\in \Omega, \\ u = v &= 0, & x &\in \partial\Omega. \end{aligned}\right\} \tag{31.2}$$

Throughout this section we assume $p, q > 1$, and we denote

$$\alpha = \frac{2(p+1)}{pq-1}, \quad \beta = \frac{2(q+1)}{pq-1}. \tag{31.3}$$

These numbers play a fundamental role in the analysis of (31.2). They represent **scaling exponents**, corresponding to the fact that, for each $\lambda > 0$, the differential equations in (31.2) are invariant under the transformation $(u, v) \mapsto (u_\lambda, v_\lambda)$, where $u_\lambda(x) = \lambda^\alpha u(\lambda x)$, $v_\lambda(x) = \lambda^\beta v(\lambda x)$, due to $\alpha + 2 = \beta p$, $\beta + 2 = \alpha q$. On the other hand we say that (u, v) is positive if $u, v > 0$ (a.e.) in Ω. Note that, of course, if (u, v) is a nontrivial nonnegative, say classical, solution of (31.2) in a domain $\Omega \subset \mathbb{R}^n$, then it is positive by the strong maximum principle.

Remarks 31.1. (i) **Other nonlinearities.** Although we shall concentrate, for simplicity, on the model case (31.2) and on a few variants, the three methods that we describe below, or their modifications, can be applied to wide varieties of systems. Let us mention systems with products or sums of powers, respectively given by

$$f = u^r v^p, \quad g = v^s u^q \tag{31.4}$$

(see [371], [454], [136], [449]), and by

$$f = u^r + v^p, \quad g = v^s + u^q$$

(see [184], [547], [449]), with $p, q, r, s > 0$. Several systems arising in physical or biological applications are also tractable by these methods. Let us mention the cooperative logistic system given by

$$f = auv + u(c - u), \quad g = buv + v(d - v)$$

with $a, b, c, d > 0$ constants (see e.g. [343] and the references therein), which arises in population dynamics, where u, v stand for the densities of two biological species. Another example is given by

$$f = uv - au, \quad g = bu \tag{31.5}$$

with $a, b > 0$ constants (see [258], [122], [449]), which arises as a model of nuclear reactor, where u and v respectively represent the neutron flux and the temperature. Each of the three methods works under different (and generally non-comparable) sets of assumptions, and its applicability depends on the problem under consideration (see Theorem 31.17 for an example in the case of (31.5)).

(ii) **Noncooperative systems.** Many interesting examples from the point of view of biological or chemical applications involve noncooperative systems or systems with balance law. Results and techniques concerning the questions of global existence and blow-up for the parabolic version of such systems are presented in Section 33 below.

(iii) **Singularities for elliptic systems.** Some results on isolated singularities for systems (31.2) and (31.1), (31.4), extending those in Section 4, can be found in [82], [80], [424]. $\square$

31.1. A priori bounds by the method of moving planes and Pohozaev-type identities

We consider the Lane-Emden system (31.2). For this system, the method described in this subsection allows to obtain complete and optimal results in the case of convex domains.

Theorem 31.2. *Assume $p, q > 1$, Ω convex and bounded, and*

$$\frac{1}{p+1} + \frac{1}{q+1} > \frac{n-2}{n}, \tag{31.6}$$

equivalently

$$\alpha + \beta > n - 2.$$

(i) *Then any positive classical solution of* (31.2) *satisfies the a priori estimate*

$$\|u\|_\infty, \|v\|_\infty \le C, \tag{31.7}$$

with C independent of (u, v).

(ii) *There exists a positive classical solution of* (31.2).

Theorem 31.3. *Assume $p, q > 1$, $n \ge 3$, Ω starshaped and bounded, and*

$$\frac{1}{p+1} + \frac{1}{q+1} \le \frac{n-2}{n}, \tag{31.8}$$

equivalently

$$\alpha + \beta \le n - 2.$$

Then (31.2) *has no positive classical solution.*

Theorems 31.2 and 31.3 are respectively due to [134] (see also [413]) and to [370]. The critical curve in the (p, q) plane:

$$\frac{1}{p+1} + \frac{1}{q+1} = \frac{n-2}{n},$$

associated with condition (31.6), is called the **Sobolev hyperbola.** Note that in the scalar case, corresponding to $p = q$, condition (31.6) reduces to $p < p_S$.

The method of proof of Theorem 31.3 is a modification of that of Section 13 in the scalar case. A common ingredient to the proofs of Theorems 31.2 and 31.3 is the following variational identity of Pohozaev-type [370], which is the analogue of Theorem 5.1 in the scalar case.

Lemma 31.4. *Assume Ω bounded.*

(i) *For any functions $u, v \in C^2(\overline{\Omega})$ such that $u = v = 0$ on $\partial\Omega$, there holds*

$$\int_\Omega \left[(x \cdot \nabla v)\,\Delta u + (x \cdot \nabla u)\,\Delta v - (n-2)\nabla u \cdot \nabla v \right] dx = \int_{\partial\Omega} (x \cdot \nu)\,\frac{\partial u}{\partial \nu}\,\frac{\partial v}{\partial \nu}\, d\sigma.$$

(ii) *For any nonnegative classical solution (u, v) of (31.2) and any $\theta \in [0, 1]$, there holds*

$$\int_\Omega \left[\left(\frac{n}{p+1} - (n-2)\theta \right) v^{p+1} + \left(\frac{n}{q+1} - (n-2)(1-\theta) \right) u^{q+1} \right] dx$$

$$= \int_{\partial\Omega} (x \cdot \nu)\,\frac{\partial u}{\partial \nu}\,\frac{\partial v}{\partial \nu}\, d\sigma. \tag{31.9}$$

Proof. (i) We compute

$$\mathrm{div}\left((x \cdot \nabla v)\,\nabla u \right) = (x \cdot \nabla v)\,\Delta u + \left(\nabla(x \cdot \nabla v) \cdot \nabla u \right)$$

$$= (x \cdot \nabla v)\,\Delta u + \sum_{i,j} \frac{\partial}{\partial x_i}\left(x_j \frac{\partial v}{\partial x_j} \right) \frac{\partial u}{\partial x_i}$$

$$= (x \cdot \nabla v)\,\Delta u + \sum_{i,j} x_j \frac{\partial^2 v}{\partial x_i \partial x_j} \frac{\partial u}{\partial x_i} + \sum_i \frac{\partial v}{\partial x_i} \frac{\partial u}{\partial x_i}.$$

Therefore

$$\mathrm{div}\left[(x \cdot \nabla v)\,\nabla u + (x \cdot \nabla u)\,\nabla v \right]$$

$$= (x \cdot \nabla v)\,\Delta u + (x \cdot \nabla u)\,\Delta v + x \cdot \nabla(\nabla u \cdot \nabla v) + 2\nabla u \cdot \nabla v.$$

On the other hand, we have

$$\mathrm{div}\left[x(\nabla u \cdot \nabla v) \right] = (\mathrm{div}\, x)\,(\nabla u \cdot \nabla v) + x \cdot \nabla(\nabla u \cdot \nabla v)$$

$$= n(\nabla u \cdot \nabla v) + x \cdot \nabla(\nabla u \cdot \nabla v).$$

By subtracting, we obtain

$$\mathrm{div}\left[(x \cdot \nabla v)\,\nabla u + (x \cdot \nabla u)\,\nabla v - x(\nabla u \cdot \nabla v) \right]$$

$$= (x \cdot \nabla v)\,\Delta u + (x \cdot \nabla u)\,\Delta v - (n-2)\nabla u \cdot \nabla v.$$

Applying the divergence theorem, it follows that

$$\int_\Omega \left[(x \cdot \nabla v)\,\Delta u + (x \cdot \nabla u)\,\Delta v - (n-2)\nabla u \cdot \nabla v \right] dx$$

$$= \int_{\partial\Omega} \left[(x \cdot \nabla v)\, \nabla u + (x \cdot \nabla u)\, \nabla v - x(\nabla u \cdot \nabla v) \right] \cdot \nu \, d\sigma.$$

Since $\nabla u = \left(\frac{\partial u}{\partial \nu}\right)\nu$, $\nabla v = \left(\frac{\partial v}{\partial \nu}\right)\nu$ on $\partial\Omega$, due to $u = v = 0$ on $\partial\Omega$, assertion (i) follows.

(ii) For a solution (u, v) of (31.2), we have

$$(x \cdot \nabla v)\, \Delta u + (x \cdot \nabla u)\, \Delta v = -(x \cdot \nabla v)\, v^p - (x \cdot \nabla u)\, u^q$$

$$= -x \cdot \nabla\left(\frac{v^{p+1}}{p+1} + \frac{u^{q+1}}{q+1} \right)$$

$$= -\mathrm{div}\left(x \left(\frac{v^{p+1}}{p+1} + \frac{u^{q+1}}{q+1} \right) \right) + n\left(\frac{v^{p+1}}{p+1} + \frac{u^{q+1}}{q+1} \right)$$

hence

$$\int_\Omega \left[(x \cdot \nabla v)\, \Delta u + (x \cdot \nabla u)\, \Delta v \right] dx = n \int_\Omega \left(\frac{v^{p+1}}{p+1} + \frac{u^{q+1}}{q+1} \right) dx. \tag{31.10}$$

On the other hand,

$$\int_\Omega \nabla u \cdot \nabla v \, dx = -\int_\Omega u\, \Delta v \, dx = \int_\Omega u^{q+1} \, dx$$

and

$$\int_\Omega \nabla u \cdot \nabla v \, dx = -\int_\Omega v\, \Delta u \, dx = \int_\Omega v^{p+1} \, dx$$

yield

$$\int_\Omega \nabla u \cdot \nabla v \, dx = \int_\Omega \left((1 - \theta)\, u^{q+1} + \theta\, v^{p+1} \right) dx. \tag{31.11}$$

In view of (i), assertion (ii) then follows by combining (31.10) and (31.11). $\quad\square$

We first prove Theorem 31.3, which follows easily from Lemma 31.4.

Proof of Theorem 31.3. In view of (31.8), by choosing $\theta = \frac{n}{(n-2)(p+1)} \in (0, 1)$, we get

$$\frac{n}{p+1} - (n-2)\theta = 0, \qquad \frac{n}{q+1} - (n-2)(1 - \theta) \le 0. \tag{31.12}$$

Identity (31.9) in Lemma 31.4 then implies $\int_{\partial\Omega} (x \cdot \nu)\, \frac{\partial u}{\partial \nu}\, \frac{\partial v}{\partial \nu}\, d\sigma \le 0$. Now since Ω is starshaped around, say, $x = 0$, we have $x \cdot \nu \ge 0$ on $\partial\Omega$, along with $\frac{\partial u}{\partial \nu}, \frac{\partial v}{\partial \nu} \le 0$, hence

$$\int_{\partial\Omega} (x \cdot \nu)\, \frac{\partial u}{\partial \nu}\, \frac{\partial v}{\partial \nu}\, d\sigma = 0. \tag{31.13}$$

If the inequality in (31.8) is strict, then so is the inequality in (31.12) and we deduce from (31.9) that $u \equiv 0$, hence $v \equiv 0$. In the equality case, then since

$x \cdot \nu \not\equiv 0$ on $\partial\Omega$, (31.13) implies $\frac{\partial u}{\partial \nu} = 0$ or $\frac{\partial v}{\partial \nu} = 0$ at some point of $\partial\Omega$. Since $-\Delta u, -\Delta v \geq 0$, $u, v \geq 0$ in Ω and $u = v = 0$ on $\partial\Omega$, we infer from Hopf's lemma that $u \equiv 0$ or $v \equiv 0$, hence $u \equiv v \equiv 0$. (Note that this last argument actually applies whenever (31.8) holds.) $\square$

Proof of Theorem 31.2. (i) It is more involved and requires several steps.

Step 1. Basic L^1_{loc} estimates. We claim that

$$\int_\Omega u\varphi_1 \, dx \leq C, \quad \int_\Omega v\varphi_1 \, dx \leq C. \tag{31.14}$$

Multiplying by φ_1, integrating by parts, and using Jensen's inequality, we obtain

$$\lambda_1 \int_\Omega u\varphi_1 \, dx = \int_\Omega v^p \varphi_1 \, dx \geq \left(\int_\Omega v\varphi_1 \, dx\right)^p$$

and

$$\lambda_1 \int_\Omega v\varphi_1 \, dx = \int_\Omega u^q \varphi_1 \, dx \geq \left(\int_\Omega u\varphi_1 \, dx\right)^q.$$

Consequently, we have

$$\left(\int_\Omega u\varphi_1 \, dx\right)^{pq} \leq \lambda_1^{p+1} \int_\Omega u\varphi_1 \, dx,$$

which yields the first inequality in (31.14). The second follows similarly.

Step 2. Estimates near $\partial\Omega$. We use the notation of Section 13 (see after Theorem 13.1). Since Ω is convex and smooth, we can find $\lambda_0, c_0 > 0$ such that

$$\Sigma'(y, \lambda) \subset \Omega, \quad \lambda \leq \lambda_0 \quad \text{and} \quad (\nu(x), \nu(y)) > c_0, \quad x \in \partial\Sigma(y, \lambda_0) \cap \partial\Omega.$$

Similarly as in Theorem 13.1, we shall apply the moving planes method (cf. [514] in the case of systems) to show that

$$u(R(y, \lambda)x) \geq u(x), \quad v(R(y, \lambda)x) \geq v(x), \qquad y \in \partial\Omega, \ x \in \Sigma(y, \lambda), \ \lambda \leq \lambda_0. \tag{31.15}$$

Without loss of generality, we may assume that $y = 0$ and that $\nu(0) = -e_1$. For each $x = (x_1, x')$, we denote $x^\lambda := R(0, \lambda)x = (2\lambda - x_1, x')$, $\Sigma_\lambda := \Sigma(0, \lambda) = \Omega \cap \{x_1 < \lambda\}$, and $\Sigma'_\lambda := \Sigma'(0, \lambda) = R(0, \lambda)\Sigma_\lambda$. Define

$$w^\lambda(x) = u(x^\lambda) - u(x), \quad z^\lambda(x) = v(x^\lambda) - v(x), \qquad \text{for } x \in \Sigma_\lambda, \ 0 < \lambda \leq \lambda_0,$$

and set

$$E := \left\{\mu \in (0, \lambda_0] : w^\lambda(x) \geq 0, \ z^\lambda(x) \geq 0 \text{ for all } x \in \Sigma_\lambda \text{ and } \lambda \in (0, \mu)\right\}.$$

Since $\frac{\partial u}{\partial x_1}(0) > 0$, $\frac{\partial v}{\partial x_1}(0) > 0$ by Hopf's lemma, we have $\lambda \in E$ for $\lambda > 0$ small. Assume for contradiction that $\bar{\lambda} := \sup E < \lambda_0$. We have

$$w^\lambda \geq 0, \quad z^\lambda \geq 0, \qquad \text{for all } x \in \Sigma_\lambda \text{ and } \lambda \in (0, \bar{\lambda}], \tag{31.16}$$

and there exists a sequence $\lambda_i \to \bar{\lambda}$, with $\bar{\lambda} < \lambda_i < \lambda_0$, such that (for instance) $\min\limits_{\overline{\Sigma_{\lambda_i}}} w^{\lambda_i} < 0$. Since $w^\lambda = 0$ on $\{x_1 = \lambda\} \cap \overline{\Omega}$ and

$$w^\lambda > 0 \quad \text{on } \{x_1 < \lambda\} \cap \partial\Omega, \quad \text{for all } \lambda < \lambda_0, \tag{31.17}$$

it follows that this minimum is attained at a point $q_i \in \Sigma_{\lambda_i}$. Therefore $\nabla w^{\lambda_i}(q_i) = 0$. On the other hand, since $\frac{\partial u}{\partial x_1} = (e_1 \cdot \nu)\frac{\partial u}{\partial \nu} \geq c > 0$ on $\{x_1 \leq \lambda_0\} \cap \partial\Omega$ and

$$w^\lambda(x) = u(2\lambda - x_1, x') - u(x_1, x') = 2(\lambda - x_1)\frac{\partial u}{\partial x_1}(\xi(x)),$$

with $|\xi(x) - x| \leq 2(\lambda - x_1)$, we see that $w^\lambda(x) \geq 0$ for x in an ε-neighborhood of $\{x_1 = \lambda\} \cap \partial\Omega$, with $\varepsilon > 0$ independent of $\lambda \in (0, \lambda_0]$. Therefore, we may assume that $q_i \to \bar{q} \in \overline{\Sigma_{\bar{\lambda}}}$, $\bar{q} \notin \{x_1 = \bar{\lambda}\} \cap \partial\Omega$, and by continuity we get

$$w^{\bar{\lambda}}(\bar{q}) = 0 \quad \text{and} \quad \nabla w^{\bar{\lambda}}(\bar{q}) = 0. \tag{31.18}$$

But (31.16) implies

$$-\Delta w^{\bar{\lambda}}(x) = v^p(x^{\bar{\lambda}}) - v^p(x) \geq 0 \quad \text{and} \quad w^{\bar{\lambda}}(x) \geq 0, \qquad x \in \Sigma_{\bar{\lambda}}.$$

By Hopf's lemma, this along with (31.18) implies $w^{\bar{\lambda}} = 0$ in $\Sigma_{\bar{\lambda}}$, contradicting (31.17). Consequently, $\bar{\lambda} = \lambda_0$, which proves (31.15). This guarantees that

$$(\nabla u(x), \nu(y)) \leq 0, \quad (\nabla v(x), \nu(y)) \leq 0, \qquad y \in \partial\Omega, \ x \in \Sigma(y, \lambda_0). \tag{31.19}$$

By Lemma 13.2 and Step 1, we deduce that $u, v \leq C$ on $\Omega_\varepsilon = \{z \in \Omega : \delta(z) < \varepsilon\}$ for some $\varepsilon, C > 0$ depending only on Ω. Using interior-boundary elliptic L^p-estimates (see Appendix A) and the embedding $W^{2,k} \hookrightarrow BUC^1$ for $k > n$, we deduce a uniform bound for $\nabla u, \nabla v$ in $\Omega_{\varepsilon/2}$. In particular, we have shown that

$$\left|\frac{\partial u}{\partial \nu}\right|, \ \left|\frac{\partial v}{\partial \nu}\right| \leq C, \quad x \in \partial\Omega. \tag{31.20}$$

Step 3. Energy estimates. We claim that

$$\int_\Omega v^{p+1} \, dx \leq C, \quad \int_\Omega u^{q+1} \, dx \leq C.$$

Since $\frac{1}{p+1} + \frac{1}{q+1} > \frac{n-2}{n}$, we may choose $\theta \in (0,1)$, such that

$$\frac{n}{p+1} - (n-2)\theta > 0, \qquad \frac{n}{q+1} - (n-2)(1-\theta) > 0.$$

Using assertion (i) of Lemma 31.4 and estimate (31.20), we deduce that

$$\int_\Omega v^{p+1}\, dx + \int_\Omega u^{q+1}\, dx \leq C \int_{\partial\Omega} (x \cdot \nu)\, \frac{\partial u}{\partial \nu}\, \frac{\partial v}{\partial \nu}\, d\sigma \leq C.$$

Step 4. Bootstrap. Pick $\rho > 1$ to be fixed later and consider the following induction hypothesis:

$$\|u\|_{(q+1)\rho^i}, \quad \|v\|_{(p+1)\rho^i} \leq C. \tag{H_i}$$

Step 3 guarantees that (H_0) is verified. Assume that (H_i) holds for some $i \in \mathbb{N}$. Then, since (u,v) solves (31.2), the linear estimate in Proposition 47.5(i) implies (H_{i+1}) provided

$$\frac{p}{(p+1)\rho^i} - \frac{1}{(q+1)\rho^{i+1}} < \frac{2}{n} \quad \text{and} \quad \frac{q}{(q+1)\rho^i} - \frac{1}{(p+1)\rho^{i+1}} < \frac{2}{n}.$$

It is thus sufficient that

$$\frac{p}{(p+1)} - \frac{1}{(q+1)\rho} < \frac{2}{n} \quad \text{and} \quad \frac{q}{(q+1)} - \frac{1}{(p+1)\rho} < \frac{2}{n},$$

i.e.

$$\frac{1}{\rho} > \max\left[(q+1)\left(\frac{n-2}{n} - \frac{1}{p+1}\right), (p+1)\left(\frac{n-2}{n} - \frac{1}{q+1}\right)\right].$$

Since, by assumption, $\frac{1}{p+1} + \frac{1}{q+1} = \frac{n-2}{n} + \varepsilon$ for some $\varepsilon > 0$, it suffices to choose

$$\frac{1}{\rho} > 1 - \varepsilon \min(p+1, q+1).$$

After a finite number of steps, we obtain $\|u\|_{\hat{q}} \leq C$, $\|v\|_{\hat{p}} \leq C$ for some $\hat{q} > nq/2$, $\hat{p} > np/2$, and a further application of Proposition 47.5(i) yields $\|u\|_\infty \leq C$, $\|v\|_\infty \leq C$.

(ii) The proof is similar to that of Corollary 10.3 (see e.g. [180] or [449, Section 4] for details). $\square$

Remarks 31.5. Limitations and extensions. (i) The above method does not extend to general systems of the form (31.1). Indeed (but for very special cases), f should not depend on u (nor g on v) because of the need of variational identities. Also, f, g cannot depend on x (at least in an arbitrary way) in order to apply the moving planes method. It can still be generalized to $f = f(v)$, $g = g(u)$, with f, g nondecreasing (in order for the system to admit a comparison principle to apply the moving planes method), provided f, g also satisfy suitable growth conditions related with the Sobolev hyperbola. These conditions can be expressed as a relation between f, g and their primitives which enables one to control $\int_\Omega v f(v)$ and $\int_\Omega u g(u)$ from the variational identities.

(ii) The method partially extends to nonconvex domains Ω (via the Kelvin transform). However, this requires additional growth restrictions if $n \geq 3$, namely $p, q \leq p_S$ in the case of (31.2). $\quad\square$

Remark 31.6. Variational methods. If the nonlinearities f, g in system (31.1) have the form $f(x, u, v) = H_v(x, u, v)$, $g(x, u, v) = H_u(x, u, v)$, then solutions of (31.1) can be found as critical points of the functional

$$\Phi(u, v) := \int_\Omega \nabla u \cdot \nabla v \, dx - \int_\Omega H(x, u, v) \, dx.$$

Considering Φ as a strongly indefinite functional in $W_0^{1,2} \times W_0^{1,2}(\Omega)$ (or, more generally, in spaces of the form $X_\alpha \times X_{1-\alpha}$, where X_α, $\alpha \in (0, 1)$, are suitable interpolation spaces between $X_0 := L^2(\Omega)$ and $X_1 := W^{2,2} \cap W_0^{1,2}(\Omega)$, see [285] or [182], for example, and cf. Section 51.1) often leads to unnecessary technical restrictions concerning the growth of the Hamiltonian H. To overcome these difficulties one can use a dual approach (see [137] in the case of systems or [24], [25] in the scalar case). In the particular case of the Lane-Emden system (31.2) we have $H = |v|^{p+1}/(p+1) + |u|^{q+1}/(q+1)$ and the dual functional has the form

$$\tilde{\Phi}(w, z) = \int_\Omega \left(\frac{|w|^{p_1}}{p_1} + \frac{|z|^{q_1}}{q_1} - \frac{1}{2}(K * w)z \right) dx,$$

where $p_1 = 1 + 1/p$, $q_1 = 1 + 1/q$, $w = |v|^{p-1}v$, $z = |u|^{q-1}u$ and K is the Green function for the negative Dirichlet Laplacian, that is $u := K * w$ is the solution of the problem

$$-\Delta u = w \quad \text{in } \Omega, \qquad u = 0 \quad \text{on } \partial\Omega.$$

(Notice that $\int_\Omega (K * w)z \, dx = \int_\Omega (K * z)w \, dx$.) The functional

$$\tilde{\Phi} : L^{p_1} \times L^{q_1}(\Omega) \to \mathbb{R}$$

possesses a mountain-pass structure and, in particular, it is easy to show that the existence result in Theorem 31.2 remains true without the assumption Ω convex. However, this approach does not provide a priori estimates of solutions.

For some particular nonlinearities f, g, system (31.1) can also be reduced to a single higher-order equation. This is for instance the case for the Lane-Emden system (31.2), which is equivalent to the problem

$$\left. \begin{array}{ll} -\Delta((-\Delta u)^{1/p}) = u^q, & x \in \Omega, \\ u = \Delta u = 0, & x \in \partial\Omega, \end{array} \right\}$$

where $u \geq 0 \geq \Delta u$. Again, this problem can solved by variational methods. $\quad\square$

31.2. Liouville-type results for the Lane-Emden system

In this subsection we state Liouville-type theorems for the Lane-Emden system (and prove some of them). These are statements about nonexistence of entire positive solutions in the whole space or in a half-space. As in the scalar case, they constitute essential pieces of information in view of the rescaling method (see next subsection).

We thus consider the following problems:

$$\left. \begin{aligned} -\Delta u &= v^p, & x \in \mathbb{R}^n, \\ -\Delta v &= u^q, & x \in \mathbb{R}^n, \end{aligned} \right\} \tag{31.21}$$

or

$$\left. \begin{aligned} -\Delta u &= v^p, & x \in \mathbb{R}^n_+, \\ -\Delta v &= u^q, & x \in \mathbb{R}^n_+, \\ u = v &= 0, & x \in \partial\mathbb{R}^n_+, \end{aligned} \right\} \tag{31.22}$$

where $p, q > 1$ and $\mathbb{R}^n_+ := \{x \in \mathbb{R}^n : x_n > 0\}$.

Conjecture 31.7. *Systems* (31.21) *and* (31.22) *do not admit any positive classical solutions if* (p, q) *lies below the Sobolev hyperbola, i.e.*

$$\alpha + \beta > n - 2.$$

Remarks 31.8. "Classical solutions" in Conjecture 31.7 means $u, v \in C^2(\mathbb{R}^n)$ and $u, v \in C^2(\mathbb{R}^n_+) \cap C(\overline{\mathbb{R}^n_+})$, respectively; no growth or decay conditions at infinity are imposed. However for the rescaling method, it is sufficient to know a Liouville-type theorem for *bounded* positive solutions. $\square$

Although the full Conjecture 31.7 has not been proved so far, it is strongly supported by the following results.

Theorem 31.9. *Let* $p, q > 1$.

(i) *Assume* $\alpha + \beta \leq n - 2$. *Then system* (31.21) *admits some radial, bounded, positive classical solution.*

(ii) *System* (31.21) *does not admit any positive classical solution in the following cases:*

(a) $\alpha + \beta > n - 2$ *and either* u, v *are radial or* $n = 3$,

(b) $\max(\alpha, \beta) \geq n - 2$,

(c) $p, q \leq p_S$, $(p, q) \neq (p_S, p_S)$.

Assertion (i) is due to [472]. As for assertion (ii), part (a) is due to [371] in the radial case. In the nonradial case, part (a) settles Conjecture 31.7 for $n = 3$. This is due to [471] for polynomially bounded solutions, and to [424] in the general case. Part (b) is actually valid for supersolutions (see Theorem 31.12 below). Part (c), which in particular recovers the (optimal) scalar case, is due to [181]. The Liouville-type result is also known in some other parts of the region $\alpha + \beta > n - 2$ (see [106]). Since most of the proofs are long and technical, we shall only prove nonexistence under assumption (b), as a consequence of Theorem 31.12 below.

As for the half-space case, we have the following reduction. The result is due to [87], generalizing the idea of Theorem 8.3 in the scalar case.

Theorem 31.10. *Let $n \geq 1$. For given $p, q > 1$, if system (31.21) does not admit any bounded, positive classical solution in dimension $n - 1$, then system (31.22) does not admit any bounded, positive classical solution in dimension n.*

Remarks 31.11. (i) Here we have made the convention that functions of $n = 0$ variables are constants (and have null Laplacian). Consequently the conclusion of Theorem 31.10 is true for $n = 1$.

(ii) If system (31.21) does not admit any bounded positive solution in dimension n, then this remains true in dimension $n - 1$, so that in particular, system (31.22) does not admit any bounded positive solution in dimension n. Indeed if (u, v) solves (31.21) in $\mathbb{R}^{n-1}$ and if we let $\tilde{u}(x) = u(x_1, \ldots, x_{n-1})$, $\tilde{v}(x) = v(x_1, \ldots, x_{n-1})$, then $(\tilde{u}, \tilde{v})$ solves (31.21) in $\mathbb{R}^n$.

(iii) On the other hand, it was shown in [424] that, for given $p, q > 1$ and n, if system (31.21) does not admit any bounded, positive classical solution, then it does not admit any positive classical solution at all, and system (31.22) does not admit any bounded, positive classical solution. $\square$

Sketch of proof of Theorem 31.10. Assume that (31.22) admits a bounded positive solution (u, v). By using the moving planes method (see the proof of Theorem 21.10 for similar arguments in the scalar case), one can show that $\frac{\partial u}{\partial x_n} \geq 0$ and $\frac{\partial v}{\partial x_n} \geq 0$ in $\mathbb{R}^n_+$. Therefore, for each $x' \in \mathbb{R}^{n-1}$,

$$U(x') := \lim_{x_n \to \infty} u(x', x_n) \quad \text{and} \quad V(x') := \lim_{x_n \to \infty} v(x', x_n)$$

are well defined and are bounded positive functions. Arguing exactly as in the proof of Theorem 8.3, we see that (U, V) is a bounded, positive classical solution of system (31.21) in $\mathbb{R}^{n-1}$. The result follows. $\square$

Case (b) of Theorem 31.9(ii) is actually true for the following system of inequalities (see [501], [372]):

$$\left.\begin{array}{ll} -\Delta u \geq v^p, & x \in \mathbb{R}^n, \\ -\Delta v \geq u^q, & x \in \mathbb{R}^n. \end{array}\right\} \qquad (31.23)$$

Theorem 31.12. *Let $p, q > 1$. System (31.23) does not admit any positive solution $u, v \in C^2(\mathbb{R}^n)$ if $\max(\alpha, \beta) \geq n - 2$.*

Proof. It is based on the rescaled test-function method. Fix $\phi \in \mathcal{D}(\mathbb{R}^n)$, $0 \leq \phi \leq 1$, such that $\phi(x) = 1$ for $|x| \leq 1$ and $\phi(x) = 0$ for $|x| \geq 2$. For each $R > 0$, put $\phi_R(x) = \phi(x/R)$. Let $m, k \geq 2$ to be fixed later. We note that $\Delta(\phi_R^m) = 0$ for $|x| \leq R$ and that

$$|\Delta(\phi_R^m)| = \left| m\phi_R^{m-1}\Delta\phi_R + m(m-1)\phi_R^{m-2}|\nabla\phi_R|^2 \right| \leq CR^{-2}\phi_R^{m-2}.$$

Multiplying the first inequality in (31.23) by ϕ_R^m and integrating by parts, we obtain

$$\int v^p \phi_R^m \leq -\int \phi_R^m \Delta u = -\int u \Delta(\phi_R^m) \leq CR^{-2} \int_{R<|x|<2R} u \phi_R^{m-2}$$

(where $\int = \int_{\mathbb{R}^n}$). Applying Hölder's inequality, it follows that

$$\int v^p \phi_R^m \leq CR^{(n/q')-2}\left(\int_{R<|x|<2R} u^q \phi_R^{(m-2)q}\right)^{1/q}.$$

Similarly, we obtain

$$\int u^q \phi_R^k \leq CR^{(n/p')-2}\left(\int_{R<|x|<2R} v^p \phi_R^{(k-2)p}\right)^{1/p}.$$

Now, since $p, q > 1$, we have $2 + (k/q) < (k - 2)p$ for k large enough, and we can then choose m such that $2 + (k/q) \leq m \leq (k - 2)p$, that is: $(k - 2)p \geq m$ and $(m - 2)q \geq k$. Therefore,

$$\left(\int v^p \phi_R^m\right)^{pq} \leq CR^{((n/q')-2)pq}\left(\int_{R<|x|<2R} u^q \phi_R^k\right)^p,$$

$$\left(\int u^q \phi_R^k\right)^p \leq CR^{((n/p')-2)p} \int_{R<|x|<2R} v^p \phi_R^m.$$

Consequently,

$$\left(\int v^p \phi_R^m\right)^{pq} \leq CR^\theta \int_{R<|x|<2R} v^p \phi_R^m, \tag{31.24}$$

where

$$\theta = pq\left(\frac{n}{q'}-2\right)+p\left(\frac{n}{p'}-2\right) = p(n(q-1)-2q)+n(p-1)-2p = (n-2)(pq-1)-2(p+1).$$

In particular,

$$\left(\int_{|x|<R} v^p\right)^{pq-1} \leq \left(\int v^p\, \phi_R^m\right)^{pq-1} \leq CR^\theta.$$

If $\alpha > n-2$, then $\theta < 0$, and by letting $R \to \infty$ we immediately obtain $v \equiv 0$. Since $u^q \leq -\Delta v = 0$, we also get $u \equiv 0$.

If $\alpha = n-2$, then $\theta = 0$, so that (31.24) implies $\int v^p < \infty$. Returning to (31.24), we then deduce

$$\left(\int v^p\, \phi_R^m\right)^{pq} \leq C \int_{R<|x|} v^p \to 0, \quad \text{as } R \to \infty,$$

hence again $v \equiv 0$ and $u \equiv 0$.

By exchanging the roles of u and v, we get the same conclusion if $\beta \geq n-2$. $\qquad\square$

31.3. A priori bounds by the rescaling method

Unlike the method based on moving planes and Pohozaev-type identity, the rescaling method allows to treat more general systems of the form (31.1). However, one has to assume, roughly speaking, that for each fixed $x \in \Omega$, f and g behave asymptotically like homogeneous functions of u, v. Several choices of homogeneity are possible. In this subsection, we shall work under the following assumptions:

$$\left.\begin{aligned}
&f(x,u,v) = a(x)v^p + f_1(x,u), \quad |f_1| \leq C(1+u^r), \quad r < \frac{p(q+1)}{p+1},\\[2mm]
&g(x,u,v) = b(x)u^q + g_1(x,v), \quad |g_1| \leq C(1+v^s), \quad s < \frac{q(p+1)}{q+1},\\[2mm]
&a,b \in C(\overline{\Omega}), \quad a,b > 0 \text{ in } \overline{\Omega}, \quad f_1,g_1 \in C(\overline{\Omega} \times \mathbb{R}).
\end{aligned}\right\} \quad (31.25)$$

Theorem 31.13. *Assume Ω bounded. For given $p,q > 1$, let (31.25) be satisfied and assume that system (31.21) does not admit any bounded, positive classical solution. Then any nonnegative classical solution of (31.1) satisfies the a priori estimate (31.7).*

Theorem 31.13 is a variant of results from [180], [184] (see also references therein). Similarly as in the scalar case (cf. Corollary 10.3), existence results can be deduced from Theorem 31.13 under suitable additional assumptions on f, g.

Proof. Let us first observe that, due to Remark 31.11(ii), the assumption of the theorem guarantees that (31.22) neither has any nontrivial solution.

Similarly as in the proof of Theorem 12.1, we proceed by contradiction. Assume that there exists a sequence (u_j, v_j) of solutions such that $\|u_j\|_\infty + \|v_j\|_\infty \to \infty$.

We may assume $\|u_j\|_\infty \geq \|v_j\|_\infty^{\alpha/\beta}$ without loss of generality. Let $x_j \in \Omega$ be such that $u_j(x_j) = \|u_j\|_\infty$ and set

$$\lambda_j := \left(\|u_j\|_\infty^{1/\alpha} + \|v_j\|_\infty^{1/\beta}\right)^{-1} \to 0, \qquad \text{as } j \to \infty.$$

By passing to a subsequence, we may assume that $x_j \to x_\infty \in \overline{\Omega}$. Setting $d_j := \text{dist}(x_j, \partial\Omega)$, we then split the proof into two cases, according to whether $d_j/\lambda_j \to \infty$ (along some subsequence) or d_j/λ_j is bounded.

Case 1: $d_j/\lambda_j \to \infty$. We rescale the solutions around x_j as follows:

$$\tilde{u}_j(y) = \lambda_j^\alpha\, u_j(x_j + \lambda_j y), \quad \tilde{v}_j(y) = \lambda_j^\beta\, v_j(x_j + \lambda_j y), \qquad y \in \Omega_j,$$

where $\Omega_j = \{y \in \mathbb{R}^n : |y| < d_j/\lambda_j\}$. Due to the definition of λ_j, it is clear that

$$\tilde{u}_j(y),\ \tilde{v}_j(y) \leq 1, \qquad y \in \Omega_j. \tag{31.26}$$

Moreover, $\tilde{u}_j^{1/\alpha}(0) = \lambda_j \|u_j\|_\infty^{1/\alpha} \geq \lambda_j \left(\|u_j\|_\infty^{1/\alpha} + \|v_j\|_\infty^{1/\beta}\right)/2 = 1/2$, hence

$$\tilde{u}_j(0) \geq 2^{-\alpha}. \tag{31.27}$$

Now, since $\alpha + 2 = \beta p$ and $\beta + 2 = \alpha q$, we find that $(\tilde{u}, \tilde{v}) = (\tilde{u}_j, \tilde{v}_j)$ satisfies the system

$$\left.\begin{array}{ll} -\Delta\tilde{u} = a(x_j + \lambda_j y)\, \tilde{v}^p + \tilde{f}_j(y), & y \in \Omega_j, \\[2mm] -\Delta\tilde{v} = b(x_j + \lambda_j y)\, \tilde{u}^q + \tilde{g}_j(y), & y \in \Omega_j. \end{array}\right\} \tag{31.28}$$

Here, $\tilde{f}_j(y) = \lambda_j^{\alpha+2} f_1(x_j + \lambda_j y, \lambda_j^{-\alpha}\tilde{u}_j(y))$ and $\tilde{g}_j(y) = \lambda_j^{\beta+2} g_1(x_j + \lambda_j y, \lambda_j^{-\beta}\tilde{v}_j(y))$. In view of our assumption (31.25) with $r < p(q+1)/(p+1) = (\alpha+2)/\alpha$, we have

$$|\tilde{f}_j| \leq C\lambda_j^{\alpha+2}(1 + \lambda_j^{-\alpha r}) \to 0, \qquad \text{as } j \to \infty. \tag{31.29}$$

Similarly we obtain

$$|\tilde{g}_j| \to 0, \qquad \text{as } j \to \infty. \tag{31.30}$$

For each fixed $R > 0$, we have $B_{2R} \subset \Omega_j$ for j sufficiently large, and $|\Delta\tilde{u}_j|, |\Delta\tilde{v}_j| \leq C(R)$ in B_{2R}, owing to (31.26), (31.28)–(31.30). It follows from interior elliptic L^p-estimates that the sequences $\tilde{u}_j, \tilde{v}_j$ are bounded in $W^{2,m}(B_R)$ for all $1 < m < \infty$. By embedding theorems, we deduce that they are bounded in $C^{1+\gamma}(\overline{B_R})$ for each $\gamma \in (0, 1)$. It follows that some subsequence of $(\tilde{u}_j, \tilde{v}_j)$ converges, locally uniformly on $\mathbb{R}^n$, to a bounded nonnegative (classical) solution of

$$\left.\begin{array}{ll} -\Delta U = a_0\, V^p, & y \in \mathbb{R}^n, \\[2mm] -\Delta V = b_0\, U^q, & y \in \mathbb{R}^n, \end{array}\right\}$$

where $a_0 = a(x_\infty) > 0$, $b_0 = b(x_\infty) > 0$. Note that a_0, b_0 can easily be scaled out to be 1. But since $U(0) \geq 2^{-\alpha}$ due to (31.27), this contradicts the Liouville-type property.

Case 2: d_j/λ_j is bounded. We may assume that $d_j/\lambda_j \to c \geq 0$. We perform the same change of coordinates $z = z(x) = (z^1, z^2, \cdots, z^n)$ as in Case 2 of the proof of Theorem 12.1. Then the solution $(\overline{u}, \overline{v}) = (\overline{u}_j(z), \overline{v}_j(z)) = (u_j(x), v_j(x))$ satisfies the following system in a half ball:

$$-\sum_{i,k} a^{ik}(z)\frac{\partial^2 \overline{u}}{\partial z^i \partial z^k} - \sum_i b^i(z)\frac{\partial \overline{u}}{\partial z^i} = a(x(z))\overline{v}^p + f_1(x(z), \overline{u}), \quad |z| < \varepsilon, \ z^1 > 0,$$

$$-\sum_{i,k} a^{ik}(z)\frac{\partial^2 \overline{v}}{\partial z^i \partial z^k} - \sum_i b^i(z)\frac{\partial \overline{v}}{\partial z^i} = b(x(z))\overline{u}^q + g_1(x(z), \overline{v}), \quad |z| < \varepsilon, \ z^1 > 0,$$

$$\overline{u} = \overline{v} = 0, \qquad\qquad\qquad\qquad\qquad |z| < \varepsilon, \ z^1 = 0.$$

Moreover, x_j becomes $z_j := z(x_j) = (d_j, 0, 0, \ldots, 0)$. Now we rescale $(\overline{u}, \overline{v})$ around z_j by setting

$$\tilde{u}_j(y) = \lambda_j^\alpha \, \overline{u}_j(z_j + \lambda_j y), \quad \tilde{v}_j(y) = \lambda_j^\beta \, \overline{v}_j(z_j + \lambda_j y), \qquad y \in \Omega_j,$$

with

$$\Omega_j = \left\{ y : \left| y - \frac{z_j}{\lambda_j} \right| < \frac{\varepsilon'}{\lambda_j}, \ y^1 > \frac{-d_j}{\lambda_j} \right\} \text{ and } \Sigma_j = \left\{ y : \left| y - \frac{z_j}{\lambda_j} \right| < \frac{\varepsilon'}{\lambda_j}, \ y^1 = \frac{-d_j}{\lambda_j} \right\}.$$

The rescaled system becomes

$$-\sum_{i,k} a^{ik}(z_j + \lambda_j y)\frac{\partial^2 \tilde{u}}{\partial y^i \partial y^k} - \lambda_j \sum_i b^i(z_j + \lambda_j y)\frac{\partial \tilde{u}}{\partial y^i}$$

$$= a(x(z_j + \lambda_j y))\, \tilde{v}^p + \tilde{f}_j(y), \quad y \in \Omega_j,$$

$$-\sum_{i,k} a^{ik}(z_j + \lambda_j y)\frac{\partial^2 \tilde{v}}{\partial y^i \partial y^k} - \lambda_j \sum_i b^i(z_j + \lambda_j y)\frac{\partial \tilde{v}}{\partial y^i}$$

$$= b(x(z_j + \lambda_j y))\, \tilde{u}^q + \tilde{g}_j(y), \quad y \in \Omega_j,$$

$$\tilde{u} = \tilde{v} = 0, \qquad\qquad\qquad\qquad y \in \Sigma_j,$$

where

$$\tilde{f}_j(y) = \lambda_j^{\alpha+2} f_1(x(z_j + \lambda_j y), \lambda_j^{-\alpha}\overline{u}_j(y)), \quad \tilde{g}_j(y) = \lambda_j^{\beta+2} g_1(x(z_j + \lambda_j y), \lambda_j^{-\beta}\overline{v}_j(y)).$$

Passing to the limit, similarly as in Case 2 of the proof of Theorem 12.1, we end up with a nonnegative solution (U, V) of

$$\left.\begin{aligned}
-\Delta U &= a_0\, V^p, & y \in \mathbb{R}^n, \ y^1 > -c, \\
-\Delta V &= b_0\, U^q, & y \in \mathbb{R}^n, \ y^1 > -c, \\
U = V &= 0, & y \in \mathbb{R}^n, \ y^1 = -c,
\end{aligned}\right\}$$

with $U(0) \geq 2^{-\alpha}$. This yields a contradiction with the Liouville-type property in a half-space mentioned at the beginning of the proof. $\square$

31.4. A priori bounds by the L^p_δ alternate bootstrap method

The method presented in this subsection relies on a specific bootstrap procedure in the scale of weighted Lebesgue spaces $L^p_\delta(\Omega)$. A simpler bootstrap argument also relying on L^p_δ-spaces has been presented for the scalar case in Section 11. Unlike the moving planes or rescaling methods, the L^p_δ bootstrap method applies to very weak solutions, and in particular it provides L^∞-regularity results for such solutions. Also, it does not suppose any monotonicity or restricted dependence, nor scale invariance properties. On the other hand, it assumes stronger growth restrictions than the previous two methods (for instance, for system (31.2) one has to assume $\max(\alpha, \beta) > n - 1$ instead of $\alpha + \beta > n - 2$). However, it will turn out that its growth conditions are optimal in the class of very weak solutions (see Theorem 31.16 below).

We consider general systems of the form (31.1), essentially under only an *upper growth* bound of the form

$$\left.\begin{array}{l} f(x, u, v) \leq C_1(1 + v^p + u^r), \\[2mm] g(x, u, v) < C_1(1 + u^q + v^s), \end{array}\right\} \qquad u, v \geq 0, \quad x \in \Omega. \qquad (31.31)$$

We also assume a standard (mild) superlinearity condition:

$$f(x, u, v) + g(x, u, v) \geq \lambda(u + v) - C_1, \qquad u, v \geq 0, \quad x \in \Omega, \qquad \text{for some } \lambda > \lambda_1. \qquad (31.32)$$

Here $f, g : \Omega \times [0, \infty)^2 \to [0, \infty)$ are Carathéodory functions, $p, q > 1$, $r, s \geq 1$, $C_1 > 0$. In what follows, we refer to the notion of L^1_δ, or very weak, solution introduced in Definition 3.1. The following result is due to [449].

Theorem 31.14. *Assume Ω bounded and (31.31), (31.32), with*

$$\max(\alpha, \beta) > n - 1 \qquad (31.33)$$

and

$$r, s < p_{BT} = \frac{n + 1}{n - 1}. \qquad (31.34)$$

Then any nonnegative very weak solution (u, v) of (31.1) belongs to $L^\infty \times L^\infty(\Omega)$ and satisfies the a priori estimate (31.7).

Similarly as in the scalar case (cf. Corollary 10.3), existence results can be deduced from Theorem 31.14 under suitable additional assumptions on f, g. Condition (31.32) can be weakened or replaced by other conditions of different form.

For instance, by applying the same method, we obtain regularity and a priori estimate for the following simple system:

$$\left.\begin{array}{ll} -\Delta u = a(x)v^p, & x \in \Omega, \\ -\Delta v = b(x)u^q, & x \in \Omega, \\ u = v = 0, & x \in \partial\Omega. \end{array}\right\} \tag{31.35}$$

Theorem 31.15. *Assume Ω bounded, $p, q > 1$, $a, b \in L^\infty(\Omega)$, $a, b \geq 0$, $a, b \not\equiv 0$ and (31.33). Then any nonnegative very weak solution of (31.35) belongs to $L^\infty \times L^\infty(\Omega)$ and satisfies the a priori estimate (31.7). Moreover, there exists a solution (u, v) of (31.35), with $u, v \in C_0 \cap W^{2,m}(\Omega)$ for all finite m, and $u, v > 0$.*

Theorem 31.15 is from [489] (see also [449]). The optimality of condition (31.33) in Theorems 31.14 and 31.15 is shown by the following result from [489], which will be proved at the end of this section (see Theorem 11.5 for the analogue in the scalar case).

Theorem 31.16. *Assume Ω bounded, $p, q > 1$ and*

$$\max(\alpha, \beta) < n - 1. \tag{31.36}$$

Then there exist functions $a, b \in L^\infty(\Omega)$, $a, b \geq 0$, $a, b \not\equiv 0$, such that system (31.35) admits a positive very weak solution (u, v) satisfying

$$u \notin L^\infty(\Omega), \quad v \notin L^\infty(\Omega).$$

In the bootstrap procedure in the proof of Theorems 31.14 and 31.15, each equation is used *alternatively*. At each step, we make use of the L_δ^p regularity theory (cf. Theorem 49.2 and Proposition 49.5 in Appendix C), and L^∞ is reached after finitely many steps. The proof of Theorem 31.15 given below presents the simplest case of application of these ideas to systems. The proof of Theorem 31.14, although based on the same basic approach, is more involved and will not be given here.

Proof of Theorem 31.15. *Step 1. Initialization.* By testing with φ_1, we obtain the basic estimate $\int_\Omega u \, dx \varphi_1$, $\int_\Omega v \, dx \varphi_1 \leq C$, i.e.

$$\|u\|_{1,\delta} + \|v\|_{1,\delta} \leq C \tag{31.37}$$

in view of (1.4). (In the case $a(x), b(x) \geq C > 0$, this is Step 1 of the proof of Theorem 31.2. For general a, b, this can be done by a simple modification using the argument in the proof of Theorem 11.3.)

We set $f := a(x)v^p$ and $g := b(x)u^q$. Then (31.37) guarantees $\|f\|_{1,\delta} + \|g\|_{1,\delta} \leq C$. Assume without loss of generality $q \geq p$ and $\beta = \frac{2(q+1)}{pq-1} > n - 1$. In particular, there holds $(p-1)(q+1) \leq pq - 1 < \frac{2(q+1)}{n-1}$ hence

$$p < p_{BT}. \tag{31.38}$$

Proposition 49.5 guarantees that

$$\|u\|_{k,\delta} + \|v\|_{k,\delta} \le C(k), \qquad \text{for all } 1 \le k < p_{BT}. \tag{31.39}$$

Note that if $n = 1$, then the growth assumptions on f, g and Theorem 49.2(i) immediately imply $\|u\|_\infty + \|v\|_\infty \le C$. We may thus assume $n \ge 2$.

We will show by a bootstrap argument that the value of k in (31.39) can be increased so as to reach $k = \infty$. Thus assume that there holds

$$\|u\|_{k,\delta} + \|v\|_{k,\delta} \le C(k) \tag{31.40}$$

for some k satisfying

$$k \ge p \quad \text{and} \quad k \ge p_{BT} - \varepsilon, \tag{31.41}$$

where $\varepsilon = \varepsilon(p, q, n) > 0$ small will be chosen below.

Step 2. Bootstrap on the first equation. Let $k_1 \in (k, \infty]$ satisfy

$$\frac{1}{k_1} > \frac{p}{k} - \frac{2}{n+1}. \tag{31.42}$$

Using Theorem 49.2(i) and the first equation, we obtain

$$\|u\|_{k_1,\delta} \le C\|\Delta u\|_{k/p,\delta} = C\|f\|_{k/p,\delta} \le C\|v^p\|_{k/p,\delta} = C\|v\|_{k,\delta}^p \le C. \tag{31.43}$$

For later use, we already note that if

$$k > \frac{(n+1)pq}{2(q+1)}, \tag{31.44}$$

then by taking $\varepsilon = \varepsilon(n, p) > 0$ in (31.41) sufficiently small, we may find

$$k_1 > \frac{(n+1)q}{2} \tag{31.45}$$

such that (31.43) is satisfied. Indeed, $\frac{p}{k} - \frac{2}{n+1} < \min\left(\frac{2}{(n+1)q}, \frac{1}{k}\right)$ and we may thus find $k_1 \in (k, \infty)$ satisfying (31.45) and (31.42), hence (31.43).

Step 3. Bootstrap on the second equation. Now assume

$$k_1 > q \tag{31.46}$$

and let $k_2 \in (k, \infty]$ satisfy

$$\frac{1}{k_2} > \frac{q}{k_1} - \frac{2}{n+1}. \tag{31.47}$$

Using Theorem 49.2(i), the second equation and (31.43), we obtain

$$\|v\|_{k_2,\delta} \le C\|\Delta v\|_{k_1/q,\delta} = C\|g\|_{k_1/q,\delta} \le C\|u^q\|_{k_1/q,\delta} = C\|u\|_{k_1,\delta}^q \le C. \tag{31.48}$$

Step 4. Fulfillment of the bootstrap conditions. Let $\rho = \rho(p, q, n) \in (0, 1)$ to be determined. Conditions (31.42), (31.46), (31.47), together with the bootstrap condition

$$\min(k_1, k_2) > \frac{k}{\rho},$$

are equivalent to

$$A := \frac{p}{k} - \frac{2}{n+1} < \frac{1}{k_1} < \min\left(\frac{\rho}{k}, \frac{1}{q}\right) \tag{31.49}$$

and

$$\frac{q}{k_1} - \frac{2}{n+1} < \frac{1}{k_2} < \frac{\rho}{k}. \tag{31.50}$$

Assume

$$k \le \frac{(n+1)pq}{2(q+1)} \tag{31.51}$$

hence, in particular, $A > 0$. Then condition (31.49) can be solved in $k_1 \in [1, \infty)$, and $1/k_1$ can be taken arbitrarily close to A, provided

$$\frac{p - \rho}{k} < \frac{2}{n+1} \tag{31.52}$$

and

$$\frac{p}{k} - \frac{2}{n+1} < \frac{1}{q}. \tag{31.53}$$

Since $k \ge p$, condition (31.52) is satisfied whenever

$$\frac{n-1}{n+1}p < \rho < 1, \tag{31.54}$$

which is allowable in view of (31.38). Due to $\beta > n - 1$, we have $(pq - 1)(n - 1) < 2q + 2$ hence $\frac{n-1}{n+1}p - \frac{2}{n+1} < \frac{1}{q}$. Taking $\varepsilon = \varepsilon(p, q, n) > 0$ small in (31.41), we thus get (31.53).

On the other hand, condition (31.50) can be solved in $k_2 \in [1, \infty)$ if

$$\frac{q}{k_1} - \frac{2}{n+1} < \frac{\rho}{k}. \tag{31.55}$$

Taking $1/k_1$ in (31.49) close enough to its lower bound A (cf. after (31.51)), (31.55) becomes equivalent to

$$\rho > 1 - \eta, \qquad \text{where } \eta := \frac{2(q+1)}{n+1}k - (pq - 1). \tag{31.56}$$

Observe that $\eta > 0$ is equivalent to $k > (n+1)/\beta$ and, since $\beta > n - 1$, this is true for $\varepsilon = \varepsilon(p, q, n) > 0$ small in (31.41). We may thus choose $\rho = \rho(p, q, n) \in (0, 1)$ satisfying (31.54) and (31.56).

Step 5. Conclusion. We deduce from Step 4 that if (31.40) holds for some k satisfying (31.41) and (31.51), then (31.40) is true with k replaced by k/ρ. Starting from (31.39), we see that some value $\bar{k} > (n+1)pq/2(q+1)$ of k is reached after a finite number of steps. It then follows from the second paragraph in Step 2 that $\|u\|_{\bar{k}_1,\delta} \leq C$ for some $\bar{k}_1 > (n+1)q/2 \geq (n+1)p/2$.

By Step 3 with $k_1 := \bar{k}_1$ and $k_2 := \infty$, it follows that $\|v\|_\infty \leq C$. We may then apply Step 2 with $k := \bar{k}_1$ and $k_1 := \infty$ to conclude that $\|u\|_\infty \leq C$. The proof is complete. $\square$

As an application of the methods in this section, one obtains the following result [449] concerning the system

$$\left.\begin{aligned}
-\Delta u &= uv - au, & x &\in \Omega, \\
-\Delta v &= bu, & x &\in \Omega, \\
u = v &= 0, & x &\in \partial\Omega,
\end{aligned}\right\} \tag{31.57}$$

mentioned in Remark 31.1(i).

Theorem 31.17. *Assume Ω bounded, $a, b > 0$, and $n \leq 4$. Then any nonnegative very weak solution of* (31.57) *belongs to* $L^\infty \times L^\infty(\Omega)$ *and satisfies the a priori estimate* (31.7)*. Moreover, there exists a classical solution of* (31.57) *with $u, v > 0$.*

Sketch of proof (see [449] for details). We use a variant of Theorem 31.14. In fact, without assuming (31.32), the growth conditions (31.31), (31.33), (31.34) alone ensure that any very weak solution satisfies $u, v \in L^\infty \cap W^{2,m}(\Omega)$ for all finite m. Moreover, if we know an a priori estimate of u and v in $L^1_\delta(\Omega)$, then this implies an a priori estimate in $L^\infty(\Omega)$ (the only role of assumption (31.32) in Theorem 31.14 is to guarantee the L^1_δ-estimate).

Take $1 < r < p_{BT}$, $p = r/(r-1)$ and $q = 1$. Using $uv \leq v^p + u^r$, and noting that $\max(\alpha, \beta) = 2(p+1)/(pq-1) = 4r - 2 > n - 1$ for r close to p_{BT} due to $n < 5$, we see that $f = uv - au$, $g = bu$ satisfy (31.31), (31.33), (31.34).

On the other hand, the L^1_δ a priori estimate can be shown as follows. We have

$$-\Delta u = -b^{-1}v\Delta v - au \geq -\frac{b^{-1}}{2}\Delta(v^2) - au.$$

Testing this inequality and the second equation in (31.57) with φ_1, we obtain

$$(\lambda_1 + a)\int_\Omega u\varphi_1\, dx \geq \frac{b^{-1}\lambda_1}{2}\int_\Omega v^2\varphi_1\, dx \geq \frac{b^{-1}\lambda_1}{2}\left(\int_\Omega v\varphi_1\, dx\right)^2$$

$$= \frac{\lambda_1^{-1}b}{2}\left(\int_\Omega u\varphi_1\, dx\right)^2.$$

This implies the desired estimate. $\square$

Remarks 31.18. Comparison with other methods. (i) The method of Section 10 based on Hardy-Sobolev inequalities has also been extended to certain systems, see [141], [258], [135], [144]. Like the L_δ^p bootstrap method, it essentially requires only upper bounds on the growth of the nonlinearities f, g. However, the growth restrictions on the nonlinearities are much stronger, unlike in the scalar case (roughly, $\min(\alpha, \beta) > n - 1$ instead of $\max(\alpha, \beta) > n - 1$; cf. [135]). The reason for this is that the bootstrap procedure in that method is based on an $H^1 \times H^1$-estimate and is carried out simultaneously on the two components. Consequently, unlike in the above proof, the possible compensation effects between the two equations are not fully exploited.

(ii) Condition (31.33) also appears in the work [85], where existence and a priori estimates are studied for system (31.2) with extra (measure) terms added in the RHS and in the boundary conditions. The method in [85] is different from that described in this section. In particular, it uses maximum principle arguments to derive comparison estimates of the form $u^{q+1} \le C(1 + v^{p+1})$. In the case of system (31.1) (without measures in the RHS), it applies typically when $0 \le f \le C_2 v^p$ and $C_1 u^q \le g \le C_2 u^q$, with $C_2 \ge C_1 > 0$ and p, q satisfying (31.33). $\quad\square$

We now turn to the proof of Theorem 31.16. Like that of Theorem 11.5, it is mainly a consequence of Lemma 49.13, where a singular solution of the linear Laplace equation with an appropriate right-hand side belonging to L_δ^1 is constructed.

Proof of Theorem 31.16. Set $\phi := |x|^{-(\alpha+2)}\chi_\Sigma$ and $\psi := |x|^{-(\beta+2)}\chi_\Sigma$, with Σ as in Lemma 49.13 and let $u, v > 0$ be the very weak solutions of (47.8) with $f = \phi, \psi$, respectively. By (49.29), we have $u \notin L^\infty$, $v \notin L^\infty$ and

$$v^p \ge C|x|^{-\beta p}\chi_\Sigma = C|x|^{-(\alpha+2)}\chi_\Sigma = C\phi,$$

$$u^q \ge C'|x|^{-\alpha q}\chi_\Sigma = C'|x|^{-(\beta+2)}\chi_\Sigma = C'\psi.$$

Setting $a(x) = \phi/v^p \ge 0$, $b(x) = \psi/u^q \ge 0$, we get $-\Delta u = \phi = a(x)v^p$, $-\Delta v = \psi = b(x)u^q$ and $a(x) \le 1/C$, $b(x) \le 1/C'$ hence $a, b \in L^\infty$. The proof is complete. $\quad\square$

Remark 31.19. Localization of singularities. The observations in Remarks 11.6 extend to the case of systems. In particular, in spite of the imposed homogeneous Dirichlet boundary condition, the singularities of the solution in Theorem 31.15 occur at a (single) boundary point. In fact, when $n - 2 < \max(\alpha, \beta) < n - 1$, system (31.1) cannot have purely interior singularities. On the contrary, for $\max(\alpha, \beta) < n - 2$, examples of similar systems which possess unbounded weak solutions with purely interior singularities can be easily constructed. Namely the pair $(u, v) = (r^{-\alpha} - 1, r^{-\beta} - 1)$, $r = |x|$, is a weak solution of system (31.1) with $f = c_1(v + 1)^p$ and $g = c_2(u + 1)^q$ for $\Omega = B_1$ and suitable constants $c_1, c_2 > 0$ (note that the right-hand sides are in L^1). $\quad\square$

32. Parabolic systems coupled by power source terms

In this section, as a simple superlinear parabolic system and an analogue of the scalar model problem (15.1), we study the system:

$$
\left.
\begin{aligned}
u_t - \Delta u &= |v|^{p-1}v, & x \in \Omega,\ t > 0,\\
v_t - \Delta v &= |u|^{q-1}u, & x \in \Omega,\ t > 0,\\
u = v &= 0, & x \in \partial\Omega,\ t > 0,\\
u(x,0) &= u_0(x), & x \in \Omega,\\
v(x,0) &= v_0(x), & x \in \Omega,
\end{aligned}
\right\}
\tag{32.1}
$$

where $p, q > 0$. We set

$$
X = L^\infty \times L^\infty(\Omega) \quad \text{and} \quad X_+ = \{(u_0, v_0) \in X : u_0, v_0 \geq 0\}.
\tag{32.2}
$$

In all this section, when $pq > 1$, the scaling exponents α, β are defined by (31.3).

Assume $p, q \geq 1$. Then problem (32.1) is locally well-posed in X (see Example 51.12). In particular,

$$
\text{if } T_{\max} < \infty, \text{ then } \lim_{t \to T_{\max}} \left(\|u(t)\|_\infty + \|v(t)\|_\infty\right) = \infty.
\tag{32.3}
$$

Also the solution satisfies

$$
u, v \in BC^{2,1}(\overline{\Omega} \times [t_1, t_2]), \qquad 0 < t_1 < t_2 < T_{\max}.
\tag{32.4}
$$

Furthermore, problem (32.1) admits a comparison principle (cf. Proposition 52.22).

Next consider the case $p, q > 0$ and $\min(p, q) < 1$. For $(u_0, v_0) \in X$, local existence can be proved easily by approximation arguments (similar to those in the proof of Proposition 51.16 for instance). Turning to the question of (non-) uniqueness, which has been studied in [171], [172], let us assume $(u_0, v_0) \in X_+$, and Ω bounded or $\Omega = \mathbb{R}^n$. Local uniqueness is true in the class of nonnegative classical solutions if either $pq \geq 1$ or $(u_0, v_0) \neq (0,0)$, but the proof is nontrivial. On the contrary, there exist infinitely many nonnegative classical solutions if $pq < 1$ and $(u_0, v_0) = (0,0)$. On the other hand, if $p, q > 0$ and (u, v) is any maximal classical solution of (32.1) with existence time denoted by $T_{\max}$, then we still have (32.3) and (32.4).

32.1. Well-posedness and continuation in Lebesgue spaces

We consider system (32.1) with initial values in the space $Y = L^{r_1} \times L^{r_2}(\Omega)$. For $(u_0, v_0) \in Y$, by a local solution of (32.1) (on $[0, T]$), we understand a function $(u, v) \in C([0, T], Y)$ which is a classical solution of (32.1) for $0 < t \leq T$ and which fulfills the initial conditions. (Actually the nonexistence result below will still hold for a weaker notion of solution, see [446] for details.)

The optimal condition for local existence/nonexistence for system (32.1) can be expressed in terms of the numbers

$$\mathcal{P} = n\left(\frac{p}{r_2} - \frac{1}{r_1}\right), \qquad \mathcal{Q} = n\left(\frac{q}{r_1} - \frac{1}{r_2}\right).$$

Theorem 32.1. (i) *(Well-posedness) Let $p, q > 1$, $r_1, r_2 > 1$ and assume*

$$\max(\mathcal{P}, \mathcal{Q}) \leq 2.$$

For all $(u_0, v_0) \in L^{r_1} \times L^{r_2}(\Omega)$, there exist $T > 0$ and a unique local solution of system (32.1) on $[0, T]$.

(ii) *(Local nonexistence) Let $p, q > 0$, $r_1, r_2 \geq 1$ and assume*

$$\max(\mathcal{P}, \mathcal{Q}) > 2.$$

Then there exists $(u_0, v_0) \in L^{r_1} \times L^{r_2}(\Omega)$, $u_0,\ v_0 \geq 0$, such that system (32.1) admits no local solution (u, v) with $u,\ v \geq 0$.

As in Section 16, it is natural to look for sufficient conditions, in terms of L^r-bounds, guaranteeing global existence.

Theorem 32.2. *(Continuation) Let $p, q \geq 1$, $pq > 1$, $n \geq 2$ and assume Ω bounded. Let (u, v) be a maximal classical solution of (32.1) and denote by T its existence time. Assume that either*

$$r_1 > \frac{n}{\alpha} = \frac{n(pq - 1)}{2(p + 1)} \qquad and \qquad \sup_{(0,T)} \|u(t)\|_{r_1} < \infty,$$

or

$$r_2 > \frac{n}{\beta} = \frac{n(pq - 1)}{2(q + 1)} \qquad and \qquad \sup_{(0,T)} \|v(t)\|_{r_2} < \infty.$$

Then $T = \infty$.

Theorems 32.1 and 32.2 are from [446] and [447], respectively. Observe that the inequality $\max(\mathcal{P}, \mathcal{Q}) < 2$ implies $r_1 > n/\alpha$ and $r_2 > n/\beta$, but that this can be true also when $\max(\mathcal{P}, \mathcal{Q}) > 2$. Therefore, the continuation property is valid under weaker assumptions on r_1, r_2 than well-posedness. This is in sharp contrast with the situation in the scalar case (cf. Theorems 15.2 and 15.3, and Corollary 24.2). Note also that an L^r-bound on a single component is enough to guarantee global existence.

Remark 32.3. The gap between conditions guaranteeing well-posedness and continuation can be heuristically explained as follows. The final profiles of a solution around a blow-up point x_0 are expected to verify the lower estimates

$$u(x,T) \ge c_1|x - x_0|^{-\alpha}, \quad v(x,T) \ge c_2|x - x_0|^{-\beta}$$

for $|x-x_0|$ small (cf. Remark 32.12(ii) for a partial result), hence $(u(\cdot,T), v(\cdot,T)) \notin L^{r_1} \times L^{r_2}$ whenever $r_1 > n/\alpha$ or $r_2 > n/\beta$. On the other hand, if a local solution exists, then u_0 *and* v_0 have to satisfy suitable integral estimates as a consequence of the variation-of-constants formula (see (32.8) below), and this leads to necessary conditions involving r_1 *and* r_2 if (32.1) is well-posed in $L^{r_1} \times L^{r_2}$. $\square$

Theorem 32.1(i) is proved in Example 51.32 of Appendix E. As for the proof of Theorem 32.1(ii), the main ingredient is the following lemma, which provides lower estimates for certain *time-space averages* of solutions of the linear heat equation with positive singular initial data.

Lemma 32.4. *Assume* $0 < p < \infty$, $1 \le r_1, r_2 < \infty$ *and*

$$n\left(\frac{p}{r_2} - \frac{1}{r_1}\right) > 2.$$

Then there exists $v_0 \in L^{r_2}(\Omega)$, $v_0 \ge 0$, *such that*

$$\left\| \int_0^t e^{-(t-s)A}\left(e^{-sA}v_0\right)^p ds \right\|_{r_1} \to \infty, \quad as \ t \to 0^+.$$

Proof. Assume $B(0, 2\rho) \subset \Omega$, $\rho > 0$, let $k \in (0, n/r_2)$, and define

$$v_0(y) = |y|^{-k}\chi_{B(0,\rho)}(y).$$

Clearly, $v_0 \in L^{r_2}(\Omega)$. Using the heat kernel estimate in Proposition 49.10, we obtain, for $s > 0$ small and $|x| \le \sqrt{s}/2$,

$$\left(e^{-sA}v_0\right)(x) = \int_{|y|<\rho} G(x,y,s)|y|^{-k}\,dy \ge c_1 s^{-n/2} \int_{\{|y-x|<\sqrt{s}\}} |y|^{-k}\,dy$$

$$\ge c_1 s^{-n/2} \int_{\{|y|<\sqrt{s}/2\}} |y|^{-k}\,dy \ge cs^{-k/2}.$$

Consequently,

$$e^{-sA}v_0 \ge cs^{-k/2}\,\chi_{B(0,\sqrt{s}/2)}, \quad for \ s > 0 \ small. \tag{32.5}$$

Next, let $t/4 \leq s \leq t/2$, with $t > 0$ small, and $|x| \leq \sqrt{s}/2$. For $|y| \leq \sqrt{s}/2$ we have $|x - y| \leq \sqrt{t - s}$, hence $G(x, y, t - s) \geq c_1(t - s)^{-n/2} \geq c_1 s^{-n/2}$ by Proposition 49.10. It follows that

$$\left(e^{-(t-s)A}\chi_{B(0,\sqrt{s}/2)}\right)(x) = \int_{|y|<\sqrt{s}/2} G(x, y, t - s)\, dy \geq c > 0.$$

Combining this with (32.5), we deduce that, for $t > 0$ small,

$$e^{-(t-s)A}\left(e^{-sA}v_0\right)^P(x) \geq cs^{-kp/2} \geq ct^{-kp/2}, \quad t/4 \leq s \leq t/2, \quad |x| \leq \sqrt{s}/2. \tag{32.6}$$

Now if $|x| \leq \sqrt{t}/4$ and t is small, it follows from (32.6) that

$$\left(\int_0^t e^{-(t-s)A}\left(e^{-sA}v_0\right)^P ds\right)(x) \geq \left(\int_{t/4}^{t/2} e^{-(t-s)A}\left(e^{-sA}v_0\right)^P ds\right)(x) \geq Ct^{1-\frac{kp}{2}}$$

(note that $s \geq t/4$ implies $\sqrt{s}/2 \geq \sqrt{t}/4 \geq |x|$). Therefore, for $t > 0$ small, we obtain

$$\left\|\int_0^t e^{-(t-s)A}\left(e^{-sA}v_0\right)^P ds\right\|_{r_1}^{r_1}$$

$$\geq \int_{\{|x|\leq\sqrt{t}/4\}} \left(\int_0^t e^{-(t-s)A}\left(e^{-sA}v_0\right)^P ds\right)^{r_1}(x)\, dx \geq Ct^{\frac{n}{2}+r_1(1-\frac{kp}{2})}.$$

Since the assumption of the lemma implies $\frac{n}{2} + r_1\left(1 - \frac{np}{2r_2}\right) < 0$, the conclusion follows by choosing k sufficiently close to n/r_2. $\quad\square$

Proof of Theorem 32.1(ii). Similarly as in Remark 15.4(iii), any nonnegative solution of (32.1) in the sense of Theorem 32.1 satisfies the variation-of-constants formula:

$$\left.\begin{aligned} u(t) &= e^{-tA}u_0 + \int_0^t e^{-(t-s)A}|v(s)|^{p-1}v(s)\, ds, \quad 0 \leq t < T, \\ v(t) &= e^{-tA}v_0 + \int_0^t e^{-(t-s)A}|u(s)|^{q-1}u(s)\, ds, \quad 0 \leq t < T. \end{aligned}\right\} \tag{32.7}$$

In particular, we have

$$\begin{aligned} u(t) &\geq e^{-tA}u_0 \geq 0, \quad 0 \leq t < T, \\ v(t) &\geq e^{-tA}v_0 \geq 0, \quad 0 \leq t < T. \end{aligned}$$

It follows that

$$\left.\begin{aligned} 0 &\leq \int_0^t e^{-(t-s)A}(e^{-sA}v_0)^p\, ds \leq u(t), \\ 0 &\leq \int_0^t e^{-(t-s)A}(e^{-sA}u_0)^q\, ds \leq v(t). \end{aligned}\right\} \tag{32.8}$$

Since $(u, v) \in C([0, T], L^{r_1} \times L^{r_2})$, the right-hand sides in (32.8) remain bounded in L^{r_1} or L^{r_2}, respectively, hence

$$\left\| \int_0^t e^{-(t-s)A} \left(e^{-sA}v_0\right)^P ds \right\|_{r_1} + \left\| \int_0^t e^{-(t-s)A} \left(e^{-sA}u_0\right)^q ds \right\|_{r_2} \leq C, \quad 0 < t < T.$$

If either $\mathcal{P} > 2$ or $\mathcal{Q} > 2$, that is,

$$n\left(\frac{p}{r_2} - \frac{1}{r_1}\right) > 2 \quad \text{or} \quad n\left(\frac{q}{r_1} - \frac{1}{r_2}\right) > 2,$$

then, by choosing $u_0 \in L^{r_1}$ or $v_0 \in L^{r_2}$ as given by Lemma 32.4, we conclude that no solution of (32.1) can exist. $\square$

Proof of Theorem 32.2. We shall prove the result only for $n \geq 4$. The proof for $n = 2, 3$ is more involved and relies on suitable interpolation spaces. (However, the proof below applies also if $n = 3$ and $p, q \geq 2$, or if $n \leq 3$ and $r_1 > q - (1/p)$, $r_2 > p - (1/q)$.)

By Propositions 48.4 and 48.5, there exists $\omega > 0$ such that

$$\left\|e^{-tA}\right\|_{\mathcal{L}(L^{m_1}, L^{m_2})} \leq C_1 t^{-\frac{n}{2}\left(\frac{1}{m_1} - \frac{1}{m_2}\right)} e^{-\omega t}, \quad 1 \leq m_1 \leq m_2 \leq \infty. \tag{32.9}$$

By a time shift, we may assume that (u, v) is smooth up to $t = 0$. In particular, it satisfies the variation-of-constants formula (32.7).

We fix $\tau \in (0, T)$ and we denote

$$|u|_m := \sup_{t \in (0, \tau)} \|u(t)\|_m < \infty, \quad 1 \leq m \leq \infty$$

(and similarly for v). In the rest of the proof, C denotes a generic constant independent of τ. Assume that

$$|v|_r \leq C \tag{32.10}$$

for some

$$r > \frac{n(pq - 1)}{2(q + 1)}. \tag{32.11}$$

Let k, l satisfy

$$1 \leq k \leq l < \infty, \quad k \geq \frac{r}{p} \quad \text{and} \quad \frac{1}{k} - \frac{1}{l} < \frac{2}{n}.$$

By the first equation in (32.7) and the smoothing property (32.9) with $m_1 = k$, $m_2 = l$ it follows that

$$|u|_l \leq C(1 + \left\||v|^p\right\|_k) = C(1 + |v|_{kp}^p)$$

hence, by (32.10) and by the interpolation inequality,

$$|u|_l \leq C(1 + |v|_\infty^{p-(r/k)}). \qquad (32.12)$$

If in addition

$$l > \frac{nq}{2},$$

then the second equation in (32.7) and (32.9) with $m_1 = l/q$, $m_2 = \infty$ imply

$$|v|_\infty \leq C\big(1 + \big||u|^q\big|_{l/q}\big) = C(1 + |u|_l^q);$$

hence, by (32.12),

$$|v|_\infty \leq C(1 + |v|_\infty^{q(p-(r/k))}).$$

It follows that $|v|_\infty \leq C$ if

$$\frac{pq-1}{qr} < \frac{1}{k}.$$

The sufficient conditions are thus

$$\max\Big(0, \frac{1}{k} - \frac{2}{n}\Big) < \frac{1}{l} < \min\Big(\frac{2}{nq}, \frac{1}{k}\Big) \qquad (32.13)$$

and

$$\frac{pq-1}{qr} < \frac{1}{k} < \min\Big(1, \frac{p}{r}\Big). \qquad (32.14)$$

Condition (32.13) can be solved in l if

$$\frac{1}{k} - \frac{2}{n} < \frac{2}{nq},$$

i.e.,

$$\frac{1}{k} < \frac{2(q+1)}{nq}.$$

Since

$$\frac{pq-1}{qr} < \frac{p}{r},$$

it then suffices to satisfy

$$\frac{pq-1}{qr} < \frac{2(q+1)}{nq} \quad \text{and} \quad \frac{pq-1}{qr} < 1,$$

that is,

$$r > \frac{n(pq-1)}{2(q+1)} \quad \text{and} \quad r > p - \frac{1}{q}.$$

Finally, note that

$$\frac{n(pq-1)}{2(q+1)} \geq p - \frac{1}{q} \quad \text{if} \quad (n-2)q \geq 2,$$

which is true for all $q \geq 1$ if $n \geq 4$, and for $q \geq 2$ if $n = 3$. The hypothesis (32.11) thus implies the solvability of (32.13)–(32.14). Consequently, $|v|_\infty \leq C$, hence $|u|_\infty \leq C$ by the first equation in (32.7). Since C is independent of τ, we deduce that u and v are uniformly bounded on Q_T, hence $T = \infty$. $\square$

32.2. Blow-up and global existence

The following result provides the conditions on the exponents p, q which imply or prevent blow-up for system (32.1) in bounded domains.

Theorem 32.5. *Assume Ω bounded, $p, q > 0$, $(u_0, v_0) \in X_+$, and set $\tilde{p} = \min(p, 1)$, $\tilde{q} = \min(q, 1)$. Let (u, v) be a maximal classical solution of (32.1) and denote by T its existence time.*

(i) If $pq > 1$, then there exists $C(p, q, \Omega) > 0$ with the following property: If $\int_\Omega (u_0^{\tilde{q}} + v_0^{\tilde{p}})\varphi_1 \, dx > C(p, q, \Omega)$, then $T < \infty$.

(ii) If $pq \leq 1$, then $T = \infty$. Moreover, if $pq < 1$, then $u(t), v(t)$ are uniformly bounded for $t \geq 0$.

Theorem 32.5 is a modification of a result from [173] (see also [224], [226] for $p, q > 1$).

Proof. (i) Denote $y = y(t) := \int_\Omega u(t)\varphi_1 \, dx$, $z = z(t) := \int_\Omega v(t)\varphi_1 \, dx$. We may assume $q = \max(p, q) > 1$ without loss of generality. Multiplying the second equation in (32.1) with φ_1, we have

$$z' = \int_\Omega v_t \varphi_1 \, dx = \int_\Omega v \Delta \varphi_1 \, dx + \int_\Omega u^q \varphi_1 \, dx.$$

Using $\Delta \varphi_1 = -\lambda_1 \varphi_1$ and Jensen's inequality yields

$$z' \geq -\lambda_1 z + y^q. \tag{32.15}$$

We first consider the easier case $p > 1$. Similarly as above, we have

$$y' \geq -\lambda_1 y + z^p.$$

Therefore $\phi := y + z$ satisfies

$$\phi' = y' + z' \geq -\lambda_1 \phi + z^p + y^q \geq -\lambda_1 \phi + z^p + y^p - y \geq -(1 + \lambda_1)\phi + 2^{1-p}\phi^p.$$

It follows that $T < \infty$ whenever $\phi(0) > C(\lambda_1, p)$.

Next consider the case $p \leq 1$. In what follows, the constants $c_i > 0$ will depend only on p, q, Ω. Recall that (u, v) satisfies the variation-of-constants formula (32.7). By (15.24), for each $0 < \sigma < s < t$, we have $e^{-(s-\sigma)A}u^{pq}(\sigma) \leq \left(e^{-(s-\sigma)A}u^q(\sigma)\right)^p$, hence

$$\int_0^s e^{-(s-\sigma)A}u^{pq}(\sigma) \, d\sigma \leq \int_0^s \left(e^{-(s-\sigma)A}u^q(\sigma)\right)^p d\sigma$$

$$\leq s^{1-p}\left(\int_0^s e^{-(s-\sigma)A}u^q(\sigma) \, d\sigma\right)^p.$$

Using (32.7), we deduce that

$$u(t) \geq e^{-tA}u_0 + \int_0^t e^{-(t-s)A}\left(\int_0^s e^{-(s-\sigma)A}u^q(\sigma)\,d\sigma\right)^p ds$$

$$\geq e^{-tA}u_0 + t^{p-1}\int_0^t e^{-(t-s)A}\left(\int_0^s e^{-(s-\sigma)A}u^{pq}(\sigma)\,d\sigma\right) ds,$$

hence

$$u(t) \geq e^{-tA}u_0 + t^{p-1}\int_0^t\int_0^s e^{-(t-\sigma)A}u^{pq}(\sigma)\,d\sigma\,ds \tag{32.16}$$

by Fubini's theorem. Put $\gamma = pq > 1$. It follows from Jensen's inequality that

$$(e^{-(t-\sigma)A}u^\gamma(\sigma), \varphi_1) = e^{-\lambda_1(t-\sigma)}(u^\gamma(\sigma), \varphi_1) \geq e^{-\lambda_1(t-\sigma)}y^\gamma(\sigma).$$

Multiplying (32.16) with φ_1, we thus obtain

$$y(t) \geq e^{-\lambda_1 t}y(0) + t^{p-1}\int_0^t\int_0^s e^{-\lambda_1(t-\sigma)}y^\gamma(\sigma)\,d\sigma\,ds.$$

Assume that $T \geq 1$. We have

$$y(t) \geq c_1 y(0) + c_1\int_0^t\int_0^s y^\gamma(\sigma)\,d\sigma\,ds =: h(t), \quad 0 < t < 1,$$

hence

$$h''(t) \geq c_1 h^\gamma, \quad 0 < t < 1. \tag{32.17}$$

To conclude, it suffices to show that this inequality cannot be satisfied whenever

$$\int_\Omega (u_0 + v_0^p)\varphi_1\,dx \geq M, \quad \text{where } M = M(p, q, \Omega) \text{ is large enough.}$$

Multiplying (32.17) by $h' \geq 0$ and integrating, we have

$$h'^2(t) \geq c_2 h^{\gamma+1}(t) - c_3 y^{\gamma+1}(0), \quad 0 < t < 1. \tag{32.18}$$

On the other hand, using (32.7) and $p \leq 1$ again, we get

$$\tilde{z}(t) := \int_\Omega v^p(t)\varphi_1\,dx \geq \int_\Omega (e^{-tA}v_0)^p\varphi_1\,dx \geq \int_\Omega (e^{-tA}v_0^p)\varphi_1\,dx = e^{-\lambda_1 t}\tilde{z}(0)$$

and next,

$$y(t) = e^{-\lambda_1 t}y(0) + \int_0^t e^{-\lambda_1(t-s)}\int_\Omega v^p(s)\varphi_1\,dx\,ds \geq c_4(y(0) + t\,\tilde{z}(0)), \quad 0 < t < 1.$$

Therefore, since $h''(t) = c_1 y^\gamma$, we have

$$h(1/2) \geq c_5(y(0) + \tilde{z}(0))^\gamma. \tag{32.19}$$

Due to $\gamma > 1$, if $y(0) + \tilde{z}(0) \geq M$ (where M is large enough), we deduce from (32.18) that $h' \geq c_6 h^{(\gamma+1)/2}$ on $(1/2, 1)$, which contradicts (32.19) for M large.

(ii) Let us first assume $pq < 1$. Let Θ be the classical solution of (19.27), and put $M = \|\Theta\|_\infty$. We observe that $(\overline{u}, \overline{v}) = (a(1 + \Theta), b(1 + \Theta))$ is a supersolution, whenever the constants $a, b > 0$ satisfy $a \geq [b(1 + M)]^p$ and $b \geq [a(1 + M)]^q$. It is thus sufficient that $(1 + M)^p b^p \leq a \leq (1 + M)^{-1} b^{1/q}$. Since $p < 1/q$, for a given (u_0, v_0), one can take a, b as above and such that $a \geq \|u_0\|_\infty$, $b \geq \|u_0\|_\infty$. The assertion then follows from the comparison principle (note that since $\overline{u} \geq a > 0$ and $\overline{v} \geq b > 0$, it applies even though p, q may be < 1 — see Remark 52.11(c)).

Finally assume $pq = 1$, and $p \geq 1$ without loss of generality. We claim that for all $a > 0$, $(\overline{u}, \overline{v}) = (a^p e^{pt}, a e^t)$ is a supersolution. Indeed, this is equivalent to $pa^p e^{pt} \geq a^p e^{pt}$ and $ae^t \geq a^{pq} e^{pqt}$, which is true due to $pq = 1$ and $p \geq 1$. It then suffices to choose $a \geq \max(\|u_0\|_\infty^{1/p}, \|v_0\|_\infty)$. $\quad\square$

Remark 32.6. Results on a priori estimates and universal bounds for global positive solutions of (32.1) can be found in [448] (see also [425, Section 6]). $\quad\square$

32.3. Fujita-type results

In this subsection we consider nonnegative solutions of the Cauchy problem associated with (32.1), i.e.:

$$\left.\begin{aligned}
u_t - \Delta u &= v^p, & x \in \mathbb{R}^n,\ t > 0, \\
v_t - \Delta v &= u^q, & x \in \mathbb{R}^n,\ t > 0, \\
u(x, 0) &= u_0(x), & x \in \mathbb{R}^n, \\
v(x, 0) &= v_0(x), & x \in \mathbb{R}^n.
\end{aligned}\right\} \tag{32.20}$$

We give a Fujita-type result for problem (32.20), i.e. we find the (optimal) conditions on p, q depending on n, which guarantee that the solution blows up in finite time for all $u_0, v_0 \geq 0$, $(u_0, v_0) \not\equiv (0, 0)$.

Theorem 32.7. *Let $p, q > 0$ satisfy $pq > 1$, and let $(u_0, v_0) \in X_+$, $(u_0, v_0) \not\equiv (0, 0)$.*

(i) *If $\max(\alpha, \beta) \geq n$, then (32.20) admits no nontrivial global solution.*

(ii) *If $\max(\alpha, \beta) < n$, then (32.20) admits global, bounded solutions for suitably small initial data.*

This result is from [171]. We will prove it only under the additional assumption $p, q \geq 1$, and we will not treat the critical value $\max(\alpha, \beta) = n$. However in this special case, the present proof, based on arguments from [192], is considerably simpler than that in [171]. We shall use Gaussian test-functions of x and differential inequalities. See [372] for a different proof in the case $p, q \geq 1$, based on rescaled test-functions of x and t.

As a preliminary to the proof, we prepare the following lemma concerning the system of differential inequalities:

$$\left. \begin{aligned} y'(t) &\geq z^p - \lambda y, & t \geq 0, \\ z'(t) &\geq y^q - \lambda z, & t \geq 0. \end{aligned} \right\} \tag{32.21}$$

Lemma 32.8. *Let $p, q > 0$ satisfy $pq > 1$ and $\lambda > 0$. Then there exists $K = K(p,q) > 0$ such that (32.21) has no global nonnegative solution $y, z \in C([0,\infty)) \cap C^1((0,\infty))$ with $y(0) \geq K\lambda^{\alpha/2}$.*

Proof. Put $\tau = \lambda^{-1}$ and assume that (y, z) exists on $[0, \tau]$. Then there exists $C_1 = C_1(q) > 0$ such that

$$y(\tau) \geq C_1 y(0) \quad \text{and} \quad z(\tau) \geq C_1 \lambda^{-1} y^q(0). \tag{32.22}$$

Indeed, we have $(ye^{\lambda t})' \geq 0$, hence $y(t) \geq y(0)e^{-\lambda t} \geq e^{-1}y(0)$ on $[0, \tau]$. This implies $(ze^{\lambda t})' \geq e^{\lambda t}y^q(t) \geq e^{-q}y^q(0)$ on $[0, \tau]$, hence $z(\tau) \geq e^{-(q+1)}\lambda^{-1}y^q(0)$, and (32.22) follows.

Next, since $pq > 1$, we may choose $A, B > 1$ depending only on p, q, such that $p(B-1) > A$ and $q(A-1) > B$. We claim that if, for some t_0, there exist $a, b > 0$ such that

$$y(t_0) > a, \quad z(t_0) > b, \quad b^p > A\lambda a, \quad \text{and} \quad a^q > B\lambda b, \tag{32.23}$$

then $(y(t), z(t))$ cannot exist globally.

To prove the claim, assume for contradiction that (y, z) exists for all $t > 0$. By a time shift, we may assume $t_0 = 0$. Let $(\tilde{y}, \tilde{z})$ be the unique, positive local solution of

$$\left. \begin{aligned} \tilde{y}'(t) &= \tilde{z}^p - \lambda \tilde{y}, & t \geq 0, \\ \tilde{z}'(t) &= \tilde{y}^q - \lambda \tilde{z}, & t \geq 0, \\ \tilde{y}(0) &= a, \quad \tilde{z}(0) = b. \end{aligned} \right\}$$

By an easy comparison argument (using the fact that $z \mapsto z^p$ and $y \mapsto y^q$ are increasing functions), it follows that $(\tilde{y}, \tilde{z})$ exists for all $t > 0$ and we have $y(t) \geq \tilde{y}(t) > 0$ and $z(t) \geq \tilde{z}(t) > 0$. Set $\phi(t) = \tilde{z}^p - A\lambda \tilde{y}$ and $\psi(t) = \tilde{y}^q - B\lambda \tilde{z}$. We have $\phi(0) > 0$ and $\psi(0) > 0$ by (32.23). Assume that $\phi, \psi > 0$ on $[0, T]$ for some $T > 0$. Then $\tilde{y}' \geq (A-1)\lambda \tilde{y}$ and $\tilde{z}' \geq (B-1)\lambda \tilde{z}$ on $(0, T]$. On the other hand, for all $t \in (0, T]$, we have

$$\phi'(t) = p\tilde{z}^{p-1}\tilde{z}' - A\lambda \tilde{y}' \geq (p(B-1) - A)\lambda \tilde{z}^p > 0$$

and

$$\psi'(t) = q\tilde{y}^{q-1}\tilde{y}' - B\lambda\tilde{z}' \geq (q(A-1) - B)\lambda\tilde{y}^q > 0.$$

We deduce that $\phi, \psi > 0$ on $[0, \infty)$. It follows that $\tilde{y}'(t) \geq c\tilde{z}^p$ and $\tilde{z}'(t) \geq c\tilde{y}^q$ with $c = 1 - \max(A^{-1}, B^{-1}) > 0$. But, as a consequence of Lemma 32.10 below, this guarantees that $(\tilde{y}, \tilde{z})$ cannot exist for all $t > 0$. This contradiction proves the claim.

Let us now show that, for suitable $K, \varepsilon, \eta > 0$ (independent of λ), $y(0) \geq K\lambda^{\alpha/2}$ guarantees that $a := \varepsilon\lambda^{\alpha/2}$ and $b := \eta\lambda^{\beta/2}$ satisfy (32.23) for $t_0 = \tau$. In view of the last claim, this will prove the lemma. The last two conditions in (32.23) are equivalent to

$$\eta^p\lambda^{\frac{p(q+1)}{pq-1}} > A\lambda\varepsilon\lambda^{\frac{p+1}{pq-1}} = A\varepsilon\lambda^{\frac{p(q+1)}{pq-1}}, \quad \varepsilon^q\lambda^{\frac{q(p+1)}{pq-1}} > B\lambda\eta\lambda^{\frac{q+1}{pq-1}} = B\eta\lambda^{\frac{q(p+1)}{pq-1}},$$

that is $\eta^p > A\varepsilon$ and $\varepsilon^q > B\eta$; such $\eta, \varepsilon > 0$ clearly exist since $pq > 1$. Due to (32.22), the first two conditions in (32.23) are satisfied if

$$\varepsilon\lambda^{\frac{p+1}{pq-1}} < C_1 K\lambda^{\frac{p+1}{pq-1}}, \quad \eta\lambda^{\frac{q+1}{pq-1}} < C_1\lambda^{-1}K^q\lambda^{\frac{q(p+1)}{pq-1}} = C_1 K^q\lambda^{\frac{q+1}{pq-1}}.$$

It thus suffices to choose $K > \max(C_1^{-1}\varepsilon, C_1^{-1/q}\eta^{1/q})$. $\quad\square$

Proof of Theorem 32.7. (i) Without loss of generality, we may assume $p \geq q$. As mentioned before, we shall prove the assertion under the stronger assumptions $p \geq q \geq 1$ ($p > 1$) and $\max(\alpha, \beta) = \alpha > n$. For each $\lambda > 0$, let $\phi_\lambda(x) = (4\pi)^{-n/2}\lambda^{n/2}e^{-\lambda|x|^2/4}$. We have

$$\partial_{x_i}\phi_\lambda = \frac{-\lambda x_i}{2}\phi_\lambda, \quad \partial^2_{x_i x_i}\phi_\lambda = \left[\frac{\lambda^2 x_i^2}{4} - \frac{\lambda}{2}\right]\phi_\lambda, \quad \text{hence} \quad \Delta\phi_\lambda \geq \frac{-n\lambda}{2}\phi_\lambda,$$

and $\int_{\mathbb{R}^n}\phi_\lambda = 1$. Multiplying the differential equations in (32.1) by ϕ_λ, integrating by parts, and using Jensen's inequality, we obtain

$$\frac{d}{dt}\int_{\mathbb{R}^n} u\phi_\lambda\,dx = \int_{\mathbb{R}^n} u\Delta\phi_\lambda\,dx + \int_{\mathbb{R}^n} v^p\phi_\lambda\,dx \geq -\frac{n\lambda}{2}\int_{\mathbb{R}^n} u\phi_\lambda\,dx + \left(\int_{\mathbb{R}^n} v\phi_\lambda\,dx\right)^p$$

and similarly

$$\frac{d}{dt}\int_{\mathbb{R}^n} v\phi_\lambda\,dx \geq -\frac{n\lambda}{2}\int_{\mathbb{R}^n} v\phi_\lambda\,dx + \left(\int_{\mathbb{R}^n} u\phi_\lambda\,dx\right)^q.$$

(the calculations can be justified similarly as in the proof of Theorem 17.1). Therefore, the functions

$$y_\lambda(t) := \int_{\mathbb{R}^n} u(t)\phi_\lambda\,dx \quad \text{and} \quad z_\lambda(t) := \int_{\mathbb{R}^n} v(t)\phi_\lambda\,dx$$

satisfy system (32.21) with λ replaced by $\tilde{\lambda} := n\lambda/2$. By shifting the time origin, we may assume $u_0 \not\equiv 0$. Consequently, since $\int_{\mathbb{R}^n} e^{-\lambda|x|^2/4} u_0\, dx \to \int_{\mathbb{R}^n} u_0\, dx \in (0, \infty]$ as $\lambda \to 0$, there exists $c_0 > 0$ such that $y_\lambda(0) \geq c_0 \lambda^{n/2}$ for $\lambda > 0$ small. Since $\alpha > n$, we have $y_\lambda(0) \geq K\tilde{\lambda}^{\alpha/2}$ for $\lambda > 0$ small, where K is given by Lemma 32.8. We then deduce from that lemma that (y_λ, z_λ), hence (u, v), cannot exist globally.

(ii) We assume $p \geq q \geq 1$, $p > 1$ and $\alpha < n$. We look for a supersolution under the form

$$\overline{u}(x,t) = \varepsilon(t+1)^a \phi(x,t), \quad \overline{v}(x,t) = \varepsilon(t+1)^b \phi(x,t),$$

with $a, b, \varepsilon > 0$ and $\phi(x,t) = (t+1)^{-n/2}\psi(x,t)$, where $\psi(x,t) = e^{-|x|^2/4(t+1)}$. Using $\phi_t - \Delta\phi = 0$ and $\psi \leq 1$, we obtain

$$\overline{u}_t - \Delta\overline{u} - \overline{v}^p = a\varepsilon(t+1)^{a-1}\phi - \varepsilon^p(t+1)^{bp}\phi^p$$

$$= [a(t+1)^{a-bp-1+n(p-1)/2} - \varepsilon^{p-1}\psi^{p-1}]\varepsilon(t+1)^{bp-pn/2}\psi \geq 0$$

for $t \geq 0$, whenever

$$a - bp \geq 1 - n(p-1)/2 \quad \text{and} \quad \varepsilon^{p-1} \leq a. \tag{32.24}$$

Symmetrically, we have $\overline{v}_t - \Delta\overline{v} - \overline{u}^q \geq 0$ whenever

$$b - aq \geq 1 - n(q-1)/2 \quad \text{and} \quad \varepsilon^{q-1} \leq b. \tag{32.25}$$

Choosing $a = \frac{n}{2} - \frac{p+1}{pq-1}$ and $b = \frac{n}{2} - \frac{q+1}{pq-1}$, the first conditions in (32.24) and (32.25) are satisfied (with equalities) and since $a, b > 0$ due to $\max(\alpha, \beta) = \alpha < n$, it suffices to choose $\varepsilon > 0$ small. It then follows from the comparison principle that (u, v) is global if $u_0 \leq \overline{u}(\cdot, 0)$ and $v_0 \leq \overline{v}(\cdot, 0)$. $\square$

32.4. Blow-up asymptotics

As compared with the scalar model problem (15.1), less is known concerning the asymptotic behavior of blowing-up solutions to system (32.1). We shall establish the following theorem concerning type I blow-up rate for monotone solutions in time. A few results concerning other aspects of the blow-up behavior will be mentioned in Remarks 32.12.

Theorem 32.9. *Consider problem (32.1) with Ω bounded, $p, q \geq 1$, $pq > 1$, and $0 \leq u_0, v_0 \in L^\infty(\Omega)$. Assume that u, v are nondecreasing in time and (u, v) is nonstationary. Then $T := T_{\max}(u_0, v_0) < \infty$ and we have*

$$\left.\begin{array}{l} C_1(T-t)^{-\alpha/2} \leq \|u(t)\|_\infty \leq C_2(T-t)^{-\alpha/2}, \\[2mm] C_3(T-t)^{-\beta/2} \leq \|v(t)\|_\infty \leq C_4(T-t)^{-\beta/2}, \end{array}\right\} \tag{32.26}$$

in $(0, T)$ for some $C_1, C_2, C_3, C_4 > 0$.

This result was proved in [158] and its proof is based on a modification of the maximum principle arguments of [219]. The assumption $u_t, v_t \geq 0$ is guaranteed if, for instance, $0 \leq u_0, v_0 \in C_0 \cap C^2(\Omega)$ and $\Delta u_0 + v_0^p \geq 0$, $\Delta v_0 + u_0^q \geq 0$ (see Remark 52.23(ii)).

For the proof, we need the following lemmas concerning the systems of differential inequalities:

$$\left. \begin{aligned} y'(t) &\geq \varepsilon z^p, \\ z'(t) &\geq \varepsilon y^q, \end{aligned} \right\} \tag{32.27}$$

and

$$\left. \begin{aligned} y'(t) &\leq z^p, \\ z'(t) &\leq y^q. \end{aligned} \right\} \tag{32.28}$$

Lemma 32.10. *Let $p, q, \varepsilon > 0$ satisfy $pq > 1$, and let $0 < T \leq \infty$. Assume that $0 \leq y, z \in C^1(0, T)$, $(y, z) \not\equiv (0, 0)$, and that (y, z) solves (32.27) on $(0, T)$. Then $T < \infty$ and there holds*

$$y(t) \leq C_1 (T - t)^{-\alpha/2}, \quad z(t) \leq C_1 (T - t)^{-\beta/2}, \quad 0 < t < T, \tag{32.29}$$

with $C_1 = C_1(p, q, \varepsilon) > 0$.

Lemma 32.11. *Let $p, q > 0$ satisfy $pq > 1$, and let $0 < T < \infty$. Assume that $0 \leq y, z$ are locally absolutely continuous and nondecreasing on $(0, T)$, and that (y, z) solves (32.28) a.e. on $(0, T)$. Assume also that $\sup_{(0,T)}(y + z) = \infty$ and that (32.29) is satisfied for some $C_1 > 0$. Then there holds*

$$y(t) \geq \eta (T - t)^{-\alpha/2}, \quad z(t) \geq \eta (T - t)^{-\beta/2}, \quad T - \eta < t < T,$$

with $\eta = \eta(p, q, C_1) > 0$.

Proof of Lemma 32.10. We have

$$\varepsilon^{-p-1} y(t) \geq \varepsilon^{-p} \int_0^t z^p(s)\, ds \geq \int_0^t \left(\int_0^s y^q(\sigma)\, d\sigma \right)^p ds =: h(t).$$

Therefore,

$$[(h')^{(p+1)/p}]' = (p+1) \left(\int_0^t y^q(s)\, ds \right)^p y^q(t) \geq (p+1)\varepsilon^{q(p+1)} h' h^q = C(h^{q+1})'.$$

Since $h(0) = h'(0) = 0$, it follows that $(h')^{(p+1)/p} \geq C h^{q+1}$. Moreover, due to $(y, z) \not\equiv (0, 0)$, we may assume $h > 0$ on (t_0, T) for some $t_0 \in (0, T)$. Putting $\gamma = p\frac{q+1}{p+1} > 1$, we get $[h^{1-\gamma}]' = -(\gamma - 1)h' h^{-\gamma} \leq -C < 0$. Integrating over (t, s)

with $t_0 < t < s < T$, we obtain $h^{1-\gamma}(t) \geq h^{1-\gamma}(s) + C(s-t) \geq C(s-t)$. It follows that $T < \infty$. By letting $s \to T$, we obtain

$$h(t) \leq C(T-t)^{-1/(\gamma-1)} = C(T-t)^{-\alpha/2}, \quad t_0 < t < T. \tag{32.30}$$

Next, fix $t_0 < t < T$ and let $\tau = (T-t)/4$. Since $y' \geq 0$, we have

$$h(t+2\tau) = \int_0^{t+2\tau} \left(\int_0^s y^q(\sigma)\, d\sigma \right)^p ds$$
$$\geq \tau \left(\int_0^{t+\tau} y^q(\sigma)\, d\sigma \right)^p \geq \tau(\tau y^q(t))^p = \tau^{p+1} y^{pq}(t).$$

In view of (32.30), we deduce

$$y^{pq}(t) \leq \tau^{-(p+1)} h(t+2\tau) \leq C\tau^{-(p+1)}(T-t-2\tau)^{-(p+1)/(pq-1)}$$
$$= C(T-t)^{-pq(p+1)/(pq-1)},$$

hence the estimate of y on (t_0, T). The estimate of z follows symmetrically. Since the constant C is independent of t_0 and $y = z = 0$ in $(0, t)$ if $h(t) = 0$, the estimates above (obtained in (t_0, T)) remain true in $(0, T)$. $\square$

Proof of Lemma 32.11. We first observe that for suitable $a, b > 0$ (depending on p, q) the functions

$$\overline{y}(t) := a(T-t)^{-\alpha/2}, \quad \overline{z}(t) := b(T-t)^{-\beta/2}$$

satisfy $\overline{y}'(t) = \overline{z}^p(t)$, $\overline{z}'(t) = \overline{y}^q(t)$ on $(0, T)$. We deduce that, for each $t \in (0, T)$,

$$\text{either } y(t) \geq \overline{y}(t) \text{ or } z(t) \geq \overline{z}(t). \tag{32.31}$$

(Indeed, if this failed for some $t \in (0, T)$, then we would have $y(t) < \overline{y}(t - \eta)$ and $z(t) < \overline{z}(t - \eta)$ for some $\eta > 0$ so that, by a simple comparison argument, $y(s) \leq \overline{y}(s - \eta)$ and $z(s) \leq \overline{z}(s - \eta)$, $t \leq s < T$, contradicting the fact that (y, z) is unbounded on $(0, T)$.)

Assume for contradiction that there exist sequences $\eta_i \to 0+$ and $t_i \to T$ such that
$$z(t_i) \leq \eta_i(T - t_i)^{-\beta/2}.$$

Fix $K > 1$ and put $t_i' := t_i - K(T - t_i)$. Then (32.31), (32.29) and $z' \geq 0$ guarantee that, for large i,

$$a(T - t_i)^{-\alpha/2} \leq y(t_i) \leq y(t_i') + \int_{t_i'}^{t_i} z^p(s)\, ds \leq C_1(T - t_i')^{-\alpha/2} + K\eta_i^p(T - t_i)^{1 - p(\beta/2)}.$$

Using $1 - p(\beta/2) = -\alpha/2$ and noting that $T - t_i' = (1 + K)(T - t_i)$, we get $a \leq C_1(1 + K)^{-\alpha/2} + K\eta_i^p$. Letting $i \to \infty$, we get a contradiction for $K = K(p, q, a)$ large enough. Consequently, there exists $\eta = \eta(p, q) > 0$ such that $z(t) \geq \eta(T - t)^{-\beta/2}$ on $[T - \eta, T)$. The estimate for y follows symmetrically. $\quad\square$

Proof of Theorem 32.9. We first prove the upper estimates. Using the maximum principle in a similar way as in the proof of Theorem 23.5, we obtain $u_t, v_t > 0$ in Q_T and $\partial_\nu u_t, \partial_\nu v_t < 0$ on S_T. Choosing $\tau \in (0, T)$ we can find $\varepsilon > 0$ such that $u_t(x, \tau) \geq \varepsilon v^p(x, \tau)$ and $v_t(x, \tau) \geq \varepsilon u^q(x, \tau)$ for all $x \in \Omega$. Set $f = f(v) := v^p$, $g = g(u) := u^q$ and $J := u_t - \varepsilon f$, $H := v_t - \varepsilon g$. Then

$$J_t - \Delta J = f' v_t - \varepsilon f' g + \varepsilon f'' |\nabla v|^2,$$

hence

$$J_t - \Delta J \geq f' H \qquad \text{in } Q^\tau \tag{32.32}$$

where $Q^\tau := \Omega \times (\tau, T)$, and symmetrically

$$H_t - \Delta H \geq g' J \qquad \text{in } Q^\tau. \tag{32.33}$$

Since $f'(v)$ and $g'(u)$ and nonnegative and locally bounded in $\overline{\Omega} \times [\tau, T)$, we may apply the maximum principle (Proposition 52.21) to system (32.32)–(32.33). As $J, H \geq 0$ on the parabolic boundary of Q^τ, we thus have $J, H \geq 0$ in Q^τ. Consequently, $u_t \geq \varepsilon v^p$, $v_t \geq \varepsilon u^q$ in Q^τ. Applying Lemma 32.10 with $y(t) = u(x, t)$, $z(t) = v(x, t)$ (for each fixed $x \in \Omega$) yields $T < \infty$ and the upper estimates in (32.26).

Let us turn to the lower estimates. We now set

$$y(t) = \max_{x \in \overline{\Omega}} u(x, t), \qquad z(t) = \max_{x \in \overline{\Omega}} v(x, t).$$

Arguing as in the (alternative) proof of Proposition 23.1, we obtain $y' \leq z^p$ and $z' \leq y^q$ a.e. in $(0, T)$. Consequently, the lower estimates in (32.26) are guaranteed by Lemma 32.11. $\quad\square$

Remarks 32.12. (i) **Blow-up rate.** The blow-up estimates in Theorem 32.9 were obtained before in [110] under stronger restrictions on p, q and the initial data. On the other hand, when $\Omega = \mathbb{R}^n$, (32.26) is valid for all nonglobal nonnegative solutions if $p, q > 1$ satisfy $\max(\alpha, \beta) \geq n$ [130]. This remains true for general domains if $\max(\alpha, \beta) \geq n + 1$ [199]. The proofs rely on rescaling arguments and Fujita-type theorems (cf. Remark 26.12). In the case $\Omega = \mathbb{R}^n$ and $\max(\alpha, \beta) > n$, the upper estimate was proved before in [29] by different arguments based on Moser-type iteration.

(ii) **Blow-up set.** Concerning the blow-up set for problem (32.1), the following has been proved recently in [493]. Let $\Omega = B_R$ and $p, q > 1$. Assume that $u, v \geq 0$

are radially symmetric and decreasing in $|x|$. If (u,v) blows up in finite time and satisfies the upper blow-up estimates in (32.26), then single-point blow-up occurs at $x=0$. If, moreover, $u_t, v_t \geq 0$, then the final blow-up profiles satisfy the lower estimates

$$u(x,T) \geq c_1 |x|^{-\alpha}, \quad v(x,T) \geq c_2 |x|^{-\beta}$$

for $|x|$ small. In the special case $p = q > 1$ and $n = 1$, an earlier result on single-point blow-up appeared in [216]. On the other hand, fine asymptotic properties of blow-up solutions of (32.1), including a classification of blow-up profiles, have been obtained in [29], [540] when $\Omega = \mathbb{R}^n$ under the assumption that $|p - q| \ll 1$.

(iii) **Nonsimultaneous blow-up.** For system (32.1), it is easy to see that blow-up is always simultaneous: If $T = T_{\max} < \infty$, then both components blow up in the sense that

$$\limsup_{t \to T} \|u(t)\|_\infty = \limsup_{t \to T} \|v(t)\|_\infty = \infty.$$

Indeed if u, say, were uniformly bounded on Q_T, then the second equation would yield a uniform bound on v, hence contradicting (32.3). For different systems with a weaker coupling, **nonsimultaneous blow-up** may occur. For instance, if the nonlinearities in (32.1) are replaced with $f(u,v) = u^r v^p$, $g(u,v) = v^s u^q$, or with $f(u,v) = u^r + v^p$, $g(u,v) = v^s + u^q$, then for suitable $p, q, r, s > 0$ and suitable nonnegative initial data, one component may blow up while the other remains bounded (see [466, pp. 467-472], [432], [494], [458]). $\square$

33. The role of diffusion in blow-up

In this section, we discuss the different possible effects of adding linear diffusion (and some boundary conditions) to a system of ODE's. It will turn out that quite opposite effects can be observed:

a. In the case of an ODE system whose solutions all exist globally, it can either happen that:

- diffusion preserves global existence (for all initial data),

or that:

- diffusion induces blow-up (for some initial data).

b. Consider the case of ODE systems for which (at least some) solutions blow up in finite time. We already know examples where the addition of diffusion (even with homogeneous Dirichlet conditions) will not prevent the blow-up of (some) solutions (cf. Theorem 32.5). Of course, we have encountered in Section 17 a similar situation in the scalar case. We will see that at the opposite, for certain such ODE systems, the addition of diffusion and homogeneous Dirichlet conditions can make all solutions global and bounded (again, a similar example in the scalar case was given in Section 19).

All the systems appearing in this section are locally well-posed under the assumptions that will be made on the data (this will be a consequence of Example 51.12). The existence time of the unique, maximal, classical solution is denoted by $T_{\max}$ or $T_{\max}(u_0, v_0)$, and the continuation and regularity properties (32.3) and (32.4) are valid. Also, we only consider nonnegative initial data and solutions. On the other hand, the systems in this section do not satisfy the comparison principle in general, and we shall need to rely on other techniques.

33.1. Diffusion preserving global existence

Let us consider the following system

$$\left.\begin{aligned}
u_t - a\Delta u &= f(u, v), & x \in \Omega, \ t > 0,\\
v_t - b\Delta v &= g(u, v), & x \in \Omega, \ t > 0,\\
u_\nu = v_\nu &= 0, & x \in \partial\Omega, \ t > 0,\\
u(x, 0) &= u_0(x), & x \in \Omega,\\
v(x, 0) &= v_0(x), & x \in \Omega,
\end{aligned}\right\} \tag{33.1}$$

where a, b are positive constants. Here Ω is either a bounded domain or the whole of $\mathbb{R}^n$ (in which case the boundary conditions are of course empty), and $(u_0, v_0) \in X_+$, defined in (32.2). We assume that $f, g : [0, \infty)^2 \to \mathbb{R}$ are C^1-functions and that they satisfy

$$f(0, v), \ g(u, 0) \geq 0, \quad \text{for } u, v \geq 0, \tag{33.2}$$

which ensures that system (33.1) preserves positivity. (Indeed, we may extend f to $\mathbb{R}^2$ by $f(u, v) = f(|u|, |v|)$, and (33.2) then implies $u_t - a\Delta u \geq b_1(x, t)u$, where $b_1 = f'_u(\theta u, v)$, $\theta = \theta(x, t) \in (0, 1)$, and similarly for v.)

In this subsection, we consider two classes of systems of the form (33.1): systems with dissipation of mass and systems of Gierer-Meinhardt type.

Systems with dissipation of mass

This class corresponds to nonlinearities satisfying the structure condition

$$f(u, v) + g(u, v) \leq 0, \quad \text{for all } u, v \geq 0. \tag{33.3}$$

In case Ω is bounded, condition (33.3) guarantees that system (33.1) possesses the so-called mass-dissipation property:

$$t \mapsto M(t) \ \text{ is nonincreasing, where } M(t) := \int_\Omega u(x, t)\, dx + \int_\Omega v(x, t)\, dx.$$

Indeed, this follows immediately by integrating the differential equations in (33.1) over Ω and using the boundary conditions. This property is natural in the context of chemical or biological applications, where systems of this form arise. If one looks at the corresponding kinetic system without diffusion, i.e. the ODE counterpart of (33.1):

$$\left.\begin{aligned} U' &= f(U,V), & t &> 0, \\ V' &= g(U,V), & t &> 0, \\ U(0) &= U_0 \geq 0, \\ V(0) &= V_0 \geq 0, \end{aligned}\right\} \tag{33.4}$$

then it is clear that solutions of (33.4) are global since $0 \leq U(t) + V(t) \leq U_0 + V_0$. A central issue is to determine whether or not the mass-dissipation structure condition still guarantees the global existence of solutions for the diffusive system (33.1). In the case of equal diffusions $a = b$, it is easy to see that the answer is yes. Indeed, the function $w = u + v$ then satisfies

$$w_t - a\Delta w = f + g \leq 0,$$

so $0 \leq u + v \leq \|u_0\|_\infty + \|v_0\|_\infty$ by the maximum principle and global existence follows. In the case of different diffusions $a \neq b$, a case often encountered in applications, this has long remained open and has motivated a large amount of work, along with related questions (see e.g. the survey article [348] and references therein). It will turn out that the answer is no in general (see Theorem 33.12 and the preceding comments). However, we shall now see that global existence is ensured if we make some additional assumption.

An important particular case is that when $f \leq 0$, which means that the first substance is absorbed by the reaction (systems with so-called "triangular" structure). Then one immediately obtains a uniform bound for u, since

$$u(x,t) \leq \|u_0\|_\infty, \quad x \in \overline{\Omega}, \; t \in (0, T_{\max}) \tag{33.5}$$

by the maximum principle. The problem is then reduced to obtaining a uniform estimate of v.

A simple case when this can be done is when $a > b$. This means that the absorbed substance diffuses faster than the other one. The following result for $\Omega = \mathbb{R}^n$ was proved in [348]. A similar result was obtained in [295] for Ω bounded, but the proof is more delicate.

Theorem 33.1. *Let $\Omega = \mathbb{R}^n$, $a > b > 0$ and assume*

$$f(u,v) \leq 0 \leq g(u,v), \quad \text{for all } u, v \geq 0, \tag{33.6}$$

along with (33.2), (33.3). Then, for all $(u_0, v_0) \in X_+$, the solution of problem (33.1) is global. Moreover, u, v are uniformly bounded in $\mathbb{R}^n \times [0, \infty)$.

The proof is based on a simple comparison property concerning the kernels associated with the operators $\partial_t - a\Delta$ and $\partial_t - b\Delta$.

Proof. Let us denote by $S_\lambda(t)$ the semigroup (say, on $L^\infty(\mathbb{R}^n)$) corresponding to the operator $\partial_t - \lambda\Delta$. We observe that for all $0 \le \phi \in L^\infty(\mathbb{R}^n)$,

$$\lambda \mapsto \lambda^{n/2}[S_\lambda(t)\phi](x) \quad \text{is nondecreasing for all } (x,t). \tag{33.7}$$

This follows from the fact that

$$[S_\lambda(t)\phi](x) = (4\pi\lambda t)^{-n/2}\int_{\mathbb{R}^n} \exp[-|y|^2/4\lambda t]\phi(x-y)\,dy.$$

Denoting

$$z_a(t) = -\int_0^t S_a(t-s)f(u(s),v(s))\,ds, \qquad z_b(t) = \int_0^t S_b(t-s)g(u(s),v(s))\,ds,$$

we have

$$u(t) + z_a(t) = S_a(t)u_0$$

hence $z_a(t) \le S_a(t)u_0 \le \|u_0\|_\infty$. Due to $f + g \le 0$, $f \le 0$ and (33.7), it follows that

$$z_b(t) \le -\int_0^t S_b(t-s)f(u(s),v(s))\,ds < (a/b)^{n/2}z_a(t) \le (a/b)^{n/2}\|u_0\|_\infty.$$

Therefore

$$v(t) = S_b(t)v_0 + z_b(t) \le \|v_0\|_\infty + (a/b)^{n/2}\|u_0\|_\infty.$$

This along with (33.5) yields the conclusion. $\square$

Still in the case $f \le 0$ but without assuming $a > b$, the answer is again positive under a polynomial growth assumption on g:

$$g(u,v) \le C(1 + u + v)^\gamma, \quad \text{for all } u,v \ge 0 \text{ and some } \gamma \ge 1. \tag{33.8}$$

Theorem 33.2. *Assume Ω bounded and let $a,b > 0$, $a \ne b$, and $\gamma \ge 1$. Assume (33.2), (33.3), (33.6) and (33.8). Then, for all $(u_0, v_0) \in X_+$, the solution of problem (33.1) is global.*

This result was proved in [281]. It can be shown in addition that u, v are uniformly bounded in $\overline{\Omega} \times [0, \infty)$.

The main ingredient of the proof is the following lemma, which guarantees that whenever $f + g \le 0$, v is controlled by u in L^p for any finite p. The proof is based on a duality argument.

Lemma 33.3. *Assume Ω bounded, $1 < p < \infty$, $a, b > 0$ and $T > 0$. There exists $C = C(T, p, a, b, \Omega) > 0$ such that, if $u, v \in C^{2,1}(\overline{\Omega} \times (0, T]) \cap C(\overline{Q_T})$ satisfy*

$$\left.\begin{aligned}
(u + v)_t - a\Delta u - b\Delta v &\leq 0, & x &\in \Omega, \ 0 < t < T, \\
u_\nu = v_\nu &= 0, & x &\in \partial\Omega, \ 0 < t < T,
\end{aligned}\right\} \tag{33.9}$$

then there holds

$$\|v_+\|_{L^p(Q_T)} \leq C\big(\|u(\cdot, 0) + v(\cdot, 0)\|_{L^p(\Omega)} + \|u\|_{L^p(Q_T)}\big). \tag{33.10}$$

Proof. Let $q = p'$. Fix $\chi \in \mathcal{D}(Q_T)$, $\chi \geq 0$, and let φ be the unique solution of the problem

$$\left.\begin{aligned}
-\varphi_t - b\Delta\varphi &= \chi, & x &\in \Omega, \ 0 < t < T, \\
\varphi_\nu &= 0, & x &\in \partial\Omega, \ 0 < t < T, \\
\varphi(x, T) &= 0, & x &\in \Omega.
\end{aligned}\right\} \tag{33.11}$$

We have $\varphi \geq 0$ by the maximum principle. Parabolic L^q-estimates (see Remark 48.3(ii)) guarantee

$$\|\varphi_t\|_{L^q(Q_T)} + \|D^2\varphi\|_{L^q(Q_T)} \leq C\|\chi\|_{L^q(Q_T)}. \tag{33.12}$$

Since $\varphi(\cdot, T) = 0$, this implies in particular

$$\|\varphi(\cdot, 0)\|_{L^q(\Omega)} \leq C\|\chi\|_{L^q(Q_T)}. \tag{33.13}$$

Multiplying the inequality in (33.9) by φ, integrating by parts, and using the boundary conditions and $\varphi(x, T) = 0$, we obtain

$$\begin{aligned}
0 \geq &\int\!\!\int_{Q_T} (u_t + v_t - a\Delta u - b\Delta v)\varphi \, dx \, dt \\
= &\int\!\!\int_{Q_T} \big(u(-\varphi_t - a\Delta\varphi) + v(-\varphi_t - b\Delta\varphi)\big) \, dx \, dt \\
&- \int_\Omega (u(x, 0) + v(x, 0))\varphi(x, 0) \, dx.
\end{aligned}$$

Therefore, by (33.12) and (33.13), we get

$$\begin{aligned}
\int\!\!\int_{Q_T} v\chi \, dx &= \int\!\!\int_{Q_T} v(-\varphi_t - b\Delta\varphi) \\
&\leq \int\!\!\int_{Q_T} u(\varphi_t + a\Delta\varphi) \, dx \, dt + \int_\Omega (u(x, 0) + v(x, 0))\varphi(x, 0) \, dx \\
&\leq C\big(\|u\|_{L^p(Q_T)} + \|u(\cdot, 0) + v(\cdot, 0)\|_{L^p(\Omega)}\big)\|\chi\|_{L^q(Q_T)}.
\end{aligned}$$

Since $\chi \geq 0$ is arbitrary in $\mathcal{D}(Q_T)$, the lemma follows. $\square$

Proof of Theorem 33.2. Fix $r > (n+2)/2$. By (33.5) and Lemma 33.3, we have

$$\|g(u,v)\|_{L^r(Q_T)}^r \leq C \int_0^T \int_\Omega (1 + u + v)^{r\gamma}\, dx\, dt \leq C(T),$$

for all finite $T \leq T_{\max}$. Using the variation-of-constants formula, it follows that

$$\|v(t)\|_\infty \leq Ct^{-n/2}\|v_0\|_1 + C \int_0^t (t-s)^{-n/2r}\|g(u(s),v(s))\|_{L^r(\Omega)}\, ds$$

$$\leq Ct^{-n/2}\|v_0\|_1 + C\Big(\int_0^t (t-s)^{-n/2(r-1)}\, ds\Big)^{(r-1)/r}\|g(u,v)\|_{L^r(Q_t)}$$

$$\leq Ct^{-n/2}\|v_0\|_1 + C(T)t^\theta,$$

for all $0 < t < T$, with $\theta = 1 - (n+2)/2r > 0$. This along with (33.5) yields $T_{\max} = \infty$. $\square$

Remark 33.4. The duality argument in the proof of Lemma 33.3 has other applications. For instance, under the assumptions of Theorem 33.2, it yields global existence for the system with inhomogeneous Neumann boundary conditions:

$$\left.\begin{aligned}
u_t - a\Delta u &= f(u,v), & x \in \Omega,\ t > 0,\\
v_t - b\Delta v &= g(u,v), & x \in \Omega,\ t > 0,\\
u_\nu &= \alpha_1(t), & x \in \partial\Omega,\ t > 0,\\
v_\nu &= \alpha_2(t), & x \in \partial\Omega,\ t > 0,\\
u(x,0) &= u_0(x), & x \in \Omega,\\
v(x,0) &= v_0(x), & x \in \Omega,
\end{aligned}\right\} \tag{33.14}$$

where α_i are arbitrary smooth functions. This works also in the case of Dirichlet boundary conditions $u = \alpha_1(t)$, $v = \alpha_2(t)$ [349]. The argument can also be used to study the case of nonlinearities of the form $g = -f = c(x)u^p v^q$, with sign-changing $c(x)$ (see [282], [417]). Another system of physical interest, the so-called Brusselator, corresponding to the choices $f = -uv^2 + Bv$, $g = uv^2 - (B+1)v + A$, can also be handled by similar techniques [281]. $\square$

If f, g do not have a sign, it is still possible to show global existence modulo the additional dissipation condition:

$$\lambda f(u,v) + g(u,v) \leq 0, \quad \text{for } u, v \geq 0, \tag{33.15}$$

with sufficiently large $\lambda > 1$, assuming also that f, g have at most polynomial (upper) growth:

$$f(u,v),\ g(u,v) \leq C(1 + u + v)^\gamma, \quad \text{for all } u, v \geq 0 \text{ and some } \gamma \geq 1. \tag{33.16}$$

Theorem 33.5. *Assume Ω bounded and let $a, b > 0$, $a \neq b$, and $\gamma \geq 1$. Assume* (33.2), (33.3), (33.15), (33.16), *with*

$$\lambda \geq \lambda_0(a, b, n, \gamma) := \left[\frac{(a+b)^2}{4ab}\right]^{m-1} \geq 1, \qquad (33.17)$$

where m in the smallest integer such that $m > n(\gamma - 1)/2$. Then, for all $(u_0, v_0) \in X_+$, the solution of problem (33.1) is global. Moreover, u, v are uniformly bounded in $\overline{\Omega} \times [0, \infty)$.

Theorem 33.5 is from [309]. A typical example (without sign condition) to which it applies is given by $f = u^p v^q - u^r v^s$, $g = u^r v^s - \lambda u^p v^q$ for any $p, q, r, s \geq 1$ and $\lambda > 1$ large enough (depending on p, q, r, s, n, a, b). Interestingly, it will turn out that the largeness assumption on λ is in some sense necessary (see Theorem 33.12 and the preceding paragraphs).

Remarks 33.6. (i) The proof of Theorem 33.5 is based on a suitable Lyapunov functional, cf. Lemma 33.7 below, whereas Theorem 33.2 was based on a duality argument. Note that (33.15) is satisfied in particular if $f + g \leq 0$ and $f \leq 0$. Therefore, Theorem 33.5 is more general than Theorem 33.2. However, the duality argument has other applications (cf. Remark 33.4) which do not seem to be tractable by the Lyapunov functional approach. Also, in the case of homogeneous Dirichlet boundary conditions, results similar to Theorem 33.5 have been obtained in [468] by duality techniques, but the largeness condition on λ is not explicit.

(ii) Note that if $\gamma < (n + 2)/n$, then $\lambda_0 = 1$ in (33.17) (with $m = 1$), so that condition (33.15) disappears. For earlier results related to Theorems 33.2 and 33.5, see [11], [12] (based on Moser's iteration), [460] (based on bootstrap) and [350] (based on a Lyapunov functional). On the other hand, the global existence result of Theorem 33.2 has been extended to f, g satisfying some exponential growth conditions, for instance for $g = -f = ue^v$. For this we refer to [270], [52] (relying on suitable Lyapunov functionals) and to [274], [294] (based on a delicate analysis using parabolic BMO estimates). $\square$

The key of the proof of Theorem 33.5 is the following Lyapunov functional.

Lemma 33.7. *Assume Ω bounded, $a, b > 0$,*

$$K \geq \frac{a + b}{2\sqrt{ab}} \qquad (33.18)$$

and let m be any positive integer. Assume that f, g satisfy (33.2), (33.3), (33.15), (33.16), for some $\lambda \geq K^{2(m-1)}$. Let (u, v) be a solution of (33.1) and let

$$L(t) = \int_\Omega H_m\left(u(x, t), v(x, t)\right) dx,$$

where

$$H_m(u,v) = \sum_{i=0}^{m} C_m^i K^{i^2 - i} u^i v^{m-i}, \qquad C_m^i = \frac{m!}{i!(m-i)!}.$$

Then $L'(t) \leq 0$ on the interval $(0, T_{\max})$.

Proof. Set $w = Kv$ and $L_1(t) = K^m L(t)$. We have

$$K^m H_m(u,v) = \sum_{i=0}^{m} C_m^i K^{i^2} u^i w^{m-i}$$

and w solves

$$w_t - b\Delta w = Kg(u,v).$$

Differentiating L_1 with respect to t yields

$$L_1'(t) = \int_\Omega \Big[\sum_{i=1}^{m} \big(iC_m^i K^{i^2} u^{i-1} w^{m-i} \big) u_t + \sum_{i=0}^{m-1} \big((m-i) C_m^i K^{i^2} u^i w^{m-i-1} \big) w_t \Big] dx$$

$$= \int_\Omega \sum_{i=1}^{m} \big(iC_m^i K^{i^2} u^{i-1} w^{m-i} \big) (a\Delta u + f(u,v))\, dx$$

$$+ \int_\Omega \sum_{i=1}^{m} \big((m-i+1) C_m^{i-1} K^{(i-1)^2} u^{i-1} w^{m-i} \big) (b\Delta w + Kg(u,v))\, dx$$

$$= \int_\Omega \Big[\sum_{i=1}^{m} aiC_m^i K^{i^2} u^{i-1} w^{m-i} \Delta u + b(m-i+1)C_m^{i-1} K^{(i-1)^2} u^{i-1} w^{m-i} \Delta w \Big] dx$$

$$+ \int_\Omega \Big[\sum_{i=1}^{m} iC_m^i K^{i^2} u^{i-1} w^{m-i} f(u,v)$$

$$+ (m-i+1)C_m^{i-1} K^{(i-1)^2+1} u^{i-1} w^{m-i} g(u,v) \Big] dx$$

$$=: I + J.$$

By using Green's formula we obtain

$$I = -\int_\Omega \Big(\mathcal{A} |\nabla u|^2 + \mathcal{B} \nabla u \nabla w + \mathcal{C} |\nabla w|^2 \Big)\, dx,$$

where

$$\mathcal{A} = \sum_{i=2}^{m} ai(i-1) C_m^i K^{i^2} u^{i-2} w^{m-i},$$

$$\mathcal{B} = \sum_{i=1}^{m-1} ai(m-i) C_m^i K^{i^2} u^{i-1} w^{m-i-1} + \sum_{i=2}^{m} b(i-1)(m-i+1) C_m^{i-1} K^{(i-1)^2} u^{i-2} w^{m-i},$$

and

$$C = \sum_{i=1}^{m-1} b(m-i)(m-i+1)C_m^{i-1}K^{(i-1)^2}u^{i-1}w^{m-i-1}.$$

Using the fact that

$$\begin{aligned} iC_m^i &= mC_{m-1}^{i-1}, & i &= 1,\ldots,m, \\ (m-i)C_m^i &= mC_{m-1}^i, & i &= 0,\ldots,m-1, \end{aligned} \tag{33.19}$$

we get

$$\mathcal{A} = am(m-1)\sum_{i=2}^{m} C_{m-2}^{i-2}K^{i^2}u^{i-2}w^{m-i},$$

$$\mathcal{B} = am(m-1)\sum_{i=1}^{m-1} C_{m-2}^{i-1}K^{i^2}u^{i-1}w^{m-i-1}$$

$$+ bm(m-1)\sum_{i=2}^{m} C_{m-2}^{i-2}K^{(i-1)^2}u^{i-2}w^{m-i}$$

$$=: \mathcal{B}_1 + \mathcal{B}_2,$$

and

$$\mathcal{C} = bm(m-1)\sum_{i=1}^{m-1} C_{m-2}^{i-1}K^{(i-1)^2}u^{i-1}w^{m-i-1}.$$

Putting $j = i - 2$, we have

$$\mathcal{A} = am(m-1)\sum_{j=0}^{m-2} C_{m-2}^{j}K^{(j+2)^2}u^{j}w^{m-j-2},$$

$$\mathcal{B}_2 = bm(m-1)\sum_{j=0}^{m-2} C_{m-2}^{j}K^{(j+1)^2}u^{j}w^{m-j-2},$$

and putting $j = i - 1$, we get

$$\mathcal{B}_1 = am(m-1)\sum_{j=0}^{m-2} C_{m-2}^{j}K^{(j+1)^2}u^{j}w^{m-j-2},$$

$$\mathcal{C} = bm(m-1)\sum_{j=0}^{m-2} C_{m-2}^{j}K^{j^2}u^{j}w^{m-j-2}.$$

Therefore,

$$I = -m(m-1) \sum_{j=0}^{m-2} C_{m-2}^j \int_\Omega u^j w^{m-j-2}$$

$$\times \left(aK^{(j+2)^2} |\nabla u|^2 + (a+b) K^{(j+1)^2} \nabla u \nabla w + bK^{j^2} |\nabla w|^2 \right) dx.$$

The quadratic forms (with respect to ∇u and ∇w) are positive since

$$\left((a+b)K^{(j+1)^2}\right)^2 - 4abK^{j^2} K^{(j+2)^2} = K^{2j^2+4j+2}\left((a+b)^2 - 4abK^2\right) \le 0$$

for $j = 0, 1, \ldots, m-2$, due to (33.18). It follows that $I \le 0$.

On the other hand, by (33.19), we have

$$J = m \sum_{i=1}^{m} C_{m-1}^{i-1} \int_\Omega \left(K^{i^2} f(u,v) + K^{(i-1)^2+1} g(u,v)\right) u^{i-1} w^{m-i} \, dx.$$

Since (33.3) and (33.15) imply $\mu f + g \le 0$ for all $\mu \in [1, \lambda]$, we obtain

$$K^{i^2} f(u,v) + K^{(i-1)^2+1} g(u,v) = K^{(i-1)^2+1}\left(K^{2(i-1)} f(u,v) + g(u,v)\right) \le 0$$

for $i = 1, \ldots, m$, hence $J \le 0$. $\square$

Proof of Theorem 33.5. In Lemma 33.7, we take $K = \frac{a+b}{2\sqrt{ab}}$ and m as in the statement of the theorem. Then $\lambda_0 = K^{2(m-1)}$ and we deduce from Lemma 33.7 that $u(t)$ and $v(t)$ are bounded in $L^m(\Omega)$. Since $m > n(\gamma - 1)/2$, by similar arguments as in the proof of Theorem 16.4, one deduces that they are bounded in $L^\infty(\Omega)$. (Alternatively one could use modifications of arguments in the proof of (15.2) in Theorem 15.2.) In particular, this implies $T_{\max} = \infty$ and the theorem is proved. $\square$

Remarks 33.8. (i) Simple modifications of the proofs of Theorems 33.2 and 33.5, show that global existence (without boundedness) is still true if the conditions $f+g$ and/or $\lambda f + g \le 0$ are replaced by $f + g$ and/or $\lambda f + g \le C(1 + u + v)$.

(ii) Under the assumptions of Theorem 33.5, if $u, v \ge 0$ and (u, v) solves (33.1) for $t \in (0, T)$, with the boundary conditions replaced by

$$u_\nu \le 0, \quad v_\nu \le 0,$$

then u, v are uniformly bounded in $\overline{\Omega} \times [0, T)$. This follows from a simple modification of the proof of Lemma 33.7 and Theorem 33.5. $\square$

Systems of Gierer-Meinhardt type

We consider the system

$$
\left.
\begin{aligned}
u_t - a\Delta u &= -\mu_1 u + u^p/v^q + \sigma, & x \in \Omega,\ t > 0, \\
v_t - b\Delta v &= -\mu_2 v + u^r/v^s, & x \in \Omega,\ t > 0, \\
u_\nu = v_\nu &= 0, & x \in \partial\Omega,\ t > 0, \\
u(x,0) &= u_0(x), & x \in \Omega, \\
v(x,0) &= v_0(x), & x \in \Omega,
\end{aligned}
\right\}
\qquad (33.20)
$$

where $p > 1, q, r, s \geq 0$, $a, b > 0$, $\mu_1, \mu_2, \sigma \geq 0$, and $u_0, v_0 \in C(\overline{\Omega})$ with $u_0, v_0 > 0$. By the maximum principle, we immediately obtain the lower bounds

$$
u(x,t) \geq \big(\min_{\overline{\Omega}} u_0\big)\, e^{-\mu_1 t}, \quad v(x,t) \geq \big(\min_{\overline{\Omega}} v_0\big)\, e^{-\mu_2 t}, \qquad x \in \overline{\Omega},\ 0 < t < T_{\max}.
$$

$$(33.21)$$

These bounds imply in particular that

$$
\text{if } T_{\max} < \infty, \text{ then } \limsup_{t \to T_{\max}} \|u(t)\|_\infty = \infty.
$$

System (33.20) (for instance with $p = r = 2$, $q = s = 4$, $n = 1$) arises in a biological model of pattern formation, due to [242]. Here u and v represent the concentrations of an activator and an inhibitor, respectively. The peaks of high concentration of activator give the positional information for the development of a structure, e.g. a tentacle in the polyp Hydra.

An essentially complete answer to the global existence/nonexistence question for system (33.20) is provided by the following theorem from [333] (see also [292], [398] for related results). Earlier partial results of global existence had been proved in [460], [352]. We note also that a large amount of literature has been devoted to the singular perturbation problem associated with the study of "spike-layers" (stationary solutions developing a finite number of concentration peaks as $a \to 0$). We refer for this to the surveys [391], [392].

Theorem 33.9. *Assume Ω bounded.*

(i) *Assume that*

$$
\frac{p-1}{r} < \min\Big(\frac{q}{s+1}, 1\Big).
\qquad (33.22)
$$

Then, for all $u_0, v_0 \in C(\overline{\Omega})$ with $u_0, v_0 > 0$, the solution (u, v) of problem (33.20) is global. If in addition $\mu_1, \mu_2, \sigma > 0$, then u, v are uniformly bounded in $\overline{\Omega} \times [0, \infty)$.

(ii) *Assume that*

$$
\frac{p-1}{r} > \min\Big(\frac{q}{s+1}, 1\Big), \qquad \frac{p-1}{r} \neq 1.
\qquad (33.23)
$$

Then there exist space-independent initial data $u_0, v_0 > 0$ such that the solution $(u, v) = (u(t), v(t))$ of problem (33.20) satisfies $T_{\max} < \infty$.

Remark 33.10. Diffusion preserving global existence. It follows from Theorem 33.9 that, for any p, q, r, s such that the solutions of the ODE system are all global, the addition of diffusion preserves global existence (except perhaps for $p - 1 = r$ and for the equality case in (33.22), which seem to be open). No example of blow-up seems to be known for system (33.20) beside the space-independent solutions. $\square$

The proof of assertion (i) relies on multiplier arguments and on the following consequence of Young's inequality.

Lemma 33.11. *Assume that p, q, r, s satisfy (33.22). For all $\eta, \alpha, \beta > 0$, there exist $C = C(\eta, \alpha, \beta) > 0$ and $\theta = \theta(\alpha) \in (0, 1)$ such that*

$$\alpha \frac{x^{p-1+\alpha}}{y^{q+\beta}} \le \beta \frac{x^{r+\alpha}}{y^{s+1+\beta}} + C \left(\frac{x^\alpha}{y^\beta} \right)^\theta, \qquad x \ge 0, \quad y \ge \eta. \tag{33.24}$$

Proof. Let $x > 0$ and $y \ge \eta$. Inequality (33.24) is equivalent to

$$\alpha \frac{x^{p-1}}{y^q} \le \beta \frac{x^r}{y^{s+1}} + C \left(\frac{y^\beta}{x^\alpha} \right)^{1-\theta}.$$

Write

$$\alpha \frac{x^{p-1}}{y^q} = \left(\frac{x^r}{y^{s+1}} \right)^{(p-1)/r} y^{(p-1)(s+1)/r - q} - C \left(\beta \frac{x^r}{y^{s+1}} \right)^\gamma y^{-m},$$

where $\gamma = (p-1)/r < 1$ and $m = q - (p-1)(s+1)/r > 0$. For each $0 < \varepsilon < \min(m/(s+1), 1 - \gamma)$, using $y \ge \eta$ and Young's inequality, we obtain

$$\alpha \frac{x^{p-1}}{y^q} = C \left(\beta \frac{x^r}{y^{s+1}} \right)^{\gamma+\varepsilon} y^{-m+(s+1)\varepsilon} x^{-r\varepsilon} \le C \left(\beta \frac{x^r}{y^{s+1}} \right)^{\gamma+\varepsilon} \left(\frac{y^\beta}{x^\alpha} \right)^{r\varepsilon/\alpha}$$

$$\le \beta \frac{x^r}{y^{s+1}} + C \left(\frac{y^\beta}{x^\alpha} \right)^{r\varepsilon/(1-\gamma-\varepsilon)\alpha},$$

and (33.24) follows by taking ε sufficiently small. $\square$

Proof of Theorem 33.9(i). We shall only prove global existence and uniform boundedness in the case $\mu_1, \mu_2, \sigma > 0$. Global existence in the general case can be shown by simple modifications (using the lower bound (33.21) on finite time intervals instead of (33.25) below).

Step 1. Lower estimates. We claim that there exists $c_1 > 0$ such that

$$u, v \ge c_1, \qquad x \in \overline{\Omega}, \quad 0 < t < T_{\max}. \tag{33.25}$$

Since u satisfies $u_t - a\Delta u > 0$ on $\{u < \sigma/\mu_1\}$ (along with homogeneous Neumann conditions), the maximum principle implies $u \ge \delta := \min(\sigma/\mu_1, \min_{\overline{\Omega}} u_0) > 0$ in

$\overline{\Omega} \times [0, \infty)$. Then, v satisfies $v_t - b\Delta v > 0$ on $\{v < (\delta^r/\mu_2)^{1/(s+1)}\}$ and the lower bound for v follows similarly.

Step 2. Bound for a quotient. We claim that, for all large $\alpha, \beta > 0$, the function

$$\phi = \phi_{\alpha,\beta}(t) := \int_\Omega \frac{u^\alpha}{v^\beta}\, dx$$

satisfies

$$\sup_{t \in (0, T_{\max})} \phi(t) < \infty. \tag{33.26}$$

By (33.20), we have

$$\phi'(t) = \int_\Omega \left(\alpha \frac{u^{\alpha-1} u_t}{v^\beta} - \beta \frac{u^\alpha v_t}{v^{\beta+1}}\right) dx$$

$$= \alpha \int_\Omega \frac{u^{\alpha-1}}{v^\beta}\left(a\Delta u - \mu_1 u + \sigma + \frac{u^p}{v^q}\right) dx - \beta \int_\Omega \frac{u^\alpha}{v^{\beta+1}}\left(b\Delta v - \mu_2 v + \frac{u^r}{v^s}\right) dx.$$

Using Green's formula, we deduce that

$$\phi'(t) = (-\alpha\mu_1 + \beta\mu_2)\phi + \int_\Omega \left(\alpha \frac{u^{p-1+\alpha}}{v^{q+\beta}} - \beta \frac{u^{r+\alpha}}{v^{s+1+\beta}} + \alpha\sigma \frac{u^{\alpha-1}}{v^\beta}\right) dx$$

$$+ \int_\Omega \left(-a\alpha(\alpha-1)\frac{u^{\alpha-2}}{v^\beta}|\nabla u|^2 - b\beta(\beta+1)\frac{u^\alpha}{v^{\beta+2}}|\nabla v|^2 \right. \tag{33.27}$$

$$\left. + (a+b)\alpha\beta \frac{u^{\alpha-1}}{v^{\beta+1}}\nabla u \cdot \nabla v\right) dx.$$

The last integrand can be rewritten as

$$Q := \left[-a\alpha(\alpha-1)v^2|\nabla u|^2 - b\beta(\beta+1)u^2|\nabla v|^2 + (a+b)\alpha\beta(v\nabla u)\cdot(u\nabla v)\right]\frac{u^{\alpha-2}}{v^{\beta+2}}.$$

Consequently we have $Q \leq 0$, provided we assume

$$\frac{\alpha\beta}{(\alpha-1)(\beta+1)} \leq \frac{4ab}{(a+b)^2}, \tag{33.28}$$

which guarantees that the discriminant $(a+b)^2(\alpha\beta)^2 - 4ab\alpha\beta(\alpha-1)(\beta+1)$ of the quadratic form Q is nonpositive. Owing to (33.25), we also have

$$\frac{u^{\alpha-1}}{v^\beta} = \left(\frac{u^\alpha}{v^\beta}\right)^{(\alpha-1)/\alpha} v^{-\beta/\alpha} \leq C\left(\frac{u^\alpha}{v^\beta}\right)^{(\alpha-1)/\alpha}. \tag{33.29}$$

Using (33.25), Lemma 33.11, (33.29) and Hölder's inequality, we obtain

$$\phi'(t) \leq (-\alpha\mu_1 + \beta\mu_2)\phi + C\int_\Omega \left(\frac{u^\alpha}{v^\beta}\right)^\theta dx + \alpha\sigma \int_\Omega \frac{u^{\alpha-1}}{v^\beta}\, dx$$

$$\leq (-\alpha\mu_1 + \beta\mu_2)\phi + C(\phi^\theta + \phi^{(\alpha-1)/\alpha}) \tag{33.30}$$

for some $\theta \in (0,1)$.

Now assume $\alpha \geq 2\max(1, \mu_2/\mu_1)$ and $\beta \leq 2ab/(a+b)^2 \leq 1$. Then (33.28) is satisfied and, since $-\alpha\mu_1 + \beta\mu_2 < 0$, the function

$$f(Y) := (-\alpha\mu_1 + \beta\mu_2)Y + C(Y^\theta + Y^{(\alpha-1)/\alpha})$$

has a largest positive zero, say $Y = K$. Since, by (33.30), $\phi'(t) < 0$ whenever $\phi(t) > K$, we deduce easily that $\sup_{t\in(0,T_{\max})} \phi(t) \leq \max(\phi(0), K)$, hence (33.26). Since v is bounded below, it is clear that (33.26) remains true if we enlarge β. The claim is proved.

Step 3. L^∞-bounds. By (33.26), we have $u^p v^{-q}$ and $u^r v^{-s} \in L^\infty((0, T_{\max}),$ $L^m(\Omega))$ for all $m < \infty$. By a simple argument using the variation-of-constants formula and the L^p-L^q-estimate (Proposition 48.4), we deduce that u and v are uniformly bounded and that $T_{\max} = \infty$. $\square$

Proof of Theorem 33.9(ii). We consider space-independent solutions of (33.20), i.e. solutions of the corresponding ODE system without diffusion. For spatially homogeneous initial data $u_0, v_0 \geq 1$ to be determined later, we assume for contradiction that $T_{\max}(u_0, v_0) > 1$. In what follows, all the positive constants $C, c, \ldots$ are independent of u_0, v_0.

For fixed $\alpha, \beta > 0$, let $\lambda = \alpha\mu_1 - \beta\mu_2$ and $w(t) = u^\alpha/v^\beta$. By direct calculation using (33.20) (cf. the first line of (33.27)), we have

$$w' + \lambda w = \alpha\frac{u^{p-1+\alpha}}{v^{q+\beta}} - \beta\frac{u^{r+\alpha}}{v^{s+1+\beta}} + \alpha\sigma\frac{u^{\alpha-1}}{v^\beta}. \qquad (33.31)$$

We consider two cases separately.

Case 1: $p - 1 > r$. We apply (33.31) with $\alpha = 1$. Taking β large enough and using (33.21) and $v_0 \geq 1$, we have for all $t \in [0,1]$,

$$\frac{\alpha}{2}\frac{u^p}{v^{q+\beta}} = \frac{\alpha}{2}\left(\frac{u^{r+1}}{v^{s+1+\beta}}\right)^{p/(r+1)} v^k \geq \beta\frac{u^{r+1}}{v^{s+1+\beta}} - C,$$

where $k = (s+1+\beta)\frac{p}{r+1} - q - \beta > 0$, and

$$\frac{\alpha}{2}\frac{u^p}{v^{q+\beta}} = \frac{\alpha}{2}\left(\frac{u}{v^\beta}\right)^p v^m \geq c\left(\frac{u}{v^\beta}\right)^p,$$

where $m = (p-1)\beta - q > 0$. It follows that

$$w' \geq cw^p - \lambda w - C.$$

Taking $w(0)$ large enough, this implies blow-up of u before $t = 1$; a contradiction.

Case 2: $p - 1 < r$, $(p-1)(s+1) > qr$. We claim that that there exist constants $C_1, C_2 > 0$ such that, if

$$u_0^{r-p+1} \geq C_1 v_0^{s+1-q}, \tag{33.32}$$

then

$$u^{r-p+1} \geq C_2 v^{s+1-q}, \quad 0 < t \leq 1. \tag{33.33}$$

To prove this, letting $z = e^{\lambda t} w$ and applying (33.31) with

$$\alpha = r - p + 1 > 0 \quad \text{and} \quad \beta = s + 1 - q > 0,$$

we see that, for all $t \in [0, 1]$,

$$z'(t) \leq 0 \implies \frac{u^{r+\alpha}}{v^{s+1+\beta}} \geq \frac{u^{p-1+\alpha}}{v^{q+\beta}} \alpha \beta^{-1} \implies z(t) \geq e^{-|\lambda|} \frac{u^{\alpha}}{v^{\beta}} \geq e^{-|\lambda|} \alpha \beta^{-1} =: C_1.$$

Consequently we have $z(t) \geq \min(C_1, w(0))$ on $[0, 1]$, and the claim follows with $C_2 = e^{-|\lambda|} C_1$.

Now assume (33.32). Using the first equation in (33.20) and (33.33), we deduce that

$$u' + \mu_1 u \geq \frac{u^p}{v^q} \geq c u^{p-q(r-p+1)/(s+1-q)} = c u^{\gamma}, \quad 0 < t \leq 1,$$

where $\gamma = 1 + \frac{(p-1)(s+1)-qr}{s+1-q} > 1$. But, taking u_0 larger, this implies blow-up of u before $t = 1$; a contradiction. $\square$

33.2. Diffusion inducing blow-up

In this subsection we show that certain parabolic systems admit a blowing-up solution for some particular initial data, although the corresponding system of ODE's has only global bounded solutions. We shall give three different examples, each of them involving a different method.

We first consider systems of the form

$$\left.\begin{array}{ll} u_t - a\Delta u = f(u, v), & x \in \Omega, \ t > 0, \\ v_t - b\Delta v = g(u, v), & x \in \Omega, \ t > 0, \\ u_\nu = \alpha_1(t), & x \in \partial\Omega, \ t > 0, \\ v_\nu = \alpha_2(t), & x \in \partial\Omega, \ t > 0, \\ u(x, 0) = u_0(x), & x \in \Omega, \\ v(x, 0) = v_0(x), & x \in \Omega, \end{array}\right\} \tag{33.34}$$

under the mass-dissipation structure condition $f + g \leq 0$. Sufficient conditions ensuring global existence for such systems were studied in the previous subsection.

Recall that, under a polynomial growth assumption on the nonlinearities, global existence of nonnegative solutions is true if $f + g \leq 0$ and if either:

- $\lambda f + g \leq 0$ with $\lambda > 1$ large enough, under homogeneous boundary conditions (or more generally $u_\nu, v_\nu \leq 0$); or

- $f \leq 0$, with arbitrary (smooth) functions α_i (or also under Dirichlet-Dirichlet boundary conditions).

The following result [417] shows that in case of unequal diffusions, the condition $f + g \leq 0$ is not sufficient to ensure global existence, even if the additional dissipation property $\lambda f + g \leq 0$ is also satisfied (with some $\lambda > 1$ not too large) and f and g have polynomial growth. We point out that the example below involves nonnegative solutions and functions $\alpha_i \leq 0$, so that the condition $f + g \leq 0$ still guarantees the mass-dissipation property $(d/dt) \int_\Omega (u(t) + v(t)) \, dx \leq 0$. The result has to be compared with Theorem 33.5, which is therefore in a sense optimal.

Theorem 33.12. *Let $\Omega = B_1 \subset \mathbb{R}^n$. There exist constants $a, b, T > 0$, $a \neq b$, functions $f, g \in C^\infty(\mathbb{R}^2, \mathbb{R})$ and $\alpha_1, \alpha_2 \in C^\infty([0,T], \mathbb{R})$, satisfying $\alpha_1, \alpha_2 \leq 0$,*

$$f + g \leq 0, \quad \lambda f + g \leq 0, \quad \text{for all } u, v \geq 0 \text{ and some } \lambda > 1, \tag{33.35}$$

$$f(u,v), \; g(u,v) \leq C(1 + u + v)^\gamma, \quad \text{for all } u, v \geq 0 \text{ and some } \gamma \geq 1, \tag{33.36}$$

and such that for some C^∞ initial data $u_0, v_0 \geq 0$, system (33.34) admits a classical nonnegative solution (u, v) on $(0, T)$, with

$$\lim_{t \to T} u(0, t) = \lim_{t \to T} v(0, t) = \infty.$$

Moreover, u and v blow up only at $x = 0$ as $t \to T$.

The proof is based on the construction of an explicit solution, of self-similar form, and involves some relatively heavy numerical computations (still doable by hand, but the construction was initially carried out with the help of the formal computation software Maple).

Sketch of proof (for $n = 10$). An explicit solution is searched under the form

$$u(x,t) = \frac{A(T-t) + B|x|^2}{(T - t + |x|^2)^{5/4}}, \qquad v(x,t) = \frac{C(T-t) + D|x|^2}{(T - t + |x|^2)^{5/4}},$$

with constants $A, B, C, D > 0$ to be determined. Note that this is actually a self-similar solution, since it can be rewritten under the form

$$u = (T - t)^{-1/4} U(y), \qquad v = (T - t)^{-1/4} V(y), \qquad y = x(T - t)^{-1/2},$$

with

$$U(y) = \frac{A + B|y|^2}{(1 + |y|^2)^{5/4}}, \qquad V(y) = \frac{C + D|y|^2}{(1 + |y|^2)^{5/4}}.$$

A direct calculation yields

$$u_t - a\Delta u = (T - t)^{-5/4} \frac{A_1 + B_1|y|^2 + C_1|y|^4}{(1 + |y|^2)^{(5/4)+2}}$$

and

$$v_t - b\Delta v = (T - t)^{-5/4} \frac{A_2 + B_2|y|^2 + C_2|y|^4}{(1 + |y|^2)^{(5/4)+2}},$$

where A_1, B_1, C_1 and A_2, B_2, C_2 are computed in terms of n, a, A, B and n, b, C, D, respectively. As for the functions f, g, one looks for polynomials, homogeneous and of total degree 5, of the form

$$f(u, v) = \sum_{i=0}^{5} \lambda_i u^{5-i} v^i, \qquad g(u, v) = \sum_{i=0}^{5} \mu_i u^{5-i} v^i.$$

The PDE's in system (33.34) then become equivalent to

$$\sum_{i=0}^{5} \lambda_i (A + B|y|^2)^{5-i} (C + D|y|^2)^i = (1 + |y|^2)^3 (A_1 + B_1|y|^2 + C_1|y|^4) \quad (33.37)$$

and

$$\sum_{i=0}^{5} \mu_i (A + B|y|^2)^{5-i} (C + D|y|^2)^i = (1 + |y|^2)^3 (A_2 + B_2|y|^2 + C_2|y|^4). \quad (33.38)$$

Choosing $n = 10$ (other choices are possible) and $a = 1$, $b = 1/10$, $A = 1/25$, $B = 1$, $C = 11/2$, $D = 1/10$, it turns out that it is possible to adjust the constants λ_i, μ_i in such a way that (33.37), (33.38) be satisfied, with moreover $\lambda_i + \mu_i < 0$, so that

$$\lambda f + g = \sum_{i=0}^{5} (\lambda \lambda_i + \mu_i) u^{5-i} v^i \leq 0$$

for λ equal or close to 1. Finally, for $r = |x| = 1$, we compute

$$\alpha_1(t) = u_\nu(x, t) = u_r(1, t) = \frac{(4B - 5A)(T - t) - B}{2(T - t + 1)^{9/4}}$$

and an analogous expression for $\alpha_2(t)$ (with C, D in place of A, B). Taking $T > 0$ small enough, it follows that $\alpha_i(t) \leq 0$ on $[0, T]$. $\square$

Remark 33.13. (i) **Other examples of blow-up.** An example similar to that of Theorem 33.12 is also constructed in [416] for nonlinearities $f(x,t,u,v) = c_1(x,t)u^p v^q$, $g(x,t,u,v) = c_2(x,t)u^p v^q$, with $n = 1$, $p,q > 1$ and sign-changing functions c_i such that $c_1 + c_2 \leq 0$. However, it remains an open problem to construct similar examples of blow-up in the case of homogeneous boundary conditions. On the other hand, beyond the special examples, there is a lack of general blow-up criteria for such systems, as well as of a description of possible singularities, in comparison with the scalar problems studied in Chapter II.

(ii) **Global weak solutions.** Consider problem (33.1) under the assumptions (33.2), $(u_0, v_0) \in X_+$, $f + g \leq 0$ and $\lambda f + g \leq 0$ for some $\lambda > 1$. Theorem 33.12 suggests that this problem need not admit a global classical solution. However, it was shown in [415] that there exists a global weak solution in some appropriate L^1-sense. Also, it is to be noted that in the example constructed in Theorem 33.12, it is possible to extend the solution across the blow-up time to a global weak solution. $\square$

Still for mass-dissipative systems of the form (33.34), here with $f \leq 0 \leq g$ and $f + g = 0$, the next result [64], [65] shows that mixed Dirichlet-Neumann conditions can lead to finite-time blow-up, even for equal diffusions. Namely we consider the one-dimensional problem

$$\left.\begin{aligned}
u_t - u_{xx} &= -uv^p, & x &\in (-1,1),\ t > 0, \\
v_t - v_{xx} &= uv^p, & x &\in (-1,1),\ t > 0, \\
u(\pm 1, t) &= 1, & t &> 0, \\
v_x(\pm 1, t) &= 0, & t &> 0, \\
u(x,0) &= u_0(x), & x &\in (-1,1), \\
v(x,0) &= v_0(x), & x &\in (-1,1).
\end{aligned}\right\} \tag{33.39}$$

Theorem 33.14. *Assume $p > 2$. Let $u_0, v_0 \in C^2([-1,1])$ be even and satisfy $u_0(\pm 1) = 1$, $(v_0)_x(\pm 1) = 0$,*

$$0 < u_0 \leq 1, \quad v_0 > 0 \quad in\ [-1,1],$$

$$(u_0 + v_0)_x, \quad (v_0)_x \geq 0 \quad in\ [0,1],$$

$$(u_0 + v_0)_{xx} \geq 0, \quad (v_0)_{xx} + u_0 v_0^p \geq 0 \quad in\ [-1,1]. \tag{33.40}$$

Then the solution of (33.39) satisfies $T_{\max} < \infty$ and $\lim_{t \to T_{\max}} v(1,t) = \infty$.

Remark 33.15. It has been shown in [349] that solutions of (33.39) exist globally if $0 < p \leq 2$. On the other hand, if the boundary conditions are replaced by $u = 0$ and $v_x = 0$, then global existence is true for any $p > 0$ (this follows from a simple modification of the proof of Theorem 33.2 and Lemma 33.3). Analogues of Theorem 33.14 in higher dimension can be found in [65]. $\square$

The proof of Theorem 33.14 is based on monotonicity and subsolution arguments — to obtain pointwise lower bounds for u, v — and on the use of a simple differential inequality.

Proof of Theorem 33.14. *Step 1. Absence of steady states.* We easily verify that (33.39) has no nonnegative stationary solution except $(U, V) \equiv (1, 0)$. Indeed, if (U, V) is a nonnegative stationary solution, then $V_{xx} = -UV^p \leq 0$ and $V_x(\pm 1) = 0$, hence $V_x \equiv 0$, so that $V \equiv 0$ (since $U \not\equiv 0$). But then $U_{xx} \equiv 0$, hence $U \equiv 1$ due to $U(\pm 1) = 1$.

Step 2. Monotonicity properties. From the assumptions on u_0, v_0, the functions u, v are symmetric in x, and we have $0 \leq u \leq 1$. We next observe that (w, v), with $w := u + v$, solves the equivalent system

$$
\left.
\begin{aligned}
w_t - w_{xx} &= 0, & x &\in (-1, 1), \ t > 0, \\
v_t - v_{xx} &= (w - v)v^p, & x &\in (-1, 1), \ t > 0, \\
w(\pm 1, t) &= 1 + v(\pm 1, t), & t &> 0, \\
v_x(\pm 1, t) &= 0, & t &> 0, \\
w(x, 0) &= (u_0 + v_0)(x), & x &\in (-1, 1), \\
v(x, 0) &= v_0(x), & x &\in (-1, 1).
\end{aligned}
\right\}
\tag{33.41}
$$

Now we claim that

$$
w_t, v_t \geq 0 \text{ in } [-1, 1] \times [0, T_{\max}) \text{ and } w_x, v_x \geq 0 \text{ in } [0, 1] \times [0, T_{\max}). \tag{33.42}
$$

Let $(y, z) = (w_t, v_t)$. By continuous dependence, it suffices to prove that $y, z \geq 0$ when the second inequality in (33.40) is assumed to be strict. By continuity, we have $z > 0$ in $[-1, 1]$ for t small. Assume for contradiction that there is a first $t_0 > 0$ such that $z(x_0, t_0) = 0$ for some $x_0 \in [-1, 1]$, and denote $Q_0 := [0, 1] \times [0, t_0]$ and $S_0 := \{-1, 1\} \times [0, t_0]$. Since

$$
y_t - y_{xx} = 0 \quad \text{in } Q_0,
$$

$y(\cdot, 0) \geq 0$ and $y = z \geq 0$ in S_0, the maximum principle implies $y \geq 0$ in Q_0. But we then have

$$
z_t - z_{xx} = v^p y + b(x, t)z \geq b(x, t)z \quad \text{in } Q_0,
$$

with $b = (pw - (p+1)v)v^{p-1}$, and $z \geq 0$ in Q_0. Therefore, $x_0 = \pm 1$ by the strong maximum principle (since $z(\cdot, 0) \not\equiv 0$). But this is impossible in view of Hopf's lemma, since $z_x = 0$ on S_0. We have thus proved the first part of (33.42).

Next, we have $w_{xx} = w_t \geq 0$ and $w_x(0,t) = 0$. Therefore $w_x \geq 0$ in $[0,1] \times [0, T_{\max})$. By differentiating the second equation of (33.41) in x, we see that $h := v_x$ satisfies

$$h_t - h_{xx} = v^p w_x + b(x,t)h \geq b(x,t)h \quad \text{in } [0,1] \times [0, T_{\max}),$$

with $h(0,t) = h(1,t) = 0$. The second part of (33.42) then follows from the maximum principle.

We now denote

$$M(t) := \max_{[-1,1] \times [0,t]} v = v(1,t)$$

and we assume for contradiction that $T_{\max} = \infty$.

Step 3. Unboundedness of v. We claim that

$$M(t) \to \infty, \quad t \to \infty.$$

Otherwise u, v are bounded (recall that $u \leq 1$) and since w, v are nondecreasing in time by Step 2, there would exist bounded functions (W, V) such that

$$W(x) = \lim_{t \to \infty} w(x,t), \quad V(x) = \lim_{t \to \infty} v(x,t).$$

But the monotonicity of w, v guarantees that (W, V) is a stationary solution of (33.41) (see Proposition 53.8). Letting $U := W - V$, (U, V) is thus a stationary solution of (33.39), hence $V \equiv 0$ by Step 1. This is a contradiction, since $V \geq v_0 > 0$.

Step 4. Pointwise lower bounds for u and v and differential inequality. For fixed $T > 0$, put $M = M(T)$, $\delta = \min_{[-1,1]} u_0 \leq 1$ and $\underline{u}(x,t) = \delta e^{M^{p/2}(x-1)}$. Then $\underline{u}$ satisfies

$$\left.\begin{array}{ll} \underline{u}_t - \underline{u}_{xx} + M^p \underline{u} = 0 \leq u_t - u_{xx} + M^p u, & x \in (-1,1),\ 0 < t < T, \\[2mm] \underline{u}(\pm 1, t) \leq 1, & 0 < t < T, \\[2mm] \underline{u}(x,0) \leq u_0(x), & x \in (-1,1). \end{array}\right\}$$

It follows from the maximum principle that $\underline{u} \leq u$ in $[-1,1] \times [0,T]$. We deduce that, for all t large,

$$u(x,t) \geq \eta := \delta/e > 0, \quad x_0(t) \leq x \leq 1, \tag{33.43}$$

with

$$x_0(t) := 1 - M^{-p/2}(t) \in (-1,1).$$

On the other hand, we have $v_{xx} = v_t - uv^p \geq -v^p \geq -M^p(t)$. Consequently, by Taylor expansion, for some $\xi \in (x,1)$, we have

$$v(x,t) = v(1,t) + v_x(1,t)(x-1) + v_{xx}(\xi,t)\frac{(x-1)^2}{2} \geq M(t) - M^p(t)\frac{(x-1)^2}{2}.$$

Therefore, for t large,

$$v(x,t) \geq M(t)/2, \quad x_0(t) \leq x \leq 1. \tag{33.44}$$

Integrating the second equation in (33.39) over $(-1,1)$ and using (33.43), (33.44), we see that $\phi(t) := \int_{-1}^{1} v(t)\, dx$ satisfies

$$\phi'(t) = \int_{-1}^{1} uv^p(t)\, dx \geq (1 - x_0(t))\eta(M/2)^p(t) = CM^{p/2}(t).$$

Since on the other hand $\phi(t) \leq 2M(t)$, we obtain

$$\phi'(t) \geq C\phi^{p/2}(t)$$

for t large. Since $p > 2$, this contradicts $T_{\max} = \infty$. $\qquad\square$

In our last example, we consider a system without the structure $f + g \leq 0$, but with homogeneous Neumann conditions (unlike in the previous two examples) and equal diffusions, and for which blowing-up solutions exist for some particular initial data, although the corresponding system of ODE's has only global bounded solutions.

Namely, we consider the system

$$\left.\begin{aligned}
u_t - du_{xx} &= h(u,v)(1+u) - \delta u, & x \in (-1,1),\ t &> 0, \\
v_t - dv_{xx} &= -h(u,v)(1+v) - \delta v, & x \in (-1,1),\ t &> 0, \\
u_x = v_x &= 0, & x = \pm 1,\ t &> 0, \\
u(x,0) &= u_0(x), & x \in (-1,1), & \\
v(x,0) &= v_0(x), & x \in (-1,1), &
\end{aligned}\right\} \tag{33.45}$$

with $d > 0$ and $\delta \geq 0$. Here the function $h : [0,\infty)^2 \to \mathbb{R}$, of class C^1, is assumed to satisfy:

$$h(u,v) = -h(v,u), \tag{33.46}$$

$$h(u,0) = h(0,v) = 0, \tag{33.47}$$

$$h(u,v) \geq 0, \qquad u \geq v \geq 0, \tag{33.48}$$

$$h(u,v) \geq k(u-v)^\gamma, \qquad u \geq v \geq 1, \tag{33.49}$$

for some $k, \gamma > 0$. These assumptions apply for instance to the function $h(u,v) = (uv)^m |u - v|^p (u - v)$, for any $m \geq 1$, $p \geq 0$. As for the initial data (u_0, v_0), we assume

$$u_0, v_0 \in C^1([-1,1]), \quad u_0, v_0 > 0 \ \text{ in } [-1,1], \quad (u_0)_x, (v_0)_x = 0,\ x = \pm 1, \tag{33.50}$$

$$v_0(x) = u_0(-x), \qquad\qquad\qquad (33.51)$$

$$u_0 \geq v_0 \quad \text{in } [0,1]. \qquad\qquad (33.52)$$

Due to (33.47) and the maximum principle, we have $u, v > 0$ in $[-1,1] \times [0,T)$.

Let us first observe that for the corresponding system of ODE's

$$\left. \begin{aligned} y' &= h(y,z)(1+y) - \delta y, \\ z' &= -h(y,z)(1+z) - \delta z, \\ y(0) &= y_0 \geq 0, \quad z(0) = z_0 \geq 0, \end{aligned} \right\}$$

all solutions are global, and that they decay exponentially to $(0,0)$ if moreover $\delta > 0$. This follows immediately from the fact that $(y + z + yz)' = y'(1+z) + (1+y)z' = -\delta(y + z + 2yz)$.

The following result essentially comes from [527], where it was given for $h(u,v) = uv(u - v)$. We here present the simplified proof from [492] (with more general nonlinearities).

Theorem 33.16. *Assume* (33.46)–(33.52). *There exists* $C = C(d, \delta, k, \gamma) > 0$ *such that, if*

$$\int_0^1 (u_0 - v_0)\sin(\pi x/2)\, dx \geq C \quad and \quad \int_{-1}^1 \log(1 + u_0)\, dx \geq C, \qquad (33.53)$$

then $T := T_{\max}(u_0, v_0) < \infty$.

The idea of the proof is to derive differential inequalities for two different functionals on some interval $(0, T_0)$. Integrating them yields upper estimates for the measures of two complementary subsets of $(0, T_0)$, whose sum is less than T_0, leading to a contradiction with existence up to $t = T_0$.

Remarks 33.17. Unequal diffusions. (i) The fact that the diffusion coefficients are equal in the two equations is used crucially in the proof (via the sign and symmetry properties of the two components). An example of a system with blow-up induced by unequal diffusions and homogeneous Neumann conditions can be found in [383]. The proof therein is more delicate. On the other hand, it is unknown whether or not Theorem 33.16 remains true in the case of homogeneous Dirichlet boundary conditions. As for the asymptotic blow-up behavior of solutions of (33.45), this is an essentially open problem.

(ii) In the fundamental article [516], it had been shown that unequal diffusions can destabilize an otherwise stable constant equilibrium (global existence being however preserved). Other results on diffusion-induced blow-up can be found in [133], [387], [259]. $\square$

Proof of Theorem 33.16. First note that, since $(\tilde{u}, \tilde{v}) := (v(-x,t), u(-x,t))$ solves the same system due to (33.46), (33.51), we have by uniqueness:

$$v(x,t) = u(-x,t). \qquad\qquad\qquad (33.54)$$

Next, we put

$$m(t) = \min_{x\in[-1,1]} u(x,t) = \min_{x\in[-1,1]} v(x,t), \quad M(t) = \max_{x\in[-1,1]} u(x,t) = \max_{x\in[-1,1]} v(x,t),$$

and we claim that

$$M(t) \geq 1, \quad 0 < t < \min(T, \delta^{-1}). \tag{33.55}$$

Indeed, by adding the equations for u and v, we get

$$(u+v)_t - d(u+v)_{xx} = h(u,v)(u-v) - \delta(u+v).$$

Integrating and using the boundary conditions, we deduce that

$$\frac{d}{dt} \int_{-1}^{1} (u+v)\, dx \geq -\delta \int_{-1}^{1} (u+v)\, dx,$$

hence

$$M(t) \geq \frac{1}{2} \max_{[-1,1]} (u+v) \geq \frac{1}{4} \int_{-1}^{1} (u+v)\, dx \geq \frac{1}{4} e^{-\delta t} \int_{-1}^{1} (u_0 + v_0)\, dx,$$

and (33.55) follows by taking $C \geq 2e$ in (33.53).

Now we derive two differential inequalities for the auxiliary functions ϕ and ψ, defined as follows:

$$\phi(t) := e^{(\delta + \lambda d)t} \int_{-1}^{1} w\varphi\, dx, \qquad \psi(t) := \int_{-1}^{1} z\, dx + 2\delta(t - T_1), \qquad 0 \leq t < T,$$

where

$$w = u - v, \qquad z = \log\left(\frac{1+u}{2}\right),$$

and

$$\lambda := \pi^2/4, \quad \varphi(x) = (\pi/4)\sin(\pi x/2), \quad T_1 = \min\left(\delta^{-1}, (\gamma(\delta + \lambda d))^{-1}\right).$$

We also put $T_0 = \min(T, T_1)$ and we set

$$E = \{t \in (0, T_0) : m(t) \geq 1\}, \qquad F = (0, T_0) \setminus E.$$

Claim 1. We have

$$\phi(t) > 0 \quad \text{and} \quad \phi'(t) \geq ke^{-1}\phi^{1+\gamma}\chi_E, \qquad 0 < t < T_0. \tag{33.56}$$

To show this, we subtract the equations for u and v to obtain

$$w_t - dw_{xx} = h(u,v)(2 + u + v) - \delta w. \tag{33.57}$$

Note in particular that since $w(0,t) = w_x(1,t) = 0$, (33.48), (33.52) and the maximum principle imply $w \geq 0$ on $[0,1] \times (0,T)$. Therefore, $\phi \geq 0$ on $(0,T)$ and $h(u,v)\varphi \geq 0$ on $[-1,1] \times (0,T)$ by (33.54) and (33.46). Multiplying by φ and integrating by parts yields

$$\frac{d}{dt} \int_{-1}^1 w\varphi \, dx = d\big[w_x\varphi - \varphi_x w\big]_{-1}^1 + \int_{-1}^1 h(u,v)(2+u+v)\varphi \, dx - (\delta+\lambda d)\int_{-1}^1 w\varphi \, dx.$$

Since $h(u,v)(2+u+v)\varphi \geq h(u,v)(u-v)|\varphi|$, by using (33.46), (33.48), (33.49) and Jensen's inequality we get

$$\phi'(t) \geq e^{(\delta+\lambda d)t} \, k \int_{-1}^1 |u-v|^{1+\gamma}|\varphi| \, dx \geq e^{-\gamma(\delta+\lambda d)t} \, k\,\phi^{1+\gamma} \geq e^{-1} \, k\,\phi^{1+\gamma}, \quad t \in E,$$

and $\phi'(t) \geq 0$ if $t \notin E$. This, along with $\phi(0) \geq C > 0$ (cf. (33.53)), proves the claim.

Claim 2. We have

$$\psi(t) > 0 \quad \text{and} \quad \psi'(t) \geq \frac{d}{8}\psi^2 \chi_F, \qquad 0 < t < T_0. \tag{33.58}$$

By a simple computation, we get

$$z_t - dz_{xx} = h(u,v) + d(z_x)^2 - \delta\frac{u}{1+u}.$$

Since $h(u(x,t),v(x,t))$ is odd due to (33.46) and (33.54), we have $\int_{-1}^1 h(u,v)\,dx = 0$ hence,

$$\psi'(t) = \frac{d}{dt}\int_{-1}^1 z\,dx + 2\delta \geq d\int_{-1}^1 (z_x)^2 \, dx \geq 0. \tag{33.59}$$

Since $\psi(0) = \int_{-1}^1 \log(1+u_0)\,dx - 2\log 2 - 2\delta T_1 > 0$ by taking $C > 2\log 2 + 2\delta T_1$ in (33.53), it follows in particular that $\psi > 0$. Now, if $t \in F$, i.e. $m(t) < 1$, then (33.55) implies the existence of $\xi(t) \in [-1,1]$ such that $u(\xi(t),t) = 1$, hence $z(\xi(t),t) = 0$. Therefore

$$\left(\int_{-1}^1 |z|\,dx\right)^2 \leq 4(\max|z(x,t)|)^2 \leq 4\left(\int_{-1}^1 |z_x|\,dx\right)^2 \leq 8\int_{-1}^1 (z_x)^2 \, dx. \tag{33.60}$$

Since $\int_{-1}^1 z\,dx \geq \psi$ on $[0,T_0)$ by the definition of ψ, (33.58) follows from (33.59) and (33.60).

To complete the proof of Theorem 1, we integrate (33.56) and (33.58), to obtain

$$\phi^{-\gamma}(0) \geq \gamma \int_0^{T_0} \phi'\phi^{-1-\gamma}\,ds \geq \gamma k e^{-1}|E|, \qquad \psi^{-1}(0) \geq \int_0^{T_0} \psi'\psi^{-2}\,ds \geq \frac{d}{8}|F|.$$

We deduce that

$$\min(T, \delta^{-1}, [\gamma(\delta+\lambda d)]^{-1}) = T_0 = |E| + |F| \leq (\gamma k)^{-1}e\phi^{-\gamma}(0) + 8d^{-1}\psi^{-1}(0).$$

We conclude that if $\phi(0)$ and $\psi(0) \geq C(d,\delta,k,\gamma) > 0$ large enough, then $T \leq (\gamma k)^{-1}e\phi^{-\gamma}(0) + 8d^{-1}\psi^{-1}(0) < \infty$. $\square$

33.3. Diffusion eliminating blow-up

In this subsection we consider the following system

$$\left.\begin{aligned}
u_t - d_1\Delta u &= f(u-v), & x \in \Omega,\ t > 0,\\
v_t - d_2\Delta v &= f(u-v) - v, & x \in \Omega,\ t > 0,\\
u = v &= 0, & x \in \partial\Omega,\ t > 0,
\end{aligned}\right\} \tag{33.61}$$

where $\Omega \subset \mathbb{R}^n$ is bounded, $d_1, d_2 > 0$, $d_1 - d_2 > 1/\lambda_1$, $f(w) = |w|^{p-1}w$, $1 < p < p_S$, and the initial data belong to $Z := H_0^1 \times H_0^1(\Omega)$. We also consider the corresponding system of ODE's

$$\left.\begin{aligned}
U_t &= f(U-V),\\
V_t &= f(U-V) - V.
\end{aligned}\right\} \tag{33.62}$$

The following theorem is due to [197].

Theorem 33.18. *Let the assumptions above be satisfied. Then:*

(i) *there exists a solution of (33.62) which blows up in finite time;*

(ii) *for all $(u_0, v_0) \in Z$, the solution of (33.61) is global and converges to the trivial solution $(0,0)$ in Z as $t \to \infty$.*

Proof. (i) Denote $W := U - V$ and assume $V(0) > 1$, $W(0) > \left(\frac{p+1}{2}V(0)^2\right)^{1/(p+1)}$. We will prove that the solution (U, V) blows up in finite time.

Since $V' = f(W) - V$ and $W' = V$, the functions W, V remain positive. Multiplying the equation $W'' + W' = W^p$ by W' we see that the function $E(t) := \frac{1}{2}(W'(t))^2 - \frac{1}{p+1}W(t)^{p+1}$ is nonincreasing, hence $E(t) \le E(0) < 0$. In particular,

$$W^{p+1} > \frac{p+1}{2}(W')^2 \ge (W')^2 = V^2,$$

hence $V' = W^p - V > V^{2p/(p+1)} - V$. Since $V(0) > 1$, the last differential inequality guarantees blow-up of V.

(ii) Similarly as in Example 51.27 we get that problem (33.61) is well-posed in $Y := L^{p+1} \times L^{p+1}(\Omega)$. In addition, if the initial data $(u_0, v_0) \in Z$ and the solution is bounded in Y, then this solution is global and its trajectory is relatively compact in Z, see Example 51.38. Finally, $u_t, v_t \in C^1((0, \infty), L^2(\Omega)) \cap C((0, \infty), H^2 \cap H_0^1(\Omega))$.

Fix $(u_0, v_0) \in Z$ and set $w := u - v$. Then (w, v) solves the problem

$$\left.\begin{aligned}
w_t - d_1\Delta w &= (d_1 - d_2)\Delta v + v, & x \in \Omega,\ t > 0,\\
v_t - d_2\Delta v &= -v + |w|^{p-1}w, & x \in \Omega,\ t > 0,\\
w = v &= 0, & x \in \partial\Omega,\ t > 0.
\end{aligned}\right\} \tag{33.63}$$

Let $-A$ denote the Dirichlet Laplacian in $L^2(\Omega)$. Due to $d_1 - d_2 > 1/\lambda_1$, $(d_1 - d_2)A - 1$ is a positive self-adjoint operator and its inverse

$$K := ((d_1 - d_2)A - 1)^{-1}$$

is compact, positive and commutes with both A and $A^{1/2}$. The first equation in (33.63) can be rewritten as

$$v = K(d_1 \Delta w - w_t).$$

Now the second equation in (33.63) guarantees

$$K(d_1 \Delta w_t - w_{tt}) = d_2 K \Delta(d_1 \Delta w - w_t) - K(d_1 \Delta w - w_t) + |w|^{p-1}w. \qquad (33.64).$$

Define the norm

$$\|\varphi\|_{-1} := \|K^{1/2}\varphi\|_{L^2(\Omega)} \qquad \text{for } \varphi \in L^2(\Omega).$$

Multiplying (33.64) by w_t and integrating in x over Ω, we have

$$-d_1 \|A^{1/2} w_t\|_{-1}^2 - \frac{1}{2}\frac{d}{dt}\|w_t\|_{-1}^2$$
$$= \frac{d_1 d_2}{2}\frac{d}{dt}\|Aw\|_{-1}^2 + d_2\|A^{1/2}w_t\|_{-1}^2 + \frac{d_1}{2}\frac{d}{dt}\|A^{1/2}w\|_{-1}^2$$
$$+ \|w_t\|_{-1}^2 + \frac{1}{p+1}\frac{d}{dt}\int_\Omega |w|^{p+1}dx,$$

which implies

$$\frac{d}{dt}L(t) = -(d_1 + d_2)\|A^{1/2}w_t\|_{-1}^2 - \|w_t\|_{-1}^2 \leq 0, \qquad (33.65)$$

where

$$L(t) := \frac{1}{2}\|w_t\|_{-1}^2 + \frac{d_1 d_2}{2}\|Aw\|_{-1}^2 + \frac{d_1}{2}\|A^{1/2}w\|_{-1}^2 + \frac{1}{p+1}\int_\Omega |w|^{p+1}dx.$$

Consequently, L is a Lyapunov functional (see Appendix G) and the function $w(t)$ stays bounded in $L^{p+1}(\Omega)$. Now the second equation in (33.63) and a simple estimate based on the variation-of-constants formula shows that $v(t)$ stays bounded in $W^{2-\varepsilon,(p+1)/p}(\Omega)$ for any $\varepsilon > 0$ and $t \geq t_0 > 0$. Since this space is embedded in $L^{p+1}(\Omega)$ for ε small due to $p < p_S$, we see that the solution $(u(t), v(t))$ remains bounded in Y. Consequently, it exists globally and is relatively compact in Z. Consequently, the ω-limit set $\omega(u_0, v_0)$ of this solution is a compact nonempty connected and invariant set in Z (see Proposition 53.3). Fix $(\tilde{u}_0, \tilde{v}_0) \in \omega(u_0, v_0)$ and let $(\tilde{u}, \tilde{v})$ be the solution of problem (33.61) with initial data $(\tilde{u}_0, \tilde{v}_0)$. Set $\tilde{w} = \tilde{u} - \tilde{v}$. Since the Lyapunov functional L is constant on $\omega(u_0, v_0)$, (33.65) guarantees $\tilde{w}_t = 0$, hence $\tilde{w}_{tt} = 0$. Now multiplying (33.64) (with w replaced by $\tilde{w}$) with $\tilde{w}$ and denoting by $(\cdot, \cdot)$ the scalar product in $L^2(\Omega)$ we obtain

$$d_1 d_2 (KA\tilde{w}, A\tilde{w}) + d_1(KA^{1/2}\tilde{w}, A^{1/2}\tilde{w}) + \int_\Omega |\tilde{w}|^{p+1}\,dx = 0,$$

which implies $\tilde{w} = 0$. Now the first equation in (33.63) shows $\tilde{v} = 0$, hence $\tilde{u} = 0$. This concludes the proof. $\quad\square$

Chapter IV

Equations with Gradient Terms

34. Introduction

In Chapter IV, we consider problems with nonlinearities depending on u and its space derivatives:

$$\left.\begin{aligned}
u_t - \Delta u &= F(u, \nabla u), & x \in \Omega,\ t > 0,\\
u &= 0, & x \in \partial\Omega,\ t > 0,\\
u(x,0) &= u_0(x), & x \in \Omega.
\end{aligned}\right\} \tag{34.1}$$

Here $F = F(u, \xi) : \mathbb{R} \times \mathbb{R}^n \to \mathbb{R}$ is a C^1-function (except for problem (34.4) with $1 < q < 2$, see below).

In Sections 35–39, we consider perturbations of the model problem (15.1) by terms involving first-order derivatives:

$$\left.\begin{aligned}
u_t - \Delta u &= u^p + g(u, \nabla u), & x \in \Omega,\ t > 0,\\
u &= 0, & x \in \partial\Omega,\ t > 0,\\
u(x,0) &= u_0(x), & x \in \Omega.
\end{aligned}\right\} \tag{34.2}$$

We will only consider nonnegative solutions of (34.2) (but u^p can be interpreted as $|u|^{p-1}u$ for definiteness). In many results, g might depend also on x, t, but we restrict to (34.2) for simplicity. Typical examples that we shall pay a particular attention to, are given by:

$$\left.\begin{aligned}
u_t - \Delta u &= u^p - \mu|\nabla u|^q, & x \in \Omega,\ t > 0,\\
u &= 0, & x \in \partial\Omega,\ t > 0,\\
u(x,0) &= u_0(x), & x \in \Omega,
\end{aligned}\right\} \tag{34.3}$$

with $p, q > 1$, $\mu > 0$ (dissipative gradient term) and

$$\left.\begin{aligned}
u_t - \Delta u &= u^p - a \cdot \nabla(u^q), & x \in \Omega,\ t > 0,\\
u &= 0, & x \in \partial\Omega,\ t > 0,\\
u(x,0) &= u_0(x), & x \in \Omega,
\end{aligned}\right\} \tag{34.4}$$

with $p > 1$, $q \geq 1$, $a \in \mathbb{R}^n$ (convective gradient term). A motivation for studying (34.3), (34.4) is to investigate the effect of a dissipative or convective gradient term on global existence or nonexistence of solutions, and on their asymptotic behavior, in finite or infinite time. We refer to [484], [490] for surveys on equations of the form (34.2). It will turn out that problems of this form reveal a number of interesting, qualitatively new phenomena, in comparison with the unperturbed model problem, such as new critical exponents, or changes in the parameters involved in the asymptotic blow-up behavior.

In Sections 40 and 41, we consider problems whose essential superlinear character comes from the gradient term. A simple model case is given by:

$$\left.\begin{aligned} u_t - \Delta u &= |\nabla u|^p, & x \in \Omega,\ t > 0, \\ u &= 0, & x \in \partial\Omega,\ t > 0, \\ u(x,0) &= u_0(x), & x \in \Omega, \end{aligned}\right\} \tag{34.5}$$

with $p > 1$. Problem (34.5) is often referred to as a viscous Hamilton-Jacobi equation. Also, (34.5) is related with the Kardar-Parisi-Zhang equation in the physical theory of growth and roughening of surfaces (see [252], [72] for details and references). Note that it is one of the simplest examples of a parabolic PDE with a nonlinearity depending on the first-order spatial derivatives, and can thus be considered as an analogue of the model problem (15.1). The case where the nonlinearity is replaced by $u^m |\nabla u|^p$ will also be studied. We will see that these equations exhibit phenomena qualitatively different from (15.1), such as (boundary or interior) gradient blow-up.

35. Well-posedness and gradient bounds

Throughout Chapter IV we denote

$$X := \{u \in BC^1(\overline{\Omega}) : u = 0 \text{ on } \partial\Omega\}, \tag{35.1}$$

equipped with the norm

$$\|w\|_X := \|w\|_\infty + \|\nabla w\|_\infty,$$

and $X_+ := \{w \in X : w \geq 0\}$. Problem (34.1), with F of class C^1, is locally well-posed in X (see Remark 51.11). As for problem (34.4), with Ω bounded or $\Omega = \mathbb{R}^n$ for simplicity, it is also locally well-posed in X for all $q \geq 1$ (see Example 51.15 and Proposition 51.16). In particular,

$$\text{if } T_{\max} = T_{\max}(u_0) < \infty, \text{ then } \lim_{t \to T_{\max}} \|u(t)\|_X = \infty. \tag{35.2}$$

Moreover the solution enjoys the regularity property

$$u \in BC^{2,1}(\overline{\Omega} \times [t_1, t_2]), \qquad 0 < t_1 < t_2 < T_{\max}(u_0) \tag{35.3}$$

(cf. the corresponding proof in Example 51.9). Furthermore, problem (34.1) admits a comparison principle, cf. Propositions 52.6, 52.10 and Remarks 52.11. In the case of problem (34.4), see Proposition 52.16. Those results will be frequently used throughout this chapter without explicit reference. In particular, if $F(0,0) \geq 0$ and $u_0 \in X_+$, then we have $u \geq 0$. On the other hand, in the case of problem (34.3) in a ball or in $\mathbb{R}^n$, if u_0 is radial (resp. radial nonincreasing), then u enjoys the same property, as a consequence of Proposition 52.17.

In the case of problems (34.3)–(34.5) well-posedness may actually hold true in some larger spaces, but this question is not our main concern in this chapter. However, in view of the study of the large time behavior, it will be very useful to know weaker continuation properties than (35.2). In the case of the general problem (34.1) this requires some structure assumptions on F. A rather sharp result in that direction is given by the following theorem. Here, for $k > 0$, we write $F \leq O(|\xi|^k)$ if $F(u,\xi) \leq C(u)(|\xi|^k + 1)$ and $F \leq o(|\xi|^k)$ if for all $\varepsilon > 0$, $F(u,\xi) \leq \varepsilon|\xi|^k + C_\varepsilon(u)$, where $C(u)$ and $C_\varepsilon(u)$ remain bounded on bounded sets of $u \geq 0$.

Theorem 35.1. *Consider problem (34.1) with $F(0,0) \geq 0$ and $F = f + g$, where $f, g \in C^1$ satisfy*

$$|f| \leq O(|\xi|^2), \qquad |f_\xi| \leq O(|\xi|), \qquad |f_u| \leq o(|\xi|^2) \tag{35.4}$$

and

$$g(0,\xi) \leq 0, \qquad g_u \leq 0, \qquad \xi \cdot \frac{\partial}{\partial \xi}\left(\frac{g}{|\xi|}\right) \leq 0, \qquad \text{for all } u \geq 0, \ \xi \in \mathbb{R}^n \setminus \{0\}. \tag{35.5}$$

Let $u_0 \in X_+$. If $T_{\max}(u_0) < \infty$, then

$$\lim_{t \to T_{\max}(u_0)} \|u(t)\|_\infty = \infty.$$

Theorem 35.1 is a consequence of the following Bernstein-type gradient estimate from [59], which provides a pointwise a priori estimate of ∇u assuming a bound on u.

Proposition 35.2. *Let $T > 0$ and assume that $F = f + g$, where $f, g \in C^1$ satisfy (35.4) and (35.5). Let $u \in C^{2,1}(Q_T) \cap C(\overline{Q_T})$, with $\nabla u \in C(\overline{Q_T}) \cap L^\infty(Q_T)$, be a solution of (34.1), such that*

$$0 \leq u \leq M \ \ in \ \overline{Q_T} \qquad and \qquad |\nabla u_0| \leq M \ \ in \ \overline{\Omega} \tag{35.6}$$

for some $0 < M < \infty$. Then there holds

$$|\nabla u| \leq C = C(M, T, F, \Omega) \ in \ \overline{Q_T}.$$

Remarks 35.3. (a) Theorem 35.1 reduces the proof of global existence to the derivation of a uniform estimate of u (on bounded time intervals). It also guarantees that finite-time blow-up, in case it occurs, takes place in the L^∞-norm.

(b) Assumptions (35.4), (35.5) in Theorem 35.1 can be viewed as one-sided quadratic growth restrictions. Theorem 35.1 applies for instance with $F(u, \nabla u) = f(u) + a|\nabla u|^m - \lambda u^r|\nabla u|^q$ with f of class C^1, $1 < m \le 2$, $r \ge 1$ or $r = 0$, $q > 1$ and $\lambda \ge 0$. This includes in particular problem (34.3) for any $p, q > 1$ and $\mu > 0$. In the special case of problem (34.3), the result was proved before in [433], [434], [498] by different techniques.

(c) As for problem (34.4), Theorem 35.1 applies when $q \ge 2$, but not if $1 < q < 2$, since the nonlinearity is then not Lipschitz. However, it is proved in Proposition 51.16 (by different arguments) that an L^∞-estimate is sufficient to prevent blow-up of solutions.

(d) The growth and sign assumptions in (35.4), (35.5) are essentially optimal. Indeed, the conclusion of Theorem 35.1 fails for problem (34.5) if $p > 2$ (cf. Section 40; see also Section 41 and [485] for other examples). The other assumptions on F can be slightly weakened. For instance, it is enough to assume F to be C^1 for ξ large.

(e) For earlier results under two-sided quadratic growth conditions on F, see e.g. [319], [469]. Note that when $g = 0$, the nonnegativity of u_0 and the assumption $F(0, 0) \ge 0$ are not needed. Like in [319], [469], the proof of Proposition 35.2 relies on the classical Bernstein technique, which consists in applying the maximum principle to the function $\partial v/\partial x_i$ (or to $|\nabla v|^2$), where $u = \phi(v)$. Gradient estimates can be obtained by various other techniques. Approaches based on elaborate test-function arguments are used in [320, Theorem V.4.1 and Lemma VI.3.1], where a two-sided quadratic growth assumption is made on F (but no assumption on the derivatives F_u, F_ξ), and in [88]. If $|F| \le O(|\xi|^m)$ with $m < 2$, results of this kind can be obtained via the variation-of-constants formula, or derived from well-posedness results in L^∞ (cf. Example 51.30, and see also [10] and [361, Lemma 5.1]). For related results in the radial case under (different) one-sided quadratic growth assumptions, see [512]. The technique used there is still different, based on Kruzhkov's idea of adding a new space variable. Results concerning sign-changing solutions under one-sided quadratic growth assumptions can also be found in [59], [512]. $\square$

In view of the proof of Proposition 35.2, we start with a preliminary result (under weaker assumptions) which provides control of the gradient on the boundary. The proof is based on a barrier argument (cf. [320, Lemma VI.3.1]).

Lemma 35.4. *Assume that $F(u, \xi) \le O(|\xi|^2)$. Let $T, M > 0$ and let $u \in C^{2,1}(Q_T) \cap C^{1,0}(\overline{Q_T})$ be a solution of (34.1) satisfying (35.6). Then there holds*

$$|\nabla u| \le C = C(M, F, \Omega) \text{ on } S_T.$$

Proof. Let U be the solution of

$$\left.\begin{aligned}
-\Delta U &= 1, & 1 < |x| < 2, \\
U &= 0, & |x| = 1, \\
U &= 1, & |x| = 2.
\end{aligned}\right\}$$

It is easily checked that

$$0 < U(x) < 1 \quad \text{and} \quad c_1(|x| - 1) \le U(x) \le c_2(|x| - 1), \qquad 1 < |x| < 2. \quad (35.7)$$

Let $x_0 \in \partial\Omega$. Since Ω is uniformly smooth, there exists $\rho_0 \in (0,1)$ depending only on Ω (independent of x_0) with the following property: For any $\rho \in (0, \rho_0]$, there exists $y = y(\rho) \in \mathbb{R}^n$ such that $\overline{B}(y, \rho) \cap \overline{\Omega} = \{x_0\}$. Next put

$$V(x) = \beta^{-1} \log\big(1 + e^{\beta M} U\big(\tfrac{x-y}{\rho}\big)\big), \quad \rho \le |x - y| \le 2\rho,$$

with $\beta \ge 1$, $\rho \in (0, \rho_0]$. (Observe that, up to affine changes of variables, this is just the usual Hopf-Cole exponential transformation $U = e^{\beta V}$.)

We want to compare u and V in the set $\widetilde{Q}_T = \widetilde{\Omega} \times (0, T]$, where

$$\widetilde{\Omega} = \{x \in \Omega : |x - y| < 2\rho\},$$

for suitably small ρ and $y = y(\rho)$. Due to $0 \le V \le M + 1$ and $F(u, \xi) \le O(|\xi|^2)$, a simple calculation shows that

$$-\Delta V \ge \beta|\nabla V|^2 + (2\rho^2\beta)^{-1} \ge F(V, \nabla V), \quad \rho < |x - y| < 2\rho, \qquad (35.8)$$

by taking $\beta \ge 1$ large and then $\rho \in (0, \rho_0]$ small (depending only on F and M). On the other hand, using (35.6), (35.7) and imposing in addition $\rho \le c_1/2M\beta$, we have

$$V(x) \ge \frac{\beta^{-1} e^{\beta M} U\big(\tfrac{x-y}{\rho}\big)}{1 + e^{\beta M} U\big(\tfrac{x-y}{\rho}\big)} \ge \frac{U\big(\tfrac{x-y}{\rho}\big)}{2\beta} \ge M(|x - y| - \rho) \ge u_0(x), \quad x \in \widetilde{\Omega},$$

and $V(x) > M \ge u(x, t)$ for $x \in \Omega \cap \{|x - y| = 2\rho\}$. In view of (35.8), and since $\partial\widetilde{\Omega} \subset (\Omega \cap \{|x - y| = 2\rho\}) \cup \partial\Omega$, we may then apply the comparison principle in $\widetilde{Q}_T$ to deduce that

$$u(x, t) \le V(x) \le \beta^{-1} e^{\beta M} U\big(\tfrac{x-y}{\rho}\big) \le c_2(\beta\rho)^{-1} e^{\beta M}(|x - y| - \rho), \quad (x, t) \in \widetilde{Q}_T,$$

where we also used (35.7). Due to $u(x_0, t) = V(x_0) = 0$, it follows that

$$|\nabla u(x_0, t)| = -\frac{\partial u}{\partial \nu}(x_0, t) \le -\frac{\partial V}{\partial \nu}(x_0, t) \le c_2(\beta\rho)^{-1} e^{\beta M}, \quad 0 < t < T. \quad \square$$

Proof of Proposition 35.2. Consider a function ϕ of class C^3 on some compact interval J, with $\phi' > 0$ and $\phi(J) \supset [0, M]$, and a constant $K > 0$ (ϕ and K will be specified later on). Let $h \in \mathbb{R}^n$, with $|h| = 1$. Given a solution u satisfying the assumptions of the proposition, we set

$$v := \phi^{-1}(u), \quad w := \partial_h v = h \cdot \nabla v, \quad z(x,t) := e^{-Kt} w.$$

We want to apply the maximum principle to the function z.

Step 1. Derivation of the equation for z. We have

$$F(u, \nabla u) = u_t - \Delta u = \phi'(v)(v_t - \Delta v) - \phi''(v)|\nabla v|^2,$$

hence

$$v_t - \Delta v = \frac{F(\phi(v), \phi'(v)\nabla v)}{\phi'(v)} + \frac{\phi''(v)}{\phi'(v)}|\nabla v|^2.$$

Note that $z \in C(\overline{Q_T}) \cap L^\infty(Q_T)$. Since F is C^1, by differentiating the equation for v in the direction h and using Remark 48.3(i), we get $z \in W^{2,1;q}_{loc}(Q_T)$ for all finite q. In addition, direct computation yields

$$w_t - \Delta w = \tilde{a}(x,t)\, w + b(x,t) \cdot \nabla w \quad \text{a.e. in } Q_T,$$

with

$$\tilde{a} = F_u + \frac{\phi''}{\phi'^2}\left(\xi \cdot F_\xi - F\right) + \frac{1}{\phi'^2}\left(\frac{\phi''}{\phi'}\right)' |\xi|^2 \quad \text{and} \quad b = F_\xi + 2\frac{\phi''}{\phi'^2}\,\xi,$$

where F and its derivatives are evaluated at $u = u(x,t)$, $\xi = \nabla u(x,t)$, while ϕ and its derivatives are evaluated at $v(x,t)$. Setting $a = \tilde{a} - K$, we obtain

$$z_t - \Delta z = a(x,t)\, z + b(x,t) \cdot \nabla z \quad \text{a.e. in } Q_T. \tag{35.9}$$

Step 2. Construction of a function ϕ such that $a \le 0$. Since $\xi \cdot g_\xi - g \le 0$, we look for a function ϕ such that $\phi'' \ge 0$. We take

$$\phi(s) = eM \int_0^s \exp(-e^{-\lambda\sigma})\, d\sigma, \quad s \in J := [0, 1],$$

where $\lambda > 0$ will be chosen below. For $s \in J$, we compute

$$\phi' = eM \exp(-e^{-\lambda s}), \quad \phi'' = \lambda e^{-\lambda s} \phi', \quad \left(\frac{\phi''}{\phi'}\right)' = -\lambda^2 e^{-\lambda s}.$$

Note that $M \le \phi' \le eM$, $s \in J$. In particular, we have $[0, M] \subset \phi(J)$. By (35.4), (35.5), there exist $a_0, a_1 > 0$ and, for each $\eta > 0$, there exists $C_\eta > 0$, such that

$$F_u \le \eta|\xi|^2 + C_\eta, \quad \xi \cdot F_\xi - F = \xi \cdot f_\xi - f + |\xi|\xi \cdot \tfrac{\partial}{\partial \xi}\left(\tfrac{g}{|\xi|}\right) \le a_0|\xi|^2 + a_1,$$

for $0 \le u \le M$, $\xi \in \mathbb{R}^n$. Take $\lambda = 2a_0 e M$, $\eta = 2a_0^2 e^{-\lambda}$, $K \ge C_\eta + \lambda a_1/M$. Using $g_u \le 0$, $0 \le u(x,t) \le M$ and $0 \le v(x,t) \le 1$, it follows that, for all $(x,t) \in Q_T$,

$$a(x,t) = F_u + \frac{\lambda e^{-\lambda v}}{\phi'(v)}\left(\xi \cdot F_\xi - F - \frac{\lambda}{\phi'(v)}|\xi|^2\right) - K$$

$$\le \eta|\xi|^2 + C_\eta + \frac{\lambda e^{-\lambda v}}{\phi'(v)}\left((a_0 - \frac{\lambda}{eM})|\xi|^2 + a_1\right) - K$$

$$\le (\eta - 2a_0^2 e^{-\lambda})|\xi|^2 + C_\eta + \frac{\lambda a_1}{M} - K \le 0.$$

Step 3. Conclusion. Due to (35.6), Lemma 35.4, and $\phi'(v) \ge M$, we have $z \le C = C(F,\Omega,M)$ on $\mathcal{P}_T$. Applying the maximum principle to equation (35.9) and using $a \le 0$, we deduce that $z \le C$ in $\overline{Q_T}$. Getting back to $\partial_h u = e^{Kt}\phi'(v)z$, and since h was arbitrary, the proposition follows. $\square$

Proof of Theorem 35.1. Assume for contradiction that $T := T_{\max} < \infty$ and $\liminf_{t\to T} \|u(t)\|_\infty < M$ for some $M \in (0,\infty)$. By (35.4), (35.5), there exists $K > 0$ such that

$$F(u,\xi) \le K(|\xi|^2 + 1), \quad 0 \le u \le M + 1, \ \xi \in \mathbb{R}^n.$$

Pick $t_0 \in [T - \frac{1}{K}, T)$ such that $\|u(t_0)\|_\infty \le M$ and let $\overline{u}(x,t) := M + K(t - t_0)$ for $(x,t) \in \overline{Q}$, where $Q := \Omega \times (t_0, T)$. For $(x,t) \in Q$, we have $0 \le \overline{u}(x,t) \le M + 1$, hence

$$\overline{u}_t - \Delta\overline{u} - F(\overline{u}, \nabla\overline{u}) = K - F(M + K(t - t_0), 0) \ge 0.$$

By the comparison principle, we deduce that $0 \le u \le \overline{u} \le M + 1$ in Q. Due to Proposition 35.2 it follows that $\sup_{t\in(0,T)} \|u(t)\|_X < \infty$: a contradiction. $\square$

36. Perturbations of the model problem: blow-up and global existence

In this section, we discuss the conditions on the perturbation terms which imply or prevent blow-up.

We start with a simple criterion for equation (34.4) in bounded domains, which is based on a modification of the eigenfunction method (see Theorem 17.1). The idea of the proof is from [215], [330].

Theorem 36.1. *Consider problem (34.4) with Ω bounded, $p > 1$, $q \ge 1$, and $u_0 \in X_+$.*

(i) Assume $p > q$ and set $m = p/(p-q)$. If $\int_\Omega u_0\varphi_1^m\,dx > C_1 = C_1(\Omega,p,q,a) > 0$, then $T_{\max}(u_0) < \infty$.

(ii) *Assume $q \geq p$. Then $T_{\max}(u_0) = \infty$ and $\sup_{t \geq 0} \|u(t)\|_\infty < \infty$.*

Proof. (i) Denote $y = y(t) := \int_\Omega u(t)\varphi_1^m \, dx$. Multiplying the differential equation in (34.4) with φ_1^m yields, for $0 < t < T := T_{\max}(u_0)$,

$$y' = \int_\Omega u_t \varphi_1^m \, dx = \int_\Omega \varphi_1^m \Delta u \, dx + \int_\Omega u^p \varphi_1^m \, dx + \int_\Omega (a \cdot \nabla(\varphi_1^m)) u^q \, dx.$$

We claim that

$$\int_\Omega \varphi_1^m \Delta u \, dx \geq -m\lambda_1 \int_\Omega u \varphi_1^m \, dx. \tag{36.1}$$

Since $\varphi_1^m \notin C^2(\overline{\Omega})$ (when $1 < m < 2$), we consider $(\varphi_1 + \varepsilon)^m$ and observe that

$$\Delta(\varphi_1 + \varepsilon)^m = m(\varphi_1 + \varepsilon)^{m-1}\Delta\varphi_1 + m(m-1)(\varphi_1 + \varepsilon)^{m-2}|\nabla\varphi_1|^2$$
$$\geq -m\lambda_1(\varphi_1 + \varepsilon)^{m-1}\varphi_1.$$

Integrating by parts, we obtain

$$\int_\Omega (\varphi_1 + \varepsilon)^m \Delta u \, dx = \int_\Omega u\Delta(\varphi_1 + \varepsilon)^m \, dx + \int_{\partial\Omega} (\varphi_1 + \varepsilon)^m \partial_\nu u \, d\sigma$$
$$\geq -m\lambda_1 \int_\Omega u(\varphi_1 + \varepsilon)^{m-1}\varphi_1 \, dx + \varepsilon^m \int_{\partial\Omega} \partial_\nu u \, d\sigma$$

and (36.1) follows upon letting $\varepsilon \to 0$.

Now by Hölder's inequality we have

$$\left| \int_\Omega (a \cdot \nabla(\varphi_1^m)) u^q \, dx \right| \leq m|a| \int_\Omega |\nabla\varphi_1|\varphi_1^{m-1} u^q \, dx \leq C\left(\int_\Omega \varphi_1^m u^p \, dx \right)^{q/p},$$

for some $C = C(\Omega, p, q, a) > 0$. Combining this with Jensen's inequality, we obtain

$$y' \geq \int_\Omega u^p \varphi_1^m \, dx - C\left(\int_\Omega u^p \varphi_1^m \, dx \right)^{1/p} - C\left(\int_\Omega u^p \varphi_1^m \, dx \right)^{q/p} \geq \frac{1}{2}y^p - \tilde{C},$$

for some $\tilde{C} = \tilde{C}(\Omega, p, q, a) > 0$. We infer that u cannot exist globally whenever $y(0) > (2\tilde{C})^{1/p}$.

(ii) Without loss of generality, we may assume that $a = |a|e_1$ and that $\Omega \subset \{x \in \mathbb{R}^n : 0 < x_1 < L\}$ for some $L > 0$. We seek for a (stationary) supersolution of (34.4) of the form $v(x) = Ke^{\alpha x_1}$, for arbitrarily large $K > 0$ (to guarantee $K \geq \|u_0\|_\infty$) and some $\alpha > 0$. The condition to ensure this is thus

$$-\alpha^2 K e^{\alpha x_1} \geq K^p e^{\alpha p x_1} - |a|\alpha q K^q e^{\alpha q x_1}, \quad 0 < x_1 < L,$$

which is satisfied if

$$\alpha q |a| K^{q-p} e^{\alpha(q-p)x_1} \geq 1 + \alpha^2 K^{1-p} e^{\alpha(1-p)x_1}, \quad 0 < x_1 < L.$$

Since $q \geq p > 1$, it is thus sufficient that $\alpha q|a|K^{q-p} \geq 1 + \alpha^2 K^{1-p}$. This is true for $\alpha = 2/q|a|$ and all large $K > 1$. It then follows from the comparison principle in Proposition 52.16, that $0 \leq u(x,t) \leq v(x) \leq Ke^{\alpha L}$ in Ω, as long as $u(t)$ exists. By Proposition 51.16, this implies global existence. $\quad\square$

We now turn to problem (34.3) (in bounded and unbounded domains). We begin with a result from [497] which shows that finite-time blow-up occurs for large initial data when $p > q$.

Theorem 36.2. *Consider problem* (34.3) *with* $p > q > 1$, $\mu > 0$. *Let* $u_0 = \lambda\phi$, *with* $\phi \in X_+$, $\phi \not\equiv 0$, $\lambda > 0$. *If* λ *is sufficiently large, then* $T_{\max}(u_0) < \infty$.

Remarks 36.3. (i) For problem (34.3) the eigenfunction method does not seem to apply, and the proof of Theorem 36.2 relies on a different technique, based on self-similar blowing-up subsolutions. For earlier results in that direction (and other methods), see for instance [129], [304], [433].

(ii) When $p > q$, by Young's inequality, we have

$$|a \cdot \nabla(u^q)| \leq q|a|u^{q-1}|\nabla u| \leq \tfrac{1}{2}u^p + \mu|\nabla u|^m, \qquad m = p/(p-q+1) < p,$$

for some $\mu = \mu(a,p,q) > 0$, so that any solution of (34.4) is a supersolution of $u_t - \Delta u = \tfrac{1}{2}u^p - \mu|\nabla u|^m$. Consequently, Theorem 36.2 implies blow-up of the solution of (34.4) for large initial data. However, the criterion in Theorem 36.1(i) is more precise.

(iii) **Blow-up for slow decay initial data.** For problems (34.3) with $p > q \geq 2p/(p+1)$, and (34.4) with $q = (p+1)/2$, there are blow-up results for slow decay initial data in $\Omega = \mathbb{R}^n$, similar to those known for the model problem (15.1). In fact, the conclusion of Theorem 17.12 remains valid in this case, with a different constant on the RHS of (17.14) [497]. The proof is based on Theorem 36.2, rescaling and comparison arguments. Such results extend to more general unbounded domains (containing a cone or a paraboloid); see [497], [461]. $\quad\square$

Proof of Theorem 36.2. We seek a (self-similar) subsolution of (34.3) of the form:

$$v(x,t) = \frac{1}{(1-\varepsilon t)^k}\, V\!\left(\frac{|x|}{(1-\varepsilon t)^m}\right), \qquad t_0 \leq t < 1/\varepsilon,$$

where V is defined by

$$V(y) = 1 + \frac{A}{2} - \frac{y^2}{2A}, \qquad y \geq 0.$$

Here $A, k, m, t_0, \varepsilon > 0$ (with $t_0 < 1/\varepsilon$) are to be determined. Set $R = \big(A(2+A)\big)^{1/2}$, so that $V(R) = 0$. Note that $v(x,t) > 0$ if and only if $(x,t) \in D$, where

$$D := \big\{(x,t) : t_0 \leq t < 1/\varepsilon,\ |x| < R(1-\varepsilon t)^m\big\},$$

and that v is smooth in D. We will verify that $Pv := v_t - \Delta v - v^P + \mu|\nabla v|^q \leq 0$ in D. We compute, setting $y = |x|/(1 - \varepsilon t)^m$ for convenience:

$$Pv = \frac{\varepsilon(kV(y) + myV'(y))}{(1 - \varepsilon t)^{k+1}} - \frac{V''(y) + \frac{n-1}{y}V'(y)}{(1 - \varepsilon t)^{k+2m}} - \frac{V^P(y)}{(1 - \varepsilon t)^{kp}} + \mu\frac{|V'(y)|^q}{(1 - \varepsilon t)^{(k+m)q}}.$$

The function V obviously satisfies

$$1 \leq V(y) \leq 1 + A/2, \quad -1 \leq V'(y) \leq 0, \quad \text{for } 0 \leq y \leq A,$$
$$0 \leq V(y) \leq 1, \quad -R/A \leq V'(y) \leq -1, \quad \text{for } A \leq y \leq R,$$
$$V''(y) + (n - 1)V'(y)/y = -n/A, \quad \text{for } 0 < y < R.$$

We first choose

$$k = \frac{1}{p - 1}, \quad 0 < m < \min\left\{\frac{1}{2}, \frac{p - q}{q(p - 1)}\right\},$$

so that $kp = k + 1 > k + 2m$ and $k + 1 > (k + m)q$, and next we choose:

$$A > k/m, \quad \varepsilon < \frac{1}{k(1 + A/2)}.$$

In the case $0 \leq y \leq A$, by using also $V' \leq 0$ and by taking $t_0 = t_0(\varepsilon, k, m, q, A, n, \mu)$ sufficiently close to $1/\varepsilon$, we obtain

$$Pv(x, t) \leq \frac{\varepsilon k(1 + A/2) - 1}{(1 - \varepsilon t)^{k+1}} + \frac{n/A}{(1 - \varepsilon t)^{k+2m}} + \frac{\mu}{(1 - \varepsilon t)^{(k+m)q}}$$
$$\leq (1 - \varepsilon t)^{-k-1}\left(\varepsilon k(1 + \tfrac{A}{2}) - 1 + \tfrac{n}{A}(1 - \varepsilon t_0)^{1-2m}\right.$$
$$\left. + \mu(1 - \varepsilon t_0)^{k+1-(k+m)q}\right) \leq 0.$$

In the case $A \leq y < R$, by taking $t_0 = t_0(\varepsilon, k, m, q, A, n, \mu)$ still closer to $1/\varepsilon$, we get

$$Pv(x, t) \leq \frac{\varepsilon(k - mA)}{(1 - \varepsilon t)^{k+1}} + \frac{n/A}{(1 - \varepsilon t)^{k+2m}} + \frac{\mu(R/A)^q}{(1 - \varepsilon t)^{(k+m)q}}$$
$$\leq (1 - \varepsilon t)^{-k-1}\left(\varepsilon(k - mA) + \tfrac{n}{A}(1 - \varepsilon t_0)^{1-2m}\right.$$
$$\left. + \mu\left(\tfrac{R}{A}\right)^q(1 - \varepsilon t_0)^{k+1-(k+m)q}\right) \leq 0.$$

Now, by translation, one can assume without loss of generality that $0 \in \Omega$ and $\phi \geq C$ in $B(0, \rho)$ for some $\rho, C > 0$. Therefore, for t_0 close to $1/\varepsilon$ and $\lambda > 0$ large enough, we have $u_0 \geq v(\cdot, t_0)$ in $B(0, R(1 - \varepsilon t_0)^m)$, hence in $\overline{\Omega}$. Moreover, we have

$v \leq 0$ on $\partial\Omega \times (t_0, 1/\varepsilon)$. If $T_{\max}(u_0) \geq 1/\varepsilon - t_0$, it follows from the comparison principle that

$$u(x, t - t_0) \geq v(x, t) \quad \text{in } D.$$

Since $v(0, t) \to \infty$ as $t \to 1/\varepsilon$, we conclude that $T_{\max}(u_0) \leq 1/\varepsilon - t_0 < \infty$. $\quad\square$

The next result from [498], [481] shows in particular that the blow-up condition $p > q$ in Theorem 36.2 is optimal for bounded domains (see [185], [434] for earlier results in that direction). However, for general unbounded domains, the issue depends in a crucial way on the geometry of the domain, through the notion of inradius $\rho(\Omega)$ (cf. Section 19 and Appendix D).

Theorem 36.4. *Consider problem* (34.3) *with* $q \geq p > 1$, $\mu > 0$.

(i) *Assume* $\rho(\Omega) < \infty$. *Then for all* $u_0 \in X_+$, *there holds* $T_{\max}(u_0) = \infty$ *and*

$$\sup_{t \geq 0} \|u(t)\|_\infty < \infty.$$

Assume in addition that $u_0 \in W_0^{1,r}(\Omega)$ *for some finite* $r > n \max(1, q - 1)$. *There exist* $\mu_0, \lambda > 0$ *(depending only on* Ω, p, q, r*) such that, if* $\mu \geq \mu_0$, *then*

$$\|u(t)\|_s \leq C(u_0)\, e^{-\lambda t}, \quad t \geq 0, \quad r \leq s \leq \infty. \tag{36.2}$$

(ii) *Assume* $\rho(\Omega) = \infty$. *Then there exists* $u_0 \in X_+$, *such that either*

$$T_{\max}(u_0) < \infty, \quad or \quad T_{\max}(u_0) = \infty \quad and \quad \lim_{t \to \infty} \|u(t)\|_\infty = \infty.$$

Furthermore, u_0 *can be taken in* $W_0^{1,r}(\Omega)$ *for* r *large.*

We start with assertion (i). The proof of global existence and boundedness is based on comparison arguments. The idea is to construct a stationary supersolution v in the exterior of a ball of small radius ε, which is radial and whose minimum is larger than $\|u_0\|_\infty$. The solution u is thus dominated by all the translates of v, centered at points y such that $B(y, \varepsilon) \subset \Omega^c$ (these supersolutions play the role of a barrier). Since $\rho(\Omega) < \infty$ and Ω is uniformly regular, any point x of Ω is at bounded distance of such a point y. This guarantees a uniform bound for u, hence global existence in view of the gradient estimates in Section 35. The decay will be proved by a multiplier argument, using multiplication by a power of u and the Poincaré inequalities (which are valid due to $\rho(\Omega) < \infty$).

Remark 36.5. Although the comparison function v below is unbounded, v and ∇v are bounded on the set $\{(x, t) \in \Omega \times [0, T] : u > v\}$ for each $T < T_{\max}(u_0)$, due to $u \in L^\infty(Q_T)$. Consequently the comparison principle can be applied in view of Remark 52.11(i). $\quad\square$

Proof of Theorem 36.4(i). Applying the finiteness assumption on $\rho(\Omega)$ and the uniform regularity of Ω, we may choose $\varepsilon \in (0, 1)$ such that for any ball B of radius $\rho(\Omega) + 1$, $B \cap \Omega^c$ contains a ball of radius ε. Let a be a fixed point in Ω, and let us pick x_a such that

$$B(x_a, \varepsilon) \subset \Omega^c$$

and

$$|x_a - a| \leq \rho(\Omega) + 1. \tag{36.3}$$

We seek for a supersolution of (34.3) of the form $v(x, t) = Ke^{\alpha r}$, $r = |x - x_a|$, $\alpha \geq 0$. The inequality $Pv := v_t - \Delta v + \mu|\nabla v|^q - v^p \geq 0$ needs to be checked only for $r \geq \varepsilon$. The condition to ensure is thus

$$-\alpha^2 Ke^{\alpha r} - \alpha \frac{n-1}{r} Ke^{\alpha r} + \mu \alpha^q K^q e^{\alpha q r} - K^p e^{\alpha p r} \geq 0, \quad r > \varepsilon,$$

which is satisfied if

$$\mu \alpha^q K^{q-1} e^{\alpha(q-1)r} \geq K^{p-1} e^{\alpha(p-1)r} + \alpha^2 + \alpha \frac{n-1}{\varepsilon}, \quad r > \varepsilon.$$

Since $q \geq p > 1$, this is achieved whenever

$$\mu \alpha^q K^{q-1} \geq 2K^{p-1} \quad \text{and} \quad \mu \alpha^q K^{q-1} \geq 2\alpha^2 + 2\alpha \frac{n-1}{\varepsilon}.$$

It thus suffices to choose $\alpha = (2/\mu)^{1/q}$ and next

$$K = \max\left\{\|u_0\|_\infty, 1, \left(\alpha^2 + \alpha(n-1)/\varepsilon\right)^{1/(q-1)}\right\}.$$

It then follows from the comparison principle that $0 \leq u(x, t) \leq v(x, t)$ in Ω, as long as $u(t)$ exists. In particular, using (36.3), we have

$$0 \leq u(a, t) \leq K \exp[(2/\mu)^{1/q}(\rho(\Omega) + 1)].$$

Since a was an arbitrary point in Ω, we deduce that $u(t)$ remains bounded in L^∞ on its existence interval. By virtue of Theorem 35.1, this implies global existence.

Let us next prove the exponential decay statement. Since we now assume that $u_0 \in W_0^{1,r}(\Omega)$, it follows from Example 51.29 that $u \in C([0, \infty), W_0^{1,r}(\Omega)) \cap C((0, \infty), W^{2,r} \cap W_0^{1,r}(\Omega)) \cap C^1((0, \infty), L^r(\Omega))$. We multiply the equation by u^{r-1} and integrate over Ω, which yields, for $t > 0$,

$$\frac{1}{r} \frac{d}{dt} \int_\Omega u^r \, dx = \int_\Omega u^{r-1} \Delta u \, dx + \int_\Omega u^{p+r-1} \, dx - \mu \int_\Omega u^{r-1}|\nabla u|^q \, dx.$$

Integrating by parts, it follows that[6]

$$\frac{1}{r} \frac{d}{dt} \int_\Omega u^r \, dx = \int_\Omega u^{p+r-1} \, dx - (r-1) \int_\Omega u^{r-2}|\nabla u|^2 \, dx - \mu \int_\Omega u^{r-1}|\nabla u|^q \, dx,$$

$$= \int_\Omega u^{p+r-1} \, dx - C_1 \int_\Omega |\nabla u^{r/2}|^2 \, dx - \mu C_2 \int_\Omega \left|\nabla\left(u^{\frac{q+r-1}{q}}\right)\right|^q \, dx.$$

---

[6]Note that we have $r > 2$ if $n \geq 2$, thus integration by parts can be carried out without difficulty. If $n = 1$ and $1 < r < 2$, this can still be done easily.

Here and in what follows, C, C_1, C_2 denote any constant depending only on p, q, r and Ω, but not on μ.

Now, due to Proposition 50.1, we may apply the Poincaré inequality in $H_0^1(\Omega)$ and in $W_0^{1,q}(\Omega)$ to get

$$\frac{1}{r}\frac{d}{dt}\int_\Omega u^r\,dx \le \int_\Omega u^{p+r-1}\,dx - C\int_\Omega u^r\,dx - \mu C\int_\Omega u^{q+r-1}\,dx. \qquad (36.4)$$

Using the inequality

$$x^{p+r-1} \le \varepsilon x^r + C(p,q)\varepsilon^{-(q-p)/(p-1)}x^{q+r-1}, \qquad x \ge 0,\ \varepsilon > 0$$

in case $q > p$, it follows from (36.4) that

$$\frac{d}{dt}\int_\Omega u^r\,dx \le -C\int_\Omega u^r\,dx$$

whenever $q \ge p$ and $\mu > \mu_0(\Omega, p, q, r)$ large enough. Consequently,

$$\int_\Omega u^r(t)\,dx \le \exp(-Ct)\int_\Omega u_0^r\,dx, \qquad t > 0. \qquad (36.5)$$

To prove exponential decay in L^∞, we use an argument of comparison with the model problem (15.1). Fix $t_0 > 0$. By (36.5), we have $\|u(t_0)\|_r \le M := \|u_0\|_r$. Therefore, since $r > n(p-1)/2$, by Theorem 15.2, the solution v of (15.1) with initial data $v(0) = u(t_0)$ exists on a time interval $[0, \tau]$ with $\tau = \tau(M)$ (independent of t_0) and satisfies $\|v(t)\|_\infty \le C\|v(0)\|_r\, t^{-n/2r}$ on $(0, \tau]$. Since $u(t_0 + t) \le v(t)$ on $[0, \tau]$ by the comparison principle, it follows from (36.5) that

$$\|u(t)\|_\infty \le C\|u(t-\tau)\|_r\,\tau^{-n/2r} \le C(M)\exp(-C\,(t-\tau)), \qquad t \ge \tau,$$

hence (36.2) with $s = \infty$. The general case $r \le s \le \infty$ follows by interpolating between $s = r$ and $s = \infty$. $\square$

The main ingredient of the proof of Theorem 36.4(ii) (and of Theorem 36.7 below) is the following lemma.

Lemma 36.6. *Let $p > 1$, $q > 2p/(p+1)$ and $\mu \ge 0$. There exist η, ε, $R > 0$ and a (radial) function $v \ge 0$, of class C^2 on $\mathbb{R}^n \times \mathbb{R}_+$, satisfying:*

$$P_\mu v := v_t - \Delta v - v^p + \mu|\nabla v|^q \le 0, \qquad x \in \mathbb{R}^n,\ t \ge 0, \qquad (36.6)$$

$$\operatorname{supp}(v(t)) \subset B(0, R + \eta t), \qquad t \ge 0, \qquad (36.7)$$

$$\|v(t)\|_\infty = v(0, t) \ge \varepsilon t, \qquad t \ge 0. \qquad (36.8)$$

$$\lim_{t \to \infty} v(x,t) = \infty, \qquad x \in \mathbb{R}^n, \tag{36.9}$$

$$v_t(x,t) \geq 0, \qquad x \in \mathbb{R}^n, \ t \geq 0, \tag{36.10}$$

and

$$\|\nabla v\|_{L^\infty(\mathbb{R}^n \times \mathbb{R}_+)} \leq 1. \tag{36.11}$$

Intuitively, the idea is to seek an unbounded global subsolution, whose gradient remains uniformly bounded, so that the damping effect of the gradient term can never become too important even for large q. This subsolution will take the form of a spherical "expanding wave", which propagates radially away from the origin with an increasing maximum at 0.

Proof of Lemma 36.6. We need two auxiliary functions. Let us first define a function $f : \mathbb{R} \to \mathbb{R}$, of class C^2, by

$$f(s) = \begin{cases} 0, & s \leq 0, \\ 4s^3(1-s), & 0 \leq s \leq 1/2, \\ s - 1/4, & s \geq 1/2. \end{cases}$$

It is easily seen that f satisfies, for some $\varepsilon > 0$,

$$0 \leq f' \leq 1, \ f'' \geq 0, \quad s \in \mathbb{R},$$

$$f'' + f^p \geq 3\varepsilon f', \ s \leq 1/2 \quad \text{and} \quad f^p \geq 3\varepsilon f', \ s \geq 1/2.$$

Next, we define $\beta : \mathbb{R}_+ \to \mathbb{R}$, as

$$\beta(s) = \begin{cases} s + \dfrac{(M-s)^3}{3M^2}, & 0 \leq s \leq M, \\ s, & s > M, \end{cases}$$

with $M = 2n/\varepsilon$. The function β is of class C^2 on $\mathbb{R}_+$, with the following properties:

$$0 \leq \beta(s) \leq M, \quad 0 \leq s \leq M,$$

$$s \leq \beta(s), \ 0 \leq \beta' \leq 1, \ 0 \leq \beta'' \leq \varepsilon/n, \quad s \in \mathbb{R}_+,$$

$$\beta(0) = M/3, \ \beta'(0) = 0.$$

Now we set

$$U(x,t) = f\left(M + \frac{1}{2} + \varepsilon t - \beta(|x|)\right), \quad x \in \mathbb{R}^n, \ t \geq 0,$$

which is of class C^2 on $\mathbb{R}^n \times \mathbb{R}_+$. We compute (omitting the argument in f, f', f'' for simplicity):

$$\nabla U = -\frac{x}{|x|}\beta'(|x|)f' \quad (0 \text{ if } x = 0), \qquad \Delta U = \beta'^2(|x|)f'' - f'\Delta\big(\beta(|x|)\big),$$

$$\Delta\big(\beta(|x|)\big) = \beta''(|x|) + \frac{n-1}{|x|}\beta'(|x|) \leq n \sup \beta'' \leq \varepsilon.$$

First taking $\mu = \varepsilon$ in (34.3), we have

$$P_\varepsilon U = \varepsilon f' - \beta'^2(|x|)f'' + \Delta\beta(|x|)f' - f^p + \varepsilon|\beta'(|x|)f'|^q \leq 3\varepsilon f' - \beta'^2(|x|)f'' - f^p.$$

If $s = 1/2 + M + \varepsilon t - \beta(|x|) \geq 1/2$, then $f^p \geq 3\varepsilon f'$ hence $P_\varepsilon U(x,t) \leq 0$. On the other hand, if $s \leq 1/2$, then $\beta(|x|) \geq M + \varepsilon t \geq M$. Hence $\beta'(|x|) = 1$ and $P_\varepsilon U(x,t) \leq 3\varepsilon f' - f'' - f^p \leq 0$. Now, for a general $\mu > 0$, replacing U by

$$U_\alpha(x,t) = \alpha^{2/(p-1)}U(\alpha x, \alpha^2 t),$$

we get

$$P_\mu U_\alpha = \alpha^{2p/(p-1)}\Big[U_t - \Delta U - U^p + \mu\alpha^{(q(p+1)-2p)/(p-1)}|\nabla U|^q\Big](\alpha x, \alpha^2 t)$$

$$\leq \alpha^{2p/(p-1)}[P_\varepsilon U](\alpha x, \alpha^2 t) \leq 0,$$

for $\alpha > 0$ sufficiently small since $q > 2p/(p+1)$, which proves (36.6) with $v = U_\alpha$. Finally, (36.7)–(36.11) are straightforward consequences of the definition of f (take $R = (M + 1/2)/\alpha$ and $\eta = \varepsilon\alpha$ and replace ε in (36.8) by $\varepsilon\alpha^{2p/(p-1)}$). $\quad\square$

Proof of Theorem 36.4(ii). Let R_j be a sequence of positive reals, $R_j \to \infty$. From the hypotheses, there is a sequence of disjoint balls $B_j = B(x_j, R_j') \subset \Omega$, with $R_j' > R_j$. We are going to construct a suitable subsolution $w = w(x,t)$ of (34.3) on Ω by taking advantage of the scaling properties of the equation. With v as in Lemma 36.6, we set:

$$w_j(x,t) = \frac{1}{j^{2/(p-1)}}v\Big(\frac{x-x_j}{j}, \frac{\gamma_j(t)}{j^2}\Big), \quad x \in \mathbb{R}^n, \ t \geq 0, \ j \in \mathbb{N}^*,$$

with $\gamma_j(t) = M_j t/(M_j + t)$, where the constants $M_j > 0$ will be adjusted later. By (ii)–(iii) in Lemma 36.6, we have:

$$\operatorname{supp}(w_j(t)) \subset B\big(x_j, j(R + \eta M_j/j^2)\big), \quad t \geq 0,$$

$$\|w_j(t)\|_\infty \geq \frac{\varepsilon\gamma_j(t)}{j^{2p/(p-1)}} \longrightarrow \frac{\varepsilon M_j}{j^{2p/(p-1)}} \quad \text{as } t \to \infty.$$

For $(x,t) \in \mathbb{R}^n \times \mathbb{R}_+$, it follows from (i) that

$$Pw_j = j^{-2p/(p-1)}\Big[\gamma_j'(t)v_t - \Delta v - v^p + \mu j^{(2p-q(p+1))/(p-1)}|\nabla v|^q\Big]\Big(\frac{x-x_j}{j}, \frac{\gamma_j(t)}{j^2}\Big)$$

$$\leq j^{-2p/(p-1)}\Big[v_t - \Delta v - v^p + \mu|\nabla v|^q\Big]\Big(\frac{x-x_j}{j}, \frac{\gamma_j(t)}{j^2}\Big) \leq 0,$$

where we have used the fact that $q \geq p > 2p/(p+1)$, $v_t \geq 0$ and $\gamma_j'(t) = M_j^2/(M_j + t)^2 \leq 1$. We now choose

$$M_j = j^{1+2p/(p-1)} \quad \text{and} \quad R_j = j(R + \eta M_j/j^2)$$

and define the function w as:

$$w = \sum_{j \geq 1} w_j.$$

Note that each w_j is supported on B_j and that the B_j are disjoint. By Lemma 36.6, it is clear that w is C^2 on $\mathbb{R}^n \times \mathbb{R}_+$, and hence is a classical subsolution of (34.3). Moreover, by the choice of γ_j, w is bounded on $\mathbb{R}^n \times [0, T]$ for each $T > 0$. We note that $w(0) \in X_+$. (Also, since

$$\|w_j(0)\|_\infty \leq j^{-2/(p-1)}\|v(0)\|_\infty \quad \text{and} \quad \|\nabla w_j(0)\|_\infty \leq j^{-(p+1)/(p-1)}\|\nabla v(0)\|_\infty,$$

it follows from the choice of R_j that $w(0) \in W_0^{1,r}(\Omega)$ for all large r.) By the comparison principle, the solution of (34.3) with initial data $w(0)$ remains above $w(t)$ as long as it exists, which implies the desired conclusion. $\square$

It is not known whether blow-up may occur in finite time when $q \geq p$ and $\rho(\Omega) = \infty$ (except, of course, for the trivial example when u solves the corresponding ODE, i.e. $u_0(x) = C$ in $\Omega = \mathbb{R}^n$). The next result from [498] shows that infinite-time blow-up can occur in the case $\Omega = \mathbb{R}^n$.

Theorem 36.7. *Consider problem (34.3) with $q \geq p > 1$, $\mu > 0$ and $\Omega = \mathbb{R}^n$.*

(i) Assume that $u_0 \in X_+$ has compact support. Then $T_{\max}(u_0) = \infty$.

(ii) There exists $u_0 \in X_+$ with compact support, such that $T_{\max}(u_0) = \infty$ and u is unbounded. Actually, it even holds

$$\lim_{t \to \infty} u(x,t) = \infty, \quad \text{for all } x \in \mathbb{R}^n.$$

Proof. (i) We shall actually prove that the following exponential decay condition (instead of compact support) is sufficient for global existence:

$$0 \leq u_0(x) \leq Ce^{-\varepsilon|x.a|}, \quad x \in \Omega, \qquad \text{for some } C > 0,\ a \in \mathbb{R}^n,\ |a| = 1,$$

where ε is any positive number if $q > p$, or $\varepsilon = \mu^{-1/p}$ if $q = p$. Without loss of generality, we may assume that a is the unit vector in the x_1-direction. We claim that, for a suitable choice of α, the functions

$$v_\pm(x,t) = C\exp(\alpha t \pm \varepsilon x_1)$$

are (traveling wave) supersolutions. If $q > p$ and $\beta > 0$, or if $q = p$ and $\beta = 1$, we have the elementary inequality

$$x^q \geq \beta^{(q-p)/(q-1)} x^p - \beta x, \quad x \geq 0.$$

Therefore,

$$\partial_t v_{\pm} - \Delta v_{\pm} + \mu |\nabla v_{\pm}|^q - v_{\pm}^p$$
$$\geq \partial_t v_{\pm} - \Delta v_{\pm} + \mu \beta^{(q-p)/(q-1)} |\nabla v_{\pm}|^p - \mu \beta |\nabla v_{\pm}| - v_{\pm}^p$$
$$= C \exp(\alpha t \pm \varepsilon x_1)\left(\alpha - \varepsilon^2 - \mu \beta \varepsilon\right)$$
$$+ C^p \exp\left[p(\alpha t \pm \varepsilon x_1)\right]\left(\mu \beta^{(q-p)/(q-1)} \varepsilon^p - 1\right).$$

It thus suffices to choose $\beta = (\mu \varepsilon^p)^{-(q-1)/(q-p)}$ and $\alpha = \varepsilon^2 + \mu \beta \varepsilon$, if $q > p$ and $\varepsilon > 0$, or $\beta = 1$ and $\alpha = \varepsilon^2 + \mu \varepsilon$, if $q = p$ and $\varepsilon = \mu^{-1/p}$. Then we get, thanks to the comparison principle

$$0 \leq u(x,t) \leq v_{\pm}(x,t), \quad x \in \mathbb{R}^n, \ 0 \leq t < T,$$

where $T = T_{\max}(u_0)$ (note that the comparison principle applies, for the same reason as in Remark 36.5). Consequently,

$$0 \leq u(x,t) \leq C \exp(\alpha t - \varepsilon |x_1|), \quad x \in \mathbb{R}^n, \ 0 \leq t < T,$$

hence in particular
$$\|u(t)\|_\infty \leq C \exp(\alpha t), \quad 0 \leq t < T.$$

By virtue of Theorem 35.1, this implies global existence.

(ii) Taking $u_0 = v(0)$, with v as in Lemma 36.6, it is an immediate consequence of that lemma and part (i). $\square$

Remarks 36.8. (i) **Blow-up set.** In Theorem 36.7(ii), we have global blow-up (in infinite time), i.e. the blow-up set is the whole of $\mathbb{R}^n$. Infinite-time blow-up for $q \geq p$ is also known to occur when Ω is a cone (see [498]). But in this case, blow-up takes place only at infinity (the solution remaining bounded for $t \geq 0$ in compact subsets).

(ii) In Theorem 36.4, the largeness assumption on μ_0 for decay is necessary in general. Indeed, when $q > 2p/(p+1)$, $p < p_S$ and Ω is a ball, there exist positive stationary solutions (see [129, Corollary 5.4]).

(iii) For problem (34.3) where the gradient term is replaced with $-\mu u^r |\nabla u|^q$, related results can be found in [497], [58]. $\square$

37. Fujita-type results

We consider the Cauchy problems associated with (34.3) and (34.4), i.e.:

$$\left.\begin{aligned}
u_t - \Delta u &= u^p - \mu|\nabla u|^q, & x \in \mathbb{R}^n,\ t > 0,\\
u(x,0) &= u_0(x), & x \in \mathbb{R}^n,
\end{aligned}\right\} \tag{37.1}$$

and

$$\left.\begin{aligned}
u_t - \Delta u &= u^p - a \cdot \nabla(u^q), & x \in \mathbb{R}^n,\ t > 0,\\
u(x,0) &= u_0(x), & x \in \mathbb{R}^n.
\end{aligned}\right\} \tag{37.2}$$

In this section, we give Fujita-type results for problems (37.1) and (37.2), i.e. we find conditions which guarantee that the solution blows up in finite time for all $u_0 \geq 0$, $u_0 \not\equiv 0$ (and not only for large initial data as in the previous section).

For (37.1), the following result is due to [373] and is based on a method of rescaled test-functions.

Theorem 37.1. *Consider problem (37.1) with $p > 1$, $q = 2p/(p+1)$. There exists $\mu_0(n,p) > 0$ such that if*

$$p < 1 + \frac{2}{n} \quad \text{and} \quad \mu \leq \mu_0,$$

then $T_{\max}(u_0) < \infty$ for any nontrivial $u_0 \in X_+$.

We can complement Theorem 37.1 with the following result.

Theorem 37.2. *Consider problem (37.1) with $p, q > 1$, and assume that at least one of the following assumptions holds:*

(i) $$p > 1 + \frac{2}{n};$$

(ii) $$q < \frac{2p}{p+1};$$

(iii) $$q = \frac{2p}{p+1} \quad \text{and} \quad \mu > \mu_1(n,p) > 0 \text{ large enough.}$$

Then $T_{\max}(u_0) = \infty$ and $\sup_{t \geq 0} \|u(t)\|_\infty < \infty$ for some nontrivial $u_0 \in X_+$.

Remarks 37.3. (i) **Critical exponents.** The value $q = p$ is critical for the blow-up and global existence properties of equation (34.3), as shown in the previous section. Another particular role is played by $q = 2p/(p+1)$. Indeed, for this value of q, the differential equation in (34.3) enjoys the same scaling properties as for $\mu = 0$. Namely, for any solution u and any $\alpha > 0$, the rescaled function $u_\alpha(x,t) := \alpha^{2/(p-1)} u(\alpha x, \alpha^2 t)$ is still a solution. This property is reflected in the existence of blowing-up self-similar solutions (cf. Remark 39.8(i)).

(ii) It seems to be unknown whether (37.1) admits any global solutions when $2p/(p+1) < q < p$ and $p \leq 1 + \frac{2}{n}$. Nonexistence of positive stationary solutions is known when $q > 2p/(p+1)$ and $p \leq n/(n-2)_+$ [470].

(iii) A stronger result than Theorem 37.1 actually holds: Under the assumptions of that theorem there exist no nontrivial nonnegative distributional solutions of $u_t - \Delta u = u^p - \mu|\nabla u|^q$ in $Q = \mathbb{R}^n \times (0,\infty)$, with $u \in L^p_{loc}(Q)$ and $\nabla u \in L^2_{loc}(Q)$. This follows from a small modification of the proof below and from similar arguments as in Step 1 of the proof of Theorem 18.1(i). $\square$

Proof of Theorem 37.1. Assume that $u \geq 0$ is a global solution of (37.1), classical for $t > 0$, with $u \in L^\infty_{loc}(\mathbb{R}^n \times [0,\infty))$.

Step 1. Let $\alpha \in (0,1)$, $a = (p-\alpha)/(p-1) > 1$ and $A_i > 0$, $i = 1, \ldots, 4$, with

$$C_1 := \alpha - A_1 - \mu A_3 \geq 0. \tag{37.3}$$

For simplicity, we shall write $\int$ for the space integral $\int_{\mathbb{R}^n}$ and $\int\int$ for the time-space integral $\int_0^\infty \int_{\mathbb{R}^n}$. We claim that for any compactly supported $\varphi \in C^1(\mathbb{R}^n \times [0,\infty))$, $\varphi \geq 0$, there holds

$$C_2 \int \int u^{p-\alpha}\varphi \leq C_3 \int \int |\nabla\varphi|^{2a}\varphi^{1-2a} + C_4 \int \int |\varphi_t|^a \varphi^{1-a}, \tag{37.4}$$

where

$$C_2 = 1 - C(p,\alpha)\big(A_1^{-1}A_2^{1-a'} + \mu A_3^{-p} + A_4\big), \quad C_3 = A_2/4A_1, \quad C_4 = C(p,\alpha)A_4^{1-a}$$

(the function φ will be later chosen such that the integrals on the RHS will be finite).

Fix $\tau, \varepsilon > 0$ and put $u_\varepsilon = u + \varepsilon$. Multiplying the equation by $u_\varepsilon^{-\alpha}\varphi$ and integrating by parts, we get

$$\int_\tau^\infty \int u^p u_\varepsilon^{-\alpha}\varphi + \alpha \int_\tau^\infty \int |\nabla u|^2 u_\varepsilon^{-1-\alpha}\varphi + \frac{1}{1-\alpha}\int u_\varepsilon^{1-\alpha}(\cdot,\tau)\varphi(\cdot,\tau)$$

$$= \int_\tau^\infty \int u_\varepsilon^{-\alpha}\nabla u \cdot \nabla\varphi + \mu \int_\tau^\infty \int |\nabla u|^q u_\varepsilon^{-\alpha}\varphi + \frac{1}{\alpha-1}\int_\tau^\infty \int u_\varepsilon^{1-\alpha}\varphi_t$$

$$=: I_1 + \mu I_2 + I_3. \tag{37.5}$$

Let us estimate I_1, I_2, I_3 in terms of the double integrals appearing on the LHS. Repeatedly using Young's inequality, we obtain

$$I_1 \leq A_1 \int_\tau^\infty \int |\nabla u|^2 u_\varepsilon^{-1-\alpha}\varphi + B_1 \int_\tau^\infty \int |\nabla\varphi|^2 u_\varepsilon^{1-\alpha}\varphi^{-1}$$

$$\leq A_1 \int_\tau^\infty \int |\nabla u|^2 u_\varepsilon^{-1-\alpha}\varphi + B_1 A_2 \int_\tau^\infty \int |\nabla\varphi|^{2a}\varphi^{1-2a} + B_1 B_2 \int_\tau^\infty \int u_\varepsilon^{p-\alpha}\varphi,$$

$$I_2 \leq A_3 \int_\tau^\infty \int |\nabla u|^2 u_\varepsilon^{-1-\alpha}\varphi + B_3 \int_\tau^\infty \int u_\varepsilon^{p-\alpha}\varphi$$

and

$$I_3 \leq A_4 \int_\tau^\infty \int u_\varepsilon^{p-\alpha}\varphi + C_4 \int_\tau^\infty \int |\varphi_t|^a \varphi^{1-a},$$

where $B_1 = (4A_1)^{-1}$, $B_2 = C(p,\alpha)A_2^{1-a'}$ and $B_3 = C(p)A_3^{-p}$. Plugging the above estimates in (37.5), we find that

$$\int_\tau^\infty \int u^p u_\varepsilon^{-\alpha}\varphi - (B_1 B_2 + \mu B_3 + A_4)\int_\tau^\infty \int u_\varepsilon^{p-\alpha}\varphi + C_1 \int_\tau^\infty \int |\nabla u|^2 u_\varepsilon^{-1-\alpha}\varphi$$
$$\leq C_3 \int_\tau^\infty \int |\nabla\varphi|^{2a}\varphi^{1-2a} + C_4 \int_\tau^\infty \int |\varphi_t|^a \varphi^{1-a}.$$

Due to assumption (37.3), the third term in the LHS can be left out. Since φ is compactly supported, with the help of the monotone convergence theorem, we may pass to the limit $\varepsilon \to 0$, and then $\tau \to 0$, in the first two terms of the LHS. This yields (37.4).

Step 2. Choose

$$0 < \alpha < 1 - n(p-1)/2, \tag{37.6}$$

$A_1 = \alpha/2$ and $A_3 = 1$. By taking A_2 large, A_4 small (depending only on n,p), and then $\mu < \mu_0(n,p)$ small, we have (37.3) and $C_2 > 0$.

Now consider φ of the form $\varphi(x,t) = \psi\left(\frac{|x|}{R}\right)\psi\left(\frac{t}{R^2}\right)$. Here $R > 0$, $\psi \in C^1([0,\infty))$ satisfies $\psi' \leq 0$ and

$$\psi(s) = \begin{cases} 1, & 0 \leq s \leq 1, \\ (2-s)^m, & 3/2 \leq s \leq 2, \\ 0, & s \geq 2, \end{cases}$$

with $m > 2a > 2$. Inequality (37.4) implies

$$C_2 \int\int_\Sigma u^{p-\alpha} \leq C_3 \int\int_{\Sigma'} |\nabla\varphi|^{2a}\varphi^{1-2a} + C_4 \int\int_{\Sigma'} |\varphi_t|^a \varphi^{1-a}, \tag{37.7}$$

where

$$\Sigma = \{(x,t) : |x| \leq R,\ 0 \leq t \leq R^2\}, \quad \Sigma' = \{(x,t) : |x| \leq 2R,\ 0 \leq t \leq 2R^2\}.$$

Observe that the integrals on the RHS are finite (the integrands are continuous, including at $|x| = 2R$, $t = 2R^2$ due to $m > 2a$). The substitutions $x = Ry$, $t = R^2 s$ into the integrals on the right-hand side of (37.7) then yield

$$C_2 \int\int_\Sigma u^{p-\alpha} \leq CR^{n+2-2a}.$$

Since $n + 2 - 2a < 0$ due to (37.6), by letting $R \to \infty$, we conclude that $u \equiv 0$. $\square$

Proof of Theorem 37.2. By virtue of Theorem 35.1, it suffices to obtain a uniform estimate of u.

If $p > 1 + (2/n)$, then u is a subsolution of the same problem with $\mu = 0$ and the same initial data. Global existence for small initial data then follows from Theorem 20.1 in view of the comparison principle.

If $q = 2p/(p+1)$ and $\mu > \mu_1(p)$ large enough we shall show that there exists a (bounded stationary) supersolution of the form $U(x) = \varepsilon(1 + |x|^2)^{-a}$, with $a = 1/(p-1)$, which will imply the desired conclusion.

We have

$$\nabla U = -2\varepsilon a x (1 + |x|^2)^{-(a+1)}, \quad -\Delta U = 2\varepsilon a \big[n + (n - 2 - 2a)|x|^2\big](1 + |x|^2)^{-(a+2)}.$$

By choosing $0 < \varepsilon, r_0 < 1$ small enough (depending only on n, p), we first guarantee that

$$-\Delta U \geq an\varepsilon \geq \varepsilon^p \geq U^p, \quad |x| \leq r_0. \tag{37.8}$$

Next, for $|x| > r_0$, there holds

$$\Delta U \leq C_1(1 + |x|^2)^{-(a+1)} = C_1(1 + |x|^2)^{-p/(p-1)}, \quad U^p \leq (1 + |x|^2)^{-p/(p-1)}$$

and

$$|\nabla U|^q \geq C_2(1 + |x|^2)^{-q(a+(1/2))} = C_2(1 + |x|^2)^{-p/(p-1)}$$

for some $C_1, C_2 > 0$ depending only on n, p. Therefore,

$$-\Delta U + \mu_1|\nabla U|^q \geq U^p, \quad |x| > r_0,$$

provided $\mu_1 = \mu_1(n, p)$ is chosen large enough. This along with (37.8) guarantees that U is a supersolution.

Finally, if $q < 2p/(p+1)$ and $\mu > 0$, let us put $V(x) = \alpha^{2/(p-1)}U(\alpha x)$. Since $|\nabla U|$ is bounded, we have $|\nabla U|^q \geq c|\nabla U|^{2p/(p+1)}$ in $\mathbb{R}^n$ for some $c > 0$. For $\alpha > 0$ sufficiently small, it follows that

$$\big(-\Delta V + \mu|\nabla V|^q - V^p\big)(x)$$
$$= \alpha^{2p/(p-1)}\big(-\Delta U + \mu\alpha^{(q(p+1)-2p)/(p-1)}|\nabla U|^q - U^p\big)(\alpha x)$$
$$\geq \alpha^{2p/(p-1)}\big(-\Delta U + \mu_1|\nabla U|^{2p/(p+1)} - U^p\big)(\alpha x) \geq 0,$$

so that V is a supersolution. $\square$

We now turn to problem (37.2). The following result from [5] shows that the critical number (for p) may depend on both n and q.

Theorem 37.4. *Consider problem (37.2) with $p, q > 1$, $a \neq 0$, and set*

$$p_1 = p_1(n, q) := \min\left(1 + \frac{2}{n}, 1 + \frac{2q}{n+1}\right).$$

(i) *If $q \leq p \leq p_1$, then $T_{\max}(u_0) < \infty$ for any nontrivial $u_0 \in X_+$.*

(ii) *If $p > p_1$, then $T_{\max}(u_0) = \infty$ for some nontrivial $u_0 \in X_+$.*

Remarks 37.5. (a) **Critical exponents.** It was also shown in [5] that when $q = 1$, the critical exponent becomes $p = 1 + 2/n$. We thus observe that the critical exponent $p_1(n, q)$ is a discontinuous function of q (since $p_1(n, q) \to 1 + 2/(n+1)$, as $q \to 1+$, in view of Theorem 37.4).

(b) It is known (see [498, Proposition 3.6] and its proof) that blow-up in finite or infinite time can occur for (37.2) whenever $q \geq p > 1$ and that this actually occurs for all nontrivial $u_0 \geq 0$ when $q > p > 1$ and $p < 1 + 2/n$ (see [498, Remark 3.4] and [5]). However it is unknown whether the blow-up time is finite or infinite.

(c) The following proof is a simplification of the original proof of [5] (especially for part (ii) in Case 1 below). Moreover, it yields uniform decay rates for suitably small data in assertion (ii). $\square$

Proof of Theorem 37.4. (i) We shall prove the result only for $p < p_1$, the equality case being more involved.

Set

$$\phi(x) = C \exp\left(-\frac{1}{n}(1 + |x|^2)^{1/2}\right),$$

where $C > 0$ is chosen so that $\int_{\mathbb{R}^n} \phi(x)\, dx = 1$. For $i = 1, \ldots, n$, we have

$$|\partial_{x_i}\phi| \leq \frac{\phi}{n}, \qquad \partial^2_{x_i x_i}\phi \geq -\frac{\phi}{n}. \tag{37.9}$$

Without loss of generality, we may assume that $a = |a|e_1$. Let $\gamma \geq 0$ to be fixed below. Let $\lambda \in (0, 1]$, and put $\phi_\lambda(x) = \lambda^{n+\gamma}\phi(\lambda^{1+\gamma}x_1, \lambda x')$, where $x = (x_1, x')$. By (37.9), we have

$$\int_{\mathbb{R}^n} \phi_\lambda(x)\, dx = 1, \qquad \Delta\phi_\lambda \geq -\lambda^2\phi_\lambda, \qquad |(\phi_\lambda)_{x_1}| \leq \lambda^{1+\gamma}\phi_\lambda.$$

Multiplying equation (37.2) by ϕ_λ and integrating on $\mathbb{R}^n$, we obtain, for $t > 0$,

$$\frac{d}{dt}\int_{\mathbb{R}^n} u\phi_\lambda = \int_{\mathbb{R}^n} \phi_\lambda \Delta u + \int_{\mathbb{R}^n} u^p\phi_\lambda - |a|\int_{\mathbb{R}^n} (u^q)_{x_1}\phi_\lambda$$

$$= \int_{\mathbb{R}^n} u\Delta\phi_\lambda + \int_{\mathbb{R}^n} u^p\phi_\lambda + |a|\int_{\mathbb{R}^n} u^q(\phi_\lambda)_{x_1}$$

$$\geq -\lambda^2\int_{\mathbb{R}^n} u\phi_\lambda + \int_{\mathbb{R}^n} u^p\phi_\lambda - |a|\lambda^{1+\gamma}\int_{\mathbb{R}^n} u^q\phi_\lambda$$

(this can be easily justified by using the exponential decay of ϕ and the fact that $u(\cdot,t) \in BC^2(\mathbb{R}^n)$).

Denote $y_\lambda(t) = \int_{\mathbb{R}^n} u(t)\phi_\lambda$. If $q < p$, by Young's inequality, we observe that

$$|a|\lambda^{1+\gamma}u^q = u^{p(q-1)/(p-1)}\left(|a|\lambda^{1+\gamma}u^{(p-q)/(p-1)}\right)$$
$$\leq \tfrac{1}{2}u^p + C\lambda^{(1+\gamma)(p-1)/(p-q)}u \tag{37.10}$$

for some $C = C(p,q,|a|) > 0$. If $q = p$, then (37.10) is obviously true with $C = 0$ for all λ small. Using $\int_{\mathbb{R}^n} u^p\phi_\lambda \geq y_\lambda^p$ (owing to Jensen's inequality) and (37.10), we deduce that

$$y_\lambda'(t) \geq \tfrac{1}{2}y_\lambda^p - (\lambda^2 + C\lambda^{(1+\gamma)(p-1)/(p-q)})y_\lambda.$$

It follows that y_λ, and hence u, cannot exist for all $t > 0$ whenever the RHS in the previous inequality is positive at $t = 0$. This is satisfied if

$$\left(\int_{\mathbb{R}^n} u_0(x)\phi(\lambda^{1+\gamma}x_1, \lambda x')\,dx\right)^{p-1} > 2\lambda^{-(n+\gamma)(p-1)}(\lambda^2 + C\lambda^{(1+\gamma)(p-1)/(p-q)}).$$
$$\tag{37.11}$$

Now, since $p < p_1$, by choosing $0 < \gamma < \gamma_+ := 2/(p-1) - n$ close to γ_+, we get

$$(n+\gamma)(p-1) < 2 \quad\text{and}\quad n+\gamma < (1+\gamma)/(p-q).$$

Since, by monotone convergence, the LHS in (37.11) converges to $\left(\phi(0)\int_{\mathbb{R}^n} u_0\right)^{p-1} \in (0,\infty]$ as $\lambda \to 0$, (37.11) holds for $\lambda > 0$ sufficiently small and we conclude that $T_{\max}(u_0) < \infty$.

(ii) By Proposition 51.16, it suffices to obtain a uniform estimate of u on bounded time intervals.

Case 1: $q > 1 + (1/n)$. This case is simple, since one can directly build a (self-similar) supersolution of (37.2) under the form

$$v(x,t) = t^\alpha \tilde{G}(x,t) \cdot$$

for some $0 < \alpha < n/2$, where $\tilde{G} = (4\pi)^{n/2}G$ and G is the Gaussian heat kernel. Indeed, setting $k = n/2 - \alpha$, the function v satisfies

$$v_t - \Delta v - v^p + a \cdot \nabla(v^q) = t^\alpha(\tilde{G}_t - \Delta\tilde{G}) + \alpha t^{\alpha-1}\tilde{G} - t^{\alpha p}\tilde{G}^p + t^{\alpha q}a \cdot \nabla(\tilde{G}^q)$$
$$= \alpha t^{-k-1}e^{-|x|^2/4t} - t^{-kp}e^{-p|x|^2/4t}$$
$$- qt^{-kq-1/2}\left(\frac{x \cdot a}{2\sqrt{t}}\right)e^{-q|x|^2/4t}$$
$$\geq \left(\alpha t^{-k-1} - t^{-kp} - Ct^{-kq-1/2}\right)e^{-|x|^2/4t},$$

where we used $se^{-qs^2} \leq Ce^{-s^2}$, $s \geq 0$. Now, since $p > p_1 = 1+2/n$ and $q > 1+1/n$, by taking $\alpha > 0$ sufficiently small, it follows that $kp > k+1$ and $kq + 1/2 > k+1$, so that $v_t - \Delta v - v^p + a \cdot \nabla(v^q) \geq 0$ in $\mathbb{R}^n$ for $t \geq t_0$, where $t_0 \geq 1$ is large enough. If $u_0(x) \leq t_0^{-n/2} \exp(-|x|^2/4t_0)$, the comparison principle in Proposition 52.16 then guarantees that $u(t) \leq v(t_0 + t)$ on $[0, T_{\max}(u_0))$ and u exists globally.

Case 2: $q \leq 1+(1/n)$. This case is more involved and requires the consideration of the auxiliary problem:

$$\begin{cases} v_t - \Delta v = -(1+t)^r a \cdot \nabla(v^q), & t > 0, \quad x \in \mathbb{R}^n, \\ v(x,0) = u_0(x), & x \in \mathbb{R}^n, \end{cases} \qquad (37.12)$$

with $r > 0$. By Proposition 51.16, for any $u_0 \in X_+$, problem (37.12) has a unique classical solution $v \geq 0$. By the maximum principle, we have

$$\|v(t)\|_\infty \leq \|u_0\|_\infty, \qquad (37.13)$$

which guarantees the global existence of v, in view of (51.39). Moreover, v satisfies

$$v \in L^\infty_{loc}((0, \infty), BC^2(\mathbb{R}^n)). \qquad (37.14)$$

If in addition $u_0 \in L^1(\mathbb{R}^n)$, then

$$v \in C([0, \infty), L^1(\mathbb{R}^n)). \qquad (37.15)$$

We shall use the following lemma:

Lemma 37.6. *For $1 < q \leq 2$ and $u_0 \in L^\infty \cap L^1(\mathbb{R}^n)$, $u_0 \geq 0$, the solution of (37.12) satisfies the estimate*

$$\|v(t)\|_\infty \leq C(\|u_0\|_1 + \|u_0\|_\infty)(1 + t)^{-(n+1+2r)/(2q)}. \qquad (37.16)$$

Proof. Assume that $a = |a|e_1$ without loss of generality.

Step 1. Set $z := v^{q-1}$ and $w := z_{x_1} = (q-1)v^{q-2}v_{x_1}$. We claim that

$$w(x,t) \leq \frac{r+1}{q|a|}t^{-r-1}, \quad x \in \mathbb{R}^n, \ t > 0. \qquad (37.17)$$

By the strong maximum principle (apply Proposition 52.7 in any bounded subdomain) we have $v > 0$ in $\mathbb{R}^n \times (0, \infty)$ (unless $v \equiv 0$). By continuous dependence, it thus suffices to establish (37.17) when u_0 also satisfies $u_0 \geq \varepsilon > 0$, hence $v \geq \varepsilon$. The function z verifies

$$z_t - \Delta z + \frac{q-2}{q-1}\frac{|\nabla z|^2}{z} = -q|a|(1+t)^r z z_{x_1}.$$

By parabolic regularity results, $w \in C^{2,1}(\mathbb{R}^n \times (0, \infty))$. Differentiating in x_1, we get

$$w_t - \Delta w + \frac{2(q-2)}{q-1} \frac{\nabla z \cdot \nabla w}{z} + \frac{2-q}{q-1} \frac{|\nabla z|^2}{z^2} w = -q|a|(1+t)^r (w^2 + z w_{x_1}). \quad (37.18)$$

Since $1 < q \le 2$, for each $t_0 \in (0, 1]$, the function $\tilde{w}(t) = \frac{r+1}{q|a|}(t + t_0)^{-r-1}$ is a supersolution of (37.18). On the other hand, for fixed $\tau > 0$, by taking t_0 small enough, we can ensure that $w(\tau) < \tilde{w}(0)$. Since, for $t \ge \tau$, $z, \nabla z$ are bounded and z is bounded away from 0 (due to (37.14) and $v \ge \varepsilon$), it follows from a small modification of the comparison principle in Proposition 52.6 that

$$w(x, \tau + t) \le \tilde{w}(x, t) \le \frac{r+1}{q|a|} t^{-r-1}, \quad x \in \mathbb{R}^n, \ t > 0.$$

Claim (37.17) follows by letting $\tau \to 0$.

Step 2. Write $x = (x_1, x')$. We claim that

$$\|h(t)\|_\infty \le \|u_0\|_1 (4\pi t)^{-(n-1)/2}, \quad \text{where } h(x', t) = \int_{-\infty}^{\infty} v(x_1, x', t) \, dx_1. \quad (37.19)$$

Formally, by integrating (37.12) on $\mathbb{R}$ with respect to x_1, we see that h solves $h_t - \Delta h = 0$ in $\mathbb{R}^{n-1} \times (0, \infty)$, so that (37.19) would follow as a consequence of the L^1-L^∞-estimate. However, integration needs to be justified and we shall proceed instead as follows. For fixed $R > 0$, letting $h_R(x', t) = \int_{-R}^{R} v(x_1, x', t) \, dx_1$ and integrating (37.12) on $(-R, R)$ with respect to x_1, we obtain

$$\partial_t h_R - \Delta_{x'} h_R = \left[\left(v_{x_1} - |a|(1+t)^r v^q \right)(x_1, x', t) \right]_{x_1 = -R}^{R}, \quad x' \in \mathbb{R}^{n-1}, \ t > 0.$$

Fix $0 < \tau < T < \infty$. It follows from (37.14) and (37.15) that

$$v(x, t), v_{x_1}(x, t) \to 0, \ |x| \to \infty, \quad\quad\quad\quad (37.20)$$

uniformly for $t \in [\tau, T]$. Therefore,

$$\partial_t h_R - \Delta_{x'} h_R \le \varepsilon(R), \quad x' \in \mathbb{R}^{n-1}, \ \tau \le t \le T,$$

where $\lim_{R \to \infty} \varepsilon(R) = 0$. For $x' \in \mathbb{R}^{n-1}$ and $t \in [\tau, T]$, it follows from the maximum principle that

$$h_R(x', t) \le (G_{t-\tau} * h_R(\tau))(x') + \varepsilon(R)(t - \tau).$$

By the L^1-L^∞-estimate, we deduce that

$$h_R(x', t) \le (4\pi(t - \tau))^{-(n-1)/2} \|h_R(\tau)\|_{L^1(\mathbb{R}^{n-1})} + \varepsilon(R)t$$
$$\le (4\pi(t - \tau))^{-(n-1)/2} \|v(\tau)\|_{L^1(\mathbb{R}^n)} + \varepsilon(R)T.$$

Letting $R \to \infty$ and then $\tau \to 0$, using (37.15), we deduce (37.19).

$Step\ 3.$ By (37.20) and (37.17), we have

$$v^q(x_1, x', t) = q/(q-1) \int_{-\infty}^{x_1} (v^{q-1})_{x_1} v(y_1, x', t)\, dy_1 \leq Ct^{-r-1} h(x', t).$$

This, combined with (37.19), yields (37.16) for $t \geq 1$, whereas (37.13) gives (37.16) for $t \leq 1$. $\square$

Completion of proof of Theorem 37.4. Let $U(x, t) = (1+t)^m v$, where v is a solution of (37.12) for $r = m(q-1)$ and $m > 0$ to be fixed later on. We shall prove that if $\|u_0\|_1 + \|u_0\|_\infty$ is small enough, then U is a supersolution of (37.2) (hence $v \leq U$ by the comparison principle in Proposition 52.16). We have

$$U_t - \Delta U = (1+t)^m (v_t - \Delta v) + m(1+t)^{m-1} v = -a \cdot \nabla(U^q) + m(1+t)^{m-1} v.$$

Therefore, it will be enough to see that $m(1+t)^{m-1} v \geq (1+t)^{mp} v^p$ or equivalently:

$$\|v(t)\|_\infty \leq m^{1/(p-1)}(1+t)^{-m-1/(p-1)}. \tag{37.21}$$

But, since $p > p_1 = 1 + 2q/(n+1)$, we may choose $m > 0$ so small that $m + 1/(p-1) \leq (n+1+2m(q-1))/2q$, and (37.21) follows from the lemma. The proof of Theorem 37.4 is complete. $\square$

38. A priori bounds and blow-up rates

The following result shows that universal bounds of the form (26.25), known for the model problem (15.1), remain valid for the perturbed problem (34.2) if the perturbation term is not too strong. In particular, this implies a (universal) a priori bound for global solutions and the usual blow-up rate estimate.

Theorem 38.1. *Let $p > 1$ and $T > 0$. Assume that either*

$$p < p_B$$

or

$$p < p_S, \qquad \Omega = \mathbb{R}^n\ or\ B_R, \qquad u = u(|x|, t), \qquad g = g(u, |\xi|).$$

Assume in addition that the function $g : \mathbb{R}_+ \times \mathbb{R}^n \to \mathbb{R}$ satisfies the growth assumption

$$|g(u, \xi)| \leq C_0(1 + |u|^{p_1} + |\xi|^q),$$

$$for\ some\ 1 \leq p_1 < p\ and\ 1 < q < 2p/(p+1). \tag{38.1}$$

$$g = u^{p\pm} |\nabla u|^q$$

$$|g| \leq C(|u|^p + |\xi|^q) \leq C(1 + |u|^p + |\xi|^q)$$

$$p < 1 + \frac{2}{N} = \frac{N+2}{N} < \frac{N+2}{N-2}$$

$$p < \frac{n+2}{n-2}$$

Then, for any nonnegative classical solution of

$$\left.\begin{array}{ll} u_t - \Delta u = u^p + g(u, \nabla u), & x \in \Omega, \ 0 < t < T, \\ u = 0, & x \in \partial\Omega, \ 0 < t < T, \end{array}\right\} \qquad (38.2)$$

there holds

$$u(x,t) + |\nabla u(x,t)|^{2/(p+1)} \le C\big(1 + t^{-1/(p-1)} + (T-t)^{-1/(p-1)}\big), \quad x \in \Omega, \ 0 < t < T,$$

with $C = C(p, p_1, q, C_0, \Omega) > 0$.

Assumption (38.1) is satisfied for instance for problems (34.3) and (34.4) when $q < 2p/(p+1)$ or $q < (p+1)/2$, respectively. The method of proof is based on rescaling and doubling arguments, already used in the proof of Theorem 26.8. Note that this method does not use any variational structure, and is thus well adapted to problem (38.2).

Proof. Since the proof is very similar to that of Theorem 26.8, we only sketch the main changes. Instead of (26.34), we define the functions M_k by

$$M_k := u_k^{(p-1)/2} + |\nabla u_k|^{(p-1)/(p+1)}.$$

Rescaling similarly as in (26.39) with again

$$\lambda_k := M_k^{-1}(x_k, t_k) \to 0,$$

the function v_k is now a solution of the equation

$$\left.\begin{array}{ll} \partial_s v_k - \Delta_y v_k = v_k^p + g_k, & (y,s) \in \tilde{D}_k, \\ v_k = 0, & y \in \lambda_k^{-1}(\partial\Omega - x_k), \ \ |y| < k/2, \ \ |s| < k^2/4, \end{array}\right\}$$

with

$$g_k(y,s) := \lambda_k^{2p/(p-1)} g\big(\lambda_k^{-2/(p-1)} v_k(y,s), \lambda_k^{-(p+1)/(p-1)} \nabla v_k(y,s)\big),$$

$$v_k^{(p-1)/2}(0) + |\nabla v_k|^{(p-1)/(p+1)}(0) = 1,$$

and

$$v_k^{(p-1)/2} + |\nabla v_k|^{(p-1)/(p+1)} \le 2, \quad (y,s) \in \tilde{D}_k.$$

The growth assumption (38.1) then implies

$$|g_k| \le C\lambda_k^m \ \text{ in } \tilde{D}_k, \quad \text{where } m := \min\left\{\frac{2(p-p_1)}{p-1}, \frac{2p - q(p+1)}{p-1}\right\} > 0.$$

Now, as in the proof of Theorem 26.8, we distinguish the cases (26.42) and (26.43). By using parabolic L^p-estimates, we obtain a subsequence of $\{v_k\}$ converging to a nonnegative solution v of (21.1) or (26.45). The difference is that we now use convergence in $C^{1+\sigma,\sigma/2}(\mathbb{R}^n \times \mathbb{R})$, which is satisfied due to the embedding (1.2). Therefore we get $v^{(p-1)/2}(0) + |\nabla v|^{(p-1)/(p+1)}(0) = 1$, so that v is nontrivial (moreover v and ∇v are bounded). As before, we reach a contradiction with a Liouville-type theorem. $\quad \square$

Remarks 38.2. Blow-up rate. (i) For problem (34.2), the lower blow-up estimate $\|u(t)\|_\infty \geq C(p)(T-t)^{1/(p-1)}$ is true whenever g satisfies $g(u,0) \leq 0$ and u blows up in L^∞-norm (see Theorem 35.1 for a sufficient condition). This follows from the proof of Proposition 23.1.

(ii) By using different arguments, based on a modification of the method of the auxiliary function J of [219], the upper blow-up estimate

$$\|u(t)\|_\infty \leq C(T-t)^{-\frac{1}{p-1}}, \quad 0 \leq t < T$$

(this time with C depending on u) was obtained in [131] for equation (34.3) with $q < 2p/(p+1)$, under different assumptions. Namely, no restriction is made on $p > 1$, but the solution is assumed to satisfy $u_t \geq 0$. $\square$

In the rest of this section, we shall see that the conclusions of Theorem 38.1 may become false for stronger perturbation terms in equation (34.2) (so that the growth restriction $q < 2p/(p+1)$ in (38.1) is not purely technical — although it is presently unknown whether it is optimal).

First, concerning a priori estimates of global solutions, we just recall Theorems 36.4 and 36.7, which already provide us with examples of global solutions of (34.3), unbounded as $t \to \infty$, whenever $q \geq p$ (in, e.g., $\Omega = \mathbb{R}^n$). A further counter-example in that direction can be found in [156] for (34.3) with $p > q = 2$, $n = 1$, $\Omega = (-1,1)$. In that example, the solution stabilizes (monotonically) in infinite time to a stationary solution singular at $x = 0$.

Next, we shall show that for stronger absorbing gradient terms, the blow-up rate may become faster, or type II [264]. Let us consider the following problem

$$\left.\begin{array}{ll}
u_t - u_{xx} = (u+1)^p - \lambda \dfrac{u_x^2}{u+1}, & -1 < x < 1,\ t > 0, \\[2mm]
u = 0, & x = \pm 1,\ t > 0, \\[2mm]
u(x,0) = u_0(x), & -1 < x < 1,
\end{array}\right\} \quad (38.3)$$

with $p > 1$ and $\lambda \geq 0$. Note that (38.3) is of the form (34.2) (with $g(u,u_x) = (u+1)^p - u^p - \lambda \frac{u_x^2}{u+1}$).

Theorem 38.3. *Consider problem (38.3) with $\lambda > p > 1$. Assume that $u_0 \in X_+$ is even and nonincreasing in $|x|$. If $T := T_{\max}(u_0) < \infty$, then*

$$(T-t)^{1/(p-1)}\|u(t)\|_\infty \to \infty \quad as\ \ t \to T. \tag{38.4}$$

Remarks 38.4. (i) **Instability of the blow-up rate.** It was moreover proved in [264] that the assumption on λ in Theorem 38.3 is optimal: If $0 < \lambda \leq p$ (and $u_t \geq 0$), then the usual blow-up rate is verified:

$$C_1 \leq (T-t)^{1/(p-1)}\|u(t)\|_\infty \leq C_2, \quad 0 < t < T,$$

for some constants $C_1, C_2 > 0$. This shows a phenomenon of strong sensitivity to gradient perturbations (with $\lambda = p$ being the threshold value for problem (38.3)).

(ii) It is unknown whether or not the value $q = 2p/(p+1)$ in Theorem 38.1 is optimal. However, observe that the PDE in (38.3), rewritten in terms of $v := u+1$, has the same scale invariance properties as that in (34.3) for $q = 2p/(p+1)$ (cf. Remark 37.3(i)). Theorem 38.3 thus suggests that the dividing line for problem (34.3) could be given by the scaling. Namely, global unbounded solutions might exist for $q > 2p/(p+1)$, and type II blow-up for $2p/(p+1) < q < p$ (or even for $q = 2p/(p+1)$ and μ large). This conjecture is also supported by Theorem 39.1 below. $\square$

Theorem 38.3 will be deduced from a result of [264] on dead-cores for the absorption problem

$$\left.\begin{aligned} w_t - w_{xx} &= -w^r, & -1 < x < 1,\ t > 0,\\ w(\pm 1, t) &= k, & t > 0,\\ w(x, 0) &= w_0(x), & -1 < x < 1, \end{aligned}\right\} \tag{38.5}$$

where $0 < r < 1$ and $k > 0$. Indeed, following [304], we notice that (38.3) is transformed into (38.5) by the change of unknown

$$u + 1 = aw^{-m}, \qquad m = \frac{1}{\lambda - 1}, \qquad a = m^{1/(p-1)}, \tag{38.6}$$

with

$$r = \frac{\lambda - p}{\lambda - 1} \in (0, 1), \qquad k = (\lambda - 1)^{(1-\lambda)/(p-1)}. \tag{38.7}$$

Now w is nondecreasing in $|x|$ and blow-up of u at $t = T$ is equivalent to the appearance of a dead-core for w, i.e. $w(T, 0) = 0$. Note that $w \geq 0$ exists for all times $t > 0$, with $w \in C^{2,1}([-1, 1] \times (0, \infty))$. The fast blow-up estimate (38.4) becomes equivalent to

$$\lim_{t \to T} (T - t)^{-\alpha} w(0, t) = 0, \qquad \alpha = \frac{1}{1 - r}. \tag{38.8}$$

The proof of (38.8) relies on backward similarity variables, a tool that we have already used in Section 25. Namely, following [244] (see also [228]) and [217], set

$$T - t = e^{-s}, \qquad y = x/\sqrt{T - t} \qquad \text{and} \qquad w(x, t) = (T - t)^\alpha v(y, s).$$

Then v satisfies the equation

$$v_s = v_{yy} - \frac{y}{2} v_y + \alpha v - v^r \quad \text{in } D, \tag{38.9}$$

where $D := \{(y, s) : -\log T < s < \infty, \ |y| < e^{s/2}\}$. Under the assumptions of Theorem 38.3, we shall actually show the more precise convergence statement

$$\lim_{s \to \infty} v(y, s) = V_1(y) := k_r |y|^{2\alpha}, \qquad k_r = \left[\frac{(1-r)^2}{2(1+r)}\right]^{\alpha}, \tag{38.10}$$

uniformly on $\{|y| < R\}$ for each $R > 0$, from which (38.8) (and hence (38.4)) readily follows.

A quick check reveals that the right-hand side $V_1(y)$ of (38.10) provides an (unbounded) stationary solution of (38.9), more precisely a solution of

$$V_{yy} - \frac{y}{2}V_y + \alpha V - V^r = 0, \quad y \in \mathbb{R}. \tag{38.11}$$

Note that each solution of (38.11) corresponds to a self-similar solution of $w_t = w_{xx} - w^r$ in $\mathbb{R} \times (-\infty, T)$ given by $w(x,t) = (T-t)^{1/(1-r)}V(x/\sqrt{T-t})$. On the other hand $V_1(x)$, restricted to $[-1, 1]$, is also a stationary solution of (38.5) with $k = k_r$.

The proof of Theorem 38.3 will then be carried out in three steps.

(i) Identify the stationary solutions of (38.9) (in a suitable set);

(ii) Prove that all the global solutions of (38.9) are attracted by the set of stationary solutions of (38.9) (in the locally compact topology);

(iii) Discard all the possible limits other than the stationary solution V_1.

We need three lemmas. In what follows, we shall use the fact that, by parabolic regularity results, $w_x \in C^{2,1}((-1, 1) \times (0, T)) \cap C([-1, 1] \times (0, T])$. We start with a lower estimate which is the key ingredient in step (iii).

Lemma 38.5. *Let u satisfy the hypotheses of Theorem 38.3, and w be defined by (38.6)–(38.7). There exists $c_1 > 0$ (depending on u) such that*

$$w(x,t) \geq \left[w^{1-r}(0,t) + c_1 x^2\right]^{\alpha}, \qquad |x| \leq 1, \ T/2 \leq t \leq T. \tag{38.12}$$

Proof. The basic idea of the proof is similar to that in Theorem 24.1 (cf. [219]), but some special care is required and an auxiliary nonlocal parabolic equation has to be considered — cf. (38.17) below. We set

$$J = w_x - \varepsilon x w^r.$$

It will be sufficient to show that, for $\varepsilon > 0$ small enough, there holds

$$J \geq 0 \quad \text{in } [0, 1] \times (T/2, T). \tag{38.13}$$

Indeed, we will then have $(w^{1-r})_x = (1-r)w^{-r}w_x \geq \varepsilon(1-r)x$ in $[0, 1] \times (T/2, T)$ and the estimate will immediately follow by integrating in space between 0 and x.

To prove (38.13), we first claim that for $\varepsilon > 0$ sufficiently small, we have

$$J(x, T/2) > 0 \text{ in } (0, 1], \qquad J_x(0, t) > 0 \text{ and } J(1, t) > 0 \text{ on } (T/2, T). \qquad (38.14)$$

We have $w_x \geq 0$ in $[0, 1] \times (0, T]$ and $w_x(0, t) = 0$ in $(0, T]$. Next, since $w(x, t) \leq k$ (due to $w_0 \leq k$), we have $w_x(1, t) > 0$ in $(0, T]$ by Hopf's lemma (cf. Proposition 52.7). Moreover, since $z := w_x$ satisfies $z_t - z_{xx} = -rw^{r-1}z$ in $[0, 1] \times (0, T)$, the strong maximum principle (Proposition 52.7) then implies

$$w_x(x, t) > 0 \quad \text{in } (0, 1] \times (0, T]. \qquad (38.15)$$

As z achieves its minimum value $z = 0$ at $x = 0$ for each $t \in (0, T)$, we also have

$$w_{xx}(0, t) = z_x(0, t) > 0 \quad \text{in } (0, T), \qquad (38.16)$$

in view of Hopf's lemma. The claim (38.14) follows from (38.15) and (38.16).

Let us now compute
$$(xw^r)_t = xrw^{r-1}w_t,$$

$$(xw^r)_x = w^r + xrw^{r-1}w_x,$$

$$(xw^r)_{xx} = 2rw^{r-1}w_x + xrw^{r-1}w_{xx} + xr(r-1)w^{r-2}(w_x)^2.$$

We get

$$\begin{aligned}
J_t - J_{xx} &= (w_t - w_{xx})_x - \varepsilon(xw^r)_t + \varepsilon(xw^r)_{xx} \\
&= -rw^{r-1}w_x + \varepsilon\big(-xrw^{r-1}w_t + 2rw^{r-1}w_x \\
&\qquad + xrw^{r-1}w_{xx} + xr(r-1)w^{r-2}(w_x)^2\big) \\
&= -rw^{r-1}\big(w_x + \varepsilon x(w_t - w_{xx})\big) + \varepsilon rw^{r-2}w_x\big(2w + x(r-1)w_x\big) \\
&= -rw^{r-1}J + \varepsilon rw^{r-2}w_x\big(2w + x(r-1)J + \varepsilon x^2(r-1)w^r\big).
\end{aligned}$$

Putting

$$a(x, t) = rw^{r-1} + \varepsilon xr(1-r)w^{r-2}w_x \quad \text{and} \quad b(x, t) = 2\varepsilon r(1-r)w^{2r-2}w_x,$$

we obtain

$$J_t - J_{xx} + aJ = \varepsilon rw^{r-2}w_x\big(2w - \varepsilon x^2(1-r)w^r\big)$$

$$= b\Big(\frac{w^{1-r}}{1-r} - \varepsilon\frac{x^2}{2}\Big).$$

On the other hand, we note that

$$\Big(\frac{w^{1-r}}{1-r} - \varepsilon\frac{x^2}{2}\Big)_x = w_x w^{-r} - \varepsilon x = w^{-r}J.$$

It follows that

$$J_t - J_{xx} + aJ = b\left(\frac{w^{1-r}(0,t)}{1-r} + \int_0^x w^{-r}J(y,t)\,dy\right). \tag{38.17}$$

We can then apply a simple nonlocal version of the maximum principle to deduce that $J \geq 0$. The key point which allows this is that the function b is positive. Let us give the details to make everything safe.

By continuity, using (38.14), we have

$$E := \{\tau \in (T/2, T) : J > 0 \text{ on } (0,1] \times [T/2, \tau)\} \neq \emptyset.$$

Assume for contradiction that $t_0 := \sup E < T$. Then there holds $J(x, t_0) \geq 0$ in $[0,1]$ and there exists $x_0 \in (0,1)$ such that $J(x_0, t_0) = 0$, $J_t(x_0, t_0) \leq 0$ and $J_{xx}(x_0, t_0) \geq 0$. Substituting this into (38.17) and noting that $b(x_0, t_0) > 0$, we obtain

$$0 \geq (J_t - J_{xx} + aJ)(x_0, t_0) = b(x_0, t_0)\left(\frac{w^{1-r}(0, t_0)}{1-r} + \int_0^{x_0} w^{-r}J(y, t_0)\,dy\right) > 0.$$

This contradiction shows that $t_0 = T$, which gives the desired conclusion. $\quad\square$

Lemma 38.6. *Let u satisfy the hypotheses of Theorem 38.3, and w be defined by* (38.6)–(38.7). *There exists $c_2 > 0$ (depending on u) such that*

$$w(x,t) \leq \left[w^{(1-r)/2}(0,t) + c_2|x|\right]^{2\alpha} \tag{38.18}$$

for all $T/2 \leq t \leq T$, $|x| \leq 1$. Moreover, the corresponding global solution v of (38.9) *satisfies*

$$v(y,s) \leq C(1 + |y|)^{2\alpha} \quad and \quad |v_y(y,s)| \leq C(1 + |y|)^{2\alpha-1} \tag{38.19}$$

for all $-\log(T/2) =: s_0 < s < \infty$, $|y| < e^{s/2}$.

Proof. We consider the function $J(x,t) := \frac{1}{2}w_x^2 - Cw^{r+1}$, where $C > 1$ is a constant to be determined later. We compute

$$\begin{aligned}
J_t - J_{xx} &= w_x(w_t - w_{xx})_x - w_{xx}^2 - C(r+1)\left[w^r(w_t - w_{xx}) - rw^{r-1}w_x^2\right] \\
&= -rw^{r-1}w_x^2 - w_{xx}^2 + C(r+1)\left[w^{2r} + rw^{r-1}w_x^2\right] \\
&= C(r+1)w^{2r} + r(C(r+1) - 1)w^{r-1}w_x^2 - w_{xx}^2
\end{aligned}$$

in $(0,1) \times (T/2, T)$. Using the relation $w_x^2 = 2(J + Cw^{r+1})$, we get

$$J_t - J_{xx} + b_1 J = C[1 - r + 2r(r+1)C]w^{2r} - w_{xx}^2,$$

with $b_1(x,t) = 2(1-C(r+1))rw^{r-1}$. Then using $w_{xx} = J_x/w_x + C(r+1)w^r$ (recall that $w_x > 0$ in $(0,1) \times (T/2,T)$) and setting $b_2(x,t) = J_x/w_x^2 + 2C(r+1)w^r/w_x$, we end up with

$$J_t - J_{xx} + b_2 J_x + b_1 J = C(1-r)[1 - (r+1)C]w^{2r} < 0.$$

Now, for $C > 0$ sufficiently large, we have $J < 0$ on the parabolic boundary of $Q := (0,1) \times (T/2,T)$. The maximum principle then yields $J \le 0$ in Q, hence

$$(w^{(1-r)/2})_x = \frac{1-r}{2} w_x w^{-(r+1)/2} \le C'$$

and the estimate (38.18) follows. Note that we get in turn the estimate

$$|w_x| \le Cw^{(r+1)/2}, \qquad |x| \le 1, \quad T/2 \le t \le T. \tag{38.20}$$

Let us next prove (38.19). Since $w_{xx}(0,t) \ge 0$, we have $w_t(0,t) \ge -w^r(0,t)$. By integrating between t and T, we easily get $w(0,t) \le C(T-t)^\alpha$. By combining this with (38.18), we obtain

$$v(y,s) = (T-t)^{-\alpha} w(y\sqrt{T-t}, t) \le (T-t)^{-\alpha}\left[w^{(1-r)/2}(0,t) + c_2|y|\sqrt{T-t}\right]^{2\alpha}$$

$$\le C(T-t)^{-\alpha}\left[\sqrt{T-t} + |y|\sqrt{T-t}\right]^{2\alpha} = C(1+|y|)^{2\alpha}.$$

The estimate of v_y then follows from (38.20). The proof of the lemma is complete. $\square$

Next, for step (i), we have the following.

Lemma 38.7. *Let* $V \in C^2(\mathbb{R})$ *be a solution of* (38.11) *such that*

$$V = V(|y|), \quad \text{with} \quad V' \ge 0, \quad V > 0 \quad \text{for all } y > 0,$$

and such that V *is polynomially bounded. Then*

$$V = V_1 := k_r|y|^{2/(1-r)} \quad \text{or} \quad V = V_2 := \kappa := (1-r)^{1/(1-r)}.$$

Proof. Let $W := V^{1-r}$ and denote $' = d/dy$. Since $V > 0$ for $y > 0$, W is smooth there. The equation for W is:

$$W'' - \frac{y}{2}W' + \frac{r}{1-r}\frac{W'^2}{W} + W = 1 - r. \tag{38.21}$$

By differentiating, we note that

$$W''' - \frac{y}{2}W'' + \frac{1}{2}W' = -\frac{r}{1-r}\left(\frac{W'^2}{W}\right)' = -\frac{r}{1-r}\frac{W'(2W''W - W'^2)}{W^2}. \quad (38.22)$$

Set $H := W - \frac{y}{2}W'$ and let $D = \{y > 0 : H(y) \neq 0\}$. For all $y \in D$, $Z := |H|^{\frac{1}{1-r}}$ is smooth and we compute

$$Z' = \frac{1}{1-r}|H|^{\frac{2r-1}{1-r}}HH', \qquad Z'' = \frac{1}{1-r}|H|^{\frac{2r-1}{1-r}}\left(HH'' + \frac{r}{1-r}H'^2\right),$$

hence

$$\frac{y}{2}Z' - Z'' = \frac{1}{1-r}|H|^{\frac{2r-1}{1-r}}\left(H\left(\frac{y}{2}H' - H''\right) - \frac{r}{1-r}H'^2\right),$$

and

$$H' = \frac{1}{2}W' - \frac{y}{2}W'', \qquad H'' = -\frac{y}{2}W'''.$$

Using (38.22), it follows that, for all $y \in D$,

$$\frac{y}{2}Z' - Z''$$

$$= \frac{1}{1-r}|H|^{\frac{2r-1}{1-r}}\left\{\frac{y}{2}\left(W - \frac{y}{2}W'\right)\left(W''' - \frac{y}{2}W'' + \frac{W'}{2}\right) - \frac{r}{1-r}\left(\frac{W'}{2} - \frac{y}{2}W''\right)^2\right\}$$

$$= -\frac{r}{4(1-r)^2}|H|^{\frac{2r-1}{1-r}}\left\{y(2W - yW')\frac{W'(2W''W - W'^2)}{W^2} + (W' - yW'')^2\right\}$$

$$= -\frac{r}{4(1-r)^2}|H|^{\frac{2r-1}{1-r}}\left\{W'^2 + y^2\left(\frac{WW'' - W'^2}{W}\right)^2 + 2yW'\left(\frac{WW'' - W'^2}{W}\right)\right\}$$

$$= -\frac{r}{4(1-r)^2}|H|^{\frac{2r-1}{1-r}}\left(W' + y\left(\frac{WW'' - W'^2}{W}\right)\right)^2 \leq 0$$

hence

$$(e^{-y^2/4}Z')' \geq 0 \quad \text{in } D. \quad (38.23)$$

We next claim that the function $Z \geq 0$ is nonincreasing in $(0, \infty)$. Indeed, otherwise, there would exist y_0 such that $Z(y_0) > 0$ and $Z'(y_0) > 0$, hence $Z' \geq Ce^{y^2/4}$ for $y \geq y_0$ by (38.23). Due to $|(y^{-2}W)'| = 2y^{-3}Z^{1-r}$, we would get $W \geq e^{\eta y^2}$ as $y \to \infty$, for some $\eta > 0$, contradicting the polynomial bound assumed on V.

Now assume for contradiction that Z is nonconstant on $(0, \infty)$. Then there is $R > 0$ such that $Z(0) > Z(R)$ and we may choose $\varepsilon > 0$ so small that $f := Z + \varepsilon e^{y^2/2}$ satisfies $f(0) > Z(0) > f(R)$. It follows that f has a local maximum at some $y_1 \in (-R, R)$ and that $Z(y_1) > 0$ (hence $y_1 \in D$). Therefore, at $y = y_1$, we get $0 \leq (y/2)f' - f'' \leq \varepsilon((y^2/2) - 1 - y^2)e^{y^2/2} < 0$, a contradiction.

We deduce that $W - (y/2)W' = C$ on $(0, \infty)$. By integration, we finally get $W = A + By^2$ and the conclusion follows easily by substituting into equation (38.21). $\square$

Now, the rest of the proof of Theorem 38.3 via the dead-core rate estimate (38.8), and in particular Step (ii), will be a consequence of energy arguments close to those from Section 25 (cf. [244]) for blow-up problems. A difference with [244], [217] is that here v is *not* uniformly bounded; and indeed it will be proved that, unlike in those works, v converges to an unbounded self-similar profile.

Proof of Theorem 38.3. Let $\rho(y) = e^{-y^2/4}$ and define $R(s) = e^{s/2}$ and

$$E(s) = \int_0^{R(s)} \left(\frac{v_y^2}{2} + \frac{v^{r+1}}{r+1} - \frac{\alpha v^2}{2} \right)(y, s)\rho(y)\, dy.$$

For $s \geq s_0 = -\log(T/2)$, we have

$$E'(s) = R'(s)\rho\left(\frac{v_y^2}{2} + \frac{v^{r+1}}{r+1} - \frac{\alpha v^2}{2} \right)(R(s), s) + \int_0^{R(s)} \left(v_y v_{ys} + (v^r - \alpha v)v_s \right)\rho\, dy$$

$$= \rho\left[R'(s)\left(\frac{v_y^2}{2} + \frac{v^{r+1}}{r+1} - \frac{\alpha v^2}{2} \right) + v_y v_s \right](R(s), s)$$

$$+ \int_0^{R(s)} \left(-(\rho v_y)_y + \rho(v^r - \alpha v) \right)v_s\, dy$$

$$= \rho\left[R'(s)\left(\frac{v_y^2}{2} + \frac{v^{r+1}}{r+1} - \frac{\alpha v^2}{2} \right) + v_y v_s \right](R(s), s) - \int_0^{R(s)} \rho v_s^2\, dy$$

$$\equiv A(s) - \int_0^{R(s)} \rho v_s^2\, dy.$$

On the other hand, by (38.19), we have

$$|E(s)| \leq \int_0^\infty Ce^{-y^2/4}(1 + |y|)^{4\alpha}\, dy = C, \qquad s \geq s_0$$

and, using $v_s(R(s), s) = \left(\alpha v - \frac{y}{2}v_y \right)(R(s), s)$ and (38.19), we obtain

$$|A(s)| \leq C \exp\left(-\frac{1}{4}e^s \right)e^{s/2}(1 + e^{s/2})^{4\alpha},$$

hence $A(s) \in L^1(s_0, \infty)$. It follows that

$$\int_{s_0}^\infty \int_0^{R(s)} \rho v_s^2\, dy\, ds < \infty.$$

Then, by arguing similarly as in the proof of Lemma 25.6(i), we deduce that, for each sequence $s_n \to \infty$, there exists a subsequence s'_n such that $v(\cdot, s'_n)$ converges to a solution V of (38.11), uniformly on $\{|y| < R\}$ for each $R > 0$.

But on the other hand, by the lower bound (38.12), for each $y \in \mathbb{R}$ and $s > 2\log|y|$, we have

$$v(y, s) = (T - t)^{-\alpha} w\left(y\sqrt{T - t}, t\right) \geq (T - t)^{-\alpha}\left(c_1 |y\sqrt{T - t}|^2\right)^{\alpha} = c_1^{\alpha} |y|^{2\alpha}.$$

In view of Lemma 38.7 and (38.19), this shows that necessarily $V = V_1$. The conclusion follows. $\square$

39. Blow-up sets and profiles

The following results show that there is a threshold $q = 2p/(p + 1)$ above which the absorbing gradient term has a strong influence on the final blow-up profile of solutions of (34.3), making it more and more singular as q increases to p (observe that $q/(p - q) > 2/(p - 1)$ for $2p/(p + 1) < q < p$). Theorems 39.1 and 39.2 are from [131] and [490], respectively.

Theorem 39.1. *Consider problem* (34.3) *with* $1 < q < p$, $\mu > 0$, *and* $\Omega = B_R$. *Let* $u_0 \in X_+$ *be radial nonincreasing and such that* $T := T_{\max}(u_0) < \infty$. *Then* 0 *is the only blow-up point. Moreover, for all* $\alpha > \alpha_0$, *there holds*

$$u(r, t) \leq C_\alpha r^{-\alpha}, \quad 0 \leq t < T, \quad 0 < r \leq R,$$

with

$$\alpha_0 = \begin{cases} 2/(p - 1) & \text{if } 1 < q \leq 2p/(p + 1), \\ q/(p - q) & \text{if } 2p/(p + 1) < q < p. \end{cases}$$

The optimality of Theorem 39.1 is shown by the following:

Theorem 39.2. *Under the hypotheses of Theorem* 39.1, *assume in addition that* $u_t \geq 0$ *in* Q_T. *Then there exist* $C, \eta > 0$ *such that*

$$u(r, T) := \lim_{t \to T} u(r, t) \geq C r^{-\alpha_0}, \quad 0 < r < \eta. \tag{39.1}$$

Remarks 39.3. (i) The assumption $u_t \geq 0$ is guaranteed if u_0 is a subsolution of the stationary problem (see Proposition 52.19), and it is not difficult to construct such initial data.

(ii) We have $\eta = \eta(u_0) > 0$ in (39.1), but we may take $C = C(p) > 0$ if $1 < q \leq 2p/(p + 1)$, $C = C(p, q, \mu) > 0$ if $2p/(p + 1) < q < p$. $\square$

The proof of Theorem 39.1 consists of two steps. The first one (Lemma 39.4) is a modification of the argument of [219] (cf. the proof of Theorem 24.1), which consists in estimating $-u_r$ from below near $r = 0$, by applying the maximum principle to an auxiliary function of the form $J = u_r + c_\varepsilon(r)F(u)$. The second step (in the case $q > 2p/(p + 1)$) is an additional bootstrap argument (on the value of γ for $F(u) = u^\gamma$), which enables one to reach the optimal exponent α_0.

Lemma 39.4. *Consider problem (34.3) with $1 < q < p$, $\mu > 0$, $\Omega = B_R$, and let u_0 be as in Theorem 39.1. Denote $f(u) = u^p$ and let $F \in C^2((0,\infty)) \cap C^1([0,\infty))$ satisfy $F, F', F'' \geq 0$, with $F'' F$ bounded near 0. Let $\delta > 0$ and set $K = (q-1)\mu$. Assume that*

$$G(y) := \int_y^\infty \frac{ds}{F(s)} < \infty, \quad y > 0$$

and that, for all sufficiently small $\varepsilon > 0$,

$$
\begin{aligned}
f'F - fF' &+ \varepsilon^2 r^{2+2\delta} F'' F^2 + \delta(n+\delta) r^{-2} F \\
&\geq 2\varepsilon(1+\delta) r^\delta F' F + 2^{q-1} K \varepsilon^q r^{q+q\delta} F^q F', \quad u > 0, \quad 0 < r \leq R.
\end{aligned}
\tag{39.2}
$$

Then 0 is the only blow-up point and there exists $\varepsilon_0 > 0$ such that

$$u(r,t) \leq G^{-1}(\varepsilon_0 r^{2+\delta}) \quad \text{in } [0, R] \times [T/2, T).$$

Remark 39.5. If u is a given solution satisfying the assumptions of Lemma 39.4, then the conclusion remains valid if (39.2) is assumed to hold for all r and all $u = u(r,t)$ such that $(r,t) \in (0,R) \times [T/2, T)$ (instead of for all $u > 0$ and $0 < r < R$). This fact will be used in the proof of Theorem 39.1(ii). $\square$

Proof of Lemma 39.4. Set

$$J = w + c_\varepsilon(r) F(u)$$

where $c_\varepsilon(r) = \varepsilon r^{1+\delta}$ and $w = u_r \leq 0$. By parabolic regularity results, we have $w \in C^{2,1}((0,R) \times (0,T)) \cap C([0,R] \times (0,T))$. Also, $u > 0$ in $[0,R) \times (0,T)$ by the strong maximum principle. Differentiating the equation

$$u_t - u_{rr} - \frac{n-1}{r} u_r = f(u) - \mu |u_r|^q$$

with respect to r yields

$$w_t - w_{rr} - \frac{n-1}{r} w_r = -\frac{n-1}{r^2} w + f'(u)w + q\mu |w|^{q-1} w_r. \tag{39.3}$$

Using (39.3) and writing f, F for $f(u), F(u)$, we compute the equation for J:

$$
\begin{aligned}
J_t - J_{rr} - \frac{n-1}{r} J_r &= -\frac{n-1}{r^2} w + f'(u)w + q\mu |w|^{q-1} w_r + c_\varepsilon[f - \mu |w|^q] F' \\
&\quad - 2w c_\varepsilon' F' - \left[\frac{n-1}{r} c_\varepsilon' + c_\varepsilon''\right] F - c_\varepsilon w^2 F''.
\end{aligned}
$$

Using the relations $w = J - c_\varepsilon F$, $w^2 = c_\varepsilon^2 F^2 + (J - 2c_\varepsilon F)J$ and

$$w_r = J_r - c_\varepsilon' F - c_\varepsilon F' w,$$

we obtain

$$J_t - J_{rr} - \left(\frac{n-1}{r} + q\mu|w|^{q-1}\right)J_r - b_0 J$$

$$= c_\varepsilon\left\{\left(-f' - \frac{c_\varepsilon'}{c_\varepsilon}q\mu|w|^{q-1} + \frac{n-1}{r^2} - \frac{c_\varepsilon''}{c_\varepsilon} - \frac{n-1}{r}\frac{c_\varepsilon'}{c_\varepsilon}\right)F \right.$$

$$\left. + (f + K|w|^q)F' + 2c_\varepsilon'F'F - c_\varepsilon^2 F''F^2\right\}$$

where

$$b_0 = f' - (n-1)r^{-2} - 2c_\varepsilon'F' - c_\varepsilon(J - 2c_\varepsilon F)F''.$$

The assumption (39.2) is equivalent to

$$f'F - fF' - 2c_\varepsilon'F'F + c_\varepsilon^2 F''F^2 - 2^{q-1}Kc_\varepsilon^q F^q F' + \left(\frac{c_\varepsilon''}{c_\varepsilon} + \frac{n-1}{r}\frac{c_\varepsilon'}{c_\varepsilon} - \frac{n-1}{r^2}\right)F \geq 0.$$

Combining this with $|w|^q \leq 2^{q-1}(|J|^q + c_\varepsilon^q F^q)$, we obtain

$$J_t - J_{rr} - \left(\frac{n-1}{r} + q\mu|w|^{q-1}\right)J_r - bJ \leq 0 \quad \text{in } (0, R) \times (0, T],$$

where $b = b_0 + 2^{q-1}Kc_\varepsilon F'|J|^{q-2}J$.

On the other hand, arguing as in the proof of Theorem 24.1, we obtain $u_r < 0$ in $(0, R] \times (0, T)$ and $u_{rr}(0, t) < 0$ in $(0, T)$. It follows that $J(\cdot, T/2) \leq 0$ in $[0, R]$ for ε small. Obviously $J(0, t) \leq 0$ and $J(R, t) < 0$ for all $t \in (0, T)$. Since b is bounded above in $((0, R) \times (0, \tau)) \cap \{(x, t) : J > 0\}$ for each $\tau \in (0, T)$, it follows from the maximum principle (cf. Proposition 52.4 and Remark 52.11(a)) that $J \leq 0$ in $[0, R] \times [T/2, T)$. Integrating this inequality between 0 and r yields the conclusion. $\square$

We shall show that condition (39.2) in Lemma 39.4 is satisfied for $F(u) = u^\gamma$ with suitable choices of $\gamma > 1$. The inequality (39.2) takes the form

$$(p - \gamma)u^{p+\gamma-1} + (\varepsilon r^{1+\delta})^2\gamma(\gamma - 1)u^{3\gamma-2} + \delta(n + \delta)r^{-2}u^\gamma$$

$$\geq 2\varepsilon\gamma(1 + \delta)r^\delta u^{2\gamma-1} + 2^{q-1}K\gamma(\varepsilon r^{1+\delta})^q u^{\gamma q+\gamma-1}.$$

In the proof of this inequality, we shall need the following elementary lemma.

Lemma 39.6. *Let n be a positive integer, $R, K, \delta > 0$ and $p > 1$.*
(i) *If $1 < \gamma < p$, then for $\varepsilon > 0$ small enough, there holds*

$$\frac{1}{2}\left((p - \gamma)u^{p+\gamma-1} + \delta(n + \delta)r^{-2}u^\gamma\right)$$

$$\geq 2\varepsilon\gamma(1 + \delta)r^\delta u^{2\gamma-1}, \quad 0 < r \leq R, \quad u \geq 0. \tag{39.4}$$

(ii) *If $1 < q < 2p/(p+1)$ and $\gamma \in (p/q, p)$, then for $\varepsilon > 0$ small enough, there holds*

$$\frac{1}{2}\left((p-\gamma)u^{p+\gamma-1} + (\varepsilon r^{1+\delta})^2 \gamma(\gamma-1)u^{3\gamma-2}\right)$$
$$\geq 2^{q-1}K\gamma(\varepsilon r^{1+\delta})^q u^{\gamma q + \gamma - 1}, \quad 0 < r \leq R, \quad u \geq 0. \tag{39.5}$$

(iii) *If $1 < q < p$ and $\gamma = p/q$, then for $\varepsilon > 0$ small enough, there holds*

$$\frac{1}{2}(p-\gamma)u^{p+\gamma-1} \geq 2^{q-1}K\gamma(\varepsilon r^{1+\delta})^q u^{\gamma q + \gamma - 1}, \quad 0 < r \leq R, \quad u \geq 0. \tag{39.6}$$

Proof. Inequalities (39.4) and (39.5) are consequences of Young's inequality

$$\frac{a^\alpha}{\alpha} + \frac{b^\beta}{\beta} \geq ab, \quad a, b \geq 0, \ \alpha, \beta > 1, \ \frac{1}{\alpha} + \frac{1}{\beta} = 1,$$

where we choose $\alpha = (p-1)/(\gamma-1)$ in the case of (39.4) and $\alpha = (2\gamma - 1 - p)/(2\gamma - 1 - \gamma q)$ in the case of (39.5). In inequality (39.6), it is sufficient to choose $\varepsilon \leq (\frac{p-\gamma}{2^q K\gamma})^{1/q} R^{-1-\delta}$. $\square$

We now continue with the proof of Theorem 39.1.

Proof of Theorem 39.1. First assume $1 < q < 2p/(p+1)$. In this case, we choose $F(u) = u^\gamma$ with $1 < \gamma < p$ and Lemma 39.6(i) and (ii) yields that (39.2) holds. Lemma 39.4 then implies

$$u(r,t) \leq Cr^{-\frac{2+\delta}{\gamma-1}}$$

and $\frac{2+\delta}{\gamma-1}$ can be made arbitrarily close to $2/(p-1)$.

Next consider the case $2p/(p+1) \leq q < p$. Now we first choose $F(u) = u^\gamma$ with $\gamma = p/q$ and Lemma 39.6(i) and (iii) yields that (39.2) holds. Lemma 39.4 implies

$$u(r,t) \leq Cr^{-\alpha}, \quad \alpha = \alpha(\delta, \gamma) = \frac{2+\delta}{\gamma-1}. \tag{39.7}$$

Inequality (39.6) is equivalent to

$$u^{\gamma q - p} \leq \frac{p-\gamma}{K\gamma}(2\varepsilon r^{1+\delta})^{-q}$$

and, due to the estimate (39.7) on u, it is also true (for $u = u(r,t)$ — cf. Remark 39.5) if γ is replaced by $\bar{\gamma} \in (\gamma, p)$ such that $(\bar{\gamma}q - p)\alpha < (1+\delta)q$, or, equivalently,

$$\bar{\gamma} < \frac{p}{q} + \frac{1+\delta}{2+\delta}(\gamma - 1).$$

If δ is chosen small enough, this reduces to

$$\overline{\gamma} < \frac{p}{q} + \frac{\gamma - 1}{2}.$$

Clearly,

$$\gamma < \frac{p}{q} + \frac{\gamma - 1}{2} \quad \text{if} \quad \gamma < \frac{2p}{q} - 1 \ (\leq p),$$

and

$$\alpha(\delta, \gamma) \to \frac{q}{p - q} \quad \text{as} \quad \delta \to 0, \ \gamma \to \frac{2p}{q} - 1.$$

Consequently, an obvious bootstrap argument implies the assertion. $\qquad\square$

Proof of Theorem 39.2. We modify the argument in the proof of Theorem 24.3.

Step 1. We claim that

$$\|u_r(t)\|_\infty \leq C_1 u^\gamma(0, t), \tag{39.8}$$

with $\gamma = \min\big((p+1)/2, p/q\big) > 1$.

On the one hand, since $u_t \geq 0$ and $u_r \leq 0$, we have

$$\frac{\partial}{\partial r}\left(\frac{1}{2} u_r^2 + \frac{1}{p+1} u^{p+1}\right) = (u_{rr} + u^p) u_r = \left(u_t + \mu |u_r|^q - \frac{n-1}{r} u_r\right) u_r \leq 0,$$

hence

$$\left(\frac{1}{2} u_r^2 + \frac{1}{p+1} u^{p+1}\right)(r, t) \leq \frac{1}{p+1} u^{p+1}(0, t).$$

Therefore, we get (39.8) with $\gamma = (p+1)/2$ (and $C_1 = C_1(p)$).

On the other hand, for each $t \in (0, T)$, at a point $r \in (0, R]$ where $|u_r(\cdot, t)|$ achieves its maximum, we have

$$\mu |u_r|^q = u^p + u_{rr} - u_t + \frac{n-1}{r} u_r \leq u^p,$$

due to $u_t \geq 0$, $u_r \leq 0$ and $u_{rr}(r, t) \leq 0$. This yields (39.8) with $\gamma = p/q$ (and $C_1 = \mu^{-1/q}$), hence the claim.

Step 2. For $0 < t < T := T_{\max}(u_0)$, let $r_0(t)$ be such that $u(r_0(t), t) = \frac{1}{2} u(0, t)$. Note that, since $u_r < 0$ for $0 < t < T$ and $0 < r \leq R$, the implicit function theorem guarantees that $r_0(t)$ is unique and is a continuous function of t. Since $T_{\max}(u_0) < \infty$, we have $u(0, t) \to \infty$ as $t \to T$, due to Theorem 35.1. Also, by Theorem 39.1, we know that 0 is the only blow-up point, hence $r_0(t) \to 0$ as $t \to T$. Now we have

$$-u_r \leq C_2 u^\gamma, \qquad 0 \leq r \leq r_0(t).$$

Integrating, we get

$$u^{1-\gamma}(r_0(t), t) \leq u^{1-\gamma}(0, t) + C_3 r_0(t) = 2^{1-\gamma} u^{1-\gamma}(r_0(t), t) + C_3 r_0(t),$$

hence $u(r_0(t), t) \geq C_4(r_0(t))^{-1/(\gamma-1)}$. Using $u_t \geq 0$, it follows that

$$u(r_0(t), T) \geq C_4(r_0(t))^{-1/(\gamma-1)}, \quad 0 < t < T.$$

Since r_0 is continuous and $r_0(t) \to 0$ as $t \to T$, we deduce that the range $r_0((0, T))$ contains an interval of the form $(0, \eta)$ and the conclusion follows. $\square$

For equation (38.3), the arguments in the proof of Theorem 38.3 provide precise information on the blow-up profile, which turns out to be slightly less singular than for the model problem (15.1) (cf. Remark 25.8).

Theorem 39.7. *Consider problem* (38.3) *with* $\lambda > p$. *Assume that* $u_0 \in X_+$ *is even and nonincreasing in* $|x|$. *If* $T := T_{\max}(u_0) < \infty$, *then for each* $x \neq 0$, $u(x, T) := \lim_{t \to T} u(x, t)$ *exists and it satisfies*

$$C_1 \leq |x|^{2/(p-1)} u(x, T) \leq C_2, \quad x \text{ small}, \ x \neq 0,$$

for some constants $C_1, C_2 > 0$.

Proof. The (globally defined) solution $w \geq 0$ of the transformed problem (38.5) (cf. formulas (38.6)–(38.7)) satisfies (38.12) and (38.18). In particular, by (38.12), parabolic estimates and standard embeddings, we have $u \in BUC^\alpha(\{\varepsilon < |x| < 1\} \times (T/2, T))$ for each $\varepsilon > 0$ and some $\alpha \in (0, 1)$. It follows that the limit $u(x, T)$ exists for $x \in [-1, 1] \setminus \{0\}$. On the other hand, since $w(0, T) = 0$, (38.12) and (38.18) imply

$$(c_1|x|^2)^\alpha \leq w(x, T) \leq (c_2|x|)^{2\alpha}, \quad -1 < x < 1.$$

The assertion concerning $u(x, T)$ follows immediately. $\square$

Remarks 39.8. (i) **Self-similar blowing-up solutions.** As mentioned in Remark 37.3(i) when $q = 2p/(p+1)$, the equation (34.3) is scale-invariant. This property has been exploited in [495] to construct backward self-similar (blow-up) solutions by ODE methods. More precisely, for each $0 < \mu < 2$ and $1 < p < p_0(n, \mu)$, there exists a solution of (34.3) of the form

$$u(x, t) = (T - t)^{-1/(p-1)} W\big(x/(T - t)^{1/2}\big), \tag{39.9}$$

for $(x, t) \in \mathbb{R}^n \times (-\infty, T)$. Here W is a positive, C^2, radially symmetric decreasing function on $\mathbb{R}^n$. Moreover, for all such solutions, W satisfies

$$\lim_{|x| \to \infty} |x|^{2/(p-1)} W(x) = C > 0.$$

This guarantees that u blows up at the single point $x = 0$ and admits a limiting profile, similar to that obtained for equation (38.3) in Theorem 39.7, given by

$$u(x, T) = C|x|^{-2/(p-1)}, \quad \text{for all } x \neq 0.$$

In contrast, recall that no nontrivial, backward, self-similar solutions exist for $\mu = 0$ and $p \le p_S$ (cf. Proposition 25.4).

Comparison of this result with Theorems 39.1–39.2 yields the interesting and a bit surprising observation that the gradient term can have opposite effects on the blow-up profile: When the perturbation is mild ($q = 2p/(p+1)$), the profile is slightly less singular; when the perturbation is strong ($2p/(p+1) < q < p$), it is more singular.

(ii) **Single-point vs. regional blow-up.** We have seen several examples of single-point blow-up for equations with dissipative gradient terms in the radial case (cf. Theorems 39.1 and 39.7 and Remark 39.8(i)). Also, examples of single-point blow-up for the convective problem (34.4) can be found in [218]. On the other hand, it was proved in [131] that if Ω is convex bounded, $\mu > 0$ and $q < 2p/(p+1)$, then the blow-up set of any solution of (34.3) is a compact subset of Ω. The situation is quite different when $\mu < 0$. Namely, for $q = 2$ one has single-point blow-up if $p > 2$, regional blow-up if $p = 2$, and global blow-up if $1 < p < 2$ (see [313], [304], [231]). The proof relies on the transformation $v = e^u - 1$, which converts (34.3) into the equation with mildly superlinear source $v_t - \Delta v = (1 + v)\log^p(1 + v)$.

The authors of [304] interpret this result in the following way. While the term u^p alone would force the solution to develop a spike at the maximum point, hence causing single-point blow-up, the gradient term now has a positive sign and tends to push up the steeper parts of the graph of $u(.,t)$. This enhances regional or even global blow-up, the influence of the gradient term becoming more important as the value of p decreases.

(iii) L^∞ **boundary blow-up for a Dirichlet problem.** For the convective problem (34.4), a surprising example was recently constructed in [202], of a solution blowing up (only) at the boundary, in spite of the imposed homogeneous Dirichlet boundary condition. More precisely, consider problem (34.4) with $n = 1$, $\Omega = (0, \infty)$, $p > 1$ and $q = (p+1)/2$. Then, for $-a > 0$ sufficiently large, there exists a positive solution u such that

$$\limsup_{t \to T} \sup_{x > 0} u(x, t) = \infty$$

and

$$u(x, t) \le C|x|^{-2/(p-1)}, \quad x > 0, \; 0 < t < T.$$

This solution is constructed in the backward self-similar form (39.9), now with $W(y) > 0$, $y > 0$, and $W(0) = 0$ (note that (34.4) is scale-invariant for $q = (p+1)/2$, similarly as (34.3) for $q = 2p/(p+1)$ — cf. Remark 39.8(i)).

(iv) **More self-similar profiles.** Concerning self-similar profiles, still in the case $\mu < 0$, $q = 2$, with $\Omega = \mathbb{R}^n$, it is proved in [231] that radial blow-up solutions to equation (34.3) behave asymptotically like a self-similar solution w of the following Hamilton-Jacobi equation without diffusion:

$$w_t = |\nabla w|^2 + w^p.$$

The function w is of the form

$$w(x,t) = (T-t)^{-1/(p-1)} W\left(x/(T-t)^m\right), \qquad m = (2-p)/2(p-1).$$

Note that this kind of self-similar behavior is quite different from that in (i) above (or from those known for $\mu = 0$ and p supercritical); indeed, m describes the range $(-\infty, 1/2)$ for $p \in (1, \infty)$. $\quad\square$

40. Viscous Hamilton-Jacobi equations and gradient blow-up on the boundary

In this section we study problem (34.5), which exhibits quite different phenomena from the model problem (15.1) or its perturbations (34.3), (34.4). For simplicity we shall again only consider nonnegative solutions (this assumption is essential in some, but not all, of the results).

40.1. Gradient blow-up and global existence

A basic fact about (34.5) is that solutions are uniformly bounded. Indeed, as a direct consequence of the maximum principle, for any $u_0 \in X_+$, there holds

$$0 \le u(x,t) \le \max_{x\in\overline{\Omega}} u_0(x), \quad x \in \Omega, \ 0 \le t < T_{\max}(u_0). \tag{40.1}$$

In view of (40.1), and since (34.5) is well-posed in the space X, a solution can cease to exist in finite time $T_{\max}(u_0) < \infty$ only if

$$\lim_{t \to T_{\max}(u_0)} \|\nabla u(t)\|_\infty = \infty. \tag{40.2}$$

This is what we call **gradient blow-up** (GBU for short).

Unlike the model problem (15.1), for which nonglobal solutions exist if and only if $p > 1$, the following two results show that the dividing line for the Dirichlet problem associated with (34.5) is given by $p = 2$.

Theorem 40.1. *Consider problem* (34.5) *with* $1 < p \le 2$. *Then* $T_{\max}(u_0) = \infty$ *for any* $u_0 \in X_+$. *Moreover, we have*

$$\sup_{t\ge 0} \|u(t)\|_X < \infty.$$

Theorem 40.2. *Consider problem (34.5) with $p > 2$ and Ω bounded, and let $1 \leq q < \infty$. There exists $C = C(p, q, \Omega) > 0$ such that, if $u_0 \in X_+$ and $\|u_0\|_q \geq C$, then $T_{\max}(u_0) < \infty$.*

Theorem 40.1 is an immediate consequence of the boundary gradient estimate from Lemma 35.4 and of the following simple result, which asserts that for problem (34.5), $|\nabla u|$ achieves its maximal values on the parabolic boundary.

Proposition 40.3. *Assume $p > 1$ and $u_0 \in X_+$. Let u be the solution of (34.5) and let $0 < T < T_{\max}(u_0)$. Then*

$$\sup_{t \in [0,T]} \|\nabla u(t)\|_\infty = \sup_{\mathcal{P}_T} |\nabla u|.$$

Proof. Fix $h \in \mathbb{R}^n$, with $|h| = 1$, and put $w := \partial_h v = h \cdot \nabla v$. We have $w \in C(\overline{Q_T}) \cap L^\infty(Q_T)$, and parabolic regularity results imply $w \in C^{2,1}(\overline{Q_T})$. Taking the space derivative of the equation in the direction h, we obtain

$$w_t - \Delta w = b(x, t) \cdot \nabla w \quad \text{in } Q_T,$$

where $b(x, t) = p|\nabla u|^{p-2}\nabla u$. By the maximum principle, we deduce that $\sup_{\overline{Q_T}} w \leq \sup_{\mathcal{P}_T} w$. Since h is arbitrary, the conclusion follows. $\square$

Proof of Theorem 40.2. Put $q_0 := 2(p-1)/(p-2)$. It is obviously sufficient to show the assertion for $q_0 \leq q < \infty$. Let thus set $k := q - 1 \in [p/(p-2), \infty)$. Multiplying (34.5) by u^k, we get

$$\frac{1}{k+1} \frac{d}{dt} \int_\Omega u^{k+1}(t) \, dx = \int_\Omega |\nabla u|^p u^k \, dx - k \int_\Omega |\nabla u|^2 u^{k-1} \, dx. \qquad (40.3)$$

On the other hand, by Poincaré's inequality, we have

$$\int_\Omega |\nabla u|^p u^k \, dx = C \int_\Omega |\nabla u^{(p+k)/p}|^p \geq C \int_\Omega u^{p+k} \, dx. \qquad (40.4)$$

Since $k \geq p/(p-2)$, by using Hölder's inequality and (40.4), we get

$$\int_\Omega |\nabla u|^2 u^{k-1} \, dx \leq \left(\int_\Omega |\nabla u|^p u^k \, dx \right)^{2/p} \left(\int_\Omega u^{k-p/(p-2)} \, dx \right)^{(p-2)/p}$$

$$\leq C \left(\int_\Omega |\nabla u|^p u^k \, dx \right)^{(k+1)/(k+p)}.$$

Combining this with (40.3), (40.4) and Hölder's inequality, we obtain

$$\frac{d}{dt} \int_\Omega u^{k+1} \, dx \geq \int_\Omega |\nabla u|^p u^k \, dx - C \geq C_1 \left(\int_\Omega u^{k+1} \, dx \right)^{(k+p)/(k+1)} - C_2.$$

The conclusion follows. $\square$

Remarks 40.4. (i) **Different methods of proof.** Theorem 40.2, whose proof relies on multiplication by powers of u, is due to [279] for $q = 2(p-1)/(p-2)$ and [500] in the general case. By a different argument, using the first eigenfunction, GBU for problem (34.5) was shown in [485] under a stronger condition on u_0 (see also [6]). The first example of GBU seems to be due to [210], where a one-dimensional problem with time-dependent Dirichlet boundary conditions was considered. The proof was based on subsolution arguments.

(ii) **Sharp condition for GBU.** A more precise growth condition for preventing GBU is known to be

$$|F(u, \nabla u)| \le C(u)(1 + |\nabla u|^2)h(|\nabla u|) \tag{40.5}$$

where h is positive nondecreasing and satisfies

$$\int^{\infty} \frac{ds}{sh(s)} = \infty, \tag{40.6}$$

and $C(u)$ is locally bounded (compare with condition (17.4) in the case of L^{∞}-blow-up); see [320], [337], [512]. There are known examples showing that condition (40.5)–(40.6) is sharp. A GBU result for general (including homogeneous) Dirichlet data can be found in [485]. The proof relies on eigenfunction and convex conjugate functions arguments. Earlier examples involving particular time-dependent boundary data, and relying on subsolution methods, were given in [337]. □

Unlike in the Dirichlet problem, global existence for the Cauchy problem holds for any $p > 1$ (cf. [402], [26]):

Proposition 40.5. *Consider problem* (34.5) *with* $\Omega = \mathbb{R}^n$ *and* $p > 1$. *Then* $T_{\max}(u_0) = \infty$ *for any* $u_0 \in X_+$. *Moreover, we have*

$$\sup_{t \ge 0} \|\nabla u(t)\|_{\infty} = \|\nabla u_0\|_{\infty}. \tag{40.7}$$

Proposition 40.5 is an immediate consequence of Proposition 40.3.

Remark 40.6. Unbounded domains. Although Theorem 40.2 is stated for bounded domains, GBU for large data when $p > 2$ occurs in any (regular) unbounded domain Ω other than $\mathbb{R}^n$. (Thus, for $p > 2$, Proposition 40.5 is true only in $\mathbb{R}^n$.)

Indeed, this follows from a simple comparison argument: Choose a ball $B \subset \Omega$ such that $\partial B \cap \partial \Omega$ consists of a single point, say x_0. Without loss of generality, we may assume that $B = B(0, \rho)$ and $x_0 = \rho e_1$. Let $0 \le \phi \in C^1(\overline{B})$ satisfy $\phi = 0$ on ∂B, ϕ radially symmetric, and let v be the solution of problem (34.5) with Ω replaced by B and initial data ϕ. If $\|\phi\|_1$ is sufficiently large, then v has GBU in a finite time T, due to Theorem 40.2. Since v is radially symmetric, it follows from

Proposition 40.3 that $\liminf_{t \to T} \frac{\partial v}{\partial x_1}(x_0, t) = -\infty$. Take any $u_0 \in X_+$ such that $u_0 \geq \phi$ in B and let u be the solution of (34.5). By the comparison principle, we have $u(x, t) \geq v(x, t)$ in B as long as u exists. Since $u(x_0, t) = v(x_0, t) = 0$, this implies $\frac{\partial u}{\partial x_1}(x_0, t) \leq \frac{\partial v}{\partial x_1}(x_0, t)$. Consequently GBU must occur for u no later than at time T. $\square$

40.2. Asymptotic behavior of global solutions

We start with the case of bounded domains. We have the following result on exponential decay. Assertions (i), (ii) follow from [146], [69], and (iii) from [485].

Theorem 40.7. *Consider problem (34.5) with $p > 1$, Ω bounded and $u_0 \in X_+$.*
(i) *Assume that*

$$T_{\max}(u_0) = \infty \quad and \quad \sup_{t \geq 0} \|u(t)\|_X < \infty. \tag{40.8}$$

Then there exists $C > 0$ (depending on u), such that

$$\|u(t)\|_X \leq Ce^{-\lambda_1 t}, \quad t \geq 0. \tag{40.9}$$

(ii) *If $1 < p \leq 2$, then assertions (40.8) and (40.9) are true for any $u_0 \in X_+$.*
(iii) *If $p > 2$, then assertions (40.8) and (40.9) are true whenever $\|u_0\|_X$ is sufficiently small.*

In the proof we shall use the following simple observation about steady states of (34.5) (cf. [341]):

Proposition 40.8. *Assume Ω bounded and let $p > 1$. Then the only solution $v \in C^2 \cap C_0(\Omega)$ of $\Delta v + |\nabla v|^p = 0$ is the trivial solution $v \equiv 0$.*

Proof. For $\varepsilon > 0$ small, let us denote $\omega_\varepsilon = \{x \in \Omega : \delta(x) > \varepsilon\}$. By the maximum principle applied to the equation $\Delta v + b(x) \cdot \nabla v = 0$ where $b(x) = |\nabla v|^{p-2} \nabla v$, we have $\max_{\overline{\omega}_\varepsilon} |v| = M(\varepsilon) := \max_{\partial \omega_\varepsilon} |v|$. But $v \in C_0(\Omega)$ implies $M(\varepsilon) \to 0$ as $\varepsilon \to 0$. Consequently $v \equiv 0$. $\square$

Proof of Theorem 40.7. (i) Let u_0 satisfy (40.8). We first claim that

$$\lim_{t \to \infty} \|u(t)\|_X = 0. \tag{40.10}$$

To see this, let us first observe that $\phi : X \to [0, \infty)$, defined by $\phi(v) = \|v\|_\infty$ is a Lyapunov functional for problem (34.5) (cf. Appendix G). Indeed, as a consequence of (40.1), the function $h(t) = \|u(t)\|_\infty$ is nonincreasing for $t \geq 0$. Moreover, if h is constant, then for each $t > 0$, $u(\cdot, t)$ achieves the value $\|u_0\|_\infty$ at some interior point. Applying the strong maximum principle (Proposition 52.7), we infer that u is constant, hence 0, in $\Omega \times (0, \infty)$. Therefore, ϕ is in fact a strict Lyapunov functional. Moreover, by (40.8) and parabolic estimates, $u(t)$ is bounded in $W^{2,q}(\Omega)$ for each finite q, so that $\{u(t) : t \geq 1\}$ is precompact in X. By Propositions 53.3 and 53.5, it follows that $\omega(u_0)$ (in the X topology) is nonempty and consists of equilibria. Property (40.10) thus follows from Proposition 40.8.

Now the exponential decay in (40.9) follows from Remark 51.20(ii).

(ii) This follows from Theorem 40.1 and assertion (i).

(iii) Let Θ be the classical solution of (19.27). By Hopf's lemma (cf. Proposition 52.1), we have $\Theta(x) \geq c_1 \delta(x)$ in Ω. Letting

$$M = \|\nabla\Theta\|_\infty \quad \text{and} \quad \phi = M^{-p/(p-1)}\Theta,$$

we find that

$$-\Delta\phi \geq |\nabla\phi|^p \quad \text{in } \Omega. \tag{40.11}$$

Now, assume that $\|u_0\|_X < c_1 M^{-p/(p-1)}$. Consequently, $u_0(x) \leq \|u_0\|_X \delta(x) \leq \phi(x)$ in Ω. It then follows from (40.11) and the comparison principle (see Proposition 52.16) that $u \leq \phi$ in $\overline{\Omega} \times [0, T)$, where $T = T_{\max}(u_0)$. Since $u = \phi = 0$ on $\partial\Omega$, we deduce that

$$|\nabla u| = -\frac{\partial u}{\partial\nu} \leq -\frac{\partial\phi}{\partial\nu} \leq C \quad \text{on } S_T.$$

The result then follows from Proposition 40.3 and assertion (i). $\quad\square$

Remarks 40.9. (a) **Boundedness of global solutions.** In the case $p > 2$ and Ω bounded, the question of boundedness of global solutions has been studied, too. For $n = 1$, it was shown in [38] that all global solutions are bounded in X. Consequently, they satisfy (40.9). For $n \geq 2$, this is an open problem.

On the other hand, if one looks at weaker norms, an L^∞ a priori estimate of global solutions is provided by (40.1), and it was shown in [500] that all global solutions actually decay in L^∞. Moreover, Theorem 40.2 implies the universal L^q-bound $\sup_{t\geq 0} \|u(t)\|_q \leq C(\Omega, p, q)$ for all finite q. Furthermore, under the stronger condition $p > \max(2, n)$, one actually has a universal L^∞-bound of the form $\|u(t)\|_\infty \leq C(\Omega, p, q)(1 + t^{-\alpha})$ for all $t > 0$ and some $\alpha = \alpha(n, p) > 0$ [500].

(b) **A priori estimates.** Similarly to the model problem (15.1) (cf. Section 22) one can consider the question, not only of boundedness but of a priori estimates of global solutions in X norm, and the related problem of continuity of the existence time. Consider problem (34.5) with $p > 2$ and $\Omega = (0, 1)$, fix a nontrivial $\phi \in X_+$ and define

$$E := \left\{ \lambda > 0 : T_{\max}(\lambda\phi) < \infty \quad \text{and} \quad \|u(t)\|_X \to 0, \text{ as } t \to \infty \right\}.$$

By Theorems 40.7(iii) and 40.2, we have $\lambda \in E$ for $\lambda > 0$ small and $\lambda \notin E$ for $\lambda > 0$ large. Therefore $\lambda^* := \sup E \in (0, \infty)$ and, due to Theorem 40.7(iii) and continuous dependence, $\lambda^* \notin E$. Since all global solutions decay in X (cf. Remark 40.9(a)), it follows that $T_{\max}(\lambda^*\phi) < \infty$. Consequently $T_{\max}$ is discontinuous. Moreover global solutions fail to satisfy an a priori estimate of the form $\sup_{t\geq 0} \|u(t)\|_X \leq C(\|u_0\|_X)$ (since, by continuous dependence, this would imply a bound for $\|u(\cdot;\lambda^*\phi)\|_X$, hence $T_{\max}(\lambda^*\phi) = \infty$). This exhibits a similar phenomenon as for the model problem in dimensions $n \geq 3$, which does not occur in dimensions $n = 1$ or 2 (cf. Theorem 22.1, Theorem 28.7 for radial solutions in a ball and $p > p_S$, and see after Theorem 22.13).

(c) **Unbounded global solutions.** If one considers the modification of problem (34.5) where a (smooth) inhomogeneous term $h(x) \geq 0$ is added on RHS, then boundedness of global solutions is still true in L^∞-norm, but may fail in the X norm. Indeed, examples of global solutions with $|\nabla u(x,t)|$ becoming unbounded on the boundary as $t \to \infty$ have been constructed in [500] for all $n \geq 1$. In [496], for a variant of problem (34.5) with $n = 1$, the grow-up rate of u_x is determined by techniques of matched asymptotics.

(d) Consider the situation of Theorem 40.7 in the limiting case $p = 1$. Then all solutions are still global and decay exponentially, but the decay exponent can be smaller than λ_1 (see [69]). On the other hand, decay no longer occurs in general for $0 < p < 1$. Indeed, if $\Omega = (0,1)$ for instance, it is easy to construct positive stationary solutions. We refer to [321] for results on the asymptotic behavior in this case. $\square$

We turn to the Cauchy problem. Recall that now all solutions are global by Proposition 40.5. The most complete results available concern the case of solutions with finite mass: Unless otherwise specified, we shall assume in the rest of this subsection that

$$u_0 \in X_+ \cap L^1(\mathbb{R}^n), \quad u_0 \not\equiv 0. \tag{40.12}$$

Under this assumption, the solution of (34.5) satisfies $u \in C([0,\infty), L^1(\mathbb{R}^n))$ (this can be shown by arguments similar to those in the proof of (51.42) in Proposition 51.16). Moreover, $\|u(t)\|_1$ is nondecreasing in time. This follows from

$$\int_{\mathbb{R}^n} u(t)\,dx = \int_{\mathbb{R}^n} u_0\,dx + \int_0^t \int_{\mathbb{R}^n} |\nabla u(y,s)|^p\,dy\,ds, \tag{40.13}$$

due to Proposition 48.4(b) and the variation-of-constants formula. We may thus define

$$I_\infty = \lim_{t\to\infty} \|u(t)\|_1 \in (0,\infty],$$

and a natural question is then to determine whether the *growth of mass* is limited or not, i.e., $I_\infty < \infty$ or $I_\infty = \infty$. It turns out that the problem involves two critical exponents

$$p = 2 \quad \text{and} \quad p = p_c := (n+2)/(n+1).$$

Recall that G_t denotes the Gaussian heat kernel, defined in (48.5).

Theorem 40.10. *Consider problem* (34.5) *with* $\Omega = \mathbb{R}^n$ *and* u_0 *satisfying* (40.12).
(i) *Assume* $p \geq 2$. *Then, for all* u_0, *there holds* $I_\infty < \infty$. *Moreover,*

$$\|u(t) - I_\infty G_t\|_1 \to 0, \quad t \to \infty. \tag{40.14}$$

(ii) *Assume* $1 < p \leq p_c$. *Then, for all* u_0, *there holds* $I_\infty = \infty$.

(iii) *Assume* $p_c < p < 2$. *Then we have* $I_\infty < \infty$ *for small data (in a suitable sense), and there also exist* u_0 *such that* $I_\infty = \infty$. *Furthermore,* (40.14) *is satisfied whenever* $I_\infty < \infty$.

Assertions (i) and (ii) are due to [322]. As for assertion (iii), the fact that $I_\infty < \infty$ under suitable smallness assumptions was proved in [145], [322], and the existence of at least one solution such that $I_\infty = \infty$ is due to [72]. This was next shown to occur under suitable largeness conditions on u_0 in [70]. We shall prove (i) and (ii) only. The proof of (iii) is more delicate and we refer for this to the above mentioned articles.

Proof of Theorem 40.10(i). First observe that in view of (40.7), u satisfies

$$u_t - \Delta u \leq a|\nabla u|^2, \quad x \in \mathbb{R}^n, \ t > 0$$

with $a = \|\nabla u_0\|_\infty^{p-2} > 0$. We use the Hopf-Cole transformation $v := e^{au} - 1$. The function v satisfies

$$v_t - \Delta v = a(u_t - \Delta u - a|\nabla u|^2)e^{au} \leq 0, \quad x \in \mathbb{R}^n, \ t > 0.$$

Therefore, $v(t) \leq e^{-tA}v_0$ by the maximum principle, where e^{-tA} denotes the heat semigroup in $\mathbb{R}^n$. Using the inequalities $x \leq e^x - 1 \leq xe^x$ for $x \geq 0$, it follows that

$$a\|u(t)\|_1 \leq \|v(t)\|_1 \leq \|v_0\|_1 \leq \|au_0e^{au_0}\|_1 \leq ae^{a\|u_0\|_\infty}\|u_0\|_1, \quad t \geq 0$$

hence $I_\infty < \infty$. Property (40.14) is then a consequence of Lemma 20.16 (and so is the last statement of assertion (iii)). $\square$

Proof of Theorem 40.10(ii).
Case 1: $n \geq 2$. Since $p < n$, by the Sobolev inequality and (40.13), there holds

$$\|u(t)\|_1 \geq C \int_0^t \|u(s)\|_{p^*}^p \, ds, \quad t \geq 0, \qquad \text{where } p^* = np/(n-p). \tag{40.15}$$

Also, for $|x| \leq \sqrt{t}$ and $t \geq t_0(u_0)$ large enough, we have

$$u(x,t) \geq (4\pi t)^{-n/2} \int_{\mathbb{R}^n} e^{-|x-y|^2/4t} u_0(y) \, dy$$

$$\geq (4\pi t)^{-n/2}e^{-1} \int_{|y|<\sqrt{t}} u_0(y) \, dy \geq Ct^{-n/2}\|u_0\|_1. \tag{40.16}$$

It follows that for all $s \geq t_0(u_0)$,

$$\|u(s)\|_{p^*}^{p^*} \geq \int_{|x|<\sqrt{s}} u^{p^*}(x,s)\, dx \geq C\|u_0\|_1^{p^*} s^{-\frac{n}{2}(p^*-1)},$$

which combined with (40.15) yields

$$\|u(t)\|_1 \geq C\|u_0\|_1^p \int_{t_0}^t s^{-\frac{np}{2}\left(1-\frac{1}{p^*}\right)}\, ds.$$

Since $p \leq p_c$, we have $\frac{np}{2}\left(1 - \frac{1}{p^*}\right) = \frac{(n+1)p-n}{2} \leq 1$, hence $I_\infty = \infty$.

Case 2: $n = 1$. We use the interpolation inequality

$$\|v\|_\infty^{2p-1} \leq C\|v\|_1^{p-1}\|v_x\|_p^p, \qquad \text{for all } p \geq 1 \text{ and } v \in L^1 \text{ such that } v_x \in L^p, \tag{40.17}$$

which is a consequence of

$$|v(x)|^{(p-1)/p}v(x) = \frac{2p-1}{p}\int_{-\infty}^x |v|^{(p-1)/p}v_x\, dy$$

and of Hölder's inequality. Since $\|u(t)\|_1$ is nondecreasing, it follows from (40.13) and (40.17) that

$$\|u(t)\|_1^p \geq \|u(t)\|_1^{p-1}\int_0^t \|u_x(s)\|_p^p\, ds$$

$$\geq \int_0^t \|u(s)\|_1^{p-1}\|u_x(s)\|_p^p\, ds \geq C\int_0^t \|u(s)\|_\infty^{2p-1}\, ds.$$

But (40.16) implies $\|u(t)\|_\infty \geq C\|u_0\|_1 t^{-1/2}$ for $t \geq t_0(u_0)$ large enough, hence

$$\|u(t)\|_1^p \geq C\|u_0\|_1^{2p-1}\int_{t_0}^t s^{-p+\frac{1}{2}}\, ds.$$

Since $p \leq p_c = 3/2$, we conclude that $I_\infty = \infty$. $\square$

Remarks 40.11. (a) **Nonlinear asymptotic behaviors.** Theorem 40.10 shows that when $p \geq 2$, or $p_c < p < 2$ and u_0 is small, then $I_\infty < \infty$ and the asymptotic behavior is dominated by the diffusion. When $I_\infty = \infty$, other behaviors are known. To describe this briefly, first observe that, since $\|u(t)\|_\infty$ is nonincreasing in time due to (40.1), we may set

$$N_\infty := \lim_{t\to\infty} \|u(t)\|_\infty \in [0, \infty).$$

It was proved in [70] that if (and only if)

$$N_\infty > 0, \tag{40.18}$$

then $I_\infty = \infty$ and u behaves like the viscosity solution z of the pure Hamilton-Jacobi equation $z_t = |\nabla z|^p$, with initial data $N_\infty \chi_{\{0\}}$. More precisely,

$$\lim_{t\to\infty} \|u(t) - z(t)\|_\infty = 0, \quad \text{where } z(x,t) = \left(N_\infty - c(p)\left(\frac{|x|}{t^{1/p}}\right)^{p/(p-1)}\right)_+.$$

In the range $1 < p \le p_c$, property (40.18) is true for all nontrivial (not necessarily integrable) $u_0 \in X_+$, see [251]. The situation is different in the range $p_c < p < 2$: property (40.18) holds under a suitable largeness condition on u_0 [70], but an example of a solution such that

$$I_\infty = \infty \quad \text{and} \quad N_\infty = 0 \tag{40.19}$$

has been constructed in [72]. This solution is self-similar, of the form

$$u(x,t) = (t+1)^{-k}V\left(\frac{x}{\sqrt{t+1}}\right), \qquad k = \frac{2-p}{2(p-1)},$$

where the profile $V \in L^1(\mathbb{R}^n)$ decays exponentially at infinity (cf. Remark 15.4(ii) for an analogue in the model problem (15.1)). This corresponds to an intermediate behavior involving a balance between the diffusion and the nonlinear term. It is unknown whether this self-similar solution is unique, nor if there exist solutions satisfying (40.19) other than self-similar.

(b) For estimates on I_∞ (if finite) or on the growth rate of $\|u(t)\|_1$ otherwise, see [322], [70], [251]. An alternative proof of Theorem 40.10(i) based on multiplier arguments (instead of the Hopf-Cole transformation) is also given in [322].

(c) Estimates similar to (40.14) are also true for other L^q-norms [70]. Namely, for every $q \in [1, \infty]$, there holds

$$t^{(n/2)(1-(1/q))}\|u(t) - I_\infty G_t\|_q \to 0, \quad t \to \infty.$$

(d) For general solutions of (34.5) (assuming only $u_0 \in X_+$ but not $u_0 \in L^1(\mathbb{R}^n)$), some results on the asymptotic behavior can be found in [71], [252], [251], [491]. In particular it was shown in [71], [252] by Bernstein-type techniques, that any solution satisfies the global gradient estimate

$$|\nabla u(x,t)| \le C(p)\|u_0\|_\infty^{1/p}t^{-1/p}, \quad x \in \mathbb{R}^n, \ t > 0$$

(hence $\|\nabla u(t)\|_\infty \to 0$ as $t \to \infty$).

(e) The exponents $p = 2$ and $p = p_c$ are also critical in the local existence-uniqueness theory of problem (34.5) with irregular initial data u_0; see [27], [71], [72]. $\quad\square$

40.3. Space profile of gradient blow-up

In this subsection we study the space profile of GBU of solutions to (34.5) for $p > 2$. We shall restrict ourselves to one space dimension.

Definition 40.12. Let $\Omega \subset \mathbb{R}^n$ and consider problem (34.5). We say that $x_0 \in \overline{\Omega}$ is a GBU point (in finite or infinite time) if there exist sequences $t_j \to T_{\max}(u_0)$ and $x_j \to x_0$ such that $|u_x(x_j, t_j)| \to \infty$. $\square$

In order to formulate our results, it is convenient to introduce the steady states of (34.5) for $n = 1$. They will be useful again in the study of the time rate of GBU; see the proof of Theorem 40.19 in the next subsection. To describe these steady states, let us denote

$$U(x) := d_p\, x^{(p-2)/(p-1)}, \qquad U'(x) = d'_p\, x^{-1/(p-1)}, \qquad x > 0, \qquad (40.20)$$

where $d_p = (p-2)^{-1}(p-1)^{(p-2)/(p-1)}$ and $d'_p = (p-1)^{-1/(p-1)}$. The function $U \in C([0,\infty)) \cap C^1((0,\infty))$ is a "singular" steady state. Namely, it is a solution of

$$V'' + V'^p = 0, \quad x > 0, \qquad V(0) = 0, \qquad (40.21)$$

which satisfies $U_x(0) = \infty$. Next, for each $\lambda > 0$, we put

$$U_\lambda(x) := U(x + \lambda) - U(\lambda). \qquad (40.22)$$

Each $U_\lambda \in C^1([0,\infty))$ also solves (40.21). Moreover we have $U'_\lambda(x) \to \infty$, as $x \to 0+$ and $\lambda \to 0+$, and $U_\lambda(x) \to U(x)$, uniformly for $x \in [0, 1]$, as $\lambda \to 0+$.

Our first result gives bounds on u_x away from $x = 0$ and 1. This shows in particular that GBU may occur only on the boundary.

Proposition 40.13. *Consider problem (34.5) with $p > 2$ and $\Omega = (0, 1)$. Let $u_0 \in X_+$ and $0 < t_0 < T := T_{\max}(u_0)$. There exists $C_1 > 0$ such that, for all $t_0 \le t < T$,*

$$u_x(x, t) \le U'(x) + C_1 x, \quad 0 < x \le 1 \qquad (40.23)$$

and

$$u_x(x, t) \ge -U'(1 - x) - C_1(1 - x), \quad 0 \le x < 1, \qquad (40.24)$$

where U is defined by (40.20). In particular $x = 0$ and $x = 1$ are the only possible GBU points.

The next result gives a precise description of the spatial profile around a GBU point. It is essentially due to [138] (where a slightly different problem, with non-homogeneous boundary value at $x = 1$, was actually studied).

Theorem 40.14. *Consider problem* (34.5) *with* $p > 2$ *and* $\Omega = (0,1)$. *Let* $u_0 \in X_+$ *and assume that* $T := T_{\max}(u_0) < \infty$.

(i) *For each* $x \in (0,1)$, *the limits*

$$u(x,T) := \lim_{t \to T} u(x,t) \quad and \quad u_x(x,T) := \lim_{t \to T} u_x(x,t) \quad exist \ and \ are \ finite.$$

Moreover, the first (resp., second) limit is uniform (resp., locally uniform) for $x \in (0,1)$.

(ii) *If* 0 *is a GBU point, then*

$$\lim_{t \to T} u_x(0,t) = \infty \tag{40.25}$$

and there exist $C_1, C_2 > 0$ *such that*

$$U(x) - C_1 x \le u(x,T) \le U(x) + C_2 x^2, \quad 0 < x \le 1/2 \tag{40.26}$$

and

$$U'(x) - C_1 \le u_x(x,T) \le U'(x) + C_2 x, \quad 0 < x \le 1/2, \tag{40.27}$$

where U *is defined by* (40.20). *Similar estimates hold if* 1 *is a GBU point.*

As a preliminary to the proofs, we need the following simple properties of the time-derivative u_t. (They are valid without restriction on n and will be used also in the next subsection.) We first note that $u_t \in C^{2,1}(Q_T)$ by parabolic regularity results, and that $u_t \in BC(\overline{\Omega} \times [t_0, t_1])$, $0 < t_0 < t_1 < T$, due to (35.3). The function $w := u_t$ satisfies

$$\left.\begin{aligned} w_t - \Delta w &= a(x,t) \cdot \nabla w, & x \in \Omega, \ 0 < t < T, \\ w &= 0, & x \in \partial\Omega, \ 0 < t < T, \end{aligned}\right\} \tag{40.28}$$

where

$$a(x,t) = p|\nabla u|^{p-2}\nabla u. \tag{40.29}$$

As an immediate consequence of (40.28) and of the maximum principle, we have:

Lemma 40.15. *Consider problem* (34.5) *with* $p > 1$ *and* $u_0 \in X_+$, *and let* $0 < t_0 < T := T_{\max}(u_0)$. *There exists* $C_1 > 0$ *such that*

$$|u_t| \le C_1, \quad x \in \Omega, \ t_0 \le t < T. \tag{40.30}$$

Proof of Proposition 40.13. Fix $0 < t_0 < t < T$ and let $y(x) = (u_x(x,t) - C_1 x)_+$, where C_1 is given by Lemma 40.15. The function y satisfies

$$y' + y^p = (u_{xx} - C_1)\chi_{\{u_x > C_1 x\}} + (u_x - C_1 x)_+^p, \quad \text{for a.e. } x \in (0,1).$$

For each x such that $u_x(x,t) > C_1 x$, we have $(y'+y^p)(x) \le (u_{xx} - C_1 + |u_x|^p)(x) \le 0$ by (40.30). Therefore, we have $y' + y^p \le 0$ a.e. on $(0,1)$. By integration, it follows that $y(x) \le ((p-1)x)^{-\frac{1}{p-1}}$, hence (40.23).

As for (40.24), it follows by applying (40.23) to the function $u(1-x,t)$, which satisfies the same equation. $\square$

For the proof of Theorem 40.14 we need the following lemma, which will provide a lower bound on the blow-up profile of u_x at $x = 0$.

Lemma 40.16. *Consider problem* (34.5) *with* $p > 2$ *and* $\Omega = (0,1)$. *Let* $u_0 \in X_+$ *and* $0 < t_0 < T := T_{\max}(u_0)$. *There exists* $C_3 > 0$ *such that, for all* $t_0 \leq t < T$,

$$\left[(u_x(x,t))_+ + C_3\right]^{1-p} \leq \left[u_x(0,t) + C_3\right]^{1-p} + (p-1)x, \quad 0 \leq x \leq 1. \quad (40.31)$$

Proof. Fix $t \in [t_0, T)$ and let

$$z(x) = (u_x(x,t))_+ + C_1^{1/p},$$

where C_1 is given by Lemma 40.15. The function z satisfies

$$z' + z^p = u_{xx}\chi_{\{u_x>0\}} + \left[(u_x(x,t))_+ + C_1^{1/p}\right]^p \geq (u_{xx} + |u_x|^p)\chi_{\{u_x>0\}} + C_1 \geq 0$$

a.e. on $(0,1)$ by (40.30). By integration, it follows that $z^{1-p}(x) \leq z^{1-p}(0)+(p-1)x$. Using $u_x(0,t) \geq 0$, we obtain (40.31) with $C_3 = C_1^{1/p}$. $\square$

Proof of Theorem 40.14. (i) It follows from Proposition 40.13 that $u_x(\cdot,t)$ is bounded in $L^1(0,1)$. This along with (40.1) implies $\{u(\cdot,t) : t \in (0,T)\}$ is precompact in $C([0,1])$. Using Lemma 40.15, we deduce that $\lim_{t\to T} u(x,t)$ exists, uniformly for $x \in [0,1]$.

Now fix $t_0 \in (0,T)$. By Proposition 40.13 and Lemma 40.15, we deduce that $u_x, u_{xx} \in L^\infty((\varepsilon, 1-\varepsilon)\times(t_0,T))$ for each $\varepsilon \in (0,1)$. Therefore, $\{u_x(\cdot,t) : t \in (t_0,T)\}$ is precompact in $C((0,1))$. Since $u_{xt} - u_{xxx} = p|u_x|^{p-2}u_x u_{xx}$, parabolic estimates imply $u_{xt} \in L^q((\varepsilon, 1-\varepsilon)\times(t_0,T))$ for each $\varepsilon \in (0,1)$ and each finite q. Consequently, $u_x \in BUC^\alpha([\varepsilon, 1-\varepsilon] \times [t_0,T))$ for each $\varepsilon \in (0,1)$ and some $\alpha \in (0,1)$. We deduce that $\lim_{t\to T} u_x(x,t)$ exists, locally uniformly for $x \in (0,1)$.

(ii) The upper estimates in (40.26), (40.27) follow from Proposition 40.13. To show the lower estimates, let us first note that, by assumption, there exist sequences $t_j \to T$ and $x_j \to 0$ such that $|u_x(x_j,t_j)| \to \infty$, hence $u_x(x_j,t_j) \to \infty$ due to (40.24). Moreover, by Lemma 40.15 and (34.5), we have $u_{xx} \leq C_1$, hence

$$u_x(0,t) \geq u_x(x,t) - C_1 x, \quad (x,t) \in [0,1] \times [t_0,T). \quad (40.32)$$

It follows that $u_x(0,t_j) \to \infty$. Put $\varepsilon_j := (u_x(0,t_j)+C_3)^{1-p} \to 0$, where C_3 is from Lemma 40.16. By that lemma, there exists $\eta \in (0,1)$ such that, for j large,

$$u_x(x,t_j) \geq (\varepsilon_j + (p-1)x)^{-1/(p-1)} - C_3, \quad 0 < x < \eta,$$

hence

$$u(x,t_j) \geq \frac{1}{p-2}\left[(\varepsilon_j + (p-1)x)^{(p-2)/(p-1)} - \varepsilon_j^{(p-2)/(p-1)}\right] - C_3 x, \quad 0 < x < \eta.$$

Letting $j \to \infty$, and since we already know that the limits in assertion (i) exist, we obtain the lower estimates in (40.26), (40.27) (η can be replaced by $1/2$ by enlarging the constant C_1).

Finally, using (40.32), we may write, for each $x \in (0,1)$,

$$\liminf_{t\to T} u_x(0,t) \geq \liminf_{t\to T} u_x(x,t) - C_1 x \geq U'(x) - C_1(x+1),$$

and (40.25) follows by letting $x \to 0$. $\square$

Remark 40.17. The analogue of the upper estimate (40.23) is still true in higher dimensions. Namely, by using Bernstein-type techniques, it is shown in [500] that any solution of (34.5) with $p > 2$ satisfies

$$|\nabla u(x,t)| \le C_1 \delta^{-1/(p-1)}(x) + C_2 \quad \text{in } \Omega \times [0, T_{\max}(u_0)),$$

with $C_1 = C_1(p,n) > 0$ and $C_2 = C_2(u) > 0$. $\quad\Box$

40.4. Time rate of gradient blow-up

We now study the time rate of GBU of solutions to (34.5) for $p > 2$, i.e.: the speed of divergence of $\|\nabla u(t)\|_\infty$. We begin with lower estimates. The following theorem is due to [263].

Theorem 40.18. *Consider problem* (34.5) *with* $p > 2$ *and* $\Omega \ne \mathbb{R}^n$*. Let* $u_0 \in X_+$ *and assume that* $T := T_{\max}(u_0) < \infty$*. Then there exists* $C > 0$ *such that*

$$\sup_{s \in [0,t]} \|\nabla u(s)\|_\infty \ge C(T-t)^{-1/(p-2)}, \quad t \to T. \tag{40.33}$$

In one space dimension, we have the following more precise result from [138] (see also Remark 40.23 below).

Theorem 40.19. *Consider problem* (34.5) *with* $p > 2$, $n = 1$ *and* $\Omega = (0,1)$*. Let* $u_0 \in X_+$ *and assume that* $T := T_{\max}(u_0) < \infty$*. Then there exists* $C > 0$ *such that*

$$\|u_x(t)\|_\infty \ge C(T-t)^{-1/(p-2)}, \quad t \to T. \tag{40.34}$$

Remarks 40.20. (i) **Non self-similar GBU rate.** The lower estimate (40.33) implies in particular that the GBU rate does not correspond to the one suggested by the self-similar invariance of the problem. Indeed, letting $k = (p-2)/(2(p-1))$, the scaling transformations

$$\mathcal{S}_\lambda : u \mapsto u_\lambda(x,t) := \lambda^{2k} u(\lambda^{-1}x, \lambda^{-2}t), \quad \lambda > 0,$$

leave invariant the equation in (34.5). This might allow for the existence of backward self-similar (classical) solutions of the form

$$w(x,t) = (T-t)^k V\left(\frac{x}{\sqrt{T-t}}\right) \tag{40.35}$$

(note that forward self-similar solutions in $\mathbb{R}^n$ exist for some $p < 2$, cf. Remark 40.11(a)). Now if there exists a nontrivial solution w of the form (40.35), say, on $\Omega = (0,\infty)$ with zero boundary condition at $x = 0$, and if $\nabla V \in L^\infty$, then

w will exhibit the GBU self-similar rate $\|\nabla w(t)\|_\infty = C(T-t)^{-1/(2(p-1))}$. However, since $1/(p-2) > 1/(2(p-1))$, Theorem 40.18 shows that no such solutions w exist and that the exponent of the GBU rate is always greater than that of the self-similar rate. A similar situation has been encountered for the supercritical model problem (cf. Section 23) and also for problem (38.3).

(ii) A rough argument involving the variation-of-constants formula would also give a lower estimate $(T-t)^{-1/(2(p-1))}$. $\square$

The upper blow-up rate estimate for problem (34.5) is still an open question. However, some results are known for the closely related one-dimensional problem:

$$\left.\begin{aligned}
u_t - u_{xx} &= |u_x|^P + \lambda, &\quad 0 < x < 1,\ t > 0, \\
u &= 0, &\quad x \in \{0,1\},\ t > 0, \\
u(x,0) &= u_0(x), &\quad 0 < x < 1,
\end{aligned}\right\} \tag{40.36}$$

with $p > 2$, $\lambda > 0$ and $u_0 \in X_+$. Note that the local solution of (40.36) is nonnegative and uniformly bounded on finite time intervals (since $\overline{u}(x,t) := \|u_0\|_\infty + \lambda t$ is a supersolution). Moreover, as a consequence of the proof of Theorem 40.2, gradient blow-up occurs whenever $\lambda > \lambda_0(p)$ or $\|u_0\|_q \geq C(p,q)$ for some $q \in [1,\infty)$, where $\lambda_0(p)$ and $C(p,q)$ are suitable positive constants.

Theorem 40.21. *Consider problem (40.36) with $p > 2$ and $\lambda > 0$. Let $u_0 \in X_+ \cap C^2([0,1])$ be symmetric with respect to $x = 1/2$ and satisfy*

$$u_{0,xx} + |u_{0,x}|^P + \lambda \geq 0 \quad in\ [0,1]. \tag{40.37}$$

If $T := T_{\max}(u_0) < \infty$, then there exists $C > 0$ such that

$$\|u_x(t)\|_\infty \leq C(T-t)^{-1/(p-2)}, \quad t \to T. \tag{40.38}$$

Theorem 40.21 is a variant of a result of [263], where the authors considered the equation in (34.5) under inhomogeneous boundary conditions for $n = 1$. For that problem the upper GBU rate estimate was first conjectured in [138] on the basis of numerical simulations.

Remarks 40.22. (i) Assumption (40.37) guarantees that the solution is nondecreasing in time. However, analogous assumption cannot be satisfied for problem (34.5). Indeed if $u_0 \in X_+ \cap C^2([0,1])$ verifies (40.37) with $\lambda = 0$, then $u_0 \equiv 0$ (this follows for instance from the maximum principle). On the other hand, it is unknown if Theorem 40.21 (and the corresponding result in [263]) remains true without assumption (40.37) or in higher space dimensions.

(ii) Estimate (40.38) is sharp. In fact, under the assumptions of Theorem 40.21, the lower estimate (40.34) follows from simple modifications of the proof of Theorem 40.18, along with (40.51) below. $\square$

We first give (a variant of) the proof of Theorem 40.18 from [263]. It relies on linear regularity estimates applied to the equation for u_t.

Proof of Theorem 40.18. Denote

$$m(t) := \sup_{\overline{\Omega} \times [T/2,t]} |\nabla u| = \max_{s \in [T/2,t]} \|\nabla u(s)\|_\infty, \qquad T/2 \le t < T \qquad (40.39)$$

(note that $m(t)$ and $\|\nabla u(t)\|_\infty$ are continuous due to (51.27)). In this proof, C will denote positive constants, independent of $t \in (T/2,T)$, which may change from line to line and may depend on u.

Step 1. We claim that $w := u_t$ satisfies

$$\|\nabla w(t)\|_\infty \le Cm^{p-1}(t), \quad T/2 < t < T. \qquad (40.40)$$

Let $t \in (T/2,T)$, $s \in (T/4,t)$, and put $K = \sup_{\sigma \in [0,t-s]} \sigma^{1/2}\|\nabla w(s+\sigma)\|_\infty$. For $\tau \in (0, t-s)$, in view of (40.28), (40.29) and the variation-of-constants formula, we have

$$w(s+\tau) = e^{-\tau A}w(s) + \int_0^\tau e^{-(\tau-\sigma)A}(a \cdot \nabla w)(s+\sigma)\,d\sigma.$$

Using Proposition 48.7, Lemma 40.15, and the fact that $\int_0^\tau (\tau-\sigma)^{-1/2}\sigma^{-1/2}\,d\sigma = \int_0^1 (1-z)^{-1/2}z^{-1/2}\,dz$, it follows that

$$\|\nabla w(s+\tau)\|_\infty \le C\tau^{-1/2}\|w(s)\|_\infty + C\int_0^\tau (\tau-\sigma)^{-1/2}\|a \cdot \nabla w(s+\sigma)\|_\infty\,d\sigma$$

$$\le C\tau^{-1/2} + Cm^{p-1}(t)K.$$

Multiplying by $\tau^{1/2}$ and taking the supremum for $\tau \in [0, t-s]$, we obtain

$$K \le C + C(t-s)^{1/2}m^{p-1}(t)K.$$

Now choosing $s = t - (1/4)\min\big(T, (Cm^{p-1}(t))^{-2}\big) \in (T/4,t)$, we obtain $K \le 2C$, hence

$$\|\nabla w(t)\|_\infty \le 2C(t-s)^{-1/2} \le 4C\max(T^{-1/2}, Cm^{p-1}(t)).$$

Since m is positive nondecreasing, this implies Claim (40.40).

Step 2. We next claim that m is locally Lipschitz on $(T/2,T)$ and that

$$m'(t) \le Cm^{p-1}(t), \quad \text{for a.e. } t \in (T/2,T). \qquad (40.41)$$

Let $T/2 < t < s < T$. For any $\tau \in [t,s]$ and $x \in \Omega$, it follows from the mean-value inequality and (40.40) that

$$|\nabla u(x,\tau) - \nabla u(x,t)| \le (\tau-t)\sup_{\Omega \times [t,\tau]} |\partial_t \nabla u| \le C(\tau-t)m^{p-1}(\tau)$$

hence

$$|\nabla u(x,\tau)| \leq |\nabla u(x,t)| + C(\tau - t)m^{p-1}(\tau) \leq m(t) + C(s-t)m^{p-1}(s).$$

Taking supremum for (x,τ) over $\Omega \times [t,s]$, we get

$$0 \leq m(s) - m(t) \leq C(s-t)m^{p-1}(s).$$

Since m is continuous, the claim follows.

Finally, integrating (40.41) over (t,s) with $T/2 < t < s < T$, and using $m(s) \to \infty$ as $s \to T$, we infer that

$$m(t) \geq C(T-t)^{-1/(p-2)}, \tag{40.42}$$

which implies estimate (40.33). $\square$

We next give the proof of Theorem 40.19, based on (a simplification of) the idea from [138]. It relies on a completely different argument, involving the intersections of the solution with the singular steady state.

Proof of Theorem 40.19. Recall that the steady states U and U_λ are defined in (40.20) and (40.22). Due to Proposition 40.3, we may assume, without loss of generality, that

$$\limsup_{t \to T} u_x(0,t) = \infty. \tag{40.43}$$

Fix $t_0 \in [T/2, T)$ and let

$$x_0 := \sup\{x \in (0,1] : u_x(\cdot, t_0) < U' \text{ in } (0,x)\}.$$

Note that, since $u_x(x,t_0) < U'(x)$ for $x > 0$ small, x_0 is well defined and $x_0 > 0$. On the other hand, by definition, we have $u_x(x,t_0) < U'(x)$ in $(0,x_0)$, hence $u(x,t_0) < U(x)$ in $(0,x_0]$. It follows that

$$u(x,t_0) \leq U_\lambda(x), \quad 0 \leq x \leq x_0, \qquad \text{for all } \lambda > 0 \text{ small.} \tag{40.44}$$

We claim that $x_0 \in (0,1)$, hence

$$u_x(x_0, t_0) = U'(x_0). \tag{40.45}$$

Indeed, otherwise $x_0 = 1$, so that (40.44) implies $u \leq U_\lambda$ in $[0,1] \times [t_0, T)$ for $\lambda > 0$ small, due to the comparison principle. Therefore $u_x(0,t) \leq U_\lambda'(0)$ in $[t_0, T)$, contradicting (40.43).

We next claim that

$$\sup_{t \in [t_0, T)} u(x_0, t) \geq U(x_0). \tag{40.46}$$

Suppose the contrary. Then, for all $\lambda > 0$ small, we have $u(x_0, t) \leq U_\lambda(x_0)$ in $[t_0, T)$. By (40.44) and the comparison principle, we deduce that $u \leq U_\lambda$ in $[0, x_0] \times [t_0, T)$, leading again to a contradiction.

Now, using Lemma 40.15 and (40.46), we get

$$C_1(T - t_0) \geq \int_{t_0}^{T} u_t(x_0, t) \geq U(x_0) - u(x_0, t_0) = \int_{0}^{x_0} (U'(x) - u_x(x, t_0))\, dx.$$

On the other hand, by (40.45), there clearly exists $x_1 \in (0, x_0]$ such that $U'(x_1) = \max_{[0,1]} u_x(\cdot, t_0)$. Since $U'(x) - u_x(x, t_0) > 0$ on $(0, x_0)$ by the definition of x_0, we obtain

$$C_1(T - t_0) \geq \int_{0}^{x_1} (U'(x) - u_x(x, t_0))\, dx$$

$$\geq U(x_1) - x_1 U'(x_1) = \frac{U'(x_1)^{2-p}}{(p-1)(p-2)} \geq \frac{\|u_x(t_0)\|_\infty^{2-p}}{(p-1)(p-2)},$$

and the conclusion follows. $\square$

Remark 40.23. More precise lower estimate. In addition to the hypotheses of Theorem 40.19, assume that $u_0 \in C^2([0, 1])$ and denote

$$C_1 = \max_{[0,1]} (u_{0,xx} + |u_{0,x}|^p).$$

Then the proof of Theorem 40.19 provides the value $C = \left((p-1)(p-2)C_1\right)^{-1/(p-2)}$ for the constant in (40.34). $\square$

Finally, we give the proof of Theorem 40.21, based on the ideas in [263], which relies on the application of the maximum principle to a suitable auxiliary function. Note that this function (cf. w below) is quite different from the function J used in the proof of Theorem 24.1.

Proof of Theorem 40.21. We consider the parabolic operator

$$\mathcal{L}\phi := \phi_t - \phi_{xx} - p|u_x|^{p-2} u_x \phi_x.$$

For $\sigma \in (0, 1)$ and $t_0 \in (0, T)$ to be chosen later, we introduce the auxiliary function

$$w(x, t) := \left(1 + \frac{1}{m^\sigma(t)}\right)\left(1 - \frac{u_x}{m(t)}\right), \qquad x \in [0, 1], \quad t \in [t_0, T),$$

where

$$m(t) := \max_{(x,t) \in [0,1] \times [t_0, t]} |u_x(x, t)| \to \infty, \quad \text{as } t \to T. \tag{40.47}$$

Step 1. We shall show that for suitable $t_0 \in (0, T)$ and $C > 0$, there holds

$$w + u \le C u_t \quad \text{in } (0, 1) \times (t_0, T). \tag{40.48}$$

We may assume $m(t) \ge 1$ without loss of generality. Also, by the proof of Theorem 40.18, m is locally Lipschitz on $(T/2, T)$ and (40.41) is satisfied. A direct computation shows that

$$\mathcal{L}w = -\frac{\sigma m'}{m^{\sigma+1}}\left(1 - \frac{u_x}{m}\right) + \left(1 + \frac{1}{m^\sigma}\right)\frac{u_x m'}{m^2}. \tag{40.49}$$

Since $m' \ge 0$ a.e., we have, in case $|u_x(x,t)| < \frac{\sigma}{\sigma+2} m^{1-\sigma}(t)$,

$$\mathcal{L}w = \frac{m'}{m^{\sigma+1}}\left(-\sigma + (\sigma+1)\frac{u_x}{m} + \frac{u_x}{m^{1-\sigma}}\right) \le \frac{m'}{m^{\sigma+1}}\left(-\sigma + (\sigma+2)\frac{|u_x|}{m^{1-\sigma}}\right) \le 0$$

(a.e. in t). On the other hand, if $|u_x(x,t)| \ge \frac{\sigma}{\sigma+2} m^{1-\sigma}(t)$, then by (40.49) and (40.41), we have

$$\mathcal{L}w \le \left(1 + \frac{1}{m^\sigma}\right)\frac{u_x m'}{m^2} \le C|u_x|\frac{m^{p-1}}{m^2} \le \frac{C}{m}\left(\frac{\sigma+2}{\sigma}\right)^{(p-2)/(1-\sigma)}|u_x|^{(p-1-\sigma)/(1-\sigma)}.$$

We now choose $\sigma = 1/(p-1)$, so that $(p - 1 - \sigma)/(1 - \sigma) = p$. Thus, taking $\tilde{C} := C(\frac{\sigma+2}{\sigma})^{(p-2)/(1-\sigma)}$ and using (40.47), we obtain, for t_0 close to T,

$$\mathcal{L}(w + u) \le \frac{\tilde{C}}{m}|u_x|^p - (p-1)|u_x|^p + \lambda \le -\frac{p-1}{2}|u_x|^p + \lambda, \quad \text{a.e. in } (0,1) \times [t_0, T).$$

If $|u_x(x,t)|^p \ge 2\lambda/(p-1)$, then $\mathcal{L}(w + u) \le 0$, whereas $w(x,t) \ge 1/2$ otherwise (for t_0 close to T). In all cases we thus have $\mathcal{L}(w + u) \le 2\lambda(w + u)$, hence

$$\mathcal{L}\left(e^{-2\lambda t}(w + u)\right) \le 0 = \mathcal{L}u_t, \quad \text{a.e. in } (0,1) \times [t_0, T). \tag{40.50}$$

Due to our assumptions on u_0, u is symmetric with respect to $x = 1/2$, and we have $u_t \ge 0$ (and $u_t \not\equiv 0$) in $(0,1) \times (0,T)$ by Proposition 52.19. In particular $t \mapsto u_x(0,t)$ is nonnegative and nondecreasing, and it follows from the proof of Proposition 40.3 that

$$m(t) = \|u_x(t)\|_\infty = u_x(0,t), \quad t_0 \le t < T, \tag{40.51}$$

by taking t_0 closer to T if necessary. Consequently,

$$[w + u](x,t) = 0 = u_t(x,t), \quad x \in \{0,1\}, \; t_0 \le t < T. \tag{40.52}$$

On the other hand, the strong and Hopf maximum principles guarantee that

$$u_t(x,t_0) > 0, \quad 0 < x < 1, \quad u_{tx}(0,t_0) > 0, \quad u_{tx}(1,t_0) < 0.$$

In particular there exists $C > 0$ such that

$$\left[e^{-2\lambda t_0}(w + u) - Cu_t\right](x, t_0) \leq 0, \quad 0 < x < 1. \tag{40.53}$$

Using (40.50), (40.52), (40.53), and the maximum principle (under the form of Proposition 52.8), we deduce $e^{-2\lambda t}(w + u) \leq Cu_t$ in $(0, 1) \times [t_0, T)$, hence (40.48).

Step 2. As a consequence of (40.48) and (40.51), we have

$$Cu_{xt}(0, t) = \lim_{x \to 0+} \frac{Cu_t(x, t)}{x} \geq \lim_{x \to 0+} \frac{[w + u](x, t)}{x}$$

$$\geq \lim_{x \to 0+} \frac{w(x, t)}{x} = w_x(0, t) = -\left(1 + \frac{1}{m^\sigma(t)}\right)\frac{u_{xx}(0, t)}{m(t)}$$

$$\geq \frac{|u_x(0, t)|^p}{m(t)} = u_x^{p-1}(0, t).$$

By integration, we obtain

$$u_x(0, t) \leq C(T - t)^{-1/(p-2)}, \quad t \to T,$$

which proves the result. $\square$

Remarks 40.24. (a) **Boundary layer.** Under the assumptions of Theorem 40.21, for any $K > 0$, there exists $c = c(K) > 0$ such that

$$u_x(x, t) \geq c(T - t)^{-1/(p-2)} \quad \text{for } 0 < x < K(T - t)^{(p-1)/(p-2)}$$

and t close to T (cf. [263]). This boundary layer estimate follows from the proof of Lemma 40.16, estimate (40.34) (cf. Remark 40.22(ii)) and (40.51).

(b) **Nonsymmetric initial data.** In Theorem 40.21, the symmetry assumption on u_0 can be removed. To show this, assuming that $x = 0$ (resp. $x = 1$) is a GBU point, one has to replace the interval $[0, 1]$ by $[0, 1/2]$ (resp. $[1/2, 1]$) in the proof of Theorem 40.21 (and in particular in the definition (40.47) of $m(t)$). One then uses the fact that u_x is bounded away from the boundary (cf. the proof of Theorem 40.14) and that estimate (40.41) remains true (this follows from simple modifications of the proof of Theorem 40.18, using $w\phi$ instead of w in Step 1, where $\phi = \phi(x)$ is a cut-off function equal to 1 on $[0, 1/2]$ and to 0 near $x = 1$).

On the other hand, by similar arguments, one can show that (under the hypotheses of Theorem 40.21 without the symmetry assumption) there holds $|u_x(x, t)| \geq C(T - t)^{-1/(p-2)}$ as $t \to T$, for each GBU point $x \in \{0, 1\}$.

(c) **Continuation after GBU.** For some results on continuation after GBU, see [193], [57], [201]. These references contain examples where *all* solutions can be continued in some sense after GBU (and not only threshold solutions, like in L^∞-blow-up — cf. Proposition 27.7 and Remark 27.8(c)). $\square$

41. An example of interior gradient blow-up

In the previous section we studied the phenomenon of gradient blow-up on the boundary. The aim of this section is to provide a simple example of a different behavior, namely: interior gradient blow-up.

Consider the following problem:

$$
\left.
\begin{aligned}
u_t - u_{xx} &= |u|^{m-1}u|u_x|^p, & -1 &< x < 1,\ t > 0,\\
u(\pm 1, t) &= A_\pm, & t &> 0,\\
u(x,0) &= u_0(x), & -1 &< x < 1,
\end{aligned}
\right\}
\qquad (41.1)
$$

where $p > 2$, $m \geq 1$, $A_- < 0 < A_+$ and

$$
u_0 \in C^1([-1,1]), \quad \text{with} \quad u_0(-1) = A_- \leq u_0 \leq A_+ = u_0(1) \text{ in } [-1,1]. \quad (41.2)
$$

Unlike in problem (34.5), the nonlinearity here changes sign, and this is the key feature that will allow for interior GBU rather than boundary GBU (see Remark 41.4 below).

The examples in Remark 51.11 guarantee that (41.1) is locally well-posed (observe for instance that (41.1) can be converted to a problem with homogeneous boundary conditions by the change of unknown $v(x,t) = u(x,t) - \phi(x)$, where ϕ is an affine function such that $\phi(\pm 1) = A_\pm$). By the maximum principle and (41.2), we immediately obtain

$$
A_- \leq u(x,t) \leq A_+, \quad -1 \leq x \leq 1,\ 0 \leq t < T_{\max}(u_0). \quad (41.3)
$$

Therefore $T_{\max}(u_0) < \infty$ guarantees that GBU occurs (i.e. (40.2)).

Theorem 41.1. *Consider problem* (41.1) *with* $p > 2$, $m \geq 1$. *There exists* $L = L(m,p) > 0$ *such that, if* $\max(A_+, |A_-|) > L$, *then* $T_{\max}(u_0) < \infty$ *for any* u_0 *satisfying* (41.2).

Theorem 41.1 is (a variant of) a result from [31]. The original proof was based on the construction of appropriate traveling wave sub- and supersolutions. We here present a simpler proof based on a multiplier argument similar to that in the proof of Theorem 40.2.

Proof. Let $k = (p + 2m)/(p - 2)$. In what follows, C and C_1 denote any positive constant depending only on p, m. For all $t \in (0, T_{\max}(u_0))$, multiplying (41.1) by $|u|^{k-1}u$ and integrating by parts, we get

$$
\frac{d}{dt}\int_{-1}^{1} \frac{|u|^{k+1}}{k+1}\,dx = \left[u_x|u|^{k-1}u\right]_{-1}^{1} - k\int_{-1}^{1}|u_x|^2|u|^{k-1}\,dx + \int_{-1}^{1}|u_x|^p\,|u|^{m+k}\,dx.
$$

$$
(41.4)
$$

Next, by Hölder's inequality, we have

$$\int_{-1}^{1} |u_x|^p |u|^{m+k}\, dx = C \int_{-1}^{1} \left|\left(|u|^{(m+k)/p}u\right)_x\right|^p \geq C \left|\int_{-1}^{1} \left(|u|^{(m+k)/p}u\right)_x\right|^p \tag{41.5}$$

$$= C\left(A_+^{(p+m+k)/p} + |A_-|^{(p+m+k)/p}\right)^p \geq CL^{p+m+k}.$$

Moreover, since $p(k-1)/2 = m+k$, Young's inequality yields

$$k \int_{-1}^{1} |u_x|^2 |u|^{k-1}\, dx \leq \frac{1}{2} \int_{-1}^{1} |u_x|^p |u|^{m+k}\, dx + C. \tag{41.6}$$

On the other hand, (41.3) implies

$$u_x(\pm 1, t) \geq 0. \tag{41.7}$$

Combining (41.4)–(41.7), and taking $L = L(p,m)$ large enough, we obtain

$$\frac{d}{dt} \int_{-1}^{1} |u|^{k+1}\, dx \geq CL^{p+m+k} - C_1 \geq 1.$$

Integrating and using (41.3), it follows that

$$t \leq \int_{-1}^{1} |u|^{k+1}\, dx \leq 2\big(\max(A_+, |A_-|)\big)^{k+1},$$

hence $T_{\max}(u_0) < \infty$. $\square$

Remark 41.2. It can be shown that if A_+ and $|A_-|$ are small enough, then there exist stationary, hence global solutions (see [31]). In this case the argument of the above proof still shows that GBU occurs for suitably large initial data. $\square$

The next result asserts that, for $m = 1$ and a suitable class of initial data, GBU occurs at a single interior point, namely $x_0 = 0$. Moreover, an upper estimate is given for the final profile. A much more general result of interior GBU was proved in [31] (see Remark 41.4 below). However the proof therein is more delicate.

Theorem 41.3. *Consider problem (41.1) with $p > 2$, $m = 1$, and $A_{\pm} = \pm A$ with $A > L$, where L is defined in Theorem 41.1. Let $u_0 \in C^2([-1,1])$ be an odd function satisfying*

$$u_{0,x} \geq 0, \quad u_{0,xx} \leq 0 \quad in\ [0,1], \qquad u_0(1) = A, \quad u_{0,xx}(1) + A|u_{0,x}(1)|^p = 0.$$

Then, there holds $T := T_{\max}(u_0) < \infty$,

$$\lim_{t \to T} u_x(0,t) = \infty \tag{41.8}$$

and

$$0 \leq u_x(x,t) \leq A|x|^{-1}, \qquad 0 < |x| < 1,\ 0 < t < T. \tag{41.9}$$

Proof. By local uniqueness, we have $u(-x,t) = -u(x,t)$. Let $v = u_x$ and $w = u_{xx}$. By parabolic regularity results, we have $v, w \in C^{2,1}(Q_T) \cap C(\overline{Q_T})$. We compute

$$v_t - v_{xx} = (u|v|^p)_x = |v|^p v + pu|v|^{p-2}vw.$$

Due to (41.7) and $u_{0,x} \geq 0$ in $[0,1]$, the maximum principle implies $v \geq 0$. Differentiating again in x, we get

$$w_t - w_{xx} = (p+1)v^p w + p(uv^{p-1}w)_x = a(x,t)w + b(x,t)w_x,$$

where $a = (2p+1)v^p + p(p-1)uv^{p-2}w$ and $b = puv^{p-1}$. Moreover $w(1,t) = -uv^p(1,t) \leq 0$ and, since $u(0,t) = 0$, we have $w(0,t) = -uv^p(0,t) = 0$. We thus infer from the maximum principle that

$$u_{xx} = w \leq 0, \quad 0 \leq x \leq 1,\ 0 \leq t < T.$$

For all $0 < x \leq 1$ and $0 \leq t < T$, it follows that $u_x(x,t) \leq u_x(0,t)$. Therefore, $u_x(0,t) = \|u_x(t)\|_\infty$, hence (41.8). On the other hand, by concavity, we have

$$A \geq u(x,t) - u(0,t) \geq xu_x(x,t), \quad 0 < x \leq 1,\ 0 \leq t < T,$$

hence (41.9). $\square$

Remark 41.4. In fact, it was proved in [31] that for any $m \geq 1$ and any initial data as in Theorem 41.1, interior GBU occurs, in the sense that u_x remains bounded close to the boundary. Moreover, GBU may occur only at points "where u changes sign"; more precisely, for $x_0 \in (-1,1)$, if u remains bounded away from 0 in a neighborhood of x_0 as $t \to T$, then u_x remains bounded near x_0. The proof is based on Bernstein-type arguments. $\square$

Chapter V

Nonlocal Problems

42. Introduction

In this chapter, we study various problems with nonlocal nonlinearities. The equations that we consider involve nonlocal terms taking the form of an integral in space, or in time. These terms may also be combined with local ones, either in an additive or in a multiplicative way.

In Sections 43–44, we consider several problems with space integrals from the point of view of global existence, blow-up and a priori estimates. In particular, we study in some detail the asymptotic behavior of blowing-up solutions. The phenomenon of global blow-up appears as a typical feature of nonlocal problems. As an example of applied interest, we discuss the thermistor problem, which arises in the modeling of Ohmic heating. Fujita-type results for problems with space integrals are next described in Section 45. Finally, Section 46 is devoted to nonlocal problems (in time) with memory terms.

Throughout this chapter we will only consider nonnegative solutions, except in Subsection 44.4. Unless otherwise specified, each of the problems below is locally well-posed for (nonnegative) L^∞ initial data, and the (nonnegative) solution enjoys the regularity property (16.2) (see Example 51.13 and cf. Example 51.9). Also, we have the usual blow-up alternative in L^∞ (cf. Proposition 16.1). As for the comparison principle, more care is necessary when considering nonlocal problems, since it may be valid for certain problems and fail for some others. This will be made precise whenever necessary.

43. Problems involving space integrals (I)

We consider the following problem

$$\left.\begin{aligned}
u_t - \Delta u &= \int_\Omega u^p(y,t)\,dy - ku^q, & x \in \Omega,\ t > 0, \\
u &= 0, & x \in \partial\Omega,\ t > 0, \\
u(x,0) &= u_0(x), & x \in \Omega,
\end{aligned}\right\} \qquad (43.1)$$

where $\Omega \subset \mathbb{R}^n$ is a bounded domain, $p > 1$, $q \geq 1$, $k \geq 0$ and $u_0 \in L^\infty(\Omega)$, $u_0 \geq 0$. Note that problem (43.1) with $k = 0$ is maybe the simplest analogue of the model problem (15.1) for which the nonlocal, superlinear source term is given by a space integral.

43.1. Blow-up and global existence

When $k > 0$, (43.1) involves a competition between nonlocal source and local damping terms. A basic question is to determine the conditions for global existence or nonexistence of solutions. An answer is provided by the following theorem [525], which in particular shows that the value $q = p$ represents a critical blow-up exponent. It also contains some information concerning the global asymptotic behavior.

Theorem 43.1. *Consider problem* (43.1) *with* Ω *bounded,* $p > 1$, $q \geq 1$, $k \geq 0$, *and* $0 \leq u_0 \in L^\infty(\Omega)$.

(i) *Assume* $p > q$ *or* $p = q$ *and* $k < |\Omega|$. *Then:*

   (i1) *there exists* u_0 *such that* $T_{\max}(u_0) < \infty$;

   (i2) *the trivial solution is locally exponentially stable, i.e.: for* $\|u_0\|_\infty$ *small enough,* u *is global and* $\|u(t)\|_\infty$ *decays exponentially to 0 as* $t \to \infty$.

(ii) *Assume* $p = q$ *and* $k \geq |\Omega|$. *Then the trivial solution is globally exponentially stable, i.e.: all solutions of* (43.1) *are global, bounded and* $\|u(t)\|_\infty$ *decays exponentially to 0 as* $t \to \infty$.

(iii) *Assume* $p < q$ *and* $k > 0$. *Then:*

   (iii1) *all solutions of* (43.1) *are global and bounded;*

   (iii2) *if* k *is sufficiently large, then the trivial solution is globally exponentially stable;*

   (iii3) *if* k *is sufficiently small, then there exist positive stationary solutions.*

Proof. (i1) We first prove the existence of blowing-up solutions in the case $p = q$ and $k < |\Omega|$. Fix a subdomain $\Omega' \subset\subset \Omega$, such that $\delta := (|\Omega'| - k)/2 > 0$. There exists $\psi \in \mathcal{D}(\Omega)$ such that

$$\psi = 1 \text{ in } \Omega' \quad \text{and} \quad 0 \leq \psi \leq 1 \text{ in } \Omega$$

and we have

$$\int_\Omega \psi \, dx \geq |\Omega'| = k + 2\delta \quad \text{and} \quad \Delta\psi \geq -K$$

for some $K > 0$. Let $y(t) = \int_\Omega u(t)\psi \, dx$. Multiplying (43.1) by ψ, integrating by parts and using Hölder's inequality, we obtain

$$y'(t) = \int_\Omega u\Delta\psi \, dx + \int_\Omega \psi \, dx \int_\Omega u^p \, dx - k \int_\Omega u^p \psi \, dx \geq -K \int_\Omega u \, dx + 2\delta \int_\Omega u^p \, dx$$

$$\geq \delta \int_\Omega u^p \, dx + \left[\delta|\Omega|^{1-p}\left(\int_\Omega u \, dx\right)^{p-1} - K\right] \int_\Omega u \, dx.$$

Setting $g(t) = \delta|\Omega|^{1-p}\left(\int_\Omega u\,dx\right)^{p-1} - K$ and $c = \delta\|\psi\|_\infty^{-p}|\Omega|^{1-p}$, we deduce that

$$y'(t) \geq cy^p(t) + g(t)\int_\Omega u\,dx, \quad 0 < t < T_{\max}(u_0). \tag{43.2}$$

Now let $u_0 = \mu\psi$ with $\mu^{p-1} \geq \frac{K}{\delta}\max\left(1, |\Omega|^{p-1}\left(\int_\Omega \psi\,dx\right)^{1-p}\right)$. On the one hand we have $g(0) \geq 0$. On the other hand, we get

$$\int_\Omega u_0^p\,dx - ku_0^p = \mu^p\left(\int_\Omega \psi^p\,dx - k\psi^p\right) \geq \mu^p(|\Omega'| - k) = 2\delta\mu^p \geq K\mu \geq -\Delta u_0,$$

so that $u \geq u_0$ on $(0, T_{\max}(u_0))$ by the comparison principle (Proposition 52.25). It follows that $g(t) \geq 0$ on $(0, T_{\max}(u_0))$ and (43.2) then implies $T_{\max}(u_0) < \infty$.

To prove the existence of blowing-up solutions in the case $p > q$, we just note that u satisfies

$$u_t - \Delta u \geq \int_\Omega u^p\,dx - \tilde{k}u^p - Cu$$

for some $0 < \tilde{k} < |\Omega|$ and $C > 0$. The result then follows by an obvious modification of the proof in the case $q = p$.

(i2) Let us prove the local exponential stability of the trivial solution. It obviously suffices to treat the case $k = 0$. Let Θ be defined in (19.27), and put $z(x,t) = \varepsilon(1 + \Theta(x))e^{-\alpha t}$. Then, for $\alpha, \varepsilon > 0$ sufficiently small, we have

$$z_t - \Delta z = \varepsilon(-\alpha(1 + \Theta) + 1)e^{-\alpha t} \geq \tfrac{\varepsilon}{2}e^{-\alpha t}$$

$$\geq \varepsilon^p e^{-p\alpha t}\int_\Omega (1 + \Theta)^p\,dx = \int_\Omega z^p\,dx, \quad t \geq 0.$$

Therefore, if $\|u_0\|_\infty \leq \varepsilon$, then z is a supersolution to (43.1), so that u is global and satisfies $u(x,t) \leq z(x,t) \leq Ce^{-\alpha t}$.

The local exponential stability of the trivial solution also follows from abstract results in Appendix E (see Remark 51.20(ii)).

(ii) Multiplying the equation by u^m $(m \geq 1)$, integrating by parts and using Hölder's inequality, we have

$$\frac{d}{dt}\int_\Omega \frac{u^{m+1}}{m+1}\,dx + m\int_\Omega u^{m-1}|\nabla u|^2\,dx = \int_\Omega u^m\,dx\int_\Omega u^p\,dx - k\int_\Omega u^{m+p}\,dx$$

$$\leq (|\Omega| - k)\int_\Omega u^{m+p}\,dx \leq 0.$$

It follows from the Poincaré inequality that

$$\frac{d}{dt}\int_\Omega u^{m+1}\,dx \leq -\frac{4m}{m+1}\int_\Omega |\nabla u^{(m+1)/2}|^2\,dx \leq -C\int_\Omega u^{m+1}\,dx,$$

hence

$$\int_\Omega u^{m+1}(t)\,dx \le M_0 e^{-Ct}, \quad 0 \le t < T_{\max}(u_0).$$

If we choose $m + 1 \ge p$, then (for different constants $M_0, C > 0$) u satisfies

$$\left.\begin{array}{ll} u_t - \Delta u \le M_0 e^{-Ct}, & x \in \Omega,\ t > 0, \\[4pt] u = 0, & x \in \partial\Omega,\ t > 0, \\[4pt] u(x,0) = u_0(x), & x \in \Omega. \end{array}\right\} \tag{43.3}$$

Let now $w(x,t) = M(1 + \Theta(x))e^{-\alpha t}$, with Θ defined in (19.27). If we choose $\alpha \le \min\big((2(1 + \|\Theta\|_\infty))^{-1}, C\big)$ and $M \ge \max(2M_0, \|u_0\|_\infty)$, we then have

$$w_t - \Delta w = (-\alpha(1 + \Theta) + 1)Me^{-\alpha t} \ge M_0 e^{-Ct},$$

and $w(0,.) \ge u_0$, so that w is a supersolution of (43.3). It follows that u is global and decays exponentially to 0 as $t \to \infty$.

(iii1) To show that all solutions of (43.1) are global and bounded, it suffices to note that for any constant

$$M > \max\big(\|u_0\|_\infty, (|\Omega|k^{-1})^{1/(q-p)}\big),$$

$\bar{u} \equiv M$ is a supersolution of (43.1).

(iii2) Next, let us show the global stability of 0 for k large. By Young's inequality, we have $|\Omega|u^p \le ku^q + \varepsilon(k)u$, where $\varepsilon(k) = C(|\Omega|, p, q)k^{-(p-1)/(q-p)}$, so that u satisfies

$$u_t - \Delta u \le \int_\Omega u^p\,dx - |\Omega|u^p + \varepsilon(k)u.$$

For $k \ge k_0(|\Omega|, p, q)$ sufficiently large (hence $\varepsilon(k)$ small), the conclusion then follows from an easy modification of the proof of assertion (ii).

(iii3) Finally, let us prove the existence of stationary solutions for k small. Let $U = \mu\Theta$, with Θ again defined in (19.27) and $\mu = 2(\int_\Omega \Theta^p\,dx)^{-1/(p-1)}$. For $k > 0$ sufficiently small we have

$$-\Delta U + kU^q = (1 + k\mu^{q-1}\Theta^q)\mu \le 2^{p-1}\mu = \int_\Omega U^p\,dx.$$

By a modification of Proposition 52.20, it follows that the solution of (43.1) with $u_0 = U$ is nondecreasing in time, and we already know that it is global and bounded. Now Example 51.39 and Proposition 53.8 guarantee that $u(t) \to V$ in $L^\infty(\Omega)$ where V is a (classical) stationary solution, $V \ge U$. $\quad\square$

43.2. Blow-up rates, sets and profiles

In this subsection, we study the blow-up asymptotics for problem (43.1). The methods and results of this subsection are from [480], except for Theorem 43.4 in the case $p \geq 2$ [487] and Theorem 43.11(ii) (which is an improvement of [487]). We refer to [60], [119], [525] for earlier results on blow-up asymptotics for problem (43.1) and its variants.

Our first result shows that blowing-up solutions to (43.1) exhibit global blow-up and can be described by a uniform blow-up profile in the interior of the domain.

Theorem 43.2. *Assume Ω bounded, $p > q \geq 1$, $k \geq 0$, and $0 \leq u_0 \in L^\infty(\Omega)$. Let u be the solution of (43.1) and assume that $T := T_{\max}(u_0) < \infty$. Then we have*

$$\lim_{t \to T} (T - t)^{\frac{1}{p-1}} u(x, t) = \lim_{t \to T} (T - t)^{\frac{1}{p-1}} \|u(t)\|_\infty = \left[(p - 1)|\Omega| \right]^{-\frac{1}{p-1}}, \qquad (43.4)$$

uniformly on compact subsets of Ω.

Since the solution vanishes on the boundary and blows up globally inside Ω, it follows that a boundary layer appears as $t \to T$. The following result describes the behavior of the solution u near the blow-up time in the boundary layer.

Theorem 43.3. *Under the assumptions of Theorem 43.2, for all $K > 0$, there exist some constants $C_2 \geq C_1 > 0$ and some $t_0 \in (0, T)$, such that u satisfies*

$$C_1 \frac{\delta(x)}{\sqrt{T - t}} \, \|u(t)\|_\infty \leq u(x, t) \leq C_2 \frac{\delta(x)}{\sqrt{T - t}} \, \|u(t)\|_\infty, \qquad (43.5)$$

for all (x, t) in $\Omega \times [t_0, T)$ such that $\delta(x) \leq K\sqrt{T - t}$,

From the right-hand side of (43.5), one deduces that the size of the boundary layer is at least of order $\sqrt{T - t}$ near the blow-up time, in the sense that $u(x, t) = o(\|u(t)\|_\infty)$, as $t \to T$ and $\delta(x)/\sqrt{T - t} \to 0$. However, estimate (43.5) is not enough to conclude that the size of the boundary layer is exactly of order $\sqrt{T - t}$, in the sense that $u(x, t)/\|u(t)\|_\infty \to 1$, as $t \to T$ and $\delta(x)/\sqrt{T - t} \to \infty$. The following theorem, though not very sharp regarding the actual behavior of the solution in the boundary layer, enables one to conclude that this is indeed true.

Theorem 43.4. *Under the assumptions of Theorem 43.2, for all $\varepsilon > 0$, there exists $C(\varepsilon) > 0$ such that*

$$u(x, t) \geq \|u(t)\|_\infty \left(1 - \varepsilon - C(\varepsilon) \frac{T - t}{\delta^2(x)} \right), \qquad (x, t) \in \Omega \times [0, T).$$

Therefore, we have

$$\frac{u(x,t)}{\|u(t)\|_\infty} \longrightarrow \begin{cases} 1, & \text{as } t \to T \text{ and } \dfrac{\delta(x)}{\sqrt{T-t}} \to \infty, \\[3mm] 0, & \text{as } t \to T \text{ and } \dfrac{\delta(x)}{\sqrt{T-t}} \to 0. \end{cases} \tag{43.6}$$

In other words, the size of the boundary layer decays like $\sqrt{T-t}$.

Remarks 43.5. (a) **Comparison with the local model problem.** For problem (15.1), we have seen that single-point blow-up occurs if $\Omega = B_R$ and $u \geq 0$ is radial nonincreasing (cf. Theorem 24.1). If moreover $p < p_S$, then $u(\cdot,t)$ behaves like its maximum in space-time parabolas based at $(0,T)$, that is: $u(x,t)/\|u(t)\|_\infty \to 1$ as $t \to T$, uniformly for $|x| \leq C\sqrt{T-t}$. In some cases (see Remark 25.8), it is even known that

$$\frac{u(x,t)}{\|u(t)\|_\infty} \longrightarrow \begin{cases} 1, & \text{as } t \to T \text{ and } \dfrac{|x|}{\sqrt{(T-t)|\log(T-t)|}} \to 0, \\[4mm] 0, & \text{as } t \to T \text{ and } \dfrac{|x|}{\sqrt{(T-t)|\log(T-t)|}} \to \infty. \end{cases} \tag{43.7}$$

At the opposite, blow-up for problem (43.1) is global and solutions behave like their maximum everywhere outside of a space-time parabolic neighborhood of $(\partial\Omega, T)$ (compare formulas (43.6) and (43.7)). Problems (15.1) and (43.1) thus exhibit in some sense dual blow-up behaviors.

(b) **Asymptotic influence of the local damping term.** It appears from Theorem 43.2 that the local damping term has no significant effect on the asymptotic behavior of solutions near the blow-up time if $q < p$. In the blow-up critical case $q = p$, $k < |\Omega|$, which was studied in [487], this is no longer so: Blow-up is still global and uniform on compact sets, but the constant in the RHS of (43.4) has to be replaced by $\left[(p-1)(|\Omega|-k)\right]^{-1/(p-1)}$. $\square$

The proof of the above results relies on the study of linear problems with spatially homogeneous blowing-up source, of the form

$$\left. \begin{aligned} u_t - \Delta u &= g(t), & x \in \Omega,\ 0 < t < T, \\ u &= 0, & x \in \partial\Omega,\ 0 < t < T, \\ u(x,0) &= u_0(x), & x \in \Omega. \end{aligned} \right\} \tag{43.8}$$

If g is a given function, locally Hölder continuous on $[0,T)$, and if $u_0 \in L^\infty(\Omega)$, then we know that (43.8) has a unique classical solution $u \in C^{2,1}(\overline{\Omega} \times (0,T))$, with $u - e^{-tA}u_0 \in C(\overline{\Omega} \times [0,T))$.

In what follows we shall use the following notation. We set

$$G(t) = \int_0^t g(s)\,ds \quad \text{and} \quad H(t) = \int_0^t G(s)\,ds. \tag{43.9}$$

We write $u \sim v$ for $\lim_{t \to T} u(t)/v(t) = 1$. As usual, λ_1 and φ_1 denote respectively the first Dirichlet eigenvalue and eigenfunction, normalized by $\int_\Omega \varphi_1 \, dx = 1$. Moreover we set

$$K_\rho = \{x \in \Omega : \delta(x) \geq \rho\}, \quad \rho > 0.$$

For problem (43.8), we shall prove the following theorem.

Theorem 43.6. *Assume Ω bounded, $0 \leq u_0 \in L^\infty(\Omega)$, $g \geq 0$ locally Hölder continuous on $[0, T)$. Let $u \geq 0$ be the solution of (43.8). Then we have*

$$\limsup_{t \to T} \|u(x, t)\|_\infty = \infty \tag{43.10}$$

if and only if

$$\int_0^T g(s) \, ds = \infty. \tag{43.11}$$

Furthermore, if (43.10) or (43.11) is fulfilled, then

$$\lim_{t \to T} \frac{u(x, t)}{G(t)} = \lim_{t \to T} \frac{\|u(t)\|_\infty}{G(t)} = 1, \tag{43.12}$$

uniformly on compact subsets of Ω.

The proof of Theorem 43.6 is based on eigenfunction arguments, one-sided estimates of Δu (obtained via the maximum principle), and the mean-value inequality for subharmonic functions. We need the following two simple lemmas.

Lemma 43.7. *(i) Under the assumptions of Theorem 43.6, we have*

$$u \leq G(t) + \|u_0\|_\infty, \qquad (x, t) \in Q_T. \tag{43.13}$$

(ii) If moreover $u_0 \equiv 0$, then

$$\Delta u \leq 0, \qquad (x, t) \in Q_T. \tag{43.14}$$

Proof. To prove (43.13) it suffices to notice that $\bar{u}(x, t) := G(t) + \|u_0\|_\infty$ is a supersolution to (43.8).

To show (43.14) one could apply the maximum principle to the equation satisfied by Δu, after showing that $\Delta u \in C([0, T), L^2(\Omega))$. Alternatively, one can use the following simple argument: For each $h \in (0, T)$, the function $v(x, t) := u(x, t + h) - u(x, t)$ satisfies

$$\left. \begin{aligned} v_t - \Delta v &= g(t + h) - g(t), & x \in \Omega, \ 0 < t < T - h, \\ v &= 0, & x \in \partial\Omega, \ 0 < t < T - h, \\ v(x, 0) &= u(x, h), & x \in \Omega. \end{aligned} \right\} \tag{43.15}$$

Since $u(\cdot, h) \le G(h)$ due to (43.13), we see that $\overline{v}(x, t) := G(t + h) - G(t)$ is a supersolution to (43.15), hence $u(x, t + h) - u(x, t) \le G(t + h) - G(t)$. Dividing by h and letting $h \to 0$, we get $u_t \le g(t)$ in Q_T, hence (43.14). $\quad\square$

As for the next lemma, a more accurate inequality will be given below to obtain precise boundary estimates, see (43.35). However this one is sufficient for the purpose of Theorem 43.6.

Lemma 43.8. *Assume Ω bounded and let $z \in C^2(\Omega)$ satisfy*

$$z \ge 0 \quad and \quad \Delta z \ge 0, \qquad x \in \Omega. \tag{43.16}$$

Then

$$z(x) \le \frac{C(\Omega)}{\delta^{n+1}(x)} \int_\Omega z(y)\varphi_1(y)\,dy, \qquad x \in \Omega.$$

Proof. Fix $x \in K_\rho\ (\ne \emptyset)$. By the mean-value inequality for subharmonic functions, we have

$$z(x) \le \frac{1}{|B(x, \rho/2)|} \int_{B(x,\rho/2)} z(y)\,dy = \frac{C(n)}{\rho^n} \int_{B(x,\rho/2)} z(y)\,dy.$$

Since $\inf_{K_\rho} \varphi_1 \ge c_1(\Omega)\rho$ and $z \ge 0$, we deduce that

$$z(x) \le \frac{C(\Omega)}{\rho^{n+1}} \int_{B(x,\rho/2)} z(y)\varphi_1(y)\,dy \le \frac{C(\Omega)}{\rho^{n+1}} \int_\Omega z(y)\varphi_1(y)\,dy$$

and the lemma follows. $\quad\square$

Proof of Theorem 43.6. We first consider the case $u_0 \equiv 0$. From (43.13), it is clear that (43.10) implies (43.11). Conversely, assume that (43.11) holds. Our aim is then to prove (43.12).

Define

$$z(x, t) = G(t) - u(x, t) \quad and \quad \beta(t) = \int_\Omega z(y, t)\,\varphi_1(y)\,dy.$$

By Green's formula, we have

$$\beta'(t) = \int_\Omega \big(g(t) - u_t(y, t)\big)\,\varphi_1(y)\,dy = -\int_\Omega \Delta u(y, t)\,\varphi_1(y)\,dy$$

$$= -\int_\Omega u(y, t)\,\Delta\varphi_1(y)\,dy = \lambda_1 \int_\Omega u(y, t)\,\varphi_1(y)\,dy = -\lambda_1\beta(t) + \lambda_1 G(t).$$

Integrating this equation and using $\beta(0) = 0$, we obtain

$$\beta(t) = \lambda_1 \int_0^t e^{\lambda_1(s-t)} G(s)\,ds \le \lambda_1 H(t), \tag{43.17}$$

where H is defined by (43.9). Since $z \geq 0$ and $\Delta z \geq 0$ by (43.13) and (43.14), Lemma 43.8 implies

$$z(x,t) \leq \frac{C(\Omega)}{\delta^{n+1}(x)} \int_\Omega z(y,t)\varphi_1(y)\,dy \leq \frac{\lambda_1 C(\Omega)H(t)}{\rho^{n+1}}, \qquad x \in K_\rho, \ t \in (0,T).$$
$$(43.18)$$

For t close enough to T, we have $G(t) > 0$ by (43.11), and (43.13) and (43.18) give us

$$0 \leq 1 - \frac{u(x,t)}{G(t)} \leq \frac{C(\Omega)}{\rho^{n+1}}\frac{H(t)}{G(t)}, \qquad x \in K_\rho. \tag{43.19}$$

On the other hand, since G is nondecreasing, for all $\varepsilon > 0$ we have

$$0 \leq \frac{H(t)}{G(t)} \leq \frac{\displaystyle\int_0^{T-\varepsilon} G(s)\,ds}{G(t)} + \varepsilon.$$

Using (43.11), we deduce that $\lim_{t \to T} H(t)/G(t) = 0$. In view of (43.19), this proves (43.12) for $u_0 \equiv 0$.

Finally, for general $u_0 \geq 0$, we write $u = U + e^{-tA}u_0$, where U is the solution of (43.8) corresponding to $u_0 \equiv 0$. By using $\|e^{-tA}u_0\|_\infty \leq \|u_0\|_\infty$, the general case easily follows from the case $u_0 \equiv 0$. $\square$

We are now in a position to prove Theorem 43.2.

Proof of Theorem 43.2. *Case 1: $k = 0$.* We apply Theorem 43.6 with

$$g(t) := \int_\Omega u^p(y,t)\,dy, \qquad G(t) = \int_0^t g(s)\,ds. \tag{43.20}$$

By (43.12) in Theorem 43.6, it follows that

$$\forall x \in \Omega, \quad \lim_{t \to T} u^p(x,t)/G^p(t) = 1.$$

Moreover, (43.13) implies $0 \leq u^p(x,t)/G^p(t) \leq 2$ in Ω for t close enough to T. By Lebesgue's dominated convergence theorem, we infer that

$$\int_\Omega u^p(y,t)\,dy \sim |\Omega|G^p(t), \quad t \to T,$$

hence

$$G'(t) = g(t) \sim |\Omega|G^p(t), \tag{43.21}$$

or $(G^{1-p})' \sim -(p-1)\,|\Omega|$. After integrating this equivalence between t and T, we obtain

$$G(t) \sim \left[(p-1)|\Omega|(T-t)\right]^{-1/(p-1)}. \tag{43.22}$$

The result finally follows by returning to (43.12).

Case 2: $k > 0$. It requires some modifications of the arguments from the case $k = 0$ and from the proof of Theorem 43.6 (in particular we no longer consider the case $u_0 \equiv 0$ separately). We only indicate the necessary changes.

We first note that (43.13) and consequently (43.11) are still valid. As an analogue of (43.14) in Lemma 43.7, we next establish the inequality

$$\Delta u \le C_1 := \|\Delta u(\cdot, T/2)\|_\infty, \qquad (x,t) \in \Omega \times [T/2, T]. \tag{43.23}$$

By the strong maximum principle, we have $u > 0$ in Q_T. Set $v = \Delta u$ and note that $v \in C^{2,1}(Q_T) \cap C(\overline{\Omega} \times (0,T))$ by parabolic regularity. Taking the Laplacian of equation (43.1) then yields

$$v_t - \Delta v = -q(u^{q-1}v + (q-1)u^{q-2}|\nabla u|^2) \le -qu^{q-1}v \quad \text{in } \Omega \times (0,T),$$

with $v(x,t) = -g(t) \le 0$ on the boundary, where g is still defined by (43.20). Therefore, by the maximum principle, v cannot achieve an interior positive maximum, hence (43.23).

Now set

$$z(x,t) = G(t) - u(x,t) + C_1 \frac{|x|^2}{2n} + \|u_0\|_\infty \quad \text{and} \quad \beta(t) = \int_\Omega z(y,t)\,\varphi_1(y)\,dy.$$

By (43.13) and (43.23), we have

$$z \ge 0 \quad \text{and} \quad \Delta z \ge 0, \qquad (x,t) \in \Omega \times (T/2, T). \tag{43.24}$$

On the other hand, arguing as in the proof of Theorem 43.6, we obtain

$$\beta'(t) = \lambda_1 \int_\Omega u(y,t)\varphi_1(y)\,dy + k \int_\Omega u^q(y,t)\varphi_1(y)\,dy.$$

Integrating and using Hölder's inequality and (43.11), we get

$$\beta(t) \le C \left(1 + \int_0^t \int_\Omega u^q(y,s)\,dy\,ds \right)$$
$$\le C + C(T|\Omega|)^{1-(q/p)} \left(\int_0^t \int_\Omega u^p(y,s)\,dy\,ds \right)^{q/p} = o(G(t)), \tag{43.25}$$

as $t \to T$. Owing to (43.24) we may apply Lemma 43.8, and using (43.25) we then conclude in a similar way as in the proof of Theorem 43.6 and Case 1. $\square$

To prove the boundary estimates in Theorems 43.3 and 43.4, we return to problem (43.8) and introduce the following definition.

Definition 43.9. We say that g is **sub-standard**, resp. **super-standard**, if it satisfies the following power-like growth assumption

$$g(t)/G(t) \leq k_1(T-t)^{-1}, \qquad \text{as } t \to T, \tag{43.26}$$

resp.

$$g(t)/G(t) \geq k_2(T-t)^{-1}, \qquad \text{as } t \to T, \tag{43.27}$$

with constants $k_1, k_2 > 0$. We say that g is **standard** if it satisfies (43.26) and (43.27). $\square$

Note that if (43.26) holds, then $g(t) \leq C_1(T-t)^{-(k_1+1)}$ as $t \to T$. If (43.27) holds, then $g(t) \geq C_2(T-t)^{-(k_2+1)}$ as $t \to T$, so that in particular $\int_0^T g(s)\,ds = \infty$. Conversely, g is standard whenever, for instance, $c_1(T-t)^{-\alpha} \leq g(t) \leq c_2(T-t)^{-\alpha}$ as $t \to T$, for some $\alpha > 1$ and $c_2 \geq c_1 > 0$.

Theorem 43.10. *Assume Ω bounded, $0 \leq u_0 \in L^\infty(\Omega)$, $g \geq 0$ locally Hölder continuous on $[0, T)$, and (43.11). Let $u \geq 0$ be the solution of (43.8).*

(i) Assume that g is super-standard. Then for all $K > 0$ there exist $C_1 > 0$ and $t_1 \in (0, T)$, such that

$$u(x, t) \geq C_1 \frac{\delta(x)}{\sqrt{T-t}}\, G(t),$$

for all (x, t) in $\Omega \times [t_1, T)$ such that $\delta(x) \leq K\sqrt{T-t}$.

(ii) Assume that g is sub-standard. Then for all $K > 0$ there exist $C_2 > 0$ and $t_2 \in (0, T)$, such that

$$u(x, t) \leq C_2 \frac{\delta(x)}{\sqrt{T-t}}\, G(t),$$

for all (x, t) in $\Omega \times [t_2, T)$ such that $\delta(x) \leq K\sqrt{T-t}$.

Theorem 43.11. *Assume Ω bounded, $0 \leq u_0 \in L^\infty(\Omega)$, $g \geq 0$ locally Hölder continuous on $[0, T)$. Let $u \geq 0$ be the solution of (43.8).*

(i) Assume that G is super-standard. Then

$$u(x, t) \geq \|u(t)\|_\infty \left(1 - C\frac{T-t}{\delta^2(x)}\right) - \|u_0\|_\infty \tag{43.28}$$

in $\Omega \times [0, T)$.

(ii) Assume that g is super-standard and nondecreasing. Then for all $\varepsilon > 0$, there exists $C(\varepsilon) > 0$ such that

$$u(x, t) \geq \|u(t)\|_\infty \left(1 - \varepsilon - C(\varepsilon)\frac{T-t}{\delta^2(x)}\right) \tag{43.29}$$

in $\Omega \times [0, T)$.

Note that estimate (43.28) is stronger than (43.29) when $\|u(t)\|_\infty$ is large. However, the assumption that G is super-standard is too restrictive in practice. For instance in case of problem (43.1) with $k = 0$, we have $g(t) := \int_\Omega u^p(t)\,dx \sim C(T-t)^{-p/(p-1)}$ and $G(t) \sim C'(T-t)^{-1/(p-1)}$ by (43.21), (43.22), so that G is super-standard only for $p < 2$, whereas g is standard for all $p > 1$.

For the proof of Theorem 43.10 we construct blowing-up sub-/supersolutions of barrier type (relative to interior or exterior tangent balls at boundary points). As for Theorem 43.11, it is based on refinements of the arguments leading to Theorem 43.6.

Proof of Theorem 43.10. *Step 1.* We first claim that we need only consider the case $u_0 \equiv 0$. Indeed, for general u_0, u may be decomposed as $u = e^{-tA}u_0 + U$, where U solves $U_t - \Delta U = g(t)$ with 0 initial and boundary values. Using $e^{-tA}u_0 \in C^{1,0}(\overline{\Omega}\times[T/2,T])$ and (43.11), we have $0 \le e^{-tA}u_0 \le C\,\delta(x) \le \varepsilon\,\delta(x)\,G(t)/\sqrt{T-t}$, for all $x \in \Omega$ and t close enough to T. The claim follows.

Step 2. We prove the lower estimate when $u_0 = 0$. The basic idea is to seek a suitable subsolution.

Since Ω is smooth, $\partial\Omega$ satisfies a uniform interior and exterior sphere condition i.e., for some $\underline{R}$, $\overline{R} > 0$ depending only on Ω, and for each point $\xi \in \partial\Omega$, there exist some balls $B_i(\xi)$ of radius $\underline{R}$ and $B_e(\xi)$ of radius $\overline{R}$ such that $\overline{B_i(\xi)} \cap \Omega^c = \{\xi\} = \overline{B_e(\xi)} \cap \overline{\Omega}$.

Now fix $x_0 \in \Omega$. Let $\xi \in \partial\Omega$ be such that $\delta(x_0) = |x_0 - \xi|$, and let B be the ball containing $B_i(\xi)$, tangent to both $B_i(\xi)$ and $B_e(\xi)$, of radius $R = \max(\underline{R}, \delta(x_0))$. By the definition of $\delta(x_0)$, it is clear that $B \subset \Omega$ and that $\delta(x_0) = \mathrm{dist}(x_0, \partial B)$, with $\underline{R} \le R \le \mathrm{diam}(\Omega)$. Without loss of generality, we may also assume that B is centered at the origin. Define the space-time domain $D = \overline{B} \times [0, T)$, and divide D into two sub-regions as follows:

$$D_1 :\ 0 \le d(x) < \frac{R}{2\sqrt{T}}\sqrt{T-t}, \qquad D_2 :\ d(x) \ge \frac{R}{2\sqrt{T}}\sqrt{T-t}, \qquad (43.30)$$

where $d(x) = \mathrm{dist}(x, \partial B) = R - r$, $r = |x|$. We next define

$$v(x,t) = \begin{cases} 4G(t)\dfrac{d(x)}{\sqrt{T-t}}\left(\dfrac{R}{\sqrt{T}} - \dfrac{d(x)}{\sqrt{T-t}}\right) & \text{in } D_1, \\[3ex] G(t)\dfrac{R^2}{T} & \text{in } D_2. \end{cases} \qquad (43.31)$$

It is clear that $v \in C^1(D)$, and $v(\cdot,t) \in H^2(B)$, $0 < t < T$. Moreover, $v(x,0) = 0$ in $\overline{B}$ and $v(x,t) = 0$ for $x \in \partial B$. One then computes:

$$v_t(x,t) = \begin{cases} \dfrac{4d(x)}{\sqrt{T-t}}\left[\dfrac{G(t)}{T-t}\left(\dfrac{R}{2\sqrt{T}} - \dfrac{d(x)}{\sqrt{T-t}}\right) + g(t)\left(\dfrac{R}{\sqrt{T}} - \dfrac{d(x)}{\sqrt{T-t}}\right)\right] & \text{in } D_1, \\[3ex] g(t)\dfrac{R^2}{T} & \text{in } D_2. \end{cases}$$

$$(43.32)$$

We have

$$v_r(x,t) = v_{rr}(x,t) = 0 \quad \text{in } D_2, \tag{43.33}$$

while in D_1 (where $r \geq R/2$), we find that

$$v_r(x,t) = \frac{4G(t)}{\sqrt{T-t}}\left(\frac{-R}{\sqrt{T}} + \frac{2(R-r)}{\sqrt{T-t}}\right)$$

and $v_{rr}(x,t) = \frac{-8G(t)}{T-t}$, so that

$$-\Delta v(x,t) = -v_{rr} - \frac{n-1}{r}v_r \leq \frac{4G(t)}{T-t}\left(2 + \frac{n-1}{r}\frac{R}{\sqrt{T}}\sqrt{T-t}\right) \leq \frac{8nG(t)}{T-t} \quad \text{in } D_1.$$

Therefore, we get

$$v_t - \Delta v \leq \begin{cases} \dfrac{G(t)}{T-t}\dfrac{R^2}{4T} + g(t)\dfrac{R^2}{T} + \dfrac{8nG(t)}{T-t} & \text{in } D_1, \\[2ex] g(t)\dfrac{R^2}{T} & \text{in } D_2. \end{cases}$$

Using the fact that g is super-standard, it follows that $v_t - \Delta v \leq C(R)g(t)$ in D, where $C(R) = R^2/T + (8n + R^2/4T)\,k_2^{-1}$. Therefore, $C(R)^{-1}v$ is a subsolution in D, and since $u \geq 0$, the maximum principle implies $u \geq C(R)^{-1}v$ in D. On the other hand, for any $K > 0$, we have

$$v(x,t) \geq \begin{cases} \dfrac{2R}{\sqrt{T}}G(t)\dfrac{d(x)}{\sqrt{T-t}}, & \text{if } d(x)/\sqrt{T-t} \leq R/2\sqrt{T}, \\[2ex] \dfrac{R^2}{T}G(t) \geq \dfrac{R^2}{TK}G(t)\dfrac{d(x)}{\sqrt{T-t}}, & \text{if } R/2\sqrt{T} \leq d(x)/\sqrt{T-t} \leq K. \end{cases}$$

Since $\delta(x_0) = d(x_0)$, we deduce that if $\delta(x_0) \leq K\sqrt{T-t}$, then

$$u(x_0,t) \geq C_1\, G(t)\frac{\delta(x_0)}{\sqrt{T-t}},$$

with

$$C_1 = \frac{\min(2R/\sqrt{T}, R^2/TK)}{R^2/T + (8n + R^2/4T)\,k_2^{-1}} \geq C(T,K,n,k_2)\frac{\min(R, R^2)}{1 + R^2}$$

$$\geq C(T,K,n,k_2)\min\left(\underline{R}^2, \mathrm{diam}^{-1}(\Omega)\right),$$

where we have used $\underline{R} \leq R \leq \mathrm{diam}(\Omega)$. Therefore, C_1 can be chosen independent of x_0, and the desired lower estimate follows.

Step 3. We prove the upper estimate when $u_0 = 0$. To do so, we show that the function v of Step 2, suitably modified and multiplied by a large constant, becomes a supersolution.

Fixing $x_0 \in \Omega$, and keeping the notation of Step 2, we now set $D = B'^c \times [0, T)$, with $B' = B_e(\xi)$ the exterior ball, of radius $\overline{R}$, associated with ξ, where $\xi \in \partial\Omega$ is such that $\delta(x_0) = |x_0 - \xi|$. It is clear that $\delta(x_0) = \mathrm{dist}(x_0, B')$ and we may again assume that B' is centered at the origin. Consider the function v defined by (43.31), where now $d(x) = \mathrm{dist}(x, B') = r - \overline{R}$ and $R = \overline{R}/n$, and where D_1, D_2 are still defined by (43.30). Formulae (43.32) and (43.33) are unchanged, whereas in D_1 we now have

$$v_r(x, t) = \frac{4G(t)}{\sqrt{T - t}}\left(\frac{R}{\sqrt{T}} - \frac{2(r - \overline{R})}{\sqrt{T - t}}\right)$$

and $v_{rr}(x, t) = \frac{-8G(t)}{T-t}$, so that

$$-\Delta v(x, t) \geq \frac{4G(t)}{T - t}\left(2 - \frac{n - 1}{\overline{R}}\frac{R}{\sqrt{T}}\sqrt{T - t}\right) \geq \frac{4G(t)}{T - t} \quad \text{in } D_1.$$

Therefore, we get

$$v_t - \Delta v \geq \begin{cases} \dfrac{4G(t)}{T - t} & \text{in } D_1, \\[2ex] \dfrac{R^2}{T}g(t) & \text{in } D_2. \end{cases}$$

Using the fact that g is sub-standard, we find that $v_t - \Delta v \geq C'(\overline{R})g(t)$ in D, where $C'(\overline{R}) = \min(4k_1^{-1}, \overline{R}^2 n^{-2}T^{-1})$. It follows that $C'(\overline{R})^{-1}v$ is a supersolution in D, hence in $\Omega \times [0, T)$, and the maximum principle implies $u \leq C'(\overline{R})^{-1}v$, so that

$$u(x_0, t) \leq C_2\, G(t)\frac{\delta(x_0)}{\sqrt{T - t}} \quad \text{in } [0, T),$$

with $C_2 = 4\overline{R}n^{-1}T^{-1/2}C'(\overline{R})^{-1}$, which proves the upper estimate. $\quad\square$

Proof of Theorem 43.11. *Step 1.* We shall show that

$$u(x, t) \geq G(t) - C(n)\frac{H(t)}{\delta^2(x)}, \qquad (x, t) \in \Omega \times [0, T). \tag{43.34}$$

In view of the maximum principle, it suffices to establish (43.34) for $u_0 \equiv 0$, which we assume in the rest of this step. Estimate (43.34) is equivalent to the following inequality, which is an improved version of Lemma 43.8:

$$\sup_{x \in K_\rho} z(x, t) \leq \frac{C(n)}{\rho^2} H(t), \tag{43.35}$$

where $z(x, t) := G(t) - u(x, t)$. Note that $z \geq 0$ due to (43.13).

We first establish (43.34) when Ω is a ball $B_R(x_0)$. We may assume $x_0 = 0$ without loss of generality. Fix $t \in (0, T)$, $x \in \Omega$ and set $\rho := R - |x|$. Since $\Delta z \geq 0$ by (43.14), the mean-value inequality for subharmonic functions implies

$$z(x, t) \leq \frac{C(n)}{\rho^n} \int_{B(x, \rho/2)} z(y, t) \, dy. \tag{43.36}$$

If $\rho \geq R/2$, then

$$z(x, t) \leq \frac{C(n) \, R^{1-n}}{\rho} \int_{K_{\rho/2}} z(y, t) \, dy. \tag{43.37}$$

Next suppose that $\rho < R/2$. Note that $u(\cdot, t)$ is radially symmetric due to $u_0 \equiv 0$. Switching to polar coordinates, with $z(y, t) = z(r, t)$, $r = |y|$, we may write

$$\int_{B(x, \rho/2)} z(y, t) \, dy = \int_{|x|-\rho/2}^{|x|+\rho/2} z(r, t) M(r) \, dr,$$

where $M(r) = \mathrm{Surf}(B(x, \rho/2) \cap S(0, r))$ and "Surf" denotes the surface measure. Observing that

$$M(r) \leq \mathrm{Surf}(S(x, \rho/2)) \leq C(n)\rho^{n-1},$$

it follows from (43.36) that

$$z(x, t) \leq \frac{C(n)}{\rho} \int_{R/4}^{R-\rho/2} z(r, t) \, dr \leq \frac{C(n) \, R^{1-n}}{\rho} \int_{R/4}^{R-\rho/2} z(r, t) r^{n-1} \, dr,$$

so that (43.37) is true in all cases.

Still assuming $\Omega = B_R$, fix $\rho \in (0, R)$ and $t \in (0, T)$. Since the RHS in (43.37) is a decreasing function of ρ and, for each $x \in K_\rho$, $\tilde{\rho} := R - |x| \geq \rho$, we see that

$$\sup_{x \in K_\rho} z(x, t) \leq \frac{C(n) \, R^{1-n}}{\rho} \int_{K_{\rho/2}} z(y, t) \, dy. \tag{43.38}$$

On the other hand, by (43.17) we have

$$\int_{B_R} z(y, t) \varphi_R(y) \, dy \leq \lambda_R H(t), \tag{43.39}$$

where λ_R is the first eigenvalue in B_R and φ_R is the corresponding eigenfunction, normalized by $\int_{B_R} \varphi_R = 1$. By straightforward scaling arguments, we have

$$\lambda_R = C(n) \, R^{-2} \quad \text{and} \quad \inf_{K_{\rho/2}} \varphi_R \geq c(n) \, R^{-(n+1)}\rho. \tag{43.40}$$

Inequality (43.35) then follows by combining (43.38), (43.39) and (43.40). Therefore (43.34) is proved when $\Omega = B_R$ (and we stress that the constant $C(n)$ does not depend on R).

To extend (43.34) to a general domain Ω, we fix $x_0 \in \Omega$ and consider $B = B(x_0, R) \subset \Omega$ with $R = \delta(x_0)$. Letting $\underline{u}$ be the solution of $\underline{u}_t - \Delta \underline{u} = g(t)$ in $B \times (0, T)$, with 0 initial and boundary conditions, the maximum principle implies $u \geq \underline{u}$. Since $\delta(x_0) = \mathrm{dist}(x_0, \partial B)$, (43.34) follows from the same inequality in B.

Step 2. Let us show assertion (i) of the theorem. Since $H(t) \leq k_2^{-1}(T-t)G(t)$ for t close to T by assumption, (43.34) and $u \geq 0$ imply

$$u(x,t) \geq G(t)\left(1 - C\frac{T-t}{\delta^2(x)}\right)_+, \qquad (x,t) \in \Omega \times [T_0, T), \qquad (43.41)$$

for some $T_0 \in (0, T)$. By taking a larger constant $C \geq (T - T_0)^{-1}\mathrm{diam}^2(\Omega)$, we see that (43.41) becomes in fact valid in $\Omega \times [0, T)$. Estimate (43.28) then follows by combining (43.41) and (43.13).

Step 3. To show assertion (ii) we shall use Step 1 to derive an estimate on u_t similar to (43.34), and then integrate over carefully chosen time intervals.

Take $u_0 \equiv 0$. Fix $h > 0$ and, for $t \in [0, T-h)$, put $v(\cdot, t) = u(\cdot, t+h) - u(\cdot, t)$ and $\tilde{g}(t) = g(t+h) - g(t)$. Note that $\tilde{g} \geq 0$ by assumption. The function v satisfies

$$\left. \begin{aligned} v_t - \Delta v &= \tilde{g}(t), & x \in \Omega,\ 0 < t < T - h, \\ v &= 0, & x \in \partial\Omega,\ 0 < t < T - h, \\ v(x,0) &= u(x,h), & x \in \Omega. \end{aligned} \right\} \qquad (43.42)$$

Applying the result of Step 1 to problem (43.42), we obtain

$$u(x, t+h) - u(x,t) \geq G(t+h) - G(t) - C(n)\frac{H(t+h) - H(t)}{\delta^2(x)}$$

in $\Omega \times [0, T - h)$. Dividing by h and letting $h \to 0$, and next using the assumption that g is super-standard and $u \geq 0$, we obtain

$$u_t(x,t) \geq g(t) - C(n)\frac{G(t)}{\delta^2(x)} \geq g(t)\left(1 - C\frac{T-t}{\delta^2(x)}\right)_+, \qquad (x,t) \in \Omega \times [T_0, T), \quad (43.43)$$

for some $T_0 \in (0, T)$. Fix $\gamma > 1$ and let $T_\gamma := T - \gamma^{-1}(T - T_0)$. Let $(x,t) \in \Omega \times [T_\gamma, T)$ and set $t_\gamma := T - \gamma(T - t) \in [T_0, T)$. Integrating (43.43) over (t_γ, t) yields

$$u(x,t) - u(x, t_\gamma) \geq \left(1 - C\frac{T - t_\gamma}{\delta^2(x)}\right)_+ \int_{t_\gamma}^{t} g(s)\, ds = \left(1 - \gamma C\frac{T-t}{\delta^2(x)}\right)_+ (G(t) - G(t_\gamma)).$$

Now, g being super-standard guarantees that $s \mapsto (T-s)^{k_2} G(s)$ is nondecreasing for s close to T. Taking T_γ closer to T, it follows that $G(t_\gamma) \leq G(t)\left(\frac{T-t}{T-t_\gamma}\right)^{k_2} = G(t)\gamma^{-k_2}$ for $t \in [T_\gamma, T)$, hence

$$u(x,t) \geq (1 - \gamma^{-k_2}) G(t)\left(1 - \gamma C \frac{T-t}{\delta^2(x)}\right)_+ , \qquad (x,t) \in \Omega \times [T_\gamma, T). \qquad (43.44)$$

By the maximum principle, (43.44) obviously remains true for $u_0 \geq 0$. Using (43.12), we get

$$u(x,t) \geq \|u(t)\|_\infty \left(1 - 2\gamma^{-k_2} - \gamma C \frac{T-t}{\delta^2(x)}\right), \qquad (x,t) \in \Omega \times [T_\gamma, T). \qquad (43.45)$$

Moreover, replacing γC by a larger constant $C(\gamma)$, we see that (43.45) becomes in fact valid in $\Omega \times [0,T)$. Estimate (43.29) finally follows by choosing $\gamma = (2/\varepsilon)^{1/k_2}$. $\square$

Proof of Theorems 43.3 and 43.4. Let $g(t) = |\Omega|^{-\frac{1}{p-1}} \left[(p-1)(T-t)\right]^{-\frac{p}{p-1}}$. It follows from Theorem 43.2 that, for all $\varepsilon \in (0,1)$, u satisfies $(1-\varepsilon)g(t) \leq u_t - \Delta u \leq (1+\varepsilon)g(t)$ in $\Omega \times [T_\varepsilon, T)$ for T_ε sufficiently close to T.

Taking, say, $\varepsilon = 1/2$, the maximum principle implies $v \leq u \leq w$ in $\Omega \times [T_{1/2}, T)$, where v and w solve $v_t - \Delta v = \frac{1}{2}g(t)$ and $w_t - \Delta w = \frac{3}{2}g(t)$ in $\Omega \times [T_{1/2}, T)$ with 0 boundary values and initial conditions $v(T_{1/2}) = w(T_{1/2}) = u(T_{1/2})$. Since g is standard, we deduce from Theorem 43.10 that v and w, hence u, satisfy the conclusion of Theorem 43.3.

Since g is standard and nondecreasing, by using the same comparison argument (from below) for each $\varepsilon \in (0,1)$, along with Theorem 43.11(ii), we obtain Theorem 43.4. $\square$

Remark 43.12. Other nonlocal problems. We refer to e.g. [160], [331] for results on systems of equations involving space integral terms. A different kind of nonlocal equations, of "localized" type, have also been studied by several authors. A typical example is:

$$u_t - \Delta u = u^p(x_0(t), t), \qquad (43.46)$$

with Dirichlet boundary conditions. Here $x_0 : [0, \infty) \to \Omega$ is a given (smooth) curve, which may be thought of as representing the location of a sensor driving the reaction in the whole domain. For equation (43.46), results on global (non-)existence can be found in [119], [479]. It is known that blow-up is global and the asymptotics of blow-up was studied in [524], [480], [487] (the last two references contain results similar to Theorems 43.2–43.4). (Un-)boundedness of global solutions was investigated in [462], [488]. For other equations involving localized terms, the blow-up set has been studied in [401], [221] (see Remark 44.4 below). Finally, results on systems of equations of localized type can be found in e.g. [411], [332]. $\square$

43.3. Uniform bounds from L^q-estimates

In this subsection we derive smoothing estimates for problem (43.1), obtained in [463], which are similar to those obtained in Sections 15 and 16 for the model problem (15.1). These estimates will be one of the main ingredients in the derivation of (universal) a priori bounds for global solutions in the next subsection. It turns out that the critical value of q for smoothing from L^q into L^∞ is smaller than for problem (15.1) with the same p.

Theorem 43.13. *Consider problem (43.1) with Ω bounded, $p > 1$, $q \geq 1$, $k \geq 0$, and $0 \leq u_0 \in L^\infty(\Omega)$. Assume*

$$q > \tilde{q}_c := \frac{np}{n+2}.$$

There exists $T = T(\|u_0\|_q) > 0$ and $C = C(\Omega, p, q) > 0$ such that $T_{\max}(u_0) > T$ and

$$\|u(t)\|_\infty \leq C\|u_0\|_q t^{-n/2q}, \quad 0 < t < T.$$

Remarks 43.14. (a) The number $\tilde{q}_c$ in Theorem 43.13 is optimal (up to the equality case). Indeed, it was shown in [463] that for $1 \leq q < \tilde{q}_c$ (hence $p > 1 + 2/n$) there exists a sequence of nonnegative initial data $\{u_{0,j}\} \in L^\infty(\Omega)$, bounded in L^q, and such that $T_{\max}(u_{0,j}) \to 0$.

(b) On the other hand, it is not difficult to modify the arguments in the proof to show a local well-posedness result in L^q for $q > \tilde{q}_c$, similar to Theorem 15.2. $\square$

Proof of Theorem 43.13. By the comparison principle (Proposition 52.25), it is sufficient to establish the result for $k = 0$. We proceed in two steps.

Step 1. We estimate the L^m-norm for $m = \max(p, q)$, by considering the quantity

$$H(t) := \sup_{s \in [0,t]} s^\alpha \|u(s)\|_m, \quad \text{where } \alpha = \frac{n}{2}\left(\frac{1}{q} - \frac{1}{m}\right).$$

Using the variation-of-constants formula, $\|e^{-tA}\chi_\Omega\|_m \leq C\|e^{-tA}\chi_\Omega\|_\infty \leq C, m \geq p$ and the L^q-L^m-estimate (cf. Proposition 48.4), we have

$$t^\alpha \|u(t)\|_m \leq t^\alpha \|e^{-tA}u_0\|_m + t^\alpha \int_0^t \|u(s)\|_p^p \|e^{-(t-s)A}\chi_\Omega\|_m \, ds$$

$$\leq C\|u_0\|_q + Ct^\alpha \int_0^t \|u(s)\|_m^p \, ds \leq C\|u_0\|_q + Ct^\alpha H^p(t) \int_0^t s^{-p\alpha} ds.$$

Since $p\alpha < 1$ due to $q > \tilde{q}_c$, by taking the supremum over $(0, \tau)$ we obtain

$$H(\tau) \leq C\|u_0\|_q + C\tau^{1-(p-1)\alpha} H^p(\tau), \quad 0 < \tau < T_{\max}(u_0).$$

Let

$$T := \min\left(1, \left((2C)^{-p}\|u_0\|_q^{1-p}\right)^{1/(1-(p-1)\alpha)}\right). \tag{43.47}$$

We claim that

$$H(\tau) \le 2C\|u_0\|_q, \quad 0 < \tau < \min(T, T_{\max}(u_0)). \tag{43.48}$$

Indeed otherwise, since $H(t)$ is continuous and $H(0) = 0$ (due to the regularity of u), there exists a first $\tau < \min(T, T_{\max}(u_0))$ such that

$$2C\|u_0\|_q = H(\tau) \le C\|u_0\|_q + C\tau^{1-(p-1)\alpha}(2C\|u_0\|_q)^p$$

hence $\tau \ge T$: a contradiction.

Step 2. For $0 < t < \min(T, T_{\max}(u_0))$, arguing as in Step 1 and using (43.47), (43.48) and $\alpha \le n/2q$, we get

$$t^{n/2q}\|u(t)\|_\infty \le t^{n/2q}\|e^{-tA}u_0\|_\infty + t^{n/2q}\int_0^t \|u(s)\|_p^p \,\|e^{-(t-s)A}\chi_\Omega\|_\infty \,ds$$

$$\le C\|u_0\|_q + Ct^{n/2q}\int_0^t \|u(s)\|_m^p \,ds$$

$$\le C\|u_0\|_q + Ct^{n/2q}H^p(t)\int_0^t s^{-p\alpha}ds$$

$$\le C\|u_0\|_q + CT^{1-p\alpha+n/2q}\|u_0\|_q^p \le C_1\|u_0\|_q.$$

It follows in particular that $T_{\max}(u_0) > T$ and the theorem is proved. $\square$

43.4. Universal bounds for global solutions

In this subsection we prove universal bounds for global solutions of problem (43.1). It turns out that such bounds are true for all $p > 1$, in sharp contrast with the model problem (15.1) (where even a priori estimates fail for $p \ge p_S$, cf. Theorem 28.7). The following result is due to [463].

Theorem 43.15. *Consider problem* (43.1) *with Ω bounded, $p > 1$ and $k = 0$. For all $\tau > 0$, there exists $C(\Omega, p, \tau) > 0$ such that any global nonnegative solution satisfies*

$$\|u(t)\|_\infty \le C(\Omega, p, \tau), \quad t \ge \tau. \tag{43.49}$$

As an important ingredient of the proof, we first establish uniform a priori estimates for global solutions. Note that the problem does not seem to admit an energy functional and that the proof, based on maximum principle arguments, is completely different from that of Theorem 22.1.

Proposition 43.16. *Consider problem* (43.1) *with* Ω *bounded,* $p > 1$ *and* $k = 0$. *For all* $M > 0$, *there exists* $K(\Omega, p, M) > 0$ *such that any global nonnegative solution with* $\|u_0\|_\infty \leq M$ *satisfies*

$$\|u(t)\|_\infty \leq K, \quad t \geq 0. \tag{43.50}$$

Proof. In this proof, we denote $g(t) := \int_\Omega u^p(t)\, dx$ and assume that $\|u_0\|_\infty \leq M$.

Step 1. We first establish a (universal) integral bound on the source term:

$$\int_t^{t+1} g(s)\, ds \leq C(\Omega, p), \quad t \geq 0. \tag{43.51}$$

We argue as in the proof of Theorem 17.1 and denote $y = y(t) := \int_\Omega u(t)\varphi_1\, dx$. Multiplying the equation with φ_1, integrating by parts and using $\int_\Omega \varphi_1\, dx = 1$, we obtain

$$y' + \lambda_1 y = \int_\Omega u^p\, dx. \tag{43.52}$$

By Hölder's inequality, we deduce that

$$y' \geq -\lambda_1 y + C_1 y^p$$

with $C_1 = \|\varphi_1\|_\infty^{-p}|\Omega|^{1-p}$. It follows that $y(t) \leq C_2 := (\lambda_1/C_1)^{1/(p-1)}$ for all $t \geq 0$, since otherwise u cannot exist globally. Integrating (43.52) in time, we deduce (43.51) with $C = (1 + \lambda_1)C_2$.

Step 2. This is the main step: We shall show that u becomes eventually monotone if $g(t)$ reaches a suitably large value.

Comparison with the solution of the ODE $y' = |\Omega|y^p$, $y(0) = M$, shows that there exists $t_0 = t_0(M) > 0$ such that

$$\|u(t)\|_\infty \leq 2M, \quad 0 < t \leq t_0. \tag{43.53}$$

Now L^p- and Schauder estimates guarantee that there exists $K_1 = K_1(M) > 0$ such that

$$\|\Delta u(t_0)\|_\infty \leq K_1. \tag{43.54}$$

We claim that:

$$\text{if } g(t_1) \geq K_1 \text{ for some } t_1 \geq t_0, \text{ then } u_t \geq 0 \text{ in } \Omega \times [t_1, \infty). \tag{43.55}$$

Thus assume $t_1 \geq t_0$ and $g(t_1) \geq K_1$, and pick $t_2 \in [t_0, t_1]$ such that

$$g(t_2) = \max_{[t_0, t_1]} g(t) \geq K_1. \tag{43.56}$$

Let $v := \Delta u$ and $w =: u_t$. By parabolic regularity results, we have $v, w \in C^{2,1}(Q_T) \cap C(\overline{\Omega} \times (0, T))$. Since v satisfies

$$\left.\begin{aligned}
v_t - \Delta v &= 0, & x \in \Omega, \ t > t_0, \\
v &= -g(t), & x \in \partial\Omega, \ t > t_0, \\
v(x, t_0) &= \Delta u(x, t_0), & x \in \Omega,
\end{aligned}\right\}$$

we deduce from the maximum principle, (43.54) and (43.56) that

$$\Delta u \geq \min\big(\min_{\overline{\Omega}} \Delta u(\cdot, t_0), -g(t_2)\big) = -g(t_2), \quad x \in \overline{\Omega}, \ t \in [t_0, t_1].$$

Consequently,

$$u_t(\cdot, t_2) = \Delta u(\cdot, t_2) + g(t_2) \geq 0.$$

Since w satisfies

$$\left.\begin{aligned}
w_t - \Delta w &= p \int_\Omega u^{p-1} w \, dy, & x \in \Omega, \ t > t_2, \\
w &= 0, & x \in \partial\Omega, \ t > t_2, \\
w(x, t_2) &\geq 0, & x \in \Omega,
\end{aligned}\right\}$$

where $u^{p-1} \geq 0$, we deduce from the maximum principle for nonlocal equations (see Proposition 52.24) that $u_t \geq 0$ in $\Omega \times [t_2, \infty)$, which implies the claim.

Step 3. We next deduce a uniform estimate on the source term: there exists $K_2 = K_2(M) > 0$ such that

$$g(t) \leq K_2, \quad t \geq 0. \tag{43.57}$$

Indeed, if $g(t_1) \geq K_1$ for some $t_1 \geq t_0$, then

$$g(t) \leq \int_t^{t+1} g(s) \, ds \leq C(\Omega, p), \quad t \geq t_1$$

by (43.55) and (43.51). Consequently, taking also (43.53) into account, we get (43.57) with $K_2 := \max(K_1, C(\Omega, p), |\Omega|(2M)^p)$.

Step 4. Conclusion. Let Θ be defined in (19.27). Owing to (43.57), we see that $\overline{u} := \|u_0\|_\infty + K_2\Theta$ is a supersolution to (43.1). Consequently, (43.50) with $K = \|u_0\|_\infty + K_2\|\Theta\|_\infty$ follows from the comparison principle (Proposition 52.25). $\square$

Proof of Theorem 43.15. By (43.51), there exists $t_0 \in (0, \tau/2)$ such that

$$\|u(t_0)\|_p \leq C(\Omega, p)\tau^{-1/p}.$$

Since $p > \tilde{q}_c = np/(n+2)$, applying Theorem 43.13, we infer the existence of $t_1 \in (t_0, \tau)$ such that

$$\|u(t_1)\|_\infty \leq C(\Omega, p, \tau). \tag{43.58}$$

Estimate (43.49) finally follows by combining (43.58) and (43.50) (taking t_1 as initial time). $\square$

44. Problems involving space integrals (II)

In this section, we consider a different class of nonlocal equations, of the form

$$u_t - \Delta u = \left(\int_\Omega g(u)\, dx \right)^m f(u), \tag{44.1}$$

with Ω bounded and $m \in \mathbb{R}$, $m \neq 0$.

44.1. Transition from single-point to global blow-up

We have seen in the previous section that purely nonlocal power nonlinearities give rise to global blow-up with a uniform profile (for all nonglobal solutions), whereas purely local power nonlinearities produce single-point blow-up in the radial non-increasing case (cf. Theorem 24.1). In order to understand the transition between these two complementary situations, it is natural to consider equation (44.1) with $f(u) = u^q$, $g(u) = u^{p-q}$, $p > 1$, $0 < q < p$ and $m = 1$, under Dirichlet boundary conditions, that is:

$$\left.\begin{aligned}
u_t - \Delta u &= \left(\int_\Omega u^{p-q}(y,t)\, dy \right) u^q, & x \in \Omega,\ t > 0, \\
u &= 0, & x \in \partial\Omega,\ t > 0, \\
u(x,0) &= u_0(x), & x \in \Omega.
\end{aligned}\right\} \tag{44.2}$$

In what follows we assume $u_0 \in L^\infty(\Omega)$, $u_0 \geq 0$, $u_0 \not\equiv 0$, and we shall denote

$$g(t) = \int_\Omega u^{p-q}(y,t)\, dy, \qquad G(t) := \int_0^t g(s)\, ds.$$

Remark 44.1. Non-Lipschitz case. Problem (44.2) is well-posed for $q \geq 1$ and $p \geq q+1$. In the non-Lipschitz cases $0 < q < 1$ and/or $0 < p-q < 1$, existence of a local classical solution can still be shown, either by using the Schauder fixed point theorem, or by an approximation procedure (replacing the initial and boundary conditions by $u_\varepsilon(x,0) = u_0(x) + \varepsilon$ and $u_\varepsilon(x,t) = \varepsilon$, respectively). However, local uniqueness seems to be unknown in this case, and assertion (i) of Theorem 44.2 applies to any maximal solution starting from u_0. By a simple modification of the proof of Theorem 17.1, any solution u starting from suitably large u_0 will blow up in a finite time $T = T(u)$, in the sense that $\limsup_{t \to T} \|u(t)\|_\infty = \infty$. $\square$

The following result shows that the occurrence of single-point vs. (uniform or nonuniform) global blow-up depends in a precise way on the values of q. Observe that the rate of the (uniform) global blow-up does not change when q varies in $[0,1)$ and that the bifurcation to single-point for q in $(1,p]$, occurs through a nonuniform global blow-up at $q = 1$. Theorem 44.2 is a variant of results combined from [157] and [331], except for the blow-up rate estimates (44.6)–(44.7) which are consequences of Proposition 44.3 below.

Theorem 44.2. *Assume Ω bounded, $p > 1$, $0 < q < p$, and $0 \leq u_0 \in L^\infty(\Omega)$. Let u be a nonglobal solution of problem (44.2) and denote by T its maximal existence time.*

(i) If $0 < q < 1$, then blow-up is global and uniform. More precisely:

$$\lim_{t \to T} (T-t)^{\frac{1}{p-1}} u(x,t) = \lim_{t \to T} (T-t)^{\frac{1}{p-1}} \|u(t)\|_\infty = \left[(p-1)|\Omega|\right]^{-\frac{1}{p-1}}, \qquad (44.3)$$

uniformly on compact subsets of Ω.

(ii) If $q = 1$, then blow-up is global and nonuniform. More precisely:

$$u(x,t) = k(t)\left(e^{-tA} u_0\right)(x), \qquad where \ k(t) \sim C(T-t)^{-1/(p-1)} \ as \ t \to T \quad (44.4)$$

for some constant $C > 0$ depending on u_0.

(iii) Assume $1 < q < p$, $\Omega = B_R$, $u_0 \in C^1(\overline{\Omega})$ radial nonincreasing, with $u_0(x) = 0$ for $|x| = R$. Then single-point blow-up occurs at $x = 0$. More precisely, for any $\alpha > 2/(q-1)$ there exists $C_\alpha > 0$ such that

$$u(x,t) \leq C_\alpha |x|^{-\alpha}, \qquad 0 < |x| < R, \ 0 < t < T. \qquad (44.5)$$

Assume in addition that $p - q < n(q-1)/2$. Then we have

$$\|u(t)\|_\infty \geq C_1 (T-t)^{-1/(p-1)}, \qquad 0 < t < T \qquad (44.6)$$

and, if in addition $q < p_S$, then

$$\|u(t)\|_\infty \leq C_2 (T-t)^{-1/(p-1)}, \qquad 0 < t < T, \qquad (44.7)$$

for some $C_1, C_2 > 0$.

Proof. (i) Set $v = \frac{1}{1-q} u^{1-q}$ and let

$$M(t) := \max_{\overline{\Omega}} u(\cdot, t), \qquad N(t) := \max_{\overline{\Omega}} v(\cdot, t).$$

By the argument in the (alternative) proof of Proposition 23.1, we have $M'(t) \leq M^q(t) g(t)$, hence $N'(t) \leq g(t)$ a.e. in $(0, T)$. Consequently,

$$N(t) \leq N(0) + G(t), \qquad 0 < t < T, \qquad (44.8)$$

hence in particular

$$\lim_{t \to T} G(t) = \infty. \qquad (44.9)$$

On the other hand, noting that $u > 0$ in Q_T by the strong maximum principle, we have

$$v_t - \Delta v = u^{-q}(u_t - \Delta u) + qu^{-q-1}|\nabla u|^2 \geq g(t).$$

By using (44.9), Theorem 43.6 and the maximum principle, it follows that, uniformly on compact subsets, $\liminf_{t\to T} v(x,t)/G(t) \geq 1$, hence

$$\lim_{t\to T} \frac{v(x,t)}{G(t)} = 1 \tag{44.10}$$

by (44.8). Arguing as in the proof of Theorem 43.2 for $k = 0$, we obtain after some calculations

$$G(t) \sim (1-q)^{-1}\big[(p-1)|\Omega|(T-t)\big]^{-(1-q)/(p-1)}.$$

Returning to (44.10), (44.8) and using $u = ((1-q)v)^{1/(1-q)}$ we obtain (44.3).

(ii) For $q = 1$, by direct calculation one checks that the solution of (44.2) can be written as

$$u(x,t) = e^{G(t)} e^{-tA} u_0,$$

and we have $G(t) \to \infty$ as $t \to T$. Consequently,

$$g(t) = e^{(p-1)G(t)} \int_\Omega \big(e^{-tA} u_0\big)^{p-1}\, dx$$

hence

$$\frac{d}{dt} e^{-(p-1)G(t)} = -(p-1)g(t)e^{-(p-1)G(t)} \to -C, \quad C := (p-1)\int_\Omega \big(e^{-TA} u_0\big)^{p-1}\, dx,$$

as $t \to T$. By integration, we obtain $e^{G(t)} \sim C^{1/(p-1)}(T-t)^{-1/(p-1)}$ and (44.4) follows.

(iii) The proof of (44.5) is very similar to that of Theorem 24.1. The variables f, f' now stand for $f = f(t,u) = g(t)u^q$, $f' = g(t)qu^{q-1}$, and J is defined by (24.3) with $1 < \gamma < q$. The main difference is that the condition $H \geq 0$ becomes equivalent to

$$g(t)(q-\gamma)u^{q-1} + (n+\delta)\delta r^{-2} \geq 2\varepsilon\gamma(1+\delta)u^{\gamma-1}r^\delta, \tag{44.11}$$

instead of (24.5). Since

$$g(t) \geq \int_\Omega \big(e^{-tA} u_0\big)^{p-q}\, dx \geq c > 0, \qquad 0 \leq t < T, \tag{44.12}$$

(44.11) is satisfied if ε is small enough.

To show the blow-up estimates (44.6)–(44.7) in Theorem 44.2(iii), we first establish the following more general result, where the upper bound will be proved by using arguments from [425] (cf. Theorem 26.8).

Proposition 44.3. *Let $T > 0$, $p > 1$, and let $a \in C([0,T))$ be nonnegative and bounded. Let $0 \le u \in C^{2,1}(B_R \times (0,T))$ be a radial nonincreasing solution of the equation*

$$u_t - \Delta u = a(t)\, u^p, \qquad x \in B_R,\ 0 < t < T,$$

such that $\lim_{t \to T} \|u(t)\|_\infty = \infty$.

(i) There exists $C_1 > 0$ such that

$$\|u(t)\|_\infty \ge C_1 (T - t)^{-1/(p-1)}, \qquad 0 < t < T.$$

(ii) Assume in addition that $\ell := \lim_{t \to T} a(t)$ exists in $(0,\infty)$ and that $p < p_S$. Then there exists $C_2 > 0$ such that

$$\|u(t)\|_\infty \le C_2 (T - t)^{-1/(p-1)}, \qquad 0 < t < T. \tag{44.13}$$

Proof. (i) Since $N(t) := \sup_{|x|<R} u(x,t) = u(0,t)$, we have $u_r(0,t) = 0$ and $u_{rr}(0,t) \le 0$. Therefore, $dN/dt \le a(t)N^p(t)$ and assertion (i) follows immediately upon integration.

(ii) We shall apply Lemma 26.11 with $D = (0,T)$, $\Sigma = X = [0,T]$ and the standard distance. We denote $d(s) = \mathrm{dist}(s, \partial D) = \min(s, T - s)$. Assume that estimate (44.13) fails and denote $M(t) = u^{p-1}(0,t)$. Then there exists a sequence $s_k \to T$ such that

$$M(s_k) > 2k(T - s_k)^{-1} = 2kd^{-1}(s_k). \tag{44.14}$$

It follows from Lemma 26.11 that there exists $t_k \in (0,T)$ such that

$$M(t_k)d(t_k) > 2k, \tag{44.15}$$

$$M(t_k) \ge M(s_k) \tag{44.16}$$

and

$$M(t) \le 2M(t_k) \quad \text{for all } t \in (0,T) \cap (t_k - kM^{-1}(t_k), t_k + kM^{-1}(t_k)). \tag{44.17}$$

Note that, by (44.14) and (44.16) we have

$$t_k \to T. \tag{44.18}$$

For k large, we deduce from (44.15) that $kM^{-1}(t_k) < d(t_k) = T - t_k$, so that (44.17) rewrites as

$$M(t) \le 2M(t_k) \quad \text{for all } t \in (t_k - kM^{-1}(t_k), t_k + kM^{-1}(t_k)). \tag{44.19}$$

Now we rescale u_k by setting

$$\lambda_k := M^{-1}(t_k) \to 0 \tag{44.20}$$

and

$$v_k(y,s) := \lambda_k^{1/(p-1)} u_k(\lambda_k^{1/2} y, t_k + \lambda_k s), \quad (y,s) \in \tilde{D}_k := \{|y| < R\lambda_k^{-1/2}\} \times (-k, k).$$

The function v_k solves

$$\partial_s v_k - \Delta_y v_k = a(t_k + \lambda_k s)v_k^p, \qquad (y,s) \in \tilde{D}_k. \tag{44.21}$$

Moreover we have $v_k(0,0) = 1$ and (44.19) implies

$$0 \leq v_k \leq C := 2^{1/(p-1)}, \quad (y,s) \in \tilde{D}_k. \tag{44.22}$$

By using (44.21), (44.22), (44.20), (44.18), interior parabolic estimates and the embedding (1.2), we deduce that some subsequence of v_k converges in $C^\alpha(\mathbb{R}^{n+1})$, $0 < \alpha < 1$, to a (bounded classical) solution $v \geq 0$ of

$$v_t - \Delta v = \ell \, v^p, \qquad x \in \mathbb{R}^n, \ s \in \mathbb{R}.$$

Moreover, v is radial nonincreasing and satisfies $v(0,0) = 1$. This contradicts Theorem 21.1. $\square$

End of proof of Theorem 44.2. In view of Proposition 44.3, to show (44.6) and (44.7), it suffices to verify that

$$g(t) \to \ell \in (0,\infty), \qquad \text{as } t \to T. \tag{44.23}$$

Due to (44.5) and $p - q < n(q-1)/2$, the function $g(t)$ is bounded on $[0,T)$. By (44.5), parabolic estimates and the embedding (1.2), it follows that, for some $\nu \in (0,1)$, $u \in BUC^\nu(\{\gamma < |x| < 1 - \gamma\} \times (T/2, T))$ for each $\gamma > 0$. Consequently, for all $x \in B(0,1) \setminus \{0\}$, $\lim_{t \to T} u(x,t)$ exists and is finite. Using (44.5) and $p - q < n(q-1)/2$ again, along with the dominated convergence theorem and (44.12), we obtain (44.23). $\square$

Remark 44.4. Problems involving localized nonlinearities. A different type of competition between local and nonlocal reaction terms has been studied in [401], [221] for the following variant of equation (43.46):

$$u_t - \Delta u = u^q(x_0, t) + u^p$$

$p > 1$, $q > 0$, $x_0 \in \Omega$, with Dirichlet boundary conditions, when Ω is a ball B_R and u_0 is radial decreasing. Interestingly, the critical condition is different depending on the location of x_0. Namely, for $x_0 = 0$, blow-up is always global if $p \leq q + 1$, while single-point blow-up occurs for some u_0 if $p > q + 1$. Next assume $x_0 \neq 0$. If $p < q$, then both global and single-point blow-ups occur, and there are no other possibilities. On the contrary, if $p > q$ (or $p = q > 2$), then only single-point blow-up occurs. $\square$

44.2. A problem with control of mass

We now consider equation (44.1) with $f(u) = g(u) = u^p$, $p > 1$, and $m = -1$, under Neumann boundary conditions, that is:

$$
\left.
\begin{aligned}
u_t - \Delta u &= \left(\int_\Omega u^p(y,t)\, dy \right)^{-1} u^p, & x \in \Omega,\ t > 0, \\
u_\nu &= 0, & x \in \partial\Omega,\ t > 0, \\
u(x,0) &= u_0(x), & x \in \Omega.
\end{aligned}
\right\}
\tag{44.24}
$$

In what follows we assume $u_0 \in L^\infty(\Omega)$, $u_0 \geq 0$, $u_0 \not\equiv 0$, and we shall denote

$$
k(t) = \left(\int_\Omega u^p(y,t)\, dy \right)^{-1}.
$$

As in Chapters I and II we shall use the notation

$$
p_{sg} =
\begin{cases}
\infty & \text{if } n \leq 2, \\
n/(n-2) & \text{if } n > 2.
\end{cases}
$$

Let us first observe that, by integrating the equation, we immediately obtain

$$
\int_\Omega u(t)\, dx = t + \int_\Omega u_0\, dx.
\tag{44.25}
$$

This means that the total "mass" is controlled.

We shall first investigate under what conditions the solutions of (44.24) blow up or exist globally. On a heuristic level one can expect that, when u becomes large in some sense, then the factor $k(t)$ might become large, too, and have a stabilizing effect which could prevent blow-up. Interestingly, whether or not this possible stabilizing effect is effective depends in a sharp way on the relation between the exponent p and the space dimension n. The following result is due to [284].

Theorem 44.5. *Consider problem (44.24) with Ω bounded, $p > 1$, and $0 \leq u_0 \in L^\infty(\Omega)$, $u_0 \not\equiv 0$.*

(i) *If $p < p_{sg}$, then $T_{\max}(u_0) = \infty$ for all u_0.*

(ii) *Assume $n \geq 3$, $p > p_{sg}$ and let $\Omega = B_1$. Then there exists u_0 such that $T_{\max}(u_0) < \infty$.*

Note that since the solution stays bounded in L^1, it is clear that for radial nonincreasing solutions, blow-up can occur only at the origin. As a corollary to the proof of Theorem 44.5, one obtains the following blow-up profile estimate.

Theorem 44.6. *Consider problem (44.24) with $n \geq 3$, $p > p_{sg}$ and $\Omega = B_1$. Then there exists $0 \leq u_0 \in L^\infty(\Omega)$, $u_0 \not\equiv 0$, radial nonincreasing, such that $T := T_{\max}(u_0) < \infty$, u exhibits single-point blow-up at $x = 0$, and u satisfies*

$$u(x,t) \leq C_\varepsilon |x|^{-\frac{2}{p-1}-\varepsilon}, \quad x \in \Omega, \ 0 < t < T, \qquad \text{for each } \varepsilon > 0. \qquad (44.26)$$

Moreover, (44.26) is optimal, in the sense that it cannot be satisfied for any $\varepsilon < 0$.

As for the blow-up rate, we have the following result, which is a consequence of Proposition 44.3.

Theorem 44.7. *Consider problem (44.24) with $p > 1$ and $\Omega = B_1$. Let $0 \leq u_0 \in L^\infty(\Omega)$, $u_0 \not\equiv 0$, be radial nonincreasing and assume that $T := T_{\max}(u_0) < \infty$.*

(i) There exists $C_1 > 0$ such that

$$\|u(t)\|_\infty \geq C_1(T - t)^{-1/(p-1)}, \quad 0 < t < T.$$

(ii) Assume in addition that $p < p_S$ and that u satisfies (44.26). Then there exists $C_2 > 0$ such that

$$\|u(t)\|_\infty \leq C_2(T - t)^{-1/(p-1)}, \quad 0 < t < T. \qquad (44.27)$$

Remarks 44.8. (a) **Global solutions.** For any $p > 1$, (44.24) admits global solutions for arbitrarily large initial data. Namely it suffices to take homogeneous initial data $u_0 = M$ (with any $M > 0$) and u is then given by solving the ODE, i.e.: $u_M(x,t) = M + |\Omega|^{-1}t$. We thus observe that it is the "shape" of u_0, rather than its size, which causes blow-up. On the other hand, all global solutions of (44.24) are unbounded, due to (44.25).

(b) **Failure of the comparison principle.** Problem (44.24) admits no comparison principle. For instance taking $\Omega = B_1$ and u a blow-up solution as in Theorem 44.6 (ii), we see that $u(\cdot, 0) < M$ for M large but u eventually intersects the solution $u_M(x,t) = M + |\Omega|^{-1}t$.

(c) **Interpretation of the critical exponent.** Observe that $k(t)$ is bounded due to (44.25) and Hölder's inequality and that $k(t)$ vanishes if and only if $\|u(t)\|_p$ blows up. This allows for a heuristic interpretation of the value of the critical exponent in Theorem 44.5 if we put problem (44.24) in parallel with the model equation $u_t - \Delta u = u^p$. Indeed the supercriticality condition for the L^p-norm is given by $p > n(p-1)/2$ that is, $p < p_{sg}$. More precisely, for the model problem (15.1), if $p < p_{sg}$, then $\|u(t)\|_p$ blows up whenever u is nonglobal (cf. Theorem 15.2), whereas if $p > p_{sg}$, then there exist solutions (in a ball) such that $\|u(t)\|_p$ remains bounded (cf. Theorem 24.1 and Corollary 24.2). The idea of the proof of Theorem 44.5(ii) below is precisely to use (a nontrivial modification of) the method in

Theorem 24.1 to construct initial data which yield a blow-up profile belonging to L^p and provide a control of $k(t)$ from below.

(d) **Critical case.** In the critical case $p = p_{sg}$, it is proved in [284] that the solution exists globally if $\int_\Omega u_0 \, dx$ is large enough.

(e) Explicit examples of initial data in Theorem 44.5(ii) are constructed in Lemma 44.10 below. Namely blow-up occurs whenever $u_0 \in C^2(\overline{B}_1)$ is radial and satisfies (44.33)–(44.37) with $\beta > 0$ small (depending on n, p) and $M > 0$ large (depending on n, p, β). $\quad\square$

Remarks 44.9. Other nonlocal problems. Results on blow-up for other nonlocal problems with control of mass, of the form $u_t - \Delta u = f(u) - \frac{1}{|\Omega|}\int_\Omega f(u)\,dy$ with Neumann boundary conditions and $f(u) = |u|^p$ or $|u|^{p-1}u$, can be found in [103], [284]. For physical motivation concerning such problems, see [465]. $\quad\square$

Proof of Theorem 44.5(i). The proof here is for $n \geq 3$. The cases $n = 1, 2$ can be obtained with obvious modifications. Fix $m > 1$. For any $0 < a < 1 < q$, we have

$$\int_\Omega u^{p+m}\,dx \leq \left(\int_\Omega u^{(p+m)aq}\,dx\right)^{1/q}\left(\int_\Omega u^{(p+m)(1-a)q'}\,dx\right)^{1/q'}.$$

We claim that we can find $0 < a < 1$ and $q > n/(n-2)$ such that

$$(p+m)aq = (m+1)n/(n-2) \quad \text{and} \quad (p+m)(1-a)q' \leq p. \tag{44.28}$$

Indeed, (44.28) is equivalent to

$$a = \frac{n}{(n-2)q}\frac{m+1}{m+p} \geq 1 - \frac{p}{m+p}\left(1 - \frac{1}{q}\right),$$

i.e.:

$$q \leq \frac{n}{n-2} + \frac{1}{m}\left(\frac{n}{n-2} - p\right) \tag{44.29}$$

and, since $p < n/(n-2)$, we can choose $q > n/(n-2)$ satisfying (44.29) and the corresponding a then belongs to $(0, 1)$.

Now using Hölder's inequality, we obtain

$$\int_\Omega u^{p+m}\,dx \leq C\left(\int_\Omega u^{(m+1)n/(n-2)}\,dx\right)^{1/q}\left(\int_\Omega u^p\,dx\right)^{(p+m)(1-a)/p}. \tag{44.30}$$

Multiplying (44.24) by u^m and integrating by parts over Ω, we obtain

$$\frac{d}{dt}\int_\Omega \frac{u^{m+1}}{m+1}\,dx + m\int_\Omega u^{m-1}|\nabla u|^2\,dx = \left(\int_\Omega u^p\,dx\right)^{-1}\int_\Omega u^{p+m}\,dx. \tag{44.31}$$

Set $v = u^{(m+1)/2}$. Since

$$\int_\Omega u^p \, dx \geq C\left(\int_\Omega u \, dx\right)^p \geq \left(\int_\Omega u_0 \, dx\right)^p \tag{44.32}$$

by (44.25), formulas (44.30), (44.31) and $(p+m)(1-a) < p$ imply

$$\frac{d}{dt}\int_\Omega v^2 \, dx + \frac{4m}{m+1}\int_\Omega |\nabla v|^2 \, dx \leq C\left(\int_\Omega v^{2n/(n-2)} \, dx\right)^{1/q}.$$

Using the Sobolev inequality $\|w\|_{2n/(n-2)} \leq C(\|w\|_2 + \|\nabla w\|_2)$ and $q > n/(n-2)$, we obtain

$$\frac{d}{dt}\int_\Omega v^2 \, dx \leq C\left(1 + \int_\Omega v^2 \, dx\right).$$

By integration, it follows that for all $m > 1$, $\tau > 0$,

$$\int_\Omega u^{m+1}(t) \, dx = \int_\Omega v^2(t) \, dx \leq C(m, \tau), \quad 0 < t < \min(\tau, T).$$

Therefore, using also (44.32), the right-hand side of (44.24) remains bounded in L^r on bounded time intervals for each $r < \infty$. The L^∞-boundedness of u on bounded time intervals then follows easily from the variation-of-constants formula and the L^p-L^q-estimates (cf. Proposition 48.4). We conclude that u exists globally. $\square$

The proof of part (ii) is more delicate. It requires carefully constructed initial data. This is achieved in the following lemma.

Lemma 44.10. *Let $\Omega = B_1$ and $p > p_{sg}$. Then, for all $M, \beta > 0$, one can find a radial function $u_0 \in C^2(\overline{\Omega})$ satisfying the following properties:*

$$u_0(0) \geq M, \quad u_0(1) = \beta, \quad u_{0,r}(1) = 0, \quad u_{0,r} < 0 \ \ on \ (0,1), \tag{44.33}$$

$$\int_\Omega u_0 \, dx \leq C\beta, \tag{44.34}$$

$$k(0) = \left(\int_\Omega u_0^p \, dy\right)^{-1} \geq A\beta^{-p}, \tag{44.35}$$

$$\Delta u_0 + \lambda u_0^p \geq 0, \qquad |x| \leq 1, \tag{44.36}$$

$$u_{0,r} + \mu r u_0^p \leq 0, \qquad 0 \leq r \leq 1/2, \tag{44.37}$$

where $\lambda = K\beta^{1-p}$, $\mu = L\beta^{1-p}$, and $C, A, K, L > 0$ depend only on n, p.

Proof. Let $\alpha = 2/(p-1)$ and fix a function $U \in C^2((0,1])$ such that

$$U(r) = r^{-\alpha} \ \ on \ (0,1/2], \quad U_r < 0 \ \ on \ (0,1), \quad U_r(1) = 0 \quad and \quad U(1) = 1. \tag{44.38}$$

Fix $\delta \in (0, 1/4)$ and $\beta > 0$. We define

$$
\phi(r) := \begin{cases} U(r), & \delta < r \leq 1, \\ \delta^{-\alpha}\left(1 + \frac{\alpha(\alpha+5)}{6} - \frac{\alpha(\alpha+3)}{2}\left(\frac{r}{\delta}\right)^2 + \frac{\alpha(\alpha+2)}{3}\left(\frac{r}{\delta}\right)^3\right), & 0 \leq r \leq \delta, \end{cases} \tag{44.39}
$$

and we set $u_0 = \beta\phi$. One can check that $u_0 \in C^2(\overline{\Omega})$, that $0 \leq u_0 \leq \beta U$ on $(0, 1]$, and that u_0 satisfies (44.33) whenever $0 < \delta \leq (M/\beta)^{-1/\alpha}$.

Since $p\alpha < n$, we have $\int_\Omega U^p\, dx < \infty$, hence (44.34) and (44.35). On the other hand, we have

$$
\Delta\phi + K\phi^p \geq \begin{cases} \Delta U + K, & 1/2 \leq |x| \leq 1, \\ \big(\alpha(\alpha + 2 - n) + K\big)r^{-\alpha-2}, & \delta \leq |x| \leq 1/2, \\ \big(-n\alpha(\alpha + 3) + K\big)\delta^{-\alpha-2}, & |x| \leq \delta. \end{cases}
$$

Since $\Delta u_0 + K\beta^{1-p}u_0^p = \beta(\Delta\phi + K\phi^p)$ this implies (44.36) for $K = K(n,p) > 0$ large.

Next we have

$$
\phi_r + Lr\phi^p \leq -\alpha r^{-\alpha-1} + Lr^{-\alpha p+1} = (L - \alpha)r^{-\alpha-1}, \quad \delta \leq r \leq 1/2,
$$

and

$$
\phi_r + Lr\phi^p \leq \delta^{-\alpha}\left(-\frac{\alpha(\alpha+3)r}{\delta^2} + \frac{\alpha(\alpha+2)r^2}{\delta^3}\right) + LC(\alpha)\delta^{-p\alpha}r \leq \delta^{-\alpha p}(LC(\alpha) - \alpha)r
$$

for $0 \leq r \leq \delta$. Since $u_{0,r} + L\beta^{1-p}ru_0^p = \beta(\phi_r + Lr\phi^p)$, this implies (44.37) for $L = L(p) > 0$ small. $\square$

Next, we consider the auxiliary problem

$$
\left. \begin{aligned} w_t - \Delta w &= 2\lambda w^p, & x \in \Omega,\ t > 0, \\ u_\nu &= 0, & x \in \partial\Omega,\ t > 0, \\ w(x, 0) &= u_0(x), & x \in \Omega. \end{aligned} \right\} \tag{44.40}
$$

We shall need the following upper estimate on the existence time of its solution.

Lemma 44.11. *Let $\Omega = B_1$ and $p > p_{sg}$. For $M, \beta > 0$, let λ and u_0 be as in Lemma 44.10. Then the existence time T_w of the solution of problem (44.40) satisfies $T_w \leq \frac{M^{1-p}}{\lambda(p-1)}$.*

Proof. We use a similar idea as in the proof of Theorem 23.5, applying the maximum principle to the auxiliary function $J := w_t - \lambda w^p$. By the maximum principle we have $w \geq \beta > 0$ in $Q := \Omega \times (0, T_w)$. On the other hand, Example 51.9 shows that $w \in C([0, T_w), W^{1,q}(\Omega))$ for any $q > n$, hence $w^p \in C([0, T_w), W^{1,2}(\Omega))$ and Theorem 51.1(v) guarantees $w_t \in C([0, T_w), L^2(\Omega))$. In addition, $w_t \in C^{2,1}(\overline{\Omega} \times (0, T_w))$ (cf. Example 51.10). Therefore, $J \in C([0, T_w), L^2(\Omega)) \cap C^{2,1}(\overline{\Omega} \times (0, T_w))$.

Now, J satisfies

$$J_t - \Delta J = (w_t - \Delta w)_t - \lambda p(w^{p-1}w_t - w^{p-1}\Delta w - (p-1)w^{p-2}|\nabla w|^2)$$
$$\geq 2\lambda p w^{p-1}w_t - \lambda p w^{p-1}2\lambda w^p = 2\lambda p w^{p-1}J$$

in Q. At $t = 0$, we have $J = \Delta u_0 + \lambda u_0^p \geq 0$ by (44.36). On $\partial\Omega$, we have $\frac{\partial J}{\partial \nu} = \frac{\partial}{\partial t}\left(\frac{\partial w}{\partial \nu}\right) - \lambda p w^{p-1}\frac{\partial w}{\partial \nu} = 0$. It thus follows from the maximum principle (cf. Remark 52.9) that $J \geq 0$ in $\Omega \times (0, T_w)$. But this implies $(w^{1-p})_t \leq -\lambda(p-1)$, hence in particular $w^{1-p}(0, t) + \lambda(p-1)t \leq u_0^{1-p}(0) \leq M^{1-p}$ on $(0, T_w)$ by (44.33). The lemma follows. $\square$

Observe that, for $\beta \leq A/4K$, we have $k(0) \geq A\beta^{-p} \geq 4K\beta^{1-p} = 4\lambda$. The idea of the proof is now to show that $k(t)$ cannot become smaller than 2λ before the time $t = \beta$, independently of M. This will be achieved via the next lemma, where u is estimated from above by employing a modification of an argument from [219] (cf. Theorem 24.1). This will guarantee that u dominates the solution w of the auxiliary problem (44.40) for $t \leq \beta$. But the blow-up time of w goes to 0 as M increases, which will imply blow-up of u if M is large.

Lemma 44.12. *Let $\Omega = B_1$ and $p > p_{sg}$. For $M > 0$ and $\beta \in (0, 1)$, let A, K, λ and u_0 be as in Lemma 44.10. Set $T_0 = \min(\beta, T)$. Assume in addition that $\beta \leq A/4K$, so that $k(0) \geq 4\lambda$, and define*

$$T_1 = \sup\{\tau \in [0, T_0) : k(t) \geq 2\lambda \ on \ [0, \tau]\} \in (0, T_0].$$

For each $1 < q < p$, we have

$$u(r, t) \leq C(n, p, q)\beta r^{-2/(q-1)}, \quad 0 < r \leq 1, \ 0 < t < T_1. \tag{44.41}$$

Proof. *Step 1.* By Example 51.13, we have $u_r \in C^{2,1}((0, 1) \times (0, T)) \cap C([0, 1] \times [0, T))$. Using (44.33) and the maximum principle (in particular Proposition 52.17), one deduces that

$$u \geq \beta \quad \text{and} \quad u_r \leq 0, \qquad 0 \leq r \leq 1, \ 0 < t < T. \tag{44.42}$$

Since $u_r \leq 0$, we have

$$u(r, t) \leq nr^{-n}\int_0^r u(\rho, t)\rho^{n-1}d\rho \leq C(n)r^{-n}\int_\Omega u(t)\, dx.$$

Using (44.25), (44.34) and $T_0 \leq \beta$, we deduce that

$$u(r, t) \leq C(n, p)\beta r^{-n}, \qquad 0 < r \leq 1, \ 0 < t < T_0. \tag{44.43}$$

We next claim that

$$u_r(1/2, t) \le -c(n, p)\beta, \quad 0 \le t < T_0. \tag{44.44}$$

To show this, observe that the function $v := \beta^{-1}u_r$ satisfies

$$\left.\begin{array}{ll} v_t - v_{rr} + \dfrac{n-1}{r^2}v - \dfrac{n-1}{r}v_r = pk(t)u^{p-1}v \le 0, & 1/4 < r < 1,\ 0 < t < T, \\[2mm] v(1/4, t) \le 0, \quad v(1, t) = 0, & 0 < t < T, \\[2mm] v(0, r) = U_r(r) < 0, & 1/4 < r < 1, \end{array}\right\}$$

where U, defined in (44.38), depends only on p for $r \in (1/4, 1)$. By the strong maximum principle, recalling that $T_0 \le \beta < 1$, it follows in particular that $v(1/2, t) \le -c(n, p) < 0$ for $0 \le t < T_0$, hence (44.44).

Step 2. Set $J = u_r + \eta r u^q$. We claim that for

$$\eta = C(n, p, q)\beta^{1-q}, \tag{44.45}$$

with $C(n, p, q) > 0$ sufficiently small, there holds

$$J \le 0 \quad \text{in } Q := (0, 1/2) \times (0, T_1). \tag{44.46}$$

We compute

$$\begin{aligned} J_t - J_{rr} &= \left(u_t - u_{rr}\right)_r + \eta\left(r(u^q)_t - (ru^q)_{rr}\right) \\ &= \left(-\dfrac{n-1}{r^2} + k(t)pu^{p-1}\right)u_r + \dfrac{n-1}{r}u_{rr} \\ &\quad + \eta\left(qru^{q-1}(u_t - u_{rr}) - 2qu^{q-1}u_r - q(q-1)ru^{q-2}(u_r)^2\right) \\ &\le \left(-\dfrac{n-1}{r^2} + k(t)pu^{p-1} + (n-3)q\eta u^{q-1}\right)u_r \\ &\quad + \dfrac{n-1}{r}\left(J - \eta ru^q\right)_r + q\eta k(t)ru^{p+q-1} \\ &= \left(-\dfrac{n-1}{r^2} + k(t)pu^{p-1} - 2q\eta u^{q-1}\right)\left(J - \eta ru^q\right) \\ &\quad + \dfrac{n-1}{r}J_r - \eta\dfrac{n-1}{r}u^q + q\eta k(t)ru^{p+q-1} \\ &= a(r, t)J + \dfrac{n-1}{r}J_r + b(r, t), \end{aligned}$$

where

$$a(r, t) = -\dfrac{n-1}{r^2} + k(t)pu^{p-1} - 2q\eta u^{q-1}$$

and

$$b(r, t) = \eta ru^{p+q-1}\left((q-p)k(t) + 2q\eta u^{q-p}\right).$$

Using the definition of T_1, (44.42) and (44.45), we obtain

$$b(r,t) \leq \eta r u^{p+q-1}\left(2q\eta\beta^{q-p} - 2(p-q)K\beta^{1-p}\right) \leq 0 \quad \text{in } Q.$$

On the other hand, for $t \in [0, T_1)$, we have $J(0, t) = 0$ and, by (44.43) and (44.44), the choice (44.45) implies $J(1/2, t) \leq 0$. Also, for $t = 0$, using (44.37), $u_0 \geq \beta$ and $p > q$, (44.45) implies $J(r, 0) \leq 0$ in $[0, 1/2]$. Since a is bounded from above in $(0, 1/2) \times (0, \tau)$ for each $\tau < T_1$, Claim (44.46) thus follows from the maximum principle (see Proposition 52.4).

By integrating (44.46), we have $(u^{1-q})_r \geq (q-1)\eta r$ in $(0, 1/2] \times (0, T_1)$. This combined with (44.43) yields (44.41). $\square$

Proof of Theorems 44.5(ii) and 44.6. For $M > 0$, let β and u_0 be as in Lemma 44.12. Since $p > p_{sg}$, we may fix q such that $1 + 2p/n < q < p$. We deduce from Lemma 44.12 that

$$\int_{|x| \leq 1} u^p(t) \leq C(n, p)\beta^p \int_0^1 r^{n-1-2p/(q-1)}\, dr = C(n, p)\beta^p, \qquad 0 < t < T_1.$$

Taking $0 < \beta \leq \beta_0(n, p)$ sufficiently small, we infer that

$$k(t) \geq C(n, p)\beta^{-p} \geq 4K(n, p)\beta^{1-p} = 4\lambda, \qquad 0 < t < T_1.$$

Consequently $T_1 = T_0 = \min(T, \beta)$. In particular, by the comparison principle (use Proposition 52.7), it follows that $u \geq w$ for $t < \min(T, \beta, T_w)$. But we have $T_w < \beta$ for M large by Lemma 44.11, and we know that w blows up in L^∞-norm. It follows that $T \leq T_w < \infty$, which proves Theorem 44.5(ii).

Since $T_1 = T$, the first part of Theorem 44.6 is now a direct consequence of Lemma 44.12.

Finally, let us show that estimate (44.26) cannot be satisfied for any $\varepsilon < 0$. Suppose the contrary. This implies

$$\sup_{t \in (0,T)} \|u(t)\|_q < \infty \quad \text{for some } q > n(p-1)/2. \tag{44.47}$$

On the other hand, u is bounded on S_T and, by (44.25) and Hölder's inequality, we have

$$k(t) \leq C, \quad 0 \leq t < T. \tag{44.48}$$

Owing to (44.47), by comparison argument with (a variant of) the model problem (14.1), it follows from Theorem 16.4 (or, alternatively, Theorem 15.2 or Example 51.27 in Appendix E) that u is uniformly bounded in Q_T: a contradiction. $\square$

Proof of Theorem 44.7. (i) Due to (44.48), the lower estimate follows from Proposition 44.3(i).

(ii) In view of Proposition 44.3(ii), to prove the upper estimate, it suffices to show that

$$k(t) \to \ell \in (0, \infty), \qquad \text{as } t \to T. \tag{44.49}$$

Using (44.48), (44.26), parabolic estimates and the embedding (1.2), for some $\nu \in (0,1)$ we have $u \in BUC^\nu(\{\gamma < |x| < 1 - \gamma\} \times (T/2, T))$ for each $\gamma > 0$. Consequently, for all $x \in B(0,1) \setminus \{0\}$, $\lim_{t \to T} u(x,t)$ exists and is finite. Since $2p/(p-1) < n$, using (44.26), the dominated convergence theorem and (44.48), we deduce (44.49). $\square$

Remark 44.13. By the methods of this subsection, problem (44.24) with $f(u) = u^p$ and $g(u) = u^q$ can be studied for more general values of $p, q > 1$ and $m \in \mathbb{R}$, under either Neumann or Dirichlet boundary conditions. $\square$

44.3. A problem with variational structure

We next consider equation (44.1) with

$$f(u) = |u|^{p-1}u, \quad g(u) = \lambda + \int_0^u f(s)\, ds = \lambda + \frac{|u|^{p+1}}{p+1},$$

where $p > 1$, $m = -q < 0$ and $\lambda > 0$, under Dirichlet boundary conditions. Taking $\lambda = |\Omega|^{-1}$ for simplicity, this leads to the problem

$$
\left.
\begin{aligned}
u_t - \Delta u &= \left(1 + \int_\Omega \frac{|u(y,t)|^{p+1}}{p+1}\, dy\right)^{-q} |u|^{p-1}u, & x \in \Omega, \ t > 0, \\
u &= 0, & x \in \partial\Omega, \ t > 0, \\
u(x,0) &= u_0(x), & x \in \Omega,
\end{aligned}
\right\} \tag{44.50}
$$

where $u_0 \in L^\infty(\Omega)$.

Problem (44.50) possesses a variational structure. Namely, the energy functional

$$E(u) = \frac{1}{2}\int_\Omega |\nabla u|^2\, dx - \frac{1}{1-q}\left(1 + \int_\Omega \frac{|u|^{p+1}}{p+1}\, dx\right)^{1-q}$$

(for $q \neq 1$, with an obvious modification if $q = 1$) is nonincreasing along any solution of (44.50). More precisely

$$\frac{d}{dt}E(u(t)) = -\int_\Omega u_t^2(t)\, dx$$

(this follows in the same way as in (17.7) and Example 51.28).

Theorem 44.14. *Consider problem* (44.50) *with* Ω *bounded,* $p > 1$ *and* $0 < q < (p-1)/(p+1)$. *There exists* $C = C(p,q) > 0$ *such that, if* $u_0 \in L^\infty \cap H_0^1(\Omega)$ *satisfies* $E(u_0) < -C$, *then* $T_{\max}(u_0) < \infty$.

Proof. Set $\psi(t) := \|u(t)\|_2^2$. Multiplying the equation in (44.50) by u we obtain

$$\frac{1}{2}\psi'(t) = \int_\Omega u u_t(t)\,dx = -\int_\Omega |\nabla u(t)|^2\,dx + \left(1 + \int_\Omega \frac{|u|^{p+1}}{p+1}\,dx\right)^{-q} \int_\Omega |u|^{p+1}\,dx$$

$$= -2E(u(t)) + \left(1 + \int_\Omega \frac{|u|^{p+1}}{p+1}\,dx\right)^{-q}$$

$$\times \left(\frac{(p-1) - q(p+1)}{(p+1)(1-q)} \int_\Omega |u|^{p+1}\,dx - \frac{2}{1-q}\right)$$

$$\geq -2E(u_0) + c_1 \left(1 + \int_\Omega \frac{|u|^{p+1}}{p+1}\,dx\right)^{1-q} - c_2,$$

where $c_1, c_2 > 0$ depend only on p, q. Applying Hölder's inequality, we obtain

$$\psi' \geq c\psi^\gamma - 2E(u_0) - c_2$$

with $c = c(p, q, \Omega) > 0$ and $\gamma := (p+1)(1-q)/2 > 1$. If $E(u_0) < -c_2/2$ (or $\psi^\gamma(0) > 2(E(u_0) + c_2)/c$), then this inequality implies $T_{\max}(u_0) < \infty$. $\square$

Remark 44.15. A priori bounds. Results on boundedness and a priori estimates of global solutions and universal bounds for global nonnegative solutions for problems of the form (44.50) have been proved in [187], [440], [464]. $\square$

44.4. A problem arising in the modeling of Ohmic heating

We finally consider equation (44.1) with $f(u) = \lambda e^{-u}$, $g(u) = e^{-u}$, $\lambda > 0$, $m = -2$, $n = 1$ and $\Omega = (-1, 1)$, under Dirichlet boundary conditions. Namely:

$$\left.\begin{aligned}
u_t - u_{xx} &= \lambda \left(\int_{-1}^1 e^{-u}(y,t)\,dy\right)^{-2} e^{-u}, && x \in (-1,1),\ t > 0,\\
u(\pm 1, t) &= 0, && t > 0,\\
u(x, 0) &= u_0(x), && x \in (-1,1),
\end{aligned}\right\} \quad (44.51)$$

where we assume $u_0 \in L^\infty(\Omega)$.

Problem (44.51) arises from a special case of the following elliptic-parabolic coupled system:

$$\left.\begin{aligned}
u_t - \Delta u &= \sigma(u)|\nabla \phi|^2, && x \in \Omega,\ t > 0,\\
\operatorname{div}(\sigma(u)\nabla\phi) &= 0, && x \in \Omega,\ t > 0,\\
u &= 0, && x \in \partial\Omega,\ t > 0,\\
\phi &= \phi_0, && x \in \partial\Omega,\ t > 0,\\
u(x, 0) &= u_0(x), && x \in \Omega.
\end{aligned}\right\} \quad (44.52)$$

Here u and ϕ respectively represent the temperature and the electric potential in a thermistor, i.e. a conductor whose electric conductivity $\sigma = \sigma(u)$ may vary with the temperature, and the RHS of the first equation in (44.52) stands for the heat production due to the Joule effect. We refer to [33] and the references therein for results on blow-up and global existence concerning system (44.52) and its variants. Now assume that the (thin) conductor can be represented by the interval $x \in (-1, 1)$ and that the potential ϕ is imposed to be 0 at $x = -1$ and a constant V at $x = 1$. The second equation in (44.52) becomes $(\sigma(u)\phi_x)_x = 0$ and can be integrated in $\sigma(u)\phi_x = I(t)$ (which represents the electric current per cross-sectional area unit). Denoting $\rho(u) = 1/\sigma(u)$ (electric resistivity), hence $\phi_x = \rho(u)I(t)$, we obtain $V = \int_{-1}^{1} \phi_x\,dx = I(t)\int_{-1}^{1} \rho(u)\,dx$. The first equation in (44.52) then rewrites as

$$u_t - u_{xx} = \sigma(u)|\phi_x|^2 = \rho(u)I^2(t) = \rho(u)V^2\left(\int_{-1}^{1} \rho(u)\,dx\right)^{-2}.$$

If the conductivity law is given by $\sigma(u) = e^u$ and thermal cooling is applied on the ends of the conductor, we thus arrive at problem (44.51).

We shall see that the global behavior of solutions to (44.51) is closely related to the properties of the stationary problem

$$w_{xx} + \lambda\left(\int_{-1}^{1} e^{-w}\,dy\right)^{-2} e^{-w} = 0, \quad x \in (-1, 1), \qquad \text{with } w(\pm 1) = 0. \quad (44.53)$$

Proposition 44.16. *Let $\lambda > 0$. Problem (44.53) has a (classical) solution if and only if $\lambda < 8$. Moreover the solution is unique and it is given by $w_\lambda = z_\alpha$, where*

$$z_\alpha(x) = 2\log\left(\frac{\cos(\alpha x)}{\cos \alpha}\right), \qquad (44.54)$$

and $\alpha \in (0, \pi/2)$ satisfies $\lambda = 8\sin^2 \alpha$. Furthermore, for $|x| < 1$, $z_\alpha(x)$ is an increasing function of α, hence $w_\lambda(x)$ is an increasing function of λ.

Proof. Setting $\mu = \lambda\left(\int_{-1}^{1} e^{-w}\,dy\right)^{-2}$, we see that w solves

$$z_{xx} + \mu e^{-z} = 0, \quad x \in (-1, 1), \qquad \text{with } z(\pm 1) = 0. \quad (44.55)$$

By direct calculation, we see that a solution of (44.55) is given by (44.54) where α is the unique number in $(0, \pi/2)$ such that $\mu = 2\alpha^2/\cos^2 \alpha$. On the other hand, the solution of (44.55) is unique. Indeed, if y and z are two solutions, by subtracting the equations for y and z and multiplying by $z - y$, we get

$$0 = \int_{-1}^{1} \left((y_{xx} - z_{xx}) + \mu(e^{-y} - e^{-z})\right)(z - y)\,dx \geq \int_{-1}^{1} (y_x - z_x)^2\,dx,$$

hence $y = z$. Finally, since $\int_{-1}^{1} e^{-z_\alpha}\,dy = \frac{2}{\alpha}\sin \alpha \cos \alpha$, the function z_α solves (44.53) with $\lambda = 8\sin^2 \alpha$, hence the necessary and sufficient condition on λ. The remaining assertion follows from $\frac{\partial}{\partial \alpha} z_\alpha(x) = 2(\tan \alpha - x\tan(\alpha x))$. $\square$

The following result is from [314].

Theorem 44.17. *Consider problem* (44.51) *with* $\lambda > 0$ *and* $u_0 \in L^\infty(\Omega)$.

(i) *If* $\lambda < 8$, *then the equilibrium* w_λ *is globally asymptotically stable. Namely, all solutions are global and converge to* w_λ *in* $L^\infty(-1, 1)$ *as* $t \to \infty$.

(ii) *If* $\lambda > 8$, *then all solutions blow up in finite time. Moreover, the blow-up is global, i.e.*

$$\lim_{t \to T_{\max}(u_0)} u(x, t) = \infty, \qquad -1 < x < 1. \tag{44.56}$$

(iii) *If* $\lambda = 8$, *then all solutions are global and unbounded. Moreover,*

$$\lim_{t \to \infty} u(x, t) = \infty, \qquad -1 < x < 1.$$

Problem (44.51) admits a comparison principle (cf. Proposition 52.25). The proof of Theorem 44.17 is based on suitable sub- and supersolutions, which will be constructed under a "quasi-stationary" form (cf. (44.59) below).

Proof of Theorem 44.17. By a time shift we may assume without loss of generality that

$$u_0 \in C^1([-1, 1]) \quad \text{and} \quad u_0(\pm 1) = 0. \tag{44.57}$$

Also, denoting by φ_1 the first eigenfunction, we observe that:

if $T_{\max}(u_0) = \infty$, then $u(x, t) \geq \varepsilon \varphi_1(x)$ for some $\varepsilon > 0$ small and all t large.
$$\tag{44.58}$$
Indeed, fixing $C > 0$ such that $u_0 \geq -C\varphi_1$, one easily checks that $\underline{u}(x, t) := \big(2\varepsilon - (C + 2\varepsilon)e^{-\lambda_1 t}\big)\varphi_1(x)$ is a subsolution to problem (44.51) for $\varepsilon > 0$ small enough.

In order to construct sub-/supersolutions, we put

$$v(x, t) = z_{\alpha(t)}(x), \tag{44.59}$$

where $\alpha(t)$ is a function to be determined. Plugging (44.54), (44.59) into equation (44.51), we compute

$$\mathcal{P}v := v_t - v_{xx} - \lambda\bigg(\int_{-1}^1 e^{-v}\, dy\bigg)^{-2} e^{-v}$$

$$= 2\big(\tan\alpha - x\tan(\alpha x)\big)\alpha' + \frac{2\alpha^2}{\cos^2(\alpha x)} - \frac{\lambda\alpha^2}{4\sin^2\alpha\,\cos^2(\alpha x)}$$

$$= 2\big(\tan\alpha - x\tan(\alpha x)\big)\alpha' + \frac{(8\sin^2\alpha - \lambda)\alpha^2}{4\sin^2\alpha\,\cos^2(\alpha x)}.$$

If $\lambda < 8$, then define $\bar\alpha \in (0, \pi/2)$ by $\lambda = 8\sin^2\bar\alpha$ (hence $w_\lambda = z_{\bar\alpha}$); otherwise set $\bar\alpha := \pi/2$.

We first assume $\lambda < 8$ and look for a decreasing supersolution $\overline{v} = v$. Thus taking $\alpha' \leq 0$, we have

$$\mathcal{P}\overline{v} \geq 2\alpha' \tan \alpha + \frac{(8 \sin^2 \alpha - \lambda)\alpha^2}{4 \sin^2 \alpha} \tag{44.60}$$

provided $8 \sin^2 \alpha(t) \geq \lambda$. Due to (44.57), we may choose some $\alpha_0 \in (\bar{\alpha}, \pi/2)$ close enough to $\pi/2$ so that $z_{\alpha_0}(x) \geq u_0(x)$. Let $\alpha(t)$ be the solution of the ODE

$$\alpha'(t) = \frac{(\lambda - 8 \sin^2 \alpha)\alpha^2 \cos \alpha}{8 \sin^3 \alpha}, \quad t \geq 0 \tag{44.61}$$

with $\alpha(0) = \alpha_0$. Since $8 \sin^2 \alpha_0 > \lambda$, it is clear that α exists globally and satisfies $\alpha' < 0$ and $\lim_{t \to \infty} \alpha(t) = \bar{\alpha}$. It follows from (44.60) that $\overline{v}$ is a supersolution to problem (44.51). Consequently

$$u \leq \overline{v}, \qquad 0 < t < T_{\max}(u_0), \tag{44.62}$$

hence in particular $T_{\max}(u_0) = \infty$. Moreover there holds

$$\lim_{t \to \infty} \overline{v}(x, t) = w_\lambda(x) \quad \text{uniformly in } [-1, 1]. \tag{44.63}$$

Now consider general λ again. Looking for an increasing subsolution $\underline{v} = v$, hence $\alpha'(t) \geq 0$, we have

$$\mathcal{P}\underline{v} \leq 2\alpha' \tan \alpha + \frac{(8 \sin^2 \alpha - \lambda)\alpha^2}{4 \sin^2 \alpha} \tag{44.64}$$

provided $8 \sin^2 \alpha \leq \lambda$. Assuming $T_{\max}(u_0) = \infty$ and using (44.58), we may choose some $\alpha_1 \in (0, \bar{\alpha})$ small enough so that $z_{\alpha_1}(x) \leq u_0(x)$. Take now $\alpha(t)$ to be the solution of (44.61) with $\alpha(0) = \alpha_1$. Since $8 \sin^2 \alpha_1 < \lambda$, it is clear that α exists globally and satisfies $\alpha' > 0$. Moreover, we have

$$\lim_{t \to \infty} \alpha(t) = \bar{\alpha}.$$

It follows from (44.64) that $\underline{v}$ is a subsolution to problem (44.51). Consequently

$$u \geq \underline{v}, \qquad 0 < t < T_{\max}(u_0).$$

If $\lambda < 8$, there holds in addition $\lim_{t \to \infty} \underline{v}(x, t) = w_\lambda(x)$, uniformly in $[-1, 1]$. This, along with (44.62), (44.63), proves assertion (i).

If $\lambda \geq 8$, we have shown that either $T_{\max}(u_0) < \infty$, or

$$T_{\max}(u_0) = \infty \quad \text{and} \quad u(x, t) \geq 2 \log \left(\frac{\cos(\alpha(t)x)}{\cos \alpha(t)} \right) \to \infty, \quad t \to \infty. \tag{44.65}$$

Assume $\lambda > 8$. We shall show by a further subsolution argument that (44.65) leads to a contradiction. We look for a modified subsolution of the form

$$v(x,t) = p \log\left(\frac{\cos(\alpha x)}{\cos \alpha}\right)$$

where the function

$$\alpha : \ [0, T_0) \rightarrow [\alpha_2, \pi/2] \tag{44.66}$$

and the numbers $p > 1$, $T_0 > 0$, $\alpha_2 \in (0, \pi/2)$ are to be determined.

We shall use the following elementary lemma:

Lemma 44.18. *For each $p > 1$ and $\varepsilon > 0$, there holds*

$$I(a) := \int_{-1}^{1} \frac{dy}{\cos^p(ay)} \leq \frac{4 + \varepsilon}{\pi(p-1)\cos^{p-1} a}, \qquad \text{as } a \rightarrow \left(\tfrac{\pi}{2}\right)_-. \tag{44.67}$$

Proof. We write

$$\frac{1}{2}I(a) = \int_0^1 \frac{dy}{\cos^p(a(1-y))} = \int_0^1 \frac{dy}{\sin^p(\frac{\pi}{2} - a + ay)}$$
$$\leq \frac{1}{\sin^p(a\eta)} + \int_0^\eta \frac{dy}{\sin^p(\frac{\pi}{2} - a + ay)} \tag{44.68}$$

for $0 < a < \frac{\pi}{2}$ and $0 < \eta < 1$. Fix $\eta = \eta(\varepsilon) > 0$ small. Taking $0 < \pi/2 - a < \eta$, and using $\sin x \sim x$ as $x \rightarrow 0$, we obtain

$$\int_0^\eta \frac{dy}{\sin^p(\frac{\pi}{2} - a + ay)} \leq (1 + \varepsilon/8) \int_0^\eta \frac{dy}{(\frac{\pi}{2} - a + ay)^p}$$
$$= \frac{1 + \varepsilon/8}{a(1-p)}\left[\left(\frac{\pi}{2} - a + ay\right)^{1-p}\right]_0^\eta \leq \frac{1 + \varepsilon/8}{a(p-1)}\left(\frac{\pi}{2} - a\right)^{1-p}.$$

Since $\cos a \sim \left(\frac{\pi}{2} - a\right)$ as $a \rightarrow \pi/2$, this combined with (44.68) and

$$\lim_{a \rightarrow \pi/2} \frac{\cos^{p-1} a}{\sin^p(a\eta)} = 0$$

yields (44.67). $\square$

Proof of Theorem 44.17 (continued). Assuming $\alpha'(t) \geq 0$, we have

$$\mathcal{P}v \leq p\alpha' \tan \alpha + \frac{p\alpha^2}{\cos^2(\alpha x)} - \frac{\lambda}{\cos^p \alpha \, \cos^p(\alpha x) I^2(\alpha)}.$$

For $p \in (1, 2)$, by using (44.66), (44.67) and taking α_2 close to $\pi/2$, we have

$$\frac{p\alpha^2}{\cos^2(\alpha x)} - \frac{\lambda}{\cos^p \alpha \cos^p(\alpha x) I^2(\alpha)} \leq \frac{p\alpha^2}{\cos^2(\alpha x)} - \frac{\lambda \pi^2 (p-1)^2 \cos^{p-2}\alpha}{(4+\varepsilon)^2 \cos^p(\alpha x)}$$
$$\leq \frac{\pi^2 \cos^{p-2}\alpha}{\cos^p(\alpha x)} \left(\frac{p}{4} - \frac{\lambda(p-1)^2}{(4+\varepsilon)^2} \right).$$

Since $\lambda > 8$, we can choose $p \in (1, 2)$ close to 2 and ε small such that

$$\gamma := \pi^2 \left(\frac{\lambda(p-1)^2}{(4+\varepsilon)^2} - \frac{p}{4} \right) > 0.$$

Taking α_2 still closer to $\pi/2$ and using $\tan a \sim (\frac{\pi}{2} - a)^{-1}$ as $a \to (\pi/2)_-$, it follows that

$$\mathcal{P}v \leq p\alpha' \tan \alpha - \gamma \frac{\cos^{p-2}\alpha}{\cos^p(\alpha x)} \leq 2p \left(\frac{\pi}{2} - \alpha \right)^{-1} \alpha' - \gamma \left(\frac{\pi}{2} - \alpha \right)^{p-2}. \qquad (44.69)$$

Take now $\alpha(t)$ to be the solution of

$$\alpha'(t) = \frac{\gamma}{2p} \left(\frac{\pi}{2} - \alpha \right)^{p-1}, \qquad \alpha(0) = \alpha_2.$$

Since $1 < p < 2$ and $\left((\frac{\pi}{2} - \alpha)^{2-p} \right)' = \gamma(p-2)/2p < 0$, it follows that $\alpha(t)$ reaches $\pi/2$ in a finite time $T_0 > 0$. On the other hand, owing to (44.65), we may assume that $u_0 \geq v(\cdot, 0)$ (after a time shift) which, along with (44.69), guarantees that v is a subsolution to problem (44.51). Since $\lim_{t \to T_0} v(x, t) = \infty$ in $(-1, 1)$, this contradicts $T_{\max} = \infty$.

Let us finally prove global blow-up, i.e. (44.56). Denoting

$$h(t) = \left(\int_{-1}^{1} e^{-u} \, dy \right)^{-2}$$

and arguing as in the (alternative) proof of Proposition 23.1, we see that $M(t) := \max_{x \in [-1,1]} u(x, t)$ satisfies

$$M'(t) \leq g(t) := \lambda h(t) e^{-M(t)}, \qquad \text{for a.e. } 0 < t < T := T_{\max}(u_0). \qquad (44.70)$$

Since $u \geq \min_{[-1,1]} u_0$ by the maximum principle, $T < \infty$ implies $\limsup_{t \to T} M(t) = \infty$. Integrating (44.70), we deduce that $\int_0^T g(t) \, dt = \infty$. Since $u_t - u_{xx} = \lambda h(t) e^{-u} \geq g(t)$, (44.56) then follows from Theorem 43.6(i). This completes the proof of assertion (ii).

As for the critical case $\lambda = 8$, global existence can be shown by a modified supersolution argument. We refer for this to [314]. $\quad \square$

Remarks 44.19. (a) Formal results concerning the blow-up rate (and the behavior in the boundary layer) for problem (44.51) are given in [315].

(b) Results on problem (44.51) with more general conductivity functions $\sigma(u)$ can be found in [315], [66]. For the analogue of problem (44.51) in dimension $n = 2$, results similar to Theorem 44.17 are proved in [518], [298] for radial solutions in a disk.

(c) For problem (44.51) with Neumann boundary conditions, it is easy to see that all solutions blow up in finite time: Indeed the solution of the ODE $y' = \frac{\lambda}{4}e^y$ with $y(0) = \inf u_0$ is a subsolution. $\square$

45. Fujita-type results for problems involving space integrals

We consider Cauchy problems with nonlocal source terms involving space integrals, of the form

$$
\left.
\begin{aligned}
u_t - \Delta u &= \left(\int_{\mathbb{R}^n} K(y) u^q(y,t)\, dy \right)^{(p-1)/q} u^{1+r}, \qquad && x \in \mathbb{R}^n,\ t > 0, \\
u(x,0) &= u_0(x), && x \in \mathbb{R}^n.
\end{aligned}
\right\}
\qquad (45.1)
$$

In what follows, we assume that

$$
p > 1,\, q \geq 1,\, r \geq 0,\, u_0 \in L^\infty(\mathbb{R}^n),\, u_0 \geq 0,
$$
$$
K \text{ is a positive, bounded continuous function.}
$$
$$
\tag{45.2}
$$

If $K \notin L^1(\mathbb{R}^n)$, then we assume in addition, that $u_0 \in L^1(\mathbb{R}^n)$. Under these assumptions, problem (45.1) is locally well-posed (see Example 51.13).

The critical exponent for problem (45.1) was studied in [227]. It will depend in a crucial way on whether or not the function K is integrable. In the integrable case we have the following result.

Theorem 45.1. *Assume* (45.2), $K \in L^1(\mathbb{R}^n)$, *and let* $p_c = 1 + \frac{2}{n} - r$.

(i) *If* $p < p_c$, *then* (45.1) *admits no nontrivial global solution.*

(ii) *If* $p > p_c$, *then* (45.1) *admits both global positive and blowing-up solutions.*

In the non-integrable case we need some additional assumptions on the asymptotic behavior of K.

Theorem 45.2. *Assume* (45.2) *and* $u_0 \in L^1(\mathbb{R}^n)$. *Assume in addition that* K *satisfies*

$$
c_1(1 + |x|)^{-\beta} \leq K(x) \leq c_2(1 + |x|)^{-\beta}, \qquad x \in \mathbb{R}^n, \tag{45.3}
$$

for some $\beta \in [0, n)$ and some $c_1, c_2 > 0$. Let $p_c = 1 + \frac{q(2-nr)}{n(q-1)+\beta}$.

(i) If $p < p_c$, then (45.1) admits no nontrivial global solution.

(ii) If $p > p_c$, then (45.1) admits both global positive and blowing-up solutions.

The proofs are exclusively based on comparison with suitable (self-similar) sub- and supersolutions. Note that (45.1) does admit a comparison principle (this follows from Proposition 52.27). For results in the critical case $p = p_c$, see [227].

Proof of Theorems 45.1 and 45.2. *1. Blow-up.* We look for a blowing-up subsolution under the form

$$\underline{u}(x,t) = A(T-t)^{-\alpha}f(\xi), \qquad \xi = \frac{x}{\sqrt{T-t}}, \qquad f(\xi) = e^{-|\xi|^2},$$

where $\alpha, T, A > 0$ are parameters. We compute

$$\underline{u}_t = A\alpha(T-t)^{-\alpha-1}f(\xi) + \frac{A}{2}(T-t)^{-\alpha-1}\xi \cdot \nabla_\xi f(\xi), \qquad \Delta\underline{u} = A(T-t)^{-\alpha-1}\Delta_\xi f(\xi).$$

Denoting

$$I(t) = \int_{\mathbb{R}^n} K(y) e^{-q|y|^2/(T-t)}\, dy,$$

the condition for $\underline{u}$ being a subsolution is thus given by

$$\alpha f + \frac{\xi}{2} \cdot \nabla_\xi f - \Delta_\xi f \leq A^{p+r-1}(T-t)^{1-(r+p-1)\alpha} I^{\frac{p-1}{q}}(t) f^{r+1},$$

that is

$$\alpha + 2n \leq 5|\xi|^2 + A^{p+r-1}(T-t)^{1-(r+p-1)\alpha} I^{\frac{p-1}{q}}(t) e^{-r|\xi|^2}, \qquad \xi \in \mathbb{R}^n,\ 0 < t < T. \tag{45.4}$$

In the case $K \in L^1$, assume without loss of generality that $K \geq c_0 \chi_{\{|y|<\rho\}}$ for some $c_0, \rho > 0$. We then have

$$I(t) = (T-t)^{n/2}\int_{\mathbb{R}^n} K(z\sqrt{T-t}) e^{-q|z|^2}\, dz \geq c_0(T-t)^{n/2}\int_{|z|<\rho/\sqrt{T}} e^{-q|z|^2}\, dz,$$

hence

$$I(t) \geq C(T-t)^{n/2}T^{-n/2}$$

for all $T \geq 1$ and some $C > 0$ (independent of T).

In the case when K satisfies (45.3), we have

$$I(t) \geq C(T-t)^{n/2}\int_{\mathbb{R}^n}(1+|z|\sqrt{T-t})^{-\beta} e^{-q|z|^2}\, dz$$

$$\geq C(T-t)^{n/2}T^{-\beta/2}\int_0^\infty (T^{-1/2}+\rho)^{-\beta} e^{-q\rho^2}\rho^{n-1}\, d\rho,$$

hence

$$I(t) \geq C(T - t)^{n/2} T^{-\beta/2}$$

for all $T \geq 1$ and some $C > 0$ (independent of T).

Let us now take

$$\alpha = \frac{2q + n(p - 1)}{2q(p + r - 1)} \quad \text{and} \quad A = BT^{\gamma}$$

with

$$\gamma = \begin{cases} \frac{n(p-1)}{2q(p+r-1)} & \text{if } K \in L^1, \\[2mm] \frac{\beta(p-1)}{2q(p+r-1)} & \text{if } K \text{ satisfies } (45.3). \end{cases}$$

A sufficient condition for (45.4) is then that

$$\alpha + 2n \leq 5|\xi|^2 + c_2 B^{p+r-1} e^{-r|\xi|^2}, \quad \xi \in \mathbb{R}^n.$$

This is satisfied for some large $B > 0$ and guarantees that $\underline{u}$ is a subsolution for all $T \geq 1$.

Finally assume for contradiction that u exists for all time. Since u is a positive supersolution of the linear heat equation, it follows that $u(x, 1) \geq \varepsilon\sigma^{-n/2} e^{-|x|^2/4\sigma}$ for some $\varepsilon, \sigma > 0$, hence $u(x, t + 1) \geq \varepsilon(\sigma + t)^{-n/2} e^{-|x|^2/4(\sigma+t)}$ for all $t > 0$ (cf. (18.12)). Now, the assumption $p < p_c$ means that $\alpha - \gamma > n/2$ in both cases. Taking $T = 4(\sigma + t)$ and $t > 0$ sufficiently large, we thus get

$$u(x, t + 1) \geq \varepsilon T^{-n/2} e^{-|x|^2/T} \geq \underline{u}(x, 0) = BT^{\gamma-\alpha} e^{-|x|^2/T}$$

and the comparison principle would then imply finite-time blow-up of u. Statement (i) of Theorems 45.1 and 45.2 follows.

2. *Global existence.* We look for a blowing-up supersolution under the form

$$\bar{u}(x, t) = (T + t)^{-\alpha} g(\xi), \qquad \xi = \frac{x}{\sqrt{T + t}}, \qquad g(\xi) = e^{-\sigma|\xi|^2},$$

where $\alpha, T, \sigma > 0$ are parameters. We compute

$$\bar{u}_t = -\alpha(T + t)^{-\alpha-1} g(\xi) - \frac{1}{2}(T + t)^{-\alpha-1}\xi \cdot \nabla_\xi g(\xi), \qquad \Delta\bar{u} = (T + t)^{-\alpha-1}\Delta_\xi g(\xi).$$

Denoting

$$J(t) = \int_{\mathbb{R}^n} K(y) e^{-q\sigma|y|^2/(T+t)} \, dy,$$

the condition for $\bar{u}$ being a supersolution is thus given by

$$-\alpha g - \frac{\xi}{2} \cdot \nabla_\xi g - \Delta_\xi g \geq (T + t)^{1-(r+p-1)\alpha} J^{\frac{p-1}{q}}(t) g^{r+1}.$$

Taking $\sigma = 1/4$, this amounts to

$$\frac{n}{2} - \alpha \geq (T + t)^{1-(r+p-1)\alpha} J^{\frac{p-1}{q}}(t) e^{-r\sigma|\xi|^2}, \quad \xi \in \mathbb{R}^n, \ t > 0. \tag{45.5}$$

In the case $K \in L^1$, there obviously holds $J(t) \leq \|K\|_{L^1}$. Taking

$$\frac{1}{p + r - 1} < \alpha < \frac{n}{2}$$

and T large, we obtain (45.5).

In the case when K satisfies (45.3), since $\beta < n$, we have

$$J(t) \leq C(T + t)^{n/2} \int_{\mathbb{R}^n} (1 + |z|\sqrt{T + t})^{-\beta} e^{-q\sigma|z|^2} \, dz$$

$$\leq C(T + t)^{(n-\beta)/2} \int_0^\infty |z|^{-\beta} e^{-q\sigma|z|^2} \, dz = C(T + t)^{(n-\beta)/2}.$$

Since $p > p_c$, we may take

$$\frac{2q + (n - \beta)(p - 1)}{2q(p + r - 1)} < \alpha < \frac{n}{2}$$

and T large yields

$$\frac{n}{2} - \alpha \geq C(T + t)^{1-\alpha(p+r-1)+(n-\beta)(p-1)/2q}, \quad t \geq 0,$$

hence (45.5)

In either case, we have thus shown that $\overline{u}$ is a supersolution, which implies the global existence of u whenever $0 \leq u(0) < \overline{u}(0)$. $\square$

46. A problem with memory term

We consider the following problem

$$\left. \begin{array}{rll} u_t - \Delta u = \displaystyle\int_0^t u^p(x, s) \, ds - ku^q, & x \in \Omega, \ t > 0, \\[2mm] u = 0, & x \in \partial\Omega, \ t > 0, \\[2mm] u(x, 0) = u_0(x), & x \in \Omega, \end{array} \right\} \tag{46.1}$$

where Ω is bounded, $p > 1$, $q \geq 1$, $k \geq 0$, and $u_0 \in L^\infty(\Omega)$, $u_0 \geq 0$. Notice that the problem is well-posed due to Example 51.14.

46.1. Blow-up and global existence

The following result [479] shows that $q = p$ constitutes a critical blow-up exponent for problem (46.1). Moreover, blow-up (in finite or infinite time) occurs for *all* positive solutions of (46.1), and not only for solutions with large initial data, unlike in problems (43.1) and (15.1) for instance.

Theorem 46.1. *Consider problem* (46.1) *with* Ω *bounded,* $p > 1$, $q \geq 1$, $k \geq 0$, *and* $0 \leq u_0 \in L^\infty(\Omega)$, $u_0 \not\equiv 0$.

(i) *If* $p > q$ *or* $k = 0$, *then all solutions of* (46.1) *blow up in finite time.*

(ii) *If* $p \leq q$ *and* $k > 0$, *then all solutions of* (46.1) *are global and unbounded, that is,* $\limsup_{t\to\infty} \|u(t)\|_\infty = \infty$.

The proof of Theorem 46.1 relies on the eigenfunction method (cf. the proof of Theorem 17.1), combined with the following lemma concerning the system of differential inequalities

$$
\left.
\begin{aligned}
z' &\geq y^p, \\
y' + \lambda y + k z'^r &\geq z.
\end{aligned}
\right\}
\tag{46.2}
$$

Lemma 46.2. *Assume* $0 < r < 1 < p$ *and* $k, \lambda \geq 0$. *Let the functions* $y, z \in C^1(0, T)$ *satisfy* $y \geq 0$, $z > 0$ *and* (46.2) *on* $(0, T)$. *Then* $T < \infty$.

Proof. By translating the origin of time, we may assume that actually y, $z \in C^1([0, T))$ and $z(0) > 0$. Fix γ such that $\max(r, 1/p) < \gamma < 1$. It follows from the first inequality in (46.2) that, for all $\varepsilon > 0$, there exists a constant $C_\varepsilon > 0$ such that

$$
C_\varepsilon z'^\gamma \geq y^{p\gamma} + (3\lambda + 1)y - \varepsilon \quad \text{and} \quad C_\varepsilon z'^\gamma \geq 3k z'^r - \varepsilon,
$$

hence

$$
2C_\varepsilon z'^\gamma + 3y' \geq 3\big(y' + \lambda y + k z'^r\big) + y^{p\gamma} + y - 2\varepsilon.
$$

By the second inequality in (46.2), we deduce that

$$
2C_\varepsilon z'^\gamma + 3y' \geq 3z + y^{p\gamma} + y - 2\varepsilon.
\tag{46.3}
$$

Next take $m \in (0, \gamma)$. By Young's inequality, we have

$$
2C_\varepsilon z'^\gamma = 2C_\varepsilon \frac{z'^\gamma}{z^m} z^m \leq \varepsilon z^{m/(1-\gamma)} + C'_\varepsilon \frac{z'}{z^{m/\gamma}},
$$

hence

$$
C''_\varepsilon (z^\theta)' + \varepsilon z^{m/(1-\gamma)} \geq 2C_\varepsilon z'^\gamma, \quad \text{where } \theta = 1 - (m/\gamma) \in (0, 1),
\tag{46.4}
$$

for some large constant $C''_\varepsilon > 0$.

Now assume further that $m < 1 - \gamma$ and define

$$\phi = C''_\varepsilon z^\theta + 3y.$$

By combining (46.3) and (46.4), for $\varepsilon < 1$, we get

$$\phi' \geq 3z + y^{p\gamma} + y - 2\varepsilon - \varepsilon z^{m/(1-\gamma)} \geq 2z + y^{p\gamma} + y - 3\varepsilon, \quad 0 \leq t < T.$$

Choosing $\varepsilon < z(0)/3$, setting $\nu = \min(p\gamma, 1/\theta) > 1$ and using the fact that z is nondecreasing, we then obtain

$$\phi' \geq z + y^{p\gamma} + y \geq \left[z(0)\right]^{1-\theta\nu} z^{\theta\nu} + y^\nu \geq C\phi^\nu$$

on $(0, T)$, for some $C > 0$. We conclude that $T < \infty$. $\quad\square$

Proof of Theorem 46.1. (i) Define the functions

$$y(t) = \int_\Omega u(x,t)\varphi_1(x)\,dx \quad \text{and} \quad z(t) = \int_0^t \int_\Omega u^p(x,s)\varphi_1(x)\,dx\,ds, \quad 0 \leq t < T.$$

Multiplying (46.1) by φ_1 and integrating by parts over Ω, we get:

$$y' + \lambda_1 y = \int_0^t \int_\Omega u^p(x,s)\varphi_1(x)\,dx\,ds - k\int_\Omega u^q(x,t)\varphi_1(x)\,dx, \quad 0 < t < T.$$

We may assume $q < p$ also if $k = 0$. Letting $r = q/p < 1$ and applying Jensen's inequality yields

$$y' + \lambda_1 y + k z'^r \geq z \quad \text{and} \quad z' \geq y^p.$$

The conclusion thus follows from Lemma 46.2.

(ii) If $p < q$, a simple calculation shows that $v(x,t) = C(1 + t)^{1/(q-p)}$ is a supersolution for all large $C > 0$. If $p = q$, the same holds with $v(x,t) = Ce^{Ct}$. Taking $C > \|u_0\|_\infty$, it follows from the comparison principle (Proposition 52.25) that u must exist globally.

Last, assume for contradiction that u is globally bounded by a constant $M > 0$. Then u satisfies

$$u_t - \Delta u \geq \int_0^t u^p(x,s)\,ds - aM^{q-1}u, \quad x \in \Omega, \ t > 0.$$

By the comparison principle, in view of part (i), this immediately implies finite-time blow-up: a contradiction. $\quad\square$

Remarks 46.3. (i) The assumption $r < 1$ in Lemma 46.2 is essential, at least if $k > 0$. Indeed, if $r = 1$, then $z(t) = Ce^{\mu t}$, $y(t) = (C\mu)^{1/p}e^{\mu t/p}$ is a global positive solution of (46.2) for $\mu = 1/k$ and any $C > 0$.

(ii) **Fujita-type results.** For problem (46.1) and related equations, Fujita-type results have been recently obtained in [112]. $\quad\square$

46.2. Blow-up rate

The following result shows a type I blow-up rate for monotone-in-time solutions and provides a sufficient condition for monotonicity.

Theorem 46.4. *Consider problem* (46.1) *with* Ω *bounded,* $p > 1$, $k = 0$. *Let* $u_0 \in C^1(\overline{\Omega})$, $u_0 \geq 0$, $u_0 \not\equiv 0$, *and* $T := T_{\max}(u_0)$.
(i) *Assume that:*

$$\text{there exists } t_0 \in [0, T) \text{ such that } u_t(x, t_0) \geq 0 \text{ for all } x \in \Omega. \tag{46.5}$$

Then $T < \infty$, $u_t \geq 0$ *in* $\Omega \times [t_0, T)$ *and* u *satisfies the blow-up estimate*

$$C_1(T - t)^{-2/(p-1)} \leq \|u(t)\|_\infty \leq C_2(T - t)^{-2/(p-1)}, \qquad \text{as } t \to T. \tag{46.6}$$

(ii) *Assume that* $\Phi \in C^2(\overline{\Omega})$ *satisfies* $\Phi > 0$ *in* Ω, $\Phi_{|\partial\Omega} = 0$ *and that there exist* $\varepsilon, \eta > 0$ *such that*

$$\Delta\Phi(x) \geq \varepsilon\delta(x) \quad \text{for all } x \in \Omega \text{ such that } \delta(x) \leq \eta. \tag{46.7}$$

Then, for all $\lambda > 0$ *large enough, the solution of* (46.1) *with initial data* $u_0 = \lambda\Phi$ *satisfies* (46.5).

Part (i) was proved in [335] (under the additional assumption $\Omega = B_R$ and u_0 radially symmetric decreasing). Part (ii) was proved in [486]. Note that (46.5) cannot be satisfied for $0 \leq t_0 \ll T$, due to $u_t(., 0) = \Delta u_0$,

Proof. (i) Let

$$J(x, t) = u_t - \varepsilon \int_0^t u^p \, ds, \quad (x, t) \in \Omega \times (t_1, T).$$

By Example 51.14 we have $J \in C^{2,1}(Q_T) \cap C(\overline{\Omega} \times (0, T))$. Pick $t_1 \in (t_0, T)$. Taking $\varepsilon > 0$ small enough, using (46.5) and arguing as in the proof of Theorem 23.5, this time using the nonlocal maximum principle in Proposition 52.24, we obtain that $J(\cdot, t_1) \geq 0$ in Ω. We compute

$$J_t - \Delta J = u_{tt} - \varepsilon u^p - \Delta u_t + \varepsilon p \int_0^t u^{p-1}\Delta u \, ds + \varepsilon p(p-1) \int_0^t u^{p-2}|\nabla u|^2 \, ds$$

$$\geq (1 - \varepsilon)u^p + \varepsilon p \int_0^t u^{p-1}\Big(u_t - \int_0^s u^p \, d\sigma\Big) \, ds$$

$$= (1 - \varepsilon)u_0^p + p \int_0^t u^{p-1}\Big(u_t - \varepsilon \int_0^s u^p \, d\sigma\Big) \, ds \geq p \int_0^t u^{p-1}J \, ds.$$

Since $J = 0$ on $\partial\Omega \times (t_1, T)$, it follows from Proposition 52.24 that $J \geq 0$ in $\Omega \times (t_1, T)$.

Now, for each fixed $x \in \Omega$, multiplying the inequality $u_t \geq \varepsilon \int_{t_1}^{t} u^p\, ds$ by u^p and integrating over (t_1, t), we obtain

$$u^p(x, t) \geq c\left(\int_{t_1}^{t} u^p\, ds\right)^{2p/(p+1)}, \quad t_1 < t < T.$$

It follows that $T < \infty$ and, by integrating over (t, T), we obtain

$$\int_{t_1}^{t} u^p\, ds \leq C(T - t)^{-(p+1)/(p-1)}, \quad t_1 < t < T. \tag{46.8}$$

Setting $t' = t + (T - t)/2$ and using $u_t \geq 0$, we deduce that

$$\frac{T - t}{2}\, u^p(x, t) \leq \int_{t}^{t'} u^p\, ds \leq C(T - t')^{-(p+1)/(p-1)} = C\left(\frac{T - t}{2}\right)^{-(p+1)/(p-1)},$$

hence

$$u(x, t) \leq C(T - t)^{-2/(p-1)}, \quad t_1 < t < T,$$

with $C = C(p, \varepsilon)$. The upper estimate in (46.6) follows.

On the other hand, letting $M(t) = \|u(t)\|_\infty$ and arguing as in the (alternative) proof of Proposition 23.1, we get

$$M'(t) \leq \int_{0}^{t} M^p(s)\, ds, \quad \text{for a.e. } 0 < t < T.$$

Proceeding similarly as for (46.8), we obtain

$$\int_{0}^{t} M^p\, ds \geq c_1(T - t)^{-(p+1)/(p-1)}, \quad 0 < t < T. \tag{46.9}$$

For $t_1 \leq \tau < t < T$, by using (46.9), the upper estimate in (46.6) and M being nondecreasing on $[t_1, T)$, we obtain

$$c_1(T - t)^{-(p+1)/(p-1)} \leq \int_{0}^{\tau} M^p\, ds + \int_{\tau}^{t} M^p\, ds$$

$$\leq C(T - \tau)^{-(p+1)/(p-1)} + (t - \tau)M^p(t).$$

For t close enough to T, taking $\tau = T - \gamma(T - t)$ with $\gamma = (2C/c_1)^{(p-1)/(p+1)} > 1$, we get,

$$M(t) \geq (c_1/2\gamma)^{1/p}(T - t)^{-2/(p-1)},$$

which proves the lower estimate.

(ii) Let $v = v_\lambda := u_t$. By Example 51.14 we have $v \in C^{2,1}(Q_T) \cap C([0,T), L^2(\Omega))$. The function v satisfies

$$\left.\begin{aligned} v_t - \Delta v &= u^p, & x \in \Omega,\ 0 < t < T, \\ v &= 0, & x \in \partial\Omega,\ 0 < t < T, \\ v(x,0) &= \Delta u_0(x), & x \in \Omega, \end{aligned}\right\}$$

hence

$$v(t) = e^{-tA}(\Delta u_0) + \int_0^t e^{-(t-s)A} u^p(s)\, ds.$$

Since $u(t) \geq e^{-tA} u_0$ and

$$e^{-(t-s)A}\big(e^{-sA} u_0\big)^p \geq \big(e^{-(t-s)A}(e^{-sA} u_0)\big)^p = \big(e^{-tA} u_0\big)^p,$$

we have

$$v(t) \geq e^{-tA}(\Delta u_0) + \int_0^t e^{-(t-s)A}\big(e^{-sA} u_0\big)^p\, ds \geq e^{-tA}(\Delta u_0) + t\big(e^{-tA} u_0\big)^p.$$

Therefore, for all $\lambda > 0$,

$$\frac{1}{\lambda} v_\lambda(t) \geq e^{-tA}(\Delta\Phi) + \lambda^{p-1} t\big(e^{-tA}\Phi\big)^p, \quad 0 \leq t < T(\lambda\Phi). \tag{46.10}$$

We claim that there exists $\eta_1 > 0$ such that

$$e^{-tA}(\Delta\Phi)(x) > 0 \quad \text{for all } (x,t) \text{ such that } \delta(x) \leq \eta_1 \text{ and } 0 \leq t \leq \eta_1. \tag{46.11}$$

To prove the claim, observe that, by the assumption (46.7), there exist $\gamma > 0$ and $\rho \in \mathcal{D}(\Omega)$, $\rho \geq 0$, such that $\Delta\Phi \geq \gamma\varphi_1 - \rho$ in Ω. Therefore, $e^{-tA}(\Delta\Phi) \geq \gamma e^{-\lambda_1 t}\varphi_1 - e^{-tA}\rho$. Using $\varphi_1 \geq c_1\delta(x)$ in Ω, $\rho \in \mathcal{D}(\Omega)$ and the continuity of $e^{-tA}\rho$ in $C^1(\overline{\Omega})$ at $t = 0$, claim (46.11) follows.

A straightforward calculation shows that $w(t) := \big(\|u_0\|_\infty^{-(p-1)/2} - kt\big)^{-2/(p-1)}$ is a supersolution of (46.1) for $k = (p-1)(2(p+1))^{-1/2}$. Since blow-up takes place in L^∞-norm if it occurs, this implies in particular that $T(u_0) \geq \frac{1}{k}\|u_0\|_\infty^{-(p-1)/2}$. Taking $t = t_\lambda := \frac{1}{2k}\|\lambda\Phi\|_\infty^{-(p-1)/2}$ in (46.10), we obtain

$$\frac{1}{\lambda} v_\lambda(t_\lambda) \geq e^{-t_\lambda A}(\Delta\Phi) + \frac{1}{2k}\|\Phi\|_\infty^{-(p-1)/2}\lambda^{(p-1)/2}\big(e^{-t_\lambda A}\Phi\big)^p.$$

On the one hand, since $\Phi \geq 0$, by (46.11) we have

$$\frac{1}{\lambda} v_\lambda(x, t_\lambda) \geq 0 \quad \text{if } \delta(x) \leq \eta_1 \text{ and } \lambda \geq \lambda_0(p, \Phi) > 0 \text{ large enough.}$$

On the other hand, since $\Phi > 0$ in Ω, we have $e^{-tA}\Phi > 0$ in $\Omega \times [0, \infty)$ by the strong maximum principle. Therefore, there exists $\alpha > 0$ such that $\left(e^{-tA}\Phi\right)(x) \geq \alpha$ for all (x, t) such that $\delta(x) \geq \eta_1$ and $t \in [0, 1]$. It follows that if $\delta(x) \geq \eta_1$ and $\lambda \geq \lambda_0(p, \Phi)$ (possibly larger), then

$$\frac{1}{\lambda}v_\lambda(x, t_\lambda) \geq -\|\Delta\Phi\|_\infty + C(p, \Phi)\alpha^p\lambda^{(p-1)/2} > 0.$$

We have thus shown that $u_t(x, t_\lambda) \geq 0$ in Ω whenever $\lambda \geq \lambda_0(p, \Phi)$ and the theorem is proved. $\quad\square$

Remark 46.5. Blow-up set and profiles. Results on single-point blow-up and on blow-up profiles for equations similar to (46.1) have been obtained in [67] by employing methods from [219]. $\quad\square$

Appendices

47. Appendix A: Linear elliptic equations

In this appendix we collect some fundamental estimates for linear elliptic equations.

47.1. Elliptic regularity

We assume that Ω is an arbitrary domain in $\mathbb{R}^n$ and we consider second-order elliptic differential operators of the form

$$Au = -\sum_{i,j=1}^{n} a_{ij} \frac{\partial^2}{\partial x_i \partial x_j} u + \sum_{i=1}^{n} b_i \frac{\partial}{\partial x_i} u + cu, \tag{47.1}$$

with measurable coefficients a_{ij}, b_i, c satisfying the ellipticity condition

$$\sum_{i,j} a_{ij}(x)\xi_i\xi_j \geq \lambda|\xi|^2 \qquad \text{for all } x \in \Omega,\ \xi \in \mathbb{R}^n, \tag{47.2}$$

with $\lambda > 0$ and a uniform bound

$$|a_{ij}|, |b_i|, |c| \leq \Lambda. \tag{47.3}$$

We consider the linear problem

$$Au = f \qquad \text{in } \Omega, \tag{47.4}$$

where $f = f(x)$ is a given function.

A **strong solution** of (47.4) is a function $u \in W^{2,1}_{loc}(\Omega)$ which satisfies (47.4) a.e. We denote by $\|u\|_{k,p;D}$ the norm in $W^{k,p}(D)$; in particular $\|u\|_{k,p;\Omega} = \|u\|_{k,p}$.

The following result (cf. [250, Theorems 9.11 and 9.13]) contains the basic interior and interior-boundary elliptic L^p-estimates.

Theorem 47.1. *Let Ω be an arbitrary domain in $\mathbb{R}^n$ and assume (47.2) and (47.3). Let $u \in W^{2,p}_{loc} \cap L^p(\Omega)$, $1 < p < \infty$, be a strong solution of (47.4), where a_{ij} are continuous and $f \in L^p(\Omega)$.*

(i) Consider a subdomain $\Omega' \subset\subset \Omega$. Then

$$\|u\|_{2,p;\Omega'} \leq C(\|u\|_p + \|f\|_p), \tag{47.5}$$

where C depends only on $n, p, \Omega, \Omega', \lambda, \Lambda$, and the moduli of continuity of the a_{ij} on Ω'.

(ii) *Let Σ be an open subset of $\partial\Omega$ of class C^2, $u \in W^{2,p}(\Omega)$ and $u = 0$ on Σ in the sense of traces. Let $a_{ij} \in C(\Omega \cup \Sigma)$ and $\Omega' \subset\subset \Omega \cup \Sigma$. Then (47.5) is true, where C depends also on Σ.*

As for interior and interior-boundary elliptic Schauder estimates, we have the following theorem (cf. [250, Corollary 6.3, Theorems 6.6, 6.19 and Lemma 6.16]).

Theorem 47.2. *Let Ω be an arbitrary bounded domain in $\mathbb{R}^n$, assume (47.2), and let f and the coefficients of A belong to $BUC^\alpha(\Omega)$, where $\alpha \in (0,1)$.*

(i) *Consider a subdomain $\Omega' \subset\subset \Omega$. If $u \in C^2(\Omega)$ is a solution of (47.4), then $u \in BUC^{2+\alpha}(\Omega')$ and*

$$\|u\|_{BUC^{2+\alpha}(\Omega')} \leq C\big(\|u\|_\infty + \|f\|_{BUC^\alpha(\Omega)}\big),$$

where C depends only on $n, \alpha, \lambda, \Omega, \Omega'$ and the norms of the coefficients of A in $BUC^\alpha(\Omega)$.

(ii) *Assume Ω of class $C^{2+\alpha}$ and let $\varphi \in BUC^{2+\alpha}(\Omega)$. If $u \in C^2(\Omega) \cap C(\overline{\Omega})$ is a solution of (47.4) satisfying $u = \varphi$ on $\partial\Omega$, then $u \in BUC^{2+\alpha}(\Omega)$ and*

$$\|u\|_{BUC^{2+\alpha}(\Omega)} \leq C\big(\|u\|_\infty + \|f\|_{BUC^\alpha(\Omega)} + \|\varphi\|_{BUC^{2+\alpha}(\Omega)}\big),$$

where C depends only on $n, \alpha, \lambda, \Omega$ and the norms of the coefficients of A in $BUC^\alpha(\Omega)$.

If we deal only with weaker type of solutions (say, variational), then the regularity assumptions $u \in W^{2,p}(\Omega)$ or $u \in C^2(\Omega)$ in the above theorems can often be verified by means of the following existence-uniqueness theorem (cf. [250, Theorems 9.15 and 6.13]). See Remark 47.4(ii) for an example.

Theorem 47.3. *Assume (47.2), (47.3), $c \leq 0$ and let Ω be a bounded domain of class C^2.*

(i) *Let $a_{ij} \in C(\overline{\Omega})$, $f \in L^p(\Omega)$ and $\varphi \in W^{2,p}(\Omega)$, where $1 < p < \infty$. Then equation (47.4) has a unique (strong) solution $u \in W^{2,p}(\Omega)$ satisfying $u - \varphi \in W_0^{1,p}(\Omega)$.*

(ii) *Let f and the coefficients of A belong to $B(\Omega) \cap C^\alpha(\Omega)$, $\alpha \in (0,1)$, and $\varphi \in C(\partial\Omega)$. Then equation (47.4) has a unique (classical) solution $u \in C^{2+\alpha}(\Omega) \cap C(\overline{\Omega})$ satisfying $u = \varphi$ on $\partial\Omega$.*

Remarks 47.4. (i) Assume that A has constant coefficients. Then the following regularity result can be deduced from Theorem 47.1(i): if $u, f \in L^p_{loc}(\Omega)$ for some $1 < p < \infty$ and $Au = f$ in $\mathcal{D}'(\Omega)$, then $u \in W^{2,p}_{loc}(\Omega)$. To show this, it suffices to apply Theorem 47.1(i) to the convolution products $u * \rho_j$, where ρ_j is a sequence of mollifiers, i.e.

$$\rho_j(x) = j^n \rho(jx), \quad 0 \le \rho \in \mathcal{D}(\mathbb{R}^n), \quad \int_{\mathbb{R}^n} \rho(x)\,dx = 1 \qquad (47.6)$$

(see the end of the proof of Proposition 47.6 below for a more detailed, similar argument). Similarly, using Theorem 47.2(i), we obtain that $u \in C^{2+\alpha}(\Omega)$ whenever $u, f \in C^\alpha(\Omega)$ for some $0 < \alpha < 1$ and $Au = f$ is satisfied in $\mathcal{D}'(\Omega)$.

(ii) For A with leading coefficients of class C^1, Theorem 47.1(i) remains true if we only assume that $u \in W^{1,p}_{loc} \cap L^p(\Omega)$, equation (47.4) being understood in the variational sense. The idea of the proof is as follows. Taking ψ a smooth cut-off function, the regularity of u and a_{ij} allows to apply Theorem 47.3(i) to the equation satisfied by the function $u\psi$ in a smooth domain $\Omega' \subset\subset \Omega$. We can then conclude by using the uniqueness of variational solutions.

(iii) As a useful consequence of the above theorems, we can prove the following property. Assume that A has constant coefficients, let $\Omega \subset \mathbb{R}^n$ be a (possibly unbounded) domain of class C^2, Σ an open subset of $\partial\Omega$, and $f \in L^p_{loc}(\Omega \cup \Sigma)$, with $p > n$. Assume that $u \in W^{2,p}_{loc}(\Omega) \cap C(\Omega \cup \Sigma)$ satisfies $Au = f$ a.e. in Ω and $u - 0$ on Σ. Then $u \in W^{2,p}_{loc}(\Omega \cup \Sigma)$. If we further assume that Ω is of class $C^{2+\alpha}$ and that $f \in C^\alpha(\Omega \cup \Sigma)$ for some $\alpha \in (0,1)$, then $u \in C^{2+\alpha}(\Omega \cup \Sigma)$.

Let us prove this in the case $A = -\Delta$ for simplicity. Let $x_0 \in \Sigma$. One can find $r > 0$ and a bounded domain ω, as smooth as Ω, such that $\Omega \cap B(x_0, r) \subset \omega \subset \Omega$ and $\partial\Omega \cap B(x_0, r) \subset \Sigma$. Let $\varphi \in \mathcal{D}(\mathbb{R}^n)$ be such that $\operatorname{supp}(\varphi) \subset B(x_0, r)$ and $\varphi = 1$ near $x = x_0$. Then $v := u\varphi$ satisfies

$$-\Delta v = \tilde{f} := f\varphi - 2\nabla u \cdot \nabla\varphi - u\Delta\varphi \quad \text{in } \mathcal{D}'(\omega). \qquad (47.7)$$

Since $\tilde{f} \in W^{-1,p}(\omega)$, there exists a unique $w \in W^{1,p}_0(\omega) \subset C_0(\omega)$, such that $-\Delta w = \tilde{f}$. Also, we have $w \in W^{2,p}_{loc}(\omega)$ due to $\tilde{f} \in L^p_{loc}(\omega)$ and part (i). By the maximum principle in Proposition 52.1(i), we deduce that $w = v$. It follows that $u \in W^{1,p}_{loc}(\Omega \cup \Sigma)$. Getting back to equation (47.7), we now have $\tilde{f} \in L^p(\omega)$. By Theorem 47.3(i) and the uniqueness of w, we deduce that $w \in W^{2,p}(\omega)$, hence $u \in W^{2,p}_{loc}(\Omega \cup \Sigma)$. Now, if also $\Omega \in C^{2+\alpha}$ and $f \in C^\alpha(\Omega \cup \Sigma)$, then $\tilde{f} \in BUC^\beta(\omega)$ with $\beta = \min(\alpha, 1 - n/p)$. By Theorem 47.2(ii), we get $v \in BUC^{2+\beta}(\omega)$, hence $u \in C^{2+\beta}(\Omega \cup \Sigma)$. Iterating, we finally obtain $\tilde{f} \in BUC^\alpha(\omega)$ and $u \in C^{2+\alpha}(\Omega \cup \Sigma)$. $\square$

47.2. L^p-L^q-estimates

The following regularity results for the Laplacian are often used in bootstrap arguments in nonlinear problems. The notion of L^1-solution of the Laplace equation

$$\left.\begin{aligned} -\Delta u &= f && \text{in } \Omega, \\ u &= 0 && \text{on } \partial\Omega \end{aligned}\right\} \qquad (47.8)$$

has been introduced in Definition 3.1.

Proposition 47.5. *Let $\Omega \subset \mathbb{R}^n$ be a bounded domain of class $C^{2+\alpha}$ for some $\alpha \in (0,1)$, and assume that $1 \le p \le q \le \infty$ satisfy*

$$\frac{1}{p} - \frac{1}{q} < \frac{2}{n}. \tag{47.9}$$

Let $f \in L^1(\Omega)$ and u be the L^1-solution of (47.8).
(i) If $f \in L^p(\Omega)$, then $u \in L^q(\Omega)$ and

$$\|u\|_q \le C(\Omega, p, q)\|f\|_p.$$

(ii) If $f_+ \in L^p(\Omega)$, then $u_+ \in L^q(\Omega)$ and

$$\|u_+\|_q \le C(\Omega, p, q)\|f_+\|_p.$$

Proposition 47.6. *Let Ω be an arbitrary bounded domain in $\mathbb{R}^n$, $\Omega' \subset\subset \Omega$, and assume that $1 \le p \le q \le \infty$ satisfy (47.9). Let $u \in L^1(\Omega)$ be such that $-\Delta u =: f \in L^1(\Omega)$ (where Δu is understood in the sense of distributions).*
(i) If $f \in L^p(\Omega)$, then $u \in L^q(\Omega')$ and

$$\|u\|_{L^q(\Omega')} \le C(\Omega, \Omega', p, q)\big(\|f\|_{L^p(\Omega)} + \|u\|_{L^1(\Omega)}\big). \tag{47.10}$$

(ii) If $f_+ \in L^p(\Omega)$, then $u_+ \in L^q(\Omega')$ and

$$\|u_+\|_{L^q(\Omega')} \le C(\Omega, \Omega', p, q)\big(\|f_+\|_{L^p(\Omega)} + \|u_+\|_{L^1(\Omega)}\big). \tag{47.11}$$

Proposition 47.5 will be proved in Appendix C, along with the analogous result in L^p_δ-spaces (Theorem 49.2). As for Proposition 47.6, for $p > 1$, inequality (47.10) with $\|u\|_{L^1(\Omega)}$ replaced by $\|u\|_{L^p(\Omega)}$ would follow from Theorem 47.3(i) and the Sobolev inequality. However, since we need the case $p = 1$ (and also (47.11)) in the applications, we have to rely on different, classical arguments, using the fundamental solution and Green's formula.

Proof of Proposition 47.6. We give the proof for $n \ge 3$ only. The cases $n = 1, 2$ can be treated similarly.

(i) We first assume that u is smooth, say C^2. Fix $x \in \Omega'$ and $0 < r < R := \min\big(1, \mathrm{dist}(\Omega', \partial\Omega)\big)$. Let $\Gamma_r(y) = c_n(|y|^{2-n} - r^{2-n})$, where $c_n = ((n-2)|S^{n-1}|)^{-1}$, be the fundamental solution of the Laplacian vanishing for $|y| = r$. It is well known that

$$u(x) = \int_{|y|<r} \Gamma_r(y) f(x + y)\, dy + (n - 2)c_n r^{1-n} \int_{|y|=r} u(x + y)\, d\sigma, \tag{47.12}$$

where σ denotes the surface measure of the sphere $\{y \in \mathbb{R}^n : |y| = r\}$. (The representation formula (47.12) follows by integrating by parts the function $\Delta u(x + y)\Gamma_r(y)$ on the annulus $\{\varepsilon < |y| < r\}$ and letting $\varepsilon \to 0$, see e.g. [250, Section 2.4].) By integrating (47.12) in r over $(R/2, R)$, we get

$$|u(x)| \le c_n R \int_{|y|<R} |y|^{2-n}|f(x+y)|\,dy + C(n)R^{1-n} \int_{R/2<|y-x|<R} |u|\,dy$$
$$\le c_n R\big(h * |\tilde{f}|\big)(x) + C(n)R^{1-n}\|u\|_{L^1(\Omega)},$$

where $h(y) := |y|^{2-n}\chi_{\{|y|<R\}}$ and $\tilde{f}$ denotes the extension of f by 0. Since $h \in L^r(\mathbb{R}^n)$ for any $r < n/(n-2)$, the Young inequality for convolutions yields

$$\|h * |\tilde{f}|\|_q \le \|h\|_r \|\tilde{f}\|_p = C\|f\|_{L^p(\Omega)}, \quad \text{with } 1 - \frac{1}{r} = \frac{1}{p} - \frac{1}{q} < \frac{2}{n}.$$

Inequality (47.10) for smooth u follows.

In the general case, let $u_j = u * \rho_j$, where ρ_j is a sequence of mollifiers defined by (47.6), and $f_j := -\Delta u_j = f * \rho_j$. Note that, for a given subdomain $\omega \subset\subset \Omega$, $u_j \in C^\infty(\omega)$ for all j large enough. Choosing $\Omega' \subset\subset \Omega'' \subset\subset \Omega$, we have

$$\|u_j\|_{L^q(\Omega')} \le C\big(\|f_j\|_{L^p(\Omega'')} + \|u_j\|_{L^1(\Omega'')}\big) \tag{47.13}$$

by the previous step. Using the facts that $u_j \to u$ in $L^1(\Omega'')$ and $f_j \to f$ in $L^p(\Omega'')$, we may pass to the limit in (47.13) with help of Fatou's lemma, and the conclusion follows.

(ii) By (47.12), we have

$$u_+(x) \le \int_{|y|<r} \Gamma_r(y)f_+(x+y)\,dy + (n-2)c_n r^{1-n} \int_{|y|=r} u_+(x+y)\,d\sigma,$$

The rest of the proof is then similar. $\qquad\square$

We conclude this subsection by the following technical lemma, which makes precise some relations between distributional and L^1-solutions.

Lemma 47.7. *Let Ω be an arbitrary bounded domain in $\mathbb{R}^n$, $\omega \subset\subset \Omega$, and let $\psi \in \mathcal{D}(\Omega)$ be such that $\psi = 1$ in ω. If $f \in L^1_{loc}(\Omega)$ and $u \in L^1_{loc}(\Omega) \cap C^1(\overline{\Omega} \setminus \omega)$ is a solution of*

$$-\Delta u = f \quad in \ \mathcal{D}'(\Omega), \tag{47.14}$$

then $w := u\psi$ is an L^1-solution of

$$\left.\begin{aligned} -\Delta w = \tilde{f} &:= f\psi - 2\nabla u \cdot \nabla \psi - u\Delta \psi &&in \ \Omega, \\ w = 0 && on \ \partial\Omega. \end{aligned}\right\}$$

Proof. Fix an open set U such that $\operatorname{supp}(\psi) \subset U \subset\subset \Omega$ and let $\varphi \in C^2(\overline{\Omega}) \cap C^\infty(U)$. We write

$$-\int_\Omega u\psi\Delta\varphi\,dx = -\int_\Omega u\Delta(\psi\varphi)\,dx + 2\int_\Omega u\nabla\psi\cdot\nabla\varphi\,dx + \int_\Omega u\varphi\Delta\psi\,dx.$$

Since $\psi\varphi \in \mathcal{D}(\Omega)$ and $u \in C^1(\overline{\Omega}\setminus\omega)$, by using (47.14) and integrating by parts we obtain

$$-\int_\Omega w\Delta\varphi\,dx = \int_\Omega f\psi\varphi\,dx - \int_\Omega (2\nabla u\cdot\nabla\psi + u\Delta\psi)\varphi\,dx = \int_\Omega \tilde{f}\varphi\,dx. \qquad (47.15)$$

Finally, since $C^2(\overline{\Omega}) \cap C^\infty(U)$ is dense in $C^2(\overline{\Omega})$, (47.15) remains true for all $\varphi \in C^2(\overline{\Omega})$ and the conclusion follows. $\square$

Remark 47.8. Proposition 47.5 remains true in case of equality in (47.9), provided $p > 1$ and $q < \infty$. Indeed, noting that $n \geq 3$, this follows from estimate (48.8) below and the Marcinkiewicz interpolation theorem. $\square$

47.3. An elliptic operator in a weighted Lebesgue space

In this part we prove some basic properties of weighted spaces L_g^2, H_g^1, H_g^2 and the elliptic operator L defined below.

Let $g(y) := e^{|y|^2/4}$, $y \in \mathbb{R}^n$,

$$L_g^q := \{f \in L^q(\mathbb{R}^n) : \int_{\mathbb{R}^n} |f(y)|^q g(y)\,dy < \infty\},$$

$$H_g^1 := \{f \in L_g^2 : \nabla f \in L_g^2\}, \quad H_g^2 := \{f \in H_g^1 : \nabla f \in H_g^1\} \text{ and}$$

$$Lv := -\Delta v - \frac{y\cdot\nabla v}{2} = -\frac{1}{g}\nabla\cdot(g\nabla v), \qquad v \in H_g^2. \qquad (47.16)$$

Estimate (47.18) below (with $u = \partial v/\partial y_i$) shows that $L : H_g^2 \to L_g^2$ is a continuous linear operator. We will consider L as an unbounded operator in the Hilbert space L_g^2 with domain of definition H_g^2. Notice that

$$(Lv, w)_g = \int_{\mathbb{R}^n} (\nabla v\cdot\nabla w)g\,dy, \qquad v, w \in H_g^2,$$

where $(u, v)_g := \int_{\mathbb{R}^n} uvg\,dy$ is the scalar product in L_g^2. Hence L is symmetric and positive. The following two lemmas show that L is a self-adjoint operator with compact inverse.

Lemma 47.9. *The space H_g^1 is compactly embedded in L_g^2.*

Proof. Assume $u \in H_g^1$ and set $v := u\sqrt{g}$. Then

$$\nabla v - \frac{y}{4}v = \sqrt{g}\,\nabla u. \tag{47.17}$$

Fix $R > 0$. By integration by parts, we have

$$\int_{B_R} g|\nabla u|^2\,dy = \int_{B_R} |\nabla v|^2\,dy + \frac{1}{16}\int_{B_R} |y|^2|v|^2\,dy - \frac{1}{2}\int_{B_R} vy \cdot \nabla v\,dy$$

$$= \int_{B_R} |\nabla v|^2\,dy + \frac{1}{16}\int_{B_R} |y|^2|v|^2\,dy$$

$$+ \frac{n}{4}\int_{B_R} |v|^2\,dy - \frac{R}{4}\int_{\partial B_R} v^2\,d\sigma.$$

Since $v \in L^2(\mathbb{R}^n)$ and $\int_{\mathbb{R}^n} v^2\,dy = \int_0^\infty \int_{\partial B_r} v^2\,d\sigma\,dr$, there exists a sequence $R_j \to \infty$ such that $R_j \int_{\partial B_{R_j}} v^2\,d\sigma \to 0$. Therefore, as $R = R_j \to \infty$, we obtain

$$\int_{\mathbb{R}^n} g|\nabla u|^2\,dy \geq \frac{1}{16}\int_{\mathbb{R}^n} |y|^2|u|^2 g\,dy. \tag{47.18}$$

Now assume $u_k \to u$ weakly in H_g^1. Then Rellich's theorem guarantees $u_k \to u$ in $L_{loc}^2(\mathbb{R}^n)$. Denoting by $\|\cdot\|_{2,g}$ the norm in L_g^2, we have

$$\|u_k - u\|_{2,g}^2 = \int_{|y|\leq R} |u_k - u|^2 g\,dy + \int_{|y|>R} |u_k - u|^2 g\,dy =: A_k + B_k,$$

where $A_k \to 0$ as $k \to \infty$ and

$$B_k \leq R^{-2}\int_{|y|>R} |u_k - u|^2|y|^2 g\,dy \leq c_1 R^{-2}\|u_k - u\|_{H_g^1}^2 \leq c_2 R^{-2}$$

due to (47.18). These estimates guarantee $u_k \to u$ in L_g^2. $\square$

Lemma 47.10. *For any $f \in L_g^2$ there exists a unique $u \in H_g^2$ such that $Lu = f$.*

Proof. We shall write $\int f$ instead of $\int_{\mathbb{R}^n} f(y)\,dy$, L^q instead of $L^q(\mathbb{R}^n)$, and we set $H^k := W^{k,2}(\mathbb{R}^n)$. Denote $F(u) := \frac{1}{2}\int |\nabla u|^2 g - \int fug$. Then F achieves its minimum in H_g^1 for a unique u satisfying

$$-\Delta u - \frac{y}{2}\cdot\nabla u = f \qquad \text{in } \mathcal{D}'(\mathbb{R}^n).$$

Standard regularity results imply $u \in H^2_{loc}$. Setting $v := u\sqrt{g}$, we have $v \in H^2_{loc} \cap H^1$ by (47.17), (47.18). Moreover,

$$-\Delta v + \left(\frac{n}{4} + \frac{|y|^2}{16}\right)v = f\sqrt{g}. \tag{47.19}$$

Multiplying this equation by $(-\Delta v)\phi_k$, where $\phi_k(y) := \phi_0(|y|/k)$ and $\phi_0 : \mathbb{R}_+ \to [0,1]$ is a smooth function satisfying $\phi_0(s) = 1$ for $s \le 1$ and $\phi_0(s) = 0$ for $s \ge 2$, we obtain

$$\int |\Delta v|^2 \phi_k + \int |\nabla v|^2 \left(\frac{n}{4} + \frac{|y|^2}{16}\right)\phi_k$$
$$= \int \sqrt{g}f(-\Delta v)\phi_k - \frac{1}{8}\int (y \cdot \nabla v)v\phi_k - \int v\nabla v \cdot \left(\frac{n}{4} + \frac{|y|^2}{16}\right)\nabla \phi_k.$$

Using Cauchy's inequality, $\sqrt{g}f, |y|v \in L^2$ and $|\nabla \phi_k| \le C/k$, we get

$$\int \left(\frac{1}{2}|\Delta v|^2 + \left(\frac{n}{8} + \frac{|y|^2}{16}\right)|\nabla v|^2\right)\phi_k$$
$$\le \frac{1}{2}\int |f|^2 g + C\int |y|^2|v|^2 + \frac{C}{k}\int |v||\nabla v| + C\int |yv||\nabla v||y||\nabla \phi_k| \tag{47.20}$$
$$=: A + B + C_k + D_k.$$

We have $|y||\nabla \phi_k(y)| \le \frac{|y|}{k}|\nabla \phi_0\left(\frac{|y|}{k}\right)| \le C$, hence

$$|yv||\nabla v||y||\nabla \phi_k| \le C|yv||\nabla v| \in L^1.$$

Since $y\nabla \phi_k \to 0$ a.e., we get $D_k \to 0$. Obviously $C_k \to 0$, hence, letting $k \to \infty$ in (47.20) we deduce

$$\int \left(\frac{1}{2}|\Delta v|^2 + \left(\frac{n}{8} + \frac{|y|^2}{16}\right)|\nabla v|^2\right) \le \frac{1}{2}\int |f|^2 g + C\int |y|^2|v|^2 < \infty. \tag{47.21}$$

In particular, $\Delta v \in L^2$. Since also $v \in L^2$, we have $v \in H^2$ by standard elliptic regularity. In addition, (47.19) implies $|y|^2 v \in L^2$ and inequality (47.21) guarantees $|y||\nabla v| \in L^2$. Let us now write

$$\frac{\partial^2 v}{\partial x_i \partial x_j} = \sqrt{g}\frac{\partial^2 u}{\partial x_i \partial x_j} + \frac{y_j}{4}\sqrt{g}\frac{\partial u}{\partial x_i} + \frac{y_i}{4}\sqrt{g}\frac{\partial u}{\partial x_j} + \frac{y_i y_j}{16}\sqrt{g}u + \frac{\delta_{ij}}{4}\sqrt{g}u$$
$$=: A_1 + A_2 + A_3 + A_4 + A_5. \tag{47.22}$$

The LHS of (47.22) belongs to L^2 since $v \in H^2$. Next we have $A_2, A_3 \in L^2$ due to $|y||\nabla v| \in L^2$, $|y|^2 v \in L^2$ and (47.17). Finally, $A_4 \in L^2$ due to $|y|^2 v \in L^2$, and $A_5 \in L^2$. Consequently, $A_1 \in L^2$, which proves $u \in H^2_g$. $\square$

We will also need the following lemma.

Lemma 47.11. $H_g^1 \hookrightarrow L_g^z$, where $z = 2^*$ if $n > 2$, $z \in [2, \infty)$ is arbitrary if $n \leq 2$.

Proof. First note that $H^1(\mathbb{R}^n) \hookrightarrow L^z(\mathbb{R}^n)$. Assume $u \in H_g^1$. Then (47.17) and (47.18) imply $\nabla(u\sqrt{g}) \in L^2$ and

$$\|\nabla(u\sqrt{g})\|_{L^2} \leq C\|u\|_{H_g^1},$$

hence $u\sqrt{g} \in L^z$ and $\|u\sqrt{g}\|_{L^z} \leq C\|u\|_{H_g^1}$. Now the inequality

$$\int |u|^z g \, dy \leq \int |u|^z g^{z/2} \, dy$$

concludes the proof. $\square$

Remarks 47.12. (i) Lemmas 47.9 and 47.11 guarantee that H_g^1 is compactly embedded in L_g^{p+1} for any $p \in [1, p_S)$.

(ii) The proofs of Lemmas 47.10 and 47.11 show that $H_g^2 \hookrightarrow L_g^z$, where $z = 2n/(n-4)$ if $n > 4$, $z \in [2, \infty)$ is arbitrary if $n \leq 4$. $\square$

Lemma 47.13. *Let* $\lambda_1^L < \lambda_2^L < \cdots$ *denote all distinct eigenvalues of* L. *Then* $\lambda_k^L = (n + k - 1)/2$, $k = 1, 2, \ldots$, *and the eigenspaces are*

$$Ker\,(L - \lambda_k^L) = Span\,\{D^\beta \phi_1 : |\beta| = k - 1\},$$

where $\phi_1(y) = e^{-|y|^2/4}$, $D^\beta = \partial_1^{\beta_1} \cdots \partial_n^{\beta_n}$, $|\beta| = \beta_1 + \cdots + \beta_n$.

Proof. Let $u \in L_g^2$ and let $\hat{u}$ denote the Fourier transform of u. Since $|\cdot|^m u \in L^2(\mathbb{R}^n)$ for any $m \geq 0$, we have $\hat{u} \in \bigcap_{m \geq 0} H^m(\mathbb{R}^n) \subset C^\infty(\mathbb{R}^n)$. Assume $Lu = \lambda u$. Applying the Fourier transform we obtain

$$|\xi|^2 \hat{u}(\xi) + \frac{n}{2}\hat{u}(\xi) + \frac{1}{2}\xi \cdot \nabla\hat{u}(\xi) = \lambda\hat{u}(\xi).$$

Set $v(\xi) = e^{|\xi|^2}\hat{u}(\xi)$. Then $v \in C^\infty(\mathbb{R}^n)$ and

$$\xi \cdot \nabla v(\xi) = (2\lambda - n)v(\xi),$$

which guarantees that v is a homogeneous function of degree $(2\lambda - n)$ (cf. the Euler identity for homogeneous functions). As $v \in C^\infty(\mathbb{R}^n)$, the degree $(2\lambda - n)$ has to be a nonnegative integer, hence $v = P_{k-1}$, where P_{k-1} is a homogeneous polynomial of degree $(k-1)$ and $k \in \{1, 2, \ldots\}$. Then $\hat{u}(\xi) = P_{k-1}(\xi)e^{-|\xi|^2}$, hence $u = cP_{k-1}(D)\phi_1$. $\square$

48. Appendix B: Linear parabolic equations

This appendix is devoted to the estimates and various notions of solutions of linear parabolic equations.

48.1. Parabolic regularity

Let Ω be an arbitrary domain in $\mathbb{R}^n$ and $T > 0$. We consider the problem

$$u_t + Au = f \qquad \text{in } Q_T, \qquad (48.1)$$

where the operator A is defined in (47.1) and its coefficients a_{ij}, b_i, c depend on $z := (x, t) \in Q_T$,

$$\sum_{i,j} a_{ij}(z)\xi_i\xi_j \geq \lambda|\xi|^2 \qquad \text{for all } z \in Q_T, \ \xi \in \mathbb{R}^n. \qquad (48.2)$$

A **strong solution** of (48.1) is a function $u \in W_{loc}^{2,1;1}(Q_T)$ satisfying (48.1) a.e.

The following result (cf. [338, Theorems 7.13, 7.15, 7.17 and Corollary 7.16]) contains the basic interior and interior-boundary parabolic L^p-estimates, and an existence-uniqueness statement. See also [320] for additional results concerning parabolic L^p-theory.

Theorem 48.1. *Let Ω be an arbitrary bounded domain in $\mathbb{R}^n$. Assume (48.2) and (47.3). Let $u \in W_{loc}^{2,1;p} \cap L^p(Q_T)$, $1 < p < \infty$, be a strong solution of (48.1), where $a_{ij} \in C(\overline{Q_T})$ and $f \in L^p(Q_T)$.*
(i) If $Q' \subset Q_T$ and $\mathrm{dist}\,(Q', \mathcal{P}_T) > 0$, then

$$\|u\|_{2,1;p;Q'} \leq C(\|u\|_{p;Q_T} + \|f\|_{p;Q_T}), \qquad (48.3)$$

where C depends only on $n, p, Q_T, Q', \lambda, \Lambda$, and the moduli of continuity of the a_{ij}.
(ii) Let Ω be of class C^2 and either Σ be an open subset of S_T or $\Sigma = \mathcal{P}_T$. Assume $u \in W^{2,1;p}(Q_T)$ and $u = 0$ on Σ. Let $Q' \subset Q_T$, $\mathrm{dist}(Q', \mathcal{P}_T \setminus \Sigma) > 0$ if $\Sigma \neq \mathcal{P}_T$. Then (48.3) is true, where C depends also on Σ.
(iii) Let Ω be of class C^2, $\varphi \in W^{2,1;p}(Q_T)$, $f \in L^p(Q_T)$. Then there exists a unique (strong) solution u of (48.1) satisfying $u = \varphi$ on $\mathcal{P}_T$. Moreover, u satisfies the estimate

$$\|u\|_{2,1;p;Q_T} \leq C\big(\|f\|_{p;Q_T} + \|\varphi\|_{2,1;p;Q_T}\big).$$

The following result (cf. [338, Theorems 4.28 and 5.14]) contains the basic interior-boundary parabolic Schauder estimate and an existence-uniqueness statement. We restrict ourselves to global estimates; local estimates can be easily derived by applying this theorem to the function $u\psi$ where ψ is a smooth cut-off function. See also [214] for additional results concerning parabolic Schauder theory.

Theorem 48.2. *Assume* (48.2). *Let* $\alpha \in (0,1)$ *and let* Ω *be a bounded domain of class* $C^{2+\alpha}$. *Assume* $a_{ij}, b_i, c, f \in BUC^{\alpha,\alpha/2}(Q_T)$, $\varphi \in BUC^{2+\alpha,1+\alpha/2}(Q_T)$.

(i) *If* $u \in BUC^{2+\alpha,1+\alpha/2}(Q_T)$ *is a solution of* (48.1) *satisfying* $u = \varphi$ *on* $\mathcal{P}_T$, *then*

$$|u|_{2+\alpha;Q_T} \le C\big(\|u\|_\infty + |f|_{\alpha;Q_T} + |\varphi|_{2+\alpha;Q_T}\big),$$

where C *depends only on* $n, \alpha, \lambda, \Omega$ *and the norms of* a_{ij}, b_i, c *in* $BUC^{\alpha,\alpha/2}(Q_T)$.

(ii) *There exists a unique solution* $u \in C(\overline{Q_T}) \cap C^{2,1}(Q_T)$ *of* (48.1) *satisfying* $u = \varphi$ *on* $\mathcal{P}_T$. *If* $\varphi_t + A\varphi = f$ *on* $\partial\Omega \times \{0\}$, *then* $u \in BUC^{2+\alpha,1+\alpha/2}(Q_T)$ *and*

$$|u|_{2+\alpha;Q_T} \le C\big(|f|_{\alpha;Q_T} + |\varphi|_{2+\alpha;Q_T}\big).$$

Remark 48.3. (i) Assume that A has constant coefficients. Then, by similar arguments as in Remark 47.4(i), one can deduce the following regularity results from Theorems 48.1(i) and 48.2(i). If $u, f \in L^p_{loc}(Q_T)$ for some $1 < p < \infty$ and $u_t + Au = f$ in $\mathcal{D}'(Q_T)$, then $u \in W^{2,1;p}_{loc}(Q_T)$. If $u, f \in C^{\alpha,\alpha/2}(Q_T)$ for some $0 < \alpha < 1$ and $u_t + Au = f$ in $\mathcal{D}'(Q_T)$, then $u \in C^{2+\alpha,1+\alpha/2}(Q_T)$.

(ii) (**Neumann boundary conditions**) Under the assumptions Ω bounded, (48.2), (47.3), $a_{ij} \in C(\overline{Q_T})$, $1 < p < \infty$ and $f \in L^p(Q_T)$, if $u \in W^{2,1;p}_{loc} \cap L^p(Q_T)$ is a strong solution of (48.1) and satisfies $\partial_\nu u = 0$ on S_T and $u = 0$ on $\Omega \times \{0\}$, then we have the estimate $\|u\|_{2,1;p;Q_T} \le C\|f\|_{p;Q_T}$. Similarly, Theorem 48.2(i) remains valid if the condition $u = \varphi$ on $\mathcal{P}_T$ is replaced by $\partial_\nu u = \partial_\nu \varphi$ on S_T and $u = \varphi$ on $\Omega \times \{0\}$. These facts follow from [337, Theorem 7.20] (see also [161, Theorem 8.2]) and [337, Theorem 4.31], respectively. For existence-uniqueness results analogous to Theorem 48.2(ii), see [337, Theorem 4.31]. $\square$

In the rest of Appendix B and in Appendix C we shall restrict ourselves to the Laplace operator for simplicity, but many results can be extended to more general uniformly elliptic divergence form operators with sufficiently smooth coefficients.

48.2. Heat semigroup, L^p-L^q-estimates, decay, gradient estimates

In this subsection we collect some useful properties of the Dirichlet heat semigroup.

Let Ω be an arbitrary domain in $\mathbb{R}^n$ and let $-A_2$ denote the Dirichlet Laplacian in $L^2(\Omega)$, that is the Laplacian on $L^2(\Omega)$ subject to homogeneous Dirichlet boundary conditions (see [154] for its precise definition and for the proof of the following statements). Then $-A_2$ is a nonnegative self-adjoint operator and it generates a C^0-semigroup e^{-tA_2} on $L^2(\Omega)$. The space $L^1 \cap L^\infty(\Omega)$ is invariant under e^{-tA_2} and e^{-tA_2} may be extended from $L^1 \cap L^\infty(\Omega)$ to a positive contraction semigroup $T_p(t)$ on $L^p(\Omega)$ for each $1 \le p \le \infty$. These semigroups are strongly continuous if $1 \le p < \infty$ and $T_\infty(t)f \to f$ as $t \to 0+$ in the weak-star topology. In addition,

$T_p(t)f = T_q(t)f$ for $f \in L^p \cap L^q(\Omega)$ and $p, q \in [1, \infty]$. If no confusion seems likely, we will denote all the semigroups T_p, $1 \le p \le \infty$, by the same symbol e^{-tA} and call them the heat semigroup in Ω (more precisely, the **Dirichlet heat semigroup in** Ω, or the heat semigroup in Ω with homogeneous Dirichlet boundary conditions). Note that $u = e^{-tA}f$ solves the heat equation $u_t - \Delta u = 0$ in $\Omega \times (0, \infty)$. In addition, if Ω is smooth enough (for instance if it satisfies an exterior cone condition at each point of $\partial\Omega$), then $u \in C(\overline{\Omega} \times (0, \infty))$ and $u = 0$ on $\partial\Omega \times (0, \infty)$.

There exists a positive C^∞-function $G_\Omega : \Omega \times \Omega \times (0, \infty) \to \mathbb{R}$ (**Dirichlet heat kernel**) such that $(e^{-tA}f)(x) = \int_\Omega G_\Omega(x, y, t) f(y)\, dy$ for any $f \in L^p(\Omega)$, $1 \le p \le \infty$ (the subscript Ω in G_Ω will be often omitted if no confusion is likely). In addition,

$$G_{\Omega_1}(x, y, t) \le G_{\Omega_2}(x, y, t) \tag{48.4}$$

whenever $\Omega_1 \subset \Omega_2$ and $x, y \in \Omega_1$, and $G_\Omega(x, y, t) = G_\Omega(y, x, t)$ for all $x, y \in \Omega$ and $t > 0$. If $\Omega = \mathbb{R}^n$, then $G_{\mathbb{R}^n}(x, y, t) = G(x - y, t)$, where

$$G(x, t) = G_t(x) := (4\pi t)^{-n/2} e^{-x^2/4t} \tag{48.5}$$

is the Gaussian heat kernel, hence $e^{-tA}f = G_t * f$. Note that the functions G_t satisfy the semigroup property under convolution:

$$G_{t+s} = G_t * G_s, \qquad s, t > 0. \tag{48.6}$$

Let us also observe that if $\lambda > \sigma(-A_2)$ and $B_\lambda := (\lambda + A_2)^{-1}$, then $B_\lambda = \int_0^t e^{-\lambda t} e^{-tA_2}\, dt$ and

$$K_{\Omega,\lambda}(x, y) := \int_0^t e^{-\lambda t} G_\Omega(x, y, t)\, dt \tag{48.7}$$

is the kernel of the operator B_λ, that is $B_\lambda f(x) = \int_\Omega K_{\Omega,\lambda}(x, y) f(y)\, dy$. Notice that for each $f \in L^2(\Omega)$, $B_\lambda f$ is the unique solution of the problem

$$\lambda u - \Delta u = f \quad \text{in } H^{-1}(\Omega), \qquad u \in H_0^1(\Omega),$$

and $K_{\Omega,\lambda}$ is the Green function of this problem. If Ω is bounded and if there is no risk of confusion, we denote simply $K(x, y) = K_\Omega(x, y) = K_{\Omega,0}(x, y)$, which is the (elliptic) Green kernel of the Dirichlet Laplacian. Moreover, for $n \ge 3$, we have

$$K_\Omega(x, y) \le C_n |x - y|^{2-n}, \tag{48.8}$$

as a consequence of (48.4), (48.5) and (48.7).

The following L^p-L^q-estimate for the heat semigroup is of fundamental importance in the study of semilinear problems.

Proposition 48.4. *Let $(e^{-tA})_{t\geq 0}$ be the heat semigroup in $\mathbb{R}^n$ and $G_t(x) = G(x,t)$ the Gaussian heat kernel. We have the following properties.*

(a) $\|G_t\|_1 = 1$ for all $t > 0$.

(b) If $\Phi \geq 0$, then $e^{-tA}\Phi \geq 0$ and $\|e^{-tA}\Phi\|_1 = \|\Phi\|_1$.

(c) If $1 \leq q \leq \infty$, then $\|e^{-tA}\Phi\|_q \leq \|\Phi\|_q$ for all $t > 0$.

(d) If $1 \leq p < q \leq \infty$ and $1/r = 1/p - 1/q$, then $\|e^{-tA}\Phi\|_q \leq (4\pi t)^{-n/(2r)}\|\Phi\|_p$ for all $t > 0$.

(e) For an arbitrary domain $\Omega \subset \mathbb{R}^n$, assertions (c) and (d) remain valid if e^{-tA} is replaced with the Dirichlet heat semigroup in Ω.

Proof. Statement (a) is well known, statement (b) follows from Fubini's theorem and part (a). Statement (c) follows from the contractivity of the semigroup $T_q(t)$ (see above); it also easily follows from the estimate $\|G_t * \Phi\|_q \leq \|G_t\|_1\|\Phi\|_q$.

Interpolating between (b) and the inequality $\|e^{-tA}\Phi\|_\infty \leq (4\pi t)^{-n/2}\|\Phi\|_1$ we obtain

$$\|e^{-tA}\Phi\|_q \leq (4\pi t)^{-(n/2)(1-1/q)}\|\Phi\|_1. \tag{48.9}$$

Interpolating between (48.9) and (c) yields (d).

To prove assertion (e), denote by e^{-tA_Ω} the Dirichlet heat semigroup in Ω. Let $\tilde{\Phi}(x) = \Phi(x)$ if $x \in \Omega$, $\tilde{\Phi}(x) = 0$ otherwise. By (48.4) we have

$$|e^{-tA_\Omega}\Phi| \leq e^{-tA_\Omega}|\Phi| \leq e^{-tA}|\tilde{\Phi}|.$$

The conclusion follows from assertions (c) and (d). $\square$

In the case of bounded domains, we have the following classical property of uniform exponential decay.

Proposition 48.5. *Let Ω be an arbitrary bounded domain and let $(e^{-tA})_{t\geq 0}$ be the Dirichlet heat semigroup in Ω. For all $1 \leq p \leq \infty$ and all $\Phi \in L^p(\Omega)$, there holds*

$$\|e^{-tA}\Phi\|_p \leq C(\Omega)e^{-\lambda_1 t}\|\Phi\|_p, \qquad t \geq 0. \tag{48.10}$$

Proof. If $0 < t < 2$, then (48.10) follows from Proposition 48.4(c). We may thus assume $t \geq 2$. It is well known that

$$\|e^{-tA}\Phi\|_2 \leq e^{-\lambda_1 t}\|\Phi\|_2, \qquad t \geq 0. \tag{48.11}$$

Using (48.11), Proposition 48.4(d) for $p = 2$, $q = \infty$, Hölder's inequality and $|\Omega|^{1/p} \leq \max(1, |\Omega|)$, we get

$$\|e^{-tA}\Phi\|_p \leq |\Omega|^{1/p}\|e^{-tA}\Phi\|_\infty \leq (4\pi)^{-n/4}|\Omega|^{1/p}\|e^{-(t-1)A}\Phi\|_2$$
$$\leq C(\Omega)e^{-\lambda_1(t-2)}\|e^{-A}\Phi\|_2.$$

The assertion then follows from $\|e^{-A}\Phi\|_2 \leq \|\Phi\|_2 \leq C(\Omega)\|\Phi\|_p$ if $p \geq 2$, and from $\|e^{-A}\Phi\|_2 \leq \|\Phi\|_p$ (owing to Proposition 48.4(d)) if $p < 2$. $\quad\square$

In the case of the whole space and integrable initial data, the asymptotic behavior is described by a multiple of the Gaussian heat kernel (see [175], or [168] for further results).

Proposition 48.6. *Let* $\Phi \in L^1(\mathbb{R}^n)$ *and put* $M = \int_{\mathbb{R}^n} \Phi\, dx$.

(i) *There holds*
$$\|e^{-tA}\Phi - M\,G_t\|_1 \to 0, \quad t \to \infty.$$

(ii) *If, in addition,* $x\,\Phi(x) \in L^1(\mathbb{R}^n)$, *then*
$$\|e^{-tA}\Phi - M\,G_t\|_1 \leq Ct^{-1/2}\|x\,\Phi(x)\|_1, \quad t > 0,$$
where $C = C(n) > 0$.

Proof. We first establish assertion (ii). Let $\Phi \in L^1\big(\mathbb{R}^n; (1 + |x|)\,dx\big)$.

$$\big(e^{-tA}\Phi - M\,G_t\big)(x) = (4\pi t)^{-n/2} \int_{\mathbb{R}^n} \big(e^{-|x-y|^2/4t} - e^{-|x|^2/4t}\big)\Phi(y)\,dy$$

$$= \frac{(4\pi t)^{-n/2}}{2\sqrt{t}} \int_{\mathbb{R}^n} \int_0^1 \frac{y \cdot (x - \theta y)}{\sqrt{t}} e^{-|x-\theta y|^2/4t}\Phi(y)\,dy\,d\theta.$$

Using $\sup_{s>0} se^{-s^2/8} < \infty$ and Fubini's theorem, we deduce that

$$\|e^{-tA}\Phi - M\,G_t\|_1 \leq Ct^{-(n+1)/2} \int_0^1 \int_{\mathbb{R}^n} \int_{\mathbb{R}^n} e^{-|x-\theta y|^2/8t}|y||\Phi(y)|\,dx\,dy\,d\theta$$

$$= Ct^{-1/2} \int_0^1 \int_{\mathbb{R}^n} |y||\Phi(y)|\,dy\,d\theta = Ct^{-1/2}\|x\,\Phi(x)\|_1.$$

Let us next prove assertion (i). Fix $\Phi \in L^1(\mathbb{R}^n)$ and pick a sequence $\{\varphi_j\} \in \mathcal{D}(\mathbb{R}^n)$ such that $\int_{\mathbb{R}^n} \varphi_j\,dx = M$ and $\varphi_j \to \Phi$ in $L^1(\mathbb{R}^n)$. For each j we write

$$\|e^{-tA}\Phi - M\,G_t\|_1 \leq \|e^{-tA}\varphi_j - M\,G_t\|_1 + \|e^{-tA}(\Phi - \varphi_j)\|_1$$

$$\leq \|e^{-tA}\varphi_j - M\,G_t\|_1 + \|\Phi - \varphi_j\|_1.$$

By assertion (ii), it follows that

$$\limsup_{t\to\infty} \|e^{-tA}\Phi - M\,G_t\|_1 \leq \|\Phi - \varphi_j\|_1.$$

and the conclusion follows by letting $j \to \infty$. $\quad\square$

We conclude with a smoothing estimate for the gradient, which we state without proof (this follows from [320, Theorem IV.16.3, p. 413]).

Proposition 48.7. *Let* Ω *be a domain of class* $C^{2+\alpha}$ *for some* $\alpha \in (0,1)$ *and let* $(e^{-tA})_{t\geq 0}$ *be the Dirichlet heat semigroup in* Ω. *For all* $\Phi \in L^\infty(\Omega)$, *there holds*
$$\|\nabla e^{-tA}\Phi\|_\infty \leq C(\Omega)(1 + t^{-1/2})\|\Phi\|_\infty, \quad t > 0.$$

48.3. Weak and integral solutions

In this subsection we compare various notions of solutions of the inhomogeneous linear heat equation. Related semigroup and smoothing properties will be described in Appendix C (Subsection 49.2).

Assume that Ω is a bounded domain of class $C^{2+\alpha}$ for some $\alpha \in (0,1)$. Similarly as in Remarks 15.4(iv) and (v), we may define integral and weak L_δ^1-solutions of the linear problem

$$\left.\begin{aligned}
u_t - \Delta u &= f, & x &\in \Omega,\ t \in (0,T), \\
u &= 0, & x &\in \partial\Omega,\ t \in (0,T), \\
u(x,0) &= u_0(x), & x &\in \Omega,
\end{aligned}\right\} \qquad (48.12)$$

as follows.

Definition 48.8. (i) Let $u_0 \in L_\delta^1(\Omega)$ and $f \in L_{loc}^1((0,T), L_\delta^1(\Omega))$. A function $u \in C([0,T), L_\delta^1(\Omega)) \cap L_{loc}^1((0,T), L^1(\Omega))$ is a **weak L_δ^1-solution** of (48.12) if $u(\cdot,0) = u_0$ and, for any $0 < \tau < t < T$,

$$\int_\tau^t \int_\Omega f\varphi = -\int_\tau^t \int_\Omega u(\varphi_t + \Delta\varphi) - \int_\Omega u(\tau)\varphi(\tau) \qquad (48.13)$$

for all $\varphi \in C^2(\overline{\Omega} \times [\tau,t])$ such that $\varphi = 0$ on $\partial\Omega \times [\tau,t]$ and $\varphi(t) = 0$.

(ii) Let u_0, f be nonnegative measurable functions and let G denote the Dirichlet heat kernel in Ω. Then

$$u(x,t) := \int_\Omega G(x,y,t)u_0(y)\,dy + \int_0^t \int_\Omega G(x,y,t-s)f(y,s)\,dy\,ds \leq \infty$$

is called the **integral solution** of (48.12). $\square$

Proposition 48.9. *Let Ω be as above and let $u_0 \in L_\delta^1(\Omega)$.*

(i) *If $f \in L_{loc}^1([0,T), L_\delta^1(\Omega))$, then problem (48.12) possesses a unique weak L_δ^1-solution. Moreover $u \in L_{loc}^1([0,T), L^1(\Omega))$ and (48.13) is also satisfied for $\tau = 0$.*

(ii) *If $f \in L_{loc}^1((0,T), L_\delta^1(\Omega))$ and problem (48.12) possesses a weak L_δ^1-solution, then $f \in L_{loc}^1([0,T), L_\delta^1(\Omega))$.*

Proof. (i) Let $f \in L_{loc}^1([0,T), L_\delta^1(\Omega))$. We first prove the uniqueness. Assume that u_1, u_2 are two weak solutions of (48.12) and set $w := u_1 - u_2$. Then w is a weak solution of the homogeneous problem (48.12) (with $f = 0$ and $u_0 = 0$). In particular,

$$\int_\tau^t \int_\Omega w(\varphi_t + \Delta\varphi) + \int_\Omega w(\tau)\varphi(\tau) = 0 \qquad (48.14)$$

whenever $0 < \tau < t < T$ and $\varphi \in C^{2,1}(\overline{\Omega} \times [\tau, t])$ satisfies $\varphi = 0$ on $\partial\Omega \times [\tau, t]$ and $\varphi(t) = 0$.

Fix $t \in (0, T)$. Let $\psi \in \mathcal{D}(Q_t)$ and let $\varphi = \varphi_\psi$ be the solution of the problem

$$
\begin{aligned}
-\varphi_t - \Delta\varphi &= \psi && \text{in } Q_t, \\
\varphi &= 0 && \text{on } S_t, \\
\varphi(t) &= 0 && \text{in } \Omega.
\end{aligned}
$$

Then passing to the limit in (48.14) as $\tau \to 0$ we obtain

$$
\int_0^t \int_\Omega w\psi = 0,
$$

hence $w = 0$ a.e.

In order to prove the existence, we may assume that $u_0 \geq 0$ and $f \geq 0$ (otherwise we decompose u_0 and f into their positive and negative parts and use the linearity of the problem (48.12)). Set $u_{0,k} := \min(u_0, k)$ and $f_k := \min(f, k)$, $k = 1, 2, \dots$. Let u_k be the (strong) solution of (48.12) with f and u_0 replaced by f_k and $u_{0,k}$, respectively. Then

$$
u_k(x, t) = \int_\Omega G(x, y, t)u_{0,k}(y)\, dy + \int_0^t \int_\Omega G(x, y, t - s)f_k(y, s)\, dy\, ds, \qquad (48.15)
$$

where G denotes the Dirichlet heat kernel in Ω. Passing to the limit in (48.15) as $k \to \infty$ we get $u_k(x, t) \nearrow u(x, t)$, where u satisfies

$$
u(x, t) = \int_\Omega G(x, y, t)u_0(y)\, dy + \int_0^t \int_\Omega G(x, y, t - s)f(y, s)\, dy\, ds. \qquad (48.16)
$$

Notice also that $u_k \in C([0, T), L^r(\Omega)) \cap W_{loc}^{2,1;r}(\overline{\Omega} \times (0, T))$ for all $r \in (1, \infty)$ (see Theorem 48.1 and Appendix E). Let $0 \leq \tau < t < T$, $q > 1$ and $\varphi \in C([\tau, t], L^q(\Omega)) \cap W_{loc}^{2,1;q}(\overline{\Omega} \times (\tau, t))$ satisfy $\varphi = 0$ on $\partial\Omega \times (\tau, t)$ and $(\varphi_t + \Delta\varphi) \in L^1(\Omega \times (\tau, t))$. Multiplying the equation for u_k by φ, integrating over $\Omega \times (\tau', t')$ with $\tau < \tau' < t' < t$ and letting $\tau' \to \tau$, $t' \to t$, we obtain

$$
\int_\tau^t \int_\Omega f_k\varphi = -\int_\tau^t \int_\Omega u_k(\varphi_t + \Delta\varphi) + \int_\Omega u_k(t)\varphi(t) - \int_\Omega u_k(\tau)\varphi(\tau). \qquad (48.17)
$$

Set $\varphi := \psi$, where ψ is the solution of the problem

$$
\begin{aligned}
-\psi_t - \Delta\psi &= 1 && \text{in } Q_t, \\
\psi &= 0 && \text{on } S_t, \\
\psi(t) &= 0 && \text{in } \Omega.
\end{aligned}
$$

Then (48.17) with $\tau = 0$ implies

$$\int_0^t \int_\Omega u_k = \int_0^t \int_\Omega f_k \psi + \int_\Omega u_{0,k} \psi(0) \le C(t) < \infty,$$

hence the sequence $\{u_k\}$ is bounded in $L^1(Q_t)$, $u_k \to u$ in $L^1(Q_t)$ and $u \in L^1_{loc}([0,T), L^1(\Omega))$.

Next set $\varphi := \chi$, where $\chi(x,s) = e^{\lambda_1(s-t)}\varphi_1(x)$, which satisfies $\chi_s + \Delta\chi = 0$ in Q_t. For $k \ge j$, it follows from (48.17) with $\tau = 0$ and (1.4) that

$$c_1 \int_\Omega (u_k - u_j)(t)\delta \le \int_\Omega (u_k - u_j)(t)\varphi_1 = \int_0^t \int_\Omega (f_k - f_j)\chi + \int_\Omega (u_{0,k} - u_{0,j})\chi(0)$$

$$\le c_2 \int_0^t \int_\Omega (f_k - f_j)\delta + c_2 \int_\Omega (u_{0,k} - u_{0,j})\delta.$$

This estimate guarantees that $\{u_k\}$ is a Cauchy sequence in $C([0,t], L^1_\delta(\Omega))$, hence $u \in C([0,T), L^1_\delta(\Omega))$.

Finally, fix $0 \le \tau < t < T$ and $\varphi \in C^2(\overline{\Omega} \times [\tau, t])$ satisfying $\varphi - 0$ on $\partial\Omega \times [\tau, t]$ and $\varphi(t) = 0$. Then passing to the limit in (48.17) as $k \to \infty$ we see that u is a weak solution of (48.12) and that (48.13) is also satisfied for $\tau = 0$.

For future reference, we note that the solution u that we have just constructed satisfies

$$\int_0^t \int_\Omega |f|\delta \le C \int_\Omega |u(t)|\delta, \qquad 0 < t < T, \tag{48.18}$$

where C remains bounded for T bounded. Indeed (still assuming $u_0, f \ge 0$ without loss of generality), (48.18) follows by passing to the limit $k \to \infty$ in (48.17) with $\tau = 0$ and $\varphi = \chi$.

(ii) Now assume that problem (48.12) possesses a weak L^1_δ-solution u. Then, for each $\tau \in (0, T)$, u coincides with the weak L^1_δ-solution of (48.12) on (τ, T) with initial data $u(\tau)$, given by part (i). For each $t \in (0, T)$, estimate (48.18) guarantees that

$$\int_\tau^t \int_\Omega |f|\delta \le C \int_\Omega |u(t)|\delta, \qquad 0 < \tau < t,$$

and the assertion follows by letting $\tau \to 0$. $\square$

Corollary 48.10. *Let Ω be as above, $u_0 \in L^1_\delta(\Omega)$, $u \in L^1_{loc}(Q_T)$, and $f : Q_T \to \mathbb{R}$ be measurable. Assume that $u_0, u, f \ge 0$.*

(i) If $f \in L^1_{loc}((0,T), L^1_\delta(\Omega))$ and u is a weak L^1_δ-solution of (48.12), then it is an integral solution of (48.12).

(ii) If u is an integral solution of (48.12), then $f \in L^1_{loc}((0,T), L^1_\delta(\Omega))$ and u is a weak L^1_δ-solution of (48.12).

Proof. If u is a weak solution, then $f \in L^1_{loc}([0,T), L^1_\delta(\Omega))$ by Proposition 48.9(ii), and the proof of Proposition 48.9(i) (cf. formula (48.16)) shows that u is an integral solution.

Let u be an integral solution of (48.12). Again, the proof of Proposition 48.9(i) guarantees that u is a weak solution provided we show $f \in L^1_{loc}([0,T), L^1_\delta(\Omega))$. Let f_k, u_k be as in the proof of Proposition 48.9(i). Let $0 < t < T' < T$ and $\psi \in \mathcal{D}(Q_T)$, $\psi \geq 0$, $\psi(\cdot, t) \not\equiv 0$. Let φ be the solution of the problem

$$
\begin{aligned}
-\varphi_t - \Delta\varphi &= \psi && \text{in } Q_{T'}, \\
\varphi &= 0 && \text{on } S_{T'}, \\
\varphi(T') &= 0 && \text{in } \Omega.
\end{aligned}
$$

Then there exists $\varepsilon > 0$ such that

$$
\varphi(x,s) \geq \varepsilon\delta(x) \qquad \text{for all } (x,s) \in Q_t.
$$

Multiplying the equation $\partial_t u_k - \Delta u_k = f_k$ by φ we obtain

$$
\varepsilon \int_0^t \int_\Omega f_k \delta \leq \int_0^{T'} \int_\Omega f_k \varphi = \int_0^{T'} \int_\Omega u_k \psi - \int_\Omega u_{0,k} \varphi(0) \leq \int_0^{T'} \int_\Omega u\psi < \infty,
$$

hence $\int_0^t \int_\Omega f\delta < \infty$, which guarantees $f \in L^1_{loc}([0,T), L^1_\delta(\Omega))$. $\square$

Corollary 48.11. *Let Ω be as above, $q \geq 1$ and let u be a mild L^q-solution of (48.12) (that is $u \in C([0,T), L^q(\Omega))$, $u(0) = u_0$, $f \in L^1_{loc}((0,T), L^1(\Omega))$ and (15.5) is true with $f(u)$ replaced by f). Then u is a weak L^1_δ-solution of (48.12).*

Proof. Fix $\tau_0 \in (0,T)$ and set $u_{0,1} := u_+(\tau_0)$, $u_{0,2} := u_-(\tau_0)$, $f_1 := f_+$, $f_2 := f_-$,

$$
v_i(t) = v_i(t; \tau_0) := e^{-(t-\tau_0)A} u_{0,i} + \int_{\tau_0}^t e^{-(t-s)A} f_i(s)\, ds, \qquad \tau_0 \leq t < T,\ i = 1,2.
$$

Then v_i, $i = 1,2$, are nonnegative integral solutions of problem (48.12) with $[0,T), u_0, f$ replaced by $[\tau_0, T), u_{0,i}, f_i$. Consequently, v_i, $i = 1,2$, are weak solutions of those problems and $v := v_1 - v_2$ is a weak solution of (48.12) on $[\tau_0, T)$ with initial data $u(\tau_0)$. On the other hand, $v_1(t; \tau_0) - v_2(t; \tau_0) = u(t)$ for any $t \in (\tau_0, T)$, hence $u = v$ is a weak solution of (48.12) on $[0,T)$. $\square$

Remark 48.12. In the case of $\Omega = \mathbb{R}^n$, for instance, and of nonnegative data u_0, f, one can also study the relations between local classical nonnegative solutions of (48.12) and integral solutions. Let $\Omega = \mathbb{R}^n$, $u_0 \in L^1_{loc}(\mathbb{R}^n)$, f be locally Hölder continuous in Q_T, with $u_0 \geq 0$ a.e. and $f \geq 0$. Assume that $0 \leq u \in C^{2,1}(Q_T) \cap C([0,T); L^1_{loc}(\mathbb{R}^n))$ is a solution of $u_t - \Delta u = f$ in Q_T, with $u(\cdot, 0) = u_0$. Then u satisfies (48.16) in Q_T, where all the integrals are in particular finite (see [482] and cf. also [526]). Such property may be useful, e.g., when considering problems of Fujita-type. $\square$

49. Appendix C: Linear theory in L_δ^p-spaces and in uniformly local spaces

In this section, we state and prove some useful properties of the Laplace and heat equations in weighted Lebesgue spaces $L_\delta^p(\Omega)$ and in uniformly local Lebesgue spaces $L_{ul}^p(\mathbb{R}^n)$. We refer to Section 1 for the definition of these spaces.

49.1. The Laplace equation in L_δ^p-spaces

Very weak, or L_δ^1, solutions of the Laplace equation (47.8) have been introduced in Definition 3.1. We have the following existence-uniqueness result (see [94]; estimate (49.1) is proved there for $q = 1$ and in the general case in [107]).

Theorem 49.1. *Let $\Omega \subset \mathbb{R}^n$ be a bounded domain of class $C^{2+\alpha}$ for some $\alpha \in (0,1)$, and let $f \in L_\delta^1(\Omega)$. Then there exists a unique $u \in L^1(\Omega)$ such that u is an L_δ^1-solution of problem (47.8). Moreover, for all $1 \le q < n/(n-1)$, we have $u \in L^q(\Omega)$ and*

$$\|u\|_q \le C(n, q, \Omega) \, \|f\|_{1,\delta}. \tag{49.1}$$

Furthermore, the maximum principle is satisfied, i.e.: $f \ge 0$ a.e. implies $u \ge 0$ a.e.

Proof. We start by proving the uniqueness. Thus assume that $u \in L^1(\Omega)$ is an L_δ^1-solution of problem (47.8) with $f = 0$. Take any $h \in \mathcal{D}(\Omega)$ and let $\varphi \in C^2(\overline{\Omega})$ be the classical solution of

$$\left.\begin{aligned} -\Delta\varphi &= h &&\text{in } \Omega, \\ \varphi &= 0 &&\text{on } \partial\Omega. \end{aligned}\right\}$$

Then $\int_\Omega uh\,dx = 0$ by (3.3). It follows that $u = 0$, hence the uniqueness assertion.

Let us show the existence. We may assume $f \ge 0$ without loss of generality (writing $f = f_+ - f_-$). Let $f_i = \min(f, i)$ and denote by u_i the strong solution of (47.8) with f replaced by f_i. Let Θ be the classical solution of (19.27). For $j \ge i$, we have $f_j \ge f_i \ge 0$, hence $u_j \ge u_i \ge 0$ by the maximum principle. Testing the equation for $u_j - u_i$ with Θ, we have

$$\|u_j - u_i\|_1 = \int_\Omega (u_j - u_i)\,dx = \int_\Omega (f_j - f_i)\Theta\,dx.$$

Since $f_i \to f$ in $L_\delta^1(\Omega)$, we deduce that $\{u_i\}$ is a Cauchy sequence in $L^1(\Omega)$, and we denote by $u \in L^1(\Omega)$ its limit. Observe that $u \ge 0$. For any $\varphi \in C^2(\overline{\Omega})$ with $\varphi = 0$ on $\partial\Omega$, we then have

$$\int_\Omega u(-\Delta\varphi)\,dx = \lim_{i\to\infty} \int_\Omega u_i(-\Delta\varphi)\,dx = \lim_{i\to\infty} \int_\Omega f_i\varphi\,dx = \int_\Omega f\varphi\,dx, \tag{49.2}$$

hence u is an L^1_δ-solution of (47.8).

Next, the choice $\varphi = \Theta$ in (49.2) yields $\int_\Omega u\,dx = \int_\Omega f\Theta\,dx$. Assuming again $f \geq 0$, this implies estimate (49.1) for $q = 1$. The case $1 < q < n/(n-1)$ will be proved along with Theorem 49.2. $\square$

The following results describe the optimal regularity of the Dirichlet Laplacian in the scale of L^p_δ-spaces (see [84], [200] for Theorem 49.2 and [489] for Theorem 49.3). The proofs will be given in Subsection 49.4 below.

Theorem 49.2. *Let $\Omega \subset \mathbb{R}^n$ be a bounded domain of class $C^{2+\alpha}$ for some $\alpha \in (0,1)$. Assume that $1 \leq p \leq q \leq \infty$ satisfy*

$$\frac{1}{p} - \frac{1}{q} < \frac{2}{n+1}. \tag{49.3}$$

Let $f \in L^1_\delta(\Omega)$ and let u be the L^1_δ-solution of (47.8).
(i) If $f \in L^p_\delta(\Omega)$, then $u \in L^q_\delta(\Omega)$ and

$$\|u\|_{q,\delta} \leq C(p,q,\Omega)\,\|f\|_{p,\delta}. \tag{49.4}$$

(ii) If $f_+ \in L^p_\delta(\Omega)$, then $u_+ \in L^q_\delta(\Omega)$ and

$$\|u_+\|_{q,\delta} \leq C(p,q,\Omega)\,\|f_+\|_{p,\delta}. \tag{49.5}$$

Theorem 49.3. *Let $\Omega \subset \mathbb{R}^n$ be a bounded domain of class $C^{2+\alpha}$ for some $\alpha \in (0,1)$. Assume that $1 \leq p < q \leq \infty$ satisfy*

$$\frac{1}{p} - \frac{1}{q} > \frac{2}{n+1}. $$

Then there exists $f \in L^p_\delta(\Omega)$ such that the L^1_δ-solution u of (47.8) satisfies

$$u \notin L^q_\delta(\Omega).$$

Remarks 49.4. (a) In Theorem 49.2 one may take in particular $q = \infty$ for $p > (n+1)/2$ and any $q < (n+1)/(n-1)$ for $p = 1$.

(b) By a density argument, it is easy to see that the L^1_δ-solution u of (47.8) is given by $u(x) = \int_\Omega K(x,y)f(y)\,dy$, where $K(x,y)$ is the Dirichlet Green kernel in Ω.

(c) By similar arguments as in the proofs of Theorems 49.2 and 49.3 (see [489] for details), one can obtain further optimal regularity properties of the solution u of (47.8). Namely, assuming $1 \leq p \leq q \leq \infty$:

- if $\frac{n+1}{p} - \frac{n}{q} < 2$, then $u \in L^q(\Omega)$ and $\|u\|_q \leq C\|f\|_{p,\delta}$;

- if $\frac{n+1}{p} - \frac{n}{q} > 2$, then there exists $f \in L_\delta^p$ such that $u \notin L^q(\Omega)$;

- if $\frac{n+1}{p} - \frac{n}{q} < 1$, then $u \in W_0^{1,q}(\Omega)$ and $\|u\|_{1,q} \leq C\|f\|_{p,\delta}$;

- if $\frac{n+1}{p} - \frac{n}{q} > 1$, then there exists $f \in L_\delta^p$ such that $u \notin W_0^{1,q}(\Omega)$.

In particular, it follows that if $f \in L_\delta^p$ for some $p > 1$, then $u \in W_0^{1,q}(\Omega)$ for $q > 1$ close to 1, so that the boundary conditions in (47.8) are also satisfied in the sense of traces. We note that in the example constructed in the proof of Theorem 49.3, the solution u possesses a singularity at a (single) boundary point $a \in \partial\Omega$ and that $u \in W_{loc}^{2,m}(\overline{\Omega} \setminus \{a\})$ for all finite m.

(d) Theorem 49.2 remains true in case of equality in (49.3) provided $p > 1$, $q < \infty$ and $n \neq 2$ (see [362], where equality cases in Remark (c) are also treated).

(e) If $f \in L_\delta^1$, then, for each $\alpha \in (0,1)$, we have $u/\delta^\alpha \in L^1(\Omega)$ and $\|u/\delta^\alpha\|_1 \leq C(\alpha)\|f\|_{1,\delta}$. This can be shown by using the singular test-function ξ from Lemma 10.4 for smooth f and the general case follows by density. $\quad\square$

We close this subsection by proving a useful, simple consequence of Theorem 49.2 (cf. [449, Proposition 2.3]).

Proposition 49.5. *Let $\Omega \subset \mathbb{R}^n$ be a bounded domain of class $C^{2+\alpha}$ for some $\alpha \in (0,1)$. Let $f \in L_\delta^1(\Omega)$ and let u be the L_δ^1-solution of (47.8). Then, for any $1 \leq k < (n+1)/(n-1)$, we have*

$$\|u\|_{k,\delta} \leq C(\Omega, k)\big(\|u_+\|_{1,\delta} + \|f_-\|_{1,\delta}\big).$$

Proof. Using (3.3) with $\varphi = \varphi_1$ and (1.4), we obtain

$$\int_\Omega |f|\varphi_1 = \int_\Omega f\varphi_1 + 2\int_\Omega (f_-)\varphi_1$$
$$= \lambda_1 \int_\Omega u\varphi_1 + 2\int_\Omega (f_-)\varphi_1 \leq C(\Omega)\big(\|u_+\|_{1,\delta} + \|f_-\|_{1,\delta}\big).$$

Applying Theorem 49.2(i) with $p = 1$ and using $\varphi_1 \geq c_1\delta$, we deduce that

$$\|u\|_{k,\delta} \leq C(\Omega, k)\|f\|_{1,\delta} \leq C(\Omega, k)\big(\|u_+\|_{1,\delta} + \|f_-\|_{1,\delta}\big). \quad\square$$

49.2. The heat semigroup in L_δ^p-spaces

We start by introducing a natural extension of the Dirichlet heat semigroup. Here we also use the spaces $L_{\varphi_1}^p(\Omega)$, which are defined similarly as $L_\delta^p(\Omega)$. Note that if Ω is C^2-smooth, then $L_{\varphi_1}^p(\Omega) \doteq L_\delta^p(\Omega)$, due to (1.4).

Proposition and Definition 49.6. *Let Ω be an arbitrary bounded domain in $\mathbb{R}^n$. The Dirichlet heat semigroup admits a unique extension to $L_{\varphi_1}^1(\Omega)$, still denoted by $(e^{-tA})_{t \geq 0}$. It is a contraction semigroup on $L_{\varphi_1}^1(\Omega)$, which satisfies*

$$\|e^{-tA}\phi\|_{1,\varphi_1} = e^{-\lambda_1 t}\|\phi\|_{1,\varphi_1}, \quad t \geq 0, \quad \phi \in L_{\varphi_1}^1(\Omega). \tag{49.6}$$

Moreover the maximum principle is satisfied, i.e.:

$$\phi \in L_{\varphi_1}^1(\Omega) \text{ and } \phi \geq 0 \text{ a.e. imply } e^{-tA}\phi \geq 0 \text{ a.e.} \tag{49.7}$$

Furthermore, for each $1 < p < \infty$, $(e^{-tA})_{t \geq 0}$ restricts to a contraction semigroup on $L_{\varphi_1}^p(\Omega)$, which satisfies

$$\|e^{-tA}\phi\|_{p,\varphi_1} \leq e^{-(\lambda_1/p)t}\|\phi\|_{p,\varphi_1}, \quad t \geq 0, \quad \phi \in L_{\varphi_1}^p(\Omega). \tag{49.8}$$

In addition, if Ω is of class C^2, then we have

$$\|e^{-tA}\phi\|_{p,\delta} \leq C(\Omega)\, e^{-(\lambda_1/p)t}\|\phi\|_{p,\delta}, \quad t \geq 0, \quad \phi \in L_\delta^p(\Omega), \quad 1 \leq p < \infty. \tag{49.9}$$

Proof. Let $\phi \in L^2(\Omega)$ with $\phi \geq 0$. Since e^{-tA} is self-adjoint on $L^2(\Omega)$ and $e^{-tA}\varphi_1 = e^{-\lambda_1 t}\varphi_1$, we have, for all $t \geq 0$,

$$\|e^{-tA}\phi\|_{1,\varphi_1} = (e^{-tA}\phi, \varphi_1) = (\phi, e^{-tA}\varphi_1) = e^{-\lambda_1 t}(\phi, \varphi_1) = e^{-\lambda_1 t}\|\phi\|_{1,\varphi_1}. \tag{49.10}$$

Writing $\phi = \phi_+ - \phi_-$ and using the linearity and the positivity preserving property of e^{-tA}, it follows that (49.10) is true for all $\phi \in L^2(\Omega)$.

Now fix $\phi \in L_{\varphi_1}^1(\Omega)$ and pick a sequence $\{\phi_i\}$ in $L^2(\Omega)$, such that $\phi_i \to \phi$ in $L_{\varphi_1}^1(\Omega)$. For each fixed $t > 0$, (49.10) implies

$$\|e^{-tA}\phi_i - e^{-tA}\phi_j\|_{1,\varphi_1} = e^{-\lambda_1 t}\|\phi_i - \phi_j\|_{1,\varphi_1},$$

thus $\{e^{-tA}\phi_i\}$ is a Cauchy sequence in $L_{\varphi_1}^1(\Omega)$. Consequently, we may define $e^{-tA}\phi := \lim_{i \to \infty} e^{-tA}\phi_i$, and it follows from (49.10) that the limit is independent of the choice of the sequence $\{\phi_i\}$, hence the uniqueness assertion. Moreover, (49.6) is satisfied. On the other hand, if $\phi \geq 0$, by choosing $\phi_i = \min(\phi, i) \geq 0$, we obtain (49.7).

Finally let $1 \le p < \infty$ and $\phi \in L^p_{\varphi_1}(\Omega)$. By using Jensen's inequality and $\int_\Omega G(x,y,t)\,dy \le 1$, we have

$$|e^{-tA}\phi|^p \le e^{-tA}(|\phi|^p), \quad \phi \in L^p_{\varphi_1}(\Omega) \tag{49.11}$$

(first assume $\phi \in L^p(\Omega)$ and then argue by density). Therefore, using (49.6), we get

$$\|e^{-tA}\phi\|^p_{p,\varphi_1} = \||e^{-tA}\phi|^p\|_{1,\varphi_1} \le \|e^{-tA}(|\phi|^p)\|_{1,\varphi_1}$$
$$= e^{-\lambda_1 t}\||\phi|^p\|_{1,\varphi_1} = e^{-\lambda_1 t}\|\phi\|^p_{p,\varphi_1},$$

hence (49.8). If Ω is smooth, then (49.9) follows from (49.8) and (1.4). $\square$

The following result provides optimal smoothing estimates for the Dirichlet heat semigroup in the scale of L^p_δ-spaces (see [200] for assertion (i) and [200], [489] for assertion (ii)). Its proof is postponed to Subsection 49.4 below.

Theorem 49.7. *Let $\Omega \subset \mathbb{R}^n$ be a bounded domain of class C^2, let $1 \le p \le q \le \infty$ and set $\beta = \frac{n+1}{2}\left(\frac{1}{p} - \frac{1}{q}\right)$.*
(i) *For all $\phi \in L^p_\delta(\Omega)$, we have*

$$\|e^{-tA}\phi\|_{q,\delta} \le C(p,q,\Omega)\,\|\phi\|_{p,\delta}t^{-\beta}, \quad t > 0. \tag{49.12}$$

(ii) *For all $\varepsilon > 0$, there exist a function $\phi \in L^p_\delta(\Omega)$ and a constant $C > 0$, such that*

$$\|e^{-tA}\phi\|_{q,\delta} \ge Ct^{-\beta+\varepsilon}, \quad \text{for } t > 0 \text{ small.}$$

Remarks 49.8. (a) The elliptic and parabolic estimates in Theorems 49.2 and 49.7 exhibit a remarkable *dimension shift phenomenon:* they are similar to those in standard L^p-spaces in $n+1$ dimensions (cf. Proposition 47.5 and Proposition 48.4).

(b) Assume Ω smooth. Recalling that $L^\infty_\delta = L^\infty$ and interpolating between (49.9) with $p = 1$ and (48.10) with $p = \infty$, we see that there exists $C = C(\Omega) > 0$ such that

$$\|e^{-tA}\phi\|_{p,\delta} \le C\,e^{-\lambda_1 t}\|\phi\|_{p,\delta}, \quad t \ge 0, \quad \phi \in L^p_\delta(\Omega), \quad 1 \le p \le \infty, \tag{49.13}$$

which is an alternative to (49.8).

(c) Assume Ω smooth and let $u_0 \in L^1_\delta(\Omega)$. By a density argument, it is easy to see that $u(t) := e^{-tA}u_0$ satisfies $u(x,t) = \int_\Omega G(x,y,t)u_0(y)\,dy$ in $\Omega \times (0,\infty)$. Moreover u is a weak L^1_δ-solution of (48.12) with $f = 0$ in the sense of Remark 15.4(v). $\square$

49.3. Some pointwise boundary estimates for the heat equation

We here state and prove some pointwise estimates for the heat (and the Laplace) equation, involving the distance to the boundary, which are essential to establish the L^p_δ-properties stated above. Some of them are also used at other places.

Proposition 49.9. *Let Ω be a bounded domain of class C^2. There exists $C = C(\Omega) > 0$ such that, for all $\phi \in L^\infty(\Omega)$,*

$$\left|\left(e^{-tA}\phi\right)(x)\right| \le C\|\phi\|_\infty \frac{\delta(x)}{\sqrt{t}}, \quad x \in \Omega, \ t > 0. \tag{49.14}$$

Proposition 49.9 can be derived as a consequence of Gaussian estimates for the gradient of the heat kernel [477] (or of the reverse of estimate (49.17) below). However, we shall give a maximum principle based, self-contained proof relying on arguments from [346], [347].

Proof. *Step 1.* We consider the auxiliary problem

$$\left.\begin{aligned} V_t - \Delta V &= 1, & x \in \Omega, \ t > 0, \\ V &= 0, & x \in \partial\Omega, \ t > 0, \\ V(x,0) &= 0, & x \in \Omega, \end{aligned}\right\} \tag{49.15}$$

and we claim that, for some $T = T(\Omega) > 0$, there holds

$$V(x,t) \le 2\sqrt{t}\,\delta(x), \quad x \in \Omega, \ 0 < t \le T. \tag{49.16}$$

To show (49.16) we use a barrier argument based on the construction of a suitable supersolution. Fix $x_1 \in \Omega$ and pick $x_2 \in \partial\Omega$ such that $\delta(x_1) = |x_1 - x_2|$. Since the domain Ω is C^2-smooth and bounded, one can find $0 < \rho < R$ independent of x_1, and $a \in \mathbb{R}^n$, such that $\Omega \subset D := \{x \in \mathbb{R}^n : \rho < |x - a| < R\}$ and $\overline{\Omega} \cap \overline{B}(a,\rho) = \{x_2\}$. For $(x,t) \in Q := D \times (0,\infty)$, we consider the function

$$\overline{V}(x,t) = t\,\varphi(y),$$

where

$$y = \frac{|x - a| - \rho}{\sqrt{t}} \quad \text{and} \quad \varphi(y) = \begin{cases} y(2 - y) & \text{if } 0 \le y < 1, \\ 1 & \text{if } y \ge 1. \end{cases}$$

The function $\overline{V}$ is C^1 in t on Q and C^2 in x on $\tilde{Q} := \{(x,t) \in Q : y \ne 1\}$. We compute

$$\overline{V}_t = \varphi(y) - \frac{y}{2}\varphi'(y) \ge \chi_{\{y \ge 1\}} - \frac{1}{2}\chi_{\{0 \le y < 1\}}$$

and, for all $(x,t) \in \tilde{Q}$,

$$-\Delta \overline{V} = -\varphi''(y) - \frac{n-1}{|x-a|}\sqrt{t}\,\varphi'(y) \geq 2\chi_{\{0 \leq y < 1\}} - 2(n-1)\rho^{-1}\sqrt{T}\,\chi_{\{0 \leq y < 1\}}.$$

Taking $T = (\rho/4(n-1))^2$, we obtain $\overline{V}_t - \Delta\overline{V} \geq 1$ in $\tilde{Q}$. On the other hand, it is easy to see that $0 \leq \overline{V} \in C(\overline{D} \times (0,\infty))$, $\overline{V} \in C^1((0,\infty), L^2(D))$ and $\overline{V}(\cdot, t) \in H^2(D)$ for each $t > 0$. Moreover, $\overline{V}(\cdot, t) \to 0$ in $L^\infty(D)$ as $t \to 0$. Consequently, $\overline{V}$ is a (weak) supersolution to (49.15). It follows from the maximum principle that $V \leq \overline{V}$ in $\Omega \times (0,T]$, hence in particular

$$V(x_1, t) \leq t\,\varphi\big((|x_1 - a| - \rho)/\sqrt{t}\big) \leq 2\sqrt{t}\,(|x_1 - a| - \rho) = 2\sqrt{t}\,\delta(x_1), \quad 0 < t \leq T.$$

Step 2. Let $U(t) = e^{-tA}\chi_\Omega$. For each $\tau > 0$, the maximum principle yields $U(0) - U(\tau) = \chi_\Omega - e^{-\tau A}\chi_\Omega \geq 0$, hence $U(t) - U(t+\tau) = e^{-tA}(U(0) - U(\tau)) \geq 0$. Therefore U is nonincreasing in time. By the variation-of-constants formula, it follows that

$$V(t) = \int_0^t U(s)\,ds \geq t\,U(t).$$

This combined with (49.16) yields

$$\big(e^{-tA}\chi_\Omega\big)(x) \leq \frac{2\delta(x)}{\sqrt{t}}, \quad x \in \Omega,\ 0 < t \leq T.$$

By the maximum principle, we deduce that (49.14) is true for $0 < t \leq T$. If $t \geq T$, using (48.10) with $p = \infty$, we obtain

$$\big(e^{-tA}\chi_\Omega\big)(x) \leq \frac{2\delta(x)}{\sqrt{T}}\|e^{-(t-T)A}\chi_\Omega\|_\infty \leq \frac{2\delta(x)}{\sqrt{T}}Ce^{-\lambda_1(t-T)}, \quad x \in \Omega.$$

The proposition follows. $\square$

Proposition 49.10. *Let Ω be an arbitrary domain in $\mathbb{R}^n$. There exist constants $c_1 > 0$ and $c_2 \geq 2$ depending only on n, such that the Dirichlet heat kernel $G(x,y,t)$ in Ω satisfies*

$$G(x,y,t) \geq c_1 t^{-n/2},$$

for all $t > 0$ and $x, y \in \Omega$ such that

$$\delta(x) \geq c_2\sqrt{t} \quad \text{and} \quad |x - y| \leq \sqrt{t}.$$

Proposition 49.10 is a consequence of the sharp estimate [545]

$$G(x,y,t) \geq C_1 \min\!\big(1, \tfrac{\delta(x)\delta(y)}{t}\big)t^{-n/2}e^{-C_2|x-y|^2/t}, \quad \text{for } t > 0 \text{ small}, \qquad (49.17)$$

but the proof of (49.17) is much more delicate. (Estimate (49.17) is in fact proved in [545] for C^2 bounded domains and $n \geq 3$; the reverse inequality, with different constants C_1, C_2, is also true [153].) Here we give an elementary and self-contained proof of Proposition 49.10 based only on the maximum principle.

Proof. Fix $y \in \Omega$, let $\rho = \delta(y)$, $B = B(y, \rho)$, and denote

$$u(x, t) = (4\pi t)^{-n/2} e^{-|x-y|^2/4t}, \quad x \in B, \ t > 0.$$

For $x \in \partial B$, we have $u(x, t) = \rho^{-n} g(t\rho^{-2})$, where $g(s) = (4\pi s)^{-n/2} e^{-1/4s}$. Let $a(n) := \sup_{s>0} g(s)$ (which is finite) and put

$$\underline{u}(x, t) = u(x, t) - M, \qquad M := a(n)\rho^{-n}.$$

Then $\underline{u}$ satisfies

$$\left.\begin{array}{ll} \underline{u}_t - \Delta\underline{u} = 0, & x \in B, \ t > 0, \\ \underline{u} \leq 0, & x \in \partial B, \ t > 0, \end{array}\right\} \tag{49.18}$$

and moreover $\underline{u}(\cdot, t) \to \delta_y - M$ in the sense of measures, as $t \to 0$, where δ_y is the Dirac measure at point y. It follows from the maximum principle that $G(x, y, t) \geq \underline{u}(x, t)$ in $B \times (0, \infty)$. (More precisely, one can easily show that inequality (52.16), with $u(x, t)$ replaced by $G(x, y, t) - u(x, t)$, is satisfied for $f = 0$ and $u_0 = 0$; so the assertion follows from Proposition 52.13(ii).) In particular, if $\delta(x) \geq c_2\sqrt{t}$ and $|x - y| \leq \sqrt{t}$, hence $\rho = \delta(y) \geq (c_2 - 1)\sqrt{t}$, we obtain

$$G(x, y, t) \geq \left((4\pi)^{-n/2} e^{-1/4} - a(n)(c_2 - 1)^{-n}\right) t^{-n/2} \geq c_1(n) t^{-n/2}$$

provided we choose $c_2 = c_2(n) > 1$ large enough. $\quad\square$

Proposition 49.11. *Let Ω be a bounded domain of class C^2.*

(i) The Dirichlet heat kernel $G(x, y, t)$ in Ω satisfies

$$G(x, y, t) \geq c(t, \Omega)\, \delta(x)\delta(y), \quad x, y \in \Omega, \ t > 0$$

where the constant $c(t, \Omega)$ is uniformly positive for t bounded and bounded away from 0.

(ii) There exists a constant $c = c(\Omega) > 0$ such that the Dirichlet Green kernel $K(x, y)$ in Ω satisfies

$$K(x, y) \geq c\, \delta(x)\delta(y), \quad x, y \in \Omega.$$

Remarks 49.12. (i) Proposition 49.11 provides quantitative versions of Hopf's lemma. Namely, let $f \in L^1_\delta(\Omega)$ satisfy $f \geq 0$ a.e. Then the $L^1_\delta(\Omega)$ solution u of the Laplace equation (47.8) satisfies

$$u(x) = \int_\Omega K(x,y)f(y)\,dy \geq c(\Omega)\,\|f\|_{1,\delta}\,\delta(x), \quad x \in \Omega.$$

Likewise, for the heat equation, we have

$$\left(e^{-tA}f\right)(x) = \int_\Omega G(x,y,t)f(y)\,dy \geq c(t,\Omega)\,\|f\|_{1,\delta}\,\delta(x), \quad x \in \Omega,\ t > 0. \tag{49.19}$$

(ii) Estimate (49.19) is sharp in the sense that, for all $f \in L^1_\delta(\Omega)$,

$$\left|(e^{-tA}f)(x)\right| \leq c(\Omega)\,t^{-(n+2)/2}\,\|f\|_{1,\delta}\,\delta(x), \quad x \in \Omega,\ t > 0.$$

This follows by writing $e^{-tA}f = e^{-(t/2)A}\bigl(e^{-(t/2)A}f\bigr)$ and combining (49.14) with (49.12) for $p = 1$, $q = \infty$. $\quad\square$

Again, Proposition 49.11 is a consequence of estimate (49.17). We give a simple proof essentially based on [345] (see also [92]).

Proof of Proposition 49.11. (i) We may assume, without loss of generality, that $B(0,4\rho) \subset \Omega$ for some $\rho > 0$. In what follows, $c(t)$ will denote any constant depending only on t and Ω (or ρ) and such that $c(t)$ is uniformly positive for t bounded and bounded away from 0. For each $y \in \mathbb{R}^n$, let us denote by $(e^{-tA_y})_{t\geq 0}$ the Dirichlet heat semigroup in $B(y,3\rho)$.

Fix $y \in B(0,\rho)$ and $t > 0$. Since $B(y,3\rho) \subset \Omega$, the maximum principle implies

$$e^{-tA}\delta_y \geq e^{-tA_y}\delta_y \quad \text{in } B(y,3\rho), \tag{49.20}$$

where δ_y is the Dirac measure at point y. Also, by the strong maximum principle, we have

$$e^{-tA_0}\delta_0 \geq c(t)\,\chi_{B(0,2\rho)}. \tag{49.21}$$

Since $\left(e^{-tA_y}\delta_y\right)(x) = \left(e^{-tA_0}\delta_0\right)(x-y)$, it follows from (49.20) and (49.21) that

$$e^{-tA}\delta_y \geq c(t)\,\chi_{B(y,2\rho)} \geq c(t)\,\chi_{B(0,\rho)}. \tag{49.22}$$

On the other hand, by Hopf's lemma (see Proposition 52.7), we have

$$e^{-tA}\chi_{B(0,\rho)} \geq c(t)\,\delta. \tag{49.23}$$

Combining (49.22) and (49.23) (with t replaced by $t/2$), we obtain

$$e^{-tA}\delta_y = e^{-(t/2)A}\bigl(e^{-(t/2)A}\delta_y\bigr) \geq c(t)\,e^{-(t/2)A}\chi_{B(0,\rho)} \geq c(t)\,\delta.$$

In other words, we have shown that

$$G(x, y, t) = (e^{-tA}\delta_y)(x) \geq c(t)\,\delta(x)\chi_{B(0,\rho)}(y), \quad x, y \in \Omega, \ t > 0. \qquad (49.24)$$

Using $G(x, y, t) = G(y, x, t) = (e^{-tA}\delta_y)(x) = (e^{-tA}\delta_x)(y)$, and (49.24), (49.23) (with t replaced by $t/2$), we then obtain

$$G(x, y, t) = \big(e^{-(t/2)A}(e^{-(t/2)A}\delta_x)\big)(y)$$
$$\geq c(t)\,\delta(x)\big(e^{-(t/2)A}\chi_{B(0,\rho)}\big)(y) \geq c(t)\,\delta(x)\delta(y),$$

hence assertion (i).

(ii) Since $K(x, y) = \int_0^\infty G(x, y, t)\,dt$, this is an immediate consequence of assertion (i). $\square$

49.4. Proof of Theorems 49.2, 49.3 and 49.7

We begin with the L_δ^p-L_δ^q-estimates. We first treat the parabolic case (Theorem 49.7(i)). The elliptic case (Theorem 49.2) will next be deduced as a consequence.

Proof of Theorem 49.7(i). In this proof, C denotes any positive constant depending only on Ω (not on p, q). Let $\phi \in L^2(\Omega)$ with $\phi \geq 0$. Since e^{-tA} is self-adjoint on $L^2(\Omega)$ we deduce from Proposition 49.9 that, for all $t > 0$,

$$\|e^{-tA}\phi\|_1 = (e^{-tA}\phi, \chi_\Omega) = (\phi, e^{-tA}\chi_\Omega) \leq Ct^{-1/2}\,(\phi, \delta),$$

hence

$$\|e^{-tA}\phi\|_1 \leq Ct^{-1/2}\,\|\phi\|_{1,\delta}, \quad t > 0. \qquad (49.25)$$

Writing $\phi = \phi_+ - \phi_-$ and using the linearity and the positivity preserving property of e^{-tA}, it follows that (49.25) is true for all $\phi \in L^2(\Omega)$. Let now $\phi \in L_\delta^1(\Omega)$ and take $\phi_i \in L^2(\Omega)$ such that $\phi_i \to \phi$ in $L_\delta^1(\Omega)$. We have $e^{-tA}\phi_i \to e^{-tA}\phi$ in $L_\delta^1(\Omega)$ by (49.6), hence a.e. (up to a subsequence). By Fatou's lemma, we infer that (49.25) is true for all $\phi \in L_\delta^1(\Omega)$.

Next using the L^1-L^∞-estimate (see Proposition 48.4), we deduce that

$$\|e^{-tA}\phi\|_\infty = \|e^{-(t/2)A}(e^{-(t/2)A}\phi)\|_\infty$$
$$\leq (2\pi t)^{-n/2}\|e^{-(t/2)A}\phi\|_1 \leq Ct^{-(n+1)/2}\|\phi\|_{1,\delta}. \qquad (49.26)$$

Now take $\phi \in L_\delta^p(\Omega)$. Using (49.11) and applying (49.26) with ϕ replaced by $|\phi|^p$, we get

$$\|e^{-tA}\phi\|_\infty^p = \||e^{-tA}\phi|^p\|_\infty \leq \|e^{-tA}(|\phi|^p)\|_\infty$$
$$\leq Ct^{-(n+1)/2}\||\phi|^p\|_{1,\delta} = Ct^{-(n+1)/2}\|\phi\|_{p,\delta}^p,$$

hence (49.12) for $q = \infty$. For $p \leq q < \infty$, combining this with (49.9) yields

$$\|e^{-tA}\phi\|^q_{q,\delta} \leq \|e^{-tA}\phi\|^{q-p}_\infty \|e^{-tA}\phi\|^p_{p,\delta} \leq C^{(q-p)/p}C^p t^{-(n+1)(q-p)/2p}\|\phi\|^q_{p,\delta}.$$

Raising to the power $1/q$, we obtain (49.12). $\square$

Proof of Theorem 49.2. (i) Let us first assume $f \in \mathcal{D}(\Omega)$. Observe that u is a solution to the inhomogeneous heat equation with initial data u and right-hand side f. Therefore,

$$u = e^{-tA}u + \int_0^t e^{-sA}f\,ds,$$

for all $t > 0$, by the variation-of-constants formula. Next, by (49.12) and (49.13), we have

$$\|e^{-sA}f\|_{q,\delta} \leq Cs^{-\beta}\|e^{-(s/2)A}f\|_{p,\delta} \leq Cs^{-\beta}e^{-\lambda_1 s/2}\|f\|_{p,\delta}.$$

Consequently,

$$\|u\|_{q,\delta} \leq Ct^{-\beta}e^{-\lambda_1 t/2}\|u\|_{p,\delta} + C\left(\int_0^t s^{-\beta}e^{-\lambda_1 s/2}\,ds\right)\|f\|_{p,\delta},$$

where the integral over $(0,t)$ is convergent due to $\beta < 1$. Estimate (49.4) for $f \in \mathcal{D}(\Omega)$ follows upon letting $t \to \infty$. (Note that if $q < \infty$ one can also use (49.9) instead of (49.13).)

Now, in the general case $f \in L^p_\delta(\Omega)$, the conclusion follows by a density argument: Take $f_i \in \mathcal{D}(\Omega)$ such that $f_i \to f$ in $L^p_\delta(\Omega)$ and let u_i be the solution of (47.8) with f replaced by f_i. By (49.1) for $q = 1$ (which we already proved), we have $u_i \to u$ in $L^1(\Omega)$, hence a.e. (up to a subsequence). Passing to the limit in $\|u_i\|_{q,\delta} \leq C\|f_i\|_{p,\delta}$ by Fatou's lemma, the conclusion follows.

(ii) Let $v \geq 0$ be the L^1_δ-solution of (47.8) with f replaced by f_+. We have $u \leq v$ by the maximum principle (cf. Theorem 49.1), hence $u_+ \leq v$. Estimate (49.5) then follows from (49.4). $\square$

Proof of Proposition 47.5. It is completely similar to that of Theorem 49.2, except that we use Propositions 48.5 and 48.4, instead of formulas (49.12) and (49.13). $\square$

Proof of (49.1) in Theorem 49.1. For any $\phi \in L^1_\delta(\Omega)$, using inequality (49.25), the L^1-L^q-estimate and (49.9), we get

$$\|e^{-tA}\phi\|_q \leq Ct^{-\frac{n}{2}(1-\frac{1}{q})}\|e^{-(t/2)A}\phi\|_1 \leq Ct^{-\frac{n+1}{2}+\frac{n}{2q}}\|e^{-(t/4)A}\phi\|_{1,\delta}$$

$$\leq Ce^{-\lambda_1 t/4}t^{-\frac{n+1}{2}+\frac{n}{2q}}\|\phi\|_{1,\delta}.$$

Arguing as in the proof of Theorem 49.2(i), we then obtain (49.1). $\square$

We now proceed to prove the optimality results, namely Theorem 49.3 and Theorem 49.7(ii). The proofs are based on the construction of an appropriate right-hand side of the Laplace equation (or initial data of the heat equation), with suitable boundary singularities. It is supported in a conical subdomain of Ω with vertex at a boundary point. The following lemma provides key lower estimates of the corresponding solutions in the same cone. This construction is used also in Sections 11 and 31 to show the existence of unbounded solutions of nonlinear elliptic equations and systems.

Lemma 49.13. *Let $n \geq 2$ and let $\Omega \subset \mathbb{R}^n$ be a bounded domain of class $C^{2+\gamma}$ for some $\gamma \in (0,1)$. Assume that $0 \in \partial\Omega$. Let $\alpha < n - 1$. There exist $R > 0$ and a revolution cone Σ_1 of vertex 0, with $\Sigma := \Sigma_1 \cap B_{2R} \subset \Omega$, such that the function*

$$\phi := |x|^{-(\alpha+2)} \chi_\Sigma \tag{49.27}$$

belongs to $L^1_\delta(\Omega)$ and enjoys the following properties.
(i) Denote $V(t) = e^{-tA}\phi$. Then

$$V(x,t) \geq Ct^{-(\alpha+2)/2} \tag{49.28}$$

for all x, t such that $x \in \Sigma$, $|x| \leq R$ and $\sigma|x| \leq \sqrt{t} \leq 2\sigma|x|$, where $\sigma > 0$ is a constant.
(ii) The L^1_δ-solution $U > 0$ of (47.8) with $f = \phi$ satisfies

$$U \geq C|x|^{-\alpha} \chi_\Sigma. \tag{49.29}$$

Proof. Write $x = (x_1, x')$, $x' = (x_2, \ldots, x_n)$. Since Ω is a C^2-domain, we may assume without loss of generality that Ω contains the (truncated) revolution cone

$$\Sigma_0 := \big\{ x : |x'| \leq 2\theta x_1, \ |x| \leq 3R \big\},$$

for some $\theta, R > 0$. Next define

$$\Sigma_1 := \big\{ x : |x'| \leq \theta x_1 \big\}, \qquad \Sigma := \Sigma_1 \cap B_{2R},$$

and let ϕ be defined by (49.27). The fact that $\phi \in L^1_\delta$ will follow from Lemma 49.14 below.

Let the constant $c_2 \geq 2$ be given by Proposition 49.10. We observe that there exists $\sigma = \sigma(\theta) \in (0, 1/c_2)$ such that

$$\delta(x) \geq \mathrm{dist}(x, \Sigma_0^c) \geq 2c_2\sigma|x|, \qquad \text{for all } x \in \Sigma. \tag{49.30}$$

(Indeed, $\mathrm{dist}(x, \{z : |z'| = 2\theta z_1\}) \geq |x|\sin(\beta - \beta')$, where $\beta = \arctan(2\theta)$, $\beta' = \arctan\theta$, and $\mathrm{dist}(x, \{z : |z| = 3R\}) \geq R \geq |x|/2$.)

Let now x, t satisfy $x \in \Sigma$, $|x| \leq R$ and $\sigma|x| \leq \sqrt{t} \leq 2\sigma|x|$. In particular, we have $t \leq (2\sigma|x|)^2 \leq R^2$ and, by (49.30), $\delta(x) \geq c_2\sqrt{t}$. By Proposition 49.10, it follows that

$$V(x,t) = \int_\Omega G(x,y,t)\phi(y)\,dy \geq c_1 t^{-n/2} \int_{|x-y|\leq\sqrt{t}} |y|^{-(\alpha+2)}\chi_\Sigma(y)\,dy.$$

Observe that, due to $x \in \Sigma$, $|x| \leq R$ and $t \leq R^2$, we have $\Sigma \cap B(x,\sqrt{t}) \supset (x+\Sigma) \cap B(x,\sqrt{t})$, hence $\mathrm{meas}(\Sigma \cap B(x,\sqrt{t})) \geq Ct^{n/2}$. Since $\sigma|x| \leq \sqrt{t} \leq 2\sigma|x|$ (with $0 < \sigma < 1/2$) and $|x-y| \leq \sqrt{t}$ imply $c\sqrt{t} \leq |y| \leq C\sqrt{t}$, we obtain

$$V(x,t) \geq Ct^{-n/2}\,t^{-(\alpha+2)/2}\,\mathrm{meas}(\Sigma \cap B(x,\sqrt{t})) \geq Ct^{-(\alpha+2)/2}. \qquad (49.31)$$

This proves (i).

Let $x \in \Sigma$. If $|x| \leq R$, by (49.28), we have

$$U(x) = \int_0^\infty V(x,t)\,dt \geq \int_{\sigma^2|x|^2}^{4\sigma^2|x|^2} Ct^{-(\alpha+2)/2}\,dt \geq C|x|^{-\alpha}.$$

If $|x| \geq R$, then $\delta(x) \geq 2c_2\sigma R$ due to (49.30). By Remark 49.12(i), it follows that $U(x) \geq C$ with $C > 0$ independent of x. Thus (ii) is proved. $\quad\square$

As for the integrability properties of the functions ϕ, U, V, we have the following simple lemma.

Lemma 49.14. *Let Ω, α, ϕ, U be as in Lemma 49.13.*

(i) *Assume $\alpha > -2$. The function $\phi \in L_\delta^p(\Omega)$ if and only if $p < (n+1)/(\alpha+2)$.*

(ii) *Assume $\alpha > 0$. If $q \geq (n+1)/\alpha$, then $U \notin L_\delta^q(\Omega)$.*

(iii) *For $1 \leq q \leq \infty$, there holds $\|V(t)\|_{q,\delta} \geq Ct^{\frac{n+1}{2q}-\frac{\alpha+2}{2}}$ for $t > 0$ small.*

Proof. (i) We have $\|\phi\|_{p,\delta}^p = C\int_\Sigma |x|^{-(\alpha+2)p}\delta(x)\,dx$. By (49.30) and $\delta(x) \leq |x|$, the last integral has the same nature (finite or infinite) as

$$\int_\Sigma |x|^{1-(\alpha+2)p}\,dx = \int_0^{2R} r^{n-(\alpha+2)p}\,dr \int_{\Sigma'} d\omega,$$

where $\Sigma' = \{x/|x| \in S^{n-1} : x \in \Sigma \setminus \{0\}\}$. Therefore, $\phi \in L_\delta^p$ if and only if $p < (n+1)/(\alpha+2)$.

(ii) In view of (49.29), this follows from assertion (i).

(iii) Due to (49.28) we may assume $q < \infty$. Let $A(t) = \{x \in \Sigma : \sigma|x| \leq \sqrt{t} \leq 2\sigma|x|\}$. For $t < \sigma^2 R^2$, we have $A(t) \subset B_R$. By (49.28) and (49.30), it follows that

$$\int_\Omega V^q(x,t)\delta(x)\,dx \geq Ct^{-\frac{\alpha+2}{2}q} \int_{A(t)} \delta(x)\,dx \geq Ct^{\frac{1}{2}-\frac{\alpha+2}{2}q} \int_{A(t)} dx$$

$$\geq Ct^{\frac{1}{2}-\frac{\alpha+2}{2}q} \int_{\sqrt{t}/2\sigma}^{\sqrt{t}/\sigma} r^{n-1}\,dr \int_{\Sigma'} d\omega = Ct^{\frac{n+1}{2}-\frac{\alpha+2}{2}q}. \qquad \square$$

After these preparations, we can now easily conclude.

Proof of Theorem 49.3. By assumption, one can choose $\alpha \in (0, n-1)$ such that $\frac{n+1}{q} < \alpha < \frac{n+1}{p} - 2$. The result then follows from Lemmas 49.13 and 49.14(i) and (ii). $\square$

Proof of Theorem 49.7(ii). Choose $\alpha > -2$ such that $\frac{n+1}{p} - 2 - 2\varepsilon < \alpha < \frac{n+1}{p} - 2$. Then $\frac{n+1}{2q} - \frac{\alpha+2}{2} < -\beta + \varepsilon$ and the result follows from Lemmas 49.13 and 49.14(i) and (iii). $\square$

49.5. The heat equation in uniformly local Lebesgue spaces

We have the following smoothing property for the linear heat equation in uniformly local spaces.

Proposition 49.15. *Let $1 \le p < \infty$.*
(i) The heat semigroup on $\mathbb{R}^n$, given by $e^{-tA}\phi = G_t * \phi$, is well defined on L^p_{ul} and $e^{-tA}(L^p_{ul}) \subset L^\infty$ for all $t > 0$.
(ii) Let $0 < T < \infty$, $p \le q \le \infty$ and $\phi \in L^p_{ul}$. Then

$$\|e^{-tA}\phi\|_{q,ul} \le C(n,p,q,T) t^{-\frac{n}{2}\left(\frac{1}{p} - \frac{1}{q}\right)} \|\phi\|_{p,ul}, \quad 0 < t \le T.$$

We use the following simple lemma.

Lemma 49.16. *Let $1 \le p < \infty$. The norms $\|\cdot\|_{p,ul}$ and $\|\cdot\|_{p,*}$, where*

$$\|\phi\|_{p,*} := \sup_{a \in \mathbb{R}^n} \left(\int_{\mathbb{R}^n} |\phi(a-y)|^p G_1(y)\, dy \right)^{1/p} = \|G_1 * |\phi|^p\|_\infty^{1/p},$$

are equivalent on L^p_{ul}.

Proof. On the one hand, we have

$$\int_{\mathbb{R}^n} |\phi(y)|^p G_1(a-y)\, dy \ge (4\pi)^{-n/2} e^{-1/4} \int_{|y-a|<1} |\phi(y)|^p\, dy,$$

hence $\|\phi\|_{p,*} \ge c\|\phi\|_{p,ul}$. On the other hand, there holds

$$\int_{\mathbb{R}^n} |\phi(y)|^p G_1(y)\, dy \le \sum_{k \in \mathbb{Z}^n} \exp\left[-\tfrac{1}{4}\left(\tfrac{2|k|}{\sqrt{n}} - 1\right)_+^2\right] \int_{|y-2k/\sqrt{n}|<1} |\phi(y)|^p\, dy$$

$$\le C \sup_{a \in \mathbb{R}^n} \left(\int_{|y-a|<1} |\phi(y)|^p\, dy \right)$$

with $C = \sum_{k \in \mathbb{Z}^n} \exp\left[-\frac{1}{4}\left(\frac{2|k|}{\sqrt{n}} - 1\right)_+^2\right] < \infty$. After a translation, this implies $\|\phi\|_{p,*} \leq C^{1/p}\|\phi\|_{p,ul}$. $\quad\square$

Proof of Proposition 49.15. Let $1 \leq p < \infty$ and $\phi \in L^p_{ul}$. We may assume $\phi \geq 0$ without loss of generality. Moreover, by the semigroup property (48.6), it is sufficient to consider $T = 1$ and $0 < t \leq 1$.

By Jensen's inequality and $\|G_t\|_{L^1} = 1$, it follows that

$$(G_t * \phi)^p \leq G_t * \phi^p. \tag{49.32}$$

On the other hand, we have

$$G_t \leq t^{-n/2}G_1, \quad 0 < t \leq 1. \tag{49.33}$$

Using (49.32) and (49.33), Lemma 49.16 implies in particular that $e^{-tA}\phi$ is well defined as an element of L^∞, hence assertion (i).

From (49.32) and $\|G_t\|_{L^1} = 1$, we deduce

$$\|G_1 * (e^{-tA}\phi)^p\|_\infty = \|G_1 * (G_t * \phi)^p\|_\infty \leq \|G_1 * G_t * \phi^p\|_\infty$$
$$= \|G_t * G_1 * \phi^p\|_\infty \leq \|G_1 * \phi^p\|_\infty$$

hence

$$\|e^{-tA}\phi\|_{p,*} \leq \|\phi\|_{p,*}, \quad t > 0. \tag{49.34}$$

On the other hand, (49.32) and (49.33) imply

$$\|e^{-tA}\phi\|_\infty \leq \|G_t * \phi^p\|_\infty^{1/p} \leq t^{-n/2p}\|G_1 * \phi^p\|_\infty^{1/p} = t^{-n/2p}\|\phi\|_{p,*}. \tag{49.35}$$

Now for $p \leq q < \infty$, it follows from (49.34) and (49.35) that

$$\|e^{-tA}\phi\|_{q,*}^q = \|G_1 * (e^{-tA}\phi)^q\|_\infty \leq \|G_1 * (e^{-tA}\phi)^p\|_\infty \|e^{-tA}\phi\|_\infty^{q-p}$$
$$= \|e^{-tA}\phi\|_{p,*}^p \|e^{-tA}\phi\|_\infty^{q-p} \leq t^{-(n/2)(q/p-1)}\|\phi\|_{p,*}^q.$$

This along with (49.35) and Lemma 49.16 yields assertion (ii). $\quad\square$

50. Appendix D: Poincaré, Hardy-Sobolev, and other useful inequalities

50.1. Basic inequalities

In this subsection we recall some basic inequalities which we frequently use.

Young's inequality. Let $1 < p < \infty$, $\varepsilon > 0$ and let $q = p' = p/(p-1)$. Then

$$xy \le \frac{\varepsilon^p x^p}{p} + \frac{\varepsilon^{-q} y^q}{q}, \qquad x, y > 0.$$

In what follows, Ω is an arbitrary domain in $\mathbb{R}^n$.

Hölder's inequality. Let $1 \le p \le \infty$ and $q = p' = p/(p-1)$. Then

$$\|uv\|_1 \le \|u\|_p \|v\|_q, \qquad u \in L^p(\Omega), \ v \in L^q(\Omega).$$

A useful consequence is the following **interpolation inequality.** Let $1 \le p < r < q \le \infty$. If $u \in L^p \cap L^q(\Omega)$, then $u \in L^r(\Omega)$ and

$$\|u\|_r \le \|u\|_p^\theta \|u\|_q^{1-\theta}, \qquad \text{where } \theta = \left(\frac{1}{r} - \frac{1}{q} \right) \left(\frac{1}{p} - \frac{1}{q} \right)^{-1} \in (0,1).$$

Jensen's inequality. Assume that $F : \mathbb{R} \to [0, \infty)$ is a convex function, and that $w : \Omega \to [0, \infty]$ is measurable and satisfies $\int_\Omega w(x)\,dx = 1$. If u is a measurable function on Ω such that $uw, F(u)w \in L^1(\Omega)$, then

$$F\left(\int_\Omega u(x)w(x)\,dx \right) \le \int_\Omega F(u(x))w(x)\,dx.$$

Sobolev's inequality. Let $1 \le p < n$ and denote $p^* = np/(n-p)$. Then

$$\|u\|_{p^*} \le C(n,p)\|\nabla u\|_p, \qquad u \in W_0^{1,p}(\Omega). \tag{50.1}$$

50.2. The Poincaré inequality

Let Ω be an arbitrary domain in $\mathbb{R}^n$ and let $1 \le q < \infty$. The Poincaré inequality in $W_0^{1,q}(\Omega)$ is the statement that

$$\|v\|_q \le C_q(\Omega)\|\nabla v\|_q, \quad \text{for all } v \in W_0^{1,q}(\Omega). \tag{50.2}$$

It is well known (see e.g. [90]) that (50.2) holds in any bounded domain, or more generally in any domain which is bounded in one direction. However, since this is a basic inequality in the study of elliptic and parabolic problems, it is important to have a characterization of those domains Ω such that (50.2) is true. It turns out that there is a simple geometric necessary condition, which is also almost sufficient. Moreover the equivalence is true for uniformly regular domains.

To this end, let us introduce the notion of **inradius** $\rho(\Omega)$ of a domain Ω:

$$\rho(\Omega) = \sup\{r > 0 : \Omega \text{ contains a ball of radius } r\} = \sup_{x \in \Omega} \operatorname{dist}(x, \partial\Omega).$$

We also define the **strict inradius** $\rho'(\Omega) \ge \rho(\Omega)$, given by:

$$\rho'(\Omega) - \inf\{R > 0 : \exists \varepsilon > 0 \text{ such that for any ball } B \text{ of radius } R,$$
$$B \cap \Omega^c \text{ contains a ball of radius } \varepsilon\}.$$

The relation between Poincaré inequalities and the inradius and strict inradius is given by the following result.

Proposition 50.1. *Let Ω be an arbitrary domain in $\mathbb{R}^n$.*

(i) *If (50.2) holds for some $q \in [1, \infty)$, then $\rho(\Omega) < \infty$.*

(ii) *If $\rho'(\Omega) < \infty$, then (50.2) holds for all $1 \le q < \infty$.*

(iii) *Assume that Ω is uniformly regular or, more generally, that Ω satisfies a uniform exterior cone condition. Then for all $1 \le q < \infty$, (50.2) holds if and only if $\rho(\Omega) < \infty$.*

Examples 50.2. Let us give some simple examples concerning inradius and strict inradius in the case of unbounded domains.

(a) If Ω is contained in a strip, then ρ and ρ' are both finite, while if Ω contains an infinite cone, then they are both infinite.

(b) If Ω is the complement of a periodic net of points, $\Omega = \mathbb{R}^n \setminus \bigcup_{z \in \mathbb{Z}^n} \{Rz\}$, for some $R > 0$, then $\rho(\Omega) = n^{1/2}R/2$, $\rho'(\Omega) = \infty$.

(c) If Ω is the complement of a periodic net of balls of constant radius, $\Omega = \mathbb{R}^n \setminus \bigcup_{z \in \mathbb{Z}^n} \overline{B}(Rz, \varepsilon)$, for some $0 < \varepsilon < R/2$, then $\rho(\Omega) = \rho'(\Omega) = n^{1/2}R/2 - \varepsilon$. $\quad\square$

Part (i) of Proposition 50.1 is easy. The idea of the proof of part (ii) is due to [4, Lemma 7.4 p. 75], where this is done for $q = 2$ (see [481] for the general case).

On the other hand, Proposition 50.1 (for all q) can be proved as a consequence of more general and difficult results, where the domain need not be uniformly regular (see [336, Corollary 2]). See also [154, Section 1.5] and the references therein for related results.

Proof of Proposition 50.1. (i) Assume $\rho(\Omega) = \infty$. This means that Ω contains some ball $B_j = B(x_j, j)$ for all $j \geq 1$. Fixing a test-function $w \in \mathcal{D}(\mathbb{R}^n)$, $w \geq 0$, $w \not\equiv 0$ with $\mathrm{supp}(w) \subset B(0,1)$, and setting $w_j(x) = w((x - x_j)/j)$, we get that $w_j \in \mathcal{D}(\Omega)$ and that

$$\|w_j\|_q = j^{n/q}\|w\|_q \quad \text{and} \quad \|\nabla w_j\|_q = j^{(n/q)-1}\|\nabla w\|_q.$$

Consequently, (50.2) is false.

(ii) By density, it obviously suffices to consider the case $v \in \mathcal{D}(\Omega)$.

Applying the definition of $\rho' := \rho'(\Omega)$, we may choose $\varepsilon \in (0,1)$ such that for any ball B of radius $\rho' + 1$, $B \cap \Omega^c$ contains a ball of radius ε. Let Q be any cube of edge $2(\rho' + 1)$, such that $Q \cap \Omega \neq \emptyset$. By translation we may assume that Q is centered at the origin. By the definition of ρ', there exists a point a such that $B(a, \varepsilon) \subset B(0, \rho' + 1) \cap \Omega^c$. In particular, $B(a, \varepsilon) \cap \Omega = \emptyset$ and $d(0, a) < \rho' + 1$ (hence $a \in Q$).

Using polar coordinates about a, denoted by (r, ω), we may represent the cube Q by the set $\tilde{Q} = \{(r, \omega) : \omega \in S^{n-1},\ 0 \leq r < R(\omega)\}$, where $R(\omega)$ is some (continuous nonnegative) function. Using $\mathrm{supp}(v) \subset \Omega$ and $B(a, \varepsilon) \cap \Omega = \emptyset$, we get

$$\int_Q |v(x)|^q\, dx = \int_{S^{n-1}} \int_\varepsilon^{R(\omega)} |v(r, \omega)|^q r^{n-1}\, dr\, d\omega.$$

Now, for all $x \in Q$, there holds

$$d(a, x) \leq d(a, 0) + d(0, x) \leq R := (1 + n^{1/2})(\rho' + 1),$$

hence $R(\omega) \leq R$. Using Hölder's inequality, we have, for all $(r, \omega) \in \tilde{Q}$,

$$|v(r, \omega)|^q = \left| \int_\varepsilon^r v_r(\sigma, \omega)\, d\sigma \right|^q \leq R^{q-1} \int_\varepsilon^{R(\omega)} |v_r(\sigma, \omega)|^q\, d\sigma$$

$$\leq \varepsilon^{1-n} R^{q-1} \int_\varepsilon^{R(\omega)} |v_r(\sigma, \omega)|^q \sigma^{n-1}\, d\sigma.$$

It follows that

$$\int_{S^{n-1}} \int_\varepsilon^{R(\omega)} |v(r, \omega)|^q r^{n-1}\, dr\, d\omega \leq \frac{R^{n+q-1}}{n\varepsilon^{n-1}} \int_{S^{n-1}} \int_\varepsilon^{R(\omega)} |v_r(r, \omega)|^q r^{n-1}\, dr\, d\omega,$$

hence

$$\int_Q |v(x)|^q \, dx \leq \frac{R^{n+q-1}}{n\varepsilon^{n-1}} \int_Q |\nabla v(x)|^q \, dx.$$

Dividing $\mathbb{R}^n$ into a periodic net of cubes of edge $2(\rho' + 1)$, and summing this inequality over all cubes yields the same inequality with $\mathbb{R}^n$ instead of Q, that is (50.2), with

$$C_q(\Omega) = (1 + n^{1/2})^{1+(n-1)/q} n^{-1/q} (\rho'(\Omega) + 1) \left(\frac{2 + \rho'(\Omega)}{\varepsilon} \right)^{(n-1)/q}.$$

(iii) This follows immediately from (i) and (ii). $\square$

50.3. Hardy and Hardy-Sobolev inequalities

The following lemma is a simple version of the Hardy inequality.

Lemma 50.3. *Let $\Omega \subset \mathbb{R}^n$ be a bounded domain of class C^1. Then there exists a positive constant $C = C(\Omega)$ such that $\|u/\delta\|_2 \leq C\|\nabla u\|_2$ for all $u \in W_0^{1,2}(\Omega)$.*

Proof. First consider the case $n = 1$, $\Omega = (0,1)$ and assume $u(x) = 0$ for $x \in (0, \varepsilon]$. Then integration by parts and the Cauchy inequality imply

$$\int_0^1 \frac{u^2}{x^2} \, dx = -\frac{1}{x} u^2(x) \Big|_\varepsilon^1 + 2 \int_0^1 \frac{1}{x} u u' \, dx \leq 2 \left(\int_0^1 \frac{u^2}{x^2} \, dx \right)^{1/2} \left(\int_0^1 (u')^2 \, dx \right)^{1/2},$$

hence

$$\left\| \frac{u}{x} \right\|_2 \leq 2\|u'\|_2. \tag{50.3}$$

If, in general, $u \in W_0^{1,2}(0,1)$, then there exist $u_k \in \mathcal{D}(0,1)$ such that $u_k \to u$ a.e. and in $W^{1,2}(0,1)$. Fatou's lemma and (50.3) imply

$$\int_0^1 \frac{u^2}{x^2} \, dx \leq \liminf_{k\to\infty} \int_0^1 \frac{u_k^2}{x^2} \, dx \leq \liminf_{k\to\infty} 4\|u_k'\|_2^2 = 4\|u'\|_2^2.$$

This inequality and the symmetric estimate $\|u/(1-x)\|_2 \leq 2\|u'\|_2$ imply the assertion in the case $\Omega = (0,1)$.

Let $n > 1$, $\Omega = (0,1)^n$ and $u \in \mathcal{D}(\Omega)$. Writing $x = (x_1, x')$, $x' = (x_2, x_3, \ldots, x_n)$, integrating the inequality

$$\int_0^1 \frac{u^2(x_1, x')}{x_1^2} \, dx_1 \leq 4 \int_0^1 \left(\frac{\partial u}{\partial x_1} (x_1, x') \right)^2 dx_1$$

over $x' \in (0,1)^{n-1}$ and using Fubini's theorem we obtain

$$\int_\Omega \frac{u^2}{x_1^2} \, dx \le 4 \int_\Omega \left(\frac{\partial u}{\partial x_1}\right)^2 dx \le 4 \int_\Omega |\nabla u|^2 \, dx.$$

Similarly as above, this implies the assertion in the case $\Omega = (0,1)^n$.

If $\Omega \subset \mathbb{R}^n$ is a C^1 bounded domain, then one can use standard localization arguments (partition of unity and flattening the boundary $\partial\Omega$) in order to prove the assertion. $\square$

A combination of Lemma 50.3 and the Sobolev inequality (50.1) with $p = 2$ yields the following Hardy-Sobolev inequality.

Lemma 50.4. *Let $\Omega \subset \mathbb{R}^n$ be a bounded domain of class C^1, $n \ge 3$, $\tau \in [0,1]$, with $1/q = 1/2^* + \tau/n$. Then there exists a positive constant $C = C(\Omega,\tau)$ such that $\|u/\delta^\tau\|_q \le C\|\nabla u\|_2$ for all $u \in W_0^{1,2}(\Omega)$.*

Proof. Due to Lemma 50.3 and the Sobolev inequality we may assume $\tau \in (0,1)$. Setting $m := 2/\tau$ and $s := 2^*/(1-\tau)$ we have $1/q = 1/m + 1/s$ and Hölder's inequality implies

$$\left\|\frac{u}{\delta^\tau}\right\|_q \le \left\|\frac{u^\tau}{\delta^\tau}\right\|_m \|u^{1-\tau}\|_s = \left\|\frac{u}{\delta}\right\|_{m\tau}^\tau \|u\|_{s(1-\tau)}^{1-\tau} \le C\|\nabla u\|_2^\tau \|\nabla u\|_2^{1-\tau} = C\|\nabla u\|_2,$$

where we used Lemma 50.3 and the Sobolev inequality again. $\square$

Remark 50.5. One can easily see that if $n = 2$ or $n = 1$, then the assertion of Lemma 50.4 remains true for any $q \ge 1$ satisfying $1/q > \tau/2$ or $1/q > \tau - 1/2$, respectively. $\square$

51. Appendix E: Local existence, regularity and stability for semilinear parabolic problems

51.1. Analytic semigroups and interpolation spaces

In this subsection we recall some basic facts on strongly continuous analytic semigroups and interpolation spaces. We refer to [77], [273], [410], [13], [14], [16], [344] and [513] for details.

Let X_0 be a Banach space endowed with the norm $|\cdot|_0$ and let A be a closed linear densely-defined operator in X_0. Denote by $D(A)$ the domain of definition of A endowed with the graph norm $\|x\|_A := |x|_0 + |Ax|_0$ and let X_1 be a Banach space endowed with the norm $|\cdot|_1$ and satisfying $X_1 \doteq D(A)$. Then $-A$ generates a C^0 analytic semigroup e^{-tA} in X_0 if and only if there exist $C > 0$ and $\omega \in \mathbb{R}$ such that $\omega + A : X_1 \to X_0$ is an isomorphism and

$$|\lambda||x|_0 + |x|_1 \leq C|(\lambda + A)x|_0 \qquad \text{for all } x \in X_1, \operatorname{Re}\lambda \geq \omega. \tag{51.1}$$

Set

$$\omega(-A) := \sup\{\operatorname{Re}\lambda : \lambda \in \sigma(-A)\},$$

where $\sigma(-A)$ denotes the spectrum of $-A$. If $-A$ generates a C^0 analytic semigroup in X_0 and $\omega > \omega(-A)$, then there exists $C > 0$ such that (51.1) is true.

Unless explicitly stated otherwise, throughout the rest of Appendix E we shall assume that

$$\left.\begin{array}{l} X_0 \text{ is a reflexive Banach space,} \\[4pt] -A \text{ generates a } C^0 \text{ analytic semigroup in } X_0, \\[4pt] \omega > \omega(-A). \end{array}\right\} \tag{51.2}$$

We will also consider the scale of spaces X_α and operators A_α, $\alpha \in [-1,1]$, defined as follows.

Let X_{-1} be the completion of X_0 endowed with the norm $|x|_{-1} := |(\omega+A)^{-1}x|_0$. Given $\theta \in (0,1)$, set $X_\theta := (X_0, X_1)_\theta$ and $X_{-1+\theta} := (X_{-1}, X_0)_\theta$, where $(\cdot,\cdot)_\theta$ is either the complex interpolation functor $[\cdot,\cdot]_\theta$ or any of the real interpolation functors $(\cdot,\cdot)_{\theta,p}$, $1 < p < \infty$. Given $\theta \in [0,1]$, let A_θ be the X_θ-realization of A (i.e. $A_\theta x = Ax$ for $x \in D(A_\theta) := \{x \in X_\theta : Ax \in X_\theta\}$) and let $A_{-1+\theta}$ be the closure of A in $X_{-1+\theta}$ (A is closable in $X_{-1+\theta}$). The following theorem is a consequence of [14, Theorems 8.1, 8.3 and Corollary 8.2], [16, Theorems II.1.2.2, III.2.5.6, III.3.4.1, III.4.10.7 and Chapter V] and the proof of [344, Proposition 4.2.1].

Theorem 51.1. *Let* $-1 \leq \beta \leq \alpha \leq 1$. *Then the following assertions are true.*

(i) *The space X_α is densely embedded in X_β; the embedding $X_\alpha \hookrightarrow X_\beta$ is compact provided A has compact resolvent and $\alpha > \beta$.*

(ii) *We have*

$$(X_\beta, X_\alpha)_{\eta_+} \hookrightarrow X_{(1-\eta)\beta+\eta\alpha} \hookrightarrow (X_\beta, X_\alpha)_{\eta_-}$$

for any $0 < \eta_- < \eta < \eta_+ < 1$ and the embeddings are dense (almost reiteration property).

(iii) *A_α is the X_α-realization of A_β and $\sigma(A_\alpha) = \sigma(A_\beta)$.*

(iv) *$-A_\alpha$ generates a C^0 analytic semigroup e^{-tA_α} in X_α. In addition,*

$$e^{-tA_\alpha} = e^{-tA_\beta}\big|_{X_\alpha},$$

and there exists $C = C(\omega, A) > 0$ such that

$$\|e^{-tA_\beta}\|_{\mathcal{L}(X_\beta, X_\alpha)} \leq Ct^{\beta-\alpha}e^{\omega t} \qquad \text{for all } t > 0. \tag{51.3}$$

468 Appendices

(v) *Let $u_0 \in X_0$, $\eta, \varepsilon > 0$, $\eta + \varepsilon < 1$, $f \in C^\varepsilon([0,T], X_\eta) + C([0,T], X_{\eta+\varepsilon})$. Then there exists a unique $u \in C([0,T], X_0) \cap C^1((0,T], X_0) \cap C((0,T], X_1)$ which solves the linear Cauchy problem*

$$\dot{u} + Au = f \quad in \ (0,T], \qquad u(0) = u_0. \tag{51.4}$$

In addition, u satisfies the variation-of-constants formula

$$u(t) = e^{-tA} u_0 + \int_0^t e^{-(t-s)A} f(s)\, ds.$$

If $u_0 \in X_\eta$, then $u \in C([0,T], X_\eta)$. If $u_0 \in X_1$, then $u \in C^1([0,T], X_0)$. If $\rho \in (0,1)$, $\theta \in [0,1]$ and $f \in C^\rho((0,T], X_\theta)$, then $u \in C^{1+\rho}((0,T], X_\theta)$.

(vi) *Let X_0 be a UMD space, $\omega(-A) < 0$,*

$$\|A^{it}\| \le M e^{\theta|t|} \quad for \ some \ M > 0, \ \theta \in [0, \pi/2) \ and \ all \ t \in \mathbb{R}, \tag{51.5}$$

$u_0 \in \mathbb{X}_0 := (X_0, X_1)_{1-1/p,p}$ and $f \in \mathbb{X}_f := L^p([0,T], X_0)$, where $1 < p < \infty$. Then the Cauchy problem (51.4) possesses a unique solution $u \in \mathbb{X} := L^p([0,T], X_1) \cap W^{1,p}([0,T], X_0)$ and

$$\|u\|_{\mathbb{X}} \le C(\|u_0\|_{\mathbb{X}_0} + \|f\|_{\mathbb{X}_f}),$$

where C does not depend on u_0, f and T.

(vii) *Let $\alpha \ge 0$, $\alpha - 1 < \gamma < \alpha$, $f \in L^\infty((0,T), X_{\alpha-1})$, and*

$$v(t) := \int_0^t e^{-(t-s)A} f(s)\, ds.$$

Then $v \in C^{\alpha-\gamma}([0,T], X_\gamma)$.

The definition and properties of UMD spaces can be found in [16, Sections III.4.4–III.4.5]. For example, the Lebesgue spaces $L^q(\Omega)$, $1 < q < \infty$, and Hilbert spaces are UMD spaces. For sufficient conditions for the boundedness of imaginary powers of A see Remark 51.5 and also [18], [161], [166], [169], [430], [476].

If $\alpha \in [-1,1]$ and no confusion seems likely, then we will shortly write A and e^{-tA} instead of A_α and e^{-tA_α}, respectively. We will also denote by $|\cdot|_\alpha$ the norm in X_α.

Remarks 51.2. (i) In [273] the author uses the fractional power spaces X^α, $\alpha \ge 0$, instead of the interpolation spaces X_α. However, if the operator A has bounded imaginary powers (that is if the estimate in (51.5) is true for some $M > 0$, $\theta \ge 0$ and all $t \in \mathbb{R}$), then the fractional power spaces are equivalent to the interpolation spaces obtained by using the complex interpolation functor $[\cdot, \cdot]_\theta$, see [16, Theorem V.1.5.4]. In the general case, we still have $X^\alpha \hookrightarrow X_\beta$ and $X_\alpha \hookrightarrow X^\beta$ whenever $1 \ge \alpha > \beta \ge 0$.

(ii) The advantage of interpolation and extrapolation spaces becomes evident in Subsection 51.5 where we deal with singular initial data. Extrapolation spaces also naturally appear if one uses semigroup approach to problems with nonlinear boundary conditions (see [14], [15], [17], [444]). $\square$

We will also need the following interpolation estimate (see [440, Proposition 2.1] and the references therein for a more general statement). We say that (E_0, E_1) is an interpolation couple of Banach spaces if E_0, E_1 are Banach spaces and there exists a locally convex space E such that $E_0, E_1 \hookrightarrow E$.

Proposition 51.3. *Let (E_0, E_1) be an interpolation couple of Banach spaces. Let $1 \le p_0, p_1 < \infty$, $\theta \in (0,1)$, $1/p_\theta = (1-\theta)/p_0 + \theta/p_1$, $s := 1-\theta$, $E_\theta := (E_0, E_1)_{\theta, p_\theta}$. Then*

$$W^{1,p_0}([0,T], E_0) \cap L^{p_1}([0,T], E_1) \hookrightarrow W^{s,p_\theta}([0,T], E_\theta)$$

and the norm of this embedding can be estimated by a constant $C(T_0)$ for all $T \in (0, T_0]$.

If E_1 is compactly embedded in E_0 and $s < 1 - \theta$, then the above embedding is compact.

If $p > 1$, $r \ge 1$ and $\Omega \subset \mathbb{R}^n$ is open, then Proposition 51.3 implies

$$W^{1,2}([0,T], L^2(\Omega)) \cap L^{(p+1)r}([0,T], L^{p+1}(\Omega)) \hookrightarrow L^\infty([0,T], L^q(\Omega)) \qquad (51.6)$$

for any $q \in [2, p+1-(p-1)/(r+1))$ (see [440] for details and see [114] for a direct proof).

Examples 51.4. (See [13] and [14].)

(i) Let $\Omega \subset \mathbb{R}^n$ be uniformly regular of class C^2, $1 < q < \infty$, $X_0 = L^q(\Omega)$, $X_1 = W^{2,q} \cap W_0^{1,q}(\Omega)$ (this choice of X_1 corresponds to Dirichlet boundary conditions). Let A be the unbounded linear operator in X_0 with domain of definition X_1 defined by

$$Au = -\sum_{i,j=1}^n a_{ij} \frac{\partial^2}{\partial x_i \partial x_j} u + \sum_{i=1}^n b_i \frac{\partial}{\partial x_i} u + cu,$$

where $a_{ij}, b_i, c \in BUC(\Omega)$ and $a_{ij} = a_{ji}$ are uniformly elliptic. Then $-A$ generates a C^0 analytic semigroup in X_0. Let $(\cdot, \cdot)_\theta$ be the complex interpolation functor if $\theta = 1/2$ and the real interpolation functor $(\cdot, \cdot)_{\theta, q}$ otherwise. Then

$$X_\theta = X_\theta(q) \doteq \begin{cases} \{u \in W^{2\theta, q}(\Omega) : u = 0 \text{ on } \partial\Omega\} & \text{if } 2\theta > 1/q, \\ W^{2\theta, q}(\Omega) & \text{if } 1/q > 2\theta \ge 0, \end{cases}$$

$X_{1/2q}(q) \hookrightarrow W^{1/q, q}(\Omega)$, and

$$X_\theta(q) \doteq \left(X_{-\theta}(q')\right)' \qquad \text{if } \theta < 0. \qquad (51.7)$$

(ii) If we set $X_1 = \{u \in W^{2,q}(\Omega) : \partial u/\partial n = 0 \text{ on } \partial\Omega\}$ (Neumann boundary conditions), then the assertions in (i) remain true with

$$X_\theta = X_\theta(q) \doteq \begin{cases} \{u \in W^{2\theta,q}(\Omega) : \partial u/\partial n = 0 \text{ on } \partial\Omega\} & \text{if } 2\theta > 1 + 1/q, \\ W^{2\theta,q}(\Omega) & \text{if } 1 + 1/q > 2\theta \geq 0, \end{cases}$$

$X_{1/2+1/2q}(q) \hookrightarrow W^{1+1/q,q}(\Omega)$, and (51.7). $\square$

Remark 51.5. Assume that Ω, A and X_α, $\alpha \in [-1, 1]$, are as in Examples 51.4. Then A satisfies (51.5) (see [17] and cf. also [161] and the references therein). If u solves (51.4), $1 < p < \infty$ and $\eta \in (1 - 1/p, 1]$, then Theorem 51.1(vi) guarantees the maximal regularity property

$$\|u\|_{W^{1,p}([0,T],L^q(\Omega))} + \|u\|_{L^p([0,T],W^{2,q}(\Omega))} \leq C\big(|u_0|_\eta + \|f\|_{L^p([0,T],L^q(\Omega))}\big), \quad (51.8)$$

where $C > 0$ does not depend on f, u_0 and T. $\square$

In what follows we will also need the following singular Gronwall inequality (see [16, Theorem 3.3.1]).

Proposition 51.6. *Let $\alpha, \beta \in [0,1)$ and $\varepsilon > 0$. Then there exists a positive constant $c := c(\alpha, \beta, \varepsilon)$ such that the following is true:*
If $A, B > 0$ and $u : [0, T) \to \mathbb{R}_+$ satisfies $[t \mapsto t^\beta u(t)] \in L^\infty_{loc}([0, T))$ and

$$u(t) \leq At^{-\beta} + B \int_0^t (t - \tau)^{-\alpha} u(\tau)\, d\tau, \qquad \text{for a.a. } t \in (0, T),$$

then
$$u(t) \leq At^{-\beta}\big(1 + cBt^{1-\alpha}e^{(1+\varepsilon)\mu t}\big) \qquad \text{for a.a. } t \in (0, T),$$

where $\mu := (\Gamma(1 - \alpha)B)^{1/(1-\alpha)}$.

51.2. Local existence and regularity for regular data

Recall that we assume (51.2) and that X_α, $\alpha \in [-1, 1]$ denote the corresponding interpolation-extrapolation scale of spaces. The proofs of the following theorem and Theorems 51.17, 51.19, 51.21, 51.25, 51.33 below are based on well-known and frequently used ideas (see [273], [344], for example).

Theorem 51.7. *Fix $1 \geq \alpha > \beta \geq 0$ and assume that $F : X_\beta \to X_{\alpha-1}$ is locally Lipschitz continuous, uniformly on bounded subsets of X_β. Let $u_0 \in X_\beta$. Then there exists $T = T(|u_0|_\beta) > 0$ such that the integral equation*

$$u(t) = e^{-tA}u_0 + \int_0^t e^{-(t-s)A}F(u(s))\, ds \qquad (51.9)$$

has a unique solution $u \in C([0,T], X_\beta)$. In addition, there exists $C = C(A) > 0$ such that $|u(t)|_\beta \leq C|u_0|_\beta + 1$ for all $t \in [0,T]$.

If $\gamma \in [\beta, \alpha)$, $\gamma > \alpha - 1$, then $u \in C^{\alpha-\gamma}((0,T], X_\gamma)$. Moreover we have the following continuous dependence property: if $\gamma \in [\beta, \alpha)$ and u and $\tilde{u}$ are two solutions with initial data u_0 and $\tilde{u}_0$, respectively, then there exist $T = T(|u_0|_\beta, |\tilde{u}_0|_\beta) > 0$ and $C > 0$ independent of the initial data such that

$$|u(t) - \tilde{u}(t)|_\gamma \leq Ct^{\beta-\gamma}|u_0 - \tilde{u}_0|_\beta \qquad \text{for all } t \in (0,T]. \tag{51.10}$$

Finally, the solution can be continued on the maximal existence interval $[0, T_{\max})$, where either $T_{\max} = \infty$ or $\lim_{t \to T_{\max}} |u(t)|_\beta = \infty$.

Proof. Due to (51.3), there exists $C_A > 0$ such that

$$\|e^{-tA}\|_{\mathcal{L}(X_{\alpha_1}, X_{\alpha_2})} \leq C_A t^{\alpha_1 - \alpha_2} \qquad \text{for all } t \in (0,1] \text{ and } -1 \leq \alpha_1 \leq \alpha_2 \leq 1. \tag{51.11}$$

Let $M > 2C_A|u_0|_\beta$. The assumptions on F guarantee the existence of $C_F = C_F(M) > 0$ such that

$$|F(u)|_{\alpha-1} \leq C_F \quad \text{and} \quad |F(u) - F(v)|_{\alpha-1} \leq C_F|u - v|_\beta \tag{51.12}$$

for all $u, v \in X_\beta$ satisfying $|u|_\beta, |v|_\beta \leq M$. Assume $T \in (0,1]$ and let $B_M = B_{M,T}$ denote the closed ball in the Banach space $Y_T := C([0,T], X_\beta)$ with center 0 and radius M. We will use the Banach fixed point theorem for the mapping $\Phi_{u_0} : B_M \to B_M$, where

$$\Phi_{u_0}(u)(t) := e^{-tA}u_0 + \int_0^t e^{-(t-s)A} F(u(s)) \, ds. \tag{51.13}$$

Let $u \in B_M$. Then

$$|\Phi_{u_0}(u)(t)|_\beta \leq \|e^{-tA}\|_{\mathcal{L}(X_\beta, X_\beta)}|u_0|_\beta + \int_0^t \|e^{-(t-s)A}\|_{\mathcal{L}(X_{\alpha-1}, X_\beta)}|F(u(s))|_{\alpha-1} \, ds$$

$$\leq C_A|u_0|_\beta + C_A C_F \int_0^t (t-s)^{\alpha-1-\beta} \, ds$$

$$\leq \frac{1}{2}M + \frac{C_A C_F}{\alpha - \beta} T^{\alpha-\beta} \leq M,$$

provided $T \leq \tau_0$, where $\tau_0 = \tau_0(M) > 0$ is small enough. Hence Φ_{u_0} maps B_M into B_M for $T \leq \tau_0$. Given $u, v \in B_M$, we have

$$|(\Phi_{u_0}(u) - \Phi_{u_0}(v))(t)|_\beta \leq \int_0^t \|e^{-(t-s)A}\|_{\mathcal{L}(X_{\alpha-1}, X_\beta)}|F(u(s)) - F(v(s))|_{\alpha-1} \, ds$$

$$\leq C_A C_F \int_0^t (t-s)^{\alpha-1-\beta}|u(s) - v(s)|_\beta \, ds$$

$$\leq C_A C_F \frac{T^{\alpha-\beta}}{\alpha - \beta}\|u - v\|_{Y_T} \leq \frac{1}{2}\|u - v\|_{Y_T},$$

provided $T \le \tau_1$, where $\tau_1 = \tau_1(M) > 0$ is small enough. Consequently, Φ_{u_0} is a contraction in $B_{M,T}$ for $T \le \tau_2 := \min(\tau_0, \tau_1)$ and possesses a unique fixed point u in $B_{M,T}$. It is easily seen that this solution of (51.9) is unique in Y_T. Notice also that $\tau_2 = \tau_2(|u_0|_\beta)$ if we fix $M = 2C_A|u_0|_\beta + 1$, for example.

Let $\gamma \in [\beta, \alpha)$. If $\gamma > \alpha - 1$, then $u \in C^{\alpha-\gamma}((0, T], X_\gamma)$ due to Theorem 51.1(vii). Next assume $u_0, \tilde{u}_0 \in X_\beta$ and fix $M > 2C_A \max(|u_0|_\beta, |\tilde{u}_0|_\beta)$. Set

$$u^0(t) := e^{-tA} u_0, \qquad \tilde{u}^0(t) := e^{-tA} \tilde{u}_0,$$
$$u^{k+1} := \Phi_{u_0}(u^k), \qquad \tilde{u}^{k+1} := \Phi_{\tilde{u}_0}(\tilde{u}^k), \qquad k = 0, 1, 2, \ldots.$$

Due to the above existence proof, u^k and $\tilde{u}^k$ converge to the solutions u and $\tilde{u}$ in B_M for T small enough. Now (51.11) implies the following inequality for $k = 0$ and all $t \in [0, T]$:

$$|u^k(t) - \tilde{u}^k(t)|_\beta \le 2C_A|u_0 - \tilde{u}_0|_\beta. \tag{51.14}$$

Assume that (51.14) is true for some $k \ge 0$. Then

$$|u^{k+1}(t) - \tilde{u}^{k+1}(t)|_\beta \le \|e^{-tA}\|_{\mathcal{L}(X_\beta, X_\beta)} |u_0 - \tilde{u}_0|_\beta$$
$$+ \int_0^t \|e^{-(t-s)A}\|_{\mathcal{L}(X_{\alpha-1}, X_\beta)} |F(u^k(s)) - F(\tilde{u}^k(s))|_{\alpha-1}\, ds$$
$$\le C_A|u_0 - \tilde{u}_0|_\beta + C_A C_F \int_0^t (t-s)^{\alpha-1-\beta} |u^k(s) - \tilde{u}^k(s)|_\beta\, ds$$
$$\le \left(C_A + 2C_A^2 C_F \frac{T^{\alpha-\beta}}{\alpha - \beta}\right) |u_0 - \tilde{u}_0|_\beta \le 2C_A|u_0 - \tilde{u}_0|_\beta,$$

provided T is small enough. Consequently, (51.14) is true for all k. Passing to the limit we obtain

$$|u(t) - \tilde{u}(t)|_\beta \le 2C_A|u_0 - \tilde{u}_0|_\beta.$$

Using this estimate we finally obtain

$$|u(t) - \tilde{u}(t)|_\gamma = |\Phi_{u_0}(u)(t) - \Phi_{\tilde{u}_0}(\tilde{u})(t)|_\gamma$$
$$\le \|e^{-tA}\|_{\mathcal{L}(X_\beta, X_\gamma)} |u_0 - \tilde{u}_0|_\beta$$
$$+ \int_0^t \|e^{-(t-s)A}\|_{\mathcal{L}(X_{\alpha-1}, X_\gamma)} |F(u(s)) - F(\tilde{u}(s))|_{\alpha-1}\, ds$$
$$\le C_A t^{\beta-\gamma} |u_0 - \tilde{u}_0|_\beta + C_A C_F \int_0^t (t-s)^{\alpha-1-\gamma} |u(s) - \tilde{u}(s)|_\beta\, ds$$
$$\le \left(C_A + 2C_A^2 C_F \frac{T^{\alpha-\beta}}{\alpha - \gamma}\right) t^{\beta-\gamma} |u_0 - \tilde{u}_0|_\beta \le 2C_A t^{\beta-\gamma} |u_0 - \tilde{u}_0|_\beta,$$

provided T is small enough.

The existence of a maximal solution follows in the same way as in the proof of Proposition 16.1. $\quad\square$

Remarks 51.8. (i) The solution u in Theorem 51.7 satisfies $u \in C^1((0, T_{\max}),$ $X_{\gamma-1})$ and

$$\dot{u} + A_{\gamma-1} u = F(u), \qquad t \in (0, T_{\max}),$$

for all $\gamma \in [\beta, \alpha)$. In fact, set $\tilde{X}_0 := X_{\gamma-1}$, $\tilde{X}_1 := X_\gamma$ and let $\tilde{X}_\eta$, $\eta \in [-1, 1]$, be the corresponding interpolation-extrapolation scale. Then $X_{\alpha-1} \hookrightarrow \tilde{X}_\eta$ for any $\eta \in (0, \alpha - \gamma)$ due to Theorem 51.1(ii), hence $F(u) \in C([0, T_{\max}), \tilde{X}_\eta)$. Now the assertion follows from Theorem 51.1(v).

(ii) It is straightforward to check that all statements in Theorem 51.7 remain true for nonautonomous nonlinearities of the form $F = F(t, u)$ provided $F : [0, \infty) \times X_\beta \to X_{\alpha-1}$ is measurable in t, locally Lipschitz continuous in u (uniformly on bounded subsets of $[0, \infty) \times X_\beta$) and $F(\cdot, 0)$ is bounded in $X_{\alpha-1}$ on bounded subsets of $[0, \infty)$.

Similarly, if we assume that $D \subset X_\beta$ is open, $F : D \to X_{\alpha-1}$ is locally Lipschitz continuous (uniformly on bounded sets $M \subset D$ satisfying $\operatorname{dist}_{X_\beta}(M, \partial D) > 0$) and $u_0 \in D$, then there exists a unique maximal solution $u \in C([0, T_{\max}), D)$ and (at least) one of the following possibilities occurs: (a) $T_{\max} = \infty$; (b) $\lim_{t \to T_{\max}} |u(t)|_\beta = \infty$; (c) $\liminf_{t \to T_{\max}} \operatorname{dist}_{X_\beta}(u(t), \partial D) = 0$.

Finally, if $\infty > r > 1/(\alpha - \beta)$, $F : C([0, T], X_\beta) \to L^r([0, T], X_{\alpha-1})$ is uniformly Lipschitz continuous on bounded sets and has the Volterra property (that is $F(u)|_{[0,t]}$ depends on $u|_{[0,t]}$ only), and $u_0 \in X_{\alpha-1/r}$, then the problem

$$\begin{aligned} u_t + Au &= F(u), \qquad t > 0, \\ u(0) &= u_0, \end{aligned} \tag{51.15}$$

has a unique maximal strong solution $u \in C([0, T_{\max}), X_\beta) \cap W_{loc}^{1,r}([0, T_{\max}), X_{\beta-1})$ due to [21, Theorem 2.3]. Strong solution means that the equation $u_t + Au = F(u)$ is satisfied for a.e. t. Notice also that $F(u) \in L_{loc}^r([0, T_{\max}), X_{\alpha-1})$ is well defined for $u \in C([0, T_{\max}), X_\beta)$ due to the Volterra property of F. Additional regularity and stability results for solutions of (51.15) can be found in [21]. In particular, $u \in C^\rho([0, T_{\max}), X_\beta) \cap W_{loc}^{1,r}([0, T_{\max}), X_{\gamma-1})$ for all $\rho < \alpha - \beta - 1/r$ and $\gamma \in (\beta, \alpha)$ and the solution u is global ($T_{\max} = T$ and $u \in C([0, T], X_\beta)$) whenever $F(u) \in L^r([0, T_{\max}), X_{\alpha-1})$.

(iii) Let α, β, γ, F and u_0 be as in Theorem 51.7, and $T_{\max} = T_{\max}(u_0)$ be the maximal existence time of the solution u of (51.9). Fix $t \in (0, T_{\max})$. Using (51.10) one can easily prove the existence of positive constants C, ε (depending on t and $\max_{0 \le s \le t} |u(s)|_\beta$) such that $T_{\max}(\tilde{u}_0) > t$ and

$$|\tilde{u}(t) - u(t)|_\gamma < C|\tilde{u}_0 - u_0|_\beta$$

for any $\tilde{u}_0 \in X_\beta$ satisfying $|\tilde{u}_0 - u_0|_\beta < \varepsilon$.

(iv) Let X_0 be a (reflexive) ordered Banach space with a total positive cone P_0 and let the semigroup e^{-tA_0} be positive (note that P_0 is total if $P_0 - P_0$ is dense in

X_0). Define positive cones P_θ in X_θ, $\theta \in [-1, 1]$ as follows: $P_\theta = P \cap X_\theta$ if $\theta > 0$, P_θ is the closure of P in X_θ if $\theta < 0$. Then X_θ become ordered Banach spaces and the semigroups e^{-tA_θ} are positive. If, in addition, F maps P_β into $P_{\alpha-1}$ and $u_0 \in P_\beta$, then the corresponding solution u is obviously nonnegative. In fact, $u = \lim u^k$, where $u^0 = e^{-tA} u_0 \geq 0$ and $u^{k+1} = \Phi_{u_0} u^k \geq 0$ whenever $u^k \geq 0$.

In particular, if e^{-tA} is positive, $u_0 \in P_\beta$ and $F : P_\beta \to P_{\alpha-1}$, then F need not be defined for $u \notin P_\beta$ (any regular extension of F to X_β leads to the same positive solution u).

(v) A simple modification of the proof of Theorem 51.7 shows that the assumption $\beta \geq 0$ can be replaced with $\beta \geq -1$. $\quad\square$

Example 51.9. Let Ω, A and X_α, $\alpha \in [-1, 1]$, be as in Examples 51.4 and $q > n$. Let $f \in C^1(\mathbb{R})$ and let F be the Nemytskii mapping associated with f, that is $F(u)(x) = f(u(x))$. Assume also that either $f(0) = 0$ or Ω is bounded. Fix $\beta = 1/2$, $\alpha = 1$ and let $u_0 \in X_\beta$. Recall from Examples 51.4 that $X_\beta = W_0^{1,q}(\Omega)$ or $X_\beta = W^{1,q}(\Omega)$ if we consider Dirichlet or Neumann boundary conditions, respectively. Since $W^{1,q}(\Omega) \hookrightarrow L^\infty \cap L^q(\Omega)$, we see that the assumptions of Theorem 51.7 are satisfied and we obtain a unique maximal solution $u \in C([0, T_{\max}), X_\beta)$. In addition, $u \in C^{1-\gamma}((0, T_{\max}), X_\gamma)$ for $\gamma \in [1/2, 1)$. Choose γ such that $\rho := 1 - \gamma = (1 - n/q)/3$. Then $X_\gamma \hookrightarrow BUC^{1+\rho}(\Omega)$, hence $u \in C^\rho((0, T_{\max}), BUC^{1+\rho}(\Omega))$.

Remark 51.8(i) implies $\dot{u} + A_{\gamma-1} u = F(u)$ in $(0, T_{\max})$. Fix $0 < \delta < T < T_{\max}$, choose $\psi \in C^\infty(\mathbb{R})$ such that $\psi(t) = 0$ for $t \leq \delta/2$, $\psi(t) = 1$ for $t \geq \delta$, and set $v(t) := \psi(t) u(t)$. Then

$$\dot{v} + A_{\gamma-1} v = \tilde{f} \quad \text{in } (0, T_{\max}), \qquad v(0) = 0, \tag{51.16}$$

where $\tilde{f}(t) := \psi(t) F(u(t)) + \psi_t(t) u(t)$. Assume that the coefficients of the operator A belong to $BUC^\rho(\Omega)$ and Ω is a bounded domain of class $C^{2+\rho}$. Since $\tilde{f}$ is also Hölder continuous, Theorem 48.2(ii) shows that there exists a classical solution w of problem (51.16). The uniqueness of solutions of (51.16) guarantees $w = v$, hence u is a classical solution for $t > 0$. Theorem 48.2(ii) also implies

$$u \in BC^{2,1}(\overline{\Omega} \times [t_1, t_2]) \quad \text{whenever } 0 < t_1 < t_2 < T_{\max}. \tag{51.17}$$

If Ω is unbounded, then (51.17) can be shown by using a smooth cut-off function in the x-variable.

This example can be straightforwardly modified for more general nonlinearities and systems (cf. also Remark 51.8(ii)). If $F(t, u)(x) = f(x, t, u(x, t), \nabla u(x, t))$, for example, then one obtains the existence of a maximal solution $u \in C([0, T_{\max}), X_\beta)$ provided Ω is bounded, the function $f = f(x, t, u, \xi)$ is C^1, its derivatives satisfy the growth condition

$$|\partial_t f| + |\partial_u f| + (1 + |\xi|)|\partial_\xi f| \leq C(|u|)(1 + |\xi|^p)$$

and $q > n \max(1, p - 1)$ (see [14] for details). Note that the regularity of f with respect to x and t can be considerably relaxed. $\quad\square$

Example 51.10. Let Ω, A and X_α, $\alpha \in [-1, 1]$, be as in Example 51.4(i), and let $p > 1$ and

$$q > \max\left(1, \frac{n(p-1)}{p+1}, \frac{np}{n+p}\right).$$

Fix $z \in (\max(1, q/p, nq/(n+q)), \min(q, nq/p(n-q)_+)]$ and assume that $F(u)(x) = f(x, u(x))$ where $f \in C^1$ satisfies $f(\cdot, 0) \in L^z(\Omega)$ and $|\partial_u f(x, u)| \le a(x) + C|u|^{p-1}$ with $a \in L^{p'z}(\Omega)$ (the regularity of f with respect to x can be relaxed). Set

$$\beta = \frac{1}{2} \quad \text{and} \quad \alpha = \frac{1}{2}\left(2 + \frac{n}{q} - \frac{n}{z}\right).$$

Then $\alpha \in (\beta, 1]$, $X_\beta = W_0^{1,q}(\Omega) \hookrightarrow L^{pz}(\Omega)$ and $L^z(\Omega) \hookrightarrow X_{\alpha-1}$ (due to $X_{1-\alpha}(q')$ $\hookrightarrow W^{2-2\alpha, q'}(\Omega) \hookrightarrow L^{z'}(\Omega)$ and (51.7)). Since F considered as a map from $L^{pz}(\Omega)$ to $L^z(\Omega)$ is locally Lipschitz continuous (uniformly on bounded sets), it has the same properties as a map $F : X_\beta \to X_{\alpha-1}$. Consequently, given $u_0 \in W_0^{1,q}(\Omega)$, Theorem 51.7 guarantees the existence of a maximal solution $u \in C([0, T_{\max}), W_0^{1,q}(\Omega))$ satisfying $u \in C((0, T_{\max}), X_\gamma)$ for all $\gamma < \alpha$.

Next assume $f = f(u)$ and notice that this restriction and our assumptions on f imply $f(0) = f'(0) = 0$ if Ω is unbounded. In fact, if Ω is unbounded, then its measure has to be infinite (since Ω is uniformly regular of class C^2), hence the spaces $L^z(\Omega)$ and $L^{p'z}(\Omega)$ do not contain nonzero constants. If $q \ge n$ or $p \le n/(n-q)$, then we may set $z = q$, hence $\alpha = 1$, and we obtain

$$u \in C((0, T_{\max}), X_\gamma) \qquad \text{for all } \gamma \in [1/2, 1). \tag{51.18}$$

Assume $q < n$, $p > n/(n-q)$, and consider $t_0 > 0$ small and $\tilde{\beta} \in (\beta, \alpha)$. Since $u(t_0) \in X_{\tilde{\beta}}$ we may repeat the considerations above with

$$\tilde{z} := \min\left(q, \frac{nq}{p(n - 2\tilde{\beta}q)_+}\right), \qquad \tilde{\alpha} := \frac{1}{2}\left(2 + \frac{n}{q} - \frac{n}{\tilde{z}}\right),$$

to obtain $u \in C((0, T_{\max}), X_\gamma)$ for all $\gamma < \tilde{\alpha}$. If $q \ge n/(2\tilde{\beta})$ or $p \le n/(n - 2\tilde{\beta}q)$, then $\tilde{z} = q$, $\tilde{\alpha} = 1$ and we obtain (51.18) again. Otherwise we notice that $\tilde{z} > z$, $\tilde{\alpha} > \alpha$, and use a bootstrap argument (enlarging $\tilde{\beta}$, $\tilde{z}$ and $\tilde{\alpha}$) to see that (51.18) is always true.

Next choose $\gamma \in (\beta, 1)$. Since

$$X_\gamma = W^{2\gamma, q} \cap W_0^{1,q}(\Omega) \hookrightarrow W_0^{1,\tilde{q}}(\Omega) \quad \text{for some } \tilde{q} > q,$$

we can repeat the arguments above with q replaced by $\tilde{q}$. An obvious bootstrap w.r.t. $\tilde{q}$ shows

$$u \in C((0, T_{\max}), W^{2\gamma, \tilde{q}} \cap W_0^{1,\tilde{q}}(\Omega)) \qquad \text{for all } \gamma < 1 \text{ and } \tilde{q} \in [q, \infty).$$

Notice also that the considerations in Example 51.9 guarantee now

$$u \in C^\rho((0, T_{\max}), BUC^{1+\rho} \cap W_0^{1,\tilde{q}}(\Omega)) \tag{51.19}$$

for some $\rho \in (0,1)$ and all $\tilde{q} \in [q, \infty)$.

Fix $t \in (0, T_{\max})$. Then the bootstrap argument used above, Remark 51.8(iii) and the embedding $W^{2\gamma,\tilde{q}}(\Omega) \hookrightarrow BUC^1(\Omega)$ for suitable $\gamma, \tilde{q}$ guarantee the existence of $\varepsilon, C > 0$ (depending on t and $\max_{0 \le s \le t} \|u(s)\|_{W^{1,q}(\Omega)}$) such that given $\tilde{u}_0 \in W_0^{1,q}(\Omega)$ satisfying $\|\tilde{u}_0 - u_0\|_{W^{1,q}(\Omega)} < \varepsilon$, we have $T_{\max}(\tilde{u}_0) > t$ and

$$\|\tilde{u}(t) - u(t)\|_{BC^1} < C\|\tilde{u}_0 - u_0\|_{W^{1,q}(\Omega)}. \tag{51.20}$$

In fact, Remark 51.26(iii) and Example 51.27 below guarantee that the RHS in (51.20) can be replaced with $C\|\tilde{u}_0 - u_0\|_r$ for any $r > n(p-1)/2, r > 1$. In addition, estimate (15.18) shows that the same is true if $r = 1 > n(p-1)/2$.

Next assume that f' is locally Hölder continuous. Then (51.19) implies the existence of $\rho > 0$ such that $F(u) \in C^\rho((0, T_{\max}), X_{1/2})$. Now Theorem 51.1(v) guarantees $u \in C^{1+\rho}((0, T_{\max}), X_{1/2}) \cap C((0, T_{\max}), X_1)$, hence

$$u \in \mathcal{W}_q := C^{1+\rho}((0, T_{\max}), W_0^{1,q}(\Omega)) \cap C((0, T_{\max}), W^{2,q}(\Omega)).$$

Since $u(t) \in W_0^{1,\tilde{q}}(\Omega)$ for all $t > 0$ and $\tilde{q} \ge q$ the arguments above imply $u \in \mathcal{W}_{\tilde{q}}$ for all $\tilde{q} \in [q, \infty)$.

Next assume that f'' is locally Hölder continuous. We will show that $u_t \in \mathcal{W}_{\tilde{q}}$ for all $\tilde{q} \in [q, \infty)$. Fix $0 < \delta < T_{\max}/2$ and choose a cut-off function $\psi \in C^\infty(\mathbb{R})$ such that $\psi(t) = 0$ for $t \le \delta$ and $\psi(t) = 1$ for $t \ge 2\delta$. Notice that the function $u_t\psi$ formally solves the linear problem

$$\begin{aligned}
w_t + Aw &= f'(u)u_t\psi + u_t\psi_t && \text{in } \Omega \times (0, T_{\max}), \\
w &= 0 && \text{on } \partial\Omega \times (0, T_{\max}), \\
w(\cdot, 0) &= 0 && \text{in } \Omega.
\end{aligned} \tag{51.21}$$

Theorem 51.1(v) guarantees that the solution w of (51.21) belongs to $\mathcal{W}_{\tilde{q}}$. Set $W(t) := \int_0^t (w(s) + u(s)\psi_t(s))\, ds$. Then is is easy to see that both W and $u\psi$ satisfy the same linear problem

$$\begin{aligned}
W_t + AW &= f(u)\psi + u\psi_t && \text{in } \Omega \times (0, T_{\max}), \\
W &= 0 && \text{on } \partial\Omega \times (0, T_{\max}), \\
W(\cdot, 0) &= 0 && \text{in } \Omega,
\end{aligned}$$

hence $W \equiv u\psi$ and $W_t(s) = u_t(s)$ for $s > 2\delta$. Since $W_t = w + u\psi_t \in \mathcal{W}_{\tilde{q}}$ and $\delta > 0$ was arbitrary, we see that $u_t \in \mathcal{W}_{\tilde{q}}$. If we only assumed that f' is locally Hölder continuous (instead of f'' locally Hölder continuous) and if the coefficients of A belong to $BUC^\rho(\Omega)$ and Ω is of class $C^{2+\rho}$, then applying L^p- and subsequently Schauder estimates to (51.21) (along with a cut-off argument if Ω is unbounded) would yield $u_t \in C^{2+\rho, 1+\rho/2}(\overline{\Omega} \times (0, T))$ for suitable $\rho > 0$. Similar regularity properties can be derived in the same way in the case of Neumann boundary conditions. $\square$

Remark 51.11. Let $\Omega \subset \mathbb{R}^n$ be uniformly regular of class C^2, let X_0 be any of the spaces $L^\infty(\Omega), BC(\overline{\Omega}), BUC(\Omega), C_*(\Omega) := \{u \in BUC(\Omega) : \lim_{|x|\to\infty} u(x) = 0\}$, and

$$\mathcal{A}u = -\sum_{i,j=1}^n a_{ij}\frac{\partial^2}{\partial x_i \partial x_j}u + \sum_{i=1}^n b_i \frac{\partial}{\partial x_i}u + cu,$$

where $a_{ij}, b_i, c \in BUC(\Omega)$ and $a_{ij} = a_{ji}$ are uniformly elliptic. Let A be the unbounded operator in X_0 defined by $Au = \mathcal{A}u$ for $u \in D(A)$, where

$$D(A) = \left\{ u \in \bigcap_{q \geq 1} W^{2,q}_{loc}(\overline{\Omega}) : u, \mathcal{A}(u) \in X_0, \ u = 0 \text{ on } \partial\Omega \right\}.$$

Note that X_0 is not reflexive and A is not densely defined, in general, since

$$\overline{D(A)}^{X_0} = \{u \in BUC(\Omega) \cap X_0 : u = 0 \text{ on } \partial\Omega\}.$$

Nevertheless, [344, Corollary 3.1.21] guarantees that $-A$ is sectorial, hence it generates an analytic semigroup e^{-tA} in X_0 (see [344] for the definition and properties of sectorial operators). Notice that this semigroup is not strongly continuous if $D(A)$ is not dense in X_0. However, its restriction to $\overline{D(A)}^{X_0}$ is strongly continuous, cf. [344, Remark 2.1.5].

Let $X_1 := D(A)$ be endowed with the graph norm and let $(X_\gamma, |\cdot|_\gamma)$, $\gamma \in (0,1)$, be Banach spaces satisfying

$$(X_0, X_1)_{\gamma,1} \hookrightarrow X_\gamma \hookrightarrow (X_0, X_1)_{\gamma,\infty}. \tag{51.22}$$

We will also assume that the spaces X_γ have the following property: if A_γ denotes the X_γ-realization of A, then

$$-A_\gamma \text{ is sectorial in } X_\gamma \quad \text{and} \quad \sigma(A_\gamma) \subset \sigma(A), \qquad \gamma \in (0,1). \tag{51.23}$$

For example, if $X_\gamma = (X_0, X_1)_\gamma$, where $(\cdot,\cdot)_\gamma$ is any of the real interpolation functors $(\cdot,\cdot)_{\gamma,p}$, $1 \leq p \leq \infty$, or the complex interpolation functor $[\cdot,\cdot]_\gamma$, then both (51.22) and (51.23) are true, see [344]. Similarly, if $X_0 = BC(\overline{\Omega})$, then the space

$$X_{1/2} := \{u \in BC^1(\overline{\Omega}) : u = 0 \text{ on } \partial\Omega\} \tag{51.24}$$

satisfies (51.22), (51.23) with $\gamma = 1/2$ due to [344, Propositions 3.1.27, 3.1.28, Theorem 3.1.25] and standard elliptic regularity theory. In general, [344, Proposition 3.1.28] and (51.22) imply

$$X_\gamma \hookrightarrow BUC^{2\gamma-\varepsilon}(\overline{\Omega}), \qquad \gamma \in (0,1], \ 0 < \varepsilon < 2\gamma. \tag{51.25}$$

The proofs of [344, Propositions 2.3.1, 2.2.9] show that the semigroup e^{-tA} satisfies estimates (51.3) for $0 \leq \beta \leq \alpha < 1$. These estimates can be used for the proof of similar existence results as above. To be more precise, assume that $1 = \alpha > \beta \geq 0$, $\tilde{u}_0 \in X_\beta$ and $F : X_\beta \to X_0$ is locally Lipschitz continuous, uniformly on bounded subsets of X_β. Then [344, Theorem 7.1.2] guarantees the existence of $r, T > 0$ such that, given $u_0 \in X_\beta$, $|u_0 - \tilde{u}_0|_\beta < r$, the integral equation (51.9) has a unique solution $u \in L^\infty((0,T), X_\beta)$. In addition, $u \in C([0,T], X_\delta)$ for $\delta \in [0, \beta)$,

$$(u - e^{-tA} u_0) \in C([0,T], X_\beta), \tag{51.26}$$

and

$$u \in C^{1-\gamma}((0,T], X_\gamma), \qquad \gamma \in (0,1). \tag{51.27}$$

These results imply the existence of a maximal solution and one can also prove similar assertions to those in Remarks 51.8(ii)–(iv). Notice also that if $\beta = 0$ and $u_0 \in \overline{D(A)}^{X_0}$, then (51.26) and the strong continuity of the restriction of e^{-tA} to $\overline{D(A)}^{X_0}$ guarantee $u \in C([0,T], X_0)$.

In particular, if $F(u)(x) = f(u(x))$ with $f \in C^1$, then $F : X_0 \to X_0$ is locally Lipschitz continuous, uniformly on bounded sets. Therefore, setting $\alpha = 1$ and $\beta = 0$ we get a solution u of (51.9) on the maximal time interval $[0, T_{\max}(u_0))$ for any $u_0 \in X_0$. In addition, the analogue of Remark 51.8(iii) and (51.25) guarantee the following: if $u_0 \in L^\infty(\Omega)$ and $t \in (0, T_{\max}(u_0))$ are fixed, then there exist $C, \varepsilon > 0$ such that $T_{\max}(\tilde{u}_0) > t$ and

$$\|\tilde{u}(t) - u(t)\|_{BC^1} \leq C \|\tilde{u}_0 - u_0\|_\infty \tag{51.28}$$

for any $\tilde{u}_0 \in L^\infty(\Omega)$ satisfying $\|\tilde{u}_0 - u_0\|_\infty < \varepsilon$.

If $F(u)(x) = f(u(x), \nabla u(x))$ with $f \in C^1$ and $X_0 = BC(\overline{\Omega})$, then F has obviously the required continuity properties as a map $F : X_{1/2} \to X_0$, where $X_{1/2}$ is defined in (51.24). Hence, setting $\alpha = 1$ and $\beta = 1/2$ we get a maximal solution

$$u \in C([0, T_{\max}), X_{1/2}) \tag{51.29}$$

provided $u_0 \in X_{1/2}$. Of course, analogous statements are true for nonlinearities of the form $F(u)(x) = f(x, u(x), \nabla u(x))$ or $F(t,u)(x) = f(x, t, u(x), \nabla u(x))$, cf. Remark 51.8(ii) and [344].

Finally, similar results are true if we consider Neumann boundary conditions instead of Dirichlet boundary conditions (that is, if we replace the condition $u = 0$ on $\partial\Omega$ in the definition of $D(A)$ by $\partial u / \partial \nu = 0$ on $\partial\Omega$) see [344, Corollary 3.1.24 and Theorem 3.1.26], for example. $\square$

Example 51.12. Let $\Omega \subset \mathbb{R}^n$ be uniformly regular of class C^2, $f, g \in C^1$, $d_1, d_2 > 0$ and consider the system

$$\left. \begin{array}{l} u_t - d_1 \Delta u = f(u,v), \\ v_t - d_2 \Delta v = g(u,v), \end{array} \right\} \qquad x \in \Omega, \ t > 0, \tag{51.30}$$

complemented with homogeneous Dirichlet boundary conditions if $\Omega \neq \mathbb{R}^n$. Consider also initial data $u_0, v_0 \in L^\infty(\Omega)$. Set $X_0 = L^\infty \times L^\infty(\Omega)$,

$$X_1 = \left\{ (u,v) : u, v \in \bigcap_{q \geq 1} W^{2,q}_{loc}(\overline{\Omega}), \ u, v, \Delta u, \Delta v \in L^\infty(\Omega), \ u|_{\partial\Omega} = v|_{\partial\Omega} = 0 \right\},$$

$X_\gamma = (X_0, X_1)_\gamma$, $\gamma \in (0,1)$, $A(u,v) = (-d_1 \Delta u, -d_2 \Delta v)$ for $(u,v) \in X_1$ and $F(u,v) = (f(u,v), g(u,v))$. Then $F : X_0 \to X_0$ is locally Lipschitz continuous (uniformly on bounded sets) and a straightforward modification of Remark 51.11 shows that the problem has a unique maximal solution satisfying $(u,v) - e^{-tA}(u_0, v_0) \in C([0, T_{\max}), X_0)$ and $(u,v) \in C^{1-\gamma}((0, T_{\max}), X_\gamma)$ for any $\gamma < 1$. Using the analogue of (51.25) we see that both u and v solve linear scalar problems with Hölder continuous right-hand sides so that one can use Schauder estimates to prove higher regularity of these solutions.

Analogous assertions as above are also true in the case of homogeneous Neumann conditions (or Dirichlet conditions for u and Neumann conditions for v). In addition, if we prescribe inhomogeneous Neumann boundary conditions of the form $\partial_\nu u = h_1(t)$, $\partial_\nu v = h_2(t)$, where h_1, h_2 are smooth, then we can find smooth functions u_h, v_h satisfying these boundary conditions and we obtain the existence results by solving the system

$$\partial_t \tilde{u} - d_1 \Delta \tilde{u} = f(\tilde{u} + u_h, \tilde{v} + v_h) + d_1 \Delta u_h - \partial_t u_h,$$
$$\partial_t \tilde{v} - d_2 \Delta \tilde{v} = g(\tilde{u} + u_h, \tilde{v} + v_h) + d_2 \Delta v_h - \partial_t v_h,$$

with homogeneous Neumann boundary conditions (by using the analogue of Remark 51.8(ii)).

Finally, using (the analogues of) Remarks 51.8(ii),(iv) one can also solve the problem if the functions f, g are defined for nonnegative (or positive) arguments only, provided the initial data are nonnegative (or positive) and either $f, g \geq 0$ or we can guarantee the positivity of the solution by other means. $\square$

Example 51.13. Let $\Omega \subset \mathbb{R}^n$ be a bounded domain of class $C^{2+\rho}$ for some $\nu \in (0,1)$, let f, g be C^1-functions and consider the equation

$$u_t - \Delta u = f\left(u, \int_\Omega g(u)\, dx\right), \qquad x \in \Omega, \ t > 0, \tag{51.31}$$

complemented with homogeneous Dirichlet boundary conditions. Assume also that the initial data $u_0 \in L^\infty(\Omega)$. Since the nonlinearity

$$F : L^\infty(\Omega) \to L^\infty(\Omega) : u \mapsto f\left(u, \int_\Omega g(u)\, dx\right)$$

is locally Lipschitz (uniformly on bounded sets), we can use Remark 51.11 in order to solve the problem. Similarly as in Example 51.12, we can also consider Neumann boundary conditions and nonlinearities defined for nonnegative or positive arguments only, for example

$$F(u) = u^p \left(\int_\Omega u^q \, dx \right)^{-m}, \tag{51.32}$$

where $p, q \geq 1$ and $m > 0$.

The same arguments apply to the equation

$$u_t - \Delta u = \left(\int_{\mathbb{R}^n} K u^p \right)^r u^q, \qquad x \in \mathbb{R}^n, \ t > 0, \tag{51.33}$$

where $p, q \geq 1$, $r > 0$, $K \in L^1(\mathbb{R}^n)$ is positive and continuous, the initial data $u_0 \in L^\infty(\mathbb{R}^n)$ are nonnegative and not identically zero. If in addition $u_0 \in L^1(\mathbb{R}^n)$, then the assumption $K \in L^1(\mathbb{R}^n)$ can be replaced by $K \in L^\infty(\mathbb{R}^n)$. In fact, the existence of a unique mild solution $u \in L^\infty((0,T), X)$, $X := L^1 \cap L^\infty(\mathbb{R}^n)$, follows by a direct application of the Banach fixed point theorem to the mapping defined in (51.13) in a ball of the space $L^\infty((0,T), X)$.

Further regularity of solutions of the above problems can be obtained by considering those problems as linear problems with bounded (or Hölder continuous) RHS, cf. Examples 51.9, 51.10. In particular, the solutions of (51.31) are classical for $t > 0$.

Finally, let us consider the homogeneous Neumann problem for the nonlinearity (51.32) with $p = q > 1$, $m = 1$ (see (44.24)). Assume that Ω is the unit ball and $u_0 \in C^2(\overline{\Omega})$ is radial and positive, $u_0(x) = U_0(|x|)$ where $U_0'(1) = 0$. Then $u(x,t) = U(|x|, t)$ for some $U = U(r,t)$. As mentioned above, u is a classical solution for $t > 0$. Set $T := T_{\max}$. Theorem 51.7 (with the choice $\alpha = 1$, $\beta = 1 - \varepsilon$, $\varepsilon > 0$ small, and $X_0 = L^r(\Omega)$, $r > n/(1 - 2\varepsilon)$) also guarantees

$$u \in C([0, \tau], W^{2-2\varepsilon, r}(\Omega)) \hookrightarrow C([0, \tau], C^1(\overline{\Omega})), \quad 0 < \tau < T.$$

The function $v(x,t) := U_r(|x|, t) \in C(\overline{\Omega} \times [0,T))$ is a weak (and, consequently strong) solution of the linear equation $v_t - \Delta v = (\int_\Omega u^p)^{-1} p u^{p-1} v$ complemented by homogeneous Dirichlet boundary conditions on S_T. Now Schauder estimates imply $v \in C^{2,1}(Q_T)$. $\square$

Example 51.14. Let $\Omega \subset \mathbb{R}^n$ be a bounded domain of class $C^{2+\rho}$ for some $\nu \in (0,1)$, let $p > 1$, $q \geq 1$, $k \geq 0$, and consider the problem

$$\left. \begin{aligned} u_t - \Delta u &= \int_0^t |u|^{p-1} u(x,s) \, ds - k|u|^q, & x \in \Omega, \ t > 0, \\ u &= 0, & x \in \partial\Omega \ t > 0, \\ u(x,0) &= u_0, & x \in \Omega. \end{aligned} \right\} \tag{51.34}$$

First notice that if $F(u) = \int_0^t F_1(u(s))\, ds + F_2(u)$ where F_1, F_2 satisfy the assumptions of Theorem 51.7, then a straightforward modification of the proof shows that the first part of that theorem remains true. More precisely, given $u_0 \in X_\beta$ there exists a unique local solution $u \in C([0,T], X_\beta)$ and $u \in C((0,T], X_\gamma)$ for all $\gamma \in [\beta, \alpha)$, $|u(t)|_\gamma \le Ct^{\beta-\gamma}$ for $t > 0$. Combining these arguments with Remark 51.11 we see that problem (51.34) is well-posed in $X_0 := L^\infty(\Omega)$ and the solution satisfies $u \in C((0,T], X_\gamma)$ and $|u(t)|_\gamma \le Ct^{-\gamma}$ for all $\gamma \in [0,1)$, where X_γ, $\gamma \in (0,1]$, are the spaces from Remark 51.11.

Fix $t_0 \in (0,T)$ and $\varepsilon > 0$ small. Since the nonlinearity $F_1(u) = |u|^{p-1}u$ satisfies $\|F_1(u)\|_{BC^1} \le C(\|u\|_\infty)\|\nabla u\|_\infty$, $\{u \in BC^1(\overline{\Omega}) : u|_{\partial\Omega} = 0\} \hookrightarrow X_{1/2-\varepsilon}$ and (51.25) is true, we see that

$$|F_1(u(s))|_{1/2-\varepsilon} \le \|F_1(u)\|_{BC^1} \le C|u(s)|_{1/2+\varepsilon} \le Cs^{-1/2-\varepsilon}$$

and, in particular,

$$F^0 := \int_0^{t_0} F_1(u(s))\, ds \in X_{1/2-\varepsilon} \hookrightarrow BUC^{1-3\varepsilon}(\overline{\Omega}). \qquad (51.35)$$

Parabolic L^p-estimates and embedding (1.2) guarantee that u is Hölder continuous in both x and t for $t \ge t_0$, hence $F(u) - F^0$ is Hölder continuous. Now (51.35) and Schauder estimates guarantee that u is a classical solution of (51.34) for $t > 0$. Obviously this remains true for the maximal solution on $(0, T_{\max})$. (Notice that the Hölder continuity of u for $t > 0$ also follows from Remark 51.8(ii) with the choice $1 > \alpha > \beta > 0$, $r > \max(1/(\alpha - \beta), n/2\beta)$ and $X_0 = L^r(\Omega)$.)

Finally assume $k = 0$. Set $T := T_{\max}$. Similar arguments as at the end of Example 51.10 show that u_t solves the linear problem

$$\left.\begin{aligned} v_t - \Delta v &= |u|^{p-1}u, & & x \in \Omega,\ 0 < t < T_{\max}, \\ v &= 0, & & x \in \partial\Omega,\ 0 < t < T_{\max}, \end{aligned}\right\} \qquad (51.36)$$

hence Schauder estimates guarantee

$$v = u_t \in C^{2,1}(Q_T) \cap C(\overline{\Omega} \times (0,T)).$$

Let $\varphi \in C^2(\overline{\Omega} \times [0,T))$, $\varphi = 0$ on $\partial\Omega \times [0,T)$. Multiplying the equation in (51.34) with φ, integrating by parts and passing to the limit as $t \to 0+$ we obtain

$$\lim_{t \to 0+} \int_\Omega u_t(x,t)\varphi(x,t)\, dx = \int_\Omega u_0(x)\Delta\varphi(x,0)\, dx.$$

In particular, if $u_0 \in H^2 \cap H_0^1(\Omega)$, then

$$\lim_{t \to 0+} \int_\Omega v(x,t)\varphi(x,t)\, dx = \int_\Omega \Delta u_0(x)\varphi(x,0)\, dx.$$

Now we infer from the uniqueness proof in Proposition 48.9 that v is (a strong) solution of (51.36) with initial data Δu_0. In particular $v = u_t \in C([0,T), L^2(\Omega))$. $\quad\square$

Example 51.15. Consider the problem

$$
\left.
\begin{aligned}
u_t - \Delta u &= c|u|^{p-1}u - a \cdot \nabla(|u|^{q-1}u), & x \in \Omega,\ t > 0,\\
u &= 0, & x \in \partial\Omega,\ t > 0,\\
u(x,0) &= u_0(x), & x \in \Omega,
\end{aligned}
\right\}
\qquad (51.37)
$$

where $\Omega \subset \mathbb{R}^n$ is uniformly regular of class $C^{2+\rho}$, $\rho \in (0,1)$, $p > 1$, $q \geq 1$, $c \geq 0$ and $a \in \mathbb{R}^n$, $a \neq 0$.

Assume first that $u_0 \in W_0^{1,r}(\Omega)$ with $r \in (n,\infty)$. Set $X_0 := L^r(\Omega)$, $X_1 := W^{2,r} \cap W_0^{1,r}(\Omega)$, $Au := -\Delta u$ for $u \in X_1$, and $F(u) := |u|^{p-1}u - a \cdot \nabla(|u|^{q-1}u)$. Let $X_\theta = X_\theta(r)$ be defined as in Example 51.4(i), in particular $X_{1/2} = W_0^{1,r}(\Omega)$. Choose $\varepsilon > 0$ such that $(1-2\varepsilon)r > n$ and set $\alpha := 1/2$ and $\beta := 1/2-\varepsilon$. Notice that F satisfies the assumptions of Theorem 51.7 since F can be viewed as $F = F_1 + F_2$, where

$$
F_1 : X_\beta \hookrightarrow L^{pr}(\Omega) \xrightarrow{\ |u|^{p-1}u\ } L^r(\Omega) \hookrightarrow X_{\alpha-1},
$$

$$
F_2 : X_\beta \hookrightarrow L^{qr}(\Omega) \xrightarrow{\ |u|^{q-1}u\ } L^r(\Omega) \xrightarrow{\ a\cdot\nabla^r\ } (W_0^{1,r'}(\Omega))' = X_{\alpha-1},
$$

and ∇^r is defined by

$$
\langle \nabla^r w, \varphi \rangle := -\int_\Omega w \nabla \varphi\, dx \qquad \varphi \in W_0^{1,r'}(\Omega),\ w \in L^r(\Omega).
$$

Consequently, (51.37) possesses a unique solution $u \in C([0,T], X_\beta)$.

Next consider the case

$$
u_0 \in X := \{u \in BC^1(\overline{\Omega}) : u|_{\partial\Omega} = 0\}. \qquad (51.38)
$$

Set $X_0 := BC(\overline{\Omega})$, $Au := -\Delta u$ for $u \in D(A)$, where $D(A)$ (and X_θ, $\theta \in (0,1]$) are as in Remark 51.11, $X_{1/2} = X$. If $q \geq 2$ or $q = 1$, then the function $f(u,\xi) := |u|^{p-1}u - q|u|^{q-1}(a \cdot \xi)$ is C^1, hence $F : X_{1/2} \to X_0$ has the required continuity properties and (51.37) possesses a unique solution $u \in L^\infty((0,T), X_{1/2})$ satisfying $u \in C([0,T], X_\delta)$ for $\delta < 1/2$, $(u - e^{-tA}u_0) \in C([0,T], X_{1/2})$ and (51.27), see Remark 51.11. In addition, this solution can be continued on the maximal existence interval $[0, T_{\max}(u_0))$ and Remark 51.35 with the choice $\alpha = 1$, $\gamma = 0$ and $\beta = 1/2$ (or Proposition 51.34 with $\alpha = 1$, $\gamma = 0$ and $\beta \in (1/2,1)$) guarantee that

$$
\text{if } T_{\max}(u_0) < \infty, \text{ then } \limsup_{t \to T_{\max}(u_0)} \|u(t)\|_\infty = \infty. \qquad (51.39)
$$

Finally, Schauder estimates show that u is a classical solution for $t > 0$ and the maximum principle guarantees that $u \geq 0$ if $u_0 \geq 0$.

Obviously, all the assertions above concerning the case (51.38) remain true for all $q \geq 1$ if we replace the nonlinearity $|u|^{q-1}$ in the definition of f with $(|u| + \varepsilon)^{q-1}$, $\varepsilon \in (0, 1]$. Let u_ε denote the corresponding solution. Then the arguments in Remark 51.35 (with $\alpha = 1$, $\gamma = 0$ and $\beta = 1/2$) guarantee that given $T < \infty$, $C_0 > 0$,

$$\|\nabla u_\varepsilon(t)\|_\infty \leq C_1, \ t \in [0, T], \quad \text{provided } \|u_\varepsilon(t)\|_\infty \leq C_0, \ t \in [0, T], \qquad (51.40)$$

where the constant $C_1 > 0$ does not depend on ε. In the following proposition we use the approximation solutions u_ε in order to show the solvability of (51.37) in $X_{1/2}$ for any $q \geq 1$. For simplicity we restrict ourselves to nonnegative solutions and to the case Ω bounded or $\Omega = \mathbb{R}^n$. $\square$

Proposition 51.16. *Let $\Omega \subset \mathbb{R}^n$ be a bounded domain of class $C^{2+\gamma}$ for some $\gamma \in (0, 1)$ or $\Omega = \mathbb{R}^n$. Consider the problem* (51.37), *where*

$$u_0 \in X_+ = \{u \in BC^1(\overline{\Omega}) : u|_{\partial\Omega} = 0, \ u \geq 0\}.$$

(i) There exists a unique, maximal classical solution $u \in C^{2,1}(\overline{\Omega} \times (0, T))$ of (51.37), *such that $u \in C([0, T), C(\overline{\Omega}))$ $(u \in C([0, T), BC(\mathbb{R}^n))$ if $\Omega = \mathbb{R}^n)$ and $\nabla u \in L^\infty_{loc}([0, T), L^\infty(\Omega))$. Moreover,* (51.39) *is true (with $T_{\max}(u_0) = T$).*
(ii) Let $\Omega = \mathbb{R}^n$. Then u also satisfies

$$u \in L^\infty_{loc}((0, T), BC^2(\mathbb{R}^n)). \qquad (51.41)$$

If in addition $u_0 \in L^1(\mathbb{R}^n)$, then

$$u \in C([0, T), L^1(\mathbb{R}^n)). \qquad (51.42)$$

Proof. The uniqueness of the solution is guaranteed by the comparison principle in Proposition 52.16.

To establish existence, we consider the approximating problem

$$\left.\begin{array}{ll} \partial_t u_\varepsilon - \Delta u_\varepsilon = c u_\varepsilon^p - q(u_\varepsilon + \varepsilon)^{q-1}(a \cdot \nabla u_\varepsilon), & x \in \Omega, \ t > 0, \\[4pt] u_\varepsilon = 0, & x \in \partial\Omega, \ t > 0, \\[4pt] u_\varepsilon(x, 0) = u_0(x), & x \in \Omega. \end{array}\right\} \qquad (51.43)$$

By Example 51.15, for each $\varepsilon \in (0, 1]$, there exists $T_\varepsilon \in (0, 1]$ and a unique classical solution $u_\varepsilon \in L^\infty([0, T_\varepsilon], X_+)$ of (51.43) satisfying $(u_\varepsilon - e^{-tA}u_0) \in C([0, T_\varepsilon], X_+)$. Moreover, u_ε can be continued as long as $\|u_\varepsilon(t)\|_\infty$ remains bounded and (51.40) is true. By comparing with the solution of the ODE $y'(t) = cy^p$, $y(0) = M := \|u_0\|_\infty$, we see that

$$0 \leq u_\varepsilon(t) \leq 2M, \qquad 0 < t \leq T_0 := C(p)M^{1-p}. \qquad (51.44)$$

In particular, $T_\varepsilon \geq T_0$ and

$$\|\nabla u_\varepsilon(t)\|_\infty \leq C_1, \qquad 0 < t \leq T_0. \tag{51.45}$$

Now, (51.45) guarantees that the RHS of (51.43) is bounded in $L^\infty(Q_{T_0})$ independently of ε. In the case Ω bounded, by parabolic L^r-estimates and the embedding (1.2), it follows that u_ε is bounded in $C^{1+\sigma,\sigma/2}(\overline{\Omega} \times (0, T_0])$ for some $\sigma \in (0, 1)$. Applying this estimate to the RHS we deduce from Schauder estimates that u_ε is bounded in $C^{2+\alpha,1+\alpha/2}(\overline{\Omega} \times (0, T_0])$ for some $\alpha \in (0, 1)$. Therefore (some subsequence of) u_ε converges to a classical solution $u \in C^{2,1}(\overline{\Omega} \times (0, T_0])$ of (34.4). In the case $\Omega = \mathbb{R}^n$, we may apply the same argument in $B(x_0, 1)$ for each $x_0 \in \mathbb{R}^n$ and we obtain a bound of u_ε in $C^{2+\alpha,1+\alpha/2}(B(x_0, 1) \times (0, T_0])$, with constant independent of x_0. This yields a solution $u \in C^{2,1}(\mathbb{R}^n \times (0, T_0])$ of (34.4), with $u_t, D^2 u$ bounded in $\mathbb{R}^n \times (\tau, T_0]$ for each $\tau > 0$. Note that this implies $u \in C((0, T_0], BC(\mathbb{R}^n))$ and (51.41). Moreover, in both cases, (51.45) and the variation-of-constants formula imply

$$\|u_\varepsilon(t) - e^{-tA}u_0\|_\infty \leq Ct\big(M^p + (M+1)^{q-1}C_1\big). \tag{51.46}$$

Passing to the limit $\varepsilon \to 0$, we get (51.46) with $u(t)$ instead of $u_\varepsilon(t)$, hence the continuity of $u(t)$ in $C(\overline{\Omega})$ (or in $BC(\mathbb{R}^n)$) at $t = 0$.

Since the solution u satisfies the variation-of-constants formula for $t > 0$, assertion (51.39) follows from Proposition 51.34 or Remark 51.35 (cf. the same argument in Example 51.15).

Finally, let us prove (51.42). Observe that for $\Omega = \mathbb{R}^n$ and $u \in C^{2,1}(\mathbb{R}^n \times (0, T))$ such that $u \in C([0, T], BC(\mathbb{R}^n)) \cap L^\infty((0, T), W^{1,\infty}(\mathbb{R}^n))$, (51.37) is equivalent to the integral equation

$$u(t) = G_t * u_0 + c \int_0^t G_{t-s} * u^p(s)\, ds + \int_0^t (a \cdot \nabla G_{t-s}) * u^q(s)\, ds. \tag{51.47}$$

When $u_0 \in X_+ \cap L^1(\mathbb{R}^n)$, one can obtain a (unique) solution of (51.47) as a fixed point in a suitable ball of the metric space $C([0, T], L^1 \cap BC(\mathbb{R}^n)) \cap L^\infty((0, T), W^{1,\infty}(\mathbb{R}^n))$ endowed with its natural norm. This can be done by using similar (and in fact simpler) arguments as in the proof of Theorem 15.3, using in particular the fact that $\|\nabla G_t * f\|_1 \leq Ct^{-1/2}\|f\|_1$, $f \in L^1(\mathbb{R}^n)$. Moreover, it is easy to see that the solution of (51.47) can be continued in this space as long as $\|u(t)\|_\infty$ remains bounded. By the already proved uniqueness statement, we deduce that the two solutions coincide, with same existence time, which completes the proof. $\qquad\square$

51.3. Stability of equilibria

استقرار التوازن

Theorem 51.17. *Let* α, β, F *be as in Theorem 51.7. Let, in addition,* $\omega(-A) < 0$ *and*

$$|F(u)|_{\alpha-1} = o(|u|_\beta) \qquad \text{as } |u|_\beta \to 0.$$

Then the zero solution of (51.9) is (locally) exponentially asymptotically stable. More precisely, given $\tilde\omega \in (\omega(-A), 0)$ *there exist* $\delta^* > 0$ *and* $C > 0$ *such that the solution* u *with initial data* u_0 *satisfying* $|u_0|_\beta < \delta^*$ *exists globally and*

$$|u(t)|_\beta \le C e^{\tilde\omega t} |u_0|_\beta \qquad \text{for all } t \ge 0. \tag{51.48}$$

Proof. Let $\tilde\omega \in (\omega(-A), 0)$. Choose $\omega \in (\omega(-A), \tilde\omega)$. Then (51.3) guarantees

$$\left.\begin{array}{l} \|e^{-tA}\|_{\mathcal{L}(X_{\alpha-1}, X_\beta)} \le C(\omega, A) t^{\alpha-1-\beta} e^{\omega t}, \\[2mm] \|e^{-tA}\|_{\mathcal{L}(X_\beta, X_\beta)} \le C(\omega, A) e^{\omega t}, \end{array}\right\} \qquad \text{for all } t > 0, \tag{51.49}$$

where $C(\omega, A) \ge 1$. Set

$$C^* = C(\omega, A) \int_0^\infty \tau^{\alpha-1-\beta} e^{(\omega-\tilde\omega)\tau}\, d\tau$$

and choose $\varepsilon > 0$ such that

$$|F(u)|_{\alpha-1} \le \frac{1}{2C^*} |u|_\beta \qquad \text{whenever } |u|_\beta \le \varepsilon. \tag{51.50}$$

Choose $\delta^* = \varepsilon / 2C(\omega, A)$ and let $|u_0|_\beta < \delta^*$. We may assume $u_0 \ne 0$. Set

$$T = \sup\{t \in (0, T_{\max}(u_0)) : |u(s)|_\beta \le 2C(\omega, A) e^{\tilde\omega s} |u_0|_\beta \text{ for all } s \in [0, t]\}$$

and notice that $T > 0$ and $|u(s)|_\beta \le \varepsilon$ for all $s \in [0, T)$. If $T = \infty$, then (51.48) is true. Hence, assume $T < \infty$. Then $T < T_{\max}(u_0)$ due to the uniform bound of $|u(s)|_\beta$ for $s \in [0, T)$, hence

$$|u(T)|_\beta = 2C(\omega, A) e^{\tilde\omega T} |u_0|_\beta. \tag{51.51}$$

On the other hand, using (51.49), (51.50), the inequality in the definition of T and the definition of C^* we obtain

$$|u(T)|_\beta \le C(\omega, A) e^{\omega T} |u_0|_\beta + C(\omega, A) \int_0^T (T-s)^{\alpha-1-\beta} e^{\omega(T-s)} |F(u(s))|_{\alpha-1}\, ds$$

$$\le C(\omega, A) e^{\omega T} |u_0|_\beta + \frac{C(\omega, A)^2}{C^*} e^{\tilde\omega T} |u_0|_\beta \int_0^T (T-s)^{\alpha-1-\beta} e^{(\omega-\tilde\omega)(T-s)}\, ds$$

$$< C(\omega, A) e^{\omega T} |u_0|_\beta + C(\omega, A) e^{\tilde\omega T} |u_0|_\beta \le 2C(\omega, A) e^{\tilde\omega T} |u_0|_\beta,$$

which contradicts (51.51) and concludes the proof. $\square$

Remarks 51.18. (i) A combination of Theorem 51.17 with estimates of the form (51.10) or (51.20) shows that the solution u in Theorem 51.17 also decays exponentially to zero in stronger norms than $|\cdot|_\beta$.

(ii) Theorem 51.17 can also be used in order to prove the stability of a non-zero equilibria. In fact, let $w \in X_\alpha$, $A_{\alpha-1} w = F(w)$. Assume that $F : X_\beta \to X_{\alpha-1}$ is Fréchet differentiable at w and set

$$F_w(v) := F(w+v) - F(w) - F'(w)v,$$

hence

$$|F_w(v)|_{\alpha-1} = o(|v|_\beta) \qquad \text{as } |v|_\beta \to 0.$$

Let us first consider the special case $\alpha = 1$. Set $\tilde{X}_1 = X_1$, $\tilde{X}_0 = X_0$, $\tilde{A} = A - F'(w)$ (with domain X_1) and assume that

$$\tilde{A} \text{ generates a } C^0 \text{ analytic semigroup in } \tilde{X}_0 \text{ and } \omega(-\tilde{A}) < 0. \tag{51.52}$$

Notice that if A has compact resolvent, then $F'(w) \in \mathcal{L}(X_\beta, X_0)$ is a compact perturbation of A, hence the first part of (51.52) is automatically satisfied. Set $v(t) = u(t) - w$ and $v_0 = u_0 - w$. Then (51.9) can be written as

$$v(t) = e^{-t\tilde{A}} v_0 + \int_0^t e^{-(t-s)\tilde{A}} F_w(v(s))\, ds$$

and one can use Theorem 51.17 with A and F replaced by $\tilde{A}$ and F_w, respectively.

If $\alpha < 1$ and $F'(w)|_{X_1} \in \mathcal{L}(X_1, X_0)$, then one can still use the same arguments as above (provided (51.52) is true). In the general case we set $\tilde{X}_0 = X_{\alpha-1}$, $\tilde{X}_1 = X_\alpha$, $\tilde{A} = A_{\alpha-1} - F'(w)$ (with domain $\tilde{X}_1$) and assume that (51.52) is true. Set also $\tilde{\alpha} = 1$ and choose $\tilde{\beta} \in (\beta+1-\alpha, 1)$. Then $\tilde{X}_{\tilde{\beta}} \hookrightarrow X_\beta$ and one can use Theorem 51.17 with A, F, α, β replaced by $\tilde{A}, F_w, \tilde{\alpha}, \tilde{\beta}$, respectively.

(iii) The conclusions of Theorem 51.17 remain true in the situation of Remark 51.11. $\square$

Theorem 51.19. *Let α, β, F be as in Theorem 51.7, $p > 1$ and*

$$|F(u)|_{\alpha-1} = O(|u|_\beta^p) \qquad \text{as } |u|_\beta \to 0. \tag{51.53}$$

Assume that $\sigma(-A) = \{\omega_1\} \cup \sigma_2$, where $\omega_1 < 0$ is a simple eigenvalue of $-A$ with eigenspace E_1 and $\sigma_2 \subset \{\lambda : \operatorname{Re}\lambda \le \omega_2\}$ for some $\omega_2 < \omega_1$. Fix $\omega \in (\max(\omega_2, \omega_1 p), \omega_1)$. Then there exist $\delta, C > 0$ and a continuous map $K : X_\beta \to E_1$ such that the solution u with initial data u_0 satisfying $|u_0|_\beta < \delta$ exists globally and

$$|u(t) - K(u_0)e^{\omega_1 t}|_\beta \le Ce^{\omega t}|u_0|_\beta \qquad \text{for all } t \ge 0. \tag{51.54}$$

Proof. Let P_1 and P_2 denote the spectral projections in X_β corresponding to the spectral sets $\{\omega_1\}$ and σ_2, respectively, and $E_2 := P_2(X_\beta)$. Then $X_\beta = E_1 \oplus E_2$, E_2 is A (and e^{-tA}) invariant, $\sigma(-A|_{E_2}) = \sigma_2$ and $-A|_{E_2}$ generates the analytic semigroup $e^{-tA}|_{E_2}$ (see [273, Section 1.5] and the references therein), hence (51.3) implies

$$|e^{-tA}P_2 u|_\beta \le Ce^{\omega t}\min(|u|_\beta, t^{\alpha-1-\beta}|u|_{\alpha-1}) \tag{51.55}$$

for all $u \in X_\beta$. Choose $\tilde{\omega} \in (\omega_1, \omega/p)$. Then Theorem 51.17 guarantees the existence of $\delta \in (0,1)$ and $C > 0$ such that

$$|u(t)|_\beta \le Ce^{\tilde{\omega}t}|u_0|_\beta \tag{51.56}$$

whenever $|u_0|_\beta < \delta$. Assume $|u_0|_\beta < \delta$ and denote $u_i(t) = P_i u(t)$, $i = 1,2$. Since $P_i e^{-tA} = e^{-tA}P_i$ we have

$$u_2(t) = e^{-tA}P_2 u_0 + \int_0^t e^{-(t-s)A}P_2 F(u(s))\,ds,$$

hence (51.53), (51.55), (51.56) and $\tilde{\omega}p < \omega$ imply

$$|u_2(t)|_\beta \le Ce^{\omega t}|u_0|_\beta + C\int_0^t e^{\omega(t-s)}(t-s)^{\alpha-1-\beta}|u(s)|_\beta^p\,ds \le Ce^{\omega t}|u_0|_\beta. \tag{51.57}$$

Set

$$K(u_0) := \lim_{t\to\infty} u_1(t)e^{-\omega_1 t} = P_1 u_0 + \int_0^\infty e^{-\omega_1 s}P_1 F(u(s))\,ds$$

(the integral is convergent since $\|P_1 F(u(s))\|_{E_1} \le Ce^{\tilde{\omega}ps}|u_0|_\beta^p$ due to (51.53), (51.56) and $\tilde{\omega}p < \omega < \omega_1$). Now the assertion follows from (51.57) and the estimate

$$\|u_1(t) - K(u_0)e^{\omega_1 t}\|_{E_1} = e^{\omega_1 t}\left\|\int_t^\infty e^{-\omega_1 s}P_1 F(u(s))\,ds\right\|_{E_1} \le Ce^{\tilde{\omega}pt}|u_0|_\beta^p. \quad \square$$

Remarks 51.20. (i) The proof of Theorem 51.19 implies

$$K(u_0) = P_1 u_0 + O(|u_0|_\beta^p).$$

(ii) Let us consider the situation from Remark 51.11 with Ω bounded, $\mathcal{A}u = -\Delta u$, $X_0 = BC(\overline{\Omega})$ and $X_{1/2}$ defined by (51.24). Set $\alpha = 1$ and $\beta = 1/2$. Then the statement in Theorem 51.19 remains true for this choice of spaces since A has the required properties, ω_1 is a simple eigenvalue of $A_{1/2}$ and $\sigma(A_{1/2}) \subset \sigma(A)$. $\quad \square$

51.4. Self-adjoint generators with compact resolvent

The proof of Theorem 51.21 below is based on an idea used in the construction of stable manifolds for general semilinear parabolic problems (see [273, Theorem 5.2.1] or [101, Lemma 4.1], for example). We will use this idea in a specific situation in order to obtain more precise information than that in [273] or [101]. In addition to (51.2) we will also assume that

$$
\left.\begin{array}{l}
X_0 \text{ is a Hilbert space,} \\[4pt]
A : X_0 \to X_0 \text{ is self-adjoint and has compact resolvent,} \\[4pt]
\omega_1 > \omega_2 > \cdots \text{ are all distinct eigenvalues of } -A, \\[4pt]
(\cdot,\cdot)_\theta \text{ is the complex interpolation functor for all } \theta \in [0,1].
\end{array}\right\}
\qquad (51.58)
$$

Then X_α, $\alpha \in [-1,1]$, are Hilbert spaces and the operators A_α are self-adjoint (see [16, Theorem V.1.5.15]). Let P_i, Q_i and R_i, $i = 1,2,\ldots$, denote the spectral projections in X_0 corresponding to the spectral sets $\{\omega_i, \omega_{i+1}, \ldots\}$, $\{\omega_1, \ldots, \omega_{i-1}\}$ and $\{\omega_i\}$, respectively. Let $P_{i,\alpha}$ denote the restriction $P_i|_{X_\alpha}$ if $\alpha > 0$ and the closure of P_i in X_α if $\alpha < 0$. Then $P_{i,\alpha}$ is the spectral projection in X_α corresponding to the spectral set $\{\omega_i, \omega_{i+1}, \ldots\}$ and analogous assertions are true for Q_i and R_i. Without fear of confusion we will write P_i, Q_i, R_i instead of $P_{i,\alpha}, Q_{i,\alpha}, R_{i,\alpha}$. Since

$$
-A = \sum_{j=1}^{\infty} \omega_j R_j \qquad \text{and} \qquad e^{-tA} = \sum_{j=1}^{\infty} e^{\omega_j t} R_j,
$$

it is easy to see that there exist $C_i > 0$, $i = 1,2,\ldots$, such that

$$
\|e^{-tA} P_i\|_{\mathcal{L}(X_\alpha, X_\alpha)} \le e^{\omega_i t}, \qquad \alpha \in [-1,1],\ t \ge 0,
$$

$$
\|e^{-tA} P_i\|_{\mathcal{L}(X_{\alpha-1}, X_\alpha)} \le \frac{C_i}{t} e^{\omega_i t}, \qquad \alpha \in [0,1],\ t > 0,
$$

and, by interpolation,

$$
\|e^{-tA} P_i\|_{\mathcal{L}(X_\beta, X_\alpha)} \le C_i t^{\beta-\alpha} e^{\omega_i t}, \qquad -1 \le \beta \le \alpha \le 1,\ t > 0. \qquad (51.59)
$$

Similarly,

$$
\|e^{-tA} Q_i\|_{\mathcal{L}(X_\beta, X_\alpha)} \le C_i e^{\omega_1 t}, \qquad \beta, \alpha \in [-1,1],\ t \ge 0,
$$

$$
\|e^{tA} Q_i\|_{\mathcal{L}(X_\beta, X_\alpha)} \le C_i e^{-\omega_{i-1} t}, \qquad \beta, \alpha \in [-1,1],\ t \ge 0, \qquad (51.60)
$$

where $e^{tA} Q_i := \sum_{j=1}^{i-1} e^{-\omega_j t} R_j$ if $t \ge 0$.

Theorem 51.21. *Assume* (51.58). *Let* α, β, F *be as in Theorem* 51.7, $p > 1$, $F(0) = 0$ *and*

$$|F'(u) - F'(v)|_{\alpha-1} \leq C_F |u - v|_\beta \left(|u|_\beta^{p-1} + |v|_\beta^{p-1} \right) \qquad \text{for } |u|_\beta, |v|_\beta \text{ small.} \quad (51.61)$$

Fix $i \geq 1$ *with* $\omega_i < 0$ *and choose* $\lambda \in [\omega_i, 0]$, $\lambda < \omega_{i-1}$ *if* $i > 1$. *Then there exist* $\rho = \rho_i > 0$ *small and* $\tilde{C}_i > 0$ *with the following properties: given* $v_0 \in P_i X_\beta$, $|v_0|_\beta \leq \rho$, *there exists a unique* $z_0 \in Q_i X_\beta$ *such that the solution of* (51.9) *with* $u_0 := v_0 + z_0$ *is global and satisfies* $|u(t)|_\beta \leq 2\rho e^{\lambda t}$ *for all* $t \geq 0$. *In addition,*

$$|u(t)|_\beta \leq 2|v_0|_\beta e^{\omega_i t} \qquad \text{for all } t \geq 0 \quad (51.62)$$

and

$$|z_0|_\beta \leq \tilde{C}_i |v_0|_\beta^p. \quad (51.63)$$

Finally, if $|R_i v_0|_\beta > \tilde{C}_i |v_0|_\beta^p$, *then there exists* $c = c(v_0) > 0$ *such that*

$$|u(t)|_\beta \geq c|R_i v_0|_\beta e^{\omega_i t} \qquad \text{for all } t \geq 0. \quad (51.64)$$

Proof. Let u be a global solution of (51.9). Then u can be written in the form $u = v + z$, where

$$v(t) = e^{-tA} v_0 + \int_0^t e^{-(t-s)A} P_i F(u(s)) \, ds,$$
$$\quad (51.65)$$
$$z(t) = e^{-tA} z_0 + \int_0^t e^{-(t-s)A} Q_i F(u(s)) \, ds,$$

$v_0 = P_i u_0$, $z_0 = Q_i u_0$ ($z_0 = 0$ and $z = 0$ if $i = 1$). Assume first that

$$|u(t)|_\beta \leq c e^{\lambda t}, \quad t \geq 0, \quad (51.66)$$

where $c > 0$ is small. If $i > 1$, then

$$|e^{tA} z(t)|_\beta = |e^{tA} Q_i z(t)|_\beta \leq e^{-\omega_{i-1} t} |z(t)|_\beta \leq c e^{(\lambda - \omega_{i-1})t} \to 0 \quad \text{as } t \to \infty,$$
$$|e^{sA} Q_i F(u(s))|_\beta \leq C_i e^{-\omega_{i-1} s} |F(u(s))|_{\alpha-1} \leq C_i C_F c^p e^{(p\lambda - \omega_{i-1})s},$$
$$\quad (51.67)$$

hence (51.65) guarantees

$$z_0 = - \int_0^\infty e^{sA} Q_i F(u(s)) \, ds \quad (51.68)$$

and $u = \Phi_{v_0}(u)$, where

$$\Phi_{v_0}(u)(t) := e^{-tA} v_0 + \int_0^t e^{-(t-s)A} P_i F(u(s)) \, ds - \int_t^\infty e^{-(t-s)A} Q_i F(u(s)) \, ds.$$
$$\quad (51.69)$$

On the other hand, if u is any function in $C([0,\infty), X_\beta)$ satisfying (51.66) and $u = \Phi_{v_0}(u)$ for some $v_0 \in P_i X_\beta$, then obviously u solves (51.9), where $u_0 = v_0 + z_0$ and z_0 is given by (51.68).

Denote

$$\|u\| = \sup_{t \geq 0} |u(t)|_\beta e^{-\lambda t}$$

and $B_\rho = \{u \in C([0,\infty), X_\beta) : \|u\| \leq 2\rho\}$. We will show that, given $v_0 \in P_i X_\beta$, $|v_0|_\beta \leq \rho$, the mapping Φ_{v_0} possesses a unique fixed point in B_ρ provided $\rho > 0$ is small enough. Given $u \in B_\rho$, we have

$$e^{-\lambda t}|\Phi_{v_0}(u)(t)|_\beta \leq e^{-\lambda t}\|e^{-tA}P_i\|_{\mathcal{L}(X_\beta, X_\beta)}|v_0|_\beta$$

$$+ \int_0^t e^{-\lambda t}\|e^{-(t-s)A}P_i\|_{\mathcal{L}(X_{\alpha-1}, X_\beta)}|F(u(s))|_{\alpha-1}\, ds$$

$$+ \int_t^\infty e^{-\lambda t}\|e^{-(t-s)A}Q_i\|_{\mathcal{L}(X_{\alpha-1}, X_\beta)}|F(u(s))|_{\alpha-1}\, ds$$

$$\leq |v_0|_\beta + C_i C_F (2\rho)^p \left[\int_0^t (t-s)^{\alpha-1-\beta} e^{-(\lambda-\omega_i)(t-s)+\lambda(p-1)s}\, ds \right.$$

$$\left. + \int_t^\infty e^{(\omega_{i-1}-\lambda)(t-s)+\lambda(p-1)s}\, ds \right]$$

$$< 2\rho,$$

provided ρ is small enough and $i > 1$ (analogous estimates are true for $i = 1$). Notice that the above estimates also imply

$$e^{-\lambda t}|\Phi_{v_0}(u)(t) - e^{-tA}v_0|_\beta \leq \tilde{C}_i \rho^p$$

for some $\tilde{C}_i > 0$ and that similar estimates guarantee

$$\|\Phi_{v_0}(u) - \Phi_{v_0}(\tilde{u})\| \leq \frac{1}{2}\|u - \tilde{u}\| \qquad \text{for } u, \tilde{u} \in B_\rho.$$

Consequently, $\Phi_{v_0} : B_\rho \to B_\rho$ is a contraction and possesses a unique fixed point in B_ρ. In addition, (51.68) and (51.67) imply $|z_0|_\beta \leq \tilde{C}_i \rho^p$. Repeating the above arguments with $\rho := |v_0|_\beta$ and $\lambda := \omega_i$ we obtain estimates (51.62), (51.63) and

$$e^{-\omega_i t}|u(t) - e^{-tA}v_0|_\beta \leq \tilde{C}_i |v_0|_\beta^p. \tag{51.70}$$

Set $w_0 := R_i v_0$, $y_0 := P_{i+1} v_0 = v_0 - w_0$ and assume $|w_0|_\beta > \tilde{C}_i |v_0|_\beta^p$. Then

$$e^{-\omega_i t}|e^{-tA}y_0|_\beta \leq e^{-(\omega_i - \omega_{i+1})t}|y_0|_\beta,$$

$$e^{-\omega_i t}|e^{-tA}w_0|_\beta = |w_0|_\beta,$$

hence (51.70) yields

$$e^{-\omega_i t}|u(t)|_\beta \geq |w_0|_\beta - e^{-(\omega_i - \omega_{i+1})t}|y_0|_\beta - \tilde{C}_i |v_0|_\beta^p > c|w_0|_\beta,$$

provided $c < 1 - \tilde{C}_i |v_0|_\beta^p / |w_0|_\beta$ and t is large enough. This concludes the proof. $\qquad \square$

Corollary 51.22. *Assume* (51.58). *Let* α, β, F, p *be as in Theorem* 51.21, *and let* u *be a global solution of* (51.9) *satisfying* $|u(t)|_\beta \to 0$ *as* $t \to \infty$. *Set*

$$\Lambda := \inf\{\lambda \le 0 : \lim_{t\to\infty} |u(t)|_\beta e^{-\lambda t} = 0\}$$

and assume $\Lambda \in (-\infty, 0)$. *Then there exist* $C_1, C_2 > 0$ *and* $i \ge 1$ *such that* $\Lambda = \omega_i$ *and*

$$C_1 e^{\omega_i t} \le |u(t)|_\beta \le C_2 e^{\omega_i t}, \qquad t \ge 0.$$

Proof. The same arguments as in the proof of [2, Corollary A.11] guarantee the existence of i such that $\Lambda = \omega_i$, $|u(t)|_\beta^{1/t} \to e^{\omega_i}$ and $\mathrm{dist}_{X_\beta}\big(u(t)/|u(t)|_\beta, S_\beta\big) \to 0$ as $t \to \infty$, where $S_\beta := \{v \in R_i X_\beta : |v|_\beta = 1\}$. Choose $\lambda \in (\omega_i, 0)$, $\lambda < \omega_{i-1}$ if $i > 1$, and let $\rho = \rho_i > 0$ be the constant from Theorem 51.21. Then $|u(t)|_\beta^{1/t} \to e^{\omega_i}$ implies $|u(t + t_0)|_\beta \le \rho e^{\lambda t}$ for $t_0 \ge 0$ large enough and all $t \ge 0$. Enlarging t_0 if necessary we may assume $|R_i u(t_0)|_\beta > \tilde{C}_i |v_0|_\beta^p$, where $v_0 := P_i u(t_0)$. Now the assertion follows from Theorem 51.21. $\square$

Remark 51.23. If $0 \notin \sigma(A)$, then the assumption $\Lambda < 0$ in Corollary 51.22 is automatically satisfied. In fact, using Theorem 51.21 with $\lambda = 0$ we obtain $\Lambda \le \omega_i$, where ω_i is the largest negative eigenvalue of $-A$.

The assumption $\Lambda > -\infty$ can be verified by an argument guaranteeing backward uniqueness (see [2, Lemma A.16 and Lemma B.4], for example, and cf. Example 51.24 below). $\square$

Example 51.24. Let L be the positive self-adjoint operator in the weighted space L_g^2 defined by (47.16). Recall that the domain of definition of L equals H_g^2, and L has compact inverse. Consider the problem

$$\begin{aligned} v_t + Av &= |v|^{p-1}v, & y \in \mathbb{R}^n,\ t > 0, \\ v(y,0) &= v_0(y), & y \in \mathbb{R}^n, \end{aligned} \tag{51.71}$$

where $Av = Lv - \lambda v$ and $p > 1$. Since L is self-adjoint and positive, it has bounded imaginary powers and $-A$ generates a strongly continuous analytic semigroup in $X_0 := L_g^2$ (see [16]). In addition, the domain of definition of A equals $X_1 := H_g^2$. Set $X_\theta = [X_0, X_1]_\theta$ for $\theta \in (0,1)$ and $X_{-1+\theta} = [X_{-1}, X_0]_\theta$, where X_{-1} is the completion of X_0 endowed with the norm $|v|_{-1} = |L^{-1}v|_0$. Then the abstract results in [16] imply $X_{-\theta} \doteq X_\theta'$ for $\theta \in (0,1]$. One can also easily verify $X_{1/2} = D(L^{1/2}) = H_g^1$ (cf. Remark 51.2(i)).

Let $p < p_S$. For simplicity assume $n \ge 3$ (the case $n \le 2$ is similar). Then $H_g^1 \hookrightarrow L_g^{2^*} \cap L_g^2$ due to Lemma 47.11 and, by interpolation, $X_\theta \hookrightarrow L_g^r \cap L_g^2$ for $\theta \in [0, 1/2]$ and $1/r = 2\theta/2^* + (1-2\theta)/2$. Using these embeddings, setting $F(v) = |v|^{p-1}v$, $z = \min(2, 2^*/p)$, $\beta = 1/2$ and $\alpha = 1 + (n/z' - n/2)/2$ (cf. Example 51.10) one obtains that $F : X_\beta \to X_{\alpha-1}$ and A satisfy the assumptions of Theorem 51.21. (Notice that we could also choose $\alpha = 1$ and β close to 1, $\beta < 1$, due to Remark 47.12(ii): in this case the assumption $p < p_S$ could be replaced by $p(n-4) < n$.)

Now assume $\lambda \notin \sigma(L)$ and let v be a global solution of (51.71) satisfying $|v(t)|_\beta \to 0$ as $t \to \infty$, $v_0 \not\equiv 0$ (such solutions exist due to Theorem 51.21). We will show the following:

(i) *There exist $\tilde{C}_1, \tilde{C}_2 > 0$ and $\omega_i \in \sigma(-A)$, $\omega_i < 0$, such that*

$$\tilde{C}_1 e^{\omega_i t} \le \|v(t)\|_\infty \le \tilde{C}_2 e^{\omega_i t}, \quad t \ge 1. \tag{51.72}$$

(ii) *Assume $\omega_1 < 0$ and let $\phi_1(y) = e^{-|y|^2/4}$ be the corresponding eigenfunction of A (see Lemma 47.13). Set $\hat{\omega} := \max(\omega_2, p\omega_1) < \omega_1$. Then there exists $M = M(v_0) \in \mathbb{R}$ such that*

$$\|v(t) - Me^{\omega_1 t}\phi_1\|_\infty \le Ce^{\hat{\omega} t}, \qquad t \ge 0. \tag{51.73}$$

Proof of (i). Set

$$\Lambda := \inf\{\kappa \le 0 : \lim_{t\to\infty} |v(t)|_\beta e^{-\kappa t} = 0\}.$$

Then $\Lambda < 0$ due to Remark 51.23. Let us show that $\Lambda > -\infty$.

Choose $\kappa < 0$ and assume

$$|v(t)|_\beta \le Ce^{\kappa t}, \qquad t \ge 0. \tag{51.74}$$

Let us prove that, given $t_0 > 0$, estimate (51.74) and $p < p_S$ guarantee

$$\|v(t)\|_\infty \le C(t_0)e^{\kappa t}, \qquad t \ge t_0. \tag{51.75}$$

Let $r > 1$, $\tilde{X}_0 = \tilde{X}_0(r) = L^r(\mathbb{R}^n)$, $\tilde{X}_1 = \tilde{X}_1(r) = W^{2,r}(\mathbb{R}^n)$ and let us rewrite (51.71) in the form

$$v_t + \tilde{A}v = \tilde{F}(v), \qquad t > 0,$$
$$v(0) = v_0, \tag{51.76}$$

where $\tilde{A}v = -\Delta v - \lambda v$ is considered as an unbounded operator in $\tilde{X}_0$ with domain $\tilde{X}_1$ and $\tilde{F}(v)(y) = |v(y)|^{p-1}v(y) + (y \cdot \nabla v(y))/2$. Let $\tilde{X}_\alpha$, $\alpha \in [-1, 1]$ be the scale of spaces constructed as in Example 51.4(i) (hence $\tilde{X}_\alpha \hookrightarrow W^{2\alpha,r}(\mathbb{R}^n)$ for $\alpha \ge 0$), and let $|\cdot|_\alpha^\sim$ denote the norm in this space.

Since $X_\beta = X_{1/2} = H_g^1$ and

$$\|y \cdot \nabla v(t)\|_2 \le C|v(t)|_\beta,$$

$$\|y \cdot \nabla v(t)\|_1 \le \int_{\mathbb{R}^n} |y| \cdot |\nabla v(t)| \, dy \le C \int_{\mathbb{R}^n} (|\nabla v(t)|g^{1/2})g^{-1/4} \, dy \le C|v(t)|_\beta,$$

$$\||v(t)|^{p-1}v(t)\|_r = \|v(t)\|_{pr}^p \le C|v(t)|_\beta^p, \qquad r \in [\max(1, 2/p), 2^*/p],$$

(51.74) guarantees

$$\|\tilde{F}(v(t))\|_r \le Ce^{\kappa t}, \qquad t \ge \tau, \tag{51.77}$$

where $r = \min(2, 2^*/p)$ and $\tau = 0$. Similar estimates as above and Lemma 47.11 imply $H_g^1 \hookrightarrow L^1 \cap L^{2^*}(\mathbb{R}^n)$, hence $H_g^1 \hookrightarrow \tilde{X}_0(r)$ for any $r \in (1, 2^*]$.

Fix $\varepsilon, T > 0$ small and assume that (51.77) is true for some $r \in [\min(2, 2^*/p), 2^*]$ and $\tau \ge 0$. Then using estimates (51.3) (with A replaced by $\tilde{A}$) we infer for all $t \ge \tau$,

$$|v(t+T)|_{1-\varepsilon}^{\sim} \le C\|e^{-T\tilde{A}}\|_{\mathcal{L}(\tilde{X}_0, \tilde{X}_{1-\varepsilon})} |v(t)|_0^{\sim}$$

$$+ \int_t^{t+T} \|e^{-(t+T-s)\tilde{A}}\|_{\mathcal{L}(\tilde{X}_0, \tilde{X}_{1-\varepsilon})} |\tilde{F}(v(s))|_0^{\sim} \, ds$$

$$\le C|v(t)|_\beta + C\int_t^{t+T} (t+T-s)^{\varepsilon-1} e^{\kappa s} \, ds \le Ce^{\kappa(t+T)},$$

where C depends on T, ε and the constants C in (51.74) and (51.77). If $r > n/2$, then $\tilde{X}_{1-\varepsilon} \hookrightarrow L^\infty(\mathbb{R}^n)$ for ε small enough and (51.75) follows. If $r \le n/2$, then $\tilde{X}_{1-\varepsilon} \hookrightarrow L^{q_1} \cap W^{1,q}(\mathbb{R}^n)$, where $1/q_1 = 1/r - (2-2\varepsilon)/n$, $1/q = 1/r - (1-2\varepsilon)/n$, hence

$$\|v(t)\|_{q_1} + \|\nabla v(t)\|_q \le Ce^{\kappa t} \qquad \text{for } t \ge \tau + T.$$

Notice that choosing ε small enough we have $q_1/p > r$ (due to $p < p_S$ and $r \ge \min(2, 2^*/p)$), $\tilde{q} := \varepsilon + (2-\varepsilon)q/2 > r$ and

$$\left.\begin{aligned}
\||v(t)|^{p-1}v(t)\|_{q_1/p} &= \|v(t)\|_{q_1}^p \le Ce^{p\kappa t}, \\
\|y \cdot \nabla v(t)\|_{\tilde{q}}^{\tilde{q}} &\le C\int_{\mathbb{R}^n} |\nabla v(t)|^{\tilde{q}-\varepsilon} |\nabla v(t)|^\varepsilon g^{\varepsilon/2} \, dy \\
&\le C\|\nabla v(t)\|_q^{\tilde{q}-\varepsilon} \|v(t)\|_{H_g^1}^\varepsilon \le Ce^{\tilde{q}\kappa t},
\end{aligned}\right\} \qquad t \ge \tau + T,$$

hence

$$\|\tilde{F}(v(t))\|_{\tilde{r}} \le Ce^{\kappa t}, \qquad t \ge \tilde{\tau},$$

where $\tilde{\tau} = \tau + T$ and $\tilde{r} = \min(q_1/p, \tilde{q}) > r$. An obvious bootstrap argument concludes the proof of (51.75).

Notice that similar estimates as above (or the choice $\alpha = 1$ mentioned above) imply $F(v) = |v|^{p-1}v \in C((0, \infty), X_0)$. Consequently, Theorem 51.1(v) (used with $\hat{X}_0 := X_{-\varepsilon}$, $\varepsilon > 0$ small) guarantees $u \in C^1((0, \infty), X_{-\varepsilon}) \cap C((0, \infty), X_{1-\varepsilon})$. We also have

$$|F(v)(t)|_{-\varepsilon} \le C|F(v)(t)|_0 \le Ch(t)|v(t)|_0 \le Ch(t)|v(t)|_{1/2-\varepsilon}, \qquad t \ge t_0,$$

where

$$h(t) := \||v|^{p-1}(t)\|_\infty \le C(t_0)^{p-1} e^{(p-1)\kappa t}$$

belongs to $L^2(t_0, \infty)$, hence [2, Lemma A.16] yields

$$|v(t)|_\beta \geq c|v(t)|_{-\varepsilon} \geq cC_1|v(t_0)|_{-\varepsilon}e^{-C_2(t-t_0)}$$

for suitable $c, C_1, C_2 > 0$. Consequently, $\Lambda > -\infty$.

Now we infer from Corollary 51.22 the existence of $C_1, C_2 > 0$ and $\omega_i \in \sigma(-A)$, $\omega_i < 0$, such that

$$C_1 e^{\omega_i t} \leq |v(t)|_\beta \leq C_2 e^{\omega_i t}, \quad t \geq 0. \tag{51.78}$$

Since (51.74) implies (51.75), we have

$$\|v(t)\|_\infty \leq \tilde{C}_2 e^{\omega_i t}, \quad t \geq 1, \tag{51.79}$$

and simple estimates based on the variation-of-constants formula also yield

$$|v(t)|_{\beta+\varepsilon} \leq \hat{C}_2 e^{\omega_i t}, \quad t \geq 1, \tag{51.80}$$

where $\varepsilon > 0$ is small. Since $X_{\beta+\varepsilon}$ is compactly embedded into X_β, given $\delta > 0$ we can find $C_\delta > 0$ such that

$$|v(t)|_\beta \leq \delta|v(t)|_{\beta+\varepsilon} + C_\delta\|v(t)\|_\infty. \tag{51.81}$$

Choosing $\delta < C_1/\hat{C}_2$, estimates (51.78), (51.80) and (51.81) imply

$$\|v(t)\|_\infty \geq \tilde{C}_1 e^{\omega_i t}, \quad t \geq 1,$$

for suitable $\tilde{C}_1 > 0$. Consequently, (51.72) is true.

Proof of (ii). Similarly as in (i), it suffices to prove (51.73) with $\|\cdot\|_\infty$ replaced by $|\cdot|_\beta$. The proof of this modified estimate is almost the same as the proof of Theorem 51.19; one just needs to replace estimate (51.55) by the more precise estimate (51.59) and estimate (51.56) by $|v(t)|_\beta \leq Ce^{\omega_1 t}$ (which follows from the proof of (i) or from Theorem 51.19). The only difference appears in the case $\hat{\omega} = p\omega_1 = \omega_2$, where one has to use a more precise estimate on the term $e^{-(t-s)A}P_2F(v(s))$. In fact, in this case Theorem 51.19 guarantees $v(s) = Me^{\omega_1 s}\phi_1 + w(s)$, where $|w(s)|_\beta \leq Ce^{(\omega_2+\varepsilon)s}$ and $\varepsilon \in (0, \omega_1 - \omega_2)$ is such that $\delta := (p-1)\omega_1 + \varepsilon < 0$. Consequently, $F(v(s)) = |M|^{p-1}Me^{p\omega_1 s}\phi_1^p + z(s)$, where $|z(s)|_{\alpha-1} \leq Ce^{(\omega_2+\delta)s}$. Let P_3, R_2 be the spectral projections introduced in the beginning of this subsection. Then $R_2\phi_1^p = 0$ due to Lemma 47.13, hence $P_2F(v(s)) = P_3(|M|^{p-1}Me^{p\omega_1 s}\phi_1^p) + P_2z(s)$ and

$$|e^{-(t-s)A}P_2F(v(s))|_\beta \leq C(t-s)^{\alpha-1-\beta}\left[e^{\omega_3(t-s)}e^{p\omega_1 s} + e^{\omega_2(t-s)}e^{(\omega_2+\delta)s}\right]$$

$$= C(t-s)^{\alpha-1-\beta}e^{\omega_2 t}\left[e^{(\omega_3-\omega_2)(t-s)} + e^{\delta s}\right].$$

This estimate is sufficient for the proof of (51.73) in the case $\hat{\omega} = p\omega_1 = \omega_2$. $\quad\square$

51.5. Singular initial data

In what follows we consider nonlinearities $F : X_\beta \to X_{\alpha-1}$ with (at most) polynomial growth and we will show that under suitable assumptions one can obtain existence for initial data $u_0 \in X_\delta$ with $\delta < \beta$. The following theorem is an abstract analogue of Theorem 15.2. In addition, it also covers the critical case (cf. Remark 15.4(i)).

We assume that $\beta > \delta$, $M, T > 0$, and we define the Banach space

$$Y_T := \{u \in L^\infty_{loc}((0,T], X_\beta) : \|u\|_{Y_T} := \sup_{t \in (0,T)} t^{\beta-\delta}|u(t)|_\beta < \infty\}.$$

We also denote by $B_M = B_{M,T}$ the closed ball in Y_T with center 0 and radius M.

Theorem 51.25. *Assume that* $p > 1$, $1 \geq \alpha > \beta > \delta \geq -1$, $\delta > \beta - 1/p$, $\beta \geq \alpha - 1$ *and* $F : X_\beta \to X_{\alpha-1}$ *satisfies*

$$|F(u) - F(v)|_{\alpha-1} \leq C_F |u - v|_\beta \big(1 + |u|_\beta^{p-1} + |v|_\beta^{p-1}\big). \tag{51.82}$$

Let $u_0 \in X_\delta$ *and let* Φ_{u_0} *be defined by* (51.13).
(i) If $\alpha > (\beta - \delta)p + \delta$, *then there exist* $M = M(|u_0|_\delta) \geq 1$ *and* $T = T(|u_0|_\delta) > 0$ *such that* Φ_{u_0} *possesses a unique fixed point in* $B_{M,T}$.
(ii) If $\alpha = (\beta - \delta)p + \delta$, *then there exist* $M = M(u_0) > 0$ *and* $T = T(u_0) > 0$ *such that* Φ_{u_0} *possesses a unique fixed point in* $B_{M,T}$.
In both cases, $u \in C([0,T], X_\delta) \cap C((0,T], X_\gamma)$ *for any* $\gamma \in [\delta, \alpha)$.

Proof. We will use the Banach fixed point theorem for the mapping $\Phi_{u_0} : B_M \to B_M$. Increasing C_F if necessary we may assume

$$|F(u)|_{\alpha-1} \leq C_F(1 + |u|_\beta^p). \tag{51.83}$$

Let $\gamma \in [\delta, \alpha)$, $0 < t \leq T \leq 1$, $M > 0$, $u \in B_M$, let C_A be the constant from (51.11) and denote $\xi_+ := \max(\xi, 0)$. Then

$$t^{\gamma-\delta}|\Phi_{u_0}(u)(t)|_\gamma \leq t^{\gamma-\delta}|e^{-tA}u_0|_\gamma + t^{\gamma-\delta}\left|\int_0^t e^{-(t-s)A}F(u(s))\,ds\right|_\gamma$$

$$\leq C_A|u_0|_\delta + C_A C_F t^{\gamma-\delta}\int_0^t (t-s)^{-(\gamma+1-\alpha)_+}\big(1 + |u(s)|_\beta^p\big)\,ds \tag{51.84}$$

$$\leq C_A|u_0|_\delta + C_A C_F t^{\gamma-\delta}\int_0^t (t-s)^{-(\gamma+1-\alpha)_+}\big(1 + M^p s^{(\delta-\beta)p}\big)\,ds < \infty$$

hence $\Phi_{u_0}(u)(t) \in X_\gamma$.

(i) Let $\alpha > (\beta - \delta)p + \delta$. Fix $M \geq \max(1, 2C_A|u_0|_\delta)$. Since $1 + M^p s^{(\delta - \beta)p} \leq 2M^p s^{(\delta - \beta)p}$ for $s \in (0,1]$, estimate (51.84) implies

$$t^{\beta - \delta}|\Phi_{u_0}(u)(t)|_\beta \leq \frac{1}{2}M + 2C_A C_F M^p B(\alpha - \beta, 1 - (\beta - \delta)p)T^{\alpha - \delta - (\beta - \delta)p} < M$$

provided T is small enough. Hence Φ_{u_0} maps B_M into B_M for such T.

Now let $u, v \in B_M$. Then, similarly as above,

$$t^{\beta - \delta}|\Phi_{u_0}(u)(t) - \Phi_{u_0}(v)(t)|_\beta \leq t^{\beta - \delta}\left|\int_0^t e^{-(t-s)A}\big(F(u(s)) - F(v(s))\big)\,ds\right|_\beta$$

$$\leq C_A C_F t^{\beta - \delta}\int_0^t (t-s)^{\alpha - 1 - \beta}|u(s) - v(s)|_\beta\big(1 + |u(s)|_\beta^{p-1} + |v(s)|_\beta^{p-1}\big)\,ds$$

$$\leq 3C_A C_F M^{p-1} B(\alpha - \beta, 1 - (\beta - \delta)p)T^{\alpha - \delta - (\beta - \delta)p}\|u - v\|_{Y_T},$$

hence

$$\|\Phi_{u_0}(u) - \Phi_{u_0}(v)\|_{Y_T} \leq \frac{1}{2}\|u - v\|_{Y_T},$$

provided T is small enough. Consequently, Φ_{u_0} is a contraction in B_M and it possesses a unique fixed point u.

Assume $0 < t_1 < t_2 \leq T$ and let either $t_1 \to t_2-$ or $t_2 \to t_1+$. Then

$$|u(t_2) - u(t_1)|_\gamma \leq \|e^{-t_1 A}\|_{\mathcal{L}(X_\delta, X_\gamma)}|(e^{-(t_2 - t_1)A} - 1)u_0|_\delta$$

$$+ \int_{t_1}^{t_2} \|e^{-(t_2 - s)A}\|_{\mathcal{L}(X_{\alpha - 1}, X_\gamma)}|F(u(s))|_{\alpha - 1}\,ds$$

$$+ \int_0^{t_1} \|e^{-(t_1 - s)A}\|_{\mathcal{L}(X_{\alpha - 1}, X_\gamma)}|(e^{-(t_2 - t_1)A} - 1)F(u(s))|_{\alpha - 1}\,ds$$

$$\to 0,$$

due to (51.11), (51.83), $u \in B_M$,

$$e^{-(t_2 - t_1)A}u_0 \to u_0 \qquad \text{in } X_\delta,$$

$$e^{-(t_2 - t_1)A}F(u(s)) \to F(u(s)) \qquad \text{in } X_{\alpha - 1}$$

and the Lebesgue theorem. Consequently, $u \in C((0,T], X_\gamma)$.

The continuity of $u : [0,T] \to X_\delta$ at $t = 0$ follows from the strong continuity of the semigroup e^{-tA} in X_δ and the estimate

$$|\Phi_{u_0}(u)(t) - e^{-tA}u_0|_\delta = \left|\int_0^t e^{-(t-s)A}F(u(s))\,ds\right|_\delta$$

$$\leq C_A C_F \int_0^t (t-s)^{-(\delta + 1 - \alpha)_+}\big(1 + |u(s)|_\beta^p\big)\,ds$$

$$\leq 2C_A C_F M^p B(\min(1, \alpha - \delta), 1 - (\beta - \delta)p)t^k,$$

where $k = \min(1, \alpha - \delta) - (\beta - \delta)p$.

(ii) Let $\alpha = (\beta - \delta)p + \delta$. First let us prove that

$$t^{\beta-\delta}|e^{-tA}u_0|_\beta \to 0 \qquad \text{as } t \to 0. \tag{51.85}$$

In fact, choose $t_k \to 0$ and set $S_k := t_k^{\beta-\delta}e^{-t_k A}$. Then $S_k \in \mathcal{L}(X_\delta, X_\beta)$ are uniformly bounded due to (51.11) and, given $w \in X_\beta$,

$$|S_k w|_\beta \le t_k^{\beta-\delta}|e^{-t_k A}w|_\beta \le t_k^{\beta-\delta}C_A|w|_\beta \to 0 \quad \text{as } k \to \infty.$$

Since X_β is dense in X_δ, we obtain $|S_k w|_\beta \to 0$ for any $w \in X_\delta$. Consequently, (51.85) is true.

Choose $M > 0$ such that $2C_A C_F M^{p-1}B(\alpha-\beta, 1-(\beta-\delta)p) \le 1/4$ and $T \in (0, 1]$ such that

$$t^{\beta-\delta}|e^{-tA}u_0|_\beta \le M/2 \qquad \text{for all } t \le T$$

(this choice is possible due to (51.85)). Let $u \in B_M = B_{M,T}$. Then

$$t^{\beta-\delta}|\Phi_{u_0}(u)(t)|_\beta \le t^{\beta-\delta}|e^{-tA}u_0|_\beta + C_A C_F t^{\beta-\delta}\int_0^t (t-s)^{\alpha-1-\beta}\left(1 + |u(s)|_\beta^p\right)ds$$

$$\le M/2 + C_A C_F \frac{T^{\alpha-\delta}}{\alpha-\beta} + C_A C_F M^p B(\alpha - \beta, 1 - (\beta - \delta)p)$$

$$\le 3M/4 + C_A C_F \frac{T^{\alpha-\delta}}{\alpha-\beta} \le M,$$

$$\tag{51.86}$$

provided T is small enough. Consequently, Φ_{u_0} maps B_M into B_M. Similar estimates (cf. (i)) show that Φ_{u_0} is a contraction and the corresponding fixed point possesses the required continuity properties. $\square$

Remarks 51.26. (i) It is easily seen from the proof that the existence time T in Theorem 51.25(ii) can be chosen uniform for initial data belonging to a compact subset of X_δ.

(ii) Theorem 51.25 guarantees that the solution u is unique in the ball B_M. However, similarly as in the proof of Theorem 15.2 one can prove the uniqueness of this solution in the space $\mathcal{C} := C([0, T], X_\delta) \cap C((0, T], X_\beta)$. In fact, let $v \in \mathcal{C}$ be any solution of (51.9) on a (small) interval $[0, \tau]$. Then $K := \{v(t) : 0 \le t \le \tau\}$ is compact in X_δ hence (i) and the proof of Theorem 51.25 guarantee the solvability of (51.9) in B_{M,T_K} for some $M, T_K > 0$ and for all initial data in K. Let $U(t)v(s)$ denote the corresponding solution starting at $v(s)$, $s \in [0, \tau]$, $t \in [0, T_K]$. Then $t^{\beta-\delta}|U(t)v(s)|_\beta \le M$. Fix $s \in (0, \min(T_K, \tau))$ and denote $u_1(t) := U(t)v(s)$ and $u_2(t) := v(t + s)$. Then $u_1 \in C([0, T_K], X_\beta)$ due to the existence part of Theorem 51.7 and the uniqueness in Theorem 51.25, and $u_2 \in C([0, \tau - s], X_\beta)$. In addition, both u_1 and u_2 solve (51.9) with initial data $v(s) \in X_\beta$. Hence

$u_1 = u_2$ on $[0, \min(T_K, \tau - s)]$, due to the uniqueness in Theorem 51.7 (see also Remark 51.8(v)). Consequently,

$$t^{\beta-\delta}|v(t+s)|_\beta = t^{\beta-\delta}|U(t)v(s)|_\beta \leq M.$$

Fix $t > 0$ small and let $s \to 0+$ in the previous estimate. Then we obtain $t^{\beta-\delta}|v(t)|_\beta \leq M$, hence $v = u$ due to the uniqueness in Theorem 51.25.

The example in Remark 15.4(iii) and Example 51.27 below show that the restriction $v \in C((0, T], X_\beta)$ in the uniqueness statement above is necessary, in general.

(iii) Let the assumptions of Theorem 51.25(i) be fulfilled. Similarly as in the case of initial data in X_β one can prove the existence of the maximal existence time $T_{\max} = T_{\max}(u_0)$, continuous dependence on initial data, positivity of the solution u if X_δ is ordered and e^{-tA_0} is positive, etc. For example, given $u_0, \tilde{u}_0 \in X_\delta$, there exists $T = T(|u_0|_\delta, |\tilde{u}_0|_\delta) > 0$ and $C > 0$ such that

$$|u(t) - \tilde{u}(t)|_\gamma \leq Ct^{\delta-\gamma}|u_0 - \tilde{u}_0|_\delta, \qquad t \leq T, \tag{51.87}$$

provided $\gamma \in [\delta, \alpha)$.

(iv) A simple modification of the proof of Theorem 51.25 shows that the assumption $\beta \geq \alpha - 1$ is superfluous (cf. also Remark 51.8(v)).

(v) Assumption (51.82) in Theorem 51.25 can be replaced with

$$|F(u) - F(v)|_{\alpha-1} \leq C_F \sum_{i=1}^{k} |u - v|_{\beta_i}\big(1 + |u|_{\beta_i}^{p_i-1} + |v|_{\beta_i}^{p_i-1}\big), \tag{51.88}$$

where (for all $i = 1, 2, \ldots, k$) $p_i > 1$, $1 \geq \alpha > \beta_i > \delta \geq -1$, $\delta > \beta_i - 1/p_i$, $\beta_i \geq \alpha - 1$, $\alpha \geq (\beta_i - \delta)p_i + \delta$ and $F : \bigcap_{i=1}^{k} X_{\beta_i} \to X_{\alpha-1}$. In this case, it is sufficient to use the fixed point argument in the space

$$Y_T := \left\{ u \in L^\infty_{loc}\big((0, T], \bigcap_{i=1}^{k} X_{\beta_i}\big) : \|u\|_{Y_T} := \max_i \sup_{t\in(0,T)} t^{\beta_i-\delta}|u(t)|_{\beta_i} < \infty \right\}.$$

For more complex generalizations of this type (and applications of such generalizations) see [446] and the references therein.

(vi) Estimate (51.84) implies

$$|u(t)|_\gamma \leq C(1 + t^{-(\gamma-\delta)}|u_0|_\delta) \quad \text{for all } \gamma \in [\delta, \alpha) \text{ and } t \in (0, T], \tag{51.89}$$

where T is from Theorem 51.25. $\quad\square$

Example 51.27. Let Ω, A and X_α, $\alpha \in [-1,1]$, be as in Example 51.4(i), $p > 1$ and $q \geq n(p-1)/2$, $q > 1$. Let $F(u)(x) = f(x, u(x))$, where $f = f(x, u)$ is a C^1-function satisfying $f(\cdot, 0) \in L^z(\Omega)$ and the growth condition $|\partial_u f(x, u)| \leq a(x) + C|u|^{p-1}$ with $a \in L^{p'z}(\Omega)$ and $z \in (\max(1, q/p), q]$, $z < q$ if $q = n(p-1)/2$ (cf. Example 51.10). Assume $u_0 \in X_0 = L^q(\Omega)$ and set

$$\beta = \frac{1}{2}\left(\frac{n}{q} - \frac{n}{pz}\right), \qquad \alpha = \frac{1}{2}\left(2 + \frac{n}{q} - \frac{n}{z}\right), \qquad \delta = 0.$$

Then $1 \geq \alpha \geq \beta p > 0$ and $\alpha > \beta p$ if $q > n(p-1)/2$, $\alpha < 1$ if $q = n(p-1)/2$. In addition, the choice of α and β guarantees $X_\beta \hookrightarrow W^{2\beta, q}(\Omega) \hookrightarrow L^{pz}(\Omega)$ and $L^z(\Omega) \hookrightarrow X_{\alpha-1}$ (since $X_{1-\alpha}(q') \hookrightarrow W^{2-2\alpha, q'}(\Omega) \hookrightarrow L^{z'}(\Omega)$). Consequently, F satisfies (51.82) and Theorem 51.25 guarantees the existence of a unique solution $u \in C([0, T], L^q(\Omega))$ in the corresponding ball B_M. In addition, $u \in C((0, T_{\max}), X_\gamma)$ for any $\gamma < \alpha$.

Let $f = f(u)$ where f' is locally Hölder continuous. If $q > n(p-1)/2$, then we may set $z = q$, hence $\alpha = 1$ and

$$u \in C((0, T_{\max}), W^{2\gamma, q} \cap W_0^{1,q}(\Omega)) \quad \text{for any } \gamma < 1. \tag{51.90}$$

If $q = n(p-1)/2$, then we may choose $z < q$ arbitrarily close to q, hence α arbitrarily close to 1, so that (51.90) remains true as well. Now Example 51.10 guarantees

$$u \in C^1((0, T_{\max}), W_0^{1, \tilde{q}}(\Omega)) \cap C((0, T_{\max}), W^{2, \tilde{q}}(\Omega)) \quad \text{for any } \tilde{q} \in [q, \infty). \tag{51.91}$$

In addition, Remark 51.26(iii) and (51.20) show

$$u(t; u_{0,k}) \to u(t; u_0) \quad \text{in } BUC^1(\Omega) \tag{51.92}$$

provided $u_{0,k} \to u_0$ in $L^q(\Omega)$ and $t \in (0, T_{\max}(u_0))$ is fixed. $\quad\square$

Example 51.28. Let us consider the situation in Example 51.27, where $Au = -\Delta u - \lambda u$, $f(u) = |u|^{p-1}$, $u_0 \in L^q \cap L^2(\Omega)$ with $q \geq q_c = n(p-1)/2$, $q > 1$. We will show that the corresponding energy function

$$E(t) := \int_\Omega \left(\frac{1}{2}|\nabla u(t)|^2 - \frac{\lambda}{2}u^2(t) - \frac{1}{p+1}|u(t)|^{p+1}\right) dx$$

is differentiable for $t > 0$. In addition, we will also prove that the problem generates a dynamical system in $H_0^1 \cap L^q(\Omega)$ provided $q \geq \max(q_c, p+1)$.

Example 51.27 and Theorem 51.25 guarantee the existence of a unique maximal solution $u \in C([0, T_{\max})), L^q(\Omega))$. In what follows we set $T := T_{\max}$. Let us first show that

$$u \in C([0, \tau], L^2(\Omega)) \quad \text{for some } \tau \in (0, T). \tag{51.93}$$

If $2 \geq n(p-1)/2$, then this assertion follows from the well-posedness in $L^2(\Omega)$ (see Example 51.27). Hence assume $2 < n(p-1)/2$. Then $q > 2$. Let $i \geq 0$ be the integer such that $2p^i < q \leq 2p^{i+1}$.

First assume $i = 0$. Observe that estimate (15.2) in Theorem 15.2 and Remark 15.4(i) remain obviously true for $\lambda \neq 0$. Applying (15.2) with $r = 2p$, we obtain

$$\|u^p(s)\|_2 = \|u(s)\|_{2p}^p \leq C\|u(s)\|_q^p s^{-\theta} \leq Cs^{-\theta}, \quad 0 < s < \tau,$$

for some $\tau > 0$, where u^p denotes $|u|^{p-1}u$ and

$$0 \leq \theta := \frac{np}{2}\left(\frac{1}{q} - \frac{1}{2p}\right) = \frac{n}{2q}(p - (q/2)) \leq \frac{p - (q/2)}{p - 1} < 1.$$

Therefore $u^p \in L^1((0,\tau), L^2(\Omega))$, hence $g(t) := \int_0^t e^{-(t-s)A}u^p(s)\,ds \in C([0,\tau], L^2(\Omega))$, and (51.93) is satisfied.

Next assume $i \geq 1$. For any $r \in [2p, qp]$, we have

$$u \in C([0,\tau], L^r(\Omega)) \Rightarrow u^p \in C([0,\tau], L^{r/p}(\Omega)) \Rightarrow g \in C([0,\tau], L^{r/p}(\Omega))$$
$$\Rightarrow u \in C([0,\tau], L^{r/p}(\Omega)), \tag{51.94}$$

due to $e^{-tA}u_0 \in C([0,\tau], L^2 \cap L^q(\Omega))$. First applying (51.94) with $r = q$, we obtain $u \in C([0,\tau], L^{q/p}(\Omega))$, hence $u \in C([0,\tau], L^{2p^i}(\Omega))$, due to $q/p \leq 2p^i \leq q$. Then applying (51.94) iteratively with $r = 2p^i, 2p^{i-1}, \ldots, 2p$, we end up with (51.93).

Due to (51.91) we know that there exists a positive constant C_∞ such that $|u| \leq C_\infty$ on $\Omega \times [\tau/2, T]$. Fix $\tilde{f} \in BC^1(\mathbb{R})$ such that $\tilde{f}(u) = f(u)$ for $|u| \leq C_\infty$. Then u is a solution of the equation $u_t - \Delta u - \lambda u = \tilde{f}(u)$ for $t \geq \tau/2$, hence estimate (51.91) (obtained with $q = 2$ and initial data $u(\tau/2)$) implies

$$u \in C^1([\tau, T), H_0^1(\Omega)) \cap C([\tau, T), H^2(\Omega)).$$

Consequently, $u \in C([0,T), L^2(\Omega)) \cap C^1((0,T), H_0^1(\Omega)) \cap C((0,T), H^2(\Omega))$. In particular, $u \in C^1((0,T), L^2(\Omega))$. Since also $u \in C^1((0,T), L^{\tilde{q}}(\Omega))$ for any $\tilde{q} \geq q$ due to (51.91), we have $u \in C^1((0,T), L^{p+1}(\Omega))$, hence $E \in C^1((0,T))$.

Next assume $u_0 \in H_0^1 \cap L^{q*}(\Omega)$, $q^* := \max(q_c, p+1)$. We already know that the solution satisfies $u \in C([0,T), L^2 \cap L^q(\Omega))$ for any $q \in [q_c, \infty)$, $q > 1$, in particular $u \in C([0,T), L^{q^*}(\Omega))$. Let us prove $u \in C([0,T), H_0^1(\Omega))$. Since $u \in C((0,T), H_0^1(\Omega))$ and $e^{-tA}u_0 \in C([0,T), H_0^1(\Omega))$ it is sufficient to show

$$\left\|\int_0^t e^{-(t-s)A}|u(s)|^{p-1}u(s)\,ds\right\|_{1,2} \to 0 \qquad \text{as } t \to 0. \tag{51.95}$$

Let $X_\theta(2)$, $\theta \in [0,1]$, be the scale of spaces from Example 51.4(i) (in particular, $X_0(2) = L^2(\Omega)$ and $X_{1/2}(2) = H_0^1(\Omega)$). If $q_c \neq p+1$, then there exists $\varepsilon > 0$ such that $L^{q^*/p}(\Omega) \hookrightarrow X_{-1/2+\varepsilon}(2)$, hence (51.95) follows from

$$\left\|e^{-(t-s)A}\right\|_{\mathcal{L}(X_{-1/2+\varepsilon}(2), X_{1/2}(2))} \leq (t-s)^{-1+\varepsilon}$$

and $\||u(s)|^{p-1}u(s)\|_{q^*/p} = \|u\|_{q^*}^p \le C$. Let $q_c = p+1$. Then $L^{q^*/p}(\Omega) \hookrightarrow X_{-1/2}(2)$. Since the estimate (51.85) is uniform for u_0 lying in a compact set of X_δ and the set $\{|u(s)|^{p-1}u(s) : s \in [0,T]\}$ is compact in $L^{q^*/p}(\Omega)$, we have

$$\|e^{-(t-s)A}|u(s)|^{p-1}u(s)\|_{1,2} = o\big((t-s)^{-1}\big) \qquad \text{as } t \to 0. \tag{51.96}$$

Now the smoothing estimate (15.2) with $q = p+1 = q_c$, $r = 2p$ guarantees

$$\||u(s)|^{p-1}u(s)\|_2 = \|u(s)\|_{2p}^p \le C\|u(s)\|_{p+1}^p s^{-1/2} \le Cs^{-1/2},$$

hence

$$\|e^{-(t-s)A}|u(s)|^{p-1}u(s)\|_{1,2} \le C(t-s)^{-1/2}s^{-1/2}. \tag{51.97}$$

Interpolation between (51.96) and (51.97) yields

$$\|e^{-(t-s)A}|u(s)|^{p-1}u(s)\|_{1,2} = o\big((t-s)^{-3/4}s^{-1/4}\big)$$

which guarantees (51.95). Consequently, $u \in C([0,T], H_0^1 \cap L^{q^*}(\Omega))$ and $E \in C([0,T))$. Similar estimates as above show the continuous dependence of solutions on initial data in $H_0^1 \cap L^{q^*}(\Omega)$, hence the problem generates a dynamical system in this space. Obviously, the same remains true for the space $H_0^1 \cap L^q(\Omega)$ with $q \in (q^*, \infty)$.

If $\lambda = 0$, then the continuity properties of E can in some cases be proved without the assumption $u_0 \in L^2(\Omega)$. For example, let $1 < p \le p_S$, $\Omega = \mathbb{R}^n$, $u_0 \in L^{p+1}(\mathbb{R}^n)$ and $\nabla u_0 \in L^2(\mathbb{R}^n)$. Set $q := p+1 \ge n(p-1)/2$. Then (51.91) shows

$$u \in C([0,T], L^{p+1}(\mathbb{R}^n)) \cap C^1((0,T), L^{p+1}(\mathbb{R}^n)) \cap C((0,T), W^{2,\tilde{q}}(\mathbb{R}^n))$$

for any $\tilde{q} \ge p+1$. In addition, estimate (51.84) implies $|u(t)|_\gamma \le C(\|u_0\|_{p+1})t^{-\gamma}$ for any $\gamma < 1$. If $p = p_S$ set $\gamma = 1/(2p)$. Otherwise fix $\gamma < 1/(2p)$ such that $X_\gamma = W^{2\gamma,p+1}(\mathbb{R}^n) \hookrightarrow L^{2p}(\mathbb{R}^n)$ and set

$$v(t) := \int_0^t e^{-(t-s)A} F(u(s))\, ds.$$

Then

$$\|v(t)\|_{1,2} \le \int_0^t (t-s)^{-1/2}\|F(u(s))\|_2\, ds \le \int_0^t (t-s)^{-1/2}\|u(s)\|_{2p}^p\, ds$$

$$\le \int_0^t (t-s)^{-1/2}|u(s)|_\gamma^p\, ds \le M \int_0^t (t-s)^{-1/2}s^{-\gamma p}\, ds < \infty$$

and

$$\|v(t)\|_{1,2} \to 0 \qquad \text{as} \quad t \to 0, \tag{51.98}$$

due to $M = M(t) \to 0$ as $t \to 0$ if $p = p_S$ (cf. estimates in (51.86)). Since also

$$\|\nabla e^{-tA} u_0\|_2 = \|e^{-tA} \nabla u_0\|_2 \le \|\nabla u_0\|_2,$$

we obtain $\nabla u(t) \in L^2(\mathbb{R}^n)$. Similar estimates show the local Hölder continuity of $v : (0,T) \to H^1(\mathbb{R}^n)$ and $\nabla(e^{-tA}u_0) \in C^1((0,T), L^2(\mathbb{R}^n))$. Since $u : (0,T) \to BUC \cap L^{2p}(\mathbb{R}^n)$ is locally Hölder continuous due to interpolation and $\nabla u : (0,T) \to L^2(\mathbb{R}^n)$ is also locally Hölder continuous, we have $F(u) \in C^\rho((0,T), H^1(\mathbb{R}^n))$ for some $\rho > 0$. Finally, (51.98) and Theorem 51.1(v) imply $v \in C^1((0,T), H^1(\mathbb{R}^n)) \cap C([0,T], H^1(\mathbb{R}^n))$. Since also $\nabla(e^{-tA}u_0) = e^{-tA}(\nabla u_0) \in C([0,T], L^2(\mathbb{R}^n))$ we see that the energy function E belongs to $C^1((0,T)) \cap C([0,T])$. $\square$

Example 51.29. Let Ω, A and X_α, $\alpha \in [-1,1]$, be as in Example 51.4(i), $F(u) = |u|^{r-1}u - \mu|\nabla u|^p$, where $\mu \in \mathbb{R}$, $p, r > 1$, $q > n(p-1)$, $1/r > 1/p - q/n$. Assume $u_0 \in X_{1/2} = W_0^{1,q}(\Omega)$, choose $z \in (\max(1, q/p), q)$ such that $1/r \ge 1/p - z/n$ and set

$$\beta = \frac{1}{2}\left(1 + \frac{n}{q} - \frac{n}{pz}\right), \qquad \alpha = \frac{1}{2}\left(2 + \frac{n}{q} - \frac{n}{z}\right), \qquad \delta = \frac{1}{2}.$$

Then $1 > \alpha > (\beta - \delta)p + \delta > 0$ and $\delta \in (\beta - 1/p, \beta)$. Since $F_2(u) := |\nabla u|^p$ and $F_1(u) := |u|^{r-1}u$ can be viewed as

$$F_2 : X_\beta \hookrightarrow W^{2\beta,q}(\Omega) \hookrightarrow W^{1,pz}(\Omega) \xrightarrow{\nabla} (L^{pz}(\Omega))^n \xrightarrow{|\cdot|^p} L^z(\Omega) \hookrightarrow X_{\alpha-1},$$

$$F_1 : X_\beta \hookrightarrow W^{2\beta,q}(\Omega) \hookrightarrow L^{rz}(\Omega) \xrightarrow{|u|^{r-1}u} L^z(\Omega) \hookrightarrow X_{\alpha-1},$$

we see that F satisfies (51.82). Now Theorem 51.26 and the same bootstrap argument as in Example 51.27 guarantee the existence of $T := T_{\max}$ and of a solution $u \in C([0,T], W_0^{1,q}(\Omega)) \cap C((0,T), W^{2\gamma,\tilde{q}}(\Omega))$, where $\gamma < 1$ and $\tilde{q} \in [q, \infty)$ are arbitrary. Choose $\gamma, \tilde{q}$ such that $(2\gamma - 1)\tilde{q} > n$. Then $W_{\tilde{q}} := W^{2\gamma-1,\tilde{q}}(\Omega) \hookrightarrow BUC(\Omega)$ and $|\nabla u| \in C((0,T), W_{\tilde{q}} \cap W_q)$. If $w \in W_{\tilde{q}} \cap W_q$, $w \ge 0$, then $w \le C$ in Ω, hence

$$|w^p(x) - w^p(y)| \le pC^{p-1}|w(x) - w(y)| \tag{51.99}$$

and using the intrinsic norm in W_q (see [13], for example) we obtain $w^p \in W_q$, $\|w^p\|_{W_q} \le pC^{p-1}\|w\|_{W_q}$. Since $|\nabla u| \in C((0,T), W_{\tilde{q}} \cap W_q)$ and $W_{\tilde{q}} \hookrightarrow BUC(\Omega)$, $W_q \hookrightarrow L^q(\Omega)$, using (51.99) we obtain $|\nabla u|^p \in C((0,T), L^q(\Omega))$. This fact, the local boundedness of $|\nabla u|^p : (0,T) \to W_q$ and interpolation yield $|\nabla u|^p \in C((0,T), W^{s,q}(\Omega))$ for $s \in (0, 2\gamma - 1)$, hence $F_2(u) \in C((0,T), X_\eta)$ for η small enough. Similar estimates show $F_1(u) \in C((0,T), X_\eta)$ for η small. Now Theorem 51.1(v) guarantees

$$u \in C^1((0,T), L^q(\Omega)) \cap C((0,T), W^{2,q} \cap W_0^{1,q}(\Omega)). \quad \square \tag{51.100}$$

Example 51.30. Let Ω, A and X_α, $\alpha \in [0,1]$, be as in Remark 51.11, $u_0 \in X_0$ and $F(u) = f(u, \nabla u)$, where $f \in C^1$,

$$|f_\xi(u,\xi)| \le M(|u|)(1 + |\xi|^{p-1}), \qquad 1 < p < 2,$$

and $M : \mathbb{R}_+ \to \mathbb{R}_+$ is locally bounded. Set $\delta = 0$, $\alpha = 1$, $\beta = 1/2 + \varepsilon$, where $\varepsilon \in (0, 1/p - 1/2)$, and fix $C_\infty > \|u_0\|_\infty$. Since $X_\beta \hookrightarrow BC^1(\overline{\Omega})$, (51.82) is true with $C_F = C(F, C_\infty)$ for all $u, v \in X_\beta$ satisfying $\|u\|_\infty, \|v\|_\infty \le C_\infty$. Now an obvious modification of Theorem 51.25 shows the well-posedness of problem (51.4) in $X_0 \in \{L^\infty(\Omega), BC(\overline{\Omega})\}$ (see also [344, Theorem 7.1.6]). $\square$

Example 51.31. Let Ω, A and X_α, $\alpha \in [-1,1]$, be as in Example 51.4(i), Ω bounded, $p \in (1, 1 + 2/n)$, $F(u) = \pm|u|^{p-1}u$ and u_0 be a bounded Radon measure in Ω. Fix $q \in (1, p)$ and choose δ such that

$$n - \frac{n}{q} < -2\delta < \frac{n+2}{p} - \frac{n}{q}.$$

Notice that $\delta \in (-1, 0)$. Set $\alpha = 1 + \delta$ and choose β such that

$$\frac{1}{2}\left(\frac{n}{q} - \frac{n}{p}\right) < \beta < \frac{1}{p} + \delta.$$

Then $u_0 \in X_\delta$ (since $X_\delta = (W_0^{-2\delta, q'}(\Omega))'$ and $W_0^{-2\delta, q'}(\Omega) \hookrightarrow C_0(\Omega)$) and $F : X_\beta \to X_{\alpha-1}$ (since $X_\beta \hookrightarrow L^p(\Omega)$ and $L^1(\Omega) \hookrightarrow X_\delta = X_{\alpha-1}$). In addition, $\alpha > (\beta - \delta)p + \delta > 0$ and $\delta \in (\beta - 1/p, \beta)$. Consequently, we can use Theorem 51.25 in order to get a solution $u \in C([0,T], X_\delta) \cap C((0,T], X_\gamma)$ for any $\gamma < 1 + \delta$. Choosing $\gamma = 0$ we obtain $u \in C((0,T], L^q(\Omega))$. Since $q > 1 > n(p-1)/2$, Examples 51.27 and 51.9 guarantee that u is a classical solution for $t > 0$.

Let us mention that the assumption $p < 1 + 2/n$ is also necessary for the solvability of (51.9) if u_0 is the Dirac distribution, see [95]. $\square$

Example 51.32. Consider the system

$$
\left.
\begin{aligned}
\partial_t u_1 - \Delta u_1 &= |u_2|^{p_1-1}u_2, && x \in \Omega,\ t > 0, \\
\partial_t u_2 - \Delta u_2 &= |u_1|^{p_2-1}u_1, && x \in \Omega,\ t > 0, \\
u_1 = u_2 &= 0, && x \in \partial\Omega,\ t > 0, \\
u_1(x,0) = u_{0,1}(x),\quad u_2(x,0) &= u_{0,2}(x), && x \in \Omega,
\end{aligned}
\right\} \quad (51.101)
$$

where $(u_{0,1}, u_{0,2}) \in X_0 := L^{r_1} \times L^{r_2}(\Omega)$, $p_1, p_2, r_1, r_2 > 1$ and $\Omega \subset \mathbb{R}^n$ is uniformly regular of class C^2. Assume

$$\max\left(\frac{n}{r_2}p_1 - \frac{n}{r_1}, \frac{n}{r_1}p_2 - \frac{n}{r_2}\right) \le 2. \qquad (51.102)$$

Set $X_1 := W^{2,r_1} \cap W_0^{1,r_1}(\Omega) \times W^{2,r_2} \cap W_0^{1,r_2}(\Omega)$, $Au := (-\Delta u_1, -\Delta u_2)$ for $u = (u_1, u_2) \in X_1$, and let X_α be defined similarly as in Example 51.4(i). We will use Remark 51.26(v) in order to prove the well-posedness of (51.101) in X_0.

Choose $\alpha \in (0,1)$ with $\alpha > \max_i(1 - n/(2r_i'))$, set $\delta = 0$, $\beta_i = \alpha/p_i$, $i = 1,2$, and define $z_i \in (1, r_i)$ by $n/z_i = n/r_i + 2 - 2\alpha$. Then $L^{z_1} \times L^{z_2}(\Omega) \hookrightarrow X_{\alpha-1}$ and (51.102) guarantees $W^{2\beta_1, r_2}(\Omega) \hookrightarrow L^{p_1 z_1}(\Omega)$, $W^{2\beta_2, r_1}(\Omega) \hookrightarrow L^{p_2 z_2}(\Omega)$. Now it is easy to verify (51.88), hence (51.101) is well-posed in X_0. Theorem 32.1(ii) shows that condition (51.102) is optimal. $\square$

Theorem 51.33. *Let $\alpha, \beta, \delta, p, F$ be as in Theorem 51.25. Assume, in addition, that $\omega(-A) < 0$ and*

$$|F(u)|_{\alpha-1} = o(|u|_\beta) \qquad as \quad |u|_\beta \to 0$$

if $\alpha > (\beta - \delta)p + \delta$,

$$|F(u)|_{\alpha-1} \le C_F |u|_\beta^p$$

if $\alpha = (\beta - \delta)p + \delta$. Then, given $\tilde\omega \in (\omega(-A), 0)$, there exists $\eta > 0$ and $C > 0$ such that the solution u with initial data u_0 satisfying $|u_0|_\delta < \eta$ exists globally and

$$|u(t)|_\beta \le C t^{\delta - \beta} e^{\tilde\omega t} |u_0|_\delta \qquad for \; all \; t \ge 0. \tag{51.103}$$

Proof. Fix $\eta_1 > 0$ and assume $|u_0|_\delta \le \eta_1$. If $\alpha > (\beta - \delta)p + \delta$, then estimate (51.87) with $\tilde u_0 = 0$ shows that

$$|u(t)|_\beta \le C_1 t^{\delta - \beta} |u_0|_\delta \qquad for \; all \; t \in (0, T_1], \tag{51.104}$$

where $T_1 = T_1(\eta_1) \in (0,1]$. Let $\delta^* > 0$ be the constant from Theorem 51.17. Choose $\eta > 0$ such that $C_1 T_1^{\delta - \beta} \eta < \delta^*$. Then the conclusion follows from (51.104) and Theorem 51.17 applied to the initial data $u(T_1)$.

If $\alpha = (\beta - \delta)p + \delta$ choose $\eta > 0$ such that $C_F(C^*)^p B(\alpha - \beta, 1 - (\beta - \delta)p)\eta^{p-1} < 1$, where $C^* := 2C_A$. Assume $|u_0|_\delta < \eta$ and set

$$T = \sup\{t \in (0, T_{\max}(u_0)) : |u(s)|_\beta \le C^* s^{\delta - \beta} |u_0|_\delta \text{ for all } s \in (0, t]\}.$$

Notice that $T > 0$, since $u \in B_{M, T(M)}$ and the constant M can be chosen arbitrarily small in the proof of Theorem 51.25(ii). If $T = \infty$, then (51.104) is true for $t \le T_1 := 1$ and we can proceed as in the case $\alpha > (\beta - \delta)p + \delta$. Assume $T < \infty$. Then $T < T_{\max}(u_0)$, hence

$$|u(T)|_\beta = C^* T^{\delta - \beta} |u_0|_\delta. \tag{51.105}$$

On the other hand,

$$|u(T)|_\beta \le C_A T^{\delta - \beta} |u_0|_\delta + C_A C_F \int_0^T (T - s)^{\alpha - 1 - \beta} |u(s)|_\beta^p \, ds$$

$$\le C_A T^{\delta - \beta} |u_0|_\delta + C_A C_F (C^*)^p B(\alpha - \beta, 1 - (\beta - \delta)p)|u_0|_\delta^p T^{\delta - \beta}$$

$$< C^* T^{\delta - \beta} |u_0|_\delta,$$

which yields a contradiction and concludes the proof. $\square$

51.6. Uniform bounds from L^q-estimates

In this part we present an abstract approach for obtaining L^∞-bounds of solutions from L^q-bounds. We will assume that (51.2) is true with $\omega < 0$ and use the scale (X_α, A_α) introduced above.

The idea of the proof of the next proposition is contained in the proof of [14, Theorem 12.8].

Proposition 51.34. *Let $0 \le \beta < \alpha \le 1$, $-1 \le \gamma < \beta$, $T \in (0, \infty]$ and $C_\gamma > 0$. Let $F : X_\beta \to X_{\alpha-1}$ be continuous and*

$$|F(u)|_{\alpha-1} \le C_F(|u|_\gamma)(1 + |u|_\beta^{1-\varepsilon}), \qquad u \in X_\beta, \tag{51.106}$$

where $\varepsilon \in (0, 1)$. Let $u_0 \in X_\beta$ and let $u \in C([0, T), X_\beta)$ solve (51.9). If $|u(t)|_\gamma \le C_\gamma$ for all $t \in [0, T)$, then $|u(t)|_\beta \le C_\beta$ for all $t \in [0, T)$, where C_β depends on C_γ and $|u_0|_\beta$ but not on T.

Proof. Let $\tilde{T} \in (0, T)$ and $t \le \tilde{T}$. Using (51.3) we obtain

$$
|u(t)|_\beta \le |e^{-tA}u_0|_\beta + \int_0^t \|e^{-(t-s)A}\|_{\mathcal{L}(X_{\alpha-1}, X_\beta)}|F(u(s))|_{\alpha-1}\, ds
$$

$$
\le c|u_0|_\beta + c\int_0^t e^{\omega(t-s)}(t-s)^{\alpha-1-\beta}C_F(C_\gamma)(1 + |u(s)|_\beta^{1-\varepsilon})\, ds
$$

$$
\le c|u_0|_\beta + cC_F(C_\gamma)\int_0^\infty e^{\omega\tau}\tau^{\alpha-1-\beta}\, d\tau\Big(1 + \sup_{0 \le s \le \tilde{T}} |u(s)|_\beta^{1-\varepsilon}\Big)
$$

and the assertion follows by choosing t such that $|u(t)|_\beta > \sup_{0 \le s \le \tilde{T}} |u(s)|_\beta - 1$ and letting $\tilde{T} \to T$. $\square$

Remark 51.35. Let the hypothesis of Proposition 51.34 be satisfied with $\varepsilon = 0$. Then the proof and the singular Gronwall inequality in Proposition 51.6 guarantee that $|u(t)|_\beta \le C_1 e^{C_2 t}$. $\square$

Lemma 51.36. *Let $p > 1$, $-1 \le \delta < (1 - 1/p)\gamma + \beta/p$ and*

$$|F(u)|_{\alpha-1} \le C\big(1 + |u|_\delta^p\big), \qquad u \in X_\beta. \tag{51.107}$$

Then the estimate (51.106) is true.

Proof. We can find $\theta \in (0, 1/p)$ such that $(1 - \theta)\gamma + \theta\beta > \delta$, hence $(X_\gamma, X_\beta)_\theta \hookrightarrow X_\delta$ and $|u|_\delta \le |u|_\gamma^{1-\theta}|u|_\beta^\theta$. Now the assertion is obvious. $\square$

As an application we first give an alternative proof of Theorem 16.4. This proof will not require Ω to be bounded.

Proof of Theorem 16.4. Let Ω, A and $X_\alpha = X_\alpha(q)$, $\alpha \in [-1,1]$, be as in Example 51.4(i). Notice that we can choose $\omega < 0$ if $Au = -\Delta u + au$, $a > 0$ (or $a = 0$ if Ω is bounded). Set $F(u) = |u|^{p-1}u + au$, $\gamma = 0$ and $\alpha = 1$. Using the assumption $q > n(p-1)/2$ it is easy to find $\beta < 1$ close to 1 and $\delta < \beta/p$ close to β/p such that $X_\delta \hookrightarrow L^{pq} \cap L^q(\Omega)$. Consequently,

$$|F(u)|_0 \le \|u\|_{pq}^p + a\|u\|_q \le C(1 + |u|_\delta^p),$$

hence (51.107) is true. Now assuming $\|u(t)\|_q \le C_0$, Lemma 51.36 and Proposition 51.34 guarantee $|u(t)|_\beta < C_\beta = C_\beta(C_0, |u_0|_\beta)$. Since $X_\beta \hookrightarrow L^{\tilde{q}}(\Omega)$ for some $\tilde{q} > q$, an obvious bootstrap argument shows $\|u(t)\|_\infty < C_\infty$ and concludes the proof. $\square$

Remarks 51.37. (i) If the assumptions of Theorem 16.4 are satisfied, then the above proof guarantees the estimate $U_\infty \le C(u_0)U_q^\rho$ for suitable $\rho \ge 1$.

(ii) If we consider the more general problem

$$\left.\begin{aligned}
u_t - \Delta u &= f(x,t,u,\nabla u), & x \in \Omega,\ t > 0, \\
u &= 0, & x \in \partial\Omega,\ t > 0, \\
u(x,0) &= u_0(x), & x \in \Omega,
\end{aligned}\right\} \qquad (51.108)$$

where $f = f(x,t,u,\xi)$ is a C^1-function satisfying the growth condition $|f| \le C(1 + |u|^p + |\xi|^r)$ with $p < 1 + 2q/n$ and $r < 1 + q/(n+q)$, then the L^q-bound for the solution of (51.108) guarantees the L^∞-bound. The same is true if one considers (51.108) with the nonlinear Neumann boundary condition $\partial_\nu u = g(x,t,u)$ instead of the homogeneous Dirichlet condition, provided $g \in C^1$ satisfies the growth condition $|g| \le C(1 + |u|^z)$, $z < 1 + q/n$ (see [14] and [435]).

(iii) Let $0 \le \beta < \gamma < \alpha \le 1$ and $F : X_\beta \to X_{\alpha-1}$ be bounded on bounded sets. Let $u_0 \in X_\beta$ and let $u \in C([0,\infty), X_\beta)$ be a global solution of (51.9). If $u(t)$ is uniformly bounded in X_β for all $t \ge 0$ and $\delta > 0$, then the estimate

$$|u(t)|_\gamma \le |e^{-tA}u_0|_\gamma + \int_0^t \|e^{-(t-s)A}\|_{\mathcal{L}(X_{\alpha-1},X_\gamma)}|F(u(s))|_{\alpha-1}\,ds$$

$$\le Ct^{\beta-\gamma}|u_0|_\beta + C\int_0^t e^{\omega(t-s)}(t-s)^{\alpha-1-\gamma}\,ds \le C(1 + t^{\beta-\gamma}),$$

implies the boundedness of $u(t)$ in X_γ for $t \ge \delta$. If A has compact resolvent, then the embedding $X_\gamma \hookrightarrow X_\beta$ is compact (see Theorem 51.1(i)), hence the trajectory of u is relatively compact in X_β. $\square$

Example 51.38. Let $\Omega \subset \mathbb{R}^n$ be bounded with C^2-boundary, $-A$ be the Dirichlet Laplacian and $F(u) = f(u)$, where $f \in C^1$, $|f'(u)| \le C(1 + |u|^{p-1})$, $1 < p < p_S$. Assume that $u_0 \in W_0^{1,2}(\Omega)$ and u is a global solution of (51.9). If $u(t)$ is uniformly bounded in $L^{p+1}(\Omega)$, then the trajectory of u is relatively compact in $W_0^{1,2}(\Omega)$.

In fact, Example 51.10 shows the existence of $T > 0$ such that $u \in C([0, T],$ $W_0^{1,2}(\Omega)) \cap C((0, T], W^{2\beta, q}(\Omega))$ for any $\beta < 1$ and $q \geq 1$. Fixing $q := p + 1 >$ $n(p-1)/2$, $\eta \in (0, T)$, and considering the solution u on the interval $[\eta, \infty)$, the proof of Theorem 16.4 above shows that $u(t)$ remains bounded in $W_0^{1,q} \cap W^{2\beta, q}(\Omega)$ for some β close to 1. In particular, the solution remains bounded in $W_0^{1,2}(\Omega)$ and the assertion follows from Remark 51.37(iii). $\square$

Example 51.39. Let $\Omega \subset \mathbb{R}^n$ be bounded with C^2-boundary, $-A$ be the Dirichlet Laplacian and $F : L^\infty(\Omega) \to L^\infty(\Omega)$ be uniformly Lipschitz continuous on bounded subsets of $L^\infty(\Omega)$. Assume that $u_0 \in L^\infty(\Omega)$ and u is a global solution of (51.9) which is uniformly bounded in $L^\infty(\Omega)$. Then the trajectory $\{u(t) : t \geq 1\}$ is relatively compact in $L^\infty(\Omega)$. This follows from Remark 51.11 and (the corresponding analogue of) Remark 51.37(iii). $\square$

52. Appendix F: Maximum and comparison principles. Zero number

Maximum and comparison principles represent a very useful tool in the study of scalar equations (and of some particular systems). Unfortunately, it is not easy to provide (or find in the literature) a general statement which would be applicable in all situations. We therefore prove — or at least formulate — various versions of these principles which we frequently use. For simplicity we have stated all the results for the case when the elliptic part of the equation is the Laplacian, but they remain true for more general operators (under suitable assumptions).

52.1. Maximum principles for the Laplace equation

We first recall the weak and strong maximum principles and the Hopf boundary lemma for strong subsolutions (cf. [250, Theorems 9.1, 9.6, and the proof of Lemma 3.4]).

Proposition 52.1. *Let Ω be an arbitrary domain in $\mathbb{R}^n$, $b \in L^\infty(\Omega, \mathbb{R}^n)$, and let $u \in W_{loc}^{2,n}(\Omega)$ satisfy*

$$-\Delta u + b \cdot \nabla u \leq 0 \quad a.e. \ in \ \Omega. \tag{52.1}$$

(i) *If $u \in C(\overline{\Omega})$, $u \leq 0$ on $\partial\Omega$, and Ω is bounded, then $u \leq 0$ in Ω.*

(ii) *If $u \leq 0$ in Ω, then either $u \equiv 0$ or $u < 0$ in Ω.*

(iii) *Let $x_0 \in \partial\Omega$. Assume that Ω satisfies an interior sphere condition at x_0 and that u is continuous at x_0. If $u \leq 0$ in Ω and $u(x_0) = 0$, then*

$$\liminf_{t \to 0+} t^{-1} u(x_0 - t\nu) < 0.$$

(Here the outer normal ν is defined in the natural way via the interior sphere at x_0). In particular, we have

$$\partial_\nu u(x_0) > 0$$

whenever this derivative exists.

Remark 52.2. Assertions (ii) and (iii) of Proposition 52.1 remain valid if the inequality in (52.1) is replaced with $-\Delta u + b \cdot \nabla u + cu \leq 0$ for some constant $c > 0$ (cf. e.g. [239]). This follows easily by applying Proposition 52.1(ii) and (iii) to the function $v(x) = e^{\alpha x_1} u(x)$ with $\alpha > 0$ large enough. $\square$

We next give a useful maximum principle under weaker regularity assumptions, namely for variational or, even, distributional subsolutions.

Proposition 52.3. *Let Ω be an arbitrary domain in $\mathbb{R}^n$ and let $u \in L^1_{loc}(\Omega)$ satisfy*

$$-\Delta u \leq 0 \quad in \ \mathcal{D}'(\Omega).$$

Assume that either:

(i) $u \in H^1(\Omega)$ and $u \leq 0$ on $\partial\Omega$ in the sense that $u_+ \in H^1_0(\Omega)$; or

(ii) Ω is bounded, u is continuous in a neighborhood of $\partial\Omega$ and $u \leq 0$ on $\partial\Omega$.

Then $u \leq 0$ a.e. in Ω.

Proof. We first assume (i). We shall use the Stampacchia truncation argument. By assumption we have

$$\int_\Omega \nabla u \cdot \nabla\varphi \, dx \leq 0, \quad \text{for all } 0 \leq \varphi \in \mathcal{D}(\Omega). \tag{52.2}$$

Fix a C^∞-function $G : \mathbb{R} \to \mathbb{R}_+$ such that $G(s) = 0$ for $s \leq 0$ and $0 < G'(s) \leq 1$ for $s > 0$. By our assumption that $u_+ \in H^1_0(\Omega)$, there exists a sequence $\psi_j \in \mathcal{D}(\Omega)$ such that $\psi_j \to u_+$ in $H^1(\Omega)$ and a.e. Let $\varphi_j = G \circ \psi_j$. We have $0 \leq \varphi_j \in \mathcal{D}(\Omega)$. Writing

$$|\nabla(G \circ \psi_j) - \nabla(G \circ u_+)| \leq G'(\psi_j)|\nabla\psi_j - \nabla u_+| + |G'(\psi_j) - G'(u_+)||\nabla u_+|,$$

we obtain $\nabla\varphi_j \to \nabla(G \circ u_+)$ in $L^2(\Omega)$ by dominated convergence. Since $\nabla u_+ = \chi_{\{u>0\}}\nabla u$, it follows from (52.2) that

$$\int_\Omega G'(u_+)|\nabla u_+|^2 \, dx = \int_\Omega \nabla u \cdot \nabla(G \circ u_+) \, dx = \lim_{j\to\infty} \int_\Omega \nabla u \cdot \nabla\varphi_j \, dx \leq 0.$$

Consequently, $\nabla(u_+)^2 = 2u_+\nabla u_+ = 0$ a.e. in Ω. Since $u_+ \in H^1_0(\Omega)$, we conclude that $u_+ = 0$ a.e. in Ω.

Let us next consider case (ii). For $\varepsilon > 0$, denote $\omega_\varepsilon = \{x \in \Omega : \delta(x) > \varepsilon\}$. By assumption, there exists $\varepsilon_0 > 0$ small, such that u is continuous on $\overline{\Omega} \setminus \omega_{\varepsilon_0}$. Now set $u_j := \rho_j * u$, where ρ_j is a sequence of mollifiers defined by (47.6), and fix $\varepsilon \in (0, \varepsilon_0)$. For $j \geq j_0(\varepsilon)$ large, we have $u_j \in C^2(\overline{\omega}_\varepsilon)$ and $\Delta u_j = \Delta u * \rho_j \geq 0$ in ω_ε. Therefore, the assertion in case (i) implies $\sup_{\omega_\varepsilon} u_j \leq \sup_{\partial\omega_\varepsilon} u_j$. Since $u_j \to u$ in $L^1(\omega_\varepsilon)$ and in $C(\partial\omega_\varepsilon)$, it follows that $\operatorname{ess\,sup}_{\omega_\varepsilon} u \leq \sup_{\partial\omega_\varepsilon} u$. The conclusion follows by letting $\varepsilon \to 0$ and using the fact that $\lim_{\varepsilon \to 0}(\sup_{\partial\omega_\varepsilon} u) \leq 0$. $\quad\square$

In the rest of Appendix F we shall only consider parabolic problems.

52.2. Comparison principles for classical and strong solutions

We start with a basic maximum principle for classical solutions. Note that unbounded and singular first-order coefficients are allowed (this will be used in Lemma 52.18 below).

Proposition 52.4. *Let Ω be an arbitrary domain in $\mathbb{R}^n$, $T > 0$, $b : Q_T \to \mathbb{R}^n$, $c : Q_T \to \mathbb{R}$, with $\sup_{Q_T} c < \infty$. Assume that $w = w(x,t) \in C^{2,1}(Q_T) \cap C(\overline{Q_T})$ satisfies $w \leq 0$ on $\mathcal{P}_T$, $\sup_{Q_T} w < \infty$, and*

$$w_t - \Delta w \leq b \cdot \nabla w + cw \qquad \text{in } Q_T. \tag{52.3}$$

If Ω is unbounded, assume in addition that either

$$\limsup_{|x| \to \infty, \ (x,t) \in Q_T} w(x,t) \leq 0, \tag{52.4}$$

or

$$|b(x,t)| \leq C_1(1 + |x - a|^{-1}), \quad x \in Q_T, \tag{52.5}$$

for some $a \in \mathbb{R}^n$ and $C_1 > 0$. Then $w \leq 0$ in Q_T.

Proof. We may assume $c < 0$ (if this is not true, then it is sufficient to consider the function $\tilde{w}(x,t) = e^{-\lambda t} w(x,t)$, where $\lambda > \sup_{Q_T} c$). Also, we may obviously assume $w \in C^{2,1}(\Omega \times (0,T])$.

Case 1: Ω bounded. Assume on the contrary that w achieves a positive interior maximum at some point $(x_0, t_0) \in \Omega \times (0, T]$. At this point we have $w > 0$, $\nabla w = 0$, $\Delta w \leq 0$, $w_t \geq 0$. Using $c < 0$ we obtain

$$0 \leq w_t - \Delta w - b \cdot \nabla w \leq cw < 0,$$

which yields a contradiction.

Case 2: Ω unbounded. If the conclusion is not true, then we have $w(x_0, t_0) > 0$ for some $(x_0, t_0) \in Q_T$. In case (52.4) is satisfied, then w achieves its positive maximum and we conclude as in case 1. In case (52.5) holds, arguing similarly as in [296], we set

$$v(x,t) = w(x,t) - \delta t - \varepsilon(1 + |x - a|^2)^{1/2},$$

where $\delta, \varepsilon > 0$ are such that $v(x_0, t_0) > 0$ and $\delta > \varepsilon(n + 2C_1)$. We compute

$$\nabla(1 + |x - a|^2)^{1/2} = (x - a)(1 + |x - a|^2)^{-1/2},$$
$$\Delta(1 + |x - a|^2)^{1/2} = (n + (n-1)|x - a|^2)(1 + |x - a|^2)^{-3/2} \leq n. \tag{52.6}$$

Since $v \leq 0$ on S_{t_0}, v attains its (positive) maximum in Q_{t_0} at some $(x_1, t_1) \in \Omega \times (0, t_0]$. At this point we have $w > v > 0$, $\nabla v = 0$, $\Delta v \leq 0$, $v_t \geq 0$. Using $c \leq 0$, it follows that

$$0 \leq v_t = w_t - \delta \leq \Delta w + b \cdot \nabla w + cw - \delta$$
$$\leq \Delta v + b \cdot \nabla v + n\varepsilon + \varepsilon |b||x - a|(1 + |x - a|^2)^{-1/2} - \delta$$
$$\leq \varepsilon(n + 2C_1) - \delta < 0,$$

which yields a contradiction and concludes the proof. $\quad\square$

Remark 52.5. The assumption $\sup_{Q_T} c < \infty$ in Proposition 52.4 is necessary (although it can be sometimes weakened). Consider for instance the simple examples $u(x, t) = t\varphi_1(x)$, $c(x, t) = \lambda_1 + t^{-1}$ (Ω bounded), or $u(x, t) = t$, $c(x, t) = t^{-1}$ ($\Omega = \mathbb{R}^n$), which satisfy $u_t - \Delta u = cu$ and $u > 0$ in Q_T, with $u \equiv 0$ on $\mathcal{P}_T$. $\quad\square$

We next give a version of the comparison principle for classical (sub-/super-) solutions.

Proposition 52.6. *Let Ω be an arbitrary domain in $\mathbb{R}^n$, $T > 0$, $u, v \in C^{2,1}(Q_T) \cap C(\overline{Q_T})$. Assume that $u \leq v$ on $\mathcal{P}_T$ and*

$$\partial_t u - \Delta u - f(x, u, \nabla u) \leq \partial_t v - \Delta v - f(x, v, \nabla v) \qquad \text{in } Q_T, \tag{52.7}$$

where $f = f(x, s, \xi) : \Omega \times \mathbb{R} \times \mathbb{R}^n \to \mathbb{R}$ is continuous in x and C^1 in s and ξ. Assume also that

$$u, v, \nabla v \in L^\infty(Q_T), \quad |u|, |v| \leq C_1, |\nabla v| \leq C_2 \tag{52.8}$$

and

$$|f_s(x, s, \xi)| + (1 + |x|)^{-1}|f_\xi(x, s, \xi)| \leq C_f \qquad \text{for all } |s| \leq C_1, |\xi| \leq C_2 + 1. \tag{52.9}$$

Then $u \leq v$ in Q_T.

Proof. Fix $\tau \in (0, T)$ such that $\tau e^{C_f \tau} < 1/8C_f$. It is sufficient to prove $u \leq v$ in Q_τ. Assume on the contrary $\delta := \sup_{Q_\tau}(u - v) > 0$ and choose $(x_0, t_0) \in Q_\tau$ such that $(u - v)(x_0, t_0) > \delta/2$. Consider $\varepsilon \in (0, 1)$ such that

$$\varepsilon < \min(C_f \delta / (n + C_f), e^{-C_f \tau}, e^{-C_f t_0} \delta / 4\psi(x_0))$$

and set

$$z(x,t) = e^{-C_f t}(u - v)(x,t) - 2C_f \delta t - \varepsilon\psi(x),$$

where $\psi(x) = (1+|x|^2)^{1/2}$. Then $z(x_0, t_0) > 0$ and z attains its maximum in $\overline{Q_\tau}$ at some $(\tilde{x}, \tilde{t}) \in \overline{Q_\tau}$, since $z(x,t) \to -\infty$ as $|x| \to \infty$, uniformly in t. Now $z(\tilde{x}, \tilde{t}) > 0$ implies $(\tilde{x}, \tilde{t}) \in \overline{Q_\tau} \setminus \mathcal{P}_\tau$, hence $z_t - \Delta z \geq 0$ and $\nabla z = 0$ at this point. Consequently,

$$|\nabla u(\tilde{x}, \tilde{t}) - \nabla v(\tilde{x}, \tilde{t})| \leq e^{C_f \tau}\varepsilon|\nabla\psi(\tilde{x})| < 1,$$

since $|\nabla\psi| \leq 1$. In addition, $z(\tilde{x}, \tilde{t}) > 0$ implies

$$\varepsilon|\tilde{x}| \leq \varepsilon\psi(\tilde{x}) < e^{-C_f \tilde{t}}\delta \leq \delta.$$

Now the mean value theorem guarantees the existence of s between $u(\tilde{x}, \tilde{t})$, $v(\tilde{x}, \tilde{t})$ and ξ between $\nabla u(\tilde{x}, \tilde{t})$, $\nabla v(\tilde{x}, \tilde{t})$ such that

$$\begin{aligned}
0 &\leq (z_t - \Delta z)(\tilde{x}, \tilde{t}) \\
&\leq e^{-C_f \tilde{t}}\big[f(\tilde{x}, u(\tilde{x}, \tilde{t}), \nabla u(\tilde{x}, \tilde{t})) - f(\tilde{x}, v(\tilde{x}, \tilde{t}), \nabla v(\tilde{x}, \tilde{t}))\big] \\
&\quad - C_f(z(\tilde{x}, \tilde{t}) + 2C_f \delta\tilde{t} + \varepsilon\psi(\tilde{x})) - 2C_f\delta + \varepsilon n \\
&= e^{-C_f \tilde{t}}\big[f_s(\tilde{x}, s, \nabla v(\tilde{x}, \tilde{t}))e^{C_f \tilde{t}}(z(\tilde{x}, \tilde{t}) + 2C_f \delta\tilde{t} + \varepsilon\psi(\tilde{x})) \\
&\quad\quad + f_\xi(\tilde{x}, u(\tilde{x}, \tilde{t}), \xi)e^{C_f \tilde{t}}(\nabla z(\tilde{x}, \tilde{t}) + \varepsilon\nabla\psi(\tilde{x}))\big] \\
&\quad - C_f(z(\tilde{x}, \tilde{t}) + 2C_f \delta\tilde{t} + \varepsilon\psi(\tilde{x})) - 2C_f\delta + \varepsilon n \\
&\leq C_f(1 + |\tilde{x}|)\varepsilon - 2C_f\delta + \varepsilon n \\
&< -C_f\delta + \varepsilon(n + C_f) < 0,
\end{aligned}$$

which yields a contradiction and concludes the proof. $\qquad \square$

The following proposition is a version of the strong comparison principle and of the Hopf boundary lemma (for strong solutions, in bounded domains). A more general version can be derived by using the maximum principles in [152] (cf. the proof).

Proposition 52.7. *Let Ω be a bounded domain in $\mathbb{R}^n$ of class C^2, $p > n+2$, and $T > 0$. Let $u, v \in W_{loc}^{2,1;p}(\overline{\Omega} \times (0,T]) \cap C([0,T], L^2(\Omega)) \cap L^\infty(Q_T)$. Assume*

$$\partial_t u - \Delta u - f(x,t,u,\nabla u) \leq \partial_t v - \Delta v - f(x,t,v,\nabla v) \qquad \text{in } Q_T,$$

where $f = f(x,t,s,\xi) : \overline{\Omega} \times [0,T] \times \mathbb{R} \times \mathbb{R}^n \to \mathbb{R}$ is continuous in x, t and C^1 in s and ξ. Assume also that $u(\cdot, 0) \leq v(\cdot, 0)$, $u(\cdot, 0) \not\equiv v(\cdot, 0)$, and either

$$u \leq v \qquad \text{on } S_T \qquad\qquad (52.10)$$

or

$$\partial_\nu u + bu \le \partial_\nu v + bv \qquad \text{on } S_T, \tag{52.11}$$

where $b \in C^1(\partial\Omega)$. Finally, if f depends on ξ, we also assume that $\nabla u, \nabla v \in L^\infty(Q_T)$. Then

$$u < v \quad \text{in } Q_T.$$

In addition, if $u(x_0, t_0) = v(x_0, t_0)$ for some $x_0 \in \partial\Omega$ and $t_0 \in (0, T)$, then

$$\partial_\nu u(x_0, t_0) > \partial_\nu v(x_0, t_0).$$

If (52.11) is true, then $u < v$ in $\overline{\Omega} \times (0, T)$.

Proof. Setting $w := v - u$, the mean value theorem implies

$$\partial w_t - \Delta w \ge g_1(x, t)w + g_2(x, t) \cdot \nabla w,$$

where $g_1(x, t) = \int_0^1 f_u(x, t, u + \theta(v - u), \nabla v)\, d\theta$ and $g_2(x, t) = \int_0^1 f_\xi(x, t, u, \nabla u + \theta(\nabla v - \nabla u))\, d\theta$.

Let us first consider the case $u, v \in W^{2,1;p}(Q_T)$ (hence in particular $u, v \in C^{1,0}(\overline{Q_T})$). Then the assertion follows from [152, Propositions 13.1, 13.2 and Theorem 13.5]. Note that the proofs in [152] use a result from [165] and the strong maximum principle for classical solutions (cf. [429] and [214]).

In the general case, since $g_1, g_2 \in L^\infty(Q_T)$ due to our assumptions, we may first apply Proposition 52.8 and Remark 52.9 below to deduce that $u \le v$ in Q_T. Since $u(\cdot, t) \not\equiv v(\cdot, t)$ for all sufficiently small $t > 0$ due to $u, v \in C([0, T], L^2(\Omega))$, the conclusion follows from the previous case. $\square$

52.3. Comparison principles via the Stampacchia method

We now give versions of the weak maximum and comparison principles which apply to $W^{2,1;2}_{loc}$ sub-/supersolutions and discontinuous initial data (as well as possibly unbounded domains).

Proposition 52.8. *Let $0 < T < \infty$. Let Ω be an arbitrary domain in $\mathbb{R}^n$, c be measurable and a.e. finite on Q_T with $\sup_{Q_T} c < \infty$, and $K \ge 0$. Assume that $w \in C(\overline{\Omega} \times (0, T)) \cap C([0, T], L^2_{loc}(\overline{\Omega}))$ satisfies*

$$\sup_{Q_T} w < \infty, \quad w_t, \nabla w, D^2 w \in L^2_{loc}(Q_T).$$

If $w \le 0$ on $\mathcal{P}_T$ and

$$w_t - \Delta w \le K|\nabla w| + cw \qquad \text{a.e. in } Q_T,$$

then

$$w \leq 0 \quad \text{in } Q_T.$$

Proof. Let $\varepsilon > 0$, $\lambda = \sup_{Q_T} c$, and set

$$z = we^{-\lambda t} - \varepsilon\psi,$$

where

$$\psi(x,t) = Mt + (1 + |x|^2)^{1/2}, \tag{52.12}$$

with $M = n + K$. We see that, a.e. in Q_T, there holds

$$\partial_t z - \Delta z - K|\nabla z| \leq e^{-\lambda t}(c - \lambda)w + \varepsilon(-\psi_t + \Delta\psi + K|\nabla\psi|)$$
$$\leq \varepsilon(-M + n + K) \leq 0. \tag{52.13}$$

We next apply the Stampacchia truncation method. Note that, for $R > 0$ large enough and for each $\tau > 0$, there exists $\eta = \eta(\tau) > 0$ such that

$$z \leq 0 \quad \text{in } \{x \in \Omega : \delta(x) \leq \eta \text{ or } |x| \geq R\} \times (\tau, T - \tau).$$

Our assumptions thus imply $z_+ \in C([0,T), L^2(\Omega)) \cap H^1_{loc}((0,T), L^2(\Omega))$, $z_+(0) = 0$ and, for a.e. $t \in (0,T)$, $z_+(t) \in H^1_0(\Omega \cap B_R)$. For a.e. $t \in (0,T)$, since $\Delta z(\cdot, t) \in L^2(\Omega \cap B_R)$, $\nabla(z_+)(\cdot, t) = \chi_{\{z>0\}}\nabla z(\cdot, t)$, it follows from (52.13) that

$$\frac{1}{2}\frac{d}{dt}\int_\Omega (z_+)^2(t)\,dx \leq -\int_\Omega |\nabla(z_+)|^2\,dx + K\int_\Omega |\nabla z| z_+\,dx$$
$$\leq -\int_\Omega |\nabla(z_+)|^2\,dx + \int_\Omega |\nabla(z_+)|^2\,dx + \frac{K^2}{4}\int_\Omega (z_+)^2\,dx$$
$$= \frac{K^2}{4}\int_\Omega (z_+)^2\,dx.$$

By integration, we conclude that $z_+ = 0$ in Q_T and the conclusion follows by letting $\varepsilon \to 0$. $\square$

Remark 52.9. Proposition 52.8 can be extended to the case of Neumann boundary conditions. For instance, assume that Ω is smooth and bounded, and that w satisfies the assumptions of Proposition 52.8 with $w \leq 0$ on $\mathcal{P}_T$ replaced by $w(\cdot, 0) \leq 0$ and $\partial_\nu w + bw \leq 0$ on S_T, where $\nabla w \in C(\overline{\Omega} \times (0,T))$ and $b \in L^\infty(\partial\Omega)$. Then we conclude that $w \leq 0$ in Q_T. This follows from simple modifications of the above proof, with $\varepsilon = 0$, using the trace inequality $\|v\|_{L^2(\partial\Omega)} \leq \eta\|\nabla v\|_2 + C(\eta)\|v\|_2$, $v \in H^1(\Omega)$, applied with $\eta > 0$ small and $v = z_+(t)$ for a.a. t. $\square$

Proposition 52.10. *Let $0 < T < \infty$, Ω be an arbitrary domain in $\mathbb{R}^n$, and let $f = f(s,\xi) : \mathbb{R} \times \mathbb{R}^n \to \mathbb{R}$, be a C^1-function. Let $u \in C(\overline{\Omega} \times (0,T))$ satisfy*

$$u \in C([0,T), L^2_{loc}(\overline{\Omega})), \quad u \in L^\infty(Q_T), \quad u_t, \nabla u, D^2 u \in L^2_{loc}(Q_T),$$

and similarly for v. If f depends on ξ, we also assume that $\nabla u, \nabla v \in L^\infty(Q_T)$. If $u \le v$ on $\mathcal{P}_T$ and

$$u_t - \Delta u - f(u, \nabla u) \le v_t - \Delta v - f(v, \nabla v) \quad \text{a.e. in } Q_T,$$

then

$$u \le v \quad \text{in } Q_T.$$

Proof. Let $w = u - v$ and set

$$M := \max\big(\operatorname*{ess\,sup}_{Q_T}(|u| + |\nabla u|), \operatorname*{ess\,sup}_{Q_T}(|v| + |\nabla v|)\big) < \infty \tag{52.14}$$

and

$$K := \sup\{|f_s(s,\xi)| + |f_\xi(s,\xi)| : |s|, |\xi| \le M\} < \infty.$$

Letting $c(x,t) = (f(u, \nabla u) - f(v, \nabla u))/(u - v)$ (defined to be 0 whenever the denominator vanishes), we have $|c| \le K$ and

$$w_t - \Delta w \le f(u, \nabla u) - f(v, \nabla v) = c(u - v) + (f(v, \nabla u) - f(v, \nabla v))$$
$$\le cw + K|\nabla w|$$

a.e. in Q_T. The result then follows from Proposition 52.8 applied to w. $\square$

Remarks 52.11. (a) In Proposition 52.4 (resp., Proposition 52.6) it is sufficient to assume that (52.3) (resp., (52.7)) holds in the set $\widetilde{Q}_T := \{(x,t) \in Q_T : w(x,t) > 0\}$ (resp., $\widetilde{Q}_T := \{(x,t) \in Q_T : u(x,t) > v(x,t)\}$). A similar remark holds for Propositions 52.8 and 52.10. Moreover any boundedness assumption on the functions $u, v, \nabla u, \nabla v$ needs to be verified only on the set $\widetilde{Q}_T$.

(b) The proof of Proposition 52.6 shows that we can assume $\nabla u \in L^\infty(\Omega)$ instead of $\nabla v \in L^\infty(\Omega)$. In addition, we do not need to assume the boundedness of ∇v (or ∇u) at all if f is independent of ξ. Similarly, the assumption $u, v \in L^\infty(\Omega)$ can be replaced by $\sup_{Q_T}(u - v) < \infty$ if f is independent of u.

(c) In Proposition 52.10, assume $f(s,\xi)$ to be only continuous (instead of C^1) at $s = 0$, and suppose in addition that $\inf_{Q_T} |u| > 0$ or $\inf_{Q_T} |v| > 0$. Then the conclusion remains valid. Indeed, assume for instance $\sigma := \inf_{Q_T} |v| > 0$ and let $K_0 := \sup\{|f(s,\xi)| : |s| \le M, |\xi| \le M\}$ and

$$K_1 := \sup\{|f_s(s,\xi)| + |f_\xi(s,\xi)| : \sigma/2 \le |s| \le M, |\xi| \le M\},$$

with M defined by (52.14). Then the function c in the proof verifies $|c(x,t)| \leq K_1$ if $|u(x,t)| \geq \sigma/2$, and $|c(x,t)| \leq 4K_0/\sigma$ if $|u(x,t)| < \sigma/2$. A similar remark holds concerning Proposition 52.22 (systems).

(d) The proof of Proposition 52.10 shows that it is sufficient to assume that u or $v \in L^\infty(Q_T)$, and that $\sup_{Q_T}(u-v) < \infty$.

(e) In Proposition 52.6, if f is of the form $f = f(u) + g(x, \nabla u)$, then the assumptions (52.8)–(52.9) can be replaced by

$$\limsup_{|x| \to \infty,\ (x,t) \in Q_T} (u-v)(x,t) \leq 0$$

and u or $v \in L^\infty(Q_T)$. This can be proved easily by using Proposition 52.4 and Remark (a) above.

(f) In Propositions 52.4 and 52.8, if $c \leq 0$ and if, instead of $w \leq 0$ on $\mathcal{P}_T$, we assume $w \leq M$ on $\mathcal{P}_T$ for some $M > 0$, then the conclusion is $w \leq M$ in Q_T (just apply the result to the function $w - M$).

(g) When comparing a solution with a sub-/supersolution, the above (and similar) results are usually applied on the time interval $(0, T)$ for each $T < T_{\max}(u_0)$, hence guaranteeing the boundedness of the solution (and possibly of its derivatives). $\square$

52.4. Comparison principles via duality arguments

We now provide "very weak" versions of the maximum and comparison principles, which are useful in particular in the study of complete blow-up (see Section 27). They can be also applied to show monotonicity of solutions in time (cf. Proposition 52.20 below).

Assume that Ω is a bounded domain of class $C^{2+\alpha}$ for some $\alpha \in (0,1)$. Let $T > 0$, $u_0 \in L^1_\delta(\Omega)$ and $f \in L^1_{loc}([0,T), L^1_\delta(\Omega))$. We say that $u \in L^1_{loc}(\overline{\Omega} \times [0,T))$ is a very weak supersolution of

$$\left.\begin{aligned}
u_t - \Delta u &= f, & x &\in \Omega,\ t \in (0,T), \\
u &= 0, & x &\in \partial\Omega,\ t \in (0,T), \\
u(x,0) &= u_0(x), & x &\in \Omega,
\end{aligned}\right\} \tag{52.15}$$

if

$$\int_0^\tau \int_\Omega \{u(\varphi_t + \Delta\varphi) + f\varphi\}\, dx\, ds + \int_\Omega u_0\varphi(0)\, dx \leq 0 \tag{52.16}$$

for any $0 < \tau < T$ and any $0 \leq \varphi \in C^{2,1}(\overline{\Omega} \times [0,\tau])$ such that $\varphi = 0$ on $\partial\Omega \times [0,\tau]$ and $\varphi(\tau) = 0$. Subsolutions are defined similarly (namely, u is a subsolution if $-u$ is a supersolution). Of course, the definition immediately carries over to the nonlinear case $f = f(u)$.

Remarks 52.12. (i) If $u \in C^{2,1}(\overline{\Omega} \times (0,T)) \cap C([0,T], L^1(\Omega))$ is a classical supersolution of (52.15) (i.e., satisfies (52.15) with = signs replaced by $\geq$), then it is easy to show that it is a very weak supersolution.

(ii) Alternatively, one could replace the integrability assumption on f near $t = 0$ by a continuity assumption on u (namely, $u \in C([0,T], L^1_\delta(\Omega))$ and just $f \in L^1_{loc}((0,T), L^1_\delta(\Omega))$ and adopt a definition more similar to that of weak L^1_δ-solution (cf. Definition 48.8). However, the present formulation seems better suited to certain applications, such as complete blow-up. $\square$

Proposition 52.13. *Let Ω be a bounded domain of class $C^{2+\alpha}$ for some $\alpha \in (0,1)$. Let $0 < T < \infty$ and $c \in L^\infty(Q_T)$.*

(i) *Assume that*

$$z \in L^q_{loc}(\overline{\Omega} \times [0,T)) \quad \text{for some } 1 < q < \infty. \tag{52.17}$$

If z is a very weak supersolution of

$$\left.\begin{aligned}
z_t - \Delta z &= cz, & x \in \Omega, \ 0 < t < T, \\
z &= 0, & x \in \partial\Omega, \ 0 < t < T, \\
z(x,0) &= 0, & x \in \Omega,
\end{aligned}\right\} \tag{52.18}$$

then $z \geq 0$ a.e. in Q_T.

(ii) *If $c = 0$, then assertion (i) remains true for $q = 1$.*

Proof. (i) Fix $m > \max(n/2, q')$ and a sequence of functions $c_j \in \mathcal{D}(Q_T)$ such that $c_j \to c$ in $L^m(Q_T)$. For given $0 < \tau < T$ and $0 \leq h \in \mathcal{D}(Q_\tau)$, let $\varphi_j \in C^{2,1}(\overline{Q_\tau})$ be the solution of

$$\left.\begin{aligned}
-\partial_t \varphi_j - \Delta \varphi_j &= c_j \varphi_j + h, & x \in \Omega, \ 0 < t < \tau, \\
\varphi_j &= 0, & x \in \partial\Omega, \ 0 < t < \tau, \\
\varphi_j(x,\tau) &= 0, & x \in \Omega.
\end{aligned}\right\} \tag{52.19}$$

By Proposition 52.8, we have $\varphi_j \geq 0$. Moreover, by using the variation-of-constants formula, the L^m-L^∞-estimate (Proposition 48.4), and $m > n/2$, one easily gets

$$\|\varphi_j\|_{L^\infty(Q_\tau)} \leq C, \quad j = 1, 2, \dots . \tag{52.20}$$

Applying the definition of z being a (very weak) supersolution of (52.18), with $\varphi = \varphi_j$ as a test-function, we obtain

$$0 \leq -\int_0^\tau \int_\Omega z(\partial_t \varphi_j + \Delta \varphi_j + c\varphi_j)\, dx\, ds = \int_0^\tau \int_\Omega \big(hz + (c_j - c)z\varphi_j\big)\, dx\, ds. \tag{52.21}$$

Since $(c_j - c)z\varphi_j \to 0$ in $L^1(Q_\tau)$ due to (52.17), (52.20) and $m > q'$, we deduce that $\int_0^\tau \int_\Omega hz\, dx\, ds \geq 0$, and the conclusion follows.

(ii) The argument is much simpler than in the previous case: It suffices to use (52.21) with $c = c_j = 0$ and φ instead of φ_j, where φ is the solution of (52.19) with $c_j = 0$. $\square$

We have the following (very weak) comparison principle for the semilinear problem (14.1).

Proposition 52.14. *Let Ω be a bounded domain of class $C^{2+\alpha}$ for some $\alpha \in (0,1)$, $0 < T < \infty$, and $u_0 \in L^\infty(\Omega)$. Assume that $f : \mathbb{R} \to \mathbb{R}$ is of class C^1. Let $u, v \in L^\infty(Q_T)$ be, respectively, very weak sub- and supersolutions to problem (14.1) on $(0,T)$. Then $u \leq v$ on $(0,T)$.*

Proof. This is an immediate consequence of Proposition 52.13 applied to $z := v - u$. $\square$

The comparison results in the previous subsections do not apply in the case of convective equations like

$$u_t - \Delta u = |u|^{p-1}u + a \cdot \nabla(|u|^{q-1}u)$$

with $1 < q < 2$, due to the fact that the nonlinearity is not C^1 at $u = 0$. For such problems, we shall rely instead on the following result, the proof of which involves a duality argument. For simplicity we restrict ourselves to the case Ω bounded or $\Omega = \mathbb{R}^n$.

Proposition 52.15. *Let Ω be a bounded domain of class C^2 or $\Omega = \mathbb{R}^n$. Let $T > 0$, $b \in L^\infty(Q_T, \mathbb{R}^n)$ and $c \in L^\infty(Q_T)$. Assume that $w = w(x,t) \in C^{2,1}(\overline{\Omega} \times (0,T)) \cap L^\infty(Q_T)$ and that $bw \in C^{1,0}(\overline{\Omega} \times (0,T))$. If $\Omega = \mathbb{R}^n$ assume in addition that $\nabla w \in L^\infty_{loc}((0,T), L^\infty(\mathbb{R}^n))$. If $w \leq 0$ on S_T, $\limsup_{t \to 0} w(x,t) \leq 0$ for all $x \in \Omega$, and*

$$w_t - \Delta w \leq \mathrm{div}(bw) + cw \qquad \text{in } Q_T, \tag{52.22}$$

then $w \leq 0$ in Q_T.

Proof. Fix $h \in \mathcal{D}(\Omega)$, $h \geq 0$, $0 < t_2 < T$.

First consider the case Ω bounded and let φ be the solution of the adjoint problem

$$\left.\begin{aligned}
-\varphi_t - \Delta\varphi &= -b \cdot \nabla\varphi + c\varphi, &\quad x \in \Omega,\ 0 < t < t_2,\\
\varphi &= 0, &\quad x \in \partial\Omega,\ 0 < t < t_2,\\
\varphi(x, t_2) &= h(x), &\quad x \in \Omega.
\end{aligned}\right\} \tag{52.23}$$

By parabolic L^r-regularity, we have $\varphi \in W^{2,1;r}(Q_T)$, $1 < r < \infty$, and $\varphi \geq 0$ by Proposition 52.8. For each $0 < t_1 < t_2$, multiplying (52.22) by φ, integrating by parts and using $w \leq 0$ on $\mathcal{P}_T$, $\partial\varphi/\partial\nu \leq 0 = \varphi$ on $\partial\Omega$, we obtain

$$\left[\int_\Omega w\varphi\, dx\right]_{t_1}^{t_2} = \int_{t_1}^{t_2}\int_\Omega (w\varphi_t + w_t\varphi)\, dx\, ds$$

$$\leq \int_{t_1}^{t_2}\int_\Omega (w\varphi_t + (\Delta w + \mathrm{div}(bw) + cw)\varphi)\, dx\, ds \tag{52.24}$$

$$\leq \int_{t_1}^{t_2}\int_\Omega (\varphi_t + \Delta\varphi - b \cdot \nabla\varphi + c\varphi)w\, dx\, ds = 0.$$

Letting $t_1 \to 0$, we obtain $\int_\Omega w(t_2) h \, dx \le 0$, hence $w \le 0$.

Next consider the case $\Omega = \mathbb{R}^n$. Observe that problem (52.23) still admits a solution $\varphi \in C([0, T], W^{1,1}(\mathbb{R}^n))$, $\varphi \ge 0$. This follows from a straightforward fixed-point argument, using the variation-of-constants formula and simple estimates involving the Gaussian heat kernel G. Moreover, given $1 < r < \infty$, we have $\varphi \in C([0, T], W^{1,r}(\mathbb{R}^n))$ due to Appendix E and $\varphi \in W^{2,1;r}_{loc}(\overline{Q_T})$ by Theorem 48.1, and a simple cut-off argument. For $R > 0$, arguing as in (52.24) with Ω replaced by B_R, we get

$$\left[\int_{B_R} w\varphi \, dx \right]_{t_1}^{t_2} \le \int_{t_1}^{t_2} \int_{\partial B_R} \left(\varphi \frac{\partial w}{\partial \nu} - w \frac{\partial \varphi}{\partial \nu} + (b \cdot \nu) w\varphi \right) d\sigma ds$$

$$\le C(t_1) \int_{t_1}^{t_2} \int_{\partial B_R} \left(\varphi + |\nabla \varphi| \right) d\sigma ds. \tag{52.25}$$

Since $\varphi \in C([0, T], W^{1,1}(\mathbb{R}^n))$ there exists a sequence $R_j \to \infty$ such that the RHS of (52.25) with $R = R_j$ decays to 0. Then letting $t_1 \to 0$, we obtain $\int_{\mathbb{R}^n} w(t_2) h \, dx = 0$, hence $w \le 0$. $\square$

As a direct consequence of Proposition 52.15, we obtain in particular:

Proposition 52.16. *Let Ω be a bounded domain of class C^2 or $\Omega = \mathbb{R}^n$. Let $T > 0$ and $f, g : (t, u) \ni [0, T] \times \mathbb{R} \to \mathbb{R}$ be such that f, f_u, g, g_u are continuous. Let $u, v \in C^{2,1}(\overline{\Omega} \times (0, T)) \cap L^\infty(Q_T)$. If $\Omega = \mathbb{R}^n$ assume in addition that $\nabla u, \nabla v \in L^\infty_{loc}((0, T), L^\infty(\mathbb{R}^n))$. If $u \le v$ on S_T, $\limsup_{t \to 0}(u - v)(x, t) \le 0$ for all $x \in \Omega$, and*

$$\partial_t u - \Delta u - f(t, u) - \operatorname{div}(g(t, u)) \le \partial_t v - \Delta v - f(t, v) - \operatorname{div}(g(t, v)) \qquad in \ Q_T,$$

then $u \le v$ in Q_T.

52.5. Monotonicity of radial solutions

Assume that Ω is a symmetric domain and that problem (34.1) is well-posed in a space of functions X on Ω. If the C^1-function $F = F(s, \xi)$ depends on ξ through $|\xi|$ only and if $u_0 \in X$ is radial, then the solution u of (34.1) is also radial. This follows immediately from the local uniqueness and the invariance of problem (34.1) by rotation. The same remains true in the case of Neumann boundary conditions. The following result provides sufficient conditions for the preservation of radial monotonicity.

Proposition 52.17. *Let $\Omega = B_R$ or $\Omega = \mathbb{R}^n$. In what follows we use the notation $T = T_{\max}(u_0)$.*

(i) *Consider problem (34.1) with a C^1-function $F = F(s, \xi) : \mathbb{R}_+ \times \mathbb{R}^n \to \mathbb{R}$ such that $F(s, \xi) = \tilde{F}(s, |\xi|)$ and $F(0, 0) \geq 0$. Assume that $u_0 \in BC^1(\overline{\Omega})$, $u_0 = 0$ on $\partial\Omega$, $u_0 \geq 0$, is radial nonincreasing. Then*

$$u \geq 0 \quad \text{and } u \text{ is radial nonincreasing in } Q_T. \tag{52.26}$$

(ii) *Consider problem (14.1) with $f \in C^1([0, \infty))$ such that $f(0) \geq 0$. If $u_0 \in L^\infty(\Omega)$, $u_0 \geq 0$, is radial nonincreasing, then (52.26) is true.*

(iii) *Consider problem (15.1) with $p > 1$, and let $1 \leq q < \infty$ satisfy $q > q_c = n(p-1)/2$ or $q = q_c > 1$. If $u_0 \in L^q(\Omega)$, $u_0 \geq 0$, is radial nonincreasing, then (52.26) is true.*

Moreover, in each case above, if $u_0 \not\equiv 0$, then

$$u_r < 0 \quad \text{in } (0, R] \times (0, T) \tag{52.27}$$

(with $(0, R]$ replaced by $(0, \infty)$ if $\Omega = \mathbb{R}^n$).

We need the following lemma:

Lemma 52.18. *Let $\Omega = (0, R)$, $0 < R \leq \infty$, $T > 0$, $f = f(s, \xi) \in C^1(\mathbb{R}_+ \times \mathbb{R})$. Assume that $u \in C^{2,1}(\overline{\Omega} \times (0, T)) \cap BC(\overline{Q_T})$ satisfies $u_r \in W^{2,1;2}_{loc}(Q_T) \cap BC(\overline{Q_T})$, $u_r \leq 0$ on $\mathcal{P}_T$ and*

$$u_t - u_{rr} - \frac{n-1}{r} u_r = f(u, u_r) \qquad \text{in } Q_T. \tag{52.28}$$

Then $u_r \leq 0$ in Q_T. If $u_r(\cdot, 0) \not\equiv 0$, then $u_r < 0$ in Q_T and, assuming $R < \infty$, $u_r(R, t) < 0$ for $t \in (0, T)$.

Proof. For $\lambda > 0$ large enough, the function $w := u_r e^{-\lambda t}$ solves the equation

$$w_t - w_{rr} = bw_r + cw \quad \text{a.e. in } Q_T,$$

with $b = f_\xi(u, u_r) + \frac{n-1}{r}$ and $c = f_u(u, u_r) - \lambda - \frac{n-1}{r^2} \leq 0$. Assume for contradiction that $\sup_{Q_T} w > 0$. Set $m = \sup_{Q_T} w/2$, $z = w - m$, and choose $r_0 \in (0, R)$ such that $w \leq m$ in $[0, r_0] \times [0, T]$. Then, for some $K > 0$, we have

$$z_t - z_{rr} = bz_r + c(z + m) \leq K|z_r| + cz \quad \text{a.e. in } (r_0, R) \times (0, T).$$

Using Proposition 52.8 we infer that $z \leq 0$ in $(r_0, R) \times [0, T]$, contradicting the definition of m. Consequently, $u_r \leq 0$.

Finally, to show that $u_r < 0$ it is sufficient to use Proposition 52.7 with $\tilde{\Omega} = (\varepsilon, \tilde{R})$, where $0 < \varepsilon < \tilde{R} < \infty$, $\tilde{R} \leq R$. $\quad \square$

Proof of Proposition 52.17. Let us first verify assertion (i). We know that u is radial and u satisfies (52.28) with $f(u, u_r) = F(u, (u_r, 0, \ldots, 0))$.

We have $u_r \in BC([0,R] \times [0,\tau])$, $0 < \tau < T$ (with $[0,R]$ replaced by $[0,\infty)$ if $\Omega = \mathbb{R}^n$). Also, by Remark 48.3(i), we have $u_r \in W^{2,1;2}_{loc}(Q_T)$. Since $u_{0,r} \leq 0$, $u_r(0,t) = 0$, and $u_r(R,t) \leq 0$ due to $u \geq 0$, the assertion is then a direct consequence of Lemma 52.18.

For the proof of assertions (ii) and (iii), we proceed in several steps.

Step 1. Define $\widetilde{\mathcal{D}}(\Omega) = \{v \in \mathcal{D}(\Omega) : v \geq 0, \ v \text{ is radial nonincreasing}\}$ and $\widetilde{L}^q(\Omega) = \{v \in L^q(\Omega) : v \geq 0, \ v \text{ is radial nonincreasing}\}$ for $1 \leq q \leq \infty$. It is not too difficult to show that if $u_0 \in \widetilde{L}^q(\Omega)$ and $v \in \widetilde{\mathcal{D}}(\Omega)$, then the convolution product $u_0 * v$ is radial nonincreasing, hence belongs to $\widetilde{L}^q(\Omega)$. The same is true if v is replaced by G_t, $t > 0$.

Step 2. By standard arguments of truncation and convolution with a mollifier, using Step 1, one shows that $\widetilde{\mathcal{D}}(\Omega)$ is dense in $\widetilde{L}^q(\Omega)$.

Step 3. We claim that for u_0 satisfying the assumption of (ii) or (iii), the function $z(x,t) := e^{-tA}u_0$ is radial nonincreasing. If $\Omega = \mathbb{R}^n$, then this follows from Step 1, since $z(t) = G_t * u_0$. If $\Omega = B_R$, then this is true for $u_0 \in \widetilde{\mathcal{D}}(\Omega)$ by assertion (i), and the general case follows from the density property of Step 2.

Step 4. First consider case (ii) with $f' \geq 0$, or case (iii). The solution u is constructed as the fixed point of a suitable contraction mapping (cf. the proof of Theorem 15.2 and Remark 51.11). Consequently, u is the limit of a sequence $u^{k+1} = \Phi_{u_0}(u^k) := e^{-tA}u_0 + \int_0^t e^{-(t-s)A}f(u(s))\,ds$. The conclusion then follows from the fact that the operator Φ_{u_0} preserves the radial nonincreasing property, due to $f' \geq 0$.

Step 5. In case (ii) for general f of class C^1, let $M := \sup_{Q_T} u$ and set $\tilde{f}(s) := f(s) + \lambda s$, where $\lambda > 0$ is chosen so large that $\tilde{f}' \geq 0$ on $[0,M]$. Noting that u solves $u_t - (\Delta - \lambda)u = \tilde{f}(u)$, the conclusion follows from the argument in Step 4, where e^{-tA} is replaced with $e^{-\lambda t}e^{-tA}$.

Finally, let $u_0 \not\equiv 0$. Since, given $\delta > 0$, $u(\delta) \in BC^1(\overline{\Omega})$ and $u = 0$ on $\partial\Omega$, inequality (52.27) follows from (i). $\square$

52.6. Monotonicity of solutions in time

We give two results which are useful to guarantee the monotonicity of solutions in time.

Proposition 52.19. *Let $\Omega \subset \mathbb{R}^n$ be a uniformly regular domain of class C^2, let $F = F(s,\xi) : \mathbb{R} \times \mathbb{R}^n \to \mathbb{R}$ be a C^1-function, and consider problem (34.1). Assume that $u_0 \in BC(\overline{\Omega}) \cap H^2_{loc}(\Omega)$ satisfies $u_0 = 0$ on $\partial\Omega$ and*

$$\Delta u_0 + F(u_0, \nabla u_0) \geq 0 \quad a.e. \ in \ \Omega.$$

If F depends on ξ, assume in addition that $u_0 \in BC^1(\overline{\Omega})$. Then $u_t \geq 0$ in Q_T, where $T := T_{\max}(u_0)$.

Proof. In the case when F depends on ξ, first recall that problem (34.1) is well-posed in $X = \{u \in BC^1(\overline{\Omega}) : u = 0 \text{ on } \partial\Omega\}$.

By comparing u with the subsolution $\underline{u}(x,t) := u_0(x)$ via Proposition 52.10, we obtain $u \geq u_0$ in Q_T.

Now fix $h \in (0, T)$ and put $v(t) := u(t + h)$. Since $v(0) = u(h) \geq u_0$, we infer from Proposition 52.6 that $v \geq u$ on $(0, T - h)$. The result then follows by dividing by h and letting $h \to 0$. $\square$

In case Ω is bounded and the nonlinearity depends only on u, the following alternative approach guarantees monotonicity of solutions in time under much weaker regularity on the initial data. We say that $u_0 \in L^\infty(\Omega)$ satisfies

$$\left.\begin{aligned} \Delta u_0 + f(u_0) \geq 0, \qquad & x \in \Omega, \\ u_0 \leq 0, \qquad & x \in \partial\Omega \end{aligned}\right\} \tag{52.29}$$

in the very weak sense if, for all $0 \leq \psi \in C^2(\overline{\Omega})$ such that $\psi = 0$ on $\partial\Omega$, there holds

$$\int_\Omega \{u_0 \Delta\psi + f(u_0)\psi\}\, dy \geq 0. \tag{52.30}$$

(Of course, this is satisfied in particular if u_0 belongs to $H^2 \cap H_0^1(\Omega)$ and verifies $\Delta u_0 + f(u_0) \geq 0$ a.e. in Ω.)

Proposition 52.20. *Assume that Ω is a bounded domain of class $C^{2+\alpha}$ for some $\alpha \in (0,1)$ and consider problem (14.1) with $f \in C^1(\mathbb{R})$. If $u_0 \in L^\infty(\Omega)$ satisfies (52.29) in the very weak sense, then $u_t \geq 0$ in Q_T, where $T := T_{\max}(u_0)$.*

Proof. *Step 1.* We claim that $u \geq u_0$ in Q_T.

For $0 \leq t < T$, set $v(t) := u(t) - u_0$, $c(x, t) = (f(u) - f(u_0))/(u - u_0)$ (defined to be 0 whenever the denominator vanishes), and notice that $c \in L^\infty(Q_T)$.

For given $0 < \tau < T$, let $0 \leq \varphi \in C^2(\overline{\Omega} \times [0, \tau])$ be such that $\varphi = 0$ on $\partial\Omega \times [0, \tau]$ and $\varphi(\tau) = 0$. For each $0 < t < \tau$, by integrating by parts and using (52.30) with $\psi = \int_t^\tau \varphi$ as a test-function, we obtain

$$-\int_\Omega (v\varphi)(t)\, dx = \int_t^\tau \int_\Omega (v\varphi_t + u_t\varphi)\, dx\, ds = \int_t^\tau \int_\Omega (v\varphi_t + (\Delta u + f(u))\varphi)\, dx\, ds$$

$$= \int_t^\tau \int_\Omega \{v(\varphi_t + \Delta\varphi) + f(u)\varphi + u_0\Delta\varphi\}\, dx\, ds$$

$$\geq \int_t^\tau \int_\Omega \{v(\varphi_t + \Delta\varphi) + (f(u) - f(u_0))\varphi\}\, dx\, ds$$

$$= \int_t^\tau \int_\Omega \{v(\varphi_t + \Delta\varphi + c\varphi)\}\, dx\, ds.$$

Letting $t \to 0$, hence $\|u(t) - u_0\|_1 \to 0$ (due to $\|u(t) - e^{-tA}u_0\|_\infty \to 0$), it follows that

$$\int_t^T \int_\Omega \{v(\varphi_t + \Delta\varphi + c\varphi)\}\, dx\, ds \leq 0.$$

By Proposition 52.13, we deduce that $v \geq 0$, hence the claim.

Step 2. As before, we fix $h \in (0, T)$ and let $v(t) := u(t + h)$. Since

$$u, v \in C^{2,1}(\overline{\Omega} \times (0, T - h)) \cap L^\infty_{loc}(\overline{\Omega} \times [0, T - h)) \cap C([0, T - h), L^1(\Omega))$$

are classical solutions of the first two equations in (14.1) on $(0, T - h)$, we deduce from Proposition 52.13(i) and Remark 52.12(i) that $v \geq u$ on $(0, T - h)$. The result follows by dividing by h and letting $h \to 0$. $\square$

52.7. Systems and nonlocal problems

We first give extensions of some of the preceding results to systems of cooperative type.

Proposition 52.21. *Let $0 < T < \infty$, Ω be an arbitrary domain in $\mathbb{R}^n$, $d_1, d_2 > 0$, and $a_{ij} \in L^\infty(Q_T)$, $i, j \in \{1, 2\}$, with $a_{12}, a_{21} \geq 0$. Assume that for $i = 1, 2$, the function w_i satisfies $w_i \in C(\overline{\Omega} \times (0, T)) \cap C([0, T), L^2_{loc}(\overline{\Omega}))$, $\sup_{Q_T} w_i < \infty$, $\partial_t w_i, \nabla w_i, D^2 w_i \in L^2_{loc}(Q_T)$. If $w_1, w_2 \leq 0$ on $\mathcal{P}_T$ and*

$$\left.\begin{array}{ll} \partial_t w_1 - d_1 \Delta w_1 \leq a_{11} w_1 + a_{12} w_2 & \text{a.e. in } Q_T, \\[4pt] \partial_t w_2 - d_2 \Delta w_2 \leq a_{21} w_1 + a_{22} w_2 & \text{a.e. in } Q_T, \end{array}\right\}$$

then

$$w_1, w_2 \leq 0 \qquad \text{in } Q_T.$$

Proof. Let $\varepsilon > 0$, $\lambda = 2 \max_{0 \leq i,j \leq 2} \sup_{Q_T} a_{ij}$, and set

$$z_i = w_i e^{-\lambda t} - \varepsilon\psi,$$

where ψ defined in (52.12) with $M = n$. Since $\Delta\psi - \psi_t \leq 0$ by (52.6), it follows that a.e. in Q_T, there holds

$$\begin{aligned} \partial_t z_1 - d_1 \Delta z_1 &= e^{-\lambda t}(\partial_t w_1 - d_1 \Delta w_1 - \lambda w_1) + \varepsilon(\Delta\psi - \psi_t) \\ &\leq e^{-\lambda t}\big((a_{11} - \lambda)w_1 + a_{12}w_2\big) \\ &\leq (a_{11} - \lambda)z_1 + a_{12}z_2 + \varepsilon(a_{11} + a_{12} - \lambda)\psi \leq (a_{11} - \lambda)z_1 + a_{12}z_2 \end{aligned}$$

and similarly,

$$\partial_t z_2 - d_2 \Delta z_2 \leq a_{21}z_1 + (a_{22} - \lambda)z_2.$$

Arguing as in the proof of Proposition 52.8, it follows that

$$\frac{1}{2}\frac{d}{dt}\int_\Omega (z_{1,+})^2(t)\,dx$$

$$\leq -d_1\int_\Omega |\nabla(z_{1,+})|^2\,dx + \int_\Omega (a_{11}-\lambda)(z_{1,+})^2\,dx + \int_\Omega a_{12}\,z_2\,z_{1,+}\,dx \qquad (52.31)$$

$$\leq \int_\Omega a_{12}\,z_{2,+}\,z_{1,+}\,dx \leq C\int_\Omega \left((z_{1,+})^2 + (z_{1,+})^2\right)dx,$$

where we used $a_{11}-\lambda \leq 0$ and $a_{12} \geq 0$. Similarly, we get

$$\frac{1}{2}\frac{d}{dt}\int_\Omega (z_{2,+})^2(t)\,dx \leq C\int_\Omega \left((z_{1,+})^2 + (z_{1,+})^2\right)dx. \qquad (52.32)$$

Adding up (52.31) and (52.32), integrating, and using $z_{1,+}(0) = z_{2,+}(0) = 0$, we infer that $z_{1,+} = z_{2,+} = 0$ in Q_T and the conclusion follows by letting $\varepsilon \to 0$. $\square$

By arguing similarly as in the proof of Proposition 52.10, one obtains a comparison principle for cooperative systems of the form

$$\partial_t u_i - d_i\Delta u_i - f_i(u_1,u_2) = 0, \qquad i = 1,2. \qquad (52.33)$$

Proposition 52.22. *Let $0 < T < \infty$, Ω be an arbitrary domain in $\mathbb{R}^n$, and let $f_i = f_i(u_1,u_2) : \mathbb{R}^2 \to \mathbb{R}$, $i = 1,2$, be C^1-functions such that*

$$\partial_{u_2} f_1 \geq 0, \quad \partial_{u_1} f_2 \geq 0. \qquad (52.34)$$

Let $u = (u_1,u_2)$, where $u_i \in C(\overline{\Omega}\times(0,T))$ satisfy $u_i \in L^\infty(Q_T)$, $u_i \in C([0,T),$ $L^2_{loc}(\overline{\Omega}))$, and $\partial_t u_i, \nabla u_i, D^2 u_i \in L^2_{loc}(Q_T)$. Finally, let v satisfy the same hypotheses as u. If, for $i = 1,2$, we have $u_i \leq v_i$ on $\mathcal{P}_T$ and

$$\partial_t u_i - d_i\Delta u_i - f_i(u_1,u_2) \leq \partial_t v_i - d_i\Delta v_i - f_i(v_1,v_2) \qquad a.e. \ in \ Q_T,$$

then

$$u_i \leq v_i \quad in \ Q_T, \ i = 1,2.$$

Remarks 52.23. (i) The cooperativity assumption (52.34) (or $a_{12}, a_{21} \geq 0$ in Proposition 52.21) is essential to ensure the order-preserving character of system (52.33), as shown by the following simple example. Consider system (52.33) under homogeneous Dirichlet boundary conditions, with $f_1(u,v) = -v$, $f_2(u,v) = 0$. If we take $u_0 = 0$ and $v_0 \geq 0$, $v_0 \not\equiv 0$, then, by the strong maximum principle, we have $v > 0$, hence $u < 0$, in $\Omega \times (0,\infty)$. Therefore the order with the solution $(0,0)$ at $t = 0$ is not preserved.

(ii) For system (52.33) with homogeneous Dirichlet boundary condition, under assumption (52.34), the analogues of Propositions 52.19 and 52.20 guaranteeing time-monotonicity of solutions can be established by simple modifications of the proofs. $\square$

We next turn to nonlocal problems (with space or time integral nonlinearities).

Proposition 52.24. *Let $0 < T < \infty$, Ω be an arbitrary bounded domain in $\mathbb{R}^n$, and $a, b, k \in L^\infty(Q_T)$, with $b, k \geq 0$. Assume that the function $w \in C(\overline{\Omega} \times (0, T)) \cap C([0, T), L^2(\Omega))$ satisfies $\sup_{Q_T} w < \infty$,*

$$\partial_t w, \nabla w, D^2 w \in L^2_{loc}(\overline{\Omega} \times (0, T)). \tag{52.35}$$

If $w \leq 0$ on $\mathcal{P}_T$ and either

$$\partial_t w - \Delta w \leq aw + b \int_\Omega k(y, \cdot) w(y, \cdot)\, dy \qquad a.e.\ in\ Q_T, \tag{52.36}$$

or

$$\partial_t w - \Delta w \leq aw + b \int_0^t k(\cdot, s) w(\cdot, s)\, ds \qquad a.e.\ in\ Q_T, \tag{52.37}$$

then

$$w \leq 0 \quad in\ Q_T.$$

Proof. Our assumptions imply $w_+ \in C([0, T), L^2(\Omega)) \cap C^1((0, T), L^2(\Omega))$, $w_+(0) = 0$ and, for a.e. $t \in (0, T)$, $w_+(t) \in H^1_0(\Omega)$. Moreover, for a.e. $t \in (0, T)$, we have $\Delta w(\cdot, t) \in L^2(\Omega)$, and $\nabla(w_+)(\cdot, t) = \chi_{\{w>0\}} \nabla w(\cdot, t)$.

In the case of (52.36), by using $b, k \geq 0$ and the Cauchy-Schwarz inequality, we obtain

$$\frac{1}{2}\frac{d}{dt}\int_\Omega (w_+)^2(t)\, dx \leq -\int_\Omega |\nabla(w_+)|^2\, dx + \int_\Omega a(w_+)^2\, dx + \int_\Omega bw_+\, dx \int_\Omega kw_+\, dy$$

$$\leq C \int_\Omega (w_+)^2\, dx.$$

In the case of (52.37), we obtain

$$\frac{1}{2}\frac{d}{dt}\int_\Omega (w_+)^2(t)\, dx \leq -\int_\Omega |\nabla(w_+)|^2\, dx + \int_\Omega a(w_+)^2\, dx$$

$$+ \int_\Omega bw_+ \left(\int_0^t kw_+\, ds\right) dx$$

$$\leq \int_\Omega a(w_+)^2\, dx + \int_\Omega b^2(w_+)^2\, dx + T \int_0^t \int_\Omega k^2(w_+)^2\, dx\, ds.$$

The function $\phi(t) := \int_0^t \int_\Omega (w_+)^2\, dx\, ds$ thus satisfies $\phi'' \leq C(\phi + \phi')$ and $\phi, \phi' \geq 0$, hence

$$[\phi^2 + (\phi')^2]' = 2(\phi + \phi'')\phi' \leq C[\phi^2 + (\phi')^2], \qquad 0 < t < T,$$

with $\phi(0) = \phi'(0) = 0$.

In both cases, by integration, we conclude that $w_+ = 0$ in Q_T. $\quad\square$

As a consequence of Proposition 52.24 we obtain in particular the following comparison principle. The proof is similar to that of Proposition 52.10.

Proposition 52.25. *Let $0 < T < \infty$, Ω be an arbitrary bounded domain in $\mathbb{R}^n$, and let $f : \mathbb{R} \to \mathbb{R}$ and $g = g(s,z) : \mathbb{R}^2 \to \mathbb{R}$ be C^1-functions, with either $f', \partial_z g \geq 0$ or $f', \partial_z g \leq 0$. Let $u \in C(\overline{\Omega} \times (0,T))$ satisfy $u \in L^\infty(Q_T)$, $u \in C([0,T), L^2_{loc}(\overline{\Omega}))$, and $u_t, \nabla u, D^2 u \in L^2_{loc}(\overline{\Omega} \times (0,T))$, and let v satisfy the same hypotheses as u. Finally, denote*

$$I(u,t) := \int_\Omega f(u(y,t))\, dy \qquad (resp., \ I(u,t) := \int_0^t f(u(y,s))\, ds).$$

If $u \leq v$ on $\mathcal{P}_T$ and

$$u_t - \Delta u - g\big(u, I(u,\cdot)\big) \leq v_t - \Delta v - g\big(v, I(v,\cdot)\big) \qquad a.e. \ in \ Q_T,$$

then

$$u \leq v \qquad in \ Q_T.$$

Remarks 52.26. (i) The positivity assumption on b, k is essential for the validity of the nonlocal maximum principle in Proposition 52.24, as shown by the following example from [525]: The function $w(x,t) = x^2 - t$ satisfies

$$\left.\begin{aligned}
w_t - w_{xx} &= -3 \geq -18\textstyle\int_{-1}^1 w(y,t)\, dy, && 1 < x < 1, \, 0 < t < 1/4, \\
w(\pm 1, t) &= 1 - t \geq 0, && 0 < t < 1/4, \\
w(x,0) &= x^2 \geq 0, && 1 < x < 1,
\end{aligned}\right\} \tag{52.38}$$

but $w(0,t) = -t < 0$.

(ii) Assumption (52.35) in Proposition 52.24 can be weakened to $\partial_t w, \nabla w, D^2 w \in L^2_{loc}(Q_T)$ (and similarly in Proposition 52.25). To see this it suffices to replace w in the proof by $z := w - \varepsilon e^{\lambda t}$ with $\lambda > 0$ large (using the fact that $z_+ = 0$ near the boundary similarly as in the proof of Proposition 52.8), and then let $\varepsilon \to 0$. $\square$

In the case of nonlocal problems in unbounded domains, we need a different statement.

Proposition 52.27. *Let $0 < T < \infty$, Ω be an arbitrary domain in $\mathbb{R}^n$, $a, b \in L^\infty(Q_T)$, and $k \in L^\infty(\Omega)$, with $b, k \geq 0$. Assume that the function $w \in C^{2,1}(Q_T) \cap C(\overline{Q_T})$ satisfies $w \leq 0$ on $\mathcal{P}_T$,*

$$\partial_t w - \Delta w \leq aw + b \int_\Omega k(y)w(y,\cdot)\, dy \qquad a.e. \ in \ Q_T, \tag{52.39}$$

and

$$w \in C([0,T), L^1(\Omega)), \qquad \int_\Omega k(y)w(y,0)\, dy < 0. \tag{52.40}$$

Then

$$w < 0 \quad \text{in } Q_T.$$

Proof. Denote $I(t) := \int_\Omega k(y) w(y, t) \, dy$ and let $0 < \tau < T$. We claim that if

$$I(t) \leq 0, \quad 0 \leq t \leq \tau, \tag{52.41}$$

then

$$w < 0, \quad x \in \Omega, \ 0 < t \leq \tau. \tag{52.42}$$

Indeed, let $z := e^{\lambda t} w$ with $\lambda := -\inf_{Q_T} a$ and $\tilde{a} := a + \lambda \geq 0$. By (52.41), (52.39), we have $z_t - \Delta z \leq \tilde{a} z$ in Q_T. Since $z \leq 0$ on $\mathcal{P}_T$, it follows from Proposition 52.8 that $z \leq 0$, hence $z_t - \Delta z \leq 0$ in Q_T. Since $z(0) \not\equiv 0$, the claim then follows from the standard strong maximum principle (see [429], or use Proposition 52.7 in any smooth bounded subdomain of Ω).

Now, by (52.40), the function $I(t)$ is continuous, with $I(0) < 0$. Therefore, (52.41) is true for small $\tau > 0$. Let

$$T_0 := \sup\{\tau \in (0, T) : (52.41) \ (\text{hence } (52.42)) \ \text{is true}\}$$

and assume for contradiction that $T_0 < T$. Then (52.41) and (52.42) hold for $\tau = T_0$, hence in particular $w(\cdot, T_0) < 0$. Consequently, $I(T_0) < 0$, so that (52.41) holds for some $\tau > T_0$, contradicting the definition of T_0. This proves the result. $\quad\square$

52.8. Zero number

Zero number arguments can be viewed as a sophisticated form of the maximum principle. Although they are restricted to one-dimensional or radially symmetric problems, they represent a very powerful tool.

The zero number of a function $\psi \in C((0, R))$ is defined as the number of sign changes of ψ in $(0, R)$;

$$z(\psi) = z_{[0,R]}(\psi) = \sup\{k \in \mathbb{N} : \text{there are } 0 < x_0 < x_1 < \cdots < x_k < R$$
$$\text{such that } \psi(x_i)\psi(x_{i+1}) < 0 \text{ for } 0 \leq i < k\}.$$

Let $B_R = \{x \in \mathbb{R}^n : |x| < R\}$, $t_1 < t_2$, $q \in L^\infty(B_R, (t_1, t_2))$, $u \in C(\overline{B_R} \times [t_1, t_2]) \cap W^{2,1;\infty}(B_R \times (t_1, t_2))$ and

$$u_t - \Delta u = qu \quad \text{a.e. in } B_R \times (t_1, t_2). \tag{52.43}$$

Assume that $q(\cdot, t)$ and $u(\cdot, t)$ are radially symmetric for all t, hence $q(x, t) = Q(|x|, t)$ and $u(x, t) = U(|x|, t)$. Then

$$U_t - U_{rr} - \frac{n-1}{r} U_r = QU, \quad r \in (0, R), \ t \in (t_1, t_2), \tag{52.44}$$

and $U_r(0, t) = 0$ for all $t \in (t_1, t_2)$.

Theorem 52.28. *Let q, u be as above, $u \not\equiv 0$, and either $U(R,t) = 0$ for all $t \in [t_1, t_2]$ or $U(R,t) \neq 0$ for all $t \in [t_1, t_2]$. Let $z = z_{[0,R]}$ denote the zero number in $(0, R)$. Then*

(i) $z(U(\cdot, t)) < \infty$ *for all $t \in (t_1, t_2)$,*

(ii) *the function $t \mapsto z(U(\cdot, t))$ is nonincreasing,*

(iii) *if $U(r_0, t_0) = U_r(r_0, t_0) = 0$ for some $r_0 \in [0, R]$ and $t_0 \in (t_1, t_2)$, then $z(U(\cdot, t)) > z(U(\cdot, s))$ for all $t_1 < t < t_0 < s < t_2$.*

$\frac{\partial \Phi}{\partial s}(A, S_1) > \frac{\partial \Phi}{\partial s}(A, S_2) \neq 0,$ then from (iii)

Proof. If $U(R,t) = 0$ for all t, then the assertion follows from [127, Theorem 2.1].

If $U(R,t) \neq 0$ for all t, then we may assume $U(R,t) > 0$ for all t. Fix $\varepsilon \in (0, R)$ such that $U(r,t) > \varepsilon$ for all $r \in [R - \varepsilon, R]$ and $t \in [t_1, t_2]$. Let $V = V(r)$ be the solution of

$$V_{rr} + \frac{n-1}{r} V_r = 0 \quad \text{in } [R - \varepsilon, R + \varepsilon], \qquad V(R + \varepsilon) = 0, \ V_r(R + \varepsilon) = -1,$$

and notice that $V(r) \geq \varepsilon$ for $r \leq R$. Choose $\varphi \in C^\infty([0, R+\varepsilon])$ such that $0 \leq \varphi \leq 1$, $\varphi \equiv 1$ on $[0, R - \varepsilon]$, $\varphi \equiv 0$ on $[R, R + \varepsilon]$, and set

$$\tilde{U}(r, t) = \varphi(r) U(r, t) + (1 - \varphi(r)) V(r), \qquad r \in [0, R + \varepsilon], \ t \in [t_1, t_2],$$

$$\tilde{Q}(r, t) = \begin{cases} Q(r, t) & \text{if } r \in [0, R - \varepsilon], \\ \frac{1}{\tilde{U}}\left(\tilde{U}_t - \tilde{U}_{rr} - \frac{n-1}{r}\tilde{U}_r\right) & \text{if } r \in (R - \varepsilon, R), \\ 0 & \text{if } r \in [R, R + \varepsilon]. \end{cases}$$

Then $\tilde{U}$ solves (52.44) with Q replaced by $\tilde{Q}$ and R by $R + \varepsilon$, $\tilde{U}(R + \varepsilon, t) = 0$ and the assertion follows from $z_{[0,R]}(U(\cdot, t)) = z_{[0,R+\varepsilon]}(\tilde{U}(\cdot, t))$ $\quad\square$

Remarks 52.29. (i) The assertion of Theorem 52.28 remains true for more general problems of the form

$$u_t - \Delta u = qu + bx \cdot \nabla u,$$

where $b \in W^{1,\infty}(B_R \times (t_1, t_2))$, $b(x, t) = B(|x|, t)$. This follows from the fact that the function $v(x, t) := e^{\frac{1}{2}\int_0^{|x|} B(\xi, t)\xi \, d\xi} u(x, t)$ solves a problem of the form (52.43).

(ii) If $n = 1$, then a more general statement (allowing Dirichlet, Neumann or periodic boundary conditions and more general coefficients of the differential operators) can be found in [30]. In particular, the arguments in [30] guarantee that if $x_1 < x_2$, $t_1 < t_2$ and $u = U \in C([x_1, x_2] \times [t_1, t_2])$ is a solution of

$$u_t - a(x, t)u_{xx} = b(x, t)u_x + c(x, t)u \qquad \text{in } (x_1, x_2) \times (t_1, t_2),$$

where

$$a > 0, \ a, a^{-1}, a_t, a_x, a_{xx}, b, b_t, b_x, c \in L^\infty,$$

and, for any $i \in \{1,2\}$,

$$\text{either} \quad u(x_i, t) = 0 \quad \text{for } t \in [t_1, t_2],$$
$$\text{or} \quad u(x_i, t) \neq 0 \quad \text{for } t \in [t_1, t_2],$$

then statements (i)–(iii) in Theorem 52.28 hold with $[0, R]$ replaced by $[x_1, x_2]$. Let us mention that in the case of Neumann boundary conditions one has to assume $a \equiv 1$ and $b \equiv 0$. $\quad\square$

53. Appendix G: Dynamical systems

In this section we collect some basic definitions and properties of dynamical systems. Since the statements are usually proved only in the dissipative case in the literature (see [267], for example), we also provide detailed proofs.

Definition 53.1. Let (X, d) be a complete metric space and $\tau : X \to (0, \infty]$ be lower semicontinuous. A mapping $\varphi : X \times [0, \infty) \to X$ defined for all (u, t) with $u \in X$ and $t \in [0, \tau(u))$ is called a (local) **dynamical system** on X if

(i) $\varphi(u, \cdot) : [0, \tau(u)) \to X$ is continuous,

(ii) $\varphi(\cdot, t) : X \to X$ is continuous at u for all $u \in X$ and $t < \tau(u)$,

(iii) $\varphi(u, 0) = u$ for all $u \in X$,

(iv) $\tau(\varphi(u, s)) = \tau(u) - s$ and $\varphi(u, t + s) = \varphi(\varphi(u, s), t)$ for all $u \in X$, $s \in [0, \tau(u))$ and $t \in [0, \tau(u) - s)$. $\quad\square$

In our applications, X is typically a Banach space in which the studied problem is well-posed (or just the positive cone of such space), $\tau(u)$ is the maximal time of existence of the solution with initial data u and $\varphi(u, t)$ is this solution at time t. Note that for most of our assertions, φ need not be continuous with respect to t at $t = 0$ so that we can also choose $X = L^\infty(\Omega)$, for example.

Given $u \in X$, the mapping $\varphi_u : [0, \tau(u)) : t \mapsto \varphi(u, t)$ is called the **trajectory emanating from** u. It is **global** if $\tau(u) = \infty$. A point $u \in X$ is called an **equilibrium** if $\tau(u) = \infty$ and $\varphi(u, t) = u$ for all $t \geq 0$. We denote by S the set of all equilibria.

If $\tau(u) = \infty$, then we define the ω-**limit set** of φ_u by

$$\omega(\varphi_u) = \omega(u) := \{v \in X : \text{there exist } t_k \to +\infty \tag{53.1}$$
$$\text{such that } \varphi(u, t_k) \to v \text{ as } k \to \infty\}.$$

It is easy to see that

$$\omega(u) = \bigcap_{s>0} \overline{\bigcup_{t \geq s} \{\varphi(u, t)\}} \tag{53.2}$$

and $\omega(\varphi(u, t)) = \omega(u)$ for all $t > 0$.

Proposition 53.2. *Let $\tau(u) = \infty$ and $v \in \omega(u)$. Then $\varphi(v,t) \in \omega(u)$ for all $t \in [0, \tau(v))$.*

Proof. There exist $t_k \to \infty$ such that $\varphi(u, t_k) \to v$. Given $t \in [0, \tau(v))$, set $\tau_k := t_k + t$. Then $\varphi(u, \tau_k) = \varphi(\varphi(u, t_k), t) \to \varphi(v, t)$, hence $\varphi(v, t) \in \omega(u)$. $\square$

Proposition 53.3. *Assume*

$$\tau(u) = \infty \quad and \quad \bigcup_{t \geq 0} \{\varphi(u,t)\} \ is \ relatively \ compact \ in \ X. \tag{53.3}$$

Then $\tau(v) = \infty$ for all $v \in \omega(u)$, $\omega(u)$ is compact, connected, nonempty, invariant (that is $\varphi(\omega(u), t) = \omega(u)$ for all $t > 0$) and $d(\varphi(u, t), \omega(u)) \to 0$ as $t \to \infty$.

Proof. The set $K := \overline{\{\varphi(u, t) : t \geq 0\}}$ is compact and $\omega(u) \subset K$. The set $\omega(u)$ is closed due to (53.2), hence compact.

Choose $t_k \to \infty$. Then $\{\varphi(u, t_k)\}$ is relatively compact, hence passing to a subsequence we may assume $\varphi(u, t_k) \to v$. Consequently, $v \in \omega(u)$ and $\omega(u)$ is nonempty.

Fix $v \in \omega(u)$. Proposition 53.2 guarantees $\varphi(v, t) \in \omega(u)$ for all $t \in [0, \tau(v))$. Assume on the contrary that $\tau(v) < \infty$. Choose $t_k \in [0, \tau(v))$, $t_k \to \tau(v)$. Then $\{\varphi(v, t_k)\}$ is relatively compact and passing to a subsequence we may assume $\varphi(v, t_k) \to v_\infty$. Then $\tau(\varphi(v, t_k)) = \tau(v) - t_k \to 0$ and $\tau(v_\infty) > 0$ which contradicts the lower semicontinuity of τ. Consequently, $\tau(v) = \infty$.

Due to Proposition 53.2, in order to show the invariance of $\omega(u)$ it is sufficient to prove

$$\omega(u) \subset \varphi(\omega(u), t) \qquad for \ all \ t > 0. \tag{53.4}$$

Fix $v \in \omega(u)$, $t > 0$ and $t_k \to \infty$ such that $\varphi(u, t_k) \to v$. Set $\tau_k := t_k - t$. Passing to a subsequence we may assume $\varphi(u, \tau_k) \to w \in \omega(u)$. Then

$$\varphi(w, t) = \varphi\big(\lim_{k \to \infty} \varphi(u, \tau_k), t\big) = \lim_{k \to \infty} \varphi(u, t_k) = v,$$

which proves (53.4).

Next we show that $d(\varphi(u, t), \omega(u)) \to 0$ as $t \to \infty$. Assume on the contrary that there exist $t_k \to \infty$ and $\varepsilon > 0$ such that $d(\varphi(u, t_k), \omega(u)) \geq \varepsilon$. Passing to a subsequence we may assume $\varphi(u, t_k) \to v \in \omega(u)$ which yields a contradiction.

For any $s > 0$, the set $\bigcup \{\varphi(u, t) : t \geq s\}$ is connected and relatively compact, hence its closure is connected and compact. Due to (53.2), $\omega(u)$ is a decreasing intersection of connected compact sets, hence $\omega(u)$ is connected. $\square$

A continuous function $E : X \to \mathbb{R}$ is called a **Lyapunov function** for φ if $E(\varphi(u, t)) \leq E(u)$ for all $u \in X$ and $t \in [0, \tau(u))$.

Proposition 53.4. *Let E be a Lyapunov function and (53.3) be true. Then the limit $e := \lim_{t\to\infty} E(\varphi(u,t))$ exists and $E(v) = e$ for all $v \in \omega(u)$.*

Proof. The function $t \mapsto E(\varphi(u,t))$ is nonincreasing and bounded since the set $\bigcup\{\varphi(u,t) : t \geq 0\}$ is relatively compact. Hence $e := \lim_{t\to\infty} E(\varphi(u,t))$ exists.

If $v \in \omega(u)$, then there exist $t_k \to \infty$ such that $\varphi(u,t_k) \to v$. Consequently, $E(\varphi(u,t_k)) \to E(v) = e$. $\quad\square$

A Lyapunov function E is called a **strict Lyapunov function** if the following condition is satisfied: If $E(\varphi(u,t)) = E(u)$ for all $t \in [0, \tau(u))$, then u is an equilibrium.

The following two useful results are versions of Lasalle's **invariance principle**.

Proposition 53.5. *Let E be a strict Lyapunov function and (53.3) be true. Then S is a closed nonempty set and $d(\varphi(u,t), S) \to 0$ as $t \to \infty$. In particular, $\omega(u)$ consists of equilibria.*

Proof. The continuity of φ guarantees that S is closed and Proposition 53.3 shows that $\omega(u) \neq \emptyset$. Fix $v \in \omega(u)$ and $t \geq 0$. Then $\tau(v) = \infty$ and $\varphi(v,t) \in \omega(u)$ due to Proposition 53.3, hence Proposition 53.4 implies $E(\varphi(v,t)) = E(v)$. Consequently, $v \in S$. $\quad\square$

Proposition 53.6. *Assume $\tau(u) = \infty$, $t_k \to \infty$ and $\varphi(u,t_k) \to v$. Let there exist a strict Lyapunov function E. Then $v \in S$.*

Proof. The proof of Proposition 53.4 shows that $e := \lim_{k\to\infty} E(\varphi(u,t_k))$ exists and $E(v) = e$. Fixing $t \in [0, \tau(v))$, the continuity of φ implies $\varphi(u, t_k + t) \to \varphi(v,t)$. As above, $\tilde{e} := \lim_{k\to\infty} E(\varphi(u, t_k + t))$ exists and $E(\varphi(v,t)) = \tilde{e}$. Fixing $k \in \mathbb{N}$ there exists $j \in \mathbb{N}$ such that $t_{k+j} \geq t_k + t \geq t_k$, hence

$$E(\varphi(u, t_{k+j})) \leq E(\varphi(u, t_k + t)) \leq E(\varphi(u, t_k)),$$

thus $\tilde{e} = e$ and $E(\varphi(v,t)) = E(v)$. Consequently, $v \in S$. $\quad\square$

Example 53.7. Consider problem (17.1) with $p > 1$ and $\lambda \in \mathbb{R}$, and let $q \in [\max(q_c, p+1), \infty)$. By Example 51.28, we know that this problem generates a dynamical system on the space $X := H_0^1 \cap L^q(\Omega)$. Moreover, by (17.7) in Lemma 17.5, the energy functional $E(u)$ defined in (17.6) is a strict Lyapunov functional. Now assume Ω bounded and let $u_0 \in L^\infty(\Omega)$ be such that $T_{\max}(u_0) = \infty$ and $\sup_{t>0} \|u(t)\|_\infty < \infty$. Then, as a consequence of parabolic estimates and Proposition 53.5, for each $\tau > 0$, the set $\{u(t) : t \geq \tau\}$ is relatively compact in X and the ω-limit set $\omega(u_0)$ (in the X-topology) is nonempty and consists of (classical) equilibria. Moreover, by smoothing effects, the convergence in the definition of $\omega(u_0)$ actually takes place (for instance) in $C^{1+\beta}(\overline{\Omega})$ for each $\beta \in (0,1)$.

By similar arguments, the above facts remain true for the more general problem (14.1) with $f \in C^1$ if we replace the last integral in (17.6) by $\int_\Omega F(u)\,dx$, where

$F(u) = \int_0^u f(s)\,ds$, and X by $H^1 \cap C_0(\Omega)$. (Note that, as far as Propositions 53.3–53.6 are concerned, one could alternatively work with $X := H^1 \cap L^\infty(\Omega)$: Although the continuity property at $t = 0$ in Definition 53.1(i) is not true, this is of no importance in those results.) $\square$

Under an additional time monotonicity assumption, one obtains the convergence of the trajectory to a single equilibrium. See Remark 19.13 for other conditions guaranteeing convergence in the case of dynamical systems generated by parabolic differential equations.

Proposition 53.8. *Let $(X, \leq)$ be an ordered Banach space with a closed positive cone $X^+ := \{u \in X : u \geq 0\}$. Assume that (53.3) is true and the trajectory φ_u is nondecreasing, that is $\varphi(u, t_1) \leq \varphi(u, t_2)$ whenever $t_1 \leq t_2$. Then $\omega(u)$ is a singleton contained in S.*

Proof. Proposition 53.3 guarantees that $\omega(u)$ is nonempty and invariant. Let $v^1, v^2 \in \omega(u)$. Then there exist $t_k^1 \to \infty$ and $t_k^2 \to \infty$ such that $\varphi(u, t_k^i) \to v^i$, $i = 1, 2$. Without loss of generality we may assume $t_k^1 < t_k^2 < t_{k+1}^1$ for all k. Then $\varphi(u, t_k^1) \leq \varphi(u, t_k^2) \leq \varphi(u, t_{k+1}^1)$ and passing to the limit we obtain $v^1 \leq v^2 \leq v^1$, hence $v^1 = v^2$. Consequently, $\omega(u)$ is a singleton. Since it is an invariant set, we have $\omega(u) \subset S$. $\square$

Remark 53.9. For many dynamical systems generated by parabolic differential equations, the compactness assumption (53.3) in Proposition 53.8 can be replaced by a weaker boundedness assumption. In fact, the monotonicity of the solution in time usually enables one to pass to the limit and conclude that the limit is a stationary solution. For example one can often use the following lemma. In the case of monotonicity in space, similar arguments are used in the proofs of Theorems 8.3 and 21.10(ii). $\square$

Lemma 53.10. *Let $f : \mathbb{R} \to \mathbb{R}$ be locally Hölder continuous, let Ω be an arbitrary domain in $\mathbb{R}^n$, and set $Q := \Omega \times (0, \infty)$. Let $u \in C^{2,1}(Q)$ satisfy*

$$u_t - \Delta u = f(u), \qquad (x, t) \in Q, \tag{53.5}$$

and

$$\sup_Q |u| < \infty.$$

Assume that $u_t \geq 0$ in Q, and let $v(x) := \lim_{t \to \infty} u(x, t)$. Then v is a bounded classical solution of

$$-\Delta v = f(v), \qquad x \in \Omega. \tag{53.6}$$

Proof. Let $w_j(x, t) = u(x, t + j)$. The L^p and Schauder interior parabolic estimates guarantee that the sequence $\{w_j\}$ is relatively compact in $C^{2,1}(K)$ for each compact subset K of Q. It follows that some subsequence of w_j converges to a classical solution $w = w(x,t)$ of (53.5). But it is clear that $v(x) = w(x,t)$ for each $(x,t) \in Q$. The conclusion follows. $\square$

Alternative proof. Let $\varphi \in \mathcal{D}(\Omega)$. For each $t > 0$, multiplying by φ and integrating by parts yields

$$\left[\int_\Omega u\varphi \, dx \right]_t^{t+1} + \int_t^{t+1} \int_\Omega \left(u\Delta\varphi + f(u)\varphi \right) dx \, ds = 0.$$

Passing to the limit by dominated convergence as $t \to \infty$, we obtain

$$\int_\Omega \left(v\Delta\varphi + f(v)\varphi \right) dx = 0.$$

It follows that v is a distributional solution of (53.6), hence a classical solution by standard elliptic regularity results (cf. Remarks 47.4). $\square$

Let $u_s \in S$. The **domain of attraction** of u_s is the set

$$D = D(u_s) := \{ u : \tau(u) = \infty \text{ and } \varphi(u,t) \to u_s \text{ as } t \to \infty \}.$$

We say that u_s is **asymptotically stable** if $D(u_s)$ contains a neighborhood of u_s. If u_s is asymptotically stable, then $D(u_s)$ is obviously open. If $u \in \partial D(u_s)$, then the continuity of φ implies $\varphi(u,t) \in \partial D(u_s)$ for all $t \in (0, \tau(u))$.

Let $u, v \in S$, $u \neq v$. A function $\psi : \mathbb{R} \to X$ is called a **connecting** (or **heteroclinic**) **orbit** between u and v if $\lim_{t \to -\infty} \psi(t) = u$, $\lim_{t \to +\infty} \psi(t) = v$ and $\psi(t) = \varphi(\psi(s), t - s)$ for all $-\infty < s < t < \infty$.

54. Appendix H: Methodological notes

In this orientation section, we shall summarize the different methods employed in this monograph for each of the main questions that we address. Of course, there exist other important methods which are frequently exploited for related questions and which are not represented in this book for various reasons. For example, in the elliptic part we only use the simplest variational methods to prove those existence results which are needed in the parabolic part and we do not pay attention to linking or concentration compactness. Notice also that our summary does not even contain all important methods used in this monograph. For example, matched asymptotics is used in Section 29 in the study of decay and grow-up rates which is not a central theme for us. On the other hand, matched asymptotics plays a crucial role in several fundamental papers devoted to our main questions (like blow-up rates or blow-up profiles, see [277], [522], [275], [276], for example) but the corresponding rigorous proofs are too long and technical and lie beyond the scope of this book.

Before providing the summary, let us briefly classify the main tools and techniques that are used throughout the book. We point out that, of course, many of these techniques are not specific to the field of superlinear elliptic and parabolic problems but are classical in various areas of PDE's.

A. Background tools

A1. Tools from Functional Analysis (functional spaces and inequalities, interpolation)

A2. Tools from ODE's (differential and integral inequalities, phase plane analysis)

A3. Linear elliptic and parabolic estimates (often used to obtain compactness properties)

A4. Tools from Dynamical Systems (ω-limits, Lyapunov functionals)

B. Main classes of techniques

B1. Comparison techniques (based on maximum principles, including moving planes and zero-number)

B2. Test-function and multiplier techniques[7] (in particular including variational and energy methods). In the parabolic case this often leads to a differential inequality for a functional of the solution

B3. Semigroup techniques (relying on the variation-of-constants formula)

B4. Rescaling procedures (often leading to the use of a nonlinear Liouville-type theorem via a contradiction argument)[8]

B5. Bootstrap and iteration procedures

C. Some other techniques

C1. Changes of dependent and/or independent variables (cf., e.g., similarity variables, Hopf-Cole transformation, conversion to a problem with absorption, ...)

C2. Differentiation of the PDE (cf., e.g., Bernstein-type techniques, auxiliary functions J, ...)

C3. Monotonicity techniques: use of monotonicity properties of solutions (in time or in space — usually obtained via a maximum principle), monotone approximation (cf. complete blow-up, threshold trajectories)

C4. Doubling arguments

---

[7]By a *test-function* technique, we usually understand the space or space-time integration of the PDE after multiplication by a function *independent* of the solution itself, such as the first eigenfunction for instance. In a *multiplier* technique, the function may depend on the solution, e.g. a power of the solution.

[8]Another aspect of the concept of scaling is the existence of self-similar solutions — cf. also similarity variables in C1.

C5. Duality arguments

We now turn to the more detailed summary of methods. For a given question, the choice of the applicable (and more appropriate) methods will depend on the particular properties of the problem: scale invariance, variational structure, monotonicity, convexity or boundedness of the domain, regularity assumed on solutions, ...

For some partial comparative discussion, see e.g. the beginning of sections 10–13; Remark 26.5 and the end of Remark 26.12; the paragraph after Theorem 27.2; Remarks 31.5, 31.18(i) and the beginning of Subsection 31.4.

We stress that the following list, which is mainly intended as a help and a guideline for readers, is necessarily schematic. Certain proofs may sometimes involve combinations of several methods, or some ad hoc arguments which do not appear in the list. On the other hand, some items below may partially overlap. The places where each method is used are mentioned in italic between brackets.

I. METHODS FOR ELLIPTIC PROBLEMS

M1. Methods to prove existence of solutions

M1.1. Variational methods
  (a) Minimisation under constraint [Section 6]
  (b) Minimax methods [Section 7]
M1.2. ODE methods [Section 9]
M1.3. A priori estimates and topological degree argument [Corollary 10.3 and cf. M4]
M1.4. Dynamical methods
  (a) Method based on a priori estimates of global solutions and threshold trajectories [Remark 28.8(ii)]
  (b) Stabilization of monotone bounded solutions [Theorem 43.1(iii)]

M2. Methods to prove nonexistence of solutions

Note: the methods which are mainly motivated by Liouville-type results appear only in M8.1–M8.3 below.

M2.1. Variational identities of Pohozaev-type [Corollary 5.2, Proposition 25.4, Theorem 31.3]
M2.2. Multiplication by the first eigenfunction [Remark 6.3]
M2.3. ODE techniques [Section 9]
  Note: the methods in M2.1 and M2.3 may be combined with symmetry results [Remark 6.9(i)]
M2.4. Maximum principle [Proposition 40.8]

M3. Methods to study regularity and singularities of solutions

M3.1. To prove regularity: bootstrap procedures using linear elliptic estimates $(W^{2,p}, L^p\text{-}L^q, L_\delta^p\text{-}L_\delta^q, \dots)$. This can be combined with test-function, cut-off or truncation arguments [Propositions 3.3, 3.5, and see M4.2 below]

M3.2. To establish pointwise singularity estimates
  (a) Integral estimates from M8.3 below, combined with a bootstrap procedure [Theorems 4.1 $(p_{sg} < p < p_S)$, 8.7]
  (b) Rescaling, Liouville-type results and doubling arguments [Remark 8.8(i)]
  (c) Combination of the following three ingredients: the characterization of nonnegative distributions with point support (the nonnegativity being obtained by truncation and test-function techniques); a comparison argument involving the Newton potential; a bootstrap procedure [Theorems 4.2, 4.1 $(1 < p < p_{sg})$]

M3.3. To produce singular solutions
  (a) Method based on the construction and pointwise estimates of a singular solution of the linear Laplace equation [Theorems 11.5, 31.16]
  (b) Explicit singular solutions [Remarks 3.6(ii), 31.19, Formula (40.20)]
  (c) ODE methods [Remark 3.6(ii)]

M4. Methods to prove a priori estimates

M4.1. Method of Hardy-Sobolev inequalities. Also: variant based on the use of a singular test-function [Section 10, Remark 31.18(i)]

M4.2. Bootstrap in L_δ^p-spaces. Alternate bootstrap in the case of systems [Section 11, Subsection 31.4]

M4.3. Method based on rescaling and Liouville-type theorems [Section 12, Subsection 31.3]

M4.4. Method of moving planes and Pohozaev-type identities [Section 10, Theorem 31.2]

II. METHODS FOR PARABOLIC PROBLEMS

M5. Methods for local well-posedness

M5.1. Local existence-uniqueness
  Note: we are mainly concerned with irregular initial data. The case of smooth data is standard.
  (a) Fixed-point in a metric space of functions of t, with a weight vanishing at $t = 0$, using $L^p\text{-}L^q$-estimates for the heat semigroup (variants: L_δ^p-spaces or uniformly local spaces, instead of L^p) [Theorems 15.2, 15.9, 32.1(i), Remark 43.14(b)]

(b) Similar to M5.1(a), with L^p-spaces replaced by a scale of interpolation-extrapolation spaces [Theorem 51.25]

(c) Improvement of the uniqueness class (without temporal weight): one shows that any solution actually belongs to the fixed-point space; this is achieved by a time-shift and continuous dependence argument [same as M5.1(a)-M5.1(b)]

M5.2. Local nonexistence of positive solutions: for suitable singular initial data, contradiction between two pointwise estimates for the "free" part $e^{-tA}u_0$ (namely: a lower estimate for the linear heat equation as $t \to 0$, and an upper a priori estimate depending on the nonlinear equation; see M8.5 for a closely related argument and more details) [Theorems 15.3, 15.10, 32.1(ii)]

M5.3. Local nonuniqueness

(a) Nonuniqueness for zero initial data: construction of a forward self-similar solution with exponential decay in space by ODE (shooting) methods [Remarks 15.4(ii), 40.11(a)]

(b) Construction of a singular stationary solution (which coexists with a classical solution for $t > 0$) [Remark 15.4(iii)]

(c) Nonuniqueness for general initial data: method based on concentrated perturbations of an initial data, continuity of the existence time and universal bounds [Proposition 28.1]

M5.4. Regularity and smoothing: bootstrap procedure using (e.g.) L^p-L^q-estimates for the heat semigroup [Theorems 15.2, 15.9, 15.11, 43.13].

M5.5. Continuation properties (in particular: uniform bounds from L^q-bounds)

(a) Consequence of well-posedness in M5.1(a)–M5.1(b) and smoothing property in M5.4. Also, a lower estimate of the norm of $u(\cdot, t)$ near the blow-up time can be directly deduced from the fixed-point argument [Remark 16.2(iii), Theorem 33.5]

(b) Moser-type iteration [Theorems 16.4, 33.5, Remark 33.6]

(c) Variation-of-constants formula combined with interpolation inequality and interpolation-extrapolation spaces [Proposition 51.34] (or just L^p-spaces [Theorem 32.2])

Note: in the case of systems, better results can be obtained by alternate use of each equation

(d) Consequence of lower estimates on the blow-up profile (cf. M12.3(a))

Note: *non*-continuation can be shown as a consequence of upper estimates on the blow-up profile [Corollary 24.2, Theorem 44.6, Remark 44.8(c)]

(e) Energy arguments [Proposition 16.3]

(f) Gradient bounds (in particular via Bernstein techniques): see details in [Section 35]

M6. Methods to prove global existence (and also boundedness, decay, stability)

M6.1. Multiplier, energy and Lyapunov functional methods

Note: the following three items partially overlap

(a) Use of powers of the solution as multiplier, in combination with various functional inequalities [Theorems 19.3(i), 33.9(i), 36.4(i), Remark 40.11(b), Theorems 43.1, 44.5(i)]

(b) Potential well method [Theorem 19.5(i)]

(c) Use of a Lyapunov functional [Theorems 33.5, 33.18(ii), 40.7(i)]

M6.2. Comparison methods

(a) Supersolutions with separated variables; spatially homogeneous supersolutions [Theorems 19.2, 32.5(ii), 43.1, 46.1(ii)]

(b) Stationary supersolutions (and families thereof); singular steady-states and their perturbations; quasi-stationary supersolutions [Theorem 19.15(ii), Remark 19.14, Theorems 20.5, 29.1 32.5(iii), 36.1(ii), 36.4(i), 37.2, 40.7(iii), 44.17(i)]

(c) Supersolutions involving the heat semigroup (possibly self-similar) [Theorem 20.2, Remark 20.4(i), Theorems 20.6, 20.11, 32.5(ii) and 37.4(ii)]

(d) Self-similar supersolutions [Theorem 20.6, Section 45]

(e) Traveling wave supersolutions [Theorem 36.7]

(f) Intersection-comparison with radial steady-states (to show global existence of a threshold solution) [Theorem 22.9]

M6.3. Variation-of-constants formula and semigroup estimates (e.g., L^p-L^q or exponential decay) [Remark 19.4(b), Theorems 20.15, 33.1, 40.7(i), 40.10(i), Subsection 51.3]

Note: sometimes combined with fixed-point arguments in spaces with temporal weight [Theorem 20.19, Corollary 20.20]

M6.4. Forward similarity variables (combined with construction of stable manifolds) [Proposition 20.13]

M6.5. Duality method [Theorem 33.2, Remarks 33.4, 33.6, 33.15]

M6.6. Gradient estimates [Proposition 40.5, Theorem 40.7(iii), Remark 40.11(d)]

M7. Methods to prove blow-up (in finite -or sometimes infinite- time)

Note 1: the methods in M7.1–M7.2 lead to a differential inequality for some functional of $u(\cdot, t)$.

Note 2: the methods which are mainly motivated by (blow-up) results of Fujita-type do not appear here (see M8.4(a), M8.5, M8.6).

M7.1. Eigenfunction method [Theorems 17.1, 17.3, 32.5(i), 33.16, 36.1(i), Remarks 40.4(i) and (ii), Theorem 46.1]

Note: other test-functions independent of u can sometimes be used [cf. Theorem 43.1(i)]

M7.2. Energy and multiplier methods

(a) Energy and Hölder's inequality (in bounded domains) [Theorems 17.6, 44.14]

(b) Energy and concavity argument (in general domains) [Theorem 17.6]

(c) Potential well method [Theorem 19.5(ii)]

(d) Use of a power of the solution as test-function, in combination with various functional inequalities [Theorems 40.2, 41.1]

M7.3. Comparison methods

(a) Blowing-up self-similar subsolutions [Theorems 36.2, Section 45]

(b) Other forms of subsolutions (perturbation of singular or regular steady states, expanding waves, traveling waves, quasi-stationary, ...) [Theorems 29.1, 36.4(ii), Lemma 36.6, Theorem 41.1, Remark 40.4(i), Theorem 44.17(ii)]

(c) Blow-up above a positive equilibrium [Theorem 17.8, Proposition 19.11]

(d) Comparison between domains [Remark 17.14]

(e) Cf. M11.2(a) [Theorem 23.5]

Note: in spatially nonlocal problems, this is sometimes combined with the method in M12.1(a) to obtain preliminary estimates on the nonlocal term [Theorem 44.5(ii)]

M7.4. Use of scaling properties of the equation (e.g. to prove blow-up for initial data with slow decay at infinity)

(a) Rescaled eigenfunctions [Theorem 17.12]

(b) Rescaled subsolutions [Remarks 17.13(i), 36.3(iii), Theorems 19.3(ii), 36.4(ii)]

M7.5. Construction of explicit blowing-up solutions (often under self-similar form, or by solving an ODE) [Theorems 33.9(ii), 33.12, 33.18(i)]

M7.6. Use of dynamical systems arguments (ω-limits via a strict Lyapunov functional, or via monotonicity) combined with the absence of steady-states (may lead to blow-up in finite or infinite time) [Remark 19.14, Theorems 28.7(iv), 33.14]

M7.7. Direct estimation via integration in space-time parabolas and Sobolev inequality (to prove growth of mass for a Cauchy problem) [Theorem 40.10(ii)]

M8. Methods to prove nonexistence in Liouville and Fujita-type results

Note: the methods in M8.1–M8.3 concern both elliptic and parabolic problems

M8.1. Rescaled test-functions

    (a) Spatial test-functions

        [Theorems 8.4, 31.12] in the elliptic case; [Remark 18.2(i), Theorems 32.7, 37.4] in the parabolic case (where this leads to differential inequalities)

    (b) Space-time test-functions [Theorems 18.1(i), 37.1]

M8.2. Moving plane methods

    (a) Via symmetry, using the Kelvin transform (elliptic; case of the whole space) [Theorem 8.1]

    (b) Via symmetry and reduction to a one-dimensional problem on a half-line (elliptic; case of a half-space) [Theorem 8.2]

    (c) Via monotonicity and reduction to an $(n-1)$-dimensional problem in the whole space (elliptic and parabolic; case of a half-space) [Theorems 8.3, 21.8, 31.10]

M8.3. Integral estimates, obtained by using Bochner's identity, power change of dependent variable, and multipliers involving powers of u and cut-offs [Propositions 8.6, 21.5]

M8.4. Comparison methods

    (a) Families of blowing-up self-similar subsolutions [Section 45]

    (b) Intersection-comparison with radial steady-states [Theorem 21.1]

M8.5. Method based on the variation-of-constants formula (for Fujita-type results) [Theorem 18.3]

More precisely, a contradiction is obtained by comparing two pointwise estimates for the "free" part $e^{-tA}u_0$ of the solution: the lower estimate from the linear heat equation as $t \to \infty$, and an upper a priori estimate depending on the nonlinear equation [Lemma 15.6]. The latter is proved by taking the action of the heat semigroup on the variation-of-constants formula.[9] In the critical case, the necessary additional information is provided by an L^1 lower bound based on convolution properties of Gaussians.

M8.6. Forward similarity variables [Lemma 18.4]

M9. Methods to prove boundedness of global solutions and parabolic a priori estimates

Note: the methods in M9.1(b), M9.2(b), M9.3(a) here yield only boundedness of global solutions

---

[9] For Fujita-type problems, the methods in M8.1(a) and M8.5 are essentially equivalent. In fact, in M8.1(a), one also compares a lower asymptotic estimate with an upper a priori bound. The latter follows from differential inequalities obtained by multiplying with rescaled Gaussian test-functions, and these Gaussians are nothing but the heat kernel with time as a parameter. However the argument in M8.1(a) requires more regularity on the solution. Alternatively, the upper a priori bound can be obtained by a subsolution argument (see Remark 15.7).

M9.1. Rescaling methods

(a) Method based on rescaling, elliptic Liouville-type theorems and energy [Theorem 22.1]

(b) Method based on rescaling and intersection-comparison, using the infinite intersection property of the singular and regular steady-states in $\mathbb{R}^n$ [Remark 23.13]

(c) Cf. M10.2(a) [Theorem 38.1]

M9.2. Energy methods

(a) Method based on energy estimates and on a bootstrap argument using interpolation and maximal regularity [Theorem 22.1, Proposition 22.11, Remark 44.15]

(b) Method based on energy estimates in forward similarity variables, and on a measure argument [Lemma 18.4]

M9.3. Methods based on the maximum principle

(a) Intersection-comparison with a backward self-similar solution and with the singular steady-state [Theorem 22.4]

(b) Use of a monotonicity property: the solution becomes increasing in time if it reaches a sufficiently high level (for a nonlocal problem) [Proposition 43.16]

M10. Methods to prove universal bounds of positive solutions and initial blow-up rates

M10.1. Methods based on smoothing estimates

(a) Smoothing in L^p_δ-spaces (using integral bounds obtained by the eigenfunction method, and possibly combined with a priori estimates) [Theorems 26.1, 26.14]

(b) Smoothing in L^p-spaces (using integral bounds obtained by using a singular test-function, or by the eigenfunction method, and combined with a priori estimates) [Theorems 26.1, 43.15]

(c) Smoothing in uniformly local Lebesgue spaces (using integral bounds obtained by the eigenfunction method) [Theorem 26.13]

M10.2. Rescaling methods

(a) Method based on a doubling lemma and parabolic Liouville-type theorems [Theorems 26.8, 26.9, 38.1]

(b) Method based on energy, measure arguments, and elliptic Liouville-type theorems. [Theorem 26.6, Remark 26.7]

M10.3. Methods based on space-time integral estimates

(a) Via Moser-type iteration or Harnack inequality [Theorem 26.13(i)]

(b) Via the method in M8.3 combined with Harnack inequality [Theorem 26.8]

M11. Methods to establish blow-up rates

M11.1. Lower estimates

    (a) Comparison with solutions of the ODE [Proposition 23.1, Remark 38.2(i)]

    (b) Differential inequality obtained by considering points of maxima of $u(\cdot,t)$ [Proposition 23.1, Theorems 32.9, 44.2(i), 44.17(ii), 46.4(i), 44.2(i), Proposition 44.3(i), Theorem 44.17(ii)]

    (c) Variation-of-constants formula and use of the doubling time of $\|u(t)\|_\infty$ [Remark 23.2(ii)]

    (d) Regularity estimates applied to the equation for u_t (for the GBU problem) [Theorem 40.18]

    (e) Method using the intersections of the solution with a steady-state, and the boundedness of u_t (for the GBU problem) [Theorem 40.19]

M11.2. Upper estimates[10]

    (a) Maximum principle applied to an auxiliary function J (for time-increasing solutions) [Theorems 23.5, 32.9, Remark 38.2(ii), Theorem 46.4]. See also [Theorem 40.21] for a different type of auxiliary function in the GBU problem.

    (b) Methods based on backward similarity variables

        (b)-1 Via rescaling, elliptic Liouville-type theorems and energy [Theorem 23.7]

        (b)-2 Via localized energy estimates and bootstrap [Remark 23.14(i)]

        Note: the last two methods are similar to M9.1(a) and M9.2(a), respectively, the question being equivalent to the boundedness of global solutions for the equation in similarity variables

    (c) Method based on rescaling and intersection-comparison, using the infinite intersection property of the singular and regular steady-states in $\mathbb{R}^n$ [Theorem 23.10]

    (d) Cf. M10.2(a) [Theorems 26.8, 38.1; see also Remarks 26.12, 32.12(i), Proposition 44.3(ii)]

    (e) Cf. M10.3(a) [Remark 32.12(i)]

    (f) Cf. M12.4 [Subsection 43.2, Theorem 44.2(i)]

M12. Methods to study blow-up sets and profiles

M12.1. Methods based on the maximum principle

    (a) Maximum principle applied to an auxiliary function J (to obtain single-point blow-up and upper profile estimates for radial nonin-

---

[10] One could alternatively classify the methods for upper blow-up estimates between: those using scaling and energy (M11.2(b)), those using scaling without energy (M11.2(c)–M11.2(d)), and those using neither energy nor scaling (M10.3(a), M11.2(a), M12.4).

creasing solutions) [Theorem 24.1, Remark 32.12(ii), Theorems 39.7, 44.2(iii), 44.6 (ii), Remark 46.5]

Note: sometimes combined with a bootstrap argument [cf. Theorem 39.1]

(b) Moving plane method (to prove compactness of the blow-up set) [Remark 24.6(iv)]

(c) Sub-/supersolutions of blowing-up barrier type, using the notion of sub-/super-standard functions (to obtain the blow-up behavior in the boundary layer for spatially nonlocal problems) [Theorem 43.10]

(d) Bernstein-type techniques (for blow-up profile estimates in the GBU problem) [Remark 40.17, 41.4]

M12.2. Method of backward similarity variables

(a) Combined with weighted energy and dynamical systems arguments (to show asymptotically self-similar blow-up behavior; to exclude blow-up at a given point and prove compactness of the blow-up set) [Theorems 25.1, 24.5]

Note: sometimes combined with comparison and cut-off arguments [Theorem 25.3]

(b) Combined with linearization and spectral techniques (to obtain refined blow-up estimates and classification of profiles) [Remark 25.8]

(c) Construction of exact backward self-similar solutions by ODE (phase plane) methods [Proposition 22.5, Remarks 39.8(i), (iii) and (iv)]

M12.3. Methods based on ODE's in space

(a) ODE energy estimate and use of the point of half-maximum of $u(\cdot, t)$ (to obtain lower profile estimates for radial nonincreasing solutions) [Theorems 24.3, 39.2, Remark 32.12(ii)]

(b) Differential inequalities in space, relying on the boundedness of the time derivative (for blow-up profile estimates in the GBU problem) [Theorem 40.14]

M12.4. Method based on eigenfunction arguments, one-sided estimates of Δu (via the maximum principle), and the mean-value inequality for subharmonic functions (to obtain the blow-up rate, set and profile for spatially nonlocal problems) [Subsection 43.2, Theorem 44.2(i)]

Bibliography

[1]   M. Abramowitz and I. Stegun, *Handbook of mathematical functions with formulas, graphs, and mathematical tables,*, Wiley, New York, 1972.

[2]   N. Ackermann and T. Bartsch, *Superstable manifolds of semilinear parabolic problems*, J. Dynam. Differential Equations **17** (2005), 115-173.

[3]   N. Ackermann, T. Bartsch, P. Kaplický and P. Quittner, *A priori bounds, nodal equilibria and connecting orbits in indefinite superlinear parabolic problems*, Trans. Amer. Math. Soc. (to appear).

[4]   S. Agmon, *Lectures on elliptic boundary value problems*, Van Nostrand, Princeton, N.J., 1965.

[5]   J. Aguirre and M. Escobedo, *On the blow up of solutions for a convective reaction diffusion equation*, Proc. Roy. Soc. Edinburgh Sect. A **123** (1993), 449-470.

[6]   N. Alaa, *Solutions faibles d'équations paraboliques quasi-linéaires avec données initiales mesures*, Ann. Math. Blaise Pascal **3** (1996), 1-15.

[7]   S. Alama and M. Del Pino, *Solutions of elliptic equations with indefinite nonlinearities via Morse theory and linking*, Ann. Inst. H. Poincaré Anal. Non Linéaire **13** (1996), 95-115.

[8]   S. Alama and G. Tarantello, *On semilinear elliptic equations with indefinite nonlinearities*, Calc. Var. Partial Differential Equations **1** (1993), 439-475.

[9]   F. Alessio, P. Caldiroli and P. Montecchiari, *Infinitely many solutions for a class of semilinear elliptic equations in $\mathbb{R}^N$*, Boll. Unione Mat. Ital. **(8) 4-B** (2001), 311-318.

[10]   L. Alfonsi and F.B. Weissler, *Blow up in $\mathbb{R}^n$ for a parabolic equation with a damping nonlinear gradient term*, Progress in nonlinear differential equations (N.G. Lloyd et al., eds.), Birkhäuser, Basel, 1992.

[11]   N.D. Alikakos, *L^p bounds of solutions of reaction-diffusion equations*, Comm. Partial Differential Equations **4** (1979), 827-868.

[12]   N.D. Alikakos, *An application of the invariance principle to reaction diffusion equations*, J. Differential Equations **33** (1979), 201-225.

[13]   H. Amann, *Existence and regularity for semilinear parabolic evolution equations*, Ann. Scuola Norm. Sup. Pisa Cl. Sci. (4) **11** (1984), 593-676.

[14]   H. Amann, *Parabolic evolution equations and nonlinear boundary conditions*, J. Differential Equations **72** (1988), 201-269.

[15] H. Amann, *Nonhomogeneous linear and quasilinear elliptic and parabolic boundary value problems*, Function Spaces, Differential Operators and Nonlinear Analysis (H.J. Schmeisser and H. Triebel, eds.), Teubner, Stuttgart, Leipzig, 1993, pp. 9-126.

[16] H. Amann, *Linear and quasilinear parabolic problems*, Volume I: Abstract linear theory, Birkhäuser, Basel, 1995.

[17] H. Amann, *Linear and quasilinear parabolic problems*, Volume II, in preparation.

[18] H. Amann, M. Hieber and G. Simonett, *Bounded H_∞-calculus for elliptic operators*, Differential Integral Equations **7** (1994), 613-653.

[19] H. Amann and J. López-Gómez, *A priori bounds and multiple solutions for superlinear indefinite elliptic problems*, J. Differential Equations **146** (1998), 336-374.

[20] H. Amann and P. Quittner, *Elliptic boundary value problems involving measures: existence, regularity, and multiplicity*, Adv. Differential Equations **3** (1998), 753-813.

[21] H. Amann and P. Quittner, *Semilinear parabolic equations involving measures and low regularity data*, Trans. Amer. Math. Soc. **356** (2004), 1045-1119.

[22] H. Amann and P. Quittner, *Optimal control problems with final observation governed by explosive parabolic equations*, SIAM J. Control Optim. **44** (2005), 1215-1238.

[23] A. Ambrosetti and P.H. Rabinowitz, *Dual variational methods in critical point theory and applications*, J. Funct. Anal. **14** (1973), 349-381.

[24] A. Ambrosetti and P.N. Srikanth, *Superlinear elliptic problems and the dual principle in critical point theory*, J. Math. Phys. Sci. **18** (1984), 441-451.

[25] A. Ambrosetti and M. Struwe, *A note on the problem $-\Delta u = \lambda u + u|u|^{2^*-2}$*, Manuscripta Math. **54** (1986), 373-379.

[26] L. Amour and M. Ben-Artzi, *Global existence and decay for viscous Hamilton-Jacobi equations*, Nonlinear Anal. **31** (1998), 621-628.

[27] D. Andreucci, *Degenerate parabolic equations with initial data measures*, Trans. Amer. Math. Soc. **349** (1997), 3911-3923.

[28] D. Andreucci and E. DiBenedetto, *On the Cauchy problem and initial traces for a class of evolution equations with strongly nonlinear sources*, Ann. Scuola Norm. Sup. Pisa Cl. Sci. (4) **18** (1991), 363-441.

[29] D. Andreucci, M.A. Herrero and J.J.L. Velázquez, *Liouville theorems and blow up behaviour in semilinear reaction diffusion systems*, Ann. Inst. H. Poincaré Anal. Non Linéaire **14** (1997), 1-53.

[30] S. Angenent, *The zero set of a solution of a parabolic equation*, J. Reine Angew. Math. **390** (1988), 79-96.

[31] S. Angenent and M. Fila, *Interior gradient blow-up in a semilinear parabolic equation*, Differential Integral Equations **9** (1996), 865-877.

[32] S. Angenent and R. van der Vorst, *A priori bounds and renormalized Morse indices of solutions of an elliptic system*, Ann. Inst. H. Poincaré Anal. Non Linéaire **17** (2000), 277-306.

[33] S.N. Antontsev and M. Chipot, *The thermistor problem: existence, smoothness, uniqueness, blow-up*, SIAM J. Math. Anal. **25** (1994), 1128-1156.

[34] G. Arioli, F. Gazzola, H.-Ch. Grunau and E. Sassone, *The second bifurcation branch for radial solutions of the Brezis-Nirenberg problem in dimension four*, NoDEA Nonlinear Differential Equations Appl. (to appear).

[35] D.G. Aronson and H.F. Weinberger, *Multidimensional nonlinear diffusions arising in population genetics*, Adv. Math. **30** (1978), 33-76.

[36] J.M. Arrieta and A. Rodríguez-Bernal, *Non well posedness of parabolic equations with supercritical nonlinearities*, Comm. Contemp. Math. **6** (2004), 733-764.

[37] J.M. Arrieta, A. Rodríguez-Bernal, J.W. Cholewa and T. Dlotko, *Linear parabolic equations in locally uniform spaces*, Math. Models Methods Appl. Sci. **14** (2004), 253-293.

[38] J.M. Arrieta, A. Rodríguez-Bernal and Ph. Souplet, *Boundedness of global solutions for nonlinear parabolic equations involving gradient blow-up phenomena*, Ann. Scuola Norm. Sup. Pisa Cl. Sci. (5) **3** (2004), 1-15.

[39] F.V. Atkinson, H. Brezis and L.A. Peletier, *Nodal solutions of elliptic equations with critical Sobolev exponents*, J. Differential Equations **85** (1990), 151-170.

[40] F.V. Atkinson and L.A. Peletier, *Emden-Fowler equations involving critical exponents*, Nonlinear Anal. **10** (1986), 755-776.

[41] F.V. Atkinson and L.A. Peletier, *Large solutions of elliptic equations involving critical exponents*, Asymptotic Anal. **1** (1988), 139-160.

[42] F.V. Atkinson and L.A. Peletier, *Oscillations of solutions of perturbed autonomous equations with an application to nonlinear elliptic eigenvalue problems involving critical Sobolev exponents*, Differential Integral Equations **3** (1990), 401-433.

[43] T. Aubin, *Problèmes isopérimétriques de Sobolev*, J. Differential Geometry **11** (1976), 573-598.

[44] P. Aviles, *On isolated singularities in some nonlinear partial differential equations*, Indiana Univ. Math. J. **32** (1983), 773-791.

[45] A. Bahri, *Topological results on a certain class of functionals and application*, J. Funct. Anal. **41** (1981), 397-427.

[46] A. Bahri and H. Berestycki, *A perturbation method in critical point theory and applications*, Trans. Amer. Math. Soc. **267** (1981), 1-32.

[47] A. Bahri and J.-M. Coron, *On a nonlinear elliptic equation involving the critical Sobolev exponent: the effect of the topology of the domain*, Comm. Pure Appl. Math. **41** (1988), 253-294.

[48] A. Bahri and P.-L. Lions, *Morse index of some min-max critical points. I. Application to multiplicity results*, Comm. Pure Appl. Math. **41** (1988), 1027-1037.

[49] A. Bahri and P.-L. Lions, *Solutions of superlinear elliptic equations and their Morse indices*, Comm. Pure Appl. Math. **45** (1992), 1205-1215.

[50] A. Bahri and P.-L. Lions, *On the existence of a positive solution of semilinear elliptic equations in unbounded domains*, Ann. Inst. H. Poincaré Anal. Non Linéaire **14** (1997), 365-413.

[51] J. Ball, *Remarks on blow-up and nonexistence theorems for nonlinear evolution equations*, Quart. J. Math. Oxford Ser. (2) **28** (1977), 473-486.

[52] A. Barabanova, *On the global existence of solutions of a reaction-diffusion equation with exponential nonlinearity*, Proc. Amer. Math. Soc. **122** (1994), 827-831.

[53] P. Baras, *Non unicité des solutions d'une équation d'évolution non linéaire*, Ann. Fac. Sci. Toulouse Math. (5) **5** (1983), 287-302.

[54] P. Baras and L. Cohen, *Complete blow-up after T_{max} for the solution of a semilinear heat equation*, J. Funct. Anal. **71** (1987), 142-174.

[55] P. Baras and R. Kersner, *Local and global solvability of a class of semilinear parabolic equations*, J. Differential Equations **68** (1987), 238-252.

[56] P. Baras and M. Pierre, *Critère d'existence de solutions positives pour des équations semi-linéaires non monotones*, Ann. Inst. H. Poincaré Anal. Non Linéaire **2** (1985), 185-212.

[57] G. Barles and F. Da Lio, *On the generalized Dirichlet problem for viscous Hamilton-Jacobi equations*, J. Math. Pures Appl. **83** (2004), 53-75.

[58] J.-Ph. Bartier, *Global behavior of solutions of a reaction diffusion equation with gradient absorption in unbounded domains*, Asympt. Anal. **46** (2006), 325-347.

[59] J.-Ph. Bartier and Ph. Souplet, *Gradient bounds for solutions of semilinear parabolic equations without Bernstein's quadratic condition*, C. R. Acad. Sci. Paris Sér. I Math. **338** (2004), 533-538.

[60] J. Bebernes, A. Bressan and A.A. Lacey, *Total blow-up versus single-point blow-up*, J. Differential Equations **73** (1988), 30-44.

[61] J. Bebernes and S. Bricher, *Final time blowup profiles for semilinear parabolic equations via center manifold theory*, SIAM J. Math. Anal. **23** (1992), 852-869.

[62] J. Bebernes and D. Eberly, *A description of self-similar blow-up for dimensions $n \geq 3$*, Ann. Inst. H. Poincaré Anal. Non Linéaire **5** (1988), 1-21.

[63] J. Bebernes and D. Eberly, *Mathematical problems from combustion theory*, Springer, New York, 1989.

[64] J. Bebernes and A.A. Lacey, *Finite-time blowup for a particular parabolic system*, SIAM J. Math. Anal. **21** (1990), 1415-1425.

[65] J. Bebernes and A.A. Lacey, *Finite time blowup for semilinear reactive-diffusive systems*, J. Differential Equations **95** (1992), 105-129.

[66] J. Bebernes and A.A. Lacey, *Global existence and finite-time blow-up for a class of nonlocal parabolic problems*, Adv. Differential Equations **2** (1996), 927-953.

[67] H. Bellout, *Blow-up of solutions of parabolic equations with nonlinear memory*, J. Differential Equations **70** (1987), 42-68.

[68] V. Benci and G. Cerami, *The effect of the domain topology on the number of positive solutions of nonlinear elliptic problems*, Arch. Rational Mech. Anal. **114** (1991), 79-93.

[69] S. Benachour, S. Dăbuleanu-Hapca and Ph. Laurençot, *Decay estimates for a viscous Hamilton-Jacobi equation with homogeneous Dirichlet boundary conditions*, Asympt. Anal. **51** (2007), 209-229.

[70] S. Benachour, G. Karch and Ph. Laurençot, *Asymptotic profiles of solutions to viscous Hamilton-Jacobi equations*, J. Math. Pures Appl. **83** (2004), 1275-1308.

[71] S. Benachour and Ph. Laurençot, *Global solutions to viscous Hamilton-Jacobi equations with irregular data*, Comm. Partial Differential Equations **24** (1999), 1999-2021.

[72] M. Ben-Artzi, Ph. Souplet and F.B. Weissler, *The local theory for viscous Hamilton-Jacobi equations in Lebesgue spaces*, J. Math. Pures Appl. **81** (2002), 343-378.

[73] R.D. Benguria, J. Dolbeault and M.J. Esteban, *Classification of the solutions of semilinear elliptic problems in a ball*, J. Differential Equations **167** (2000), 438-466.

[74] H. Berestycki, I. Capuzzo-Dolcetta and L. Nirenberg, *Superlinear indefinite elliptic problems and nonlinear Liouville theorems*, Topol. Methods Nonlinear Anal. **4** (1994), 59-78.

[75] H. Berestycki, I. Capuzzo-Dolcetta and L. Nirenberg, *Variational methods for indefinite superlinear homogeneous elliptic problems*, NoDEA Nonlinear Differential Equations Appl. **2** (1995), 553-572.

[76] H. Berestycki and P.-L. Lions, *Nonlinear scalar field equations. I. Existence of a ground state*, Arch. Rational Mech. Anal. **82** (1983), 313-345.

[77] J. Bergh and J. Löfström, *Interpolation spaces. An introduction*, Springer, Berlin - Heidelberg - New York, 1976.

[78] G. Bianchi, *Non-existence of positive solutions to semilinear elliptic equations on $\mathbb{R}^n$ or $\mathbb{R}^n_+$ through the method of moving planes*, Comm. Partial Differential Equations **22** (1997), 1671-1690.

[79] M.-F. Bidaut-Véron, *Initial blow-up for the solutions of a semilinear parabolic equation with source term*, Equations aux dérivées partielles et applications, articles dédiés à Jacques-Louis Lions, Gauthier-Villars, Paris, 1998, pp. 189-198.

[80] M.-F. Bidaut-Véron, *Local behaviour of the solutions of a class of nonlinear elliptic systems*, Adv. Differential Equations **5** (2000), 147-192.

[81] M.-F. Bidaut-Véron, A.C. Ponce and L. Véron, *Boundary singularities of positive solutions of some nonlinear elliptic equations*, C. R. Math. Acad. Sci. Paris **344** (2007), 83-88.

[82] M.-F. Bidaut-Véron and Th. Raoux, *Asymptotics of solutions of some nonlinear elliptic systems*, Comm. Partial Differential Equations **21** (1996), 1035 -1086.

[83] M.-F. Bidaut-Véron and L. Véron, *Nonlinear elliptic equations on compact Riemannian manifolds and asymptotics of Emden equations*, Invent. Math. **106** (1991), 489-539.

[84] M.-F. Bidaut-Véron and L. Vivier, *An elliptic semilinear equation with source term involving boundary measures: the subcritical case*, Rev. Mat. Iberoamericana **16** (2000), 477-513.

[85] M.-F. Bidaut-Véron and C. Yarur, *Semilinear elliptic equations and systems with measure data: existence and a priori estimates*, Adv. Differential Equations **7** (2002), 257-296.

[86] P. Biler, M. Guedda and G. Karch, *Asymptotic properties of solutions of the viscous Hamilton-Jacobi equation*, J. Evol. Equ. **4** (2004), 75-97.

[87] I. Birindelli and E. Mitidieri, *Liouville theorems for elliptic inequalities and applications*, Proc. Roy. Soc. Edinburgh Sect. A **128** (1998), 1217-1247.

[88] L. Boccardo, F. Murat and J.-P. Puel, *Existence results for some quasilinear parabolic equations*, Nonlinear Anal. **13** (1989), 373-392.

[89] M. Bouhar and L. Véron, *Integral representation of solutions of semilinear elliptic equations in cylinders and applications*, Nonlinear Anal. **23** (1994), 275-296.

[90] H. Brezis, *Analyse fonctionnelle*, Masson, Paris, 1983.

[91] H. Brezis, unpublished manuscript.

[92] H. Brezis and X. Cabré, *Some simple nonlinear PDE's without solutions*, Boll. Unione Mat. Ital. **(8) 1-B** (1999), 223-262.

[93] H. Brezis and T. Cazenave, *A nonlinear heat equation with singular initial data*, J. Anal. Math. **68** (1996), 277-304.

[94] H. Brezis, T. Cazenave, Y. Martel and A. Ramiandrisoa, *Blow up for $u_t - \Delta u = g(u)$ revisited*, Adv. Differential Equations **1** (1996), 73-90.

[95] H. Brezis and A. Friedman, *A nonlinear parabolic equations involving measures as initial conditions*, J. Math. Pures Appl. **62** (1983), 73-97.

[96] H. Brezis and T. Kato, *Remarks on the Schrödinger operator with singular complex potentials*, J. Math. Pures Appl. **58** (1979), 137-151.

[97] H. Brezis and P.-L. Lions, *A note on isolated singularities for linear elliptic equations*, Mathematical analysis and applications, Part A, Adv. in Math. Suppl. Stud., 7a,, Academic Press, New York-London, 1981, pp. 263-266.

[98] H. Brezis and L. Nirenberg, *Positive solutions of nonlinear elliptic equations involving critical Sobolev exponent*, Comm. Pure Appl. Math. **36** (1983), 437-477.

[99] H. Brezis and R.E.L. Turner, *On a class of superlinear elliptic problems*, Comm. Partial Differential Equations **2** (1977), 601-614.

[100] J. Bricmont and A. Kupiainen, *Universality in blow-up for nonlinear heat equations*, Nonlinearity **7** (1994), 539-575.

[101] P. Brunovský and B. Fiedler, *Number of zeros on invariant manifolds in reaction-diffusion equations*, Nonlinear Anal. **10** (1986), 179-193.

[102] P. Brunovský, P. Poláčik and B. Sandstede, *Convergence in general periodic parabolic equations in one space dimension*, Nonlinear Anal. **18** (1992), 209-215.

[103] C. Budd, B. Dold and A. Stuart, *Blowup in a partial differential equation with conserved first integral*, SIAM J. Appl. Math. **53** (1993), 718-742.

[104] C. Budd and J. Norbury, *Semilinear elliptic equations and supercritical growth*, J. Differential Equations **68** (1987), 169-197.

[105] C. Budd and Y.-W. Qi, *The existence of bounded solutions of a semilinear elliptic equation*, J. Differential Equations **82** (1989), 207-218.

[106] J. Busca and R. Manasevich, *A Liouville-type theorem for Lane-Emden system*, Indiana Univ. Math. J. **51** (2002), 37-51.

[107] X. Cabré and Y. Martel, *Weak eigenfunctions for the linearization of extremal elliptic problems*, J. Funct. Anal. **156** (1998), 30-56.

[108] L.A. Caffarelli, B. Gidas and J. Spruck, *Asymptotic symmetry and local behavior of semilinear elliptic equations with critical Sobolev growth*, Comm. Pure Appl. Math. **42** (1989), 271-297.

[109] G. Cai, *On the heat flow for the two-dimensional Gelfand equation*, Nonlinear Anal. (to appear).

[110] G. Caristi and E. Mitidieri, *Blow-up estimates of positive solutions of a parabolic system*, J. Differential Equations **113** (1994), 265-271.

[111] F. Catrina and Z.-Q. Wang, *On the Caffarelli-Kohn-Nirenberg inequalities: Sharp constants, existence (and nonexistence), and symmetry of extremal functions*, Comm. Pure Appl. Math. **54** (2001), 229-258.

[112] T. Cazenave, F. Dickstein and F.B. Weissler, *An equation whose Fujita critical exponent is not given by scaling*, Nonlinear Anal. (to appear).

[113] T. Cazenave and A. Haraux, *Introduction aux problèmes d'évolution semi-linéaires*, Ellipses, Paris, 1990, English translation: The Clarendon Press, Oxford University Press, New York, 1998.

[114] T. Cazenave and P.-L. Lions, *Solutions globales d'équations de la chaleur semi linéaires*, Comm. Partial Differential Equations **9** (1984), 955-978.

[115] T. Cazenave and F.B. Weissler, *Asymptotically self-similar global solutions of the nonlinear Schrödinger and heat equations*, Math. Z. **228** (1998), 83-120.

[116] C. Celik and Z. Zhou, *No local L^1 solution for a nonlinear heat equation*, Comm. Partial Differential Equations **28** (2003), 1807-1831.

[117] G. Cerami, S. Solimini and M. Struwe, *Some existence results for superlinear elliptic boundary value problems involving critical exponents*, J. Funct. Anal. **69** (1986), 289-306.

[118] K. Cerqueti, *A uniqueness result for a semilinear elliptic equation involving the critical Sobolev exponent in symmetric domains*, Asymptot. Anal. **21** (1999), 99-115.

[119] J.M. Chadam, A. Peirce and H.-M. Yin, *The blow-up property of solutions to some diffusion equations with localized nonlinear reactions*, J. Math. Anal. Appl. **169** (1992), 313-328.

[120] C.-Y. Chan, *Recent advances in quenching phenomena*, Proceedings of Dynamic Systems and Applications, Vol. 2, Dynamic, Atlanta, GA, 1996, pp. 107-113.

[121] K.-C. Chang and M.-Y. Jiang, *Dirichlet problem with indefinite nonlinearities*, Calc. Var. Partial Differential Equations **20** (2004), 257-282.

[122] H. Chen, *Positive steady-state solutions of a non-linear reaction-diffusion system*, Math. Methods Appl. Sci. **20** (1997), 625-634.

[123] W. Chen and C. Li, *Classification of solutions of some nonlinear elliptic equations*, Duke Math. J. **63** (1991), 615-622.

[124] W. Chen and C. Li, *Indefinite elliptic problems in a domain*, Discrete Contin. Dyn. Syst. **3** (1997), 333-340.

[125] X. Chen, M. Fila and J.-S. Guo, *Boundedness of global solutions of a supercritical parabolic equation*, Nonlinear Anal. (to appear).

[126] X.-Y. Chen and H. Matano, *Convergence, asymptotic periodicity, and finite-point blow-up in one-dimensional semilinear heat equations*, J. Differential Equations **78** (1989), 160-190.

[127] X.-Y. Chen and P. Poláčik, *Asymptotic periodicity of positive solutions of reaction diffusion equations on a ball*, J. Reine Angew. Math. **472** (1996), 17-51.

[128] M. Chipot and P. Quittner, *Equilibria, connecting orbits and a priori bounds for semilinear parabolic equations with nonlinear boundary conditions*, J. Dynam. Differential Equations **16** (2004), 91-138.

[129] M. Chipot and F.B. Weissler, *Some blow up results for a nonlinear parabolic problem with a gradient term*, SIAM J. Math. Anal. **20** (1989), 886-907.

[130] M. Chlebík and M. Fila, *From critical exponents to blowup rates for parabolic problems*, Rend. Mat. Appl. (7) **19** (1999), 449-470.

[131] M. Chlebík, M. Fila and P. Quittner, *Blow-up of positive solutions of a semilinear parabolic equation with a gradient term*, Dyn. Contin. Discrete Impuls. Syst. Ser. A Math. Anal. **10** (2003), 525-537.

[132] K.-S. Chou, S.-Z. Du and G.-F. Zheng, *On partial regularity of the border-line solution of semilinear parabolic problems*, Calc. Var. Partial Differential Equations (to appear).

[133] V. Churbanov, *An example of a reaction system with diffusion in which the diffusion terms lead to blowup*, Dokl. Akad. Nauk SSSR **310** (1990), 1308-1309, English translation in: Soviet Math. Dokl. 41 (1990), 191-192.

[134] Ph. Clément, D.G. de Figueiredo and E. Mitidieri, *Positive solutions of semilinear elliptic systems*, Comm. Partial Differential Equations **17** (1992), 923-940.

[135] Ph. Clément, D.G. de Figueiredo and E. Mitidieri, *A priori estimates for positive solutions of semilinear elliptic systems via Hardy-Sobolev inequalities*, Nonlinear partial differential equations, Pitman Res. Notes Math. Ser. 343 (A. Benkirane et al., eds.), Harlow: Longman, 1996, pp. 73-91.

[136] Ph. Clément, J. Fleckinger, E. Mitidieri and F. de Thélin, *Existence of positive solutions for nonvariational quasilinear system*, J. Differential Equations **166** (2000), 455-477.

[137] Ph. Clément and R.C.A.M. van der Vorst, *On a semilinear elliptic system*, Differential Integral Equations **8** (1995), 1317-1329..

[138] G. Conner and C. Grant, *Asymptotics of blowup for a convection-diffusion equation with conservation*, Differential Integral Equations **9** (1996), 719-728.

[139] M. Conti, L. Merizzi and S. Terracini, *Radial solutions of superlinear equations on $\mathbb{R}^N$. Part I: A global variational approach*, Arch. Rational Mech. Anal. **153** (2000), 291-316.

[140] M. Conti and S. Terracini, *Radial solutions of superlinear equations on $\mathbb{R}^N$. Part II: The forced case*, Arch. Rational Mech. Anal. **153** (2000), 317-339.

[141] C. Cosner, *Positive solutions for a superlinear elliptic systems without variational structure*, Nonlinear Anal. **8** (1984), 1427-1436.

[142] M.G. Crandall and P.H. Rabinowitz, *Some continuation and variational methods for positive solutions of nonlinear elliptic eigenvalue problems*, Arch. Rational Mech. Anal. **58** (1975), 207-218.

[143] M.G. Crandall, P.H. Rabinowitz and L. Tartar, *On a Dirichlet problem with a singular nonlinearity*, Comm. Partial Differential Equations **2** (1977), 193-222.

[144] M. Cuesta, D.G. de Figueiredo and P.N. Srikanth, *On a resonant-superlinear elliptic problem*, Calc. Var. Partial Differential Equations **17** (2003), 221-233.

[145] S. Cui, *Local and global existence of solutions to semilinear parabolic initial value problems*, Nonlinear Anal. **43** (2001), 293-323.

[146] S. Dabuleanu, *Problèmes aux limites pour les équations de Hamilton-Jacobi avec viscosité et données initiales peu régulières*, Doctoral Thesis, University of Nancy 1, 2003.

[147] L. Damascelli, M. Grossi and F. Pacella, *Qualitative properties of positive solutions of semilinear elliptic equations in symmetric domains via the maximum principle*, Ann. Inst. H. Poincaré Anal. Non Linéaire **16** (1999), 631-652.

[148] E.N. Dancer, *The effect of domain shape on the number of positive solutions of certain nonlinear equations*, J. Differential Equations **74** (1988), 120-156.

[149] E.N. Dancer, *A note on an equation with critical exponent*, Bull. London Math. Soc. **20** (1988), 600-602.

[150] E.N. Dancer, *Some notes on the method of moving planes*, Bull. Austral. Math. Soc. **46** (1992), 425-434.

[151] E.N. Dancer, *Superlinear problems on domains with holes of asymptotic shape and exterior problems*, Math. Z. **229** (1998), 475-491.

[152] D. Daners and P. Koch Medina, *Abstract evolution equations, periodic problems and applications*, Longman, Harlow, 1992.

[153] E.B. Davies, *The equivalence of certain heat kernel and Green function bounds*, J. Funct. Anal. **71** (1987), 88-103.

[154] E.B. Davies, *Heat kernels and spectral theory*, Cambridge University Press, Cambridge, 1989.

[155] M. Del Pino, M. Musso and F. Pacard, *Boundary singularities for weak solutions of semilinear elliptic problems*, J. Funct. Anal. (to appear).

[156] K. Deng, *Stabilization of solutions of a nonlinear parabolic equation with a gradient term*, Math. Z. **216** (1994), 147-155.

[157] K. Deng, *Nonlocal nonlinearity versus global blow-up*, Math. Applicata **8** (1995), 124-129.

[158] K. Deng, *Blow-up rates for parabolic systems*, Z. Angew. Math. Phys. **47** (1996), 132-143.

[159] K. Deng and H.A. Levine, *The role of critical exponents in blow-up theorems: The sequel*, J. Math. Anal. Appl. **243** (2000), 85-126.

[160] W. Deng, Y. Li and C.-H. Xie, *Semilinear reaction-diffusion systems with nonlocal sources*, Math. Comput. Modelling **37** (2003), 937-943.

[161] R. Denk, M. Hieber and J. Prüss, *R-boundedness, Fourier multipliers and problems of elliptic and parabolic type*, Mem. Amer. Math. Soc. 788, 2003.

[162] G. Devillanova and S. Solimini, *Concentration estimates and multiple solutions to elliptic problems at critical growth*, Adv. Differential Equations **7** (2002), 1257-1280.

[163] C. Dohmen and M. Hirose, *Structure of positive radial solutions to the Haraux-Weissler equation*, Nonlinear Anal. **33** (1998), 51-69.

[164] B. Dold, V.A. Galaktionov, A.A. Lacey and J.L. Vázquez, *Rate of approach to a singular steady state in quasilinear reaction-diffusion equations*, Ann. Scuola Norm. Sup. Pisa Cl. Sci. (4) **26** (1998), 663-687.

[165] G.C. Dong, *Nonlinear partial differential equations of second order*, Amer. Math. Soc., Transl. Math. Monographs 95, Providence, RI, 1991.

[166] G. Dore and A. Venni, H^∞ *functional calculus for an elliptic operator on a half-space with general boundary conditions*, Ann. Scuola Norm. Sup. Pisa Cl. Sci. (5) **1** (2002), 487-543.

[167] Y. Du and S. Li, *Nonlinear Liouville theorems and a priori estimates for indefinite superlinear elliptic equations*, Adv. Differential Equations **10** (2005), 841-860.

[168] J. Duoandikoetxea and E. Zuazua, *Moments, masses de Dirac et décomposition de fonctions*, C. R. Acad. Sci. Paris Sér. I Math. **315** (1992), 693-698.

[169] X.T. Duong and G. Simonett, H_∞*-calculus for elliptic operators with non-smooth coefficients*, Differential Integral Equations **10** (1997), 201-217.

[170] L. Dupaigne and A. Ponce, *Singularities of positive supersolutions in elliptic PDEs*, Selecta Math. (N.S.) **10** (2004), 341-358.

[171] M. Escobedo and M.A. Herrero, *Boundedness and blow up for a semilinear reaction-diffusion system*, J. Differential Equations **89** (1991), 176-202.

[172] M. Escobedo and M.A. Herrero, *A uniqueness result for a semilinear reaction diffusion system*, Proc. Amer. Math. Soc. **112** (1991), 175-185.

[173] M. Escobedo and M.A. Herrero, *A semilinear parabolic system in a bounded domain*, Ann. Mat. Pura Appl. (4) **165** (1993), 315-336.

[174] M. Escobedo and O. Kavian, *Variational problems related to self-similar solutions of the heat equation*, Nonlinear Anal. **11** (1987), 1103-1133.

[175] M. Escobedo and E. Zuazua, *Large time behavior for convection-diffusion equations in* $\mathbb{R}^N$, J. Funct. Anal. **10** (1991), 119-161.

[176] M.J. Esteban, *On periodic solutions of superlinear parabolic problems*, Trans. Amer. Math. Soc. **293** (1986), 171-189.

[177] M.J. Esteban, *A remark on the existence of positive periodic solutions of superlinear parabolic problems*, Proc. Amer. Math. Soc. **102** (1988), 131-136.

[178] A. Farina, *Liouville-type results for solutions of* $-\Delta u = |u|^{p-1}u$ *on unbounded domains of* $\mathbb{R}^N$, C. R. Math. Acad. Sci. Paris **341** (2005), 415-418.

[179] P.C. Fife, *Mathematical aspects of reacting and diffusing systems*, Lecture Notes in Biomathematics 28, Springer-Verlag, Berlin - New York, 1979.

[180] D.G. de Figueiredo, *Semilinear elliptic systems*, Nonlinear Functional Analysis and Applications to Differential Equations, World Sci. Publishing, River Edge, N.J., 1998, pp. 122-152.

[181] D.G. de Figueiredo and P. Felmer, *A Liouville-type theorem for elliptic systems*, Ann. Scuola Norm. Sup. Pisa Cl. Sci. (4) **21** (1994), 387-397.

[182] D.G. de Figueiredo and P. Felmer, *On superquadratic elliptic systems*, Trans. Amer. Math. Soc. **343** (1994), 99-116.

[183] D.G. de Figueiredo, P.-L. Lions and R.D. Nussbaum, *A priori estimates and existence of positive solutions of semilinear elliptic equations*, J. Math. Pures Appl. **61** (1982), 41-63.

[184] D.G. de Figueiredo and J. Yang, *A priori bounds for positive solutions of a non-variational elliptic system*, Comm. Partial Differential Equations **26** (2001), 2305-2321.

[185] M. Fila, *Remarks on blow up for a nonlinear parabolic equation with a gradient term*, Proc. Amer. Math. Soc. **111** (1991), 795-801.

[186] M. Fila, *Boundedness of global solutions of nonlinear diffusion equations*, J. Differential Equations **98** (1992), 226-240.

[187] M. Fila, *Boundedness of global solutions of nonlocal parabolic equations*, Nonlinear Anal. **30** (1997), 877-885.

[188] M. Fila, *Blow-up of solutions of supercritical parabolic equations*, Handbook of Differential Equations, Evolutionary equations, Vol. II (C.M. Dafermos et al., eds.), Elsevier/North-Holland, Amsterdam, 2005, pp. 105-158.

[189] M. Fila, J.R. King, M. Winkler and E. Yanagida, *Optimal lower bound of the grow-up rate for a supercritical parabolic equation*, J. Differential Equations **228** (2006), 339-356.

[190] M. Fila, J.R. King, M. Winkler and E. Yanagida, *Grow-up rate of solutions of a semilinear parabolic equation with a critical exponent*, Adv. Differential Equations **12** (2007), 1-26.

[191] M. Fila, J.R. King, M. Winkler and E. Yanagida, *Linear behaviour of solutions of a superlinear heat equation* (2006), Preprint.

[192] M. Fila, H.A. Levine and Y. Uda, *Fujita-type global existence–global non-existence theorem for a system of reaction diffusion equations with differing diffusivities*, Math. Methods Appl. Sci. **17** (1994), 807-835.

[193] M. Fila and G. Lieberman, *Derivative blow-up and beyond for quasilinear parabolic equations*, Differential Integral Equations **7** (1994), 811-821.

[194] M. Fila, H. Matano and P. Poláčik, *Immediate regularization after blow-up*, SIAM J. Math. Anal. **37** (2005), 752-776.

[195] M. Fila and N. Mizoguchi, *Multiple continuation beyond blow-up*, Differential Integral Equations **20** (2007), 671-680.

[196] M. Fila and H. Ninomiya, *Reaction versus diffusion: blow-up induced and inhibited by diffusivity*, Russian Math. Surveys **60** (2005), 1217-1235.

[197] M. Fila, H. Ninomiya and J.L. Vázquez, *Dirichlet boundary conditions can prevent blow-up in reaction-diffusion equations and systems*, Discrete Contin. Dyn. Syst. **14** (2006), 63-74.

[198] M. Fila and P. Poláčik, *Global solutions of a semilinear parabolic equation*, Adv. Differential Equations **4** (1999), 163-196.

[199] M. Fila and Ph. Souplet, *The blow-up rate for semilinear parabolic problems on general domains*, NoDEA Nonlinear Differential Equations Appl. **8** (2001), 473-480.

[200] M. Fila, Ph. Souplet and F.B. Weissler, *Linear and nonlinear heat equations in L_δ^p spaces and universal bounds for global solutions*, Math. Ann. **320** (2001), 87-113.

[201] M. Fila, J. Taskinen and M. Winkler, *Convergence to a singular steady-state of a parabolic equation with gradient blow-up*, Appl. Math. Letters **20** (2007), 578-582.

[202] M. Fila and M. Winkler, *Single-point blow-up on the boundary where the zero Dirichlet boundary condition is imposed*, J. Eur. Math. Soc. (to appear).

[203] M. Fila, M. Winkler and E. Yanagida, *Grow-up rate of solutions for a super-critical semilinear diffusion equation*, J. Differential Equations **205** (2004), 365-389.

[204] M. Fila, M. Winkler and E. Yanagida, *Slow convergence to zero for a parabolic equation with supercritical nonlinearity*, Math. Ann. (to appear).

[205] M. Fila, M. Winkler and E. Yanagida, *Convergence to self-similar solutions in a semilinear parabolic equation* (2007), Preprint.

[206] S. Filippas, M.A. Herrero and J.J.L. Velázquez, *Fast blow-up mechanisms for sign-changing solutions of a semilinear parabolic equation with critical nonlinearity*, R. Soc. Lond. Proc. Ser. A Math. Phys. Eng. Sci. **456** (2000), 2957-2982.

[207] S. Filippas and R. Kohn, *Refined asymptotics for the blowup of $u_t - \Delta u = u^p$*, Comm. Pure Appl. Math. **45** (1992), 821-869.

[208] S. Filippas and W.-X. Liu, *On the blowup of multidimensional semilinear heat equations*, Ann. Inst. H. Poincaré Anal. Non Linéaire **10** (1993), 313-344.

[209] S. Filippas and F. Merle, *Compactness and single-point blowup of positive solutions on bounded domains*, Proc. Roy. Soc. Edinburgh Sect. A **127** (1997), 47-65.

[210] A. Filippov, *Conditions for the existence of a solution of a quasi-linear parabolic equation (Russian)*, Dokl. Akad. Nauk SSSR **141** (1961), 568-570.

[211] R.A. Fisher, *The wave of advance of advantageous genes*, Ann. Eugenics **7** (1937), 335-369.

[212] D. Fortunato and E. Jannelli, *Infinitely many solutions for some nonlinear elliptic problems in symmetrical domains*, Proc. Roy. Soc. Edinburgh Sect. A **105** (1987), 205-213.

[213] R.H. Fowler, *Further studies of Emden's and similar differential equations*, Quart. J. Math. **2** (1931), 259-288.

[214] A. Friedman, *Partial differential equations of parabolic type*, Prentice Hall, 1964.

[215] A. Friedman, *Blow up of solutions of nonlinear parabolic equations*, Nonlinear diffusion equations and their equilibrium states, I (W.-M. Ni et al., eds.), Springer, 1988, pp. 301-318.

[216] A. Friedman and Y. Giga, *A single point blow-up for solutions of semilinear parabolic systems*, J. Fac. Sci. Univ. Tokyo Sect. IA Math. **34** (1987), 65-79.

[217] A. Friedman and M. Herrero, *Extinction properties of semilinear heat equations with strong absorption*, J. Math. Anal. Appl. **124** (1987), 530-546.

[218] A. Friedman and A.A. Lacey, *Blowup of solutions of semilinear parabolic equations*, J. Math. Anal. Appl. **132** (1988), 171-186.

[219] A. Friedman and B. McLeod, *Blow-up of positive solutions of semilinear heat equations*, Indiana Univ. Math. J. **34** (1985), 425-447.

[220] H. Fujita, *On the blowing up of solutions of the Cauchy problem for $u_t = \Delta u + u^{1+\alpha}$*, J. Fac. Sci. Univ. Tokyo Sec. IA Math. **13** (1966), 109-124.

[221] I. Fukuda and R. Suzuki, *Blow-up behavior for a nonlinear heat equation with a localized source in a ball*, J. Differential Equations **218** (2005), 273-291.

[222] V.A. Galaktionov, *Geometric Sturmian theory of nonlinear parabolic equations and applications*, Applied Mathematics and Nonlinear Science Series, 3, Chapman & Hall/CRC, Boca Raton, FL, 2004.

[223] V.A. Galaktionov and J.R. King, *Composite structure of global unbounded solutions of nonlinear heat equations with critical Sobolev exponents*, J. Differential Equations **189** (2003), 199-233.

[224] V.A. Galaktionov, S.P. Kurdyumov and A.A. Samarskii, *A parabolic system of quasilinear equations, I*, Differential Equations **19** (1983), 2133-2143.

[225] V.A. Galaktionov, S.P. Kurdyumov and A.A. Samarskii, *Asymptotic stability of invariant solutions of nonlinear heat-conduction equation with sources*, Differentsial'nye Uravneniya **20** (1984), 614-632, (English translation *Differential Equations* **20** (1984), 461-476).

[226] V.A. Galaktionov, S.P. Kurdyumov and A.A. Samarskii, *A parabolic system of quasilinear equations, II*, Differential Equations **21** (1985), 1544-1559.

[227] V.A. Galaktionov and H.A. Levine, *A general approach to critical Fujita exponents in nonlinear parabolic problems*, Nonlinear Anal. **34** (1998), 1005-1027.

[228] V.A. Galaktionov and S. Posashkov, *The equation $u_t = u_{xx} + u^\beta$. Localization, asymptotic*, Akad. Nauk SSSR Inst. Prikl. Mat. Preprint no. 97 (1985).

[229] V.A. Galaktionov and S. Posashkov, *Application of new comparison theorems to the investigation of unbounded solutions of nonlinear parabolic equations (Russian)*, Differentsial'nye Uravneniya **22** (1986), 1165-1173, (English translation: *Differential Equations* **22** (1986), 809-815).

[230] V.A. Galaktionov and J.L. Vázquez, *Regional blow up in a semilinear heat equation with convergence to a Hamilton Jacobi equation*, SIAM J. Math. Anal. **24** (1993), 1254-1276.

[231] V.A. Galaktionov and J.L. Vázquez, *Blowup for quasilinear heat equations described by means of nonlinear Hamilton-Jacobi equations*, J. Differential Equations **127** (1996), 1-40.

[232] V.A. Galaktionov and J.L. Vázquez, *Continuation of blow-up solutions of nonlinear heat equations in several space dimensions*, Comm. Pure Appl. Math. **50** (1997), 1-67.

[233] Th. Gallouët, F. Mignot and J.-P. Puel, *Quelques résultats sur le problème $-\Delta u = \lambda e^u$*, C. R. Acad. Sci. Paris Sér. I Math. **307** (1988), 289-292.

[234] F. Gazzola and H.-Ch. Grunau, *On the role of space dimension $n = 2 + 2\sqrt{2}$ in the semilinear Brezis-Nirenberg eigenvalue problem*, Analysis **20** (2000), 395-399.

[235] F. Gazzola, J. Serrin and M. Tang, *Existence of ground states and free boundary problems for quasilinear elliptic operators*, Adv. Differential Equations **5** (2000), 1-30.

[236] F. Gazzola and T. Weth, *Finite time blow-up and global solutions for semilinear parabolic equations with initial data at high energy level*, Differential Integral Equations **18** (2005), 961-990.

[237] J. Giacomoni, J. Prajapat and M. Ramaswamy, *Positive solution branch for elliptic problems with critical indefinite nonlinearity*, Differential Integral Equations **18** (2005), 721-764.

[238] B. Gidas, *Symmetry properties and isolated singularities of positive solutions of nonlinear elliptic equations*, Nonlinear partial differential equations in engineering and applied science, Lecture Notes in Pure and Appl. Math. 54, Dekker, New York, 1980, pp. 255-273.

[239] B. Gidas, W.-M. Ni and L. Nirenberg, *Symmetry and related properties via the maximum principle*, Comm. Math. Phys. **68** (1979), 209-243.

[240] B. Gidas and J. Spruck, *Global and local behavior of positive solutions of nonlinear elliptic equations*, Comm. Pure Appl. Math. **34** (1981), 525-598.

[241] B. Gidas and J. Spruck, *A priori bounds for positive solutions of nonlinear elliptic equations*, Comm. Partial Differential Equations **6** (1981), 883-901.

[242] A. Gierer and H. Meinhardt, *A theory of biological pattern formation*, Kybernetik **12** (1972), 30-39.

[243] Y. Giga, *A bound for global solutions of semilinear heat equations*, Comm. Math. Phys. **103** (1986), 415-421.

[244] Y. Giga and R.V. Kohn, *Asymptotically self-similar blow-up of semilinear heat equations*, Comm. Pure Appl. Math. **38** (1985), 297-319.

[245] Y. Giga and R.V. Kohn, *Characterizing blowup using similarity variables*, Indiana Univ. Math. J. **36** (1987), 1-40.

[246] Y. Giga and R.V. Kohn, *Nondegeneracy of blowup for semilinear heat equations*, Comm. Pure Appl. Math. **42** (1989), 845-884.

[247] Y. Giga, S. Matsui and S. Sasayama, *Blow up rate for semilinear heat equation with subcritical nonlinearity*, Indiana Univ. Math. J. **53** (2004), 483-514.

[248] Y. Giga, S. Matsui and S. Sasayama, *On blow-up rate for sign-changing solutions in a convex domain*, Math. Methods Appl. Sc. **27** (2004), 1771-1782.

[249] Y. Giga and N. Umeda, *On blow-up at space infinity for semilinear heat equations*, J. Math. Anal. Appl. **316** (2006), 538-555.

[250] D. Gilbarg and N.S. Trudinger, *Elliptic partial differential equations of second order*, Springer, Berlin, 1998.

[251] B. Gilding, *The Cauchy problem for $u_t = \Delta u + |\nabla u|^q$, large-time behaviour*, J. Math. Pures Appl. **84** (2005), 753-785.

[252] B. Gilding, M. Guedda and R. Kersner, *The Cauchy problem for $u_t = \Delta u + |\nabla u|^q$*, J. Math. Anal. Appl. **284** (2003), 733-755.

[253] J. Ginibre and G. Velo, *The Cauchy problem in local spaces for the complex Ginzburg Landau equation. II. Contraction methods*, Comm. Math. Phys. **187** (1997), 45-79.

[254] P. Groisman and J. Rossi, *Dependence of the blow-up time with respect to parameters and numerical approximations for a parabolic problem*, Asympt. Anal. **37** (2004), 79-91.

[255] P. Groisman, J. Rossi and H. Zaag, *On the dependence of the blow-up time with respect to the initial data in a semilinear parabolic problem*, Comm. Partial Differential Equations **28** (2003), 737-744.

[256] M. Grossi, *A uniqueness result for a semilinear elliptic equation in symmetric domains*, Adv. Differential Equations **5** (2000), 193-212.

[257] M. Grossi, P. Magrone and M. Matzeu, *Linking type solutions for elliptic equations with indefinite nonlinearities up to the critical growth*, Discrete Contin. Dyn. Syst. **7** (2001), 703-718.

[258] Y. Gu and M.-X. Wang, *Existence of positive stationary solutions and threshold results for a reaction-diffusion system*, J. Differential Equations **130** (1996), 177-291.

[259] M. Guedda and M. Kirane, *Diffusion terms in systems of reaction diffusion equations can lead to blow up*, J. Math. Anal. Appl. **218** (1998), 325-327.

[260] C. Gui, W.-M. Ni and X. Wang, *On the stability and instability of positive steady states of a semilinear heat equation in $\mathbb{R}^n$*, Comm. Pure Appl. Math. **45** (1992), 1153-1181.

[261] C. Gui, W.-M. Ni and X. Wang, *Further study on a nonlinear heat equation*, J. Differential Equations **169** (2001), 588-613.

[262] C. Gui and X. Wang, *Life span of solutions of the Cauchy problem for a semilinear heat equation*, J. Differential Equations **115** (1995), 166-172.

[263] J.-S. Guo and B. Hu, *Blowup rate estimates for the heat equation with a nonlinear gradient source term*, Discrete Contin. Dyn. Syst. (to appear).

[264] J.-S. Guo and Ph. Souplet, *Fast rate of formation of dead-core for the heat equation with strong absorption and applications to fast blow-up*, Math. Ann. **331** (2005), 651-667.

[265] J.K. Hale and G. Raugel, *Convergence in gradient-like systems with applications to PDE*, Z. Angew. Math. Phys. **43** (1992), 63-124.

[266] F. Hamel and N. Nadirashvili, *Travelling fronts and entire solutions of the Fisher-KPP equation in $\mathbb{R}^N$*, Arch. Ration. Mech. Anal. **157** (2001), 91-163.

[267] A. Haraux, *Systèmes dynamiques dissipatifs et applications*, Recherches en Mathématiques Appliquées 17, Masson, Paris, 1991.

[268] A. Haraux and P. Poláčik, *Convergence to a positive equilibrium for some nonlinear evolution equations in a ball*, Acta Math. Univ. Comenian. (N.S.) **61** (1992), 129-141.

[269] A. Haraux and F.B. Weissler, *Non-uniqueness for a semilinear initial value problem*, Indiana Univ. Math. J. **31** (1982), 167-189.

[270] A. Haraux and A. Youkana, *On a result of K. Masuda concerning reaction-diffusion equations*, Tohoku Math. J. (2) **40** (1988), 159-163.

[271] K. Hayakawa, *On nonexistence of global solutions of some semilinear parabolic differential equations*, Proc. Japan Acad. **49** (1973), 503-505.

[272] E. Heinz, *Über die Eindeutigkeit beim Cauchyschen Anfangswertproblem einer elliptischen Differentialgleichung zweiter Ordnung*, Nachr. Akad. Wiss. Göttingen **IIa** (1955), 1-12.

[273] D. Henry, *Geometric theory of semilinear parabolic equations*, Lecture Notes in Mathematics 840, Springer, Berlin - Heidelberg - New York, 1981.

[274] M.A. Herrero, A.A. Lacey and J.J.L. Velázquez, *Global existence for reaction-diffusion systems modelling ignition*, Arch. Rational Mech. Anal. **142** (1998), 219-251.

[275] M.A. Herrero and J.J.L. Velázquez, *Blow-up profiles in one-dimensional, semilinear parabolic problems*, Comm. Partial Differential Equations **17** (1992), 205-219.

[276] M.A. Herrero and J.J.L. Velázquez, *Blow-up behaviour of one-dimensional semilinear parabolic equations*, Ann. Inst. H. Poincaré Anal. Non Linéaire **10** (1993), 131-189.

[277] M.A. Herrero and J.J.L. Velázquez, *Explosion de solutions d'équations paraboliques semilinéaires supercritiques*, C. R. Acad. Sci. Paris Sér. I Math. **319** (1994), 141-145.

[278] M.A. Herrero and J.J.L. Velázquez, *A blow up result for semilinear heat equations in the supercritical case* (1994), Preprint.

[279] M. Hesaaraki and A. Moameni, *Blow-up of positive solutions for a family of nonlinear parabolic equations in general domain in $\mathbb{R}^N$*, Michigan Math. J. **52** (2004), 375-389.

[280] N. Hirano and N. Mizoguchi, *Positive unstable periodic solutions for superlinear parabolic equations*, Proc. Amer. Math. Soc. **123** (1995), 1487-1495.

[281] S. Hollis, R. Martin and M. Pierre, *Global existence and boundedness in reaction-diffusion systems*, SIAM J. Math. Anal. **18** (1987), 744-761.

[282] S. Hollis and J. Morgan, *Interior estimates for a class of reaction-diffusion systems from L^1 a priori estimates*, J. Differential Equations **98** (1992), 260-276.

[283] B. Hu, *Remarks on the blowup estimate for solutions of the heat equation with a nonlinear boundary condition*, Differential Integral Equations **9** (1996), 891-901.

[284] B. Hu and H.-M. Yin, *Semilinear parabolic equations with prescribed energy*, Rend. Circ. Mat. Palermo (2) **44** (1995), 479-505.

[285] J. Hulshof and R. van der Vorst, *Differential systems with strongly indefinite variational structure*, J. Functional Analysis **114** (1993), 32-58.

[286] J. Húska, *Periodic solutions in superlinear parabolic problems*, Acta Math. Univ. Comenian. (N.S.) **71** (2002), 19-26.

[287] R. Ikehata and T. Suzuki, *Stable and unstable sets for evolution equations of parabolic and hyperbolic type*, Hiroshima Math. J. **26** (1996), 475-491.

[288] K. Ishige and T. Kawakami, *Asymptotic behavior of solutions for some semilinear heat equations in* $\mathbb{R}^N$ (2006), Preprint.

[289] K. Ishige and H. Yagisita, *Blow-up problems for a semilinear heat equation with large diffusion*, J. Differential Equations **212** (2005), 114-128.

[290] H. Ishii, *Asymptotic stability and blowing up of solutions of some nonlinear equations*, J. Differential Equations **26** (1977), 291-319.

[291] M.A. Jendoubi, *A simple unified approach to some convergence theorems of L. Simon*, J. Funct. Anal. **153** (1998), 187-202.

[292] H. Jiang, *Global existence of solutions on an activator-inhibitor model*, Discrete Contin. Dyn. Syst. **14** (2006), 737-751.

[293] D.D. Joseph and T.S. Lundgren, *Quasilinear Dirichlet problems driven by positive sources*, Arch. Rational Mech. Anal. **49** (1973), 241-269.

[294] I. Kanel and M. Kirane, *Global solutions of reaction-diffusion systems with a balance law and nonlinearities of exponential growth*, J. Differential Equations **165** (2000), 24-41.

[295] I. Kanel and M. Kirane, *Global existence and large time behavior of positive solutions to a reaction diffusion system*, Differential Integral Equations **13** (2000), 255-264.

[296] S. Kaplan, *On the growth of solutions of quasi-linear parabolic equations*, Comm. Pure Appl. Math. **16** (1963), 305-330.

[297] T. Kato, *The Cauchy problem for quasi-linear symmetric hyperbolic systems*, Arch. Rational Mech. Anal. **58** (1975), 181-205.

[298] N. Kavallaris, A.A. Lacey and D. Tzanetis, *Global existence and divergence of critical solutions of a non-local parabolic problem in Ohmic heating process*, Nonlinear Anal. **58** (2004), 787-812.

[299] O. Kavian, *Remarks on the large time behaviour of a nonlinear diffusion equation*, Ann. Inst. H. Poincaré Analyse Non Linéaire **4** (1987), 423-452.

[300] O. Kavian, *Introduction à la théorie des points critiques*, Springer, Paris, 1993.

[301] T. Kawanago, *Asymptotic behavior of solutions of a semilinear heat equation with subcritical nonlinearity*, Ann. Inst. H. Poincaré Anal. non linéaire **13** (1996), 1-15.

[302] T. Kawanago, *Existence and behavior of solutions for* $u_t = \Delta(u^m) + u^\ell$, Adv. Math. Sci. Appl. **7** (1997), 367-400.

[303] H. Kawarada, *On solutions of initial-boundary problem for* $u_t = u_{xx} + 1/(1-u)$, Publ. Res. Inst. Math. Sci. **10** (1974/75), 729-736.

[304] B. Kawohl and L. Peletier, *Observations on blow up and dead cores for nonlinear parabolic equations*, Math. Z. **202** (1989), 207-217.

[305] J.P. Keener and H.B. Keller, *Positive solutions of convex nonlinear eigenvalue problems*, J. Differential Equations **16** (1974), 103-125.

[306] J.R. King, Personal communication 2006.

[307] K. Kobayashi, T. Sirao and H. Tanaka, *On the blowing up problem for semilinear heat equations*, J. Math. Soc. Japan **29** (1977), 407-424.

[308] A.N. Kolmogorov, I.G. Petrovsky and N.S. Piskunov, *Etude de l'équation de la diffusion avec croissance de la quantité de matière et son application à un problème biologique*, Bulletin Université d'Etat à Moscou (Bjul. Moskowskogo Gos. Univ.), Série Internationale, Section A **1** (1937), 1-26.

[309] S. Kouachi, *Existence of global solutions to reaction-diffusion systems via a Lyapunov functional*, Electron. J. Differential Equations **2001, 88** (2001), 1-13.

[310] M.K. Kwong, *Uniqueness of positive solutions of $\Delta u - u + u^p = 0$ in $\mathbb{R}^n$*, Arch. Rational Mech. Anal. **105** (1989), 243-266.

[311] A.A. Lacey, *Mathematical analysis of thermal runaway for spatially inhomogeneous reactions*, SIAM J. Appl. Math. **43** (1983), 1350-1366.

[312] A.A. Lacey, *The form of blow-up for nonlinear parabolic equations*, Proc. Roy. Soc. Edinburgh Sect. A **98** (1984), 183-202.

[313] A.A. Lacey, *Global blow-up of a nonlinear heat equation*, Proc. Roy. Soc. Edinburgh Sect. A **104** (1986), 161-167.

[314] A.A. Lacey, *Thermal runaway in a non-local problem modelling Ohmic heating. I. Model derivation and some special cases*, European J. Appl. Math. **6** (1995), 127-144.

[315] A.A. Lacey, *Thermal runaway in a non-local problem modelling Ohmic heating. II. General proof of blow-up and asymptotics of runaway*, European J. Appl. Math. **6** (1995), 201-224.

[316] A.A. Lacey and D. Tzanetis, *Global existence and convergence to a singular steady state for a semilinear heat equation*, Proc. Roy. Soc. Edinburgh Sect. A **105** (1987), 289-305.

[317] A.A. Lacey and D. Tzanetis, *Complete blow-up for a semilinear diffusion equation with a sufficiently large initial condition*, IMA J. Appl. Math. **41** (1988), 207-215.

[318] A.A. Lacey and D. Tzanetis, *Global, unbounded solutions to a parabolic equation*, J. Differential Equations **101** (1993), 80-102.

[319] O.A. Ladyženskaja, *Solution of the first boundary problem in the large for quasi-linear parabolic equations*, Trudy Moskov. Mat. Obšč. **7** (1958), 149-177.

[320] O.A. Ladyženskaja, V.A. Solonnikov and N.N. Ural'ceva, *Linear and quasilinear equations of parabolic type*, Amer. Math. Soc., Transl. Math. Monographs, Providence, R.I., 1968.

[321] Ph. Laurençot, *Convergence to steady states for a one-dimensional viscous Hamilton-Jacobi equation with Dirichlet boundary conditions*, Pacific J. Math. **230** (2007), 347-364.

[322] Ph. Laurençot and Ph. Souplet, *On the growth of mass for a viscous Hamilton-Jacobi equation*, J. Anal. Math. **89** (2003), 367-383.

[323] M. Lazzo, *Solutions positives multiples pour une équation elliptique non linéaire avec l'exposant critique de Sobolev*, C. R. Acad. Sci. Paris Sér. I Math. **314** (1992), 61-64.

[324] T. Lee and W.-M. Ni, *Global existence, large time behaviour and life span of solutions of a semilinear parabolic Cauchy problem*, Trans. Amer. Math. Soc. **333** (1992), 365-378.

[325] L.A. Lepin, *Countable spectrum of eigenfunctions of a nonlinear heat-conduction equation with distributed parameters*, Differentsial'nye Uravneniya **24** (1988), 1226-1234, (English translation: *Differential Equations* **24** (1988), 799-805).

[326] L.A. Lepin, *Self-similar solutions of a semilinear heat equation*, Mat. Model. **2** (1990), 63-74, (in Russian).

[327] II.A. Levine, *Some nonexistence and instability theorems for solutions of formally parabolic equations of the form $Pu_t = -Au + F(u)$*, Arch. Rational Mech. Anal. **51** (1973), 371-386.

[328] H.A. Levine, *The role of critical exponents in blowup theorems*, SIAM Rev. **32** (1990), 262-288.

[329] H.A. Levine, *Quenching and beyond: a survey of recent results*, Nonlinear mathematical problems in industry, II, GAKUTO Internat. Ser. Math. Sci. Appl., 2, Gakkōtosho, Tokyo, 1993, pp. 501-512.

[330] H.A. Levine, P. Sacks, B. Straughan and L. Payne, *Analysis of a convective reaction-diffusion equation (II)*, SIAM J. Math. Anal. **20** (1989), 133-147.

[331] F. Li, S.-H. Huang and C.-H. Xie, *Global existence and blow-up of solutions to a nonlocal reaction-diffusion system*, Discrete Contin. Dyn. Syst. **9** (2003), 1519-1532.

[332] F. Li and M.-X. Wang, *Properties of blow-up solutions to a parabolic system with nonlinear localized terms*, Discrete Contin. Dyn. Syst. **13** (2005), 683-700.

[333] M. Li, S. Chen and Y. Qin, *Boundedness and blow up for the general activator-inhibitor model*, Acta Math. Appl. Sinica (English Ser.) **11** (1995), 59-68.

[334] Y. Li, *Asymptotic behavior of positive solutions of equation $\Delta u + K(x)u^p = 0$ in $\mathbb{R}^n$*, J. Differential Equations **95** (1992), 304-330.

[335] Y. Li and C.-H. Xie, *Blow-up for semilinear parabolic equations with nonlinear memory*, Z. Angew. Math. Phys. **55** (2004), 15-27.

[336] E. Lieb, *On the lowest eigenvalue of the Laplacian for the intersection of two domains*, Invent. Math. **74** (1983), 441-448.

[337] G.M. Lieberman, *The first initial-boundary value problem for quasilinear second order parabolic equations*, Ann. Scuola Norm. Sup. Pisa Cl. Sci. (4) **13** (1986), 347-387.

[338] G.M. Lieberman, *Second order parabolic differential equations*, World Scientific, Singapore, 1996.

[339] Z. Lin, *Blowup estimates for a mutualistic model in ecology*, Electron. J. Qual. Theory Differential Equ. **2002, 8** (2002), 1-14.

[340] P.-L. Lions, *Isolated singularities in semilinear problems*, J. Differential Equations **38** (1980), 441-450.

[341] P.-L. Lions, *Résolution de problèmes elliptiques quasilinéaires*, Arch. Rational Mech. Anal. **74** (1980), 335-353.

[342] P.-L. Lions, *Asymptotic behavior of some nonlinear heat equations*, Phys. D **5** (1982), 293-306.

[343] Y. Lou, *Necessary and sufficient condition for the existence of positive solutions of certain cooperative system*, Nonlinear Anal. **26** (1996), 1079-1095.

[344] A. Lunardi, *Analytic semigroups and optimal regularity in parabolic problems*, Progress in Nonlinear Differential Equations and their Applications 16, Birkhäuser, Basel, 1995.

[345] Y. Martel, *Complete blow up and global behaviour of solutions of $u_t - \Delta u = g(u)$*, Ann. Inst. H. Poincaré Anal. Non Linéaire **15** (1998), 687-723.

[346] Y. Martel and Ph. Souplet, *Estimations optimales en temps petit et près de la frontière pour les solutions de l'équation de la chaleur avec données non-compatibles*, C. R. Acad. Sci. Paris Sér. I Math. **79** (1998), 575-580.

[347] Y. Martel and Ph. Souplet, *Small time boundary behavior for parabolic equations with noncompatible data*, J. Math. Pures Appl. **79** (2000), 603-632.

[348] R. Martin and M. Pierre, *Nonlinear reaction-diffusion systems*, Nonlinear equations in the applied sciences, Math. Sci. Engrg., 185, Academic Press, Boston, MA, 1992, pp. 363-398.

[349] R. Martin and M. Pierre, *Influence of mixed boundary conditions in some reaction-diffusion systems*, Proc. Roy. Soc. Edinburgh Sect. A **127** (1997), 1053-1066.

[350] K. Masuda, *On the global existence and asymptotic behavior of solutions of reaction-diffusion equations*, Hokkaido Math. J. **12** (1983), 360-370.

[351] K. Masuda, *Analytic solutions of some nonlinear diffusion equations*, Math. Z. **187** (1984), 61-73.

[352] K. Masuda and K. Takahashi, *Reaction-diffusion systems in the Gierer-Meinhardt theory of biological pattern formation*, Japan J. Appl. Math. **4** (1987), 47-58.

[353] H. Matano, *Convergence of solutions of one-dimensional semilinear parabolic equations*, J. Math. Kyoto Univ. **18** (1978), 221-227.

[354] H. Matano, *Existence of nontrivial unstable sets for equilibriums of strongly order-preserving systems*, J. Fac. Sci. Univ. Tokyo, Sect. 1A Math. **30** (1983), 645-673.

[355] H. Matano and F. Merle, *On nonexistence of type II blowup for a supercritical nonlinear heat equation*, Comm. Pure Appl. Math. **57** (2004), 1494-1541.

[356] H. Matano and F. Merle, in preparation.

[357] J. Matos, *Blow up of critical and subcritical norms in semilinear heat equations*, Adv. Differential Equations **3** (1998), 497-532.

[358] J. Matos, *Convergence of blow-up solutions of nonlinear heat equations in the supercritical case*, Proc. Roy. Soc. Edinburgh Sect. A **129** (1999), 1197-1227.

[359] J. Matos, *Unfocused blow up solutions of semilinear parabolic equations*, Discrete Contin. Dyn. Syst. **5** (1999), 905-928.

[360] J. Matos, *Self-similar blow up patterns in supercritical semilinear heat equations*, Comm. Appl. Anal. **5** (2001), 455-483.

[361] J. Matos and Ph. Souplet, *Universal blow-up rates for a semilinear heat equation and applications*, Adv. Differential Equations **8** (2003), 615-639.

[362] P.J. McKenna and W. Reichel, *A priori bounds for semilinear equations and a new class of critical exponents for Lipschitz domains*, J. Funct. Anal. **244** (2007), 220-246.

[363] P. Meier, *On the critical exponent for reaction-diffusion equations*, Arch. Rational Mech. Anal. **109** (1990), 63-72.

[364] F. Merle, *Solution of a nonlinear heat equation with arbitrarily given blow-up points*, Comm. Pure Appl. Math. **45** (1992), 263-300.

[365] F. Merle and L.A. Peletier, *Positive solutions of elliptic equations involving supercritical growth*, Proc. Roy. Soc. Edinburgh Sect. A **118** (1991), 49-62.

[366] F. Merle and H. Zaag, *Stability of the blow-up profile for equations of the type $u_t = \Delta u + |u|^{p-1}u$*, Duke Math. J. **86** (1997), 143-195.

[367] F. Merle and H. Zaag, *Optimal estimates for blowup rate and behavior for nonlinear heat equations*, Comm. Pure Appl. Math. **51** (1998), 139-196.

[368] F. Merle and H. Zaag, *Refined uniform estimates at blow-up and applications for nonlinear heat equations*, Geom. Funct. Anal. **8** (1998), 1043-1085.

[369] F. Mignot and J.-P. Puel, *Sur une classe de problèmes non linéaires avec une non-linéarité positive, croissante, convexe*, Comm. Partial Differential Equations **5** (1980), 791-836.

[370] E. Mitidieri, *A Rellich type identity and applications*, Comm. Partial Differential Equations **18** (1993), 125-151.

[371] E. Mitidieri, *Nonexistence of positive solutions of semilinear elliptic systems in $\mathbb{R}^n$*, Differential Integral Equations **9** (1996), 465-479.

[372] E. Mitidieri and S.I. Pohozaev, *A priori estimates and blow-up of solutions of nonlinear partial differential equations and inequalities*, Proc. Steklov Inst. Math. **234** (2001), 1-362.

[373] E. Mitidieri and S.I. Pohozaev, *Fujita-type theorems for quasilinear parabolic inequalities with a nonlinear gradient*, Dokl. Akad. Nauk **386** (2002), 160-164.

[374] N. Mizoguchi, *On the behavior of solutions for a semilinear parabolic equation with supercritical nonlinearity*, Math. Z. **239** (2002), 215-229.

[375] N. Mizoguchi, *Blowup rate of solutions for a semilinear heat equation with the Neumann boundary condition*, J. Differential Equations **193** (2003), 212-238.

[376] N. Mizoguchi, *Blowup behavior of solutions for a semilinear heat equation with supercritical nonlinearity*, J. Differential Equations **205** (2004), 298-328.

[377] N. Mizoguchi, *Type-II blowup for a semilinear heat equation*, Adv. Differential Equations **9** (2004), 1279-1316.

[378] N. Mizoguchi, *Boundedness of global solutions for a supercritical semilinear heat equation and its application*, Indiana Univ. Math. J. **54** (2005), 1047-1059.

[379] N. Mizoguchi, *Various behaviors of solutions for a semilinear heat equation after blowup*, J. Funct. Anal. **220** (2005), 214-227.

[380] N. Mizoguchi, *Multiple blowup of solutions for a semilinear heat equation*, Math. Ann. **331** (2005), 461-473.

[381] N. Mizoguchi, *Growup of solutions for a semilinear heat equation with supercritical nonlinearity*, J. Differential Equations **227** (2006), 652-669.

[382] N. Mizoguchi, *Multiple blowup of solutions for a semilinear heat equation II*, J. Differential Equations **231** (2006), 182-194.

[383] N. Mizoguchi, H. Ninomiya and E. Yanagida, *Diffusion induced blow-up in a nonlinear parabolic system*, J. Dynam. Differential Equations **10** (1998), 619-638.

[384] N. Mizoguchi and E. Yanagida, *Critical exponents for the blow-up of solutions with sign changes in a semilinear parabolic equation*, Math. Ann. **307** (1997), 663-675.

[385] N. Mizoguchi and E. Yanagida, *Critical exponents for the blowup of solutions with sign changes in a semilinear parabolic equation. II*, J. Differential Equations **145** (1998), 295-331.

[386] N. Mizoguchi and E. Yanagida, *Blow-up and life span of solutions for a semilinear parabolic equation*, SIAM J. Math. Anal. **29** (1998), 1434-1446.

[387] J. Morgan, *On a question of blow-up for semilinear parabolic systems*, Differential Integral Equations **3** (1990), 973-978.

[388] C.E. Mueller and F.B. Weissler, *Single point blow-up for general semilinear heat equation*, Indiana Univ. Math. J. **34** (1985), 881-913.

[389] Y. Naito, *Non-uniqueness of solutions to the Cauchy problem for semilinear heat equations with singular initial data*, Math. Ann. **329** (2004), 161-196.

[390] Y. Naito, *An ODE approach to the multiplicity of self-similar solutions for semi-linear heat equations*, Proc. Roy. Soc. Edinburgh Sect. A **136** (2006), 807-835.

[391] W.-M. Ni, *Diffusion, cross-diffusion, and their spike-layer steady states*, Notices Amer. Math. Soc. **45** (1998), 9-18.

[392] W.-M. Ni, *Qualitative properties of solutions to elliptic problems*, Handbook of Differential Equations, Stationary partial differential equations. Vol. I (M. Chipot et al., eds.), Elsevier/North-Holland, Amsterdam, 2004, pp. 157-233.

[393] W.-M. Ni and R. Nussbaum, *Uniqueness and nonuniqueness for positive radial solutions of* $\Delta u + f(u, r) = 0$, Comm. Pure Appl. Math. **38** (1985), 67-108.

[394] W.-M. Ni and P. Sacks, *Singular behavior in nonlinear parabolic equations*, Trans. Amer. Math. Soc. **287** (1985), 657-671.

[395] W.-M. Ni and P. Sacks, *The number of peaks of positive solutions of semilinear parabolic equations*, SIAM J. Math. Anal. **16** (1985), 460-471.

[396] W.-M. Ni, P. Sacks and J. Tavantzis, *On the asymptotic behavior of solutions of certain quasilinear parabolic equations*, J. Differential Equations **54** (1984), 97-120.

[397] W.-M. Ni and J. Serrin, *Existence and nonexistence theorems for ground states of quasilinear partial differential equations. The anomalous case*, Atti Accad. Naz. Lincei **77** (1986), 231-257.

[398] W.-M. Ni, K. Suzuki and I. Takagi, *The dynamics of a kinetic activator-inhibitor system*, J. Differential Equations **229** (2006), 426-465.

[399] S.M. Nikolskii, *Approximation of functions of several variables and imbedding theorems*, Springer, Berlin - Heidelberg - New York, 1975.

[400] R.D. Nussbaum, *Positive solutions of nonlinear elliptic boundary value problems*, J. Math. Anal. Appl. **51** (1975), 461-482.

[401] A. Okada and I. Fukuda, *Total versus single point blow-up of solutions of a semilinear parabolic equation with localized reaction*, J. Math. Anal. Appl. **281** (2003), 485-500.

[402] O. Oleinik and S. Kruzkov, *Quasi-linear parabolic second order equations with several independent variables*, Uspehi Mat. Nauk **16** (1961), 115-155.

[403] M. Otani, *Existence and asymptotic stability of strong solutions of nonlinear evolution equations with a difference term of subdifferentials*, Colloq. Math. Soc. Janos Bolyai 30, North-Holland, Amsterdam-New York, 1981, pp. 795-809.

[404] F. Pacard, *Existence and convergence of positive weak solutions of* $-\Delta u = u^{\frac{n}{n-2}}$ *in bounded domains of* $\mathbb{R}^n$, $n \geq 3$, Calc. Var. Partial Differential Equations **1** (1993), 243-265.

[405] C.-V. Pao, *Nonlinear parabolic and elliptic equations*, Plenum Press, New York, 1992.

[406] D. Passaseo, *Nonexistence results for elliptic problems with supercritical nonlinearity in nontrivial domains*, J. Funct. Anal. **114** (1993), 97-105.

[407] D. Passaseo, *New nonexistence results for elliptic equations with supercritical nonlinearity*, Differential Integral Equations **8** (1995), 577-586.

[408] D. Passaseo, *Multiplicity of positive solutions of nonlinear elliptic equations with critical Sobolev exponent in some contractible domains*, Manuscripta Math. **65** (1989), 147-165.

[409] L.E. Payne and D.H. Sattinger, *Saddle points and instability of nonlinear hyperbolic equations*, Israel J. Math. **22** (1975), 273-303.

[410] A. Pazy, *Semigroups of linear operators and applications to partial differential equations*, Springer, New York, 1983.

[411] M. Pedersen and Z. Lin, *Coupled diffusion systems with localized nonlinear reactions*, Comput. Math. Appl. **42** (2001), 807-816.

[412] L.A. Peletier, D. Terman and F.B. Weissler, *On the equation* $\Delta u + x \cdot \nabla u + f(u) = 0$, Arch. Rational Mech. Anal. **94** (1986), 83-99.

[413] L.A. Peletier and R. van der Vorst, *Existence and nonexistence of positive solutions of nonlinear elliptic systems and the biharmonic equation*, Differential Integral Equations **5** (1992), 747-767.

[414] M. Pierre (2003), Personal communication.

[415] M. Pierre, *Weak solutions and supersolutions in* L^1 *for reaction-diffusion systems*, J. Evol. Equ. **3** (2003), 153-168.

[416] M. Pierre and D. Schmitt, *Global existence for a reaction-diffusion system with a balance law*, Semigroups of linear and nonlinear operations and applications Math. Sci. Engrg., 185, Kluwer Acad. Publ., Dordrecht, 1993, pp. 251-258.

[417] M. Pierre and D. Schmitt, *Blowup in reaction-diffusion systems with dissipation of mass*, SIAM J. Math. Anal. **28** (1997), 259-269.

[418] P. Plecháč and V. Šverák, *Singular and regular solutions of a nonlinear parabolic system*, Nonlinearity **16** (2003), 2083-2097.

[419] S.I. Pohozaev, *Eigenfunctions of the equation $\Delta u + \lambda f(u) = 0$*, Soviet Math. Dokl. **5** (1965), 1408-1411.

[420] P. Poláčik, *Morse indices and bifurcations of positive solutions of $\Delta u + f(u) = 0$ on $\mathbb{R}^N$*, Indiana Univ. Math. J. **50** (2001), 1407-1432.

[421] P. Poláčik, *Parabolic equations: Asymptotic behavior and dynamics on invariant manifolds*, Handbook of dynamical systems, Vol. 2 (B. Fiedler, ed.), Elsevier, Amsterdam, 2002, pp. 835-883.

[422] P. Poláčik and P. Quittner, *A Liouville-type theorem and the decay of radial solutions of a semilinear heat equation*, Nonlinear Anal. **64** (2006), 1679-1689.

[423] P. Poláčik and P. Quittner, in preparation.

[424] P. Poláčik, P. Quittner and Ph. Souplet, *Singularity and decay estimates in superlinear problems via Liouville-type theorems. Part I: elliptic equations and systems*, Duke Math. J. **139** (2007) (to appear).

[425] P. Poláčik, P. Quittner and Ph. Souplet, *Singularity and decay estimates in superlinear problems via Liouville-type theorems. Part II: parabolic equations*, Indiana Univ. Math. J. **56** (2007), 879-908.

[426] P. Poláčik and K.P. Rybakowski, *Nonconvergent bounded trajectories in semilinear heat equations*, J. Differential Equations **124** (1996), 472-494.

[427] P. Poláčik and F. Simondon, *Nonconvergent bounded solutions of semilinear heat equations on arbitrary domains*, J. Differential Equations **186** (2002), 586-610.

[428] P. Poláčik and E. Yanagida, *On bounded and unbounded global solutions of a supercritical semilinear heat equation*, Math. Ann. **327** (2003), 745-771.

[429] M. Protter and H. Weinberger, *Maximum principles in differential equations*, Prentice Hall, Englewood Cliffs, N.J., 1967.

[430] J. Prüss and H. Sohr, *Imaginary powers of elliptic second order differential operators in L^p-spaces*, Hiroshima Math. J. **23** (1993), 161-192.

[431] P. Pucci and J. Serrin, *A general variational identity*, Indiana Univ. Math. J. **35** (1986), 681-703.

[432] F. Quiros and J. Rossi, *Non-simultaneous blow-up in a semilinear parabolic system*, Z. Angew. Math. Phys. **52** (2001), 342-346.

[433] P. Quittner, *Blow-up for semilinear parabolic equations with a gradient term*, Math. Methods Appl. Sci. **14** (1991), 413-417.

[434] P. Quittner, *On global existence and stationary solutions for two classes of semilinear parabolic equations*, Comment. Math. Univ. Carolin. **34** (1993), 105-124.

[435] P. Quittner, *Global existence of solutions of parabolic problems with nonlinear boundary conditions*, Banach Center Publ. **33** (1996), 309-314.

[436] P. Quittner, *Signed solutions for a semilinear elliptic problem*, Differential Integral Equations **11** (1998), 551-559.

[437] P. Quittner, *A priori bounds for global solutions of a semilinear parabolic problem*, Acta Math. Univ. Comenian. (N.S.) **68** (1999), 195-203.

[438] P. Quittner, *Universal bound for global positive solutions of a superlinear parabolic problem*, Math. Ann. **320** (2001), 299-305.

[439] P. Quittner, *A priori estimates of global solutions and multiple equilibria of a parabolic problem involving measure*, Electron. J. Differential Equations **2001, 29** (2001), 1-17.

[440] P. Quittner, *Continuity of the blow-up time and a priori bounds for solutions in superlinear parabolic problems*, Houston J. Math. **29** (2003), 757-799.

[441] P. Quittner, *Multiple equilibria, periodic solutions and a priori bounds for solutions in superlinear parabolic problems*, NoDEA Nonlinear Differential Equations Appl. **11** (2004), 237-258.

[442] P. Quittner, *Complete and energy blow-up in superlinear parabolic problems*, Recent Advances in Elliptic and Parabolic Problems (Chiun-Chuan Chen, Michel Chipot and Chang-Shou Lin, eds.), World Scientific Publ., Hackensack, NJ, 2005, pp. 217-229.

[443] P. Quittner, *The decay of global solutions of a semilinear parabolic equation*, Discrete Contin. Dyn. Syst. (to appear).

[444] P. Quittner, *Qualitative theory of semilinear parabolic equations and systems*, Lectures on evolutionary partial differential equations, Lecture notes of the Jindřich Nečas Center for Mathematical Modeling, vol. 2, Matfyzpress, Praha, 2007 (to appear).

[445] P. Quittner and A. Rodríguez-Bernal, *Complete and energy blow-up in parabolic problems with nonlinear boundary conditions*, Nonlinear Anal. **62** (2005), 863-875.

[446] P. Quittner and Ph. Souplet, *Admissible L_p norms for local existence and for continuation in semilinear parabolic systems are not the same*, Proc. Roy. Soc. Edinburgh Sect. A **131** (2001), 1435-1456.

[447] P. Quittner and Ph. Souplet, *Global existence from single-component L^p estimates in a semilinear reaction-diffusion system*, Proc. Amer. Math. Soc. **130** (2002), 2719-2724.

[448] P. Quittner and Ph. Souplet, *A priori estimates of global solutions of superlinear parabolic problems without variational structure*, Discrete Contin. Dyn. Syst. **9** (2003), 1277-1292.

[449] P. Quittner and Ph. Souplet, *A priori estimates and existence for elliptic systems via bootstrap in weighted Lebesgue spaces*, Arch. Rational Mech. Anal. **174** (2004), 49-81.

[450] P. Quittner, Ph. Souplet and M. Winkler, *Initial blow-up rates and universal bounds for nonlinear heat equations*, J. Differential Equations **196** (2004), 316-339.

[451] P.H. Rabinowitz, *Some aspects of nonlinear eigenvalue problems*, Rocky Mountain J. Math. **3** (1973), 161-202.

[452] P.H. Rabinowitz, *Multiple critical points of perturbed symmetric functionals*, Trans. Amer. Math. Soc. **272** (1982), 753-770.

[453] M. Ramos, S. Terracini and C. Troestler, *Superlinear indefinite elliptic problems and Pohožaev type identities*, J. Funct. Anal. **159** (1998), 596-628.

[454] W. Reichel and H. Zou, *Non-existence results for semilinear cooperative elliptic systems via moving spheres*, J. Differential Equations **161** (2000), 219-243.

[455] F. Rellich, *Darstellung der Eigenwerte von $\Delta u + \lambda u = 0$ durch ein Randintegral*, Math. Z. **46** (1940), 635-636.

[456] O. Rey, *A multiplicity result for a variational problem with lack of compactness*, Nonlinear Anal. **13** (1989), 1241-1249.

[457] F. Ribaud, *Analyse de Littlewood Paley pour la résolution d'équations paraboliques semi-linéaires*, Doctoral Thesis, University of Paris XI, 1996.

[458] J. Rossi and Ph. Souplet, *Coexistence of simultaneous and nonsimultaneous blow-up in a semilinear parabolic system*, Differential Integral Equations **18** (2005), 405-418.

[459] F. Rothe, *Uniform bounds from bounded L_p-functionals in reaction-diffusion equations*, J. Differential Equations **45** (1982), 207-233.

[460] F. Rothe, *Global solutions of reaction-diffusion systems*, Lecture Notes in Mathematics 1072, Springer-Verlag, Berlin - Heidelberg - New York, 1984.

[461] P. Rouchon, *Blow-up of solutions of nonlinear heat equations in unbounded domains for slowly decaying initial data*, Z. Angew. Math. Phys. **52** (2001), 1017-1032.

[462] P. Rouchon, *Boundedness of global solutions for nonlinear diffusion equations with localized source*, Differential Integral Equations **16** (2003), 1083-1092.

[463] P. Rouchon, *Universal bounds for global solutions of a diffusion equation with a nonlocal reaction term*, J. Differential Equations **193** (2003), 75-94.

[464] P. Rouchon, *Universal bounds for global solutions of a diffusion equation with a mixed local-nonlocal reaction term*, Acta Math. Univ. Comenian. (N.S.) **75** (2006), 63-74.

[465] J. Rubinstein and P. Sternberg, *Nonlocal reaction-diffusion equations and nucleation*, IMA J. Appl. Math. **48** (1992), 249-264.

[466] A.A. Samarskii, V.A. Galaktionov, S.P. Kurdyumov and A.P. Mikhailov, *Blow-up in quasilinear parabolic equations*, Nauka, Moscow, 1987, English translation: Walter de Gruyter, Berlin, 1995.

[467] D.H. Sattinger, *On global solution of nonlinear hyperbolic equations*, Arch. Rational Mech. Math. **30** (1968), 148-172.

[468] D. Schmitt, *Existence globale ou explosion pour les systèmes de réaction-diffusion avec contrôle de masse*, Doctoral Thesis, University of Nancy 1, 1995.

[469] J. Serrin, *Gradient estimates for solutions of nonlinear elliptic and parabolic equations*, Contributions to nonlinear functional analysis, Academic Press, New York, 1971, pp. 565-601.

[470] J. Serrin and H. Zou, *Existence and non-existence for ground states of quasilinear elliptic equations*, Arch. Rational Mech. Anal. **121** (1992), 101-130.

[471] J. Serrin and H. Zou, *Non-existence of positive solutions of Lane-Emden systems*, Differential Integral Equations **9** (1996), 635-653.

[472] J. Serrin and H. Zou, *Existence of positive solutions of the Lane-Emden system*, Atti Sem. Mat. Fis. Univ. Modena **46** (1998, suppl.), 369-380.

[473] J. Serrin and H. Zou, *Cauchy-Liouville and universal boundedness theorems for quasilinear elliptic equations and inequalities*, Acta Math. **189** (2002), 79-142.

[474] L. Simon, *Asymptotics for a class of non-linear evolution equations, with applications to geometric problems*, Ann. Math. **118** (1983), 525-571.

[475] S. Snoussi, S. Tayachi and F.B. Weissler, *Asymptotically self-similar global solutions of a general semilinear heat equation*, Math. Ann. **321** (2001), 131-155.

[476] S. Sohr, *Beschränkter H_∞-Funktionalkalkül für elliptische Randwertsysteme*, Dissertation, Kassel 1999.

[477] V. Solonnikov, *Green matrices for parabolic boundary value problems*, Zap. Naučn. Sem. Leningrad **14** (1969), 132-150.

[478] Ph. Souplet, *Sur l'asymptotique des solutions globales pour une équation de la chaleur semi-linéaire dans des domaines non bornés*, C. R. Acad. Sci. Paris Sér. I Math. **323** (1996), 877-882.

[479] Ph. Souplet, *Blow-up in nonlocal reaction-diffusion equations*, SIAM J. Math. Anal. **28** (1998), 1301-1334.

[480] Ph. Souplet, *Uniform blow-up profiles and boundary behavior for diffusion equations with nonlocal nonlinear source*, J. Differential Equations **153** (1999), 374-406.

[481] Ph. Souplet, *Geometry of unbounded domains, Poincaré inequalities and stability in semilinear parabolic equations*, Comm. Partial Differential Equations **24** (1999), 951-973.

[482] Ph. Souplet, *A note on the paper by Qi S. Zhang: "A priori estimates and the representation formula for all positive solutions to a semilinear parabolic problem"*, J. Math. Anal. Appl. **243** (2000), 453-457.

[483] Ph. Souplet, *Decay of heat semigroups in L^∞ and applications to nonlinear parabolic problems in unbounded domains*, J. Funct. Anal. **173** (2000), 343-360.

[484] Ph. Souplet, *Recent results and open problems on parabolic equations with gradient nonlinearities*, Electron. J. Differential Equations **2001, 20** (2001), 1-19.

[485] Ph. Souplet, *Gradient blow-up for multidimensional nonlinear parabolic equations with general boundary conditions*, Differential Integral Equations **15** (2002), 237-256.

[486] Ph. Souplet, *Monotonicity of solutions and blow-up for semilinear parabolic equations with nonlinear memory*, Z. Angew. Math. Phys. **55** (2004), 28-31.

[487] Ph. Souplet, *Uniform blow-up profile and boundary behaviour for a non-local reaction-diffusion equation with critical damping*, Math. Methods Appl. Sci. **27** (2004), 1819-1829.

[488] Ph. Souplet, *Infinite time blow-up for superlinear parabolic problems with localized reaction*, Proc. Amer. Math. Soc. **133** (2005), 431-436.

[489] Ph. Souplet, *Optimal regularity conditions for elliptic problems via L^p_δ spaces*, Duke Math. J. **127** (2005), 175-192.

[490] Ph. Souplet, *The influence of gradient perturbations on blow-up asymptotics in semilinear parabolic problems: a survey*, Progress in Nonlinear Differential Equations and Their Applications, Vol. 64, Birkhaüser, 2005, pp. 473-496.

[491] Ph. Souplet, *A remark on the large-time behavior of solutions of viscous Hamilton-Jacobi equations*, Acta Math. Univ. Comenian. (N.S.) **76** (2007), 11-13.

[492] Ph. Souplet, *A note on diffusion-induced blow-up*, J. Dynam. Differential Equations **19** (2007), to appear.

[493] Ph. Souplet, *Single point blow-up for a semilinear parabolic system*, J. Eur. Math. Soc. (to appear).

[494] Ph. Souplet and S. Tayachi, *Optimal condition for non-simultaneous blow-up in a reaction-diffusion system*, J. Math. Soc. Japan **56** (2004), 571-584.

[495] Ph. Souplet, S. Tayachi and F.B. Weissler, *Exact self-similar blow-up of solutions of a semilinear parabolic equation with a nonlinear gradient term*, Indiana Univ. Math. J. **45** (1996), 655-682.

[496] Ph. Souplet and J.L. Vázquez, *Stabilization towards a singular steady state with gradient blow-up for a convection-diffusion problem*, Discrete Contin. Dyn. Syst. **14** (2006), 221-234.

[497] Ph. Souplet and F.B. Weissler, *Self-similar subsolutions and blowup for nonlinear parabolic equations*, J. Math. Anal. Appl. **212** (1997), 60-74.

[498] Ph. Souplet and F.B. Weissler, *Poincaré's inequality and global solutions of a nonlinear parabolic equation*, Ann. Inst. H. Poincaré Anal. Non Linéaire **16** (1999), 337-373.

[499] Ph. Souplet and F.B. Weissler, *Regular self-similar solutions of the nonlinear heat equation with initial data above the singular steady state*, Ann. Inst. H. Poincaré Anal. Non Linéaire **20** (2003), 213-235.

[500] Ph. Souplet and Q.S. Zhang, *Global solutions of inhomogeneous Hamilton-Jacobi equations*, J. Anal. Math. **99** (2006), 355-396.

[501] M.A.S. Souto, *A priori estimates and existence of positive solutions of nonlinear cooperative elliptic systems*, Differential Integral Equations **8** (1995), 1245-1258.

[502] R. Sperb, *Growth estimates in diffusion-reaction problems*, Arch. Rational Mech. Anal. **75** (1980), 127-145.

[503] P.N. Srikanth, *Uniqueness of solutions of nonlinear Dirichlet problems*, Differential Integral Equations **6** (1993), 663-670.

[504] B. Straughan, *Explosive instabilities in mechanics*, Springer, Berlin, 1998.

[505] M. Struwe, *Variational methods. Applications to nonlinear partial differential equations and Hamiltonian systems*, Springer, Berlin, 2000.

[506] M. Struwe, *Infinitely many critical points for functionals which are not even and applications to nonlinear boundary value problems*, Manuscripta Math. **32** (1980), 335-364.

[507] R. Suzuki, *Asymptotic behavior of solutions of quasilinear parabolic equations with slowly decaying initial data*, Adv. Math. Sci. Appl. **9** (1999), 291-317.

[508] G. Talenti, *Best constants in Sobolev inequality*, Ann. Mat. Pura Appl. (4) **110** (1976), 353-372.

[509] S. Taliaferro, *Local behavior and global existence of positive solutions of $au^\lambda \le -\Delta u \le u^\lambda$*, Ann. Inst. H. Poincaré Anal. Non Linéaire **19** (2002), 889-901.

[510] J.I. Tello, *Stability of steady states of the Cauchy problem for the exponential reaction-diffusion equation*, J. Math. Anal. Appl. **324** (2006), 381-396.

[511] E. Terraneo, *Non-uniqueness for a critical non-linear heat equation*, Comm. Partial Differential Equations **27** (2002), 185-218.

[512] Al. Tersenov and Ar. Tersenov, *Global solvability for a class of quasilinear parabolic problems*, Indiana Univ. Math. J. **50** (2001), 1899-1913.

[513] H. Triebel, *Interpolation theory, function spaces, differential operators*, North-Holland, Amsterdam - New York - Oxford, 1978.

[514] W.C. Troy, *Symmetry properties in systems of semilinear elliptic equations*, J. Differential Equations **42** (1981), 400-413.

[515] M. Tsutsumi, *On solutions of semilinear differential equations in a Hilbert space*, Math. Japon. **17** (1972), 173-193.

[516] A.M. Turing, *The chemical basis of morphogenesis*, Philos. Trans. Roy. Soc. London Ser. B **237** (1952), 37-72.

[517] R.E.L. Turner, *A priori bounds for positive solutions of nonlinear elliptic equations in two variables*, Duke Math. J. **41** (1974), 759-774.

[518] D. Tzanetis, *Blow-up of radially symmetric solutions of a non-local problem modelling Ohmic heating*, Electron. J. Differential Equations (2002, 11), 1-26.

[519] J.J.L. Velázquez, *Local behaviour near blow-up points for semilinear parabolic equations*, J. Differential Equations **106** (1993), 384-415.

[520] J.J.L. Velázquez, *Estimates on the $(n-1)$-dimensional Hausdorff measure of the blow-up set for a semilinear heat equation*, Indiana Univ. Math. J. **42** (1993), 445-476.

[521] J.J.L. Velázquez, *Blow up for semilinear parabolic equations*, Recent advances in partial differential equations (Res. Appl. Math. 30) (M.A. Herrero, E. Zuazua, eds.), Masson, Paris, 1994, pp. 131-145.

[522] J.J.L. Velázquez, V.A. Galaktionov and M.A. Herrero, *The space structure near a blow-up point for semilinear heat equations: a formal approach*, Zh. Vychisl. Mat. i Mat. Fiziki **31** (1991), 399-411.

[523] L. Véron, *Singularities of solutions of second order quasilinear equations*, Pitman Research Notes in Mathematics Series, 353, Longman, Harlow, 1996.

[524] L. Wang and Q. Chen, *The asymptotic behaviour of blow-up solution of localized nonlinear equation*, J. Math. Anal. Appl. **200** (1996), 315-321.

[525] M.-X. Wang and Y. Wang, *Properties of positive solutions for non-local reaction-diffusion problems*, Math. Methods Appl. Sci. **19** (1996), 1141-1156.

[526] N.A. Watson, *Parabolic equations on an infinite strip*, Marcel Dekker, Boston, 1989.

[527] H. Weinberger, *An example of blowup produced by equal diffusions*, J. Differential Equations **154** (1999), 225-237.

[528] F.B. Weissler, *Semilinear evolution equations in Banach spaces*, J. Funct. Anal. **32** (1979), 277-296.

[529] F.B. Weissler, *Local existence and nonexistence for semilinear parabolic equations in L^p*, Indiana Univ. Math. J. **29** (1980), 79-102.

[530] F.B. Weissler, *Existence and non-existence of global solutions for a semilinear heat equation*, Israel J. Math. **38** (1981), 29-40.

[531] F.B. Weissler, *Single point blow-up for a semilinear initial value problem*, J. Differential Equations **55** (1984), 204-224.

[532] F.B. Weissler, *Asymptotic analysis of an ordinary differential equation and non-uniqueness for a semilinear partial differential equation*, Arch. Rational Mech. Anal. **91** (1985), 231-245.

[533] F.B. Weissler, *Rapidly decaying solutions of an ordinary differential equation with applications to semilinear elliptic and parabolic partial differential equations*, Arch. Rational Mech. Anal. **91** (1985), 247-266.

[534] F.B. Weissler, *An L^∞ blow-up estimate for a nonlinear heat equation*, Comm. Pure Appl. Math. **38** (1985), 291-295.

[535] F.B. Weissler, *L^p-energy and blow-up for a semilinear heat equation*, In Nonlinear functional analysis and its applications, Proc. Symp. Pure Math. **45/2** (1986), 545-551.

[536] B. Wollenmann, *Uniqueness for semilinear parabolic problems, PhD Thesis, Universität Zürich*, 2001.

[537] R. Xing, *A priori estimates for global solutions of semilinear heat equations in $\mathbb{R}^n$*, Nonlinear Anal. (to appear).

[538] E. Yanagida, *Uniqueness of rapidly decaying solutions to the Haraux-Weissler equation*, J. Differential Equations **127** (1996), 561-570.

[539] X.-F. Yang, *Nodal sets and Morse indices of solutions of super-linear elliptic PDEs*, J. Funct. Anal. **160** (1998), 223-253.

[540] H. Zaag, *A Liouville theorem and blow-up behavior for a vector-valued nonlinear heat equation with no gradient structure*, Comm. Pure Appl. Math. **54** (2001), 107-133.

[541] H. Zaag, *On the regularity of the blow-up set for semilinear heat equations*, Ann. Inst. H. Poincaré Anal. Non Linéaire **19** (2002), 505-542.

[542] H. Zaag, *Determination of the curvature of the blow-up set and refined singular behavior for a semilinear heat equation*, Duke Math. J. **133** (2006), 499-525.

[543] T.I. Zelenyak, *Stabilization of solutions of boundary value problems for a second order parabolic equation with one space variable*, Differential Equations **4** (1968), 17-22.

[544] L. Zhang, *Uniqueness of positive solutions of $\Delta u + u + u^p = 0$ in a ball*, Comm. Partial Differential Equations **17** (1992), 1141-1164.

[545] Q.S. Zhang, *The boundary behavior of heat kernels of Dirichlet Laplacians*, J. Differential Equations **182** (2002), 416-430.

[546] P. Zhao and C. Zhong, *On the infinitely many positive solutions of a super-critical elliptic problem*, Nonlinear Anal. **44** (2001), 123-139.

[547] H. Zou, *A priori estimates for a semilinear elliptic systems without variational structure and their applications*, Math. Ann. **323** (2002), 713-735.

List of Symbols

Standard function spaces are defined in Preliminaries.

Index